# ELECTROMAGNETIC FIELDS
# AND INTERACTIONS

## RICHARD BECKER

### EDITED BY PROFESSOR FRITZ SAUTER PH.D
### UNIVERSITY OF COLOGNE

DOVER PUBLICATIONS, INC.
NEW YORK

Published in Canada by General Publishing Company, Ltd.,
30 Lesmill Road, Don Mills, Toronto, Ontario.
Published in the United Kingdom by Constable and Company, Ltd., 10 Orange Street, London WC2H 7EG.

This Dover edition, first published in 1982, is an unabridged
republication in one volume of the work originally published
in two volumes in 1964 by Blaisdell Publishing Company, N.Y.,
simultaneously with Blackie and Son Limited, Glasgow. This
is the only authorized translation of the original German
*Theorie der Elektrizität,* published by the B.G. Teubner Ver-
lagsgesellschaft, Stuttgart. The present edition is published by
special arrangement with Blackie and Son Limited.

Manufactured in the United States of America
Dover Publications, Inc.
180 Varick Street
New York, N.Y. 10014

**Library of Congress Cataloging in Publication Data**

Becker, Richard, 1887-1955.
    Electromagnetic fields and interactions.

    Translation of: Theorie der Elektrizität.
    Reprint. Originally published: New York : Blaisdell Pub.
Co., c1964. (A Blaisdell book in the pure and applied
sciences).
    Includes indexes.
    Contents: v. 1. Electromagnetic theory and relativity —
v. 2. Quantum theory of atoms and radiation / translated
by Ivor de Teisser ; rev. by Günther Leibfried and Wilhelm
Brenig.
    1. Electromagnetic fields. 2. Relativity (Physics) 3. Elec
trons. 4. Quantum theory. I. Sauter, Fritz, 1906-    . II.
Title. III. Series: Blaisdell book in the pure and applied
sciences.
QC665.E4B413 1982          530.1'41          81-19451
ISBN 0-486-64290-9                            AACR2

# VOLUME I

# *Electromagnetic Theory and Relativity*

# FOREWORD

August Föppl's *Introduction to Maxwell's Theory* appeared in the year 1894. A completely revised second edition followed ten years later, this being the first volume of Max Abraham's *Theory of Electricity*. This, in turn, was followed a year later by a second volume on the electron theory. From the year 1930 onward, with the appearance of the eighth edition, Richard Becker took over the further editing of the work which, in succeeding years, underwent several basic changes.

With the sixteenth edition of the first volume and the eighth edition of the second, a thorough-going revision of the work, combined with an expansion to three volumes was planned by Becker. His sudden death, however, took him from the midst of his labours on this new revision.

In carrying on the work it was evident to me that Becker's plan should be continued. In remodelling—particularly in the first volume —there were many places in the previous first volume (Maxwell theory) and in the previous second volume (relativity theory) which were taken over practically unchanged. On the one hand, so as not to allow the new first volume to become too bulky, it was necessary in places to condense further and to make changes—on the other hand, however, it appeared necessary to write certain sections anew or to put them into a new form. Examples here are the sections on dipoles and quadrupoles and their radiation fields, and also the section on the forces on dipoles and quadrupoles in external fields. The sections on the energy relations and force effects in static fields were basically transformed, and with this in particular the Lorentz force, as an experimental formula, was placed at the forefront of the proceedings concerning force effects in the magnetic field.

As previously, so in the remodelling, the Gaussian CGS system of units is used. In remarks appended to certain sections, however, the more important formulae have been transcribed to the Giorgi MKSA system. In this way I hope to have suited in some measure the wishes of both physicists and engineers. The reason for preferring the Gaussian system to the volt-ampere system (which is employed more in practice) is that in the use of the first-mentioned system the formulae of both relativity theory and quantum mechanics can be written appreciably more simply. Moreoever, if the Giorgi system had been employed, the beautiful symmetry between the electric and magnetic-field quantities in the four-dimensional Minkowski formalism of special relativity would no longer be so evident.

I have attempted to bring out clearly the mutual relationship of the two measure systems. The transition from the four basic units of the MKSA system—units which at first sight seem to appear naturally—to the three basic units of the Gaussian CGS system, requires that Coulomb's law be not only considered as an experimental law, but that at the same time it be regarded as the defining equation for the unit of charge in the Gaussian system. This is because the resulting constant of proportionality, having the dimensions of force times length squared, divided by charge squared, is arbitrarily assumed to be dimensionless and equal to 1. This procedure, violently criticized by certain advocates of the MKSA system, corresponds exactly with today's custom in high-energy physics of combining the length and time dimensions with one another through the arbitrary assumption that the velocity of light in a vacuum is dimensionless and equal to 1.

Since, in the following, the equations of electrodynamics are on the one hand to be understood as quantity equations and not as relationships between the numerical values of physical quantities in special units, and since on the other hand these equations are transcribed part by part in different ways in the two unit systems employed here, only a few of all the symbols employed represent the same physical quantity in both measure systems. Examples are the symbols of kinematic and dynamic quantities, and of electric charge, electric polarization, and electric field strength. In opposition to this (though a fact often overlooked) the symbols of electric displacement as well as those of total magnetic-field quantities, have different meanings in the two systems. Table 6 gives a summary of the relationships between such quantities customarily described by the same symbol, and it also gives the conversion factors between them. I hope that by means of this table, and also through the presentation of the corresponding relationships in the text, my readers—and in particular those who are students—will be helped through the customarily troublesome process of going over from one unit system to the other.

The translation is the work of Mr. A. W. Knudsen.

F. SAUTER

COLOGNE, 1964

# CONTENTS

# PART C.   Electric current and the magnetic field

# PART D. The general fundamental equations of the electromagnetic field

# PART E.  The theory of relativity

## Chapter EI.  The physical basis of relativity theory and its mathematical aids

## Chapter EII.  The relativistic electrodynamics of empty space

## Chapter EIII.  The relativistic electrodynamics of material bodies

## Chapter EIV.  Relativistic mechanics

# PART F.  Exercise problems and solutions

## Chapter FI.  Exercises

## Chapter FII.  Solutions

Contents

**A**

---

Introduction
to
vector
and
tensor
calculus

# CHAPTER A I

## Vectors

### §1. Definition of a vector

The science of electric and magnetic phenomena prior to the appearance of Maxwell's theory was based on the concept of action at a distance between bodies which are electrified, magnetized, or traversed by electric currents. Only the ideas of Faraday differed in this respect from those of other physicists in that he conceived all electric and magnetic actions of one body upon another separated from it as the effect of an electric and/or magnetic field existing between the bodies. Although his manner of interpreting and describing the phenomena was basically a mathematical one, Faraday was nevertheless unable to give his interpretation sufficient completeness and freedom from contradiction to have it raised to the rank of a theory. In this, success was first achieved by Maxwell, who gave Faraday's ideas rigorous mathematical form and thereby created a theoretical structure which, as a *field-action theory*, is essentially different in conception from the *action-at-a-distance theory*.

Maxwell himself formulated his equations in Cartesian coordinates and only incidentally made use of quaternion theory. A general overall view of the interconnection of all formulae is however considerably facilitated by the use of *vector calculus*. This method of calculating would appear as if created for the task of representing in the best way possible the ideas of Faraday. The labour expended in becoming familiar with the method is certainly outweighed by the advantages gained. At the beginning of this work, therefore, we give an account of the theory of vectors and vector fields, together with a short section on tensors which occasionally appear in Maxwell's theory.

The simplest physical quantities are those which are completely determined in known units by specifying a single measure-number. They will be called *scalars*. Examples are mass and temperature. In general we shall represent them by Latin or Greek letters.

In addition there are physical quantities whose establishment

3

requires the employment in a known way of three specifying numbers, as for example the displacement of a point from a given initial position. We could characterize this displacement by specifying the three Cartesian coordinates of the end-position with respect to an origin placed at the initial point, from there on calculating with these scalar displacement components. If we did this, however, we should in the first place be neglecting the fact that, physically speaking, a displacement is a single idea; and secondly we should be bringing a foreign element into the question, namely the coordinate system, which has nothing to do with the displacement itself. For the description of displacements, therefore, we shall with advantage introduce quantities of a new type, definable without reference to a coordinate system, and we shall establish appropriate rules for their use. Only in the numerical evaluation of formulae will it be necessary to introduce a definite coordinate system.

The rectilinear displacement of a point, as well as all other physical quantities which, like displacements, are uniquely established by specifying their direction in space and their magnitude, and which obey the same addition rule as displacements, we call *vectors*. In the following we shall always designate vectors by bold-face letters. Thus **A** stands for a vector, graphically representable as an arrow, and $|\mathbf{A}| = A$ its magnitude, designated by the length of the arrow.

## §2. Addition and subtraction of vectors

For adding two displacements **A** and **B** we imagine a movable point which is initially at position 1 (figure 1). To this point is first assigned the

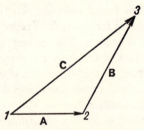

Fig. 1.—Vector addition

displacement **A** from 1 to 2; then the point is moved through displacement **B** corresponding to the interval from 2 to 3. Now the rectilinear displacement going directly from 1 to 3 is called the resultant or geometric sum **C** of the two displacements **A** and **B**:

$$C = A + B \qquad (2.1)$$

By first taking the displacement corresponding to vector **B**, and then that belonging to vector **A**, the movable point describes (figure 2) the

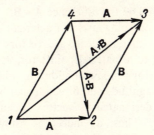

Fig. 2.—Addition and subtraction of vectors

path 143, which, with path 123 forms a parallelogram. Thus the resultant **B**+**A**, and also **A**+**B**, represents the diagonal of this parallelogram. Vector addition therefore obeys the *commutative law*

$$A + B = B + A \qquad (2.2)$$

Similarly (figure 3) the validity of the *associative law* for the sum of three or more vectors is demonstrated:

$$(A + B) + C = A + (B + C) \qquad (2.3)$$

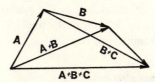

Fig. 3.—The associative law in vector addition

What meaning shall be assigned to the difference of two vectors **A** and **B**? We define the vector −**B** so that, just as for a scalar, the relationship

$$B - B = 0 \qquad (2.4)$$

holds. The vector −**B** should therefore correspond to a displacement which annuls displacement **B**, bringing the movable point back to its initial position. Thus vector $-\mathbf{B}$ has the same magnitude as $\mathbf{B}$, but the opposite direction.

As the difference of the vectors $\mathbf{A}$ and $\mathbf{B}$ we now define the sum $\mathbf{A}$ and $-\mathbf{B}$ thus:

$$\mathbf{A} - \mathbf{B} = \mathbf{A} + (-\mathbf{B}) \tag{2.5}$$

In the parallelogram of figure 2 the diagonal 13 represents the sum $\mathbf{A} + \mathbf{B}$, and the diagonal 42 the difference $\mathbf{A} - \mathbf{B}$.

Obedience to the foregoing laws characterizes those directed quantities which, according to our ideas, are vectors. Here, for example, belong forces applied to a material point; for, as statics teaches, these forces obey the parallelogram law of addition. There are however other quantities to which magnitude and direction can be ascribed, but which nevertheless are not designated as vectors because their composition follows another law. Thus, as indeed is known from kinematics, infinitesimally small rotations of a rigid system about a fixed point can be represented by vectors because the addition of two such rotations obeys the parallelogram law; not, however, finite rotations because they add in a much more complicated way.

## §3. Unit vectors, base vectors, components

By the product $\mathbf{A}$ of a scalar $\alpha$ and a vector $\mathbf{a}$

$$\mathbf{A} = \alpha \mathbf{a} \tag{3.1}$$

we understand a vector whose magnitude $A$ is equal to product of the magnitudes of the scalar and vector, and whose direction is the same as that of $\mathbf{a}$, or opposite, according as $\alpha$ is positive or negative.

In accordance with its definition, the multiplication of vectors with scalars follows the commutative law

$$\alpha \mathbf{a} = \mathbf{a} \alpha$$

The distributive law also holds, i.e.

$$(\alpha + \beta)\mathbf{a} = \alpha \mathbf{a} + \beta \mathbf{a} \qquad \alpha(\mathbf{a} + \mathbf{b}) = \alpha \mathbf{a} + \alpha \mathbf{b}$$

If, in particular, $\mathbf{a}$ is a vector in the direction of $\mathbf{A}$, but of magnitude 1, then

$$\mathbf{A} = A\mathbf{a} \tag{3.2}$$

A vector whose magnitude equals 1 will be called a *unit vector*. We shall agree to assign to a vector the dimension of its magnitude; thus a unit vector always has the dimension of a pure number.

Let there be given a fixed unit vector **s** (figure 4), as well as an arbitrary vector **A** which forms with **s** the angle $\phi \equiv \langle \mathbf{A}, \mathbf{s} \rangle$. Then we

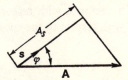

Fig. 4.—The component $A_s$ of **A** in the direction **s**

designate as the *component* of **A** with respect to the vector **s** the length of the projection of **A** on the line of the unit vector, i.e. the quantity

$$A_s = A \cos \phi = A \cos \langle \mathbf{A}, \mathbf{s} \rangle \qquad (3.3)$$

This component $A_s$ is a non-directed quantity and thus is a scalar; it will accordingly be designated by a Roman (Latin) symbol.

Summation and resolution into components of several vectors are mutually exchangeable operations; i.e. the component of the sum of

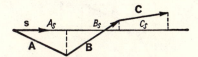

Fig. 5.—The component of **A** + **B** + **C** with respect to **s** is equal to the sum of the individual components

vectors with respect to a unit vector is equal to the algebraic sum of the corresponding components of the individual vectors. Figure 5 shows this for the case of three vectors.

Vectors of arbitrary direction and arbitrary magnitude can be described by their components with respect to three fixed non-planar unit vectors. We choose three pairwise-perpendicular unit vectors **x**, **y**, **z** as "base vectors"; their directions must therefore coincide with the axes of a Cartesian coordinate system.

As is well known, there are two kinds of Cartesian axis systems, distinguished as *right-hand* and *left-hand*. In the first kind the *x*-, *y*-, and *z*-axes (in that order) can be represented by the thumb, index finger, and middle finger of the right hand, while the second kind, in like manner, requires the left hand for this. By means of rotations, various right-hand systems can be brought into coincidence with one another; various left-hand systems likewise; but not a right-hand system into a left-hand system. Through a reflection in a coordinate plane or at a coordinate origin, however, a right-hand system will produce a left-hand system, and vice versa. In the following work we shall always make use of the right-hand system as did Maxwell.

The components of a vector **A** with respect to the base vectors **x**, **y**, **z**, or, as may be said, with respect to the coordinate axes are, according to (3.3),

$$A_x = A\cos\langle\mathbf{A},\mathbf{x}\rangle \quad A_y = A\cos\langle\mathbf{A},\mathbf{y}\rangle \quad A_z = A\cos\langle\mathbf{A},\mathbf{z}\rangle \quad (3.4)$$

They therefore follow uniquely from the magnitude $A$ and the direction cosines. Conversely the vector **A** (see figure 6) is the diagonal of length

$$A = \sqrt{(A_x^2 + A_y^2 + A_z^2)} \quad (3.5)$$

of a parallelopiped whose edge lengths are equal to the three components $A_x$, $A_y$, $A_z$. It is therefore uniquely determined, as previously

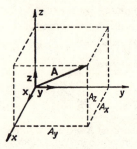

Fig. 6.—The vector **A** and its components $A_x$, $A_y$, $A_z$

mentioned, by these components. As is seen from figure 6, it is given by the vector relationship

$$\mathbf{A} = A_x\mathbf{x} + A_y\mathbf{y} + A_z\mathbf{z} \quad (3.6)$$

This equation permits calculation of the component $A_s$ of the vector **A** with respect to an arbitrary unit-vector **s**. By termwise component formation (the **x**-component in the direction of **s** is $\cos\langle\mathbf{x},\mathbf{s}\rangle$) we obtain the relationship

$$A_s = A_x\cos\langle\mathbf{x},\mathbf{s}\rangle + A_y\cos\langle\mathbf{y},\mathbf{s}\rangle + A_z\cos\langle\mathbf{z},\mathbf{s}\rangle \quad (3.7)$$

which, by using the components $s_x$, $s_y$, $s_z$ of the vector **s**, can be written in the form

$$A_s = A_x s_x + A_y s_y + A_z s_z \quad (3.8)$$

While definition (3.3) of $A_s$ is completely independent of the co-ordinate system, definition (3.8) makes use of the components of **A** and **s** in a special Cartesian coordinate system. As here, so also later in the vector calculus, we shall often come across parallel definitions, one independent of any coordinate system, the other equivalent but involving components. For general considerations the first is usually the more practical, while for carrying out calculations in concrete examples the component representation is generally to be preferred.

## §4. The scalar or inner product

From the vectors **A** and **B** we form a scalar quantity **A.B** according to the following definition which is independent of any coordinate system:

$$AB \equiv \mathbf{A} \cdot \mathbf{B} = AB \cos \langle \mathbf{A}, \mathbf{B} \rangle \qquad (4.1)$$

We call **A.B** the *inner product* or *scalar product* of the two vectors **A** and **B**.

Forming this product is suggested by observing the mechanical work performed by a force **F** in the displacement of a point through the (small) distance **s**. The quantity of work is given by the product: magnitude $F \cos \langle \mathbf{F}, \mathbf{s} \rangle$ (the projection of **F** along the direction of the displacement) times the distance travelled, **s**; this, according to (4.1) is just **F.s**.

According to the defining equation (4.1), the scalar product remains unaltered in an exchange of the order of the two factors; thus the *commutative law* holds here. That the *distributive law* holds as well follows from the definition of the scalar product and also from the result of §3 concerning the projection in a fixed direction of a sum of vectors.

According to (4.1) the scalar product of a vector with itself is equal to the square of its magnitude:

$$\mathbf{A} \cdot \mathbf{A} \equiv \mathbf{A}^2 = A^2$$

The scalar product of two vectors perpendicular to one another vanishes because $\cos \langle \mathbf{A}, \mathbf{B} \rangle = 0$. Thus, for the base vectors

$$\mathbf{x}^2 = \mathbf{y}^2 = \mathbf{z}^2 = 1 \qquad \mathbf{x} \cdot \mathbf{y} = \mathbf{y} \cdot \mathbf{z} = \mathbf{z} \cdot \mathbf{x} = 0 \qquad (4.2)$$

Further, from (3.3), the projection of **A** on the direction of the unit vector **s** can also be written as a scalar product:

$$A_s = \mathbf{A} \cdot \mathbf{s} \qquad (4.3)$$

Hence equation (3.8) can be obtained by termwise multiplication of (3.6) by **s**. Analogously, the component representation of a scalar product

$$\mathbf{A}.\mathbf{B} = A_x B_x + A_y B_y + A_z B_z \tag{4.4}$$

is found from (3.6) by scalar multiplication of this by **B**.

By considering here (4.1) and (3.4), we arrive at the familiar formula of analytic geometry.

$$\cos \langle \mathbf{A}, \mathbf{B} \rangle = \cos \langle \mathbf{A}, \mathbf{x} \rangle \cos \langle \mathbf{B}, \mathbf{x} \rangle + \cos \langle \mathbf{A}, \mathbf{y} \rangle \cos \langle \mathbf{B}, \mathbf{y} \rangle +$$
$$\cos \langle \mathbf{A}, \mathbf{z} \rangle \cos \langle \mathbf{B}, \mathbf{z} \rangle$$

### §5. The outer or vector product

We arrive at another type of product of the vectors **A** and **B** by considering the parallelogram (figure 7) which they form. This product is characterized by: (1) the contained area

$$S = AB \sin \langle \mathbf{A}, \mathbf{B} \rangle \tag{5.1}$$

of the parallelogram; (2) the sense of travel round the parallelogram, determined by the established sequence of the vectors; and (3) the orientation in space of the plane determined by **A** and **B**. We unite these three properties (surface, vector sequence, and orientation) in the concept of the *outer product* **A** × **B**.

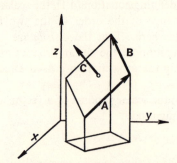

Fig. 7.—The vector product **A** × **B** as vector **C**

With every outer product so defined we can associate uniquely and reversibly a vector **C** by means of the following prescription: **C** is perpendicular to the surface of **A** × **B** and is directed so that an advance in the direction of **C** together with a rotation determined by the sense of travel given by **A** × **B** is the same as the advance of a right-hand screw. The magnitude of **C** is set equal to the contained area of the parallelogram. The vector **C**, so defined, is called the *vector product* of **A** and **B** and is set equal to the above-defined outer product

$$C = A \times B \tag{5.2}$$

From this definition it follows that since $C_z = C\cos\langle C,z\rangle$, the $z$-component of $A \times B$ is to be understood to be the projection of the parallelogram area $S \equiv C$ on the $xy$-plane. Moreover, the sum of two outer products represents a directed quantity whose $z$-component is equal to the sum of the projections of the two parallelogram areas on the $xy$-plane.

Further, from the definition of the vector product there follows the *invalidity of the commutative law*. Thus we have

$$A \times B + B \times A = 0 \tag{5.3}$$

On the other hand the *distributive law* stands:

$$(A+B) \times D = A \times D + B \times D \tag{5.4}$$

This can be shown by considering the component of this vector equation with respect to an arbitrary unit vector **s**, i.e. by considering the projections of the corresponding parallelograms on a plane perpendicular to **s**.

For the base vectors we have, in particular,

$$x \times y = -y \times x = z \qquad y \times z = -z \times y = x$$
$$z \times x = -x \times z = y \tag{5.5}$$

while, according to (5.1), products of the form $x \times x$, and also in general vector products of two parallel vectors, vanish.

We arrive at a coordinate representation of the vector product by multiplying out $A \times B$, thus:

$$A \times B = (A_x x + A_y y + A_z z) \times (B_x x + B_y y + B_z z)$$

Having regard for (5.5) we find that

$$A \times B = (A_y B_z - A_z B_y)x + (A_z B_x - A_x B_z)y + (A_x B_y - A_y B_x)z \tag{5.6}$$

or, in convenient determinant form,

$$A \times B = \begin{vmatrix} x & y & z \\ A_x & A_y & A_z \\ B_x & B_y & B_z \end{vmatrix} \tag{5.7}$$

The turning moment (torque) of a force **F** is an example of the outer product of two vectors. If **r** extends from the reference point to the point of application of the force **F**, then the turning moment **T** of the force **F** around the reference point is given by

$$\mathbf{T} = \mathbf{r} \times \mathbf{F} \tag{5.8}$$

From (5.1) its magnitude is equal to $rF\sin\langle \mathbf{r},\mathbf{F}\rangle$; it is equal, therefore, to the product: force $F$ times radius arm $r\sin\langle \mathbf{r},\mathbf{F}\rangle$.

The kinematics of rigid bodies furnishes another example (figure 8). A rigid body rotates around an axis OA. The vector **w** is placed along this axis extending from O.

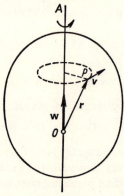

Fig. 8.—Rotation of a rigid body around the axis OA with angular velocity **w**.

The length of **w** represents the magnitude of the angular velocity and its direction that of the rotation, in the sense of a right-hand screw. If now **r** be the radius vector extending from O to any point P of the rigid body, then the velocity of P is given by

$$\mathbf{v} = \mathbf{w} \times \mathbf{r} \tag{5.9}$$

for, first of all, point P moves perpendicular to the plane containing **w** and **r**, that is, as figure 8 shows, in the direction of the vector product. And second, the magnitude of the velocity is given by the product: perpendicular distance of point P from the axis times angular velocity $w \equiv \omega$. This, however, is just the magnitude $\omega r\sin\langle \mathbf{w},\mathbf{r}\rangle$ of the vector product, which proves the assertion.

## §6.  Products of three and four vectors

(*a*) *Product of a vector and the scalar product of two other vectors:* **A.(B.C)**

Since **B.C** is a scalar, **A.(B.C)** is a vector parallel to **A**. From this it is clear that, for example, **(A.B).C** is an entirely different vector from the foregoing in so far as **A** and **C** are not parallel to one another.

(b) *Scalar product of a vector and the vector product of two other vectors:* $\mathbf{A} \cdot (\mathbf{B} \times \mathbf{C})$

We have here the important relation

$$\mathbf{A} \cdot (\mathbf{B} \times \mathbf{C}) = \mathbf{B} \cdot (\mathbf{C} \times \mathbf{A}) = \mathbf{C} \cdot (\mathbf{A} \times \mathbf{B}) \tag{6.1}$$

From the elementary rule "surface area times height", every such expression (6.1) represents the volume content of the parallelopiped formed by the vectors $\mathbf{A}$, $\mathbf{B}$, $\mathbf{C}$. Indeed, all three expressions in (6.1) give this volume with a positive sign if the vectors $\mathbf{A}$, $\mathbf{B}$, $\mathbf{C}$, in this order, form a right-hand system.

By means of (4.4) and (5.6) or (5.7) we obtain a component representation of this *spar product*; we have

$$\mathbf{A} \cdot (\mathbf{B} \times \mathbf{C}) = \begin{vmatrix} A_x & A_y & A_z \\ B_x & B_y & B_z \\ C_x & C_y & C_z \end{vmatrix} \tag{6.2}$$

(c) *Vector product of a vector and the vector product of two other vectors:* $\mathbf{A} \times (\mathbf{B} \times \mathbf{C})$

This vector is perpendicular to vector $\mathbf{A}$, and to the vector product $\mathbf{B} \times \mathbf{C}$. It therefore lies in the plane determined by vectors $\mathbf{B}$ and $\mathbf{C}$. Thus it can be expressed in the form

$$\mathbf{A} \times (\mathbf{B} \times \mathbf{C}) = \beta\mathbf{B} + \gamma\mathbf{C} \tag{6.3}$$

the coefficients $\beta$ and $\gamma$ being so far undetermined. From the requirement that the vector $\beta\mathbf{B} + \gamma\mathbf{C}$ must be perpendicular to $\mathbf{A}$ we find that

$$\beta\mathbf{A} \cdot \mathbf{B} + \gamma\mathbf{A} \cdot \mathbf{C} = 0, \quad \text{i.e.} \quad \beta = \lambda\mathbf{A} \cdot \mathbf{C}, \quad \gamma = -\lambda\mathbf{A} \cdot \mathbf{B}$$

$\lambda$ being a factor as yet undetermined. Equation (6.3) then becomes

$$\mathbf{A} \times (\mathbf{B} \times \mathbf{C}) = \lambda\{\mathbf{B}(\mathbf{A} \cdot \mathbf{C}) - \mathbf{C}(\mathbf{A} \cdot \mathbf{B})\} \tag{6.4}$$

To evaluate $\lambda$ we now go over to component representation, imagining the coordinate system specially chosen, however, so that the $x$-axis lies in the direction of the vector $\mathbf{A}$. This vector then has the three components $\{A, 0, 0\}$. Forming now, say, the $z$-component of (6.4) we find, on the one hand, for the left side on account of (5.6), the $y$-component of the vector product $\mathbf{B} \times \mathbf{C}$ multiplied by $A$. On the other hand the right side of (6.4) is equal to $\lambda A(B_z C_x - C_z B_x)$; therefore, to within the factor $\lambda A$ it is equal to the $y$-component of $\mathbf{B} \times \mathbf{C}$. Thus we must have that $\lambda = 1$, and we finally obtain

$$\mathbf{A} \times (\mathbf{B} \times \mathbf{C}) = \mathbf{B}(\mathbf{A} \cdot \mathbf{C}) - \mathbf{C}(\mathbf{A} \cdot \mathbf{B}) \qquad (6.5)$$

By working out the individual terms with the help of (6.5) we easily find that

$$\mathbf{A} \times (\mathbf{B} \times \mathbf{C}) + \mathbf{B} \times (\mathbf{C} \times \mathbf{A}) + \mathbf{C} \times (\mathbf{A} \times \mathbf{B}) = 0 \qquad (6.6)$$

(*d*) *Scalar product of two vector products:* $(\mathbf{A} \times \mathbf{B}) \cdot (\mathbf{C} \times \mathbf{D})$

This is a product of type (*b*) above in which the first vector there is replaced by the vector product of two other vectors. Applying rule (6.1) we obtain

$$(\mathbf{A} \times \mathbf{B}) \cdot (\mathbf{C} \times \mathbf{D}) = \mathbf{C} \cdot (\mathbf{D} \times (\mathbf{A} \times \mathbf{B}))$$

Now by rule (6.5) we have

$$\mathbf{D} \times (\mathbf{A} \times \mathbf{B}) = \mathbf{A}(\mathbf{B} \cdot \mathbf{D}) - \mathbf{B}(\mathbf{A} \cdot \mathbf{D})$$

It therefore follows that

$$(\mathbf{A} \times \mathbf{B}) \cdot (\mathbf{C} \times \mathbf{D}) = (\mathbf{A} \cdot \mathbf{C})(\mathbf{B} \cdot \mathbf{D}) - (\mathbf{A} \cdot \mathbf{D})(\mathbf{B} \cdot \mathbf{C}) \qquad (6.7)$$

From this we have for the special case of the square of a vector product, setting $\mathbf{C} = \mathbf{A}$ and $\mathbf{D} = \mathbf{B}$

$$(\mathbf{A} \times \mathbf{B})^2 = A^2 B^2 - (\mathbf{A} \cdot \mathbf{B})^2 \qquad (6.8)$$

The correctness of this relationship is immediately demonstrated when (4.1) and (5.1) are considered, for then

$$(\mathbf{A} \times \mathbf{B})^2 = A^2 B^2 \sin^2 \langle \mathbf{A}, \mathbf{B} \rangle, \quad \text{and} \quad (\mathbf{A} \cdot \mathbf{B})^2 = A^2 B^2 \cos^2 \langle \mathbf{A}, \mathbf{B} \rangle$$

so that (6.8) represents nothing more than the well-known relation $\sin^2 \phi = 1 - \cos^2 \phi$.

## §7. Differentiation of vectors with respect to a parameter

Let the vector $\mathbf{A}$ depend upon a scalar parameter, say the time $t$. The differential quotient, or derivative, of $\mathbf{A}$ with respect to $t$ is then defined as the limiting value of the difference quotient

$$\frac{d\mathbf{A}(t)}{dt} = \lim_{\Delta t \to 0} \frac{\mathbf{A}(t + \Delta t) - \mathbf{A}(t)}{\Delta t} \qquad (7.1)$$

Now in the division of a vector by a scalar, vectorial properties persist; therefore the derivative of a vector with respect to a scalar variable is itself a vector. If **r** is the position vector drawn from a fixed point O to a moving point P, we have (figure 9) that

$$\mathbf{v} = \frac{d\mathbf{r}}{dt} \tag{7.2}$$

where **v** is the velocity vector of the point P.

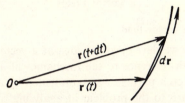

Fig. 9.—The displacement $d\mathbf{r} = \mathbf{v}\,dt$ of a moving point in time $dt$

As can be seen from (7.1), the calculation rules of differential calculus can be immediately adopted for the differentiation of the sum of vectors

$$\frac{d(\mathbf{A}+\mathbf{B})}{dt} = \frac{d\mathbf{A}}{dt} + \frac{d\mathbf{B}}{dt} \tag{7.3}$$

as well as for the product of a scalar and a vector

$$\frac{d(\alpha\mathbf{A})}{dt} = \frac{d\alpha}{dt}\mathbf{A} + \alpha\frac{d\mathbf{A}}{dt} \tag{7.4}$$

and for the inner product of two vectors

$$\frac{d(\mathbf{A}.\mathbf{B})}{dt} = \frac{d\mathbf{A}}{dt}.\mathbf{B} + \mathbf{A}.\frac{d\mathbf{B}}{dt} \tag{7.5}$$

The appropriate rule for the differentiation of the outer product is also valid, but care must be taken to write the factors in the correct sequence, since by exchanging the factors of the vector product its sign is changed:

$$\frac{d(\mathbf{A}\times\mathbf{B})}{dt} = \frac{d\mathbf{A}}{dt}\times\mathbf{B} + \mathbf{A}\times\frac{d\mathbf{B}}{dt} \tag{7.6}$$

# CHAPTER A II

## Vector Fields

### §8. Definition of a vector field

In chapter A I we developed the idea of a vector and the rules of vector algebra with reference to the mechanics of a material particle. The velocity of such a particle was represented by a single vector. In this chapter we turn to the problem of analysing the state of motion of an extended body or a liquid. Here the velocities of the various particles are in general different from one another, each particle requiring its own particular velocity vector. We have already become acquainted with an example of this in §5 where the motion of a rigid body in rotation around a fixed axis was described (equation (5.9)) by means of a position-dependent velocity vector.

In mathematical physics we speak of the *field* of a physical quantity when we consider the value of the quantity from the point of view of its dependence upon location in space. With the possible exception of certain surfaces, lines, or points, this dependence is assumed to be continuous. There are *scalar fields*, as for example density fields or temperature fields; and there are also *vector fields*, such as the velocity field in a liquid or the force field in gravitation.

The theory of vector fields was greatly advanced by the study of fluid motion, particularly by Helmholtz's fundamental researches on vortex motion. It was on this foundation that Maxwell built when he undertook to place the Faraday idea of the force field on a mathematical basis. To Maxwell, hydrodynamical analogies were more than purely mathematical pictures; hydrodynamic concepts of field mechanisms led him to the formulation of the "near-action" laws of the electromagnetic field.

We are therefore following the path of historical development when we allow ourselves to be guided by hydrodynamical analogies in our development of the mathematical theory of vector fields. In addition we shall also become acquainted with conceptual pictures borrowed from the mechanics of mass points in force fields.

## §9. The space derivative of a field quantity. The gradient

We imagine some physical quantity to be given as a field function defined for all points of space. We then go from a point described by the position vector **r**, with components $x$, $y$, $z$, to a neighbouring point having the position vector $\mathbf{r} + d\mathbf{r}$, and we inquire into the change of our field quantity for a position change $d\mathbf{r}$ having components $dx$, $dy$, $dz$.

If the field quantity under consideration is a *scalar* $\phi(\mathbf{r}) \equiv \phi(x, y, z)$, given as a continuous and differentiable function of position, then, in first approximation, the sought-for increase $d\phi = \phi(\mathbf{r} + d\mathbf{r}) - \phi(\mathbf{r})$ is equal to

$$d\phi = \frac{\partial \phi}{\partial x} dx + \frac{\partial \phi}{\partial y} dy + \frac{\partial \phi}{\partial z} dz \tag{9.1}$$

Since on the left side here we have a scalar quantity, the right side can be conceived as the scalar product of the vector $d\mathbf{r}$ with a second vector whose components are equal to the three derivatives $\dfrac{\partial \phi}{\partial x}, \dfrac{\partial \phi}{\partial y}, \dfrac{\partial \phi}{\partial z}$. We call this vector the *gradient* of the scalar $\phi$, and write

$$\operatorname{grad} \phi \equiv \mathbf{x} \frac{\partial \phi}{\partial x} + \mathbf{y} \frac{\partial \phi}{\partial y} + \mathbf{z} \frac{\partial \phi}{\partial z} \tag{9.2}$$

In a simple form, independent of any coordinate system, equation (9.1) then reads

$$d\phi = \operatorname{grad} \phi \cdot d\mathbf{r} \tag{9.3}$$

It shows that $\operatorname{grad} \phi$ stands perpendicular to the equipotential surface $\phi = \text{constant}$, and that it therefore has the direction of the maximum space rate of increase of $\phi$—for only in this case would the right-hand side of (9.3) vanish for every vector $d\mathbf{r}$ lying in the equipotential surface; that is, where $d\phi = 0$.

From the relationship

$$\mathbf{u} = \operatorname{grad} \phi, \quad \text{i.e.} \quad u_x = \frac{\partial \phi}{\partial x}, \quad u_y = \frac{\partial \phi}{\partial y}, \quad u_z = \frac{\partial \phi}{\partial z} \tag{9.4}$$

we obtain a vector field **u** from any arbitrary scalar field $\phi$. In the converse, however, it is not always possible to derive in the form (9.4), any arbitrary vector field from a scalar field. Rather, as can be shown from the component representation (9.4), the *conditions of integrability*

$$\frac{\partial u_x}{\partial y} = \frac{\partial u_y}{\partial x} \qquad \frac{\partial u_y}{\partial z} = \frac{\partial u_z}{\partial y} \qquad \frac{\partial u_z}{\partial x} = \frac{\partial u_x}{\partial z} \qquad (9.5)$$

must be fulfilled. In this case we say, on grounds later to be made manifest (see §11), that the **u**-field is irrotational.* Therefore, *the field of the gradient of a scalar is always irrotational.* Since, conversely, to every **u**-field whose components satisfy equations (9.5) there can be associated and explicitly assigned a scalar function $\phi$ in the sense of (9.4), we have the theorem that *every irrotational field can be expressed as the gradient of a scalar.*

By way of example, from the scalar field

$$\phi = -\frac{e}{r} \quad \text{with} \quad r = \sqrt{(x^2+y^2+z^2)} \qquad (9.6)$$

(with which later we shall often be concerned), we obtain from (9.4) the irrotational vector field $\mathbf{v} = \text{grad } \phi$, with

$$v_x = \frac{ex}{r^3} \qquad v_y = \frac{ey}{r^3} \qquad v_z = \frac{ez}{r^3} \qquad (9.7)$$

In hydromechanics (for $e > 0$) this describes the outflow of an incompressible fluid from a source located at the origin. Moreover, as the gradient of the "potential", $\phi$ naturally satisfies the integrability conditions (9.5).

These equations (9.5) are not satisfied for the velocity field

$$v_x = -\omega y \qquad v_y = \omega x \qquad v_z = 0 \qquad (9.8)$$

which, according to (5.9) represents the rigid rotation of a substance, such as a liquid, around the $z$-axis with the angular velocity $\omega$. This field is not irrotational. Such is, however, the case for the velocity field

$$v_x = -\frac{\omega a^2 y}{x^2+y^2} \qquad v_y = \frac{\omega a^2 x}{x^2+y^2} \qquad v_z = 0 \qquad (9.9)$$

which, according to hydromechanics, describes the streaming of an incompressible fluid around a cylinder of radius $a$ which rotates on its axis (the $z$-axis) with angular velocity $\omega$. Hence this field can be represented as the gradient of a scalar function $\phi$:

$$\mathbf{v} = \text{grad } \phi, \quad \text{with} \quad \phi = \omega a^2 \tan^{-1}\frac{y}{x} \qquad (9.10)$$

The presence of the inverse tangent makes this function multi-valued. We shall return to this question in §11.

Returning now to the initial problem of this paragraph, we ask what change a vector field quantity $\mathbf{u(r)} \equiv \mathbf{u}(x,y,z,)$ experiences in a change

---

* German: wirbelfrei = "eddy-free", or "vortex-free".—Tr.

represented by the vector $d\mathbf{r}$. The answer is provided by applying formula (9.1) or (9.3) to each one of the three components:

$$du_x = \operatorname{grad} u_x \cdot d\mathbf{r} = \frac{\partial u_x}{\partial x}dx + \frac{\partial u_x}{\partial y}dy + \frac{\partial u_x}{\partial z}dz$$

$$du_y = \operatorname{grad} u_y \cdot d\mathbf{r} = \frac{\partial u_y}{\partial x}dx + \frac{\partial u_y}{\partial y}dy + \frac{\partial u_y}{\partial z}dz \qquad (9.11)$$

$$du_z = \operatorname{grad} u_z \cdot d\mathbf{r} = \frac{\partial u_z}{\partial x}dx + \frac{\partial u_z}{\partial y}dy + \frac{\partial u_z}{\partial z}dz$$

These three equations can be combined into a single equation for the vector $d\mathbf{u}$ by making use of a differential operator first introduced by Hamilton. Its symbol is $\nabla$ (pronounced "del" or "nabla"); it is

$$\nabla \equiv \mathbf{x}\frac{\partial}{\partial x} + \mathbf{y}\frac{\partial}{\partial y} + \mathbf{z}\frac{\partial}{\partial z} \qquad (9.12)$$

This vector operator with the three components $\dfrac{\partial}{\partial x}, \dfrac{\partial}{\partial y}, \dfrac{\partial}{\partial z}$ makes possible a remarkably simple synoptical handling of many considerations of vector analysis if we calculate with it like an ordinary vector. Special care has to be taken, however, in the matter of sequence of the factors occurring in the formulae, because these involve differentiations. As a differential operator the $\nabla$-vector by itself can be characterized neither by magnitude nor by direction; it only assumes a definite value when it is applied to (i.e. operates on) some field quantity.

If it be applied to a scalar $\phi$, then by (9.2) the gradient

$$\nabla\phi \equiv \operatorname{grad} \phi \qquad (9.13)$$

is obtained. Accordingly (9.3) can be written in the form

$$d\phi = d\mathbf{r} \cdot \nabla\phi \equiv (d\mathbf{r} \cdot \nabla)\phi \qquad (9.14)$$

in which $d\mathbf{r} \cdot \nabla$ obviously represents a scalar differential operator, namely, the change with respect to the vector interval $d\mathbf{r}$ of the field quantity to be differentiated. In a similar way the three equations (9.11) can now be condensed into the vector equation

$$d\mathbf{u} = (d\mathbf{r} \cdot \nabla)\mathbf{u} \qquad (9.15)$$

in which the vector character of the right-hand side is required by **u**, while $d\mathbf{r}.\nabla$ means, as above, a scalar operator.

## §10.  The strength* of a source field and its divergence.  Gauss's theorem and Green's theorem

A concept fundamental to the whole of mathematical physics is the *surface integral*

$$\Phi = \iint_S u_n\, dS \tag{10.1}$$

in which $dS$ is an element of the surface $S$, and its normal has the direction of the unit vector **n**.

If, for example, in place of **u** above we have the velocity field **v** representing the flow of an *incompressible* fluid, then $\Phi$ is the volume of fluid which in unit time flows through $S$ by way of the side determined by **n**—for, in the time-interval $dt$, a volume of liquid $dS.v_n.dt$ passes through, which at time $t$ occupies a small skew cylinder of base $dS$, with axis in the direction of **v**, and having a height $v_n dt$. Carrying over to arbitrary vector fields **u** the concept of the quantity of fluid passing through $S$, we call the quantity $\Phi$ defined by (10.1) the "strength of the **u**-flow through $S$", or simply the *flux* of **u**.

In the following we shall often be concerned with the flux through a closed area, for which we shall write

$$\Phi = \oiint u_n\, dS \tag{10.2}$$

Herewith we establish once and for all that in this case the vector **n** shall always be understood to be the *outward-extending normal* of the enclosed region $V$.

In the special case considered above, namely, the flow of an incompressible fluid through a closed surface, there always flows from within the region outward exactly as much fluid as flows from the outside in. The total flux therefore vanishes, so long at least as at known places within $V$ no new liquid is created or destroyed. Places of the first kind we shall call (positive) *sources*; those of the second kind *sinks* (or negative sources). The total flux $\Phi$ through the closed surface $S$ can be considered as the appropriate measure of the productiveness or *strength* of the source system contained within $S$.

* German:  " Ergiebigkeit".—Tr.

As a special example we examine the flow going out from a single isolated source. Owing to symmetry, the flow of the fluid from the point source is radial and uniform in all directions. The same quantity of fluid therefore flows per second through any surface enclosing the source, and, in particular, through the surface of any sphere centred on the source. Since in the latter case $v_n$ is equal to the value $v$ of the velocity, which only depends upon $r$, we have

$$\Phi = \oiint v_n \, dS = 4\pi r^2 v = \text{constant}$$

Calling the constant $4\pi e$ (for reasons to be seen later), we have that $v = e/r^2$. For the velocity vector of the source stream we therefore obtain

$$\mathbf{v} = \frac{e\mathbf{r}}{r^3} \quad \text{i.e.} \quad v_x = \frac{ex}{r^3} \quad v_y = \frac{ey}{r^3} \quad v_z = \frac{ez}{r^3} \quad (10.3)$$

in agreement with formula (9.7) given above.

We thus have for the velocity field of the flow of an incompressible fluid from a source that

$$\Phi = \oiint v_n \, dS = 4\pi e \quad \text{or} \quad 0 \quad\quad (10.4)$$

depending upon whether the source lies inside or outside the closed region $S$. This result also holds for every other field of the form (10.3), and in particular for the *Coulomb field* of a point charge (compare §19)—for the physical consideration which has connected (10.3) with (10.4) is equivalent to a mathematical evaluation of the surface integral (10.4) for the field (10.3).

Now, however, we come to the case of a field $\mathbf{u}$ with spatially distributed sources. There exists here the task of finding a measure of *specific strength*. In the following we call this the *divergence* of the **u**-field. We designate it

$$\text{div } \mathbf{u}$$

and define it conceptually as the quotient of the total flux of **u** through a small closed surface by the volume $V$ thus enclosed, for the limiting case $V \to 0$:

$$\text{div } \mathbf{u} = \lim_{V \to 0} \frac{1}{V} \oiint u_n \, dS \quad\quad (10.5)$$

It is significant that the quantity thus defined is independent of the shape or form of the volume considered. This can be shown by direct evaluation of the right-hand side of (10.5). Rewriting the surface integral by means of (3.7), we have

$$\oiint u_n \, dS = \oiint \{u_x \cos \langle \mathbf{x}, \mathbf{n} \rangle + u_y \cos \langle \mathbf{y}, \mathbf{n} \rangle + u_z \cos \langle \mathbf{z}, \mathbf{n} \rangle\} \, dS \quad (10.6)$$

For finding the integral over the first summand on the right we imagine the enclosed volume $V$ divided into little prisms (figure 10) extending in the $x$-direction, and having a rectangular cross-section $dy \, dz$. One

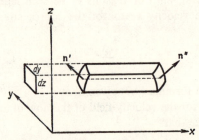

Fig. 10.—For the proof of Gauss's theorem

particular such volume element cuts from the surface of $V$ two surface elements $dS'$ and $dS''$, with normals $\mathbf{n}'$ and $\mathbf{n}''$. Moreover,

$$dS' \cos \langle \mathbf{x}, \mathbf{n}' \rangle = -dy \, dz, \quad \text{and} \quad dS'' \cos \langle \mathbf{x}, \mathbf{n}'' \rangle = +dy \, dz$$

For the prism considered, its contribution to the first summand on the right side of (10.6) is therefore

$$\{(u_x)_{x=x''} - (u_x)_{x=x'}\} \, dy \, dz = \int_{x'}^{x''} \frac{\partial u_x}{\partial x} \, dx \, dy \, dz$$

where $x'$ and $x''$ mean the $x$-coordinates of the end faces of the prism (with $x' < x''$). If, now, we sum over all prisms into which we imagine $V$ to be divided, we obtain from the first summand of (10.6) the relation

$$\oiint u_x \cos \langle \mathbf{x}, \mathbf{n} \rangle \, dS = \iiint \frac{\partial u_x}{\partial x} \, dx \, dy \, dz = \iiint \frac{\partial u_x}{\partial x} \, dv$$

which remains correct even when some of the prisms considered cut out of $S$ four, six, or even more surface elements. The volume element is designated for brevity here by $dv$. Carrying out the corresponding transformations for the other two summands of (10.6) we obtain the very important result called *Gauss's theorem:*

$$\oiint u_n \, dS = \iiint \left\{ \frac{\partial u_x}{\partial x} + \frac{\partial u_y}{\partial y} + \frac{\partial u_z}{\partial z} \right\} dv \qquad (10.7)$$

Thus from (10.5) we immediately have as the component form of div **u** the expression

$$\operatorname{div} \mathbf{u} = \frac{\partial u_x}{\partial x} + \frac{\partial u_y}{\partial y} + \frac{\partial u_z}{\partial z} \qquad (10.8)$$

independent of any relationship to the surface $S$. Corresponding to its derivation from (10.5), the divergence of the field **u** is a scalar quantity independent of any coordinate system. This also follows from the fact that div **u** can obviously be written as the scalar product of the ∇-operator (9.12) with the vector **u**:

$$\operatorname{div} \mathbf{u} = \nabla . \mathbf{u} \qquad (10.9)$$

From equation (10.8) Gauss's theorem can be written

$$\oiint u_n \, dS = \iiint \operatorname{div} \mathbf{u} \, dv \qquad (10.10)$$

This form, incidentally, permits obtaining div **u** directly from the defining equation (10.5) without the necessity of going by way of the component representation. For this the volume $V$ is broken up into small volumes $v_1, v_2, \ldots$ so that, for each of these volumes $v_j$, with surface $S_j$, the formula

$$\oiint_{S_j} u_n \, dS = \operatorname{div} \mathbf{u} \, (v_j)$$

valid in the limit $v_j \to 0$, is written down, and this is summed over the volume. The sum on the right-hand side is then just the volume integral of (10.10) if for the small volumes $v_j$ the differential $dv$ be substituted. On the left side all contributions corresponding to a surface element shared by two neighbouring volume elements cancel, for the simple reason that they always occur twice but with mutually opposite sign. Thus, in the addition of the contributions on the left side only the surface integral (10.10) over the whole surface $S$ remains.

We summarize the foregoing results of this section as follows: for continuously distributed sources the total flux of a vector field **u** over a closed surface can be calculated by Gauss's theorem as the volume integral of the divergence of **u**. The latter is therefore the proper measure of the strength or productiveness of the **u**-field. If div **u** vanishes over all space, the total flux through every closed volume is equal to zero. A finite value of total flux can however be obtained for div **u** = 0 throughout, under the assumption of one or more point sources at which the flow field is singular. The same also holds in the case for which an otherwise divergence-free flow possesses line-distributed or surface-distributed sources.

Given, for example, a plane, which we choose as the $xy$-plane, uniformly spread over with sources of an incompressible fluid, these sources will then discharge perpendicularly to the plane, and from both sides, with constant velocity

$$u_x = 0 \qquad u_y = 0 \qquad u_z = \begin{cases} +u & \text{for} \quad z > 0 \\ -u & \text{for} \quad z < 0 \end{cases} \quad (10.11)$$

Evidently div **u** = 0 for this field in all space, even up to the plane. On the other hand the source system possesses a finite surface divergence which, analogously to (10.5), we define by finding the total flux through a closed surface cutting the plane, and dividing this by the quantity of cut surface $S$, and finally, keeping the form of the included volume $V$ the same, going over to the limit $V \to 0$, thus

$$\lim_{V \to 0} \frac{1}{S} \oiint u_n \, dS \qquad (10.12)$$

If, as closed surface, we choose say a flat cylinder (figure 11) whose height is small compared with the dimensions of its two ends (these being

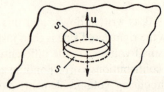

Fig. 11.—For the definition of the surface divergence

parallel to the plane and each of area $S$), then the total flux through this box is equal to $uS - (-u)S = 2uS$, and the surface divergence, from (10.12), equals $2u$. Similarly, for an arbitrary surface containing sources, the surface divergence follows the discontinuity on the surface possessed by the normal component of **u**. This is because even in any arbitrary **u**-field only the two end-surfaces of the cylinder give a finite contribution to the surface divergence; the curved surface owing to its small height furnishes a contribution that, in the limit, is vanishingly small.

Finally we consider several important transformations of Gauss's theorem (10.10) given by Green. If $\mathbf{u} = \psi\mathbf{w}$, $\psi$ being a scalar and **w** a second vector, then, from the product rule of differentiation,

$$\operatorname{div}\mathbf{u} = \operatorname{div}(\psi\mathbf{w}) = \psi\operatorname{div}\mathbf{w} + \operatorname{grad}\psi \cdot \mathbf{w}$$

and thus, on account of (10.10),

$$\oiint \psi w_n \, dS = \iiint \{\psi\operatorname{div}\mathbf{w} + \operatorname{grad}\psi \cdot \mathbf{w}\}\, dv \qquad (10.13)$$

If, further, the vector **w** is representable as the gradient of a second scalar $\phi$, then

$$\mathbf{w} = \operatorname{grad}\phi$$

and

$$\operatorname{div}\mathbf{w} = \operatorname{div}\operatorname{grad}\phi = \frac{\partial^2\phi}{\partial x^2} + \frac{\partial^2\phi}{\partial y^2} + \frac{\partial^2\phi}{\partial z^2}$$

This sum of second derivatives of a function, making its appearance here, is called the Laplace operator, $\nabla^2$:

$$\frac{\partial^2}{\partial x^2} + \frac{\partial^2}{\partial y^2} + \frac{\partial^2}{\partial z^2} \equiv \nabla \cdot \nabla = \nabla^2 \qquad (10.14)$$

Equation (10.13) thus contains the relation

$$\oiint \psi \frac{\partial\phi}{\partial n}\, dS = \iiint \{\psi\nabla^2\phi + \operatorname{grad}\psi \cdot \operatorname{grad}\phi\}\, dv \qquad (10.15)$$

valid for two arbitrary functions of position, $\phi$ and $\psi$, which are continuous within $V$, finite, and twice-differentiable. To simplify the writing in this equation we have set

$$(\operatorname{grad}\phi)_n \equiv \frac{\partial\phi}{\partial n} \qquad (10.16)$$

If, finally, from (10.15) we subtract a similar equation obtained by interchanging $\phi$ and $\psi$ throughout, we obtain

$$\oiint\left(\psi\frac{\partial\phi}{\partial n}-\phi\frac{\partial\psi}{\partial n}\right)dS = \iiint\{\psi\nabla^2\phi-\phi\nabla^2\psi\}\,dv \quad (10.17)$$

Equations (10.15) and (10.17) are called *Green's theorems*. In electrodynamics we shall often make use of them.

## §11.  The line integral and the curl.  Stokes's theorem

The line integral in a vector field **u** is fully as important in mathematical physics as the surface integral (10.1).  We connect two arbitrary points 1 and 2 by an arbitrary curve which we imagine to consist of individual line elements $d\mathbf{r}$ each having the direction of the local tangent line (in the sense from 1 to 2).  For each line element we form the scalar product $\mathbf{u}.d\mathbf{r}$ at the corresponding location and we sum over all $d\mathbf{r}$. In the limiting case when $d\mathbf{r}\to 0$ this gives the line integral

$$\int_1^2 \mathbf{u}.d\mathbf{r} \quad (11.1)$$

An important example of this procedure is the integral

$$\int_1^2 \mathbf{F}.d\mathbf{r}$$

which, in particle mechanics, represents the work performed by the force field **F** in the displacement of a mass point from 1 to 2.  This is equal to the increase of kinetic energy of the mass point on the path from 1 to 2; for, from the equation of motion

$$m\frac{d\mathbf{v}}{dt} = \mathbf{F} \quad (11.2)$$

there follows by scalar multiplication by **v** that

$$\frac{d}{dt}\left(\frac{m\mathbf{v}^2}{2}\right) = \mathbf{F}.\mathbf{v} = \mathbf{F}.\frac{d\mathbf{r}}{dt} \quad \text{therefore} \quad \left(\frac{m\mathbf{v}^2}{2}\right)_2 - \left(\frac{m\mathbf{v}^2}{2}\right)_1 = \int_1^2\mathbf{F}.\frac{d\mathbf{r}}{dt}\,dt = \int_1^2\mathbf{F}.d\mathbf{r}$$

In the special case in which the vector **u** in (11.1) can be expressed as the gradient of a scalar $\phi$ which is everywhere finite, continuous, and single-valued, the line integral (11.1) becomes independent of the choice of actual path between 1 and 2; for in this case, from (9.3) we have

$$\mathbf{u}.d\mathbf{r} = \text{grad } \phi.d\mathbf{r} = d\phi$$

so that

$$\int_1^2 \mathbf{u}.d\mathbf{r} = \int_1^2 d\phi = \phi(\mathbf{r}_2)-\phi(\mathbf{r}_1) \quad (11.3)$$

In this case the line integral has the same value for two paths having the same beginning point and end point. Thus for every closed path the line integral of the gradient vanishes:

$$\oint \mathbf{u} \cdot d\mathbf{r} = \oint d\phi = 0 \qquad (11.4)$$

This, for example, is the case for static fields in particle mechanics, where the field can be written in the form $\mathbf{F} = -\text{grad } \phi$. The work integral is then independent of path, being $\phi(\mathbf{r}_1) - \phi(\mathbf{r}_2)$, and vanishes for every closed path. The energy theorem (11.2) then has the form

$$\left(\frac{m\mathbf{v}^2}{2}\right)_1 + \phi(\mathbf{r}_1) = \left(\frac{m\mathbf{v}^2}{2}\right)_2 + \phi(\mathbf{r}_2) = \text{constant}$$

For (11.4) to be valid it is necessary that $\mathbf{u}$ be derivable from a *single-valued* scalar function $\phi$. For multi-valued $\phi$-functions the integral $\oint \mathbf{u} \cdot d\mathbf{r}$ can have a finite value everywhere.

As an example of this case we consider the flow field of a rotating cylinder given by (9.9), and described in (9.10) by

$$\mathbf{v} = \text{grad } \phi, \quad \text{with} \quad \phi = a^2\omega \tan^{-1}\frac{y}{x}$$

Here $\oint \mathbf{v} \cdot d\mathbf{r}$ vanishes for every closed path not enclosing the cylinder, for in this case the initial value and the terminal value are the same. If, however, the path of integration leads once around the cylinder in the positive sense, then, owing to the multi-valued character of the inverse-tangent function, the terminal value of $\phi$ becomes larger by the factor $2\pi a^2\omega$ than the initial value. In this case the line integral around the closed path acquires the constant value

$$\oint \mathbf{v} \cdot d\mathbf{r} = 2\pi a^2\omega$$

This is at once evident for an integration path parallel to the $xy$-plane and lying right on the surface of the cylinder—for in this case by (9.9) $\mathbf{v}$ has the direction of $d\mathbf{r}$, and the constant magnitude $v = a\omega$.

Since the integral $\oint \mathbf{v} \cdot d\mathbf{r}$ can be looked upon as providing a measure of the flow around the cylinder, it is given the name *circulation* and written

$$Z = \oint \mathbf{v} \cdot d\mathbf{r} \qquad (11.5)$$

We shall make general use of this designation in the theory of vector fields.

From the example considered we readily recognize the correctness of the following theorem: When the circulation is constant over a path surrounding a body (or, indeed, surrounding only a certain singular line), i.e. when this circulation is independent of the shape of the path chosen for its evaluation, then, for every integration path not surrounding the body or the singular line, the circulation vanishes. This theorem represents the counterpart of the law given in (10.4) for source fields. We shall meet it again in the theory of magnetic fields produced by currents.

As an example, the circulation of the flow field given by (9.8) for the rigid rotation of a liquid around the $z$-axis, with angular velocity $\omega$, is not constant. For this case we have

$$Z = \oint v_x dx + \oint v_y dy = \omega \left\{ -\oint y\,dx + \oint x\,dy \right\} = 2\omega S$$

since both $-\oint y\,dx$ and $\oint x\,dy$ represent the area of the surface $S$ which the projection of the path of integration on the $xy$-plane (in the positive sense) would enclose. The circulation therefore increases here with the magnitude of the surface $S$.

In the general case of an arbitrary **u**-field, *for sufficiently small surface area $S$*, the circulation around the periphery of $S$ is proportional to $S$. We show this now for a plane area (figure 12) parallel to the $xy$-plane

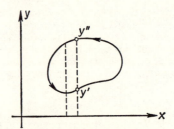

Fig. 12.—The line-integral for a plane area

and of area $S$. We move round it in the positive sense, i.e. in such sense that the circuiting, together with an advance in the positive $z$-direction, corresponds to the motion of a right-hand screw. The line integral we obtain is

$$\oint \mathbf{u}\cdot d\mathbf{r} = \oint (u_x\,dx + u_y\,dy)$$

We calculate the first summand $\oint u_x dx$ in such a way that both path elements (i.e. the one at $y''$ and the one at $y'$, where $y'' > y'$) associated with a given section $dx$ are considered jointly. Since in the journey around, $dx$ at $y'$ is to be taken positive, while $dx$ at $y''$ is negative, the contribution of these two path portions to $\oint u_x dx$ is equal to

$$dx\left\{(u_x)_{y=y'}-(u_x)_{y=y''}\right\} = -dx\int_{y'}^{y''}\frac{\partial u_x}{\partial y}dy$$

In all, therefore,

$$\oint u_x\,dx = -\iint\frac{\partial u_x}{\partial y}\,dx\,dy = -\iint\frac{\partial u_x}{\partial y}\,dS$$

in which the right side is to be integrated over the whole surface $S_z$. By an analogous treatment of the second summand $\oint u_y dy$ we thus obtain

$$\oint \mathbf{u}.\,d\mathbf{r} = \iint\left(\frac{\partial u_y}{\partial x}-\frac{\partial u_x}{\partial y}\right)dS \tag{11.6}$$

If now $S_z$ is so small that within $S_z$ the integrand does not markedly change, then $\oint \mathbf{u}.\,d\mathbf{r}$ becomes proportional to $S_z$. As a measure of *small-scale circulation*, then, it would be reasonable to take the limiting value of the quotient: circulation divided by enclosed surface $S_z$—in the limit, that is, when $S_z \to 0$. We call this limiting value the *z-component of the curl of* **u**, and write

$$(\mathrm{curl}\,\mathbf{u})_z = \lim_{S_z\to 0}\frac{1}{S_z}\oint \mathbf{u}.\,d\mathbf{r} \tag{11.7}$$

On account of (11.6) we have also that

$$(\mathrm{curl}\,\mathbf{u})_z = \frac{\partial u_y}{\partial x}-\frac{\partial u_x}{\partial y} \tag{11.8$a$}$$

By a similar procedure we obtain corresponding quantities $(\mathrm{curl}\,\mathbf{u})_x$ and $(\mathrm{curl}\,\mathbf{u})_y$ when we orient the section of area to be circuited so that its normal, corresponding to the right-hand screw situation, is parallel to, not the $z$-direction, but the $x$-direction, and then the $y$-direction:

$$(\mathrm{curl}\,\mathbf{u})_x = \frac{\partial u_z}{\partial y}-\frac{\partial u_y}{\partial z}\qquad (\mathrm{curl}\,\mathbf{u})_y = \frac{\partial u_x}{\partial z}-\frac{\partial u_z}{\partial x} \tag{11.8$b$}$$

The three quantities $(\mathrm{curl}\,\mathbf{u})_x$, $(\mathrm{curl}\,\mathbf{u})_y$, $(\mathrm{curl}\,\mathbf{u})_z$ represent just the components of a vector curl **u**, or "the curl of **u**"; and the three equations (11.8) can be assembled by means of the del-operator into a single-vector equation

$$\mathrm{curl}\,\mathbf{u} = \nabla \times \mathbf{u} \tag{11.9}$$

which can also be looked upon as the defining equation of curl **u**.

As an example, we have, according to (11.8), for the field (9.8) representing the rigid rotation of a liquid around the $z$-axis with constant angular velocity $\omega$, that

$$(\text{curl } \mathbf{v})_x = 0 \qquad (\text{curl } \mathbf{v})_y = 0 \qquad (\text{curl } \mathbf{v})_z = 2\omega$$

We call such a field for which the curl does not vanish identically, a *vortex field* or eddy field.   In contrast to this the field (9.9) representing the flow of an incompressible liquid around a rotating cylinder gives

$$\text{curl } \mathbf{v} = 0$$

In spite of its finite circulation around the cylinder this field is *vortex-free*.

For a sufficiently small arbitrarily oriented surface area $dS$ with normal $\mathbf{n}$, we have now in first approximation, owing to the vector character of $\mathbf{u}$, that

$$\oint \mathbf{u} \cdot d\mathbf{r} = (\text{curl } \mathbf{u})_n \, dS \qquad\qquad (11.10)$$

in which the line integral around the boundary curve of $dS$ is to be taken in the sense, with respect to $\mathbf{n}$, of a right-hand screw.   Summing this formula (see figure 13) over all elements $dS_1, dS_2, \ldots$ of an arbitrary surface $S$ on which curl $\mathbf{u}$ exists and is finite, we obtain *Stokes's theorem*

$$\oint \mathbf{u} \cdot d\mathbf{r} = \iint (\text{curl } \mathbf{u})_n \, dS \qquad\qquad (11.11)$$

in which the line integral is to extend around the boundary curve of the surface $S$—for, in the summation over the circuital integrals of the

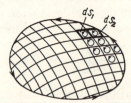

Fig. 13.—For the proof of Stokes's theorem

individual surface elements, the contributions from a border portion shared between two elements (for example the border portion between $dS_1$ and $dS_2$) cancel because they occur in pairs, but always with mutually opposite signs. In the summation, therefore, only the contributions of the boundary curve $S$ remain.

Thus, according to Stokes, the circulation $Z$ of $\mathbf{u}$ around a given closed curve is equal to the flux $\Phi$ of curl $\mathbf{u}$ through a surface $S$ bounded by the curve. It is of no importance how the surface through the boundary curve lies. For two different surfaces $S_1$ and $S_2$ having the same boundary curve we have

$$\Phi_1 = \Phi_2, \quad \text{i.e.} \quad \iint_{S_1} (\text{curl } \mathbf{u})_n \, dS = \iint_{S_2} (\text{curl } \mathbf{u})_n \, dS \quad (11.12)$$

Both surfaces together, however, completely bound a region of space $V$. If in the last equation we invert the direction of the surface normal on one of the surfaces, say $S_2$, it follows from (11.12) that

$$\oiint (\text{curl } \mathbf{u})_n \, dS = 0$$

in which the surface integral is now taken over the total surface of $V$. For the vector curl $\mathbf{u}$, then, the total flux emerging from any arbitrary region of space is equal to zero; by Gauss's theorem this vector is therefore source-free, i.e.

$$\text{div curl } \mathbf{u} = 0 \qquad (11.13)$$

This also follows immediately from the component representation (11.8), or even more simply from the vector form (11.9):

$$\text{div curl } \mathbf{u} = \nabla . (\nabla \times \mathbf{u}) = (\nabla \times \nabla) . \mathbf{u} = 0, \quad \text{because} \quad \nabla \times \nabla = 0$$

Conversely, for every vector $\mathbf{v}$ whose divergence vanishes there exists a vector field $\mathbf{u}$ whose curl is $\mathbf{v}$:

$$\boxed{\text{From } \text{div } \mathbf{v} = 0 \text{ it follows that } \mathbf{v} = \text{curl } \mathbf{u}} \quad (11.14)$$

The proof of this statement will be given in §12(*b*) by direct calculation of the vector field $\mathbf{u}$ from the given eddy distribution (or vorticity) of $\mathbf{v}$.

If, further, $\mathbf{u}$ is the gradient of a scalar $\phi$ which is single-valued and determinate within the region of integration, then, for every closed curve within the region of integration, the circulation vanishes. Consequently this field, by Stokes's theorem, is vortex-free and irrotational, i.e.

$$\text{curl grad } \phi = 0 \qquad (11.15)$$

This relationship also follows immediately for $\mathbf{u} = \text{grad}\,\phi$ either from the coordinate representation (11.8) or from the vector form (11.9). As, incidentally, a comparison of (11.8) with the relationships (9.5) shows, the equation

$$\text{curl}\,\mathbf{u} = 0$$

is exactly the requirement that $\mathbf{u}$ shall be expressible as the gradient of a scalar:

$$\boxed{\text{From}\;\text{curl}\,\mathbf{u} = 0\;\text{it follows that}\;\mathbf{u} = \text{grad}\,\phi} \qquad (11.16)$$

In §12(*a*) this function will be calculated from the given sources of $\mathbf{u}$.

For later use we calculate here the curl of the curl of a vector $\mathbf{u}$. By using the $\nabla$-form and relationship (6.5) we find that

$$\text{curl}\,\text{curl}\,\mathbf{u} = \nabla \times (\nabla \times \mathbf{u}) = \nabla(\nabla.\mathbf{u}) - (\nabla.\nabla)\mathbf{u}$$

i.e. in the customary formulation

$$\text{curl}\,\text{curl}\,\mathbf{u} = \text{grad}\,\text{div}\,\mathbf{u} - \nabla^2\mathbf{u} \qquad (11.17)$$

Finally, we shall later also need the relationship

$$\text{div}(\mathbf{u} \times \mathbf{v}) = \mathbf{v}.\text{curl}\,\mathbf{u} - \mathbf{u}.\text{curl}\,\mathbf{v} \qquad (11.18)$$

which is valid for arbitrary vectors $\mathbf{u}$ and $\mathbf{v}$, and may also be easily verified by means of the $\nabla$-form and relationship (6.1).

## §12.  Calculation of a vector field from its sources and vortices

In the two preceding sections we have seen how the scalar $\text{div}\,\mathbf{u}$ and the vector $\text{curl}\,\mathbf{u}$ can be obtained from a known vector field $\mathbf{u}$ by differentiation processes. We now treat the reverse problem: to determine a vector field $\mathbf{u}$ from its prescribed sources and vortices. We therefore seek the field for which the expressions

$$\text{div}\,\mathbf{u} = 4\pi\rho \qquad \text{curl}\,\mathbf{u} = 4\pi\mathbf{c} \qquad (12.1)$$

hold, with a given scalar function $\rho \equiv \rho(\mathbf{r})$ and vector function $\mathbf{c} \equiv \mathbf{c}(\mathbf{r})$, both different from zero in any finite region of space. Here, however, $\mathbf{c}$ cannot be chosen entirely arbitrarily. On account of (11.13) we must have

$$\text{div}\,\mathbf{c} = 0 \qquad (12.2)$$

The factors $4\pi$ in (12.1) have been introduced to bring the expressions into agreement with corresponding expressions in Gaussian units.

In order to integrate equations (12.1) we separate the desired vector field $\mathbf{u}$ into an irrotational field $\mathbf{u}_1$ with the prescribed sources, and a source-free field $\mathbf{u}_2$ having the prescribed vortices, thus:

$$\mathbf{u} = \mathbf{u}_1 + \mathbf{u}_2 \tag{12.3}$$

with

$$\operatorname{div} \mathbf{u}_1 = 4\pi\rho \qquad \operatorname{curl} \mathbf{u}_1 = 0 \tag{12.4}$$

$$\operatorname{div} \mathbf{u}_2 = 0 \qquad \operatorname{curl} \mathbf{u}_2 = 4\pi\mathbf{c} \tag{12.5}$$

We have thus divided the original problem into two partial problems: the determination of an irrotational field from its prescribed sources, and the determination of a source-free field from its prescribed vortices.

*(a) Calculation of an irrotational vector field from its source field.*
Since $\mathbf{u}_1$ is irrotational, we can write

$$\mathbf{u}_1 = -\operatorname{grad} \phi \tag{12.6}$$

in which the minus sign has been introduced to bring (12.6) into agreement with a corresponding formula in electrodynamics. Following, then, from the first equation (12.4) for the "scalar potential", we have *Poisson's equation*

$$\nabla^2 \phi = -4\pi\rho \tag{12.7}$$

For its solution we make use of Green's theorem (10.17) in which we take $\phi$ equal to the desired potential function, and $\psi$ we set equal to $1/r$, i.e. equal to the reciprocal of the distance $r$ from the first point of integration to the field point P at which we wish to evaluate $\phi$:

$$\oiint \left( \frac{1}{r} \frac{\partial \phi}{\partial n} - \phi \frac{\partial (1/r)}{\partial n} \right) dS = \iiint \left\{ \frac{1}{r} \nabla^2 \phi - \phi \nabla^2 \frac{1}{r} \right\} dv \tag{12.8}$$

For bounding the region of integration we choose on the one hand a closed surface at infinity, and on the other hand a small spherical surface around the field point whose radius $r_0$ we wish subsequently to let go to zero. We now assume that for infinitely increasing $r$, the solution we are seeking falls off at least like $1/r$. The larger outer surface furnishes no contribution to the left side of (12.8) since in this case the entire integrand falls off at least as $1/r^3$, while the surface grows only as $r^2$. A finite contribution to the left side of (12.8) is however furnished by the spherical surface around the field points. Here, since **n** points outward from the region of integration and therefore has the direction towards the field point,

$$\frac{\partial \phi}{\partial n} = -\left(\frac{\partial \phi}{\partial r}\right)_{r=r_0} \qquad \frac{\partial (1/r)}{\partial n} = -\left(\frac{\partial (1/r)}{\partial r}\right)_{r=r_0} = \frac{1}{r_0^2}$$

Since $\phi$ and grad $\phi$ certainly do not become infinite at the field point, the contribution of the first term on the left in (12.8) coming from the spherical surface around the field point does indeed vanish in the limit $r_0 \to 0$. From the second term however we obtain in the limit the finite value $-4\pi\phi(\mathrm{P})$.

On the right side of (12.8), $\nabla^2\phi$ is prescribed by (12.7), while $\nabla^2(1/r)$ vanishes everywhere within the region of integration. We therefore obtain

$$-4\pi\phi(\mathrm{P}) = -\iiint \frac{4\pi\rho}{r}\, dv$$

or, if we designate the position vector of the field point by **r**, and that of the integration point by **r′**, so that, instead of the distance $r$ and the volume $dv$, we now substitute

$$|\mathbf{r}-\mathbf{r}'| = \sqrt{[(x-x')^2+(y-y')^2+(z-z')^2]} \quad \text{and} \quad dv' = dx'\,dy'\,dz'$$

we obtain

$$\phi(\mathbf{r}) = \iiint \frac{\rho(\mathbf{r}')}{|\mathbf{r}-\mathbf{r}'|}\, dv' \tag{12.9}$$

Introducing this expression in (12.6), we obtain the desired vector $\mathbf{u}_1$ simply by differentiating with respect to the components of **r**.

We observe that, according to equation (12.9), the $\phi$-function we have found actually falls off for large $r$ at least like $1/r$; thus for very large $r$ we can neglect **r′** in the denominator $|\mathbf{r}-\mathbf{r}'|$ in the first approximation to **r**, and then, for the whole denominator, place $1/r$ outside the integral. With this the above assumption is justified in retrospect and, as a consequence, the expression (12.9) is shown to be a solution of the differential equation (12.7).

Although equation (12.9) was derived for the case of a spatially extended source field, we can also obtain from it the field of point sources, line sources, or surface-distributed sources. If, for example, there were a point source of strength $e_0$ at the point $\mathrm{P}_0$, with position-vector $\mathbf{r}_0$, we could then proceed on the basis that only in the immediate neighbourhood of $P_0$ has $\rho$ a value different from zero. We may then write the denominator in (12.9) as $1/|\mathbf{r}-\mathbf{r}_0|$, place it before the integral and, with

$$\iiint \rho(\mathbf{r}')\,dv' = e_0 \qquad (12.10)$$

obtain as the scalar potential

$$\phi(\mathbf{r}) = \frac{e_0}{|\mathbf{r}-\mathbf{r}_0|} \qquad (12.11)$$

This result, which is not changed by a subsequent passage to the limiting case of vanishingly small extension of the charge distribution, agrees (to within a change in the sign required by the minus sign, and the different location of the point source) with formula (9.6) for the potential of a point source. Conversely, since the fields superpose according to the rules of vector addition on account of the linearity of equation (12.4) in $\mathbf{u}$ and $\rho$, we can obtain equation (12.9) from (12.11) by regarding the sources contained in the individual volume elements as actually point sources of strength $\rho dv$, calculating their contributions to the potential according to (12.11), and then summing over all volume elements.

In a similar way the fields of line or surface-distributed sources can also be treated. Examples of this kind will be met later.

Incidentally, according to (12.6), we could have added to the potential (12.9) a constant without changing the field $\mathbf{u}_1$. For all problems however in which all charges are at a finite distance, we wish to establish the potential through the stipulation already realized in (12.9), that the potential shall vanish at infinity.

### (b) Calculation of a source-free vector field from its vortex field

For the integration of equation (12.5) we can represent $\mathbf{u}_2$, owing to its source-free character (11.14), as the curl of a new vector $\mathbf{A}$:

$$\mathbf{u}_2 = \operatorname{curl} \mathbf{A} \qquad (12.12)$$

We call the vector $\mathbf{A}$ so defined the *vector potential* of $\mathbf{u}_2$. Just as we can add an arbitrary constant to $\phi$ without changing $\mathbf{u}_1$, so, similarly, we may add an arbitrary irrotational vector to $\mathbf{A}$ without changing the value of $\mathbf{u}_2$. Thus we can dispose of this additional field so that the resultant A-field becomes source-free:

$$\operatorname{div} \mathbf{A} = 0 \qquad (12.13)$$

By introducing (12.12) into the second equation (12.5) we obtain

$$\operatorname{curl} \operatorname{curl} \mathbf{A} = 4\pi\mathbf{c}$$

or, on account of calculation rule (11.17) and the prescription (12.13)

$$\nabla^2 \mathbf{A} = -4\pi\mathbf{c} \qquad (12.14)$$

in complete analogy with the Poisson equation (12.7) for the scalar potential $\phi$. This analogy allows us, by considering (12.9), to write the solution of (12.14) directly:

$$\mathbf{A}(\mathbf{r}) = \iiint \frac{\mathbf{c}(\mathbf{r}')}{|\mathbf{r}-\mathbf{r}'|} \, dv' \qquad (12.15)$$

From this, in accordance with (12.12), $\mathbf{u}_2$ follows by a single differentiation.

We still have to convince ourselves that the vector field given by $\mathbf{A}$ is actually source-free, as was required by (12.13). Now obviously

$$\operatorname{div} \mathbf{A} = \iiint \mathbf{c}(\mathbf{r}') \cdot \operatorname{grad} \frac{1}{|\mathbf{r}-\mathbf{r}'|} \, dv' = -\iiint \mathbf{c}(\mathbf{r}') \cdot \operatorname{grad}' \frac{1}{|\mathbf{r}-\mathbf{r}'|} \, dv'$$

in which grad′ means the gradient taken with respect to the vector $\mathbf{r}'$, the following fact having been used:

$$\frac{\partial}{\partial x}\left(\frac{1}{|\mathbf{r}-\mathbf{r}'|}\right) = -\frac{\partial}{\partial x'}\left(\frac{1}{|\mathbf{r}-\mathbf{r}'|}\right), \quad \dots, \quad \dots$$

We have now

$$\mathbf{c}(\mathbf{r}') \cdot \operatorname{grad}' \frac{1}{|\mathbf{r}-\mathbf{r}'|} = \operatorname{div}'\left(\frac{\mathbf{c}(\mathbf{r}')}{|\mathbf{r}-\mathbf{r}'|}\right) - \frac{1}{|\mathbf{r}-\mathbf{r}'|} \operatorname{div}' \mathbf{c}(\mathbf{r}')$$

The second term on the right-hand side vanishes here because of (12.2), so that by Gauss's theorem we have

$$\operatorname{div} \mathbf{A} = -\iiint \operatorname{div}'\left(\frac{\mathbf{c}(\mathbf{r}')}{|\mathbf{r}-\mathbf{r}'|}\right) dv' = -\oiint \frac{c_n(\mathbf{r}')}{|\mathbf{r}-\mathbf{r}'|} \, dS'$$

Since, by hypothesis, the entire vortex system must remain within finite bounds, $c_n$ vanishes on the infinitely remote surface of the region of integration. Thus **A** is in fact source-free.

## §13. Orthogonal curvilinear coordinates

Many calculations in electrodynamics can be simplified by choosing, instead of a Cartesian coordinate system, another kind of system which takes advantage of the symmetry relations involved in the particular problem under study. Thus, for example, the calculation of the field of an electrically charged ellipsoid is certainly considerably easier to carry out in elliptical coordinates than in Cartesians (see §22).

Let the new coordinates $\xi_1$, $\xi_2$, $\xi_3$ be defined so that the Cartesian coordinates $x$, $y$, $z$ are known functions of $\xi_1$, $\xi_2$, $\xi_3$:

$$x = x(\xi_1, \xi_2, \xi_3) \qquad y = y(\xi_1, \xi_2, \xi_3) \qquad z = z(\xi_1, \xi_2, \xi_3)$$

We restrict ourselves to the case for which the three families of curves $\xi_1 = $ constant, $\xi_2 = $ constant, $\xi_3 = $ constant are mutually *orthogonal*. We then obtain for the distance between two neighbouring points, that is, for the line-element $ds$, the relation

$$ds^2 = dx^2 + dy^2 + dz^2 = h_1^2 \, d\xi_1^2 + h_2^2 \, d\xi_2^2 + h_3^2 \, d\xi_3^2$$

in which $h_1$, $h_2$, $h_3$ are, in general, functions of $\xi_1$, $\xi_2$, $\xi_3$. We require in addition that the new system, like the old, shall be a right-hand system.

We consider now an infinitesimal parallelopiped (figure 14) whose bounding planes coincide with the surfaces $\xi_1 = $ constant, $\xi_2 = $ constant, $\xi_3 = $ constant. Let the length of its edges be $h_1 d\xi_1$, $h_2 d\xi_2$, $h_3 d\xi_3$. Further, let $\phi(\xi_1, \xi_2, \xi_3)$ be a scalar function, and $\mathbf{u}(\xi_1, \xi_2, \xi_3)$ be a vector field having three components $u_1$, $u_2$, $u_3$ along the three co-ordinate directions $\xi_1$, $\xi_2$, $\xi_3$. We have now to investigate how the operations of vector analysis appear in the new coordinates.

Referring to figure 14, for the $\xi_1$-component of the gradient of $\phi$, we have immediately from its definition (9.3) that

$$(\text{grad } \phi)_1 = \lim_{d\xi_1 \to 0} \frac{\phi(A) - \phi(0)}{h_1 \, d\xi_1} = \frac{1}{h_1} \frac{\partial \phi}{\partial \xi_1} \tag{13.1}$$

and correspondingly for directions 2 and 3.

For finding the value of the divergence of **u** from definition (10.5), we calculate the total flux through the surface of the infinitesimal block (figure 14), and divide this by the volume $h_1 h_2 h_3 d\xi_1 d\xi_2 d\xi_3$. The flux through the surface OBHC in the direction of the outward-drawn

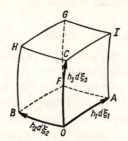

Fig. 14.—Volume element in orthogonal curvilinear coordinates

normal is $-(u_1 h_2 h_3)_{\xi_1} d\xi_2 d\xi_3$ while that through the surface AFGI is $(u_1 h_2 h_3)_{\xi_1 + d\xi_1} d\xi_2 d\xi_3$. Their difference is therefore

$$\frac{\partial}{\partial \xi_1}(u_1 h_2 h_3) d\xi_1 d\xi_2 d\xi_3$$

With the corresponding expressions for the other two surface pairs we therefore obtain

$$\operatorname{div} \mathbf{u} = \frac{1}{h_1 h_2 h_3}\left\{\frac{\partial}{\partial \xi_1}(u_1 h_2 h_3) + \frac{\partial}{\partial \xi_2}(u_2 h_3 h_1) + \frac{\partial}{\partial \xi_3}(u_3 h_1 h_2)\right\} \quad (13.2)$$

The 1-component of *curl* $\mathbf{u}$ is obtained, corresponding to definition (11.7), by applying Stokes's theorem to surface OBHC. We find that

$$(\operatorname{curl} \mathbf{u})_1 = \frac{1}{h_2 h_3}\left\{\frac{\partial}{\partial \xi_2}(u_3 h_3) - \frac{\partial}{\partial \xi_3}(u_2 h_2)\right\} \quad (13.3)$$

and correspondingly (making a cyclic exchange of the indices) for directions 2 and 3.

Finally, by combining (13.1) with (13.2), we obtain for $\nabla^2 \phi = \operatorname{div} \operatorname{grad} \phi$

$$\nabla^2 \phi = \frac{1}{h_1 h_2 h_3}\left\{\frac{\partial}{\partial \xi_1}\left(\frac{h_2 h_3}{h_1}\frac{\partial \phi}{\partial \xi_1}\right) + \frac{\partial}{\partial \xi_2}\left(\frac{h_3 h_1}{h_2}\frac{\partial \phi}{\partial \xi_2}\right) + \frac{\partial}{\partial \xi_3}\left(\frac{h_1 h_2}{h_3}\frac{\partial \phi}{\partial \xi_3}\right)\right\} \quad (13.4)$$

In a similar way the quantities $\nabla^2 u_1$, $\nabla^2 u_2$, $\nabla^2 u_3$ can naturally be found for a vector field **u**. It should however be noted that in general these quantities are different from the components of the vector

$$\nabla^2 \mathbf{u} \equiv \operatorname{grad} \operatorname{div} \mathbf{u} - \operatorname{curl} \operatorname{curl} \mathbf{u}$$

given in equation (11.17).

We now apply the foregoing formulae to two particularly important special cases:

(*a*) *Cylindrical coordinates:*    $\xi_1 = r$    $\xi_2 = \alpha$    $\xi_3 = z$

Since

$$x = r \cos \alpha \qquad y = r \sin \alpha \qquad z = z$$

we have

$$ds^2 = dr^2 + r^2 \, d\alpha^2 + dz^2$$

and therefore

$$h_1 = 1 \qquad h_2 = r \qquad h_3 = 1$$

It follows then from (13.1) to (13.4) that:

$$(\operatorname{grad} \phi)_r = \frac{\partial \phi}{\partial r} \quad (\operatorname{grad} \phi)_\alpha = \frac{1}{r} \frac{\partial \phi}{\partial \alpha} \quad (\operatorname{grad} \phi)_z = \frac{\partial \phi}{\partial z} \quad (13.1a)$$

$$\operatorname{div} \mathbf{u} = \frac{1}{r} \frac{\partial (r u_r)}{\partial r} + \frac{1}{r} \frac{\partial u_\alpha}{\partial \alpha} + \frac{\partial u_z}{\partial z} \qquad (13.2a)$$

$$(\operatorname{curl} \mathbf{u})_r = \frac{1}{r} \frac{\partial u_z}{\partial \alpha} - \frac{\partial u_\alpha}{\partial z} \qquad (\operatorname{curl} \mathbf{u})_\alpha = \frac{\partial u_r}{\partial z} - \frac{\partial u_z}{\partial r}$$

$$(\operatorname{curl} \mathbf{u})_z = \frac{1}{r} \frac{\partial (r u_\alpha)}{\partial r} - \frac{1}{r} \frac{\partial u_r}{\partial \alpha} \qquad (13.3a)$$

$$\nabla^2 \phi = \frac{1}{r} \frac{\partial}{\partial r} \left( r \frac{\partial \phi}{\partial r} \right) + \frac{1}{r^2} \frac{\partial^2 \phi}{\partial \alpha^2} + \frac{\partial^2 \phi}{\partial z^2} \qquad (13.4a)$$

(*b*) *Spherical polar coordinates:*    $\xi_1 = r$    $\xi_2 = \theta$    $\xi_3 = \alpha$

Since

$$x = r \sin \theta \cos \alpha \qquad y = r \sin \theta \sin \alpha \qquad z = r \cos \theta$$

we have

$$ds^2 = dr^2 + r^2 \, d\theta^2 + r^2 \sin^2 \theta \, d\alpha^2$$

and therefore

$$h_1 = 1 \qquad h_2 = r \qquad h_3 = r \sin \theta$$

From (13.1) to (13.4) there follow then:

$$(\text{grad }\phi)_r = \frac{\partial \phi}{\partial r} \qquad (\text{grad }\phi)_\theta = \frac{1}{r}\frac{\partial \phi}{\partial \theta} \qquad (\text{grad }\phi)_\alpha = \frac{1}{r\sin\theta}\frac{\partial \phi}{\partial \alpha} \qquad (13.1b)$$

$$\text{div }\mathbf{u} = \frac{1}{r^2}\frac{\partial(r^2 u_r)}{\partial r} + \frac{1}{r\sin\theta}\frac{\partial(\sin\theta\, u_\theta)}{\partial \theta} + \frac{1}{r\sin\theta}\frac{\partial u_\alpha}{\partial \alpha} \qquad (13.2b)$$

$$(\text{curl }\mathbf{u})_r = \frac{1}{r\sin\theta}\frac{\partial(\sin\theta\, u_\alpha)}{\partial \theta} - \frac{1}{r\sin\theta}\frac{\partial u_\theta}{\partial \alpha} \qquad (\text{curl }\mathbf{u})_\theta = \frac{1}{r\sin\theta}\frac{\partial u_r}{\partial \alpha} - \frac{1}{r}\frac{\partial(r u_\alpha)}{\partial r}$$

$$(\text{curl }\mathbf{u})_\alpha = \frac{1}{r}\frac{\partial(r u_\theta)}{\partial r} - \frac{1}{r}\frac{\partial u_r}{\partial \theta} \qquad (13.3b)$$

$$\nabla^2\phi = \frac{1}{r^2}\frac{\partial}{\partial r}\left(r^2\frac{\partial \phi}{\partial r}\right) + \frac{1}{r^2\sin\theta}\frac{\partial}{\partial \theta}\left(\sin\theta\frac{\partial \phi}{\partial \theta}\right) + \frac{1}{r^2\sin^2\theta}\frac{\partial^2\phi}{\partial \alpha^2} \qquad (13.4b)$$

# CHAPTER A III

## Tensors

### §14. Definition of a tensor. The anti-symmetric tensor

Along with scalar quantities and vectors, there enter into the mechanics of continua and into electrodynamics yet other directed quantities which are called *tensors*. The original concept of a tensor (and therewith the explanation of its name) is the condition of stress or tension within a solid body.

Through a point P of a stressed body we place a surface element $dS$ and assign to it a normal direction **n**. If we imagine a cut to have been made in the material adjacent to $dS$ and on the same side as **n** then, if the remaining material is not to experience any displacement, we must provide a force distribution on the surface element. Relating this force to the unit of surface area by dividing by $dS$, we obtain the stress $\mathbf{T}_n$ acting on the material at the position P of the surface element. To every location of the surface element given by the unit vector **n** there corresponds a stress vector $\mathbf{T}_n$ with components $T_{nx}, T_{ny}, T_{nz}$. Thus the component $T_{nm}$ of $\mathbf{T}_n$ in the direction of a second unit-vector **m** is given, according to the rules of vector calculus (see (3.8)), by

$$T_{nm} = \mathbf{T}_n \cdot \mathbf{m} = T_{nx} m_x + T_{ny} m_y + T_{nz} m_z \qquad (14.1)$$

When $\mathbf{m} = \mathbf{n}$ we have the component $T_{nn}$ which is called the *normal stress* or *tensile stress*. If **m** is perpendicular to **n** we have components of the *shearing stress*.

We now ask how the stress vectors corresponding to different directions **n** are related to one another. For two mutually opposite directions **n** and $-\mathbf{n}$ it follows from the law: action = reaction, that

$$\mathbf{T}_{-n} = -\mathbf{T}_n \qquad (14.2)$$

Further, we choose a right-angled coordinate system with P as origin, and we cut off from the apex of the first octant an infinitesimally small tetrahedron (figure 15), whose base (i.e. the cut surface) is the face $dS$ with outer normal **n**. It has components $n_x = \cos\alpha$, $n_y = \cos\beta$,

41

$n_z = \cos\gamma$, and the stress $\mathbf{T}_n$ acts on it. The remaining faces are then $dS\cos\alpha$, $dS\cos\beta$, $dS\cos\gamma$, and on them the stresses $\mathbf{T}_{-x}$, $\mathbf{T}_{-y}$, $\mathbf{T}_{-z}$ act. (The outward normals of these tetrahedral surfaces extend in directions

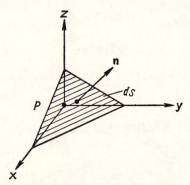

Fig. 15.—Equilibrium of the tetrahedron

which are negative with respect to the axes.) The forces on the tetra-hedron are in equilibrium when

$$\mathbf{T}_n \, dS + \mathbf{T}_{-x} \, dS \cos\alpha + \mathbf{T}_{-y} \, dS \cos\beta + \mathbf{T}_{-z} \, dS \cos\gamma = 0$$

From this, with (14.2) the desired relation follows:

$$\mathbf{T}_n = \mathbf{T}_x \cos\alpha + \mathbf{T}_y \cos\beta + \mathbf{T}_z \cos\gamma = \mathbf{T}_x \, n_x + \mathbf{T}_y \, n_y + \mathbf{T}_z \, n_z \qquad (14.3)$$

Written out in components, it is

$$\left.\begin{array}{l} T_{nx} = T_{xx} \, n_x + T_{yx} \, n_y + T_{zx} \, n_z \\ T_{ny} = T_{xy} \, n_x + T_{yy} \, n_y + T_{zy} \, n_z \\ T_{nz} = T_{xz} \, n_x + T_{yz} \, n_y + T_{zz} \, n_z \end{array}\right\} \qquad (14.4)$$

Putting these expressions into (14.1), we obtain the following trans-formation formula:

$$\left.\begin{array}{l} T_{nm} = T_{xx} \, n_x \, m_x + T_{yx} \, n_y \, m_x + T_{zx} \, n_z \, m_x + \\ \quad + T_{xy} \, n_x \, m_y + T_{yy} \, n_y \, m_y + T_{zy} \, n_z \, m_y + \\ \quad + T_{xz} \, n_x \, m_z + T_{yz} \, n_y \, m_z + T_{zz} \, n_z \, m_z \end{array}\right\} \qquad (14.5)$$

This gives the component $T_{nm}$, along an arbitrary direction $\mathbf{m}$, of a stress vector $\mathbf{T}_n$ which acts on a surface whose normal has the arbitrary direction $\mathbf{n}$.

Physically, $T_{nm}$ is a quantity in the stress field independent of the coordinate system, in contrast to the "components" $T_{xx}, T_{xy}, \ldots, T_{zz}$ referred to specific coordinate directions.

The quantity described by nine such components and having the property of yielding a scalar when combined with the two vectors **n** and **m** in the manner of (14.5), is called a *tensor*, or more accurately, a *tensor of the second rank*, since it is characterized by doubly-indexed components. In physics there also occur occasionally tensors of the third or higher rank, and they have correspondingly indexed components. In product formations like (14.5), but involving three or more vectors, they yield scalar quantities; but we shall not investigate them here. In this mode of description, vectors are tensors of the first rank.

The formulae in tensor calculations are considerably simplified if instead of indices $x, y, z$, we write the indices 1, 2, 3. We shall arrange the nine components of a tensor of the second rank in the convenient form of a square matrix, thus:

$$\begin{pmatrix} T_{xx} & T_{xy} & T_{xz} \\ T_{yx} & T_{yy} & T_{yz} \\ T_{zx} & T_{zy} & T_{zz} \end{pmatrix} \equiv \begin{pmatrix} T_{11} & T_{12} & T_{13} \\ T_{21} & T_{22} & T_{23} \\ T_{31} & T_{32} & T_{33} \end{pmatrix} \equiv (T_{ik}) \qquad (14.6)$$

By writing the components of a vector **v** in an analogous way as $v_1, v_2, v_3$, we can, for example, write formula (14.5) simply as a double sum

$$T_{nm} = \sum_i \sum_k T_{ik}\, n_i\, m_k \qquad (14.7)$$

where the sums here (and in following) are to extend over the three values (1, 2, 3) of the given indices.

In matrix notation the diagonal extending from upper left to lower right and containing the matrix elements with two equal indices is called the *principal diagonal*. If the elements which are images of one another with respect to the principal diagonal are in each case *equal*, the matrix and the tensor described by it are said to be *symmetric*. Thus, for a symmetric tensor, we have

$$T_{ik} = T_{ki} \qquad (14.8)$$

As, for example, elasticity theory shows the stress tensor we have considered above is symmetric. If, however, the elements reflected about the principal diagonal are in each case of opposite sign, i.e.

$$T_{ik} = -T_{ki} \qquad (14.9)$$

while the elements of the principal diagonal vanish, the tensor is called *anti-symmetric* or also *skew-symmetric*.

Every tensor $\mathbf{T}$ can be expressed as the sum of a symmetric and an anti-symmetric tensor. Accordingly, the sum $\mathbf{T}^{(c)}$ of two tensors $\mathbf{T}^{(a)}$ and $\mathbf{T}^{(b)}$ is defined as the tensor whose components $T_{ik}^{(c)}$ are equal to the sum of the corresponding components $T_{ik}^{(a)}$ and $T_{ik}^{(b)}$. This sum definition holds, in particular, for the stress tensor and is taken over for all tensors as an essential part of the definition of a tensor. (Compare the corresponding relationships in the combination of vectors!)

From the relationship

$$T_{ik} = \tfrac{1}{2}(T_{ik} + T_{ki}) + \tfrac{1}{2}(T_{ik} - T_{ki}) \tag{14.10}$$

we can therefore decompose any arbitrary tensor $\mathbf{T}$ into a symmetric part $\mathbf{T}^{(s)}$ with components $T_{ik}^{(s)} = \tfrac{1}{2}(T_{ik} + T_{ki})$ and an anti-symmetric part $\mathbf{T}^{(a)}$ with components $T_{ik}^{(a)} = \tfrac{1}{2}(T_{ik} - T_{ki})$.

We wish to pursue further this decomposition of a tensor by an example in the flow of an incompressible fluid. We again describe the motion by the velocity field $\mathbf{v}(\mathbf{r})$ and consider (figure 16) two fluid

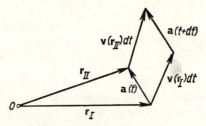

Fig. 16.—Change with time of the vector $\mathbf{a}$ between two mass-points I and II in the flow-field $\mathbf{v}(\mathbf{r})$

elements which at time $t$ are located at $\mathbf{r}_I$ and $\mathbf{r}_{II}$, and at time $t + dt$ are at $\mathbf{r}_I + \mathbf{v}(\mathbf{r}_I)dt$ and $\mathbf{r}_{II} + \mathbf{v}(\mathbf{r}_{II})dt$. Their mutual separation at time $t$ is the magnitude and direction given by the vector $\mathbf{a}(t) = \mathbf{r}_{II} - \mathbf{r}_I$, and at time $t + dt$ by the vector $\mathbf{a}(t + dt) = \mathbf{a}(t) + [\mathbf{v}(\mathbf{r}_{II}) - \mathbf{v}(\mathbf{r}_I)]dt$. Accordingly, the change of $\mathbf{a}(t)$ per second is equal to

$$\frac{d\mathbf{a}(t)}{dt} = \lim_{dt \to 0} \frac{\mathbf{a}(t + dt) - \mathbf{a}(t)}{dt}$$

$$= \mathbf{v}(\mathbf{r}_{II}) - \mathbf{v}(\mathbf{r}_I) \tag{14.11}$$

If the two fluid elements are infinitely close to one another, we can write that $r_{II} = r_I + a$, and we can consider $a$ as a small quantity. If we develop the right side of (14.11) in components of $a$ in the manner of (9.11), and if further we designate the components by index numbers, writing

$$x = x_1 \qquad y = x_2 \qquad z = x_3 \qquad (14.12)$$

we obtain from (14.11) the component equations

$$\frac{da_i}{dt} = \sum_k \frac{\partial v_i}{\partial x_k} a_k \quad \text{with} \quad i = 1, 2, 3 \qquad (14.13)$$

Since these equations, after multiplying by the components $b_i$ of an arbitrary vector $v$ and summing over the three $i$-values as permitted by the left side, give a scalar quantity

$$\sum_i \frac{da_i}{dt} b_i \equiv \frac{da}{dt} \cdot b = \sum_i \sum_k \frac{\partial v_i}{\partial x_k} b_i a_k$$

corresponding to (14.7), we can collect the nine derivatives

$$\frac{\partial v_i}{\partial x_k} = T_{ik} \qquad (14.14)$$

as the components of a tensor $T$.

We now decompose this tensor, in accordance with (14.10), into a symmetrical part with the components

$$T_{ik}^{(s)} = \tfrac{1}{2} \left( \frac{\partial v_i}{\partial x_k} + \frac{\partial v_k}{\partial x_i} \right) \qquad (14.15)$$

which we shall consider more fully in §15, and into an anti-symmetric part with components

$$T_{ik}^{(a)} = \tfrac{1}{2} \left( \frac{\partial v_i}{\partial x_k} - \frac{\partial v_k}{\partial x_i} \right) \qquad (14.16)$$

The latter we can assemble in the matrix

$$(T_{ik}^{(a)}) = \begin{pmatrix} 0 & -w_z & w_y \\ w_z & 0 & -w_x \\ -w_y & w_x & 0 \end{pmatrix} \qquad (14.17)$$

in which, having regard for (11.8), we introduce for brevity the vector

$$w = \tfrac{1}{2} \operatorname{curl} v \qquad (14.18)$$

Substituting now in (14.13), instead of the full complement of tensor components, only those of the anti-symmetric part (14.16), we obtain the vector equation

$$\left\{\frac{da}{dt}\right\}_{\text{antisymm.}} = \mathbf{w} \times \mathbf{a} \tag{14.19}$$

as can be immediately shown by writing out the individual components of this equation. A comparison with (5.9) shows that equation (14.19) describes nothing more than a rotation of the fluid around the fluid element designated in figure 16 by I, and participating in the flow. Here the vector $\mathbf{w}$ establishes the axis of rotation and the angular velocity.

From (14.17) and (14.18) we see that the angular velocity $\mathbf{w}$ and curl $\mathbf{v}$ are really not vectors, but, with their three components, belong to an anti-symmetric tensor. For such quantities the name *axial vector* or *pseudo vector* is sometimes employed to distinguish from a normal or *polar* vector. A further important example of an axial vector is the vector product $\mathbf{c} = \mathbf{a} \times \mathbf{b}$ of two polar vectors $\mathbf{a}$ and $\mathbf{b}$. This would actually have to be described as an anti-symmetric tensor with components

$$c_{ik} = a_i b_k - a_k b_i$$

The representation given by (14.17) of an anti-symmetric tensor as a vector is only possible in three-dimensional space, and even here only so long as the calculation is carried out consistently with a right-hand (or with a left-hand) system. In going from a right-hand to a left-hand system, as by reflection about the origin, all components of a polar vector change their sign, while all tensor components, being doubly-indexed quantities—and thus all components of axial vectors—retain their signs unchanged.

In the foregoing paragraphs the introduction of axial vectors showed that for their explanation we needed the concept of the right-hand screw. This concept and the consequent restriction to right-handed coordinate systems can, however, be completely avoided when, instead of axial vectors, the corresponding antisymmetric tensors are introduced.

It should also be mentioned that the vector product of a polar vector with an axial vector, as well as the curl of an axial vector, are again polar vectors. Moreover, the scalar product of a polar with an axial vector is a *pseudo scalar*, and, like a scalar, it remains unchanged in a rotation of the coordinate system; in a reflection about the origin, however, its sign changes. Such a pseudo scalar is, for example, the spar product (6.1) of three polar vectors; representing the volume of the parallelopiped determined by these vectors, its sign changes in going from a right-hand to a left-hand system; i.e. the signs of all components in (6.2) change.

## §15. The symmetric tensor and its invariants. The deviator

While the anti-symmetric part of a tensor can always be regarded as an (axial) vector, and thus leads to known concepts and calculations, the symmetric part of a tensor represents something basically new. In these paragraphs we shall therefore concern ourselves only with symmetric tensors.

Next, we consider again the example of fluid flow, restricting ourselves at the outset, however, to the symmetric part of tensor (14.14). Instead of (14.13), we therefore write

$$\frac{da_i}{dt} = \sum_k T_{ik} a_k \quad \text{with} \quad T_{ik} = \tfrac{1}{2}\left(\frac{\partial v_i}{\partial x_k} + \frac{\partial v_k}{\partial x_i}\right) \tag{15.1}$$

or for the following, more conveniently,

$$b_i \equiv a_i(t+dt) = a_i(t) + \sum_k T_{ik} a_k(t)\, dt \tag{15.2}$$

Owing to the smallness of the summation term, we can easily solve this expression for $a_i(t)$, so long as we restrict ourselves to terms of the first order in $T_{ik} dt$:

$$a_i = b_i - \sum_k T_{ik} b_k\, dt \tag{15.3}$$

We now consider in the description of figure 16 the totality of all fluid elements II which at time $t$ have a fixed separation $a$ from fluid element I. They lie on the spherical surface given by

$$\sum_i a_i^2 = a^2 \tag{15.4}$$

Owing to the inhomogeneity of the fluid flow, at time $t + dt$ this spherical surface will have become somewhat deformed. We obtain the form of the new surfaces as an equation between the components $b_i$ by substitution of (15.3) in (15.4); we find, again in first approximation, that

$$\sum_i b_i^2 - 2 \sum_i \sum_k T_{ik} b_i b_k\, dt = a^2 \tag{15.5}$$

This is the equation of an almost-spherical ellipsoid.

We now imagine the coordinate system chosen so that the new coordinate axes coincide with the principal axes of the ellipsoid. In this coordinate system equation (15.5) must be the principal-axis equation of the ellipsoid, i.e. in this coordinate system the mixed tensor components $T_{ik}$, with $i \neq k$, must vanish, so that (15.5) can be written as

$$\sum_i b_i^2(1 - 2T_{ii}\, dt) = a^2$$

From this, in the same approximation as above, we obtain for the length of the semi-axes of the ellipse

$$a(1+T_{11}\,dt) \qquad a(1+T_{22}\,dt) \qquad a(1+T_{33}\,dt)$$

and thus, as the volume of the ellipsoid (since $T_{ii} = \partial v_i/\partial x_i$), we have

$$\tfrac{4}{3}\pi a^3(1+[T_{11}+T_{22}+T_{33}]\,dt) = \tfrac{4}{3}\pi a^3(1+\operatorname{div}\mathbf{v}\,dt)$$

During the time interval $dt$ the included volume increases by

$$dV = V(T_{11}+T_{22}+T_{33})\,dt = V\operatorname{div}\mathbf{v}\,dt \qquad (15.6)$$

Consequently $dV$ is different from zero when there are sources within $V$. When sources are absent, the volume of the mass of liquid remains constant, as may the shape of the liquid mass as well.

Just as for the tensor (15.1), we can give for every symmetric tensor a special orientation of the coordinate system such that in this system only the components in the principal diagonal possess finite values. The coordinate transformation (rotation) by which this particular orientation is attained is called the *principal-axis transformation*, and a tensor possessing components different from zero only in the principal diagonal is said to be *transformed to principal axes*.

In order to find this special orientation of axes we could, in the above example, have sought for the direction of vector **a** for which vector **b**, according to (15.2), would be parallel to **a**—for which therefore $b_i = \kappa a_i$ could be substituted. Analogously, for the general symmetric tensor with components $T_{ik}$ we seek that direction of the vector **a** for which

$$\sum_i T_{ik}a_k = \lambda a_i \quad \text{for} \quad i = 1,2,3 \qquad (15.7)$$

holds. This is a system of linear homogeneous equations in the three quantities $a_1$, $a_2$, $a_3$, having, according to a theorem of algebra, a finite solution only when the determinant of the coefficients vanishes; that is, when

$$\begin{vmatrix} T_{11}-\lambda & T_{12} & T_{13} \\ T_{21} & T_{22}-\lambda & T_{23} \\ T_{31} & T_{32} & T_{33}-\lambda \end{vmatrix} = 0 \qquad (15.8)$$

This is an algebraic equation of the third degree in $\lambda$. It therefore possesses three roots $\lambda^{\mathrm{I}}$, $\lambda^{\mathrm{II}}$, $\lambda^{\mathrm{III}}$, which are called the eigenvalues of the tensor **T**, or of the coefficient matrix $(T_{ik})$. From each of these roots there follows from (15.7) a definite direction of the vector **a** so that in $\mathbf{a}^{\mathrm{I}}$, $\mathbf{a}^{\mathrm{II}}$, $\mathbf{a}^{\mathrm{III}}$ we have the three directions in which we must take the coordinates so that the tensor **T** is transformed to principal axes.

We show first that for a symmetric tensor the three roots of the $\lambda$-equation (15.8) are real. Assuming this were not the case, we should obtain complex values for the corresponding $a_i$. Thus, multiplying (15.7) by the complex conjugate quantities $a_i^*$ and summing over $i$ we should have

$$\sum_i \sum_k T_{ik} a_i^* a_k = \lambda \sum_i a_i^* a_i$$

The double sum here is real since $T_{ik} = T_{ki}$; in this case the inconsequential exchanging of the summation indices $i$ and $k$ leads formally to the complex conjugate value. Since the sum on the right side is real, this also holds for $\lambda$, contrary to the above assumption.

We show further that the three principal axes are perpendicular to one another whenever the three $\lambda$-values are all different. For this (15.7) gives for the first solution

$$\sum_k T_{ik} a_k^{\mathrm{I}} = \lambda^{\mathrm{I}} a_i^{\mathrm{I}} \tag{15.9}$$

Next we multiply this expression by $a_i^{\mathrm{II}}$ and sum over $i$:

$$\sum_i \sum_k T_{ik} a_i^{\mathrm{II}} a_k^{\mathrm{I}} = \lambda^{\mathrm{I}} \sum_i a_i^{\mathrm{II}} a_i^{\mathrm{I}}$$

Finally, we extract from this the equation which we obtain by interchanging I and II. Then, since $T_{ik} = T_{ki}$, the two double sums cancel one another, so that we must have

$$(\lambda^{\mathrm{I}} - \lambda^{\mathrm{II}}) \sum_i a_i^{\mathrm{II}} a_i^{\mathrm{I}} = 0$$

Thus if $\lambda^{\mathrm{I}}$ and $\lambda^{\mathrm{II}}$ are different, the scalar product $\mathbf{a}^{\mathrm{II}} \cdot \mathbf{a}^{\mathrm{I}}$ vanishes, and the two vectors are therefore perpendicular to one another. If it had happened that $\lambda^{\mathrm{I}} = \lambda^{\mathrm{II}}$, then $\mathbf{a}^{\mathrm{I}}$ and $\mathbf{a}^{\mathrm{II}}$, as well as any linear combination of these two vectors, would have been a solution of (15.7), and we could always choose two mutually-perpendicular base vectors $\mathbf{a}^{\mathrm{I}}$ and $\mathbf{a}^{\mathrm{II}}$. This is the case occurring when the more general triaxial ellipsoid degenerates into an ellipsoid of revolution.

As with a vector's magnitude, the value of which is always independent of the particular location of the coordinate system, we have the three eigenvalues $\lambda^{\mathrm{I}}$, $\lambda^{\mathrm{II}}$, $\lambda^{\mathrm{III}}$, which are three definitive parts of the symmetric tensor **T**, independent of coordinate system. We arrive further at *invariants of the tensor* derivable from them by writing out (15.8):

$$-\lambda^3 + \lambda^2 Sp(\mathbf{T}) - \lambda C(\mathbf{T}) + D(\mathbf{T}) = 0 \qquad (15.10)$$

Here the quantity

$$Sp(\mathbf{T}) = T_{11} + T_{22} + T_{33} \qquad (15.11)$$

is called the *spur* of the component matrix and consists of the sum of the terms of the principal diagonal. $C(\mathbf{T})$ in (15.10) is the quantity

$$C(\mathbf{T}) = T_{22} T_{33} + T_{33} T_{11} + T_{11} T_{22} - T_{23}^2 - T_{31}^2 - T_{12}^2$$

but has no special name. $D(\mathbf{T})$ in the same expression is the determinant of the components of **T**. Since equation (15.10) must have the form

$$-(\lambda - \lambda^I)(\lambda - \lambda^{II})(\lambda - \lambda^{III}) = 0$$

in order to lead to the roots $\lambda^I$, $\lambda^{II}$, $\lambda^{III}$, we must have, for example,

$$Sp(\mathbf{T}) = \lambda^I + \lambda^{II} + \lambda^{III}$$

The spur of a tensor, like the quantity $C(\mathbf{T})$ and a tensor's determinant, is an invariant of the tensor.

We must still convince ourselves that all the $T_{ik}$, with $i \neq k$, actually vanish whenever the directions of the principal axes coincide with the directions of the coordinate axes If the vector $\mathbf{a}^I$ has the direction of, say, the x-axis, then $a_1^I = a^I \neq 0$, $a_2^I = a_3^I = 0$. Taking this into equation (15.9) we obtain the three equations

$$T_{11} a^I = \lambda^I a^I \qquad T_{21} a^I = 0 \qquad T_{31} a^I = 0$$

$T_{21}$ and $T_{31}$ therefore vanish, while $T_{11} = \lambda^I$. Carrying through similar considerations for the vectors $\mathbf{a}^{II}$ and $\mathbf{a}^{III}$ lying in the y- and z-directions, we find finally for the component matrix the simple relationship

$$(T_{ik}) \equiv \begin{pmatrix} T_{11} & 0 & 0 \\ 0 & T_{22} & 0 \\ 0 & 0 & T_{33} \end{pmatrix} = \begin{pmatrix} \lambda^I & 0 & 0 \\ 0 & \lambda^{II} & 0 \\ 0 & 0 & \lambda^{III} \end{pmatrix} \qquad (15.12)$$

which shows particularly clearly the significance of the three eigenvalues of a tensor.

Finally, attention should be drawn to a development often seen in tensor calculus, the *deviator*. This is a symmetrical tensor with vanishing spur. We can understand its name from the example in hydrodynamics considered above. We found there that in the time interval $dt$ an initially spherical region of liquid deformed into an ellipsoid and simultaneously, in general, also changed its volume. Now according to (15.6), this volume change is given by the spur of the deformation tensor. If it vanishes, the tensor describes a change of form, without however a change of volume.

We can thus pass from an arbitrary symmetrical tensor with co-efficients $T_{ik}$ to a tensor with vanishing spur, i.e. to the associated deviator, by subtracting from the components $T_{ii}$ the third part of the spur, viz. $\frac{1}{3}(T_{11} + T_{22} + T_{33})$. For the deviator associated with the tensor, we thus arrive at the following component matrix:

$$\begin{pmatrix} \frac{1}{3}(2T_{11} - T_{22} - T_{33}) & T_{12} & T_{13} \\ T_{21} & \frac{1}{3}(2T_{22} - T_{11} - T_{33}) & T_{23} \\ T_{31} & T_{32} & \frac{1}{3}(2T_{33} - T_{11} - T_{22}) \end{pmatrix} \quad (15.13)$$

The procedure specified for the construction of the deviator holds independently of the coordinate system, for it means diminishing the tensor **T** by $\frac{1}{3}(T_{11} + T_{22} + T_{33})$ of the *unit tensor* **E**, in whose component matrix there are 1's throughout the principal diagonal and 0's at all other places:

$$E_{ik} = \delta_{ik} = \begin{cases} 1 & \text{for} \quad i = k \\ 0 & \text{for} \quad i \neq k \end{cases} \quad (15.14)$$

This *unit tensor* preserves its form in a rotation of the coordinate system, as can be easily verified by means of (14.7). This also follows because for it the tensor ellipsoid degenerates into a sphere.

**B**

---

The
electrostatic
field

# CHAPTER B I

## Electric Charge and the Electrostatic Field in Vacuum

### §16. Electric charge

If a stick of sealing-wax is rubbed with cat's fur, both bodies are put into a peculiar condition in which light bodies in their neighbourhood are set in motion. We say that by rubbing, bodies become *electrified* and that they carry an *electric charge*. Experimentally, the charges do not cling permanently to the sealing-wax or to the cat's fur. They can be transferred to other bodies by contact. The origin of the charged condition is not exclusively bound up with the process of rubbing; a piece of metal can, for example, be charged by a momentary contact with the pole of an electric battery.

Two charged bodies exert a force upon one another. This force can be used to measure the charge, as for example by means of an *electrometer*. From results which have been obtained in such measurements, in particular in the simultaneous transfer of different charges to such an instrument, the existence of *charges of different sign* has been concluded, these charges when in combination being added algebraically. The charge on the rubbed cat's fur has been arbitrarily called positive, and that on the stick of sealing-wax negative.

For electric charges we have the *conservation law: the total charge of a closed system remains constant*. Thus electric charges can be neither created nor destroyed. It therefore follows that simultaneously an equal quantity of charge of each sign (+ and −) is produced or disappears. In rubbing the sealing-wax with the cat's fur there must be exactly as much positive charge on the fur as there is negative charge on the sealing-wax. In the transfer of charge by contact of one body with another, the sum of the charges of both bodies before and after contact must remain the same.

### §17. The elementary electrical quantum

For a fuller understanding of electric and magnetic processes in matter, and for an understanding of the phenomena of charged-particle radiations, there is the decisive fact learned from many different experiments that, just as in matter, electric charge possesses *atomicity of structure*, and that there is a smallest, not further divisible, quantity of electric charge. This smallest charge quantity is called the elementary electrical quantum. In the following we shall always designate it by $e_0$.

The first hint of the existence of this elementary quantum was given by one of *Faraday's laws of electrolysis:* In the passage of an electric current through an aqueous solution of a salt, e.g. silver nitrate, silver is continually deposited. In this process, the mass of deposited silver is at any time proportional to the quantity of electricity which has passed through the electrolyte; indeed, according to Faraday, in the deposition of exactly one mole from a singly-ionized electrolyte, a quantity of electricity $F = 96,500$ coulombs (C), (the "Faraday constant") flows through the solution. Since the number of atoms or ions in a mole is equal to the Avogadro number $N_0 = 6.025 \times 10^{23}$, it can be concluded from the Faraday result that, in electrolysis at least, electricity, just like matter, is divided into atomic units, and that every singly-ionized positive ion arriving at the cathode carries with it a charge

$$e_0 = F/N_0 = 1.60 \times 10^{-19} \text{C}$$
$$= 4.80 \times 10^{-10} \text{ Gaussian units of charge*} \qquad (17.1)$$

These considerations were first clearly stated by G. J. Stoney and H. v. Helmholtz in 1881. The evaluation of the elementary quantum $e_0$ from electrolysis assumes, along with the (relatively easily determined) Faraday constant, a knowledge of the Avogadro number $N_0$. Concerning the various methods of determing the Avogadro number, the reader is referred to textbooks of physical chemistry.

A method for directly determining the elementary electrical quantum independently of electrolysis has been devised by R. A. Millikan. The Millikan technique is based upon the following principle: The charge found on an electrically isolated body is directly measured. The charge of the body is then somehow altered (say, by irradiating with ultra-violet light), and the charge is again measured, etc. If it is found here that all charge values obtained are always integral multiples of a smallest charge, this charge can be regarded as corresponding to the elementary quantum. Obviously only a small number of elementary quanta are allowed to be on the body, so that it is possible to conclude the existence of a whole-number multiple with reasonable accuracy.

---

* Concerning the units employed, see §§18 and 19.

For this reason it is necessary to make the body very small—so small, in fact, that it can only be observed ultra-microscopically.

In the Millikan apparatus (figure 17) there is a droplet of some substance (mercury, for example) hovering in air between the plates of a horizontally disposed capacitor having plate separation $d$. When this drop, of mass $m$, is uncharged, it

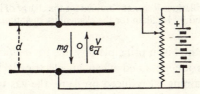

Fig. 17.—Scheme of Millikan's method for determining $e_0$

is pulled downward by gravity with the force $mg$. If the charge has the value $e$ the gravitational force can just be compensated by appropriately charging the capacitor (to, say, voltage $V$) and, since the droplet within the capacitor experiences an upward force $eV/d = eE$ (see §18), we have the condition for hovering that

$$mg = eV/d$$

The evaluation of the charge $e$ depends upon the determination of the mass $m$ of the droplet. Since for an ultra-microscopic particle a direct measurement of the particle radius is not possible, it is the mass measurement here which really becomes the experimental problem in measuring $e_0$. Millikan solved this problem by measuring the velocity $v$ with which a particle descends under the influence of gravity alone, the voltage $V$ being switched off. The frictional resistance of the surrounding air is so great that the descent of the drop takes place with constant velocity. By using Stokes's formula $v = mg/6\pi\eta a$ for the velocity of descent of a spherical drop of radius $a$ in a medium of viscosity $\eta$, and by taking account of the relationship $m = \frac{4}{3}\pi a^3(\rho - \rho_0)$, in which $\rho$ is the density of the substance forming the drop, and $\rho_0$ that of the surrounding air, we succeed in determining the radius $a$ of the drop, and therewith the mass $m$. (We do not discuss the corrections which, for very small droplets, have to be applied both to Stokes's formula and to the mass relationship, the latter owing to the possible absorption of air molecules by the droplet.) Over all his experiments, the average found by Millikan had the value $e_0 = 4.77 \times 10^{-10}$ CGS units.

In recent years the accuracy of Millikan's method of determining $e_0$ has been surpassed by many other methods in which measurements are made of $e_0$ in combination with various other atomic constants. The present best value averaged from all these methods is given in (17.1). Several of these methods of measurement are noteworthy in that they more or less directly reveal the atomistic character of electricity and its carrier.

Especially impressive in this respect is the method employed by E. Regener for *determining $e_0$ with the help of $\alpha$-particles*. According to Rutherford the $\alpha$-particles emitted in radioactive decay (of RaC, for example) consist of charged helium atoms, i.e. helium ions in which each α-particle carries a charge $2e_0$. For determining the elementary quantum, we have to measure on the one hand the totality of charge emitted by a radioactive preparation in the course of, say, one second, and on the other hand to determine the number of α-particles emitted during the same interval. Such counting is actually possible because every α-particle impinging upon a zinc sulphide screen produces a microscopically visible light flash (scintillation). By counting these scintillations Regener obtained for the charge of individual α-particles twice the value found by Millikan for the elementary quantum.

The same method is in principle possible for *artificially-produced ion rays*. Such rays are obtained when an electric field is applied to a glass vessel containing, at reduced pressure, the gas to be investigated. The ions of the gas, some of which are always present while others are continually being produced anew in impacts, are thereby accelerated toward one of the two electrodes. By means of "canals" (holes) in the electrode these rays can be admitted to the region of investigation, and thus, following the discoverer (Goldstein), we speak of *canal rays*. In an electrolyte carrying a current they are bound up with the transport of matter and with the familiar precipitation of matter ("deposition" or "plating out") at the appropriate collecting electrode.

In contrast to this, in a highly evacuated glass vessel with two electrodes, one of which (the cathode) is heated to a sufficiently high temperature, upon application of voltage, a current flows which can be maintained without producing the slightest chemical change at the cathode or anode. The further investigation of this electricity coming from the cathode (*cathode rays*) is made possible by its ability to be deflected in an electric and/or magnetic field (see §18 and §44). It turns out here that the current is carried by particles with a charge $-e_0$ and a mass which is about 1840 times smaller than the mass of the lightest atoms, the atoms of hydrogen. These entities are called *electrons*.

These electrons play a very important part in the whole of atomic physics. According to Rutherford, every atom consists of: (1) a heavy *nucleus* in which almost the whole of the atom's mass resides, (2) a definite number of electrons which orbit round this nucleus. The number $Z$ of these electrons is called the *atomic number* of the atom, since the physical and chemical properties of the atom depend essentially upon this number. It is just the number reached in counting through the periodic system of the elements, beginning with hydrogen, for which $Z = 1$. Since the atom as a whole is electrically neutral, the nucleus of an atom of atomic number $Z$ must have a nuclear charge $Ze_0$ so as to compensate the charge $-Ze_0$ of the orbital electrons. Thus the helium nucleus, corresponding to the atomic number 2 of helium, has a charge $2e_0$, in agreement with the investigation of Regener mentioned above.

Since a full exposition of atomic physics will be given in the second volume of this series, we shall let these brief indications of the structure of atoms suffice here. They place us in the position, already in this first volume, of obtaining a deeper insight into the electric and magnetic behaviour of individual atoms and of the ponderable matter formed out of them.

## §18. Electric field strength and the electric potential

Electric charges exert forces upon one another. This force effect, in its dependence upon the magnitude of the charges and their mutual separation, can be made the sole object of the investigation as was done in the usual theory of action-at-a-distance prior to Faraday and Maxwell. Going beyond this, however, we can also speak in the sense of Faraday and Maxwell of an *electric force field* in the neighbourhood of every charged body. This field, like the gravitational field of the earth, is assumed to exist even at points at which there is no charge upon which it could act. The investigation of this force field, which is looked upon primarily as the origin of the force action, is the object of a *field-action theory*.

The Faraday-Maxwell field concept is suggested and supported by the following experimental fact: The force $\mathbf{F}(\mathbf{r})$ experienced by a small test charge, as for example a goldleaf-covered pith-ball, at a point having the position vector $\mathbf{r}$ in a system of different and arbitrarily distributed charged bodies, is directly proportional to the charge $e$ of the test body. We can therefore always write

$$\mathbf{F} = e\mathbf{E} \qquad (18.1)$$

in which $\mathbf{E} = \mathbf{E}(\mathbf{r})$ is a function of position in the system considered, but is independent of the test body. This quantity is called the *electric field strength*. It is characteristic of the Maxwell theory that a reality independent of the existence of the probe charge is immediately ascribed to this vector field $\mathbf{E}$, and that it forms the actual object of the investigation.

Expression (18.1) for the force on a test charge in an electric field is not unconditionally valid. It ceases to be accurately valid as soon as the test body is brought too close to a charged or uncharged body, and the greater the value of the test charge the more is this so. It is also inaccurate when the field strength varies too strongly with position, and the more so the greater the dimensions of the test body. Later on we shall learn the causes of these discrepancies, and thus understand more fully our expression (18.1) for the force. For the present, however, we must employ a sufficiently small and sufficiently weak charged body if we wish to evaluate the strength of an electric field on the basis of equation (18.1).

If now a charge $e$ is moved by a force $\mathbf{F}$ through the distance $d\mathbf{r}$, the field, according to the basic rules of mechanics, performs a quantity of work $\mathbf{F} \cdot d\mathbf{r} = e\mathbf{E} \cdot d\mathbf{r}$. In the displacement of the charge from point 1 to point 2, the work done is equal to the line integral

$$A_{12} = \int_1^2 \mathbf{F} \cdot d\mathbf{r} = e \int_1^2 \mathbf{E} \cdot d\mathbf{r} \qquad (18.2)$$

For an electric field, just as for the static force-fields of mechanics, this work must be independent of the shape of the path going from 1 to 2; in particular, for a closed path it must vanish:

$$\oint \mathbf{F} \cdot d\mathbf{r} = e \oint \mathbf{E} \cdot d\mathbf{r} = 0$$

In mechanics it follows from this special statement of the *energy theorem* that a static force-field must be representable as the (negative) gradient of a function of position, the *potential energy*. Similarly we can conclude that for every electrostatic field the electric field strength must be expressible in the form

$$\mathbf{E} = -\operatorname{grad} \phi \qquad (18.3)$$

The scalar function of position is called the *electrostatic potential $\phi$*, and the work integral obtained for unit charge,

$$V_{12} = \int_1^2 \mathbf{E} \cdot d\mathbf{r} = \phi_1 - \phi_2 \qquad (18.4)$$

is called the *potential difference* or also the *voltage* between points 1 and 2.

We first remark that equation (18.3) contains an important assertion concerning the field strength $\mathbf{E}$ in the electrostatic field: Since $\mathbf{E}$ can be expressed as the gradient of the potential function $\phi$, it must (see §11) be *irrotational:*

$$\oint \mathbf{E} \cdot d\mathbf{r} = 0 \quad \text{or} \quad \text{curl}\,\mathbf{E} = 0 \qquad (18.5)$$

These relationships represent, in integral and differential form respectively, the first of the four *Maxwell equations* for the electromagnetic field. They are for the special case of a static field. We shall become acquainted with the general form of these equations in §45.

We remark further that $eV_{12} = e(\phi_1 - \phi_2)$ is equal to the work performed by the field in the movement of a small probe body of charge $e$ from point 1 to point 2, and thus it is also equal to the decrease of the potential energy of this charge $e$ in the electrostatic field. If the charge moves freely under the influence of the field, we have then, corresponding to (11.2), the equation of motion of the test charge (of mass $m$) as

$$m\frac{d\mathbf{v}}{dt} = e\mathbf{E} \qquad (18.6)$$

and, in consequence of the theorem of conservation of energy, this is equal to the increase in the kinetic energy of the body

$$(\tfrac{1}{2}mv^2)_2 - (\tfrac{1}{2}mv^2)_1 = e\int_1^2 \mathbf{E} \cdot \mathbf{v}\,dt = e\int_1^2 \mathbf{E} \cdot d\mathbf{r}$$

$$= e(\phi_1 - \phi_2) = eV_{12} \qquad (18.7)$$

This is thus the kinetic energy which any charged particle initially at rest (as in cathode rays or ion beams) attains in traversing a given portion of an electric field (the acceleration path) through the falling potential.

If $\mathbf{E} = \mathbf{E}(\mathbf{r})$ is known, the motion of charged particles in this field can be determined by integrating equation (18.6). Thus, for an **E**-field constant in space, charged particles move in a parabolic trajectory, just like point-masses in the gravitational field of the earth. By twice integrating (18.6) with respect to time we obtain

$$\mathbf{v} = \mathbf{v}_0 + \frac{e}{m}\mathbf{E}t \qquad \mathbf{r} = \mathbf{r}_0 + \mathbf{v}_0 t + \frac{e}{m}\mathbf{E}\frac{t^2}{2}$$

in which $\mathbf{r}_0$ and $\mathbf{v}_0$ are respectively the initial position and the initial velocity. If we imagine the $z$-axis of a Cartesian coordinate system placed in the direction of the field, and the $xz$-plane parallel to $\mathbf{v}_0$, then $v_y = 0$, $y = y_0$, and we have

$$x = x_0 + v_{0x}\,t \qquad z = z_0 + v_{0z}\,t + \frac{eEt^2}{2m}$$

If the time is eliminated here, a trajectory in the form of a parabola opening in the direction of positive $z$ is obtained.

*Remarks:* It seems appropriate that we now occupy ourselves with the question of the system of units to be employed for the electric and magnetic quantities—for, while in other branches of physics equations can always be written in the same form independent of the system of units employed, and differing only in the numerical value of the physical quantities appearing, in electrodynamics, in going from one system of units to another, different formulae are written for the different systems. This is why the establishment of units is so important here. While all formulae even here could doubtless be written in so general a way that the constants appearing in them would serve for any system of units, the formulae would then become very cumbersome and complicated. Therefore, in the different systems of units, definite simplifications are effected by arbitrarily establishing constants such that, in the particular domain of application, the formulae which appear assume their simplest possible form.

Particularly suited to the requirements of basic physics as well as of atomic physics and relativity are the (absolute) *Gaussian CGS units.* In this system, in which the electrical units are those of the electrostatic CGS system, while the magnetic units are from the electromagnetic CGS system, the basic units: the centimetre (cm), the gramme (g) and the second (s) are employed, and all electric and magnetic units are linked to these. Thus, for example, the unit of electric charge is established through the observation that a small body carrying the unit of charge experiences a force $F = 1$ cm g s$^{-2} = 1$ dyne when it is distant 1 cm from an equally charged second body (see § 19).

Suited to the requirements of applied electrodynamics and electrical technology, and since 1948 internationally recognized, is the *Giorgi system*, now generally known as the MKSA system. In this system, along with the three non-electromagnetic basic units: the metre (m), kilogramme (kg) and second (s), a fourth specifically electromagnetic measure unit, the ampere (A), has been introduced as the unit of current strength. Here the ("absolute") ampere is defined by saying that two parallel infinitely long thin wires, in each of which there flows in the same direction a current of strength 1 A, when separated by 1 m, attract each other with a force of $2 \times 10^{-7}$ m kg/s$^2$ per metre of wire length. (This definition is equivalent to establishing (see § 46, page 178) the permeability of a vacuum (constant of induction) at $\mu_0 = 4\pi \times 10^{-7}$ Vs/Am.) The absolute volt, as the unit of electrical potential, is then traced back to the four basic units of the MKSA system, by saying that the electromagnetic unit of energy, the joule = watt-second = volt-ampere-second, shall be equal to m$^2$ kg s$^{-2}$, i.e. 1 volt = 1 m$^2$ kg/A s$^3$.

Later in the text we shall regularly make use of the Gaussian CGS system of units. At the conclusion of certain paragraphs, however, as remarks in smaller print, we shall give the more important of the derived formulae in the MKSA system.

(See also the tables on p. 427 and following.)

## §19. Coulomb's law. The flux of electric force

In §18 we were concerned with the force effects of an electric field on charges. We wish now to consider the field-producing effect of charges.

One of the most important results of quantitative electrical theory prior to Faraday was *Coulomb's law:*

*The force which two charged bodies, 1 and 2, whose dimensions are small compared to their separation, exert on one another has the direction of the line joining the charges and is inversely proportional to the square of their separation distance r.*

Since we can look upon either charge at pleasure as the test charge, in the sense of §18, we can write

$$F = \frac{e_1 e_2}{r^2} f \qquad (19.1)$$

in which the factor $f$ is independent not only of the condition of the bodies but of their relative positions as well. The quantity $f$ has the dimensions of force $\times$ (length)$^2$/(charge)$^2$. Furthermore, $\mathbf{F}$ acts in the sense of a repulsion when $e_1$ and $e_2$ have the same sign, and as an attraction when the two signs are opposite.

We now introduce the *Gaussian system of units* in which, by arbitrarily establishing the hitherto undetermined dimension of charge, we make $f$ dimensionless, and we set it $= 1$. The unit quantity of electricity (electric charge) is thus defined so that upon a quantity of electricity equal to itself, at a distance of 1 cm, it exerts a force of 1 dyne. Hereby, according to (18.1), the unit of $\mathbf{E}$ is also established. In the language of Faraday and Maxwell, we can now describe the results of the Coulomb measurements in the following manner: *An electric point charge produces in its vicinity an electric field* $\mathbf{E}$ *which, in direction and magnitude is given by*

$$\mathbf{E} = \frac{e}{r^2} \frac{\mathbf{r}}{r} = \frac{e\mathbf{r}}{r^3} \qquad (19.2)$$

Here $\mathbf{r}$ is the vector extending from the charge to the field point.

As a comparison of (19.2) with (9.7) or (10.3) will show, the field $\mathbf{E}$ of a point charge corresponds formally to the velocity field $\mathbf{v}$ of an incompressible fluid issuing from a point source of strength $e$. Corresponding to the results of §18, $\mathbf{E}$, like $\mathbf{v}$, is therefore irrotational, and, according to (18.3), can be represented as the gradient of the "Coulomb potential"

$$\phi = \frac{e}{r} \tag{19.3}$$

We have, further, as in (10.4), that

$$\oiint E_n \, dS = 4\pi e \quad \text{or} \quad = 0 \tag{19.4}$$

according as the charge $e$ lies within or without the closed surface over which the integral is taken. The integral $\iint_S E_n \, dS$ is called the *flux* of electric force through the surface $S$.

Further, since experiment shows that in the simultaneous action of several charges $e_1$, $e_2$, $e_3$, ... their field contributions as well as their force actions on a test charge add according to the vector law of addition, we can generalize the electric-flux theorem (19.4) to the statement:

*The total electric flux $\oiint E_n \, dS$ through a closed surface is equal to $4\pi$ times the total charge in the region $V$ enclosed by the surface.*

In place of (19.4), therefore, we have the expression

$$\oiint E_n \, dS = 4\pi \sum_{(e_j \, \text{in} \, V)} e_j \tag{19.5}$$

where the summation is only to extend over the charges within $V$.

There may be so many densely distributed point charges with which to reckon that it is expedient to collect the charges within the volume element $dv$ into a total charge $\rho \, dv$, where $\rho$ represents the *volume density of charge*. The electric-flux theorem is then

$$\oiint E_n \, dS = 4\pi \iiint \rho \, dv \tag{19.6}$$

in which the volume integration is to extend over the region enclosed by the surface. With the help of Gauss's theorem (10.10) we obtain as the differential form of the electric-flux theorem

$$\operatorname{div} \mathbf{E} = 4\pi\rho \tag{19.7}$$

This is the *second Maxwell equation*, for the special case of volume charges in a vacuum. We shall meet it in its general form in §27.

If a large number of point charges are distributed in the immediate neighbourhood of a surface so that the charges on the surface element $dS$ can be considered as a surface charge $\omega\,dS$, with *surface density* $\omega$, then, according to §10, we have on the surface a source of the normal component of **E** ("surface divergence"):

$$(E_n)_1 - (E_n)_2 = 4\pi\omega \tag{19.8}$$

in which the index 1 refers to the same side of the surface as that indicated by the normal **n**.

*If the charge distribution is given*, the electrostatic potential $\phi$ is calculated, and with it the field **E** according to the procedures of §12, or, more simply, by superposing the Coulomb fields of the individual charge elements. For a system of point charges $e_1, e_2, \ldots e_h$ located at points having position vectors $\mathbf{r}_1, \mathbf{r}_2, \ldots \mathbf{r}_h$, we have, corresponding to (19.3) that

$$\phi(\mathbf{r}) = \sum_{j=1}^{h} \frac{e_j}{|\mathbf{r} - \mathbf{r}_j|} \tag{19.9}$$

From (19.7), and the fact that $\mathbf{E} = -\operatorname{grad}\phi$, we have for volume charge-distributions the *Poisson equation*

$$\nabla^2\phi = -4\pi\rho \tag{19.10}$$

and it follows from (12.9), or immediately from (19.9) when $\rho(\mathbf{r}')dv'$ is substituted for $e_j$ and $\mathbf{r}'$ for $\mathbf{r}_j$ that

$$\phi(\mathbf{r}) = \iiint \frac{\rho(\mathbf{r}')}{|\mathbf{r} - \mathbf{r}'|} dv' \tag{19.11}$$

Correspondingly, for surface distribution of charges, we have

$$\phi(\mathbf{r}) = \iint \frac{\omega(\mathbf{r}')}{|\mathbf{r} - \mathbf{r}'|} dS' \tag{19.12}$$

*Remark:* While in the Gaussian system of units the constant $f$ in Coulomb's law (19.1) is arbitrarily chosen equal to 1 and the units of charge and electric intensity thereby established, in the MKSA system we start with an arbitrary definition of the unit of current strength (A) and from this we deduce the units of the remaining electromagnetic quantities, as for example the unit of voltage, as watts/amperes, or the unit of charge the coulomb (C) as ampere-seconds. But then, however, the constant $f$ of Coulomb's law becomes a quantity to be determined experimentally, having dimensions of voltage times length divided by charge. In the MKSA system we say that

$$f = 1/4\pi\varepsilon_0$$

where, according to experiment, the new constant $\varepsilon_0$ is equal to

$$\varepsilon_0 = 8 \cdot 854 \times 10^{-12} \text{ As/Vm} \approx \frac{1}{4\pi \times 9 \times 10^9} \text{ As/Vm}$$

In order to arrive at a comparison of the units for the two measure-systems employed in this book, we take as our starting point that the symbols $F$, $e$ and $r$ in equation (19.1) shall, in the two systems, represent the same physical quantities. This must also hold then for the symbol $f$. In the Gaussian system this quantity thus becomes dimensionless and simply equal to 1, and this is because we have set the unit of charge (which we wish temporarily to designate here by $e_g$) equal to $1 \sqrt{(\text{dyne cm}^2)} = 1 \sqrt{(\text{erg cm})}$. If we did not do this, then, in these units, we should have that $f = 1 \text{ erg cm}/e_g{}^2$. On the other hand we have in the MKSA system that $f = 1/4\pi\varepsilon_0 \approx 9 \times 10^9$ Vm/As. Setting these expressions equal to one another, and observing that $1 \text{VAs} = 1 \text{VC} = 1 \text{Ws} = 10^7 \text{ ergs}$, and that $1 \text{ m} = 10^2 \text{ cm}$, we obtain the relationship

$$f = 1 \frac{\text{erg cm}}{e_g{}^2} \approx 9 \times 10^9 \frac{\text{Vm}}{\text{As}} = 9 \times 10^{18} \frac{\text{erg cm}}{\text{C}^2}$$

From this it follows that $1 \text{C} \approx 3 \times 10^9 \, e_g$; correspondingly, the unit of current, the ampere, is about $3 \times 10^9$ larger than the Gaussian unit of current.

In order to find out how big $1 \text{V}$ of potential difference is in Gaussian units, we make use of the defining equation of the volt:

$$1 \text{V} = 1 \frac{\text{W}}{\text{A}} \approx \frac{1 \times 10^7}{3 \times 10^9} \frac{\text{erg}}{e_g} = \frac{1}{300} \sqrt{\frac{\text{erg}}{\text{cm}}}$$

Correspondingly, the unit of field strength $1 \text{ V/m}$ is about $1/30{,}000$ of the corresponding Gaussian unit.

Finally, we wish further to remark that the physical quantity $\varepsilon_0$ (whose value in the MKSA system was given above) is, like $f$, dimensionless in the Gaussian system. The value of $\varepsilon_0$ in the Gaussian system is $1/4\pi$. (See the table on p. 431.)

We now wish to transcribe to the MKSA system the formulae of §19 (which are written in the Gaussian system), retaining throughout the factor $f = 1/4\pi\varepsilon_0$ of Coulomb's law.

Corresponding to (19.2) and (19.3), Coulomb's law now reads

$$\mathbf{E} = \frac{e\mathbf{r}}{4\pi\varepsilon_0 r^3} = - \text{ grad } \phi \quad \text{with} \quad \phi = \frac{e}{4\pi\varepsilon_0 r}$$

With this, from (19.5) and (19.6) we have for the electric flux theorem that

$$\oiint E_n \, dS = \frac{1}{\varepsilon_0} \text{ times the enclosed volume.}$$

From this, in place of (19.7), there follows the differential law that

$$\text{div } \mathbf{E} = \frac{\rho}{\varepsilon_0}$$

while equation (19.10) assumes the form

$$\nabla^2 \phi = - \frac{'\rho}{\varepsilon_0}$$

Finally, in formulae (19.9), (19.11) and (19.12) the factor $1/4\pi\varepsilon_0$ of the Coulomb field enters, so that instead of (19.11), for example, we have for the potential of a volume distribution of charge the expression

$$\phi(\mathbf{r}) = \frac{1}{4\pi\varepsilon_0} \int\int\int \frac{\rho(\mathbf{r}')\,dv'}{|\mathbf{r}-\mathbf{r}'|}$$

For simplifying the formulation of the electric-flux theorem it was first suggested by Heaviside that instead of the Gaussian system of units another system be introduced by assigning to the factor $f$ in Coulomb's law the value $1/4\pi$. The formulae in Heaviside units are therefore obtained from formulae of the technical system of units by letting $\varepsilon_0$ everywhere equal 1. However, in spite of the factor-free form thus obtained for the flux-theorem expression, this system of units has not found significant acceptance either in pure or in applied physics.

## §20. The distribution of electricity on conductors

In problems of electrostatics, matters are seldom so simple that the distribution of the electricity is given, and the potential then evaluated by means of (19.9), (19.11), or (19.12). The distribution of electricity on metal bodies is itself determined by particular requirements, and we turn now to the establishment of these.

Metals, in their simultaneous contact with two other differently charged bodies, possess the property of conducting a certain amount of charge from one of the bodies to the other. Bodies possessing this property are called *electrical conductors*, and others without this property are called *insulators*. These classes of bodies are not always easy to distinguish.

The decision whether to describe an object as a conductor or an insulator depends essentially on the duration of the observation. If the object is brought into an electrostatic field, there first ensues within such an object a field, and this field has in every case an electric current as a consequence. This current has the tendency to produce a charge distribution on the surface of the body so that everywhere within the body the external field is just compensated. When this condition is reached we have again before us an electrostatic condition in which within the object a null field obtains everywhere. Now there are two extreme cases possible. Either the time-interval for reaching this final condition is small compared to the time of observation (e.g. $10^{-6}$ s); in this case we shall always find a zero field within the object and thus describe it as a conductor. Or the time is very long (days or months); then any current will be so small that during the usual time of observation no appreciable influence on our observation has taken place. In this case we speak of an insulator. Pure electrostatics, then, deals only with idealized bodies, namely bodies for which every time is either infinitely short ("metals"), or else infinitely long ("insulators"). Thus in electrostatics metals are characterized by the field **E** within the conductor being everywhere zero. Or, in other words:

*The electrostatic potential $\phi$ within a conductor is constant.*

The field of charged metal bodies placed in an otherwise charge-free and matter-free region is to be described as follows:
In the whole external space

$$\operatorname{div}\mathbf{E} = -\operatorname{div}\operatorname{grad}\phi = -\nabla^2\phi = 0 \qquad (20.1)$$

Within the region occupied by metal there is no field, and thus there, and on the surface of each conductor, the potential $\phi$ has a constant value

$$\phi = \text{constant} \qquad \mathbf{E} = 0 \text{ within metal} \qquad (20.2)$$

Inside the metal there are therefore no net charges; there are however charges on the surface. From these surface charges of density $\omega$ there emanates a flux of electric force for which we have

$$E_n = 4\pi\omega \qquad (20.3)$$

**n** being the outward-drawn normal. The conductor then carries the total charge

$$e = \oiint \omega\, dS = \frac{1}{4\pi} \oiint E_n\, dS = -\frac{1}{4\pi} \oiint \frac{\partial\phi}{\partial n}\, dS \qquad (20.4)$$

Now in general we know either the field **E** from which the charge distribution on the surfaces could be calculated by (20.3), or this distribution itself is known and from it the field can be calculated by (19.12). Essentially, the basic problem of electrostatics, where conductors are present, consists in solving Laplace's equation (20.1) with the auxiliary condition (20.2) such that either the potential of the individual conductors agrees with the given values of $\phi$, or their total charge agrees with the given value of $e$. Thus, for a metallic body either $\phi$ or $e$ can be specified beforehand; the value of the unspecified quantity then follows from the solution of the problem.

That with the above requirements this solution is uniquely determined follows from Green's theorem (10.15). For the region bounded by the metal surface this is

$$\sum_i \oiint_{S_i} \phi\, \frac{\partial\phi}{\partial n}\, dS = -\iiint (\operatorname{grad}\phi)^2\, dv$$

Assuming that $\phi_1$ and $\phi_2$ are two solutions of the problem, then for $\phi = \phi_1 - \phi_2$, we should have on each surface $S_i$ either that $\phi = 0$, or that

$$\oiint \frac{\partial \phi}{\partial n} dS = 0$$

Consequently, in the whole space $(\text{grad} \, \phi)^2 = 0$; this means, however, that $\phi_1$ and $\phi_2$ differ by at most an additive constant, when, that is, for each conductor the charge $e_i$ has been previously specified. If, however, for just one of the conductors the potential $\phi_i$ has been specified, the absolute magnitude of the potential everywhere is then determined.

## §21. The capacitance of spherical and parallel-plate capacitors

The problem of electrostatics has been rigorously solved in only a few cases. The simplest case is that of a charged metal sphere. Let $e$ be its charge and $a$ its radius. We infer at once from the symmetry that the distribution of charge is uniform, so we can write for the surface density $\omega$

$$\omega = \frac{e}{4\pi a^2}$$

The potential of this irrotational field is

$$\phi = \frac{e}{r} + k \tag{21.1}$$

On the sphere it has the constant value

$$\phi_a = \frac{e}{a} + k$$

In order to obtain an electrostatic field that is physically possible we must now assign a terminus or sink for the flux of force issuing from the charged sphere. We shall say that a second concentric hollow sphere of internal radius $b$ encloses the first, and that its surface is charged with negative electricity. Since the charge $-e$ is distributed uniformly over this sphere, the surface density has the value

$$\omega = -\frac{e}{4\pi b^2}$$

and at $r = b$ the potential (21.1) is

$$\phi_b = \frac{e}{b} + k$$

This arrangement is called a *spherical capacitor*. The quotient of the positive charge $e$ by the "voltage" or potential difference $\phi_a - \phi_b = V$ between the positively and negatively charged conductors is called the *capacitance* of the capacitor.* For the spherical capacitor we have

$$V = \phi_a - \phi_b = e\left(\frac{1}{a} - \frac{1}{b}\right) = e\frac{b-a}{ab}$$

and its capacitance is then

$$C = \frac{e}{\phi_a - \phi_b} = \frac{ab}{b-a} \tag{21.2}$$

By diminishing the distance $b - a$ between the two spheres very large capacitances can be obtained.

When the capacitance of a single sphere by itself is spoken of, it is implied that the outer sphere which carries the opposite charge is situated at a very great distance; in this case the capacitance of the sphere is equal to its radius $a$. In laboratory experiments the total quantity of electricity associated with a field is always nil. It is therefore necessary in each case to specify the position of the corresponding charge of the opposite sign, i.e. the place where the flux of force from the sphere ends. In laboratory experiments the flux terminates on the walls of the room, or on the surface of some conductor in the room. If these terminating objects are situated at distances which are great compared with the radius of the sphere, the capacitance of the sphere is practically equal to its radius.

A *parallel-plate capacitor* consists of two metallic plates, the planes of which are parallel, and the distance between them small compared with their lateral dimensions. Neglecting the fringing of the lines of force near the edge, we have between the plates a uniform field

$$E = \frac{\phi_1 - \phi_2}{d}$$

and consequently we have a surface density $\omega$ of electricity given by

$$\omega = \frac{\phi_1 - \phi_2}{4\pi d}$$

* Physicists often speak of the "capacity" of a "condenser" rather than of the "capacitance" of a "capacitor".

The capacitance of a parallel-plate capacitor in which the separation of the plates is $d$ and the area of each plate is $S$ is therefore

$$C = \frac{S\omega}{\phi_1 - \phi_2} = \frac{S}{4\pi d} \qquad (21.3)$$

This formula can be regarded as a special case of (21.2). If, in fact, the two radii $a$ and $b$ are very nearly equal, the spherical capacitor may be looked upon as a plate capacitor with plate separation $d = b - a$, and plate area $S = 4\pi ab$.

*Remark:* In the MKSA system the voltage of a spherical capacitor is

$$V = \phi_a - \phi_b = \frac{e}{4\pi\varepsilon_0}\left(\frac{1}{a} - \frac{1}{b}\right) = \frac{e(b-a)}{4\pi\varepsilon_0 ab}$$

and its capacitance is therefore

$$C = \frac{4\pi\varepsilon_0 ab}{b-a}$$

For the parallel-plate capacitor we have in place of (21.3)

$$C = \frac{\varepsilon_0 S}{d}$$

The adherent of the "rationalized" system of units, such as Heaviside's system of the MKSA system mentioned in §19, sees in the system an especial advantage in that here the factor $4\pi$ appears for the spherical capacitor and not, as in the Gaussian system, for the parallel-plate capacitor. Also it is to the advantage of the MKSA system (in contrast to the Gaussian system) that the capacitance no longer has the dimension of a length—unsatisfactory from an electrodynamic standpoint— but the dimensions $A\,s/V \equiv$ farad F, which is immediately understandable from the general definition of capacitance. The new unit, the farad, is $9 \times 10^{11}$ times larger than the corresponding Gaussian unit.

## §22. The prolate ellipsoid of revolution

We wish now to consider an electrically charged conducting prolate ellipsoid of revolution. What is its field, and what is the value of its capacitance? Since for conductors of arbitrary form there is no general solution to the fundamental problem of electrostatics as formulated in §20, we wish to calculate the capacitance of the prolate ellipsoid of revolution by a particular method practicable only for this form of conductor. We imagine the line which joins the focal points of the ellipsoid to possess a uniform distribution of sources. We first show that the equipotential surfaces of the accompanying irrotational field are confocal ellipsoids of revolution. The strength of the entire source-line, of length $2c$, we set equal to $e$. The potential produced by it is given by

$$\phi = \frac{e}{2c} \int_{-c}^{+c} \frac{d\zeta}{r}$$

in which $\zeta$ is the distance of the point of integration from the mid-point of the source line, and $r$ is the distance of the field point from the source "point". Obviously this $\phi$-function satisfies Laplace's equation (20.1). Placing the $z$-axis along the source line and the origin of coordinates at its mid-point, we have $r = \sqrt{[(z-\zeta)^2 + x^2 + y^2]}$ and we find that

$$\phi = -\frac{e}{2c} \left\{ \ln(z - \zeta + r) \right\}_{\zeta = -c}^{\zeta = +c} = \frac{e}{2c} \ln \frac{z + c + r_1}{z - c + r_2} \qquad (22.1)$$

where $r_1$, $r_2$ are the distances of the field point from the end points designated $\zeta = -c$ and $\zeta = +c$ of the source line.

For discussing equation (22.1) we introduce elliptical coordinates $u$ and $v$ in the $zh$-plane:

$$u = \tfrac{1}{2}(r_1 + r_2) \qquad v = \tfrac{1}{2}(r_1 - r_2)$$

From figure 18 we immediately see that

$$r_1^2 = (z + c)^2 + h^2 \qquad r_2^2 = (z - c)^2 + h^2$$

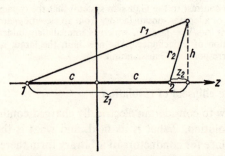

Fig. 18.—For calculating the potential of a charged line of length $2c$

From these equations it follows that

$$r_1 = u + v \qquad r_2 = u - v$$

and

$$cz = uv$$

Upon introducing these expressions into (22.1), the fraction in the logarithm can be simplified, and we obtain

$$\phi = \frac{e}{2c} \ln \frac{u+c}{u-c}$$

The potential is therefore constant on the surface $u = $ constant. These are the ellipsoids of revolution with the end points of the charged line as foci. At very great distances ($u \gg c$) we have

$$\frac{u+c}{u-c} \approx 1 + \frac{2c}{u} \qquad \ln\left(1 + \frac{2c}{u}\right) \approx \frac{2c}{u}$$

In this extreme case $u$ is practically equal to the distance from the origin, and $\phi$ therefore asymptotically passes over to the potential of a point charge $e$.

If we imagine any one of the family of prolate ellipsoids as being a conductor, then the field of this surface satisfies all the conditions of the electrostatic problem for the very distant sphere on the one hand, and those of the boundary surface on the other. The field here is irrotational and contains no sources; the total flux of force issuing from the ellipsoid is equal to the strength $e$ of the charged line; and lastly, the two conducting surfaces bounding the field are equipotential surfaces. Since, according to §20, these conditions uniquely determine the electrostatic field, $\phi$ then is the potential of the desired field.

The semi-major axis $a$ of the imaginary conducting ellipsoid is equal to the value of $u$ on its surface. We have, therefore, on this ellipsoid

$$\phi_1 = \frac{e}{2c} \ln \frac{a+c}{a-c}$$

and on the infinitely distant sphere $\phi_2 - 0$. Accordingly, for the capacitance $C$ of the prolate ellipsoid of revolution with semi-major axis $a$ and eccentricity $c$ we have

$$\frac{1}{C} = \frac{\phi_1}{e} = \frac{1}{2c} \ln \frac{a+c}{a-c}$$

In terms of the semi-minor axis $b = \sqrt{(a^2 - c^2)}$, we have

$$\frac{1}{C} = \frac{1}{\sqrt{(a^2 - b^2)}} \ln \frac{a + \sqrt{(a^2 - b^2)}}{b}$$

For ellipsoids prolate in the extreme, i.e. for small values of the quotient $b/a$, we have

$$\frac{1}{C} = \frac{1}{a} \ln \frac{2a}{b}$$

This expression gives at once the approximate capacitance of a straight wire of length $2a$ and diameter $2b$.

The distribution of charge on the surface of the ellipsoid is determined by a simple rule: a slice of the ellipsoid perpendicular to the $z$-axis and of thickness $dz$ carries a charge $e\,dz/2a$. If then we imagine the whole charge of the ellipsoid projected perpendicularly on to its axis of symmetry, this axis would be uniformly charged.

For proof we recall the fact that everywhere the lines of electric force are perpendicular to the equipotential surfaces $u = $ constant (these are confocal ellipsoids of revolution), and therefore always lie on the surfaces $v = $ constant (which are confocal hyperboloids of revolution). Making use of the electric-flux theorem (19.4) for a region bounded by the two hyperboloids belonging to $v$ and to $v+dv$ and the ellipsoid having the parameter value $u_0$, we find that the hyperboloid surfaces contribute nothing to the surface integral because on them $E_n = 0$. Thus in the integral of the flux of electric force only the surface element of the ellipsoid actually contributes. Consequently the surface charge between the two hyperboloids $v$ and $v+dv$ must always have the same value independent of the length $a < u_0$ of the semi-major axis of the charge-bearing ellipsoid. It must therefore be equal to the charge $e\,dv/2c$ belonging to the interval $dv$ on the axis, the latter being uniformly charged for, since $z = uv/c$, on the axis $u = c$ we have $z = v$ and therefore $dz = dv$. On the other hand, on the ellipsoid $u = a$ we obviously have $dz = a\,dv/c$, so that the charge $e\,dz/2a$ must also, as asserted, be carried by the portion of surface belonging to $dv$.

From elementary geometry and the equation of the ellipse

$$\frac{z^2}{a^2} + \frac{h^2}{b^2} = 1$$

we have for the area of the excised surface-piece

$$dS = 2\pi h \sqrt{(dz^2 + dh^2)} = 2\pi b\,dz \sqrt{[1 - (cz/a^2)^2]}$$

The surface charge density is therefore

$$\omega = \frac{e}{4\pi ab \sqrt{[1 - (cz/a^2)^2]}}$$

The density has its smallest value $\omega_{\min} = e/4\pi ab$ on the equator of the ellipsoid (where $z = 0$), and its greatest value $\omega_{\max} = e/4\pi b^2$ at the poles ($z = \pm a$), in agreement with the well-known fact that the field strength $E = 4\pi\omega$ on the surface of a charged metallic body increases with increasing curvature ("sharp-point effect").

## §23. Induced charges

### (a) Point charge opposite a conducting plane

We suppose the field to be bounded on one side by an infinite plane which forms the surface of a conductor. At a point A, distant $a$ from this plane, let a small body be placed, charged with $e$ units of electricity

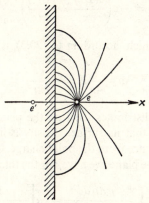

Fig. 19.—Pattern of the field lines of a point charge opposite a conducting plane

(figure 19). The dimensions of this body are so small that its electric field, in the absence of the conducting plane, would be derivable from the potential

$$\phi = \frac{e}{r}$$

The question now arises: how is the field affected by the presence of the conducting plane? The above potential obviously in no way satisfies the requirement of being constant on the surface equivalent to that of the conducting plane. We can, however, obtain a field for which that plane is an equipotential surface by taking, along with the point A, another point B which is the image of A with respect to the plane, and supposing a charge of opposite sign $-e$ to be placed at B. If $r'$ is the distance of the field point from the image point, then

$$\phi = \frac{e}{r} - \frac{e}{r'} \tag{23.1}$$

represents the potential of the combined field in the half-space considered. On the same side of the plane as point A this irrotational field is source-free, with the exception of point A itself; from here a flux of force issues, the strength being $4\pi e$.

On the bounding plane the electric field, which is there normal to the plane, has the value

$$E = -\frac{\partial \phi}{\partial x} = -e\left(\frac{\partial(1/r)}{\partial x} - \frac{\partial(1/r')}{\partial x}\right)_{x=0} = -\frac{2ea}{r^3}$$

The surface density which, according to (20.3), is proportional to it is

$$\omega = \frac{1}{4\pi}E = -\frac{ea}{2\pi r^3} \tag{23.2}$$

Hence the electricity distributes itself on the plane surface of the conductor in such a way that its surface density is inversely proportional to the cube of the distance from the point charge. On introducing polar coordinates $(\rho, \psi)$ in the plane, we have, for the total charge on the plane,

$$\int \omega \, dS = -\frac{ea}{2\pi}\int_0^\infty \int_0^{2\pi} \frac{\rho \, d\rho \, d\psi}{(\rho^2 + a^2)^{3/2}} = ea\left[\frac{1}{\sqrt{(\rho^2 + a^2)}}\right]_{\rho=0}^{\rho=\infty} = -e$$

Accordingly, the whole of the flux of force which begins at A terminates on the plane surface of the conductor.

The field strength which produces this surface charge in the surrounding space and also at the location of charge $e$, is identical with that which would be produced by the image charge $e' = -e$. There acts on $e$, therefore, the *image force* $e^2/4a^2$, directed toward the metallic surface. The corresponding potential energy of the charge is equal to $-e^2/4a$.

The phenomenon in which an electrically charged body calls forth a charge of opposite sign on the surface of a neighbouring conductor, originally uncharged, is called *electrical influence* or *electrostatic induction*. This phenomenon may be regarded as a consequence of the fact that the field cannot penetrate into the interior of the conductor. If the conductor is of finite dimensions and has no conducting connection with other bodies, then, seeing that its total charge continues to be nil, the flux of electric force which reaches it on the side facing the exciting point must leave the conductor again on the other side. In the case discussed above, namely that of a conductor extending to infinity and cutting out the field on one side, we must regard the charge $+e$, which was produced at the same time as the induced charge $-e$ when the exciting point was brought up, as having been removed to infinity.

When a point charge (say a small charged body used as a test charge for exploring a field) is brought into the neighbourhood of a charged conductor, its field, as influenced by the presence of the conductor, is

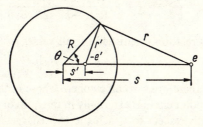

Fig. 20.—Location of the point of zero potential for two charges of opposite sign

superimposed upon the original field of the conductor. Hence the actual force on the test body will not correspond to the original charge distribution on the conductor, but to the distribution as changed by the presence of the testing body itself. It follows that the force will not give an exact measure of the original field, and that its measure will be less and less correct the greater the charge on the testing body and the nearer it is brought to the conductor. In the immediate neighbourhood of the conducting surface the method of finding the vector **E** given in §18 is only correct when the charge on the test body can be made infinitely small. Strictly speaking, the vector **E** is determined only as the limiting value of the ratio of the force on the test charge to the magnitude of this charge as this magnitude is indefinitely decreased.

### (b) *Point charge and conducting sphere*

To investigate the *induced charge on a conducting sphere* we first consider the following problems (figure 20): Let two charges $e$ and $-e'$, and their distance of separation, be given. It is required to find the surface on which the potential

$$\phi = \frac{e}{r} - \frac{e'}{r'}$$

has the value zero. Let $e'$ be, absolutely, the smaller of the two charges.

We take as origin of polar coordinates $(R, \theta)$ a point beyond $e'$ on the prolongation of the line $e \rightarrow e'$ and denote its distance from the two charges by $s$ and $s'$. We have then that

$$r^2 = R^2 + s^2 - 2Rs\cos\theta \qquad r'^2 = R^2 + s'^2 - 2Rs'\cos\theta$$

The potential is therefore equal to zero if

$$\frac{e^2}{e'^2} = \frac{r^2}{r'^2} = \frac{s}{s'} \frac{R^2/s + s - 2R\cos\theta}{R^2/s' + s' - 2R\cos\theta}$$

We see that this relation is satisfied for all values of $\theta$ if

$$R^2 = ss' \quad \text{and} \quad s/s' = e^2/e'^2$$

The potential is therefore zero on a sphere whose centre divides the line joining the two point charges externally in the ratio of the squares of the charges, and with respect to which the two charges occupy conjugate points.

We now consider a charge $e$ which is distant $s$ from the centre of a charged sphere of radius $R$. Let the sphere at first be held at potential zero (by being in conductive contact with the earth). A glance at figure 20 allows us to write down the solution at once: The sphere is imagined to have been removed, and instead of this a point charge

$$-e' = -e\sqrt{(s'/s)} = -eR/s$$

is placed a distance $s' = R^2/s$ from the (former) centre of the sphere. This charge then, together with the given point charge, produces a field whose potential has the value zero at the originally given spherical surface, and, external to this surface, exhibits only the single source $e$. The potential outside the earthed sphere is therefore given by

$$\phi = \frac{e}{r} - \frac{e'}{r'}$$

If, as another problem, we consider the sphere to be insulated and uncharged prior to the approach of the point charge, then it naturally remains uncharged (overall) at all later times. In order to describe its field we have therefore to imagine a further charge $+e'$ placed within it such that the constancy of the potential on the surface is not thereby

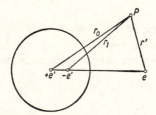

Fig. 21.—A point charge opposite an insulated metal sphere

disturbed; i.e. we think of a charge $+e'$ in the centre of the sphere (figure 21). The potential of the arrangement consisting of the point charge and the insulated uncharged sphere thus becomes

$$\phi = \frac{e}{r} - \frac{e'}{r'} + \frac{e'}{r_0}$$

in which $r_0$ denotes the distance of the field point from the centre of the sphere. On the surface of the sphere the potential is $e'/R = e/s$, i.e. it is the same as would prevail at the centre of the sphere in the absence of the latter.

In virtue of the induced charges, the force with which the charge $e$ is attracted toward the electrically isolated sphere is equal to

$$ee'\left(\frac{1}{(s-s')^2} - \frac{1}{s^2}\right) = \frac{e^2 R}{s^3}\left(\frac{s^4}{(s^2-R^2)^2} - 1\right)$$

If the distance $s - R = x$ of the charge from the spherical surface is small compared to the sphere radius $R$, the force is approximately equal to $e^2/4x^2$, and is therefore equal to the image force due to an infinite plane metallic surface. If, however, $s$ is large compared to the radius, we obtain the attractive force $2e^2 R^3/s^5$, corresponding to potential energy $-e^2 R^3/2s^4$ of a point charge.

It is interesting to see what happens if we send the point charge $e$ off to infinity, while at the same time increasing the value of its charge in such a way that the field which it produces in the neighbourhood of the sphere, namely $E = e/s^2$, retains the finite value $E_0$. In this process the image point $-e'$ goes, of course, to the centre of the sphere, and in such manner that

$$e's' = eR^3/s^2$$

retains the finite value $R^3 E_0$. Thus, for the insulated metallic sphere we obtain a double source, or, as we say (see §24), an *electric dipole* at the centre of the sphere, whose *moment* is given vectorially by the relation

$$\mathbf{p} = R^3 \mathbf{E}_0 \qquad (23.3)$$

The field $\mathbf{E}_0$ of the infinitely distant and infinitely large point charge is, of course, uniform in the neighbourhood of the sphere. Thus, *in a uniform electric field an insulated conducting sphere becomes "polarized" through induction in such a way that its surface charge acts in the exterior space like a dipole of moment $\mathbf{E}_0 R^3$, supposed to be situated at the centre of the sphere.*

### §24. The electric field at a great distance from field-producing charges. The dipole and quadrupole field

In §19 we calculated the potential of a given distribution of charge, and in so doing we found the relations* (19.11) and (19.9):

$$\phi(\mathbf{r}) = \int \frac{\rho(\mathbf{r}')}{|\mathbf{r} - \mathbf{r}'|} \, dv' \quad \text{and/or} \quad = \sum \frac{e_i}{|\mathbf{r} - \mathbf{r}_i|} \qquad (24.1)$$

We consider now the special case of a charge system lying everywhere within finite bounds, and we seek the potential and field pattern at distances which are large compared with the separation of the respective charges. We accordingly place the origin of the coordinate system in the neighbourhood of the charge system so that we shall always have $|\mathbf{r}| \gg |\mathbf{r}'|$ or $|\mathbf{r}| \gg |\mathbf{r}_i|$, and we develop $1/|\mathbf{r} - \mathbf{r}'|$ or $1/|\mathbf{r} - \mathbf{r}_i|$ in a Taylor series in increasing powers of the components of $\mathbf{r}'$ or $\mathbf{r}_i$. Since

$$\left( \frac{\partial}{\partial x'} \frac{1}{|\mathbf{r} - \mathbf{r}'|} \right)_{\mathbf{r}'=0} = - \left( \frac{\partial}{\partial x} \frac{1}{|\mathbf{r} - \mathbf{r}'|} \right)_{\mathbf{r}'=0} = - \frac{\partial}{\partial x} \frac{1}{r}, \quad \text{etc.} \qquad (24.2)$$

---

* Here and subsequently, for brevity, we shall usually write volume integrals with a single integral sign.

we have that

$$\frac{1}{|\mathbf{r}-\mathbf{r}'|} = \frac{1}{r} - \left( x'\frac{\partial(1/r)}{\partial x} + y'\frac{\partial(1/r)}{\partial y} + z'\frac{\partial(1/r)}{\partial z} \right) +$$

$$+ \frac{1}{2}\left( x'^2\frac{\partial^2(1/r)}{\partial x^2} + y'^2\frac{\partial^2(1/r)}{\partial y^2} + z'^2\frac{\partial^2(1/r)}{\partial z^2} \right) +$$

$$+ \left( x'y'\frac{\partial^2(1/r)}{\partial x\,\partial y} + x'z'\frac{\partial^2(1/r)}{\partial x\,\partial z} + y'z'\frac{\partial^2(1/r)}{\partial y\,\partial z} \right) + \dots \quad (24.3)$$

Thus we have, for example,

$$-\frac{\partial(1/r)}{\partial x} = \frac{1}{r^2}\frac{x}{r} \qquad \frac{\partial^2(1/r)}{\partial x^2} = \frac{1}{r^3}\frac{3x^2-r^2}{r^2} \qquad \frac{\partial^2(1/r)}{\partial x\,\partial y} = \frac{1}{r^3}\frac{3xy}{r^2} \qquad (24.4)$$

Since $x/r$, $y/r$, $z/r$ are trigonometric functions independent of $r$, the first expression in (24.3) in brackets decreases like $1/r^2$ with increasing $r$, and the next two like $1/r^3$, while further terms go with even higher powers of $r$. This $r$-dependence also carries over to the potential evaluated in (24.1), which we write in the form

$$\phi = \phi_0 + \phi_1 + \phi_2 + \dots \qquad (24.5)$$

in which all terms of development (24.3) occurring in the potential expression* and which, in going outward, fall off like $1/r^{n+1}$ are assembled in $\phi_n(\mathbf{r})$. We wish now to calculate and discuss the quantities $\phi_0$, $\phi_1$, and $\phi_2$.

### (a) The Coulomb field

As the first term of the development (24.5) we obtain immediately from (24.1) and (24.3)

$$\phi_0(\mathbf{r}) = \frac{1}{r}\int \rho(\mathbf{r}')\,dv \quad \text{or} \quad = \frac{1}{r}\sum_i e_i \qquad (24.6)$$

---

* The most general solution vanishing at infinity, of Laplace's equation $\nabla^2\phi = 0$, can be written in the form

$$\phi(\mathbf{r}) = \sum_{n=0}^{\infty} \frac{1}{r^{n+1}} Y_n(\theta, \alpha)$$

in conformity with (24.5). The quantities $Y_n$, which depend only on the polar angles, are called *generalized spherical harmonics*. As is shown in appropriate textbooks, for every index $n$ there are in all $2n+1$ independent spherical harmonics; the $Y_n$ therefore contain $2n+1$ arbitrary parameters which can be subsequently determined from the charge distribution producing the field.

In the first approximation the charge distribution acts at a great distance as if it consisted of a point charge of strength $\int \rho \, dv$ or $\sum_i e_i$, placed at the origin.

### (b) The dipole field

If the total charge of the system vanishes, the series development (24.5) begins with the term $\phi_1$. This can be given as

$$\phi_1(\mathbf{r}) = \frac{p_x x + p_y y + p_z z}{r^3} = \frac{\mathbf{p} \cdot \mathbf{r}}{r^3} \qquad (24.7)$$

in which vector $\mathbf{p}$, with its components $p_x$, $p_y$, $p_z$, is given according to (24.3) and (24.4) by

$$\mathbf{p} = \int \mathbf{r}' \rho(\mathbf{r}') \, dv' \quad \text{or} \quad = \sum_i \mathbf{r}_i \, e_i \qquad (24.8)$$

From (24.7) we obtain the field belonging to $\phi_1$:

$$\mathbf{E}_1(\mathbf{r}) = -\frac{\mathbf{p}}{r^3} + \frac{3(\mathbf{p} \cdot \mathbf{r})\mathbf{r}}{r^5} \qquad (24.9)$$

Figure 22, which represents the case of a horizontally oriented dipole, illustrates the field lines and the lines of constant potential, the latter provided with numbers, but to an arbitrary scale.

Such a field is called a *dipole field*, because the simplest charge distribution which for large distances leads to this field consists of two point charges (poles) of equal strength but of opposite sign. This arrangement is called a (physical) *dipole*. If $\mathbf{a}$ is the vector extending from the negative charge $-e$ to the positive charge $+e$, the *dipole moment* $\mathbf{p}$, defined by (24.8), is equal to

$$\mathbf{p} = e(\mathbf{r}_+ - \mathbf{r}_-) = e\mathbf{a} \qquad (24.10)$$

As follows from its derivation, the field of a physical dipole with finite pole separation $\mathbf{a}$ conforms to the field of (24.9) only for large distances, i.e. for $r \gg a$. If now, however, as we did in the preceding paragraph, we *simultaneously let e tend to infinity and let* $\mathbf{a}$ *tend to zero*, in such a way that $e\mathbf{a}$ retains the constant value $\mathbf{p}$, we arrive at an arrangement of charge which is called an ideal (or mathematical) dipole, the field of which is identical with the dipole field (24.9) down to $r = 0$.

Because of the importance of the field of an ideal dipole, we wish

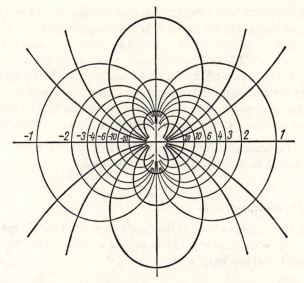

Fig. 22.—Field lines (drawn heavy) and equipotential lines (light)
for a horizontally oriented dipole

to derive its field in another way (figure 23). We start with a physical
dipole of finite pole separation **a**. For itself alone, the point charge $+e$
produces at the field point P the potential $\phi_+ = e/r$. Except for a

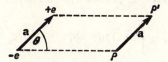

Fig. 23.—For defining a dipole

change of sign, the potential contribution of $-e$ at P is the same as
that of $+e$ at P′ which is displaced from P by the amount **a**. For the
total potential at P we therefore have

$$\phi(P) = \phi_+(P) + \phi_-(P) = \phi_+(P) - \phi_+(P')$$

According to the definition of the gradient, however, for sufficiently
small **a** this is

$$\phi = -\mathbf{a}\operatorname{grad}\phi_+ = -e\mathbf{a}\operatorname{grad}\frac{1}{r} = \mathbf{p}\cdot\frac{\mathbf{r}}{r^3}$$

in which again we have substituted $\mathbf{p} = e\mathbf{a}$. In the limits $\mathbf{a} \to 0$, $e \to \infty$ this formula holds exactly for all $\mathbf{r}$, and it exactly agrees with (24.7).

To avoid misunderstanding it should be pointed out that according to development (24.3) or (24.5), *a single point charge not at the origin leads to a dipole part of the field*. As can be immediately seen from (24.8) for the case of non-vanishing total charge, the dipole moment of the charge system can always be made to vanish by displacing the origin an amount $\sum_i \mathbf{r}_i e_i / \sum_i e_i$. The field $\phi_0$ is then so altered that the field $\phi_1$ is compensated away. In the case of vanishing total charge, i.e. for $\phi_0 = 0$, such a compensation is not possible.

### (c) The quadrupole field

If the total charge as well as the dipole moment of the charge system vanishes, then development (24.5) begins with the term $\phi_2$. From (24.3) and (24.4) this has the form

$$\left.\begin{aligned}
\phi_2(\mathbf{r}) = Q_{xx} \frac{3x^2 - r^2}{2r^5} + Q_{yy} \frac{3y^2 - r^2}{2r^5} + Q_{zz} \frac{3z^2 - r^2}{2r^5} \\
+ Q_{xy} \frac{3xy}{r^5} + Q_{xz} \frac{3xz}{r^5} + Q_{yz} \frac{3yz}{r^5}
\end{aligned}\right\} \quad (24.11)$$

with the abbreviations (the primes will be omitted in the following)

$$\left.\begin{aligned}
Q_{xx} = \int x^2 \rho(\mathbf{r})\, dv \quad &\text{or} \quad = \sum_i x_i^2 e_i \\
Q_{xy} = \int xy \rho(\mathbf{r})\, dv \quad &\text{or} \quad = \sum_i x_i y_i e_i \\
\cdots \cdots \cdots \cdots \cdots \cdots \cdots
\end{aligned}\right\} \quad (24.12)$$

The field thus given is called a *quadrupole field*. If it is required to produce the general field (24.11) with arbitrarily given $Q_{xx}$, $Q_{xy}$, $\ldots$ by means of a system of point charges having vanishing total charge ($\sum e_i = 0$) and vanishing dipole moment ($\sum \mathbf{r}_i e_i = 0$), at least four charges are in general required whose value and position are however not uniquely determined by the given $Q$-values. But a unique problem does result when, for example, we require additionally that all four charges shall have the same magnitude and shall occupy the four corners of a rhombus (figure 24a). The *elongated quadrupole* (figure 24b) represents an important special case. Here the rhombus collapses to a straight line which carries charges $+e$ on its ends, and a charge $-2e$ in the middle.

Clearly, the models of figure 24 can be made up of two equal dipoles rotated 180° with respect to one another and placed at the appropriate dipole separation. Thus, now, as with the dipole, we can pass from the real quadrupole having finite charges and charge separations, to the

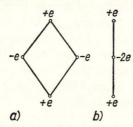

Fig. 24.—(a) General quadrupole (b) Stretched quadrupole

ideal quadrupole in which all separations are allowed to go to zero in a constant ratio; and, inverse to the square of this ratio, all charges are allowed to go to infinity—all this in such a way that the $Q$-values remain constant. Equation (24.11) then rigorously represents the quadrupole field for all **r**-values, and not only approximately for large $r$.

We must now briefly concern ourselves with the $Q$-values which, according to (24.12), are the components of a symmetric tensor, the *quadrupole moment* tensor. By §15 we can transform this tensor to principal axes, i.e. place the coordinate system so that all components having two different indices vanish. We have in addition further freedom in the choice of the components of the quadrupole moment. As is immediately clear from (24.11) the quadrupole field is not changed when the same constant $C$ is added to each of the three quantities $Q_{xx}$, $Q_{yy}$, $Q_{zz}$. We find the deeper reason for this by going back to the development (24.3): If there we add to $x'^2$, $y'^2$, $z'^2$ the same quantity $c$, we change the expression inside the second brackets by

$$c\left(\frac{\partial^2(1/r)}{\partial x^2}+\frac{\partial^2(1/r)}{\partial y^2}+\frac{\partial^2(1/r)}{\partial z^2}\right) = c\nabla^2\frac{1}{r} = 0$$

If therefore, for example, $Q_{xx} < Q_{yy} < Q_{zz}$, we can choose $C = -Q_{yy}$ so that, in the system of principal axes, the moment tensor takes the form

$$\begin{pmatrix} Q_{xx}-Q_{yy} & 0 & 0 \\ 0 & 0 & 0 \\ 0 & 0 & Q_{zz}-Q_{yy} \end{pmatrix} \equiv \begin{pmatrix} Q'_{xx} & 0 & 0 \\ 0 & 0 & 0 \\ 0 & 0 & Q_{zz} \end{pmatrix} \quad (24.13)$$

with $Q'_{xx} < 0$, $Q'_{zz} > 0$. By an appropriately chosen axis-ratio of the rhombus, the quadrupole shown in figure 24a leads immediately to such a tensor. Thus the assertion made above is additionally proved, namely that every quadrupole field can be produced by a suitably oriented quadrupole of rhombic form.*

In general however the diagonal terms of the moment tensor are not normalized as in (24.13); rather, by subtracting one-third of the spur we go to the corresponding *deviator* with vanishing spur (see §15). For the components of the *reduced quadrupole moment* thus obtained we have, in the principal diagonal,

$$\left. \begin{aligned} Q'_{xx} &= \tfrac{1}{3}\int(2x^2-y^2-z^2)\rho\,dv \quad \text{or} \quad = \tfrac{1}{3}\int\sum_i(2x_i^2-y_i^2-z_i^2)e_i \\ Q'_{yy} &= \tfrac{1}{3}\int(2y^2-x^2-z^2)\rho\,dv \quad \text{or} \quad = \tfrac{1}{3}\int\sum_i(2y_i^2-x_i^2-z_i^2)e_i \\ Q'_{zz} &= \tfrac{1}{3}\int(2z^2-x^2-y^2)\rho\,dv \quad \text{or} \quad = \tfrac{1}{3}\int\sum_i(2z_i^2-x_i^2-y_i^2)e_i \end{aligned} \right\} \quad (24.14)$$

while the other components remain unchanged. The advantage of this normalization is that for a spherically symmetric charge distribution $\rho = \rho(r)$ all components of the quadrupole moment vanish; for, on symmetry grounds,

$$\int x^2\rho(r)\,dv = \int y^2\rho(r)\,dv = \int z^2\rho(r)\,dv = \tfrac{1}{3}\int r^2\rho(r)\,dv$$

$$\int xy\rho(r)\,dv = \int xz\rho(r)\,dv = \int yz\rho(r)\,dv = 0$$

---

* The most general representation of a quadrupole field contains altogether five arbitrary parameters, namely (as an example) the quantities $Q'_{xx}$ and $Q'_{zz}$, as well as the three parameters which fix the orientation of the coordinate system. This agrees with the statement made in the footnote on p. 81 that, in the potential, the general spherical harmonic $Y_2(\theta, \alpha)$ in the quadrupole term which falls off like $1/r^3$ contains in all five arbitrary constants.

In this normalization the $Q'$-components can be regarded as a measure of the amount by which the charge system deviates from spherical symmetry.

We now consider the important special case in which the charges are distributed with rotational symmetry about an axis, say the $z$-axis. This axis is naturally then the principal axis of the moment tensor, so that we have $Q'_{xy} = Q'_{xz} = Q'_{yz}$. Designating further, as in §15, the three reduced principal quadrupole moments by $Q^{\mathrm{I}} = Q'_{xx}$, $Q^{\mathrm{II}} = Q'_{yy}$, $Q^{\mathrm{III}} = Q'_{zz}$, we have from rotational symmetry that $Q^{\mathrm{I}} = Q^{\mathrm{II}}$ and, because of the vanishing spur of the components matrix,

$$Q^{\mathrm{III}} = -2Q^{\mathrm{I}} = -2Q^{\mathrm{II}} \qquad (24.15)$$

Thus the quadrupole of a rotationally symmetric charge distribution is characterized by a single quantity. It can therefore be represented by a quadrupole elongated in the $z$-direction (figure 24$b$) whose non-reduced moment tensor, according to (24.12), possesses only the non-vanishing component $Q_{zz} = Q = 2ea^2$. For its reduced moment tensor, however, according to (24.14) we have

$$Q^{\mathrm{III}} = -2Q^{\mathrm{I}} = -2Q^{\mathrm{II}} = \tfrac{2}{3}Q \qquad (24.16)$$

Looking back at (24.14), we generally give as the quadrupole moment $Q$ of the rotationally symmetric charge distribution the quantity*

$$Q = \tfrac{1}{2} \int (2z^2 - x^2 - y^2)\rho\, dv = \int r^2\rho\, \frac{3\cos^2\eta - 1}{2}\, dv \quad (24.17)$$

where $\eta$ is the angle between the line to $dv$ and the $z$-axis. As an example, for an ellipsoid of revolution filled homogeneously with a volume distribution of charge, $a$ and $c$ being semi-axes, we have (with $x^2 + y^2 = \xi^2$)

$$Q = \tfrac{1}{2}\rho_0 \int_{-c}^{+c} dz \int_0^{a\sqrt{[1-(z/c)^2]}} 2\pi\xi\, d\xi(2z^2 - \xi^2)$$

$$= \tfrac{4}{15}\pi\rho_0\, a^2 c(c^2 - a^2) = \tfrac{1}{5}e(c^2 - a^2)$$

where $e$ is the total charge $\tfrac{4}{3}\pi\rho_0 a^2 c$ of the ellipsoid. Thus $Q$ is positive for a prolate ellipsoid of revolution ($c > a$) and negative for an oblate ellipsoid ($c < a$).

---

* In atomic physics the quantity $Q/e_0$, $e_0$ being the elementary charge quantum, or $2Q/e_0$ is usually designated as the quadrupole moment $Q$. (See H. Kopfermann, *Kernmomente*, 2nd ed., Leipzig 1956.)

Finally, it follows from (24.11) that the potential of the field of a rotationally symmetric quadrupole is given by

$$\phi_2(\mathbf{r}) = Q\frac{3z^2 - r^2}{2r^5} = \frac{Q}{r^3}\frac{3\cos^2\theta - 1}{2} \tag{24.18}$$

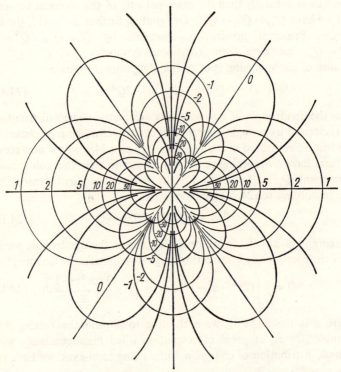

Fig. 25.—Field lines (drawn heavy) and equipotential lines (light) for a horizontally oriented stretched quadrupole

in which $\theta$ is the angle between the line to the field point and the quadrupole axis.* Figure 25 shows the accompanying field lines and potential lines.

---

* The $\theta$-factor in $\phi_2$ (also the $\eta$-factor in the $Q$-formula (24.17)) is the second zonal harmonic $P_2 (\cos\theta)$.

# CHAPTER B II

## Electrostatics of Dielectrics

### §25. The parallel-plate capacitor with dielectric insulation

We have so far restricted ourselves to the electric field in a vacuum. When we have spoken of the field in air there has been a slight inaccuracy which, as we shall shortly see, is in most cases of no actual importance. Accordingly, we now assert that the formulae of the foregoing chapters relate to a vacuum and to metals bounding a vacuum.

The fundamental discovery was made by Faraday that the capacitance of a capacitor changes when the space between its conductors is occupied by an insulator such as glass, sulphur, or petroleum. With any known material so used, the capacitance is in fact increased. The factor $\varepsilon$, by which $C$ is thus increased, proves to be a constant characteristic of the interposed substance. It is called the *permittivity* or *dielectric constant**
of the material in question. From §21 we therefore have for the parallel-plate capacitor

$$C = \frac{\varepsilon S}{4\pi d} \tag{25.1}$$

Numerical values of $\varepsilon$ at $20°\mathrm{C}$ and standard atmospheric pressure are, for example:

| | | | |
|---|---|---|---|
| Air | 1·0005 | Porcelain | 6 |
| Sulphur dioxide | 1·003 | Alcohol | 26 |
| Petroleum | 2·1 | Water | 81 |
| Glass | 5 to 8 | | |

In a vacuum $\varepsilon$ has by definition the value 1.

On the basis of the processes taking place in a parallel-plate capacitor (figure 26), we wish now to obtain a clear picture of the nature of Faraday's discovery.

Let the two opposing capacitor plates (each of area $S$, separated by distance $d$) be maintained at the constant potential difference

---

* Formerly "specific inductive capacity".

$V = \phi_1 - \phi_2$ by means, let us say, of an electric battery. Then, in a vacuum the electric field between the plates (directed from above to below in figure 26) has everywhere the constant value $E_0 = V/d$ and, correspondingly, on each plate there is a surface charge density of value $\omega_0$, thus

$$4\pi\omega_0 = E_0 = \frac{V}{d}$$

If now we insert into the capacitor an insulating plate of thickness $d$ and permittivity $\varepsilon$, we have in the part of the capacitor containing the dielectric the same field strength $E_0$ as before, since the line integral

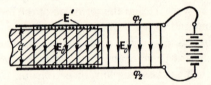

Fig. 26.—Insertion of dielectric into a parallel-plate capacitor

$\int_1^2 \mathbf{E} . d\mathbf{r}$ extending from one capacitor plate to the other has the given value $V$. With the insertion of the insulating material, however, the capacitance is increased, and also the surface charge density, the latter by the factor $\varepsilon$:

$$\omega = \varepsilon\omega_0$$

Accordingly, while the plate is being inserted in the capacitor, the battery must supply a quantity of electricity

$$\omega_P = \omega - \omega_0 = (\varepsilon - 1)\omega_0 = \frac{\varepsilon - 1}{4\pi} E_0 \qquad (25.2)$$

for every square centimetre covered by the insulator. That the battery really does this can easily be shown with the help of a ballistic galvanometer. In order that this effect may take place it is not at all necessary that the insulator (say a glass plate) be in contact with the plates of the capacitor. It happens just the same if we allow a narrow space to exist between the metal and the insulator, so long as the width of this space is small compared with the plate separation $d$. Now, experimentally, such a space does not influence the capacitance nor therefore the surface density of charge; consequently the narrow space must have an electric field strength of value

$$4\pi\omega = E' = \varepsilon E_0$$

for the metal surface borders on a vacuum. If we pass from the region of the narrow space into the insulator, we find that the field strength undergoes a discontinuity from $\varepsilon E_0$ to $E_0$. A discontinuity in the normal component of the field strength is always, however, synonymous with the existence of a surface charge. The effect of the insulator on the electrostatic field is therefore as though its surface carried a charge of density $\omega_P$, in accordance with (25.2).

*Remark:* In the MKSA system the "relative" permittivity $\varepsilon$ is also defined as the increase (by the factor $\varepsilon$) which is made in the capacitance of a capacitor through the introduction of a dielectric. Thus here also $\varepsilon$ is a pure number, in contrast to the quantity $\varepsilon_0 = 8.86 \times 10^{-12}$ As/Vm mentioned in the concluding remark of §19. This is usually (though not always happily) described as the permittivity of a vacuum.

The capacitance of a parallel-plate capacitor having a material dielectric is accordingly given by

$$C = \varepsilon\varepsilon_0 S/d$$

and the polarization charge $\omega_P$ is connected with the produced field by the relation

$$\omega_P = \omega - \omega_0 = (\varepsilon - 1)\omega_0 = (\varepsilon - 1)\varepsilon_0 E_0$$

## §26. Dielectric polarization

We give the name *polarizability* to the property described in §25, wherein an insulator, as a whole uncharged, influences a field applied to it. We speak of the insulator as having been *polarized* by the field. In order to understand this property we must assume that every material body contains positive and negative electric charges and, if it is electrically neutral, that it has each kind of charge in like amount. In conductors at least one of the two kinds of electricity is free to move (conduction electrons in metals; ions in electrolytes). In an insulator both kinds of electricity are quasi-elastically bound to fixed locations. Under the influence of the electric field the charges are displaced a short distance, the positives in the direction of the field and the negatives counter thereto; the amount of this displacement is proportional to the field strength. This mutual displacement of the two kinds of charge is called *polarization* and is measured by the vector **P** for which we can give two different but entirely equivalent definitions as follows:

Starting with the concept of polarization-displaced charges in insulators, we *first* define **P** so that $P_n dS$ means the quantity of charge which passes through a surface element $dS$ in the direction of the normal $\mathbf{n}$ when the insulator, initially unpolarized, is brought into a state of polarization by applying the field.

From this definition of $\mathbf{P}$ it follows immediately that a polarized insulator carries on its surface the surface charge

$$\omega_P = P_n \qquad (26.1)$$

If we consider a flat cylinder ("pill-box") of cross-sectional area $dS$, with one base surface lying outside the insulator and the other entirely within it, then a quantity of charge $P_n \, dS$ enters this cylinder during the act of polarizing the dielectric. In a similar way, for the charge which, during polarization, passes out of a closed volume within the body, we find the value

$$e_P = - \oiint P_n \, dS$$

in which the integral extends over the boundary surface of the enclosed volume, and $\mathbf{n}$ is the outward-drawn normal. By Gauss's theorem this expression can be written in the form

$$e_P = - \iiint \rho_P \, dv$$

with

$$\rho_P = - \operatorname{div} \mathbf{P} \qquad (26.2)$$

We see here that for the case of non-uniform polarization an additional volume distribution of charge $\rho_P$ has to be reckoned with.

We arrive at a *second* definition of $\mathbf{P}$ on considering the idea that in a polarized medium, owing to the charge displacement, every atom becomes the carrier of an *electric dipole* (see below). Since in their field effects these atomic dipoles combine additively, it would appear reasonable to combine all dipoles contained within a volume $dv$ into a single dipole. We call its moment $\mathbf{P} \, dv$, thus defining $\mathbf{P}$ as *the electric dipole moment per unit volume*.

We have now to show that the two definitions of $\mathbf{P}$ are in all ways equivalent. For this we calculate, from the second definition, the potential $\phi_P$ that is produced by a polarized body. From (24.7) the potential of a dipole of moment $\mathbf{p}$ is given by

$$\phi = \mathbf{p} \frac{\mathbf{r}}{r^3} = \mathbf{p} \cdot \operatorname{grad}' \frac{1}{r}$$

in which **r** is the vector from the source point (dipole) to the field point, and the prime on the gradient indicates differentiation with respect to the coordinates of the source point. Accordingly we have

$$\phi_P = \iiint \mathbf{P} \cdot \left( dv' \operatorname{grad}' \frac{1}{r} \right)$$

Since

$$\operatorname{div}' \frac{\mathbf{P}}{r} = \frac{1}{r} \operatorname{div}' \mathbf{P} + \mathbf{P} \cdot \operatorname{grad}' \frac{1}{r}$$

by using Gauss's theorem we obtain

$$\phi_P = \oiint \frac{P_n}{r} dS' - \iiint \frac{\operatorname{div}' \mathbf{P}}{r} dv'$$

This equation, however, says nothing other than that the insulator carries a charge of surface density $\omega_P = P_n$ and a space-distributed charge of density $\rho_P = -\operatorname{div} \mathbf{P}$, in agreement with formulae (26.1) and (26.2) coming from the first definition of **P**.

In the case of the parallel-plate capacitor (figure 26) examined above, we are obviously having to deal with a homogeneous polarization of the inserted dielectric in which the vector **P** is directed from above to below, its magnitude being given by the polarization charge

$$\omega_P = P = \frac{\varepsilon - 1}{4\pi} E$$

which comes out on the surface of the plate. Thus, also, with the help of the permittivity $\varepsilon$, the numerical connection between the vectors **P** and **E** is established as

$$\mathbf{P} = \frac{\varepsilon - 1}{4\pi} \mathbf{E} = \chi \mathbf{E} \tag{26.3}$$

The factor of proportionality $\chi$ thus defined is called the *electric susceptibility* of the material considered.

Let it be here noted, however, that this simple relationship between the polarization **P** and the field strength **E** by which the polarization is produced is by no means valid for all dielectrics. Thus in single crystals, for example, the direction of **P** does not in general coincide with that of $\mathbf{E}$. Rather, in this case, instead of the scalar quantities $\varepsilon$ and $\chi$, we have to deal with the tensors $\varepsilon_{ik}$ and $\chi_{ik}$:

$$P_i = \sum_{k=1}^{3} \chi_{ik} E_k \quad \text{with} \quad \begin{cases} \varepsilon_{ii} = 1 + 4\pi\chi_{ii} \\ \varepsilon_{ik} = \quad\; 4\pi\chi_{ik} \quad \text{for} \quad i \neq k \end{cases} \quad (26.4)$$

A linear interdependence of the components of $\mathbf{P}$ and $\mathbf{E}$ still holds here. There are, however, substances (e.g. Rochelle salt, barium titanate, etc.) in which the components of $\mathbf{P}$ are related to those of $\mathbf{E}$ in a much more complicated and non-linear way. In addition they show a dependence upon the prior treatment of the material similar to ferromagnetism, where no simple connection exists between the magnetization and the magnetic field producing it. In the following, however, unless otherwise noted, we shall always restrict ourselves to the simple case in which (26.3) is valid.

The linear relationship given by (26.3) between $\mathbf{P}$ and $\mathbf{E}$ can be understood from the behaviour of the individual atoms or molecules of the substance in the electric field. A distinction has to be made here between the case where the individual building blocks of the material do not possess a dipole moment in the unpolarized state, and the case where, already in the field-free condition, the molecules have a "permanent" dipole moment like, say, the ionic molecules HCl, HBr, and HI.

In the first case an applied electric field $\mathbf{F}$ produces in the individual atoms a dipole whose moment, for not too strong fields, is proportional to $\mathbf{F}$:

$$\mathbf{p} = \alpha\mathbf{F} \quad (26.5)$$

Here $\alpha$ is called the *polarizability*. Its value can be determined by calculation for any model representation of atomic features ("atom model"). Thus, from every atomic model so far developed, approximately the same value $\alpha \approx R^3$ has been obtained, $R$ being the radius of the atom. Long before the existence of the electron theory Mosotti had proposed to describe the dielectric property of matter by the assumption that the individual atoms behave like perfectly conducting spheres. In this case, by (23.3), we obtain exactly the result

$$\alpha = R^3 \quad (26.6)$$

If, in an analogous fashion, molecules be regarded as long stretched conducting structures (e.g. ellipsoids) $\alpha$ acquires an anisotropy depending upon molecular orientation, and this, among other things, means for $\chi$ a tensor structure corresponding to (26.4).

If, however, the individual molecules already possess a permanent dipole moment $p_0$ in the unpolarized state, then, in addition to the field-induced dipole moment as already described, there enters a partial ordering of the dipoles in the direction of the field, for the field attempts to rotate the dipoles into an orientation along its own direction. Against this orienting action of the field, however, there are the disorienting actions of thermal motion in the material and the persistent irregular and fluctuating interactions of the individual molecules. The calculation of the average component $\bar{p}$ of $\mathbf{p}$ in the direction of the applied field thus becomes a problem in statistical mechanics. For dipoles capable of free rotation, e.g. "dipolar gases", where the field strength is not too strong,

$$\bar{\mathbf{p}} = \frac{p_0^2}{3kT}\mathbf{F} \quad \text{i.e.} \quad \alpha = \frac{p_0^2}{3kT} \tag{26.7}$$

in which $T$ is the absolute temperature, and $k$ is Boltzmann's constant ($k = 1 \cdot 38 \times 10^{-16}$ erg/deg). (See for this and the following the full discussion in Section A of Volume III.)

In order to be able to obtain the macrophysical susceptibility from the foregoing considerations of atomic dipole moments, the connection with, and the distinction between, the *macrophysical field strength* E in matter and the *effective field strength* F introduced in (26.5) and (26.7) have to be known. We therefore arrive at the following decisive problem: In the space between the individual building blocks of matter (atomic nuclei, electrons) there certainly exists an electromagnetic field having extremely rapid variation with respect both to space and to time. Let us call the atomic electric field $\mathbf{e} \equiv \mathbf{e}(\mathbf{r}, t)$. In the neighbourhood of the $j$th point charge $e_j$ it tends to become infinite like $e_j \dfrac{\mathbf{r}-\mathbf{r}_j}{|\mathbf{r}-\mathbf{r}_j|^3}$. The field $\mathbf{F}_j$ acting on this point charge is therefore obtained from e by deducting its own field, which hitherto has furnished no contribution, thus

$$\mathbf{F}_j = \lim_{\mathbf{r} \to \mathbf{r}_j}\left(\mathbf{e}(\mathbf{r},t) - e_j\frac{\mathbf{r}-\mathbf{r}_j}{|\mathbf{r}-\mathbf{r}_j|^3}\right)$$

From the $\mathbf{F}_j$ so constructed the *effective field strength* F introduced above is obtained by taking the time-average over the positions of the charges taking part in the production of the atomic dipole moment. In contrast, the *macrophysical field strength* E is immediately obtained from the atomic field strength e by space- and time-averaging. Thus F would coincide with E only for the case of rarified gases in which the individual atoms can occupy arbitrary positions in space. In general, however, F and E are different from one another. As will be shown in Volume III, for example, for liquids without any permanent dipole moment, and for cubic crystals,

$$\mathbf{F} = \mathbf{E} + \frac{4\pi}{3}\mathbf{P} \tag{26.8}$$

For $n$ atoms we have in this case for the polarizability $\alpha$ per unit volume, from (26.5),

$$\mathbf{P} = n\alpha\mathbf{F} \quad \text{therefore} \quad \mathbf{F}\left(1 - \frac{4\pi}{3}n\alpha\right) = \mathbf{E} \tag{26.9}$$

Following from (26.3) we have then

$$\varepsilon - 1 = 4\pi\chi = \frac{4\pi n\alpha}{1 - \dfrac{4\pi}{3}n\alpha} \quad \text{i.e.} \quad \frac{\varepsilon-1}{\varepsilon+2} = \frac{4\pi}{3}n\alpha \tag{26.10}$$

This is the well-established formula of Clausius and Mosotti. According to this formula the quotient $(\varepsilon-1)/(\varepsilon+2)$, and not $\varepsilon-1$, is proportional to the density of the material. Only for low densities, i.e. where $n\alpha \ll 1$, we have, approximately

$$\varepsilon - 1 = 4\pi\chi \approx 4\pi n\alpha \tag{26.11}$$

*Remark:* In going over to the MKSA system, formulae (26.1) and (26.2) for the polarization charges remain unchanged. Instead of (26.3), however, we now have

$$\mathbf{P} = (\varepsilon - 1)\varepsilon_0\mathbf{E} = \chi\varepsilon_0\mathbf{E}$$

Here too the susceptibility $\chi$ is introduced as a dimensionless constant of the material. Its value is greater, however, by the factor $4\pi$ than the susceptibility $\chi$ defined by (26.3). (In the Gaussian system we have that $\varepsilon_0 = 1/4\pi$.)

## §27. The fundamental equations of electrostatics for insulators. The Maxwell displacement vector

The results of the last two sections now allow us to formulate the basic equations of electrostatics for the case in which metallic bodies are imbedded in dielectrics. First, we have as before that

$$\operatorname{curl} \mathbf{E} = 0, \quad \text{i.e.} \quad \mathbf{E} = -\operatorname{grad} \phi \tag{27.1}$$

for, here also, in carrying a charge around a closed curve no work can be performed. Therefore, at the interface between two dielectrics *the tangential component of* $\mathbf{E}$ *is continuous*. Otherwise, taking a test charge along a bounding interface would involve different amounts of work according as the path was by way of the first or the second insulator. We should recall here that even for metal surfaces the tangential component of $\mathbf{E}$ remains continuous, inasmuch as both inside the metal and outside, the tangential component of $\mathbf{E}$ vanishes.

In contrast to (27.1) the second fundamental equation of electrostatics—the electric flux theorem—has to be changed because of the possibility of *polarization charges* $\rho_P$ and $\omega_P$ making their appearance along with the hitherto considered *true* charges $\rho$ and $\omega$. Thus, instead of (19.7) we now have within insulators

$$\operatorname{div} \mathbf{E} = 4\pi(\rho + \rho_P) = 4\pi(\rho - \operatorname{div} \mathbf{P}) \tag{27.2}$$

while on the bounding surface of two insulators 1 and 2, from (19.8) and (26.1), we have that

$$(E_n)_1 - (E_n)_2 = 4\pi(\omega + \omega_P) = 4\pi\left[\omega - (P_n)_1 + (P_n)_2\right] \tag{27.3}$$

in which the normal $\mathbf{n}$ at the boundary surface points from medium 2 to medium 1, and for this reason the polarization charge in the latter medium is to be inserted with a minus sign.

The distinction between true charges and polarization charges is that the former can be removed while the latter are irremovably bound to matter. If, for example, two small flat metal plates lying one on the other be placed inside a parallel-plate capacitor so that the plane of the small plates is perpendicular to the field lines, induced charges then exist on the plates, and these persist when the plates are separated from one another in the capacitor field and are then removed. By measuring these charges, conclusions can be drawn with respect to the field in the capacitor. Now for two small juxtaposed glass plates similarly placed in a capacitor field, surface charges are again produced. These, however, vanish upon removal of the glass plates from the capacitor regardless of whether beforehand they were in mutual contact or not.

When later we speak simply of charges, we shall always mean true charges. Indeed, especially for large values of $\varepsilon$, distinguishing between true charges and polarization charges is often possible only with difficulty.

In equations (27.2) and (27.3) the quantities $\mathbf{E}$ and $\mathbf{P}$ appear only in the combination $\mathbf{E}+4\pi\mathbf{P}$. For the new vector

$$\mathbf{D} = \mathbf{E}+4\pi\mathbf{P} \qquad (27.4)$$

we introduce the special designation *electric displacement*. Within an insulator we have for it

$$\operatorname{div}\mathbf{D} = 4\pi\rho \qquad (27.5)$$

and on the boundary between two media

$$(D_n)_1 - (D_n)_2 = 4\pi\omega \qquad (27.6)$$

The sources of $\mathbf{D}$ are therefore true charges, while the sources of $\mathbf{E}$ are given as the sum of true charges and polarization charges—in the older literature lumped together as "free" charges.

From (26.3) and (27.4), in a normal isotropic dielectric,

$$\mathbf{D} = \varepsilon\mathbf{E} \qquad (27.7)$$

in which $\varepsilon$ can be a function of position. In many treatments $\mathbf{D}$ is straightway defined by (27.7). In view of the results of §26, however, this definition is much more specialized than the generally valid definition (27.4). Nevertheless, as remarked, we shall in the following mostly restrict ourselves to the simple special case where (27.7) is valid.

In this case, for two uncharged insulators, the applicable boundary conditions (continuity of the tangential component of $\mathbf{E}$ and of the normal component of $\mathbf{D}$) have as a consequence (figure 27) a unique *refraction law of force lines:* Letting $\alpha_1$, and $\alpha_2$ be the angles between the force lines and the normal on either side of the boundary surfaces, from

$$E_1 \sin\alpha_1 = E_2 \sin\alpha_2 \qquad D_1 \cos\alpha_1 = D_2 \cos\alpha_2$$

in combination with (27.7), we have immediately that

$$\tan\alpha_1 : \tan\alpha_2 = \varepsilon_1 : \varepsilon_2$$

Thus force lines, on entering an insulator with large $\varepsilon$, are refracted away from the normal.

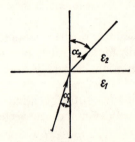

Fig. 27.—Refraction of field lines at the boundary between two dielectrics

*Remark:* In the MKSA system, for which (27.2) would be written as

$$\operatorname{div}\mathbf{E} = \frac{1}{\varepsilon_0}(\rho + \rho_P) = \frac{1}{\varepsilon_0}(\rho - \operatorname{div}\mathbf{P})$$

the displacement vector $\mathbf{D}$ is defined by

$$\mathbf{D} = \varepsilon_0\mathbf{E} + \mathbf{P}, \text{ or for special cases, } \mathbf{D} = \varepsilon\varepsilon_0\mathbf{E}$$

Hence for $\mathbf{D}$, within a charged medium,

$$\operatorname{div}\mathbf{D} = \rho$$

and on a charged boundary surface

$$(D_n)_1 - (D_n)_2 = \omega$$

In the Gaussian system of units $\mathbf{D}$ and $\mathbf{E}$ are equal for a vacuum, and for polarized media they differ only by a numerical factor. In the MKSA system, however, they possess different dimensions: $\mathbf{E}$ is measured as a force quantity, in V/m; $\mathbf{D}$ as a charge quantity, measured in As/m². Thus, at the surface of metals, the normal component of $\mathbf{D}$ in the surrounding space and therewith the magnitude $D$, are dimensionally and numerically equal to the density $\omega$ of the true surface charge.

It should however be observed that, with respect to the symbol $\mathbf{D}$ (and in contrast to $\mathbf{E}$, $\mathbf{P}$, and $\varepsilon_0$), two different amounts of the same physical entity are denoted in the two measure-systems considered. Thus, if temporarily we designate by $\mathbf{D}^*$ the $\mathbf{D}$ of the technical measure-system, the above defining equation for this quantity, after multiplication by $4\pi$, assumes the form

$$4\pi\mathbf{D}^* = 4\pi\varepsilon_0\mathbf{E} + 4\pi\mathbf{P}$$

In going over to the Gaussian system, since $\varepsilon_0 = 1/4\pi$, the right side becomes identical to the right side of (27.4). So we have that

$$\mathbf{D} = 4\pi\mathbf{D}^*$$

## §28. Point charge opposite a semi-infinite dielectric

We imagine that a point charge A is a distance $a$ away from the plane surface of a dielectric, and we inquire into the change in the Coulomb field brought about by the presence of the dielectric. The problem corresponds to that solved in §23 for the conducting plane. There we had only to consider the field in the air space, since beyond the conducting plane there was no field; now, however, the field within the insulator has also to be considered. We shall assume that the insulator is semi-infinite. We let its permittivity be $\varepsilon_2$, and that of the air $\varepsilon_1$.

We endeavour to solve this problem by the method of electrical images. Let B be the image point of A with respect to the surface of the dielectric, and let $r$ and $r'$ be the distances of a field point from A and B respectively. For the potential in air we have

$$\phi_1 = \frac{e}{\varepsilon_1 r} - \frac{e'}{\varepsilon_1 r'}$$

Let the field in air correspond to the true charge $e$ at A and the imagined true charge $-e'$ at B. This statement satisfies the requirement that sources of electrical displacement within the air space exist only at A; for the image point B lies outside the air space. We seek to represent the field within the dielectric through the statement that

$$\phi_2 = \frac{e''}{\varepsilon_2 r}$$

The field in the insulator shall therefore be constructed as if the insulator occupied all space, and as if the true charge $e''$ were located at A. This statement corresponds to the requirement that within the dielectric no sources or sinks of electrical displacement shall appear. We show that the field will actually be represented by these statements by proving that the boundary conditions at the air-dielectric interface will be fulfilled through appropriate choice of the as yet undetermined quantities $e'$ and $e''$. First, concerning the normal components of **D**, we have

$$D_{n1} = \frac{ea}{r^3} + \frac{e'a}{r'^3} \qquad D_{n2} = \frac{e''a}{r^3}$$

in which **n** points in the direction from the air space to the insulator.

From the requirement that the discontinuity of **D** across the surface vanishes on the boundary surface we have, since $r = r'$, that

$$e + e' - e'' = 0$$

On the other hand, the tangential component of **E** has to have the same value on both sides of the plane of separation; and this will be the case when $\phi_1 = \phi_2$ is satisfied along the plane, since **E** is the negative gradient of $\phi$.

We therefore require secondly that

$$\frac{e - e'}{\varepsilon_1} = \frac{e''}{\varepsilon_2}$$

From the two equations, both linear in $e$, $e'$, $e''$, we have

$$\frac{e - e'}{e + e'} = \frac{\varepsilon_1}{\varepsilon_2} \qquad e' = e\frac{\varepsilon_2 - \varepsilon_1}{\varepsilon_2 + \varepsilon_1} \qquad e'' = e\frac{2\varepsilon_2}{\varepsilon_2 + \varepsilon_1}$$

and through these the imaginary "true" charges $-e'$ at B and $+e''$ at A are determined. The force lines within the dielectric extend radially outward from A, while in the air space the field is represented by the superposition of two fields, one coming from the source point A and the other from the sink at B.

If a conductor be substituted for the dielectric, then according to §23, in order to determine the potential in the air space, a charge of value $-e$ must be given to the image point B. The disturbing action of the dielectric on the field strength in air, compared with the disturbing action of the conductor, is given by

$$e' : e = (\varepsilon_2 - \varepsilon_1) : (\varepsilon_2 + \varepsilon_1)$$

The dielectric always therefore exerts less influence than the conductor. In the extreme case in which the permittivity $\varepsilon_2$ of the insulator is very large compared with that of air, $e'$ becomes equal to $e$, i.e. the conductor influences the field in the air space just like an insulator of infinitely large permittivity.

The field strength inside the dielectric, compared to the case of no dielectric, appears reduced by the factor

$$\frac{e''}{\varepsilon_2} : \frac{e}{\varepsilon_1} = 2\varepsilon_1 : (\varepsilon_2 + \varepsilon_1)$$

In the extreme case of infinite permittivity, the electric field strength inside an insulator is zero, just as within a conductor.

### §29. Dielectric sphere in a uniform field

We consider a sphere of radius $a$ and permittivity $\varepsilon_1$, embedded in another dielectric $\varepsilon_2$ occupying all the rest of space. In the latter dielectric we suppose a homogeneous field $\mathbf{E}_0$, in the direction of the positive $x$-axis, to have existed prior to the introduction of the sphere. How is this field modified by the introduction of the sphere? In order to answer this question we determine a potential $\phi$ having the following properties:

(1) At great distances from the sphere ($\lim r \to \infty$) $\phi$ must tend to $-E_0 x$.

(2) At the surface of the sphere the normal component of the gradient of $\phi$ experiences a discontinuity such that $\varepsilon \partial \phi / \partial r$ has the same value on both sides.

(3) $\phi$ itself, and therefore the tangential component of grad $\phi$, are continuous on passing through the spherical surface.

(4) $\phi$ everywhere satisfies Laplace's equation $\nabla^2 \phi = 0$.

We denote by $\phi_1$ and $\phi_2$ the values of the potential inside and outside the sphere respectively, and we make the assertions that

$$\phi_1 = -E_i x \quad \text{and} \quad \phi_2 = -E_0 x + E_0 \frac{kx}{r^3} \qquad (29.1)$$

This means: Inside the sphere the homogeneous field $E_i$ prevails; the sphere is therefore homogeneously polarized in the $x$-direction. In the external region, however, the sphere acts as if a dipole of moment $E_0 k$ were located at its centre.

Of our four requirements, the first and the fourth are satisfied by our assertions concerning $\phi$. We have now to show that by suitable choice of the constants $E_i$ and $k$ still at our disposal, the boundary conditions (2) and (3) can also be satisfied. With polar coordinates ($x = r \cos \theta$), we have

$$\phi_1 = -E_i r \cos \theta \qquad \phi_2 = -E_0 \cos \theta \left( r - \frac{k}{r^2} \right)$$

Our boundary conditions now require that

$$\phi_1 = \phi_2 \quad \text{and} \quad \varepsilon_1 \frac{\partial \phi_1}{\partial r} = \varepsilon_2 \frac{\partial \phi_2}{\partial r} \quad \text{for} \quad r = a$$

This furnishes the two equations

$$E_i = E_0\left(1 - \frac{k}{a^3}\right) \qquad \varepsilon_1 E_i = \varepsilon_2 E_0\left(1 + \frac{2k}{a^3}\right)$$

out of which we have

$$\frac{k}{a^3} = \frac{\varepsilon_1 - \varepsilon_2}{\varepsilon_1 + 2\varepsilon_2} \qquad E_i = E_0\frac{3\varepsilon_2}{\varepsilon_1 + 2\varepsilon_2} = E_0\left(1 - \frac{\varepsilon_1 - \varepsilon_2}{\varepsilon_1 + 2\varepsilon_2}\right) \quad (29.2)$$

We consider the special case $\varepsilon_1 = \varepsilon$, $\varepsilon_2 = 1$, i.e. the case of the dielectric sphere in a vacuum. In its interior the field is reduced by the factor $3/(\varepsilon + 2)$ compared to the homogeneous field. In the space outside the sphere, the field acts like that of a dipole of moment

$$\mathbf{p} = \mathbf{E}_0 k = \mathbf{E}_0 a^3 \frac{\varepsilon - 1}{\varepsilon + 2} \qquad (29.3)$$

Its polarization $\mathbf{P}$ (moment per unit volume) has the value

$$\mathbf{P} = \mathbf{E}_0 \cdot \frac{3}{4\pi} \frac{\varepsilon - 1}{\varepsilon + 2} = \mathbf{E}_i \cdot \frac{\varepsilon - 1}{4\pi} \qquad (29.4)$$

and the internal field $\mathbf{E}_i$ producing this polarization is represented by

$$\mathbf{E}_i = \mathbf{E}_0 - \frac{4\pi}{3}\mathbf{P} \qquad (29.5)$$

A comparison with our observations on the conducting sphere (§23) shows that the latter, in its disturbing action on a homogeneous field, behaves like an insulator of infinitely great permittivity.

## §30. The homogeneously polarized ellipsoid

If we place a body of permittivity $\varepsilon = 1 + 4\pi\chi$ in a uniform electric field $\mathbf{E}_0$, we do not (as we have seen in the example of the sphere) obtain in general a polarization $\mathbf{P} = \chi\mathbf{E}_0$, but rather the polarization $\mathbf{P} = \chi\mathbf{E}_i$—for the polarization has just the consequence that the field $\mathbf{E}_i$ obtaining within the material is essentially different from $\mathbf{E}_0$. However, the equation $\mathbf{P} = \chi\mathbf{E}_i$ has something to say about the effect of the field existing only at the particular place considered. For an arbitrary shape of the body to be polarized, the additional field $\mathbf{E}' = \mathbf{E}_i - \mathbf{E}_0$ produced within the material by $\mathbf{P}$ itself becomes a complicated function of position, and thus a homogeneous $\mathbf{E}_0$ will not in general yield a homogeneous **P**. A homogeneous **P** results only when the body has the shape of an *ellipsoid*.

We can understand directly the correctness of this assertion on the basis of earlier results for two special cases, namely for a *very long wire*, and for a *sphere*. We can look upon the long *wire* as a very much stretched ellipsoid of revolution. If the axis of rotation has the direction of the external field $\mathbf{E}_0$, then, since the tangential component of **E** is continuous, we have inside the wire that $\mathbf{E} = \mathbf{E}_i = \mathbf{E}_0$; for a long polarized wire we therefore obtain that $\mathbf{P} = \chi\mathbf{E}_0$. For the sphere, by (29.5), the field

$$\mathbf{E}_i = \mathbf{E}_0 - \tfrac{4}{3}\pi\mathbf{P} = \mathbf{E}_0 + \mathbf{E}' \qquad (30.1)$$

exists inside. From $\mathbf{P} = \chi\mathbf{E}_i$, this determines the polarization. Thus, for the relation between **P** and $\mathbf{E}_0$ we obtain

$$\mathbf{P} = \frac{1}{1/\chi + \tfrac{4}{3}\pi}\mathbf{E}_0 \qquad (30.2)$$

in agreement with (29.4).

We see from (30.1) that, owing to polarization, the external field is partially shielded. The numerical factor multiplying **P** in (30.1), a factor which depends only on the shape of the body, is called the *depolarization factor*. As we have just seen, it is equal to zero for a wire, and $4\pi/3$ for a sphere. When the reciprocal susceptibility is considerably smaller than the depolarization factor, the observable polarization is, from (30.2), determined essentially by the depolarization factor above. Herein resides the difficulty of measuring the susceptibility of short pieces of easily polarizable material.

A simple description is then only possible when, as is the case in (30.1), the vector $\mathbf{E}_i$ has the same direction as $\mathbf{E}_0$. For the general case of the *tri-axial ellipsoid* we shall prove the following: If such an ellipsoid is homogeneously polarized, and if $P_x$, $P_y$, $P_z$ are the components of **P** in the direction of the principal axes, then within the ellipsoid this polarization produces a uniform field $\mathbf{E}'$ whose components are given by

$$E'_x = -AP_x \qquad E'_y = -BP_y \qquad E'_z = -CP_z \qquad (30.3)$$

The three numbers $A$, $B$, $C$ are called the *depolarization factors* of the ellipsoid for its three axes. Their determination is the essential problem of these paragraphs. When they are known, the polarization belonging to an external field $\mathbf{E}_0$ and coming from the equation $\mathbf{P} = \chi\mathbf{E}_i = \chi(\mathbf{E}_0 + \mathbf{E}')$ is given by

$$P_x = \frac{E_{0x}}{1/\chi + A} \qquad P_y = \frac{E_{0y}}{1/\chi + B} \qquad P_z = \frac{E_{0z}}{1/\chi + C} \qquad (30.4)$$

In order to derive (30.3) we shall calculate as if the polarization consisted of a displacement of two equal and opposite charge densities $+\rho$ and $-\rho$. Generally speaking, we therefore determine the field $\mathbf{E}'$ from $\text{curl}\,\mathbf{E}' = 0$ and $\text{div}\,\mathbf{E}' = -4\pi\,\text{div}\,\mathbf{P}$.

We consider first the potential of an ellipsoid

$$\frac{x^2}{a^2} + \frac{y^2}{b^2} + \frac{z^2}{c^2} = 1 \qquad (30.5)$$

which is uniformly filled with charge of density $\rho$. For the potential $\phi$ of this ellipsoid we have different formulae according as we consider a point within the ellipsoid ($\phi = \phi_i$) or without ($\phi = \phi_o$). We wish to show that this problem is solved by the following theorem due to Dirichlet:

$$\left.\begin{array}{l} \phi_i = \pi abc\rho \displaystyle\int_0^\infty \left(1 - \frac{x^2}{a^2+\lambda} - \frac{y^2}{b^2+\lambda} - \frac{z^2}{c^2+\lambda}\right)\frac{d\lambda}{D(\lambda)} \\[3mm] \phi_o = \pi abc\rho \displaystyle\int_u^\infty \left(1 - \frac{x^2}{a^2+\lambda} - \frac{y^2}{b^2+\lambda} - \frac{z^2}{c^2+\lambda}\right)\frac{d\lambda}{D(\lambda)} \end{array}\right\} \quad (30.6)$$

Here $D(\lambda)$ is the abbreviation for

$$D(\lambda) = \sqrt{[(a^2+\lambda)(b^2+\lambda)(c^2+\lambda)]} \qquad (30.7)$$

The lower integration limit $u$ in the expression for $\phi_o$ is a function of $x$, $y$, $z$. It is defined as the largest positive root (in the whole external space) of the equation

$$\frac{x^2}{a^2+u} + \frac{y^2}{b^2+u} + \frac{z^2}{c^2+u} = 1 \qquad (30.8)$$

From this definition of $u$ it follows first that $\phi_i$ and $\phi_o$ for $u = 0$, i.e. on the surface (30.5), go over in a continuous way from one to the other. In order to justify (30.6) we have further to show that the first derivatives on the surface (30.5) are continuous, and also that $\nabla^2\phi_i = -4\pi\rho$, and $\nabla^2\phi_o = 0$. It then has to be shown that for large distances $\phi_o$ vanishes like $1/r$.

For proof, from (30.6) we write

$$\frac{\partial \phi_i}{\partial x} = -2\pi abc\rho \int_0^\infty \frac{x\, d\lambda}{(a^2+\lambda)D(\lambda)}$$

$$\frac{\partial^2 \phi_i}{\partial x^2} = -2\pi abc\rho \int_0^\infty \frac{d\lambda}{(a^2+\lambda)D(\lambda)}$$

$$\frac{\partial \phi_o}{\partial x} = -2\pi abc\rho \int_u^\infty \frac{x\, d\lambda}{(a^2+\lambda)D(\lambda)}$$

$$\frac{\partial^2 \phi_o}{\partial x^2} = -2\pi abc\rho \left\{ \int_u^\infty \frac{d\lambda}{(a^2+\lambda)D(\lambda)} - \frac{x\, \partial u/\partial x}{(a^2+u)D(u)} \right\}$$

The first derivatives are therefore continuous. Further, it follows from (30.7) that

$$\frac{d\ln D(\lambda)}{d\lambda} = \frac{1}{2}\left( \frac{1}{a^2+\lambda} + \frac{1}{b^2+\lambda} + \frac{1}{c^2+\lambda} \right) \tag{30.9}$$

Thus we have that

$$\nabla^2 \phi_i = -4\pi abc\rho \int_0^\infty \frac{d\lambda}{D(\lambda)} \frac{d\ln D(\lambda)}{d\lambda} = -\frac{4\pi abc\rho}{D(0)} = -4\pi\rho$$

On the other hand, we have

$$\nabla^2 \phi_o = -\frac{4\pi abc\rho}{D(u)} \left\{ 1 - \frac{1}{2}\left( \frac{x}{a^2+u}\frac{\partial u}{\partial x} + \frac{y}{b^2+u}\frac{\partial u}{\partial y} + \frac{z}{c^2+u}\frac{\partial u}{\partial z} \right) \right\}$$

In order to show that this expression vanishes, we differentiate (30.8) with respect to $x$:

$$\frac{2x}{a^2+u} = \left( \frac{x^2}{(a^2+u)^2} + \frac{y^2}{(b^2+u)^2} + \frac{z^2}{(c^2+u)^2} \right)\frac{\partial u}{\partial x}$$

If this equation is multiplied by $x/(a^2+u)$, and if to it are added the correspondingly formed equations for $y$ and $z$, it follows that the expression for $\nabla^2\phi_o$ is actually equal to zero.

For *large distances* $r = \sqrt{(x^2+y^2+z^2)}$, from (30.8) we have approximately for the lower limit $u \approx r^2$. In the integrands of $\phi_o$ the quantities $a^2$, $b^2$, $c^2$, compared to $\lambda$, can therefore be neglected, giving easily

$$\phi_o \approx \frac{4\pi abc\rho}{3r}$$

As it ought, therefore, $\phi_o$ for large $r$ goes over to the Coulomb potential of the total charge imagined to be all assembled at the origin.

We obtain the potential $\phi'$ of a *polarized ellipsoid* whose boundary is given by (30.5) when, first, we bring into coincidence with the ellipsoid considered above (whose charge density is $\rho$) a second identical ellipsoid with charge density $-\rho$; and then we displace the positively charged ellipsoid the small amount $\delta\mathbf{r} = \{\delta x, \delta y, \delta z\}$. This process is equivalent to the production of a uniform polarization $\mathbf{P} = \rho\,\delta\mathbf{r}$. The potential $\phi'$ thus comes out as the difference of two potentials for the full ellipsoid:

$$\phi'(\mathbf{r}) = \phi(\mathbf{r} - \delta\mathbf{r}) - \phi(\mathbf{r}) = -\delta\mathbf{r}.\operatorname{grad}\phi(\mathbf{r})$$

With $\mathbf{P} = \rho\,\delta\mathbf{r}$, and using (30.6), we therefore have within the ellipsoid that

$$\phi' = AxP_x + ByP_y + CzP_z$$

in which the numbers $A, B, C$ are

$$A = 2\pi abc \int_0^\infty \frac{d\lambda}{(a^2 + \lambda)D(\lambda)} \qquad B = 2\pi abc \int_0^\infty \frac{d\lambda}{(b^2 + \lambda)D(\lambda)}$$

$$C = 2\pi abc \int_0^\infty \frac{d\lambda}{(c^2 + \lambda)D(\lambda)} \tag{30.10}$$

It is the negative gradient of $\phi'$ which actually now yields the depolarization field $\mathbf{E}'$ given in (30.3). The depolarization factors for the directions of the three principal axes of the ellipsoid (30.5) are therefore given by (30.10).

From (30.9) it is easily shown that we can always write

$$A + B + C = 4\pi$$

This equation permits the depolarization factor to be found, for some cases, directly on symmetry grounds. Thus we have $4\pi/3$ for the sphere, $4\pi$ for a plate in the direction of the normal, and $2\pi$ for a long circular cylinder in the direction perpendicular to the axis. For other forms the integrals (30.10) have actually to be evaluated.

We wish as an example of such evaluation to work out the important practical case of a *prolate ellipsoid of revolution*. We therefore say that

$$a \geqq b = c$$

We have then that

$$A = 2\pi ab^2 \int_0^\infty \frac{d\lambda}{(a^2+\lambda)^{3/2}(b^2+\lambda)}$$

From the substitution

$$\lambda = \xi^2(a^2-b^2)-a^2$$

and by introducing the dimensionless relation

$$\delta = a/b$$

of the ellipsoid, we find that

$$A = \frac{4\pi\delta}{(\delta^2-1)^{3/2}} \int_{\delta/\sqrt{(\delta^2-1)}}^\infty \frac{d\xi}{\xi^2(\xi^2-1)}$$

$$= \frac{4\pi}{\delta^2-1}\left(\frac{\delta}{2\sqrt{(\delta^2-1)}} \ln\frac{\delta+\sqrt{(\delta^2-1)}}{\delta-\sqrt{(\delta^2-1)}}-1\right)$$

For $\delta = 1$ this formula naturally furnishes the value $4\pi/3$ of the sphere. For an almost spherical ellipsoid, i.e. for small $\delta^2-1$ we obtain by development

$$A = \tfrac{4}{3}\pi[1-\tfrac{2}{5}(\delta^2-1)+\ldots]$$

For a very long prolate ellipsoid, since $\delta \gg 1$, we have

$$A \approx \frac{4\pi}{\delta^2}(\ln 2\delta - 1)$$

# CHAPTER B III

## Force Effects and Energy Relations in the Electrostatic Field

### §31. Systems of point charges in free space

For a known field strength $E(r)$ the forces and torques on an electrical system can be immediately derived from their meaning as force per unit charge. But here, as in mechanics, it is often more convenient and more intuitive to evaluate these force effects by way of the energy contained in the system.

We consider first the simplest case of a system of point charges $e_1, e_2, \ldots, e_h$, located in a vacuum at positions $r_1, r_2, \ldots, r_h$, and we inquire concerning the stored electrical energy of this system. We determine this energy from the work necessary to bring the individual charges from an infinite distance to their respective locations. We first bring charge $e_1$ to its proper place. No work is required here since the remaining charges, owing to their distance, exert no force upon $e_1$. We now bring $e_2$ to its place, distant $|r_1 - r_2| = r_{12}$ from $e_1$, and here, because of Coulomb repulsion, work $e_1 e_2 / r_{12}$ is required. Next we bring in $e_3$ and for this we have to expend energy $e_1 e_3 / r_{13} + e_2 e_3 / r_{23}$. We continue in this way until all charges are at their places, and for this the total work

$$W = \frac{e_1 e_2}{r_{12}} + \frac{e_1 e_3}{r_{13}} + \frac{e_2 e_3}{r_{23}} + \ldots = \sum_{1 \leq j < k \leq h} \sum \frac{e_j e_k}{r_{jk}} \qquad (31.1)$$

being the work of assembling or "charging", has been performed.

The energy contributed here must in some way be stored up in the system. Even though the question concerning the localization of this energy cannot be answered by electrostatics, the *action-at-a-distance theory* has very different ideas on this question from *Maxwell's field theory*. The former makes the assumption that the work of charging somehow attaches to the individual charges as potential energy. The latter regards the field as the carrier of the electrical energy and asserts that every volume-element $dv$ in empty space, if it is in the region of an electric field, contains the energy $u\,dv$, for which the *energy density* is given by

$$u = \frac{1}{8\pi}\mathbf{E}^2 \qquad (31.2)$$

Accordingly, an extended region of space should contain the energy

$$U = \frac{1}{8\pi}\int \mathbf{E}^2\,dv \qquad (31.3)$$

We justify this statement by proving that in the above-described mutual assemblage of $h$ point charges, the expended work $W$ in (31.1) is identical with the increase of field energy $U$ involved in the process.

For this purpose we designate by $\mathbf{E}_1, \mathbf{E}_2, \ldots, \mathbf{E}_h$ the field strengths at an arbitrary field point due to the charges $e_1, e_2, \ldots, e_h$. We then have that

$$\mathbf{E} = \mathbf{E}_1 + \mathbf{E}_2 + \ldots + \mathbf{E}_h$$

and therefore

$$\mathbf{E}^2 = \{\mathbf{E}_1^2 + \mathbf{E}_2^2 + \ldots + \mathbf{E}_h^2\} + 2\{\mathbf{E}_1.\mathbf{E}_2 + \mathbf{E}_1.\mathbf{E}_3 + \ldots + \mathbf{E}_{h-1}.\mathbf{E}_h\}$$

If with this expression we form the field energy $U$, we see that the contributions

$$U_j = \frac{1}{8\pi}\int \mathbf{E}_j^2\,dv$$

coming from the quadratic terms $\mathbf{E}_j{}^2$ are not modified by the approach or presence of other charges; $U_j$ is the energy belonging to the charge $e_j$ alone. (It corresponds to the electrical part of the work which, if we start with an infinitely diffuse charge cloud, would have to be expended in assembling the charge $e_j$. For a true point charge this work would be infinitely great.) We need therefore consider only the contributions of the mixed terms $\mathbf{E}_1.\mathbf{E}_2$, etc. Writing $\mathbf{r}_1 = \mathbf{r}, \mathbf{r}_2 = \mathbf{r} - \mathbf{r}_{12}$, $\angle(\mathbf{r}, \mathbf{r}_{12}) = \theta$, we have

$$\frac{1}{4\pi}\int \mathbf{E}_1.\mathbf{E}_2\,dv = \frac{e_1 e_2}{4\pi}\iiint \frac{\mathbf{r}}{r^3}.\frac{\mathbf{r}-\mathbf{r}_{12}}{|\mathbf{r}-\mathbf{r}_{12}|^3}\,r^2\,dr\,d\Omega$$

$$= \frac{e_1 e_2}{4\pi}\iiint \frac{r-r_{12}\cos\theta}{(r^2+r_{12}^2-2rr_{12}\cos\theta)^{3/2}}\,dr\,d\Omega$$

The $r$-integral here can be worked out immediately; indefinitely integrated, it gives the value $-1/\sqrt{(r^2+r_{12}^2-2rr_{12}\cos\theta)}$ and, if taken between the limits $r = 0$ and $r = \infty$ gives $1/r_{12}$. Since the solid-angle integration leads to the factor $4\pi$, the contribution of the energy term considered is equal to $e_1 e_2/r_{12}$. Thus

$$U - (U_1 + U_2 + \ldots + U_h) = \frac{e_1 e_2}{r_{12}} + \frac{e_1 e_3}{r_{13}} + \ldots$$

is exactly equal to the work $W$ of charging in (31.1).

The representation (31.3) of the work of charge assembly, or a corresponding representation—one valid for dielectrics—will prove its productivity and utility in the subsequent paragraphs treating fields produced by continuous charge distributions. Meanwhile, however, representation (31.3) is especially convenient for point charges. We show this in two examples.

### (a) The dipole in an electrostatic field

We ask how much work is required to bring a dipole with *fixed moment p* to a definite position and to a definite orientation in an electrostatic field **E**. In order to answer this question with the help of equation (31.1), we set $e_1 = e$ and $e_2 = -e$, and $\mathbf{r}_1 = \mathbf{r} + \mathbf{s}$, $\mathbf{r}_2 = \mathbf{r}$ with $e\mathbf{s} = \mathbf{p}$, and we imagine the **E**-field as having been produced by suitably placed point charges $e_3, e_4, \ldots, e_h$:

$$\mathbf{E}(\mathbf{r}) = -\operatorname{grad} \phi(\mathbf{r}) \quad \text{with} \quad \phi(\mathbf{r}) = \sum_{k=3}^{h} \frac{e_k}{|\mathbf{r} - \mathbf{r}_k|}$$

Since the dipole separation $\mathbf{r}_{12}$ is not changed, the work required to bring in the dipole is equal to

$$W = e_1 \sum_{k=3}^{h} \frac{e_k}{|\mathbf{r}_1 - \mathbf{r}_k|} + e_2 \sum_{k=3}^{h} \frac{e_k}{|\mathbf{r}_2 - \mathbf{r}_k|} = e[\phi(\mathbf{r} + \mathbf{s}) - \phi(\mathbf{r})]$$

and therefore, for sufficiently small dipole separation,

$$W = e\mathbf{s} \cdot \operatorname{grad} \phi(\mathbf{r}) = -\mathbf{p} \cdot \mathbf{E}$$

We then obtain an energy contribution $-pE$ when we orient the dipole parallel to the field, and we must expend the work $pE$ in orienting the dipole against the field. The work of bringing in the dipole, and therewith the potential energy $U$ of the dipole, depends upon the angle $\theta$ between the dipole and the field direction:

$$U = -\mathbf{p} \cdot \mathbf{E} = -pE \cos \theta \tag{31.4}$$

A torque of value

$$T = -\frac{\partial U}{\partial \theta} = -pE\sin\theta$$

must exist, against which we must do work in increasing $\theta$. It is in fact elementary that the dipole concept gives (figure 28) for this turning moment

$$\mathbf{T} = e\mathbf{s} \times \mathbf{E} = \mathbf{p} \times \mathbf{E} \tag{31.5}$$

For a non-uniform E-field we have an additional force

$$\mathbf{F} = e\left[\mathbf{E}(\mathbf{r}+\mathbf{s}) - \mathbf{E}(\mathbf{r})\right]$$
$$= p_x\frac{\partial \mathbf{E}}{\partial x} + p_y\frac{\partial \mathbf{E}}{\partial y} + p_z\frac{\partial \mathbf{E}}{\partial z} = (\mathbf{p}.\nabla)\mathbf{E} \tag{31.6}$$

corresponding to the dependence upon position of the potential energy:

$$\mathbf{F} = -\operatorname{grad} U = \operatorname{grad}(\mathbf{p}.\mathbf{E}(\mathbf{r}))$$

Since

$$\nabla(\mathbf{p}.\mathbf{E}) - (\mathbf{p}.\nabla)\mathbf{E} = \mathbf{p} \times (\nabla \times \mathbf{E}) = 0$$

these two expressions involving static fields (for which $\operatorname{curl}\mathbf{E} = 0$) actually agree with one another. Consequently, if $0 \leqq \theta < \pi/2$, the dipole is drawn into the field, otherwise it is expelled from it.

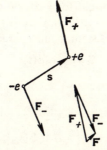

Fig. 28.—Force on a dipole in a non-uniform field

A particular consideration is the important case in which *the dipole* **p** *is produced by the field itself.* We have that

$$\mathbf{p} = \alpha\mathbf{E} \tag{31.7}$$

$\alpha$ being the field-independent polarizability; and, from (31.6), the force on the dipole in a non-uniform field is equal to

$$\mathbf{F} = (\alpha \mathbf{E} \cdot \nabla) \mathbf{E} = \frac{\alpha}{2} \operatorname{grad}(\mathbf{E}^2) \tag{31.8}$$

the latter because

$$\mathbf{E} \cdot \nabla E_x = E_x \frac{\partial E_x}{\partial x} + E_y \frac{\partial E_x}{\partial y} + E_z \frac{\partial E_x}{\partial z}$$

$$= E_x \frac{\partial E_x}{\partial x} + E_y \frac{\partial E_y}{\partial x} + E_z \frac{\partial E_z}{\partial x} = \frac{1}{2} \frac{\partial}{\partial x}(\mathbf{E}^2)$$

In the case of a variable dipole described by (31.7), the force is not derivable from the potential energy (31.4), but from the energy

$$U' = -\tfrac{1}{2} \alpha \mathbf{E}^2 = -\tfrac{1}{2} \mathbf{p} \cdot \mathbf{E} \tag{31.9}$$

The difference is required by the fact that now energy must be expended in the production of the dipole; that is, the charge separation in the dipole takes place against an opposing force which in any case is not of electrical origin, and this leads to a positive potential energy of value

$$U' - U = \mathbf{p} \cdot \frac{\mathbf{E}}{2}$$

### (b) The quadrupole in an electrostatic field

In a manner similar to that for a dipole of moment $p$, we find in general for the energy involved in introducing a group of charges into the field $\mathbf{E} = -\operatorname{grad} \phi$ of distant point charges, the value

$$W = \sum e_j \, \phi(\mathbf{r}_j)$$

or, developing $\phi$ with respect to the coordinates of the charges $e_j$,

$$W = e\phi(0) + \mathbf{p} \cdot (\operatorname{grad} \phi)_{\mathbf{r}=0}$$

$$+ \frac{1}{2} \left[ Q_{xx} \left( \frac{\partial^2 \phi}{\partial x^2} \right)_0 + Q_{yy} \left( \frac{\partial^2 \phi}{\partial y^2} \right)_0 + Q_{zz} \left( \frac{\partial^2 \phi}{\partial z^2} \right)_0 \right] +$$

$$+ \left[ Q_{xy} \left( \frac{\partial^2 \phi}{\partial x \, \partial y} \right)_0 + Q_{xz} \left( \frac{\partial^2 \phi}{\partial x \, \partial z} \right)_0 + Q_{yz} \left( \frac{\partial^2 \phi}{\partial y \, \partial z} \right)_0 \right] + \dots \tag{31.10}$$

The first two terms on the right-hand side give the energy contributions necessary for bringing in the assemblage of point charges for the case in which the assemblage possesses a total charge $e = \sum e_j$ and a resultant dipole moment $\mathbf{p} = \sum e_j \mathbf{r}_j$. If these contributions vanish, there remain, as the next non-vanishing terms, the contributions containing the second derivative of the potential. In these the $Q$-quantities mean, in the first instance, the quadrupole components according to (24.12). Since however $\nabla^2 \phi = (\nabla^2 \phi)_0 = 0$, we can subtract one-third of the quantity spur $(Q) . (\nabla^2 \phi)_0$ in the first square brackets of (31.10), so that the $Q_{jk}$ can also be understood to be the reduced quadrupole moments as in (24.14). In the following we shall calculate with these $Q'_{jk}$.

To simplify the discussion we shall consider as a *special case* that for which both the charge distribution producing the quadrupole moment, and the potential field $\phi$, are rotationally symmetric. We choose the symmetry axis of the field as the $z$-axis of our coordinate system so that in (31.10) the three mixed second derivatives of $\phi$ vanish, and thus, since $\nabla^2 \phi = 0$, we have

$$\left(\frac{\partial^2 \phi}{\partial x^2}\right)_0 = \left(\frac{\partial^2 \phi}{\partial y^2}\right)_0 = -\frac{1}{2}\left(\frac{\partial^2 \phi}{\partial z^2}\right)_0$$

Accordingly, the quadrupole term in the energy simplifies to the expression

$$U = \frac{1}{4}\left(\frac{\partial^2 \phi}{\partial z^2}\right)_0 \{2Q'_{zz} - Q'_{xx} - Q'_{yy}\} = \frac{3}{4}\left(\frac{\partial^2 \phi}{\partial z^2}\right)_0 Q'_{zz} \quad (31.11)$$

since the spur $Q'_{xx} + Q'_{yy} + Q'_{zz}$ for the reduced quadrupole moment vanishes. The axis of symmetry of the charge distribution, and therewith the third principal axis of the $Q$-tensor, moreover makes an angle $\theta$ with the $z$-axis. As in §15, we designate this axis by $\mathbf{a}^{III}$, and those corresponding to the directions of the other principal axes by $\mathbf{a}^{I}$ and $\mathbf{a}^{II}$. From (24.16) we have for the three principal quadrupole moments

$$Q^{III} = -2Q^{I} = -2Q^{II} = 2Q/3 \quad (31.12)$$

We have now to express the $Q'_{zz}$ in (31.11) in terms of $Q$. For this we draw upon equation (14.7). If we choose the $z$-axis (with unit-vector $\mathbf{z}$) as the direction of $\mathbf{n}$ and $\mathbf{m}$, and the three principal axes of the $Q$-tensor as the coordinate directions on the right-hand side, equation (14.7) reads

$$Q'_{zz} = Q^{I}(\mathbf{z} . \mathbf{a}^{I})^2 + Q^{II}(\mathbf{z} . \mathbf{a}^{II})^2 + Q^{III}(\mathbf{z} . \mathbf{a}^{III})^2$$

From (31.12) we therefore obtain

$$(\mathbf{z}\cdot\mathbf{a}^{\mathrm{I}})^2+(\mathbf{z}\cdot\mathbf{a}^{\mathrm{II}})^2+(\mathbf{z}\cdot\mathbf{a}^{\mathrm{III}})^2 = 1 \qquad \mathbf{z}\cdot\mathbf{a}^{\mathrm{III}} = \cos\theta$$

and after brief rewriting

$$U = \frac{1}{2}\left(\frac{\partial^2\phi}{\partial z^2}\right)_0 Q\,\frac{3\cos^2\theta-1}{2} \tag{31.13}$$

Thus here, as for the field of a rotationally symmetric quadrupole (equation 24.18), the angular dependence is given by the second zonal spherical harmonic $P_2(\cos\theta) = (3\cos^2\theta-1)/2$.

It is seen from (31.13) that a quadrupole in a uniform electric field experiences neither a force nor a torque. In a non-uniform field it first of all experiences a torque

$$T = -\frac{\partial U}{\partial\theta} = \frac{3}{4}\left(\frac{\partial^2\phi}{\partial z^2}\right)_0 Q\sin 2\theta \tag{31.14}$$

If the "field gradient" $\dfrac{\partial E_z}{\partial z} = -\dfrac{\partial^2\phi}{\partial z^2}$ is positive, $T$ attempts to rotate the quadrupole in the field direction. If the field gradient is negative, the quadrupole has its energy minimum when it is perpendicular to the direction of the field. This behaviour can be made clear with the example of a stretched quadrupole in the sense of figure 28b.

### §32.  Field energy when conductors and insulators are present.  Thomson's theorem

We now wish to determine the work necessary to charge an arbitrary system made up of insulators and conductors, and we would like to see if, in the sense of Maxwell, an energy density of the electric field can be given such that the total field energy reflects the work of charging.

As a beginning we consider first the *simple case of a parallel-plate capacitor* of known capacitance. We ask how much work must be done in charging it. We imagine the charging process carried out by conveying small charge elements $de'$ from the negative to the positive plate until the charging $e = CV$ is attained. If at some time there is already a charge $e' = CV'$, then, in carrying over a further charge element $de'$, we have to do an amount of work $V' de'$, so we obtain for the total work of charging the value

$$W = \int_0^e \frac{e' \, de'}{C} = \frac{e^2}{2C} = \tfrac{1}{2} eV = \tfrac{1}{2} CV^2 \qquad (32.1)$$

Obviously, this expression is independent of the presence of a dielectric in the capacitor. In the first instance it is assumed only that, if there is a dielectric, the expression $\mathbf{D} = \varepsilon \mathbf{E}$ holds for it, the permittivity $\varepsilon$ being independent of the field.

Since $V = Ed$, and $e = S\omega = SD/4\pi$, the work of charging can be written

$$W = \tfrac{1}{2} eV = \tfrac{1}{2} \cdot Ed \cdot \frac{SD}{4\pi} = \frac{1}{8\pi} \cdot ED \cdot Sd$$

Since (insignificant edge effects being neglected) $E$ and $D$ are constant inside the capacitor and vanish outside, and since $Sd$ represents the space (volume) between the plates, it would appear to make sense in our example here to speak of an energy density of the electric field given by

$$u = \frac{1}{8\pi} \mathbf{E} \cdot \mathbf{D} = \frac{\varepsilon}{8\pi} E^2 \qquad (32.2)$$

If the permittivity $\varepsilon$ be disregarded, this formula agrees with expression (31.2).

We would like now to establish formula (32.2) for the case of an arbitrary electrostatic array of conductors and insulators. The latter may carry any arbitrary volume or surface distribution of charges. We imagine a small quantity of charge $\delta e$ carried from a first metal surface having potential $\phi_1$ to a second metal surface having potential $\phi_2$. Since $\delta e_1 = -\delta e$, $\delta e_2 = +\delta e$, the work necessary for the transfer is given, corresponding to the general definition of potential, by

$$\delta W = \delta e_1 \phi_1 + \delta e_2 \phi_2 = \delta e(\phi_2 - \phi_1) \qquad (32.3)$$

We now seek to transform this expression into a volume integral over the whole region of the field. We therefore introduce the change $\delta \mathbf{D}$ in the electrical displacement occasioned by the transport of the charge just mentioned. From the electric-flux theorem we have

$$4\pi \, \delta e_1 = \oiint_{S_1} \delta D_n \, dS \qquad 4\pi \, \delta e_2 = \oiint_{S_2} \delta D_n \, dS$$

where $S_1$ and $S_2$ are the surfaces, and **n** the outward-drawn normals of the two conductors considered. Now, for all other possible metal surfaces yet remaining, as well as for the surface infinitely distant, we have $\oiint \delta D_n \, dS = 0$. Imagining the direction of all normals as reversed, we can therefore write

$$\delta W = -\frac{1}{4\pi} \oiint \phi \, \delta D_n \, dS$$

where the integration is now to extend over all metallic boundaries of the space which contains the field, including the infinitely distant surface, and **n** signifies the outward-drawn normal of this space. Applying Gauss's theorem to this expression, we obtain the volume integral

$$\delta W = -\frac{1}{4\pi} \int \operatorname{div}(\phi \, \delta \mathbf{D}) \, dv$$

extending over all space containing the field. With this transformation we have no need to regard the boundary surfaces of the dielectric as surfaces of discontinuity, since we can think of all such surfaces of discontinuity as replaced by a continuous crossing from one medium to the other. Since

$$\operatorname{div}(\phi \, \delta \mathbf{D}) = \operatorname{grad} \phi . \delta \mathbf{D} + \phi \operatorname{div} \delta \mathbf{D}$$

and since any possible remaining true charges at hand have not, by hypothesis, been altered or displaced ($\operatorname{div} \delta \mathbf{D} = 4\pi \delta \rho = 0$), we have finally that

$$\delta W = \frac{1}{4\pi} \int \mathbf{E} . \delta \mathbf{D} \, dv \qquad (32.4)$$

The small work contribution $\delta W$ which we have required in the transfer of the charge $\delta e$ from the first conductor to the second appears again completely as a change in the energy density of the electric field:

$$\delta W = \int \delta u \, dv \quad \text{with} \quad \delta u = \frac{1}{4\pi} \mathbf{E} . \delta \mathbf{D} \qquad (32.5)$$

If now we imagine the whole field as having been built up step-wise from small charge transfers, we arrive at the energy density

$$u = \frac{1}{4\pi} \int_0^{\mathbf{D}} \mathbf{E} . d\mathbf{D} \qquad (32.6)$$

*

in which **E** in the integral is to be regarded as a function of the **D**-vector which itself is variable. For the usual case where **D** = ε**E** we obviously have

$$u = \frac{1}{8\pi} \frac{\mathbf{D}^2}{\varepsilon} = \frac{\varepsilon}{8\pi} \mathbf{E}^2 = \frac{1}{8\pi} \mathbf{E} \cdot \mathbf{D} \qquad (32.7)$$

in agreement with equation (32.2).

Thus the theorem has been proved that all work expended in charging up a system can be interpreted as an increase in the electrostatic field energy. That this field energy is to be regarded (in the thermodynamic sense) not as internal energy, but as free energy, will be shown in subsequent paragraphs.

Let us first refer, however, following Thomson, to a remarkable minimal property of the field energy: when the charges in an electrostatic field move under the influence of the field strength, the energy of the field diminishes by the quantity of available work. The charges will therefore, so far as they are free to move, seek to arrange themselves in such a way that the field energy will have the least possible value. If, in particular, metallic conductors be given having charges whose distribution is in the first instance arbitrary, these charges will displace themselves until the field energy has reached a minimum. We know on the other hand that in electrostatic equilibrium the potential of a conductor is constant, and that the whole charge resides on its surface. We therefore conjecture that this charge distribution actually corresponds to a minimum of the field energy.

In order to demonstrate the correctness of this conjecture we consider, along with the field **E** and **D** of a given arrangement, a second field **E**′ and **D**′ which, like the first, must satisfy the fundamental equations

$$\text{curl}\,\mathbf{E} = 0 \qquad \text{div}\,\mathbf{D} = 4\pi\rho \qquad \mathbf{D} = \varepsilon\mathbf{E}$$

excepting that in the metallic conductors **E**′ is not constant. We prove that $U < U'$ where

$$U = \frac{1}{8\pi} \int \mathbf{E} \cdot \mathbf{D}\, dv \quad \text{and} \quad U' = \frac{1}{8\pi} \int \mathbf{E}' \cdot \mathbf{D}'\, dv$$

represent the energy of the two fields. For proof, we write

$$\mathbf{E}' = \mathbf{E} + \mathbf{E}'' \quad \text{and} \quad \mathbf{D}' = \mathbf{D} + \mathbf{D}''$$

and form the expression

$$U' - U = \frac{1}{8\pi} \int \{(\mathbf{E} + \mathbf{E}'') \cdot (\mathbf{D} + \mathbf{D}'') - \mathbf{E} \cdot \mathbf{D}\}\, dv$$

Since **D** = ε**E**, **D**′ = ε**E**′, **D**″ = ε**E**″, we have that **E**″ . **D** = **E** . **D**″ and therefore

$$U' - U = \frac{1}{4\pi} \int \mathbf{E} \cdot \mathbf{D}''\, dv + \frac{1}{8\pi} \int \mathbf{E}'' \cdot \mathbf{D}''\, dv \qquad (32.8)$$

Since the first integral, owing to the presence of the factor **E** in the integrand, is only to extend over the space between the conductors where we have div **D**″ = div **D**′ − div **D** = 0, it follows, since

$$\text{div}(\phi\mathbf{D}'') = \text{grad}\,\phi \cdot \mathbf{D}'' + \phi\,\text{div}\,\mathbf{D}'' = -\mathbf{E} \cdot \mathbf{D}''$$

by using Gauss's theorem that

$$\int \mathbf{E} . \mathbf{D}'' dv = - \int \phi D''_n dS = 0;$$

for the surface integral here is to extend over all metal surfaces, and on these the electrostatic potential $\phi$ is constant, while the remaining surface integral over $D_n''$ vanishes. Thus, with Thomson, we have

$$U' - U = \frac{1}{8\pi} \int \mathbf{E}'' . \mathbf{D}'' dv > 0$$

as soon as at any one place in space $\mathbf{E}'$ is different from $\mathbf{E}$.

Thomson's theorem therefore deduces from a minimum principle the field corresponding to the equilibrium distribution of electricity. This minimum principle corresponds exactly to the condition of equilibrium which holds for bodies having mass in the field of gravity. Such bodies are in equilibrium, in fact in stable equilibrium, when the gravitational potential energy has its smallest value for the configuration in question. We see here that, in the same way, the equilibrium of electricity on the surface of fixed conductors is characterized by a minimum value of the electrical energy. Electrical energy therefore plays the same role here as potential energy does in ordinary mechanics.

*Remark:* The chain of equations (32.1) shows that the work of charging a capacitor (and therewith its field energy) is also given in the MKSA system by $\frac{1}{2}eV$. Only now the energy obtained is no longer in ergs, but in VAs = Ws = J (joules). The expressions (32.6) and (32.7) for the energy density $u$ require formal modification, however, and in the MKSA system are given by

$$u = \int_0^{\mathbf{D}} \mathbf{E} . d\mathbf{D} \quad \text{and} \quad u = \tfrac{1}{2}\mathbf{E} . \mathbf{D} = \frac{\varepsilon\varepsilon_0}{2} \mathbf{E}^2$$

## §33.  Thermodynamical considerations of the field energy

Through §32 we have arrived at the concept of *field energy*, in that the *work of charging* an electrical system has been transformed into a volume integral over the whole space filled by field; and thus with Maxwell we could have the opinion that the total energy contribution, arrived at in the form of work done in charging the system, is completely retained by the system, stored in the form of field energy.

The internal energy of the system in the sense of thermodynamics (we wish to designate it here by $U_{th}$) could now be imagined as being always increased by the magnitude of the field energy, temporarily designated here by $U_{el}$ for distinction. This assumption is certainly correct so long as during the charging of the system no other energy is involved than the work of charging—when, in particular, the charging takes place *adiabatically*, i.e. without any introduction or removal of heat.

Since in this case, however, during the charging up or polarization, the system is in general heated, and since the polarization depends in general not only on the electric field strength but also upon the temperature, a rather obscure situation is usually attained. For this reason the experiments for testing the Maxwell theory are usually not carried out *adiabatically* but, if possible, *isothermally*, i.e. under complete compensation of any eventuating temperature differences within the system or between the system and its surroundings. But then, however, we have during the charging process also to reckon with the passage of a quantity of heat $dQ$:

$$dU_{th} = dW + dQ \tag{33.1}$$

Here $dW$ means the work of charging necessary to make a small change in the system. It can, for example, also include a contribution of mechanical work as by the motion of a metallic body or an insulator in the electric field, or by compressing the material.

If the change of state takes place *reversibly*, the process is then not one to be described by a hysteresis loop as for a ferromagnet, and the added heat $dQ$ increases the *entropy* $S$ of the system by

$$dS = dQ/T \tag{33.2}$$

in accordance with the second law of thermodynamics.

Since $d(TS) = TdS + SdT$, equation (33.1) permits the formulation

$$d(U_{th} - TS) = dW - SdT \tag{33.3}$$

If the process is carried out isothermally and reversibly, the free energy changes

$$F = U_{th} - TS$$

by just the value of the work done on the system. Thus for a charging process not involving mechanical work

$dU_{el} = dU_{th}$ for a process carried out adiabatically,
$dU_{el} = dF$ for a process carried out isothermally and reversibly.

In this respect the free energy $F$ of the latter process plays the same role as the internal energy $U_{th}$ in the former. And in general isothermal-reversible processes we have not to deal with an energy balance but with a balance of the free energy in the general form

$$dF = dU_{el} + dW_{mech} \tag{33.4}$$

The concept of electrical field energy and its availability has now been explained from the standpoint of thermodynamics.

However we now inquire further concerning the quantity of heat $dQ$ which is necessary to hold the temperature of a system constant for a pure charging process not involving other work. To calculate this we consider a cubic centimetre of a dielectric and, using small letters for the energy density and entropy, we write for a reversible change of state with input of charging energy, according to (33.1), (33.2), and (32.5),

$$T ds = du_{th} - du_{el} = du_{th} - \frac{1}{4\pi} \mathbf{E} . d\mathbf{D}$$

In order to simplify the formulation we assume further that at all times $\mathbf{E}$ and $\mathbf{D}$ are parallel to one another: ($\mathbf{E} . \mathbf{D} = ED$), and that $s$, $u_{th}$ and $D$ are unique functions of only the temperature $T$ and the field strength $E$. We have then

$$ds = \frac{1}{T}\left(\frac{\partial u_{th}}{\partial T} - \frac{E}{4\pi}\frac{\partial D}{\partial T}\right)dT + \frac{1}{T}\left(\frac{\partial u_{th}}{\partial E} - \frac{E}{4\pi}\frac{\partial D}{\partial E}\right)dE \tag{33.5}$$

Since the right-hand side is the total differential of the function, the identity

$$\frac{\partial}{\partial E}\left(\frac{\partial s}{\partial T}\right) = \frac{\partial}{\partial T}\left(\frac{\partial s}{\partial E}\right)$$

must be satisfied, where $\partial s/\partial T$ and $\partial s/\partial E$ are given through the factors of $dT$ and $dE$ on the right-hand side of (33.5). By calculating this "condition of integrability" we obtain

$$\frac{\partial u_{th}}{\partial E} = \frac{1}{4\pi} E \frac{\partial D}{\partial E} + \frac{1}{4\pi} T \frac{\partial D}{\partial T} \tag{33.6}$$

and thus, through integration with respect to $E$, since $\frac{\partial D}{\partial E} dE = dD$, for fixed $T$

$$u_{th}(T, E) - u_{th}(T, 0) = \frac{1}{4\pi} \int_0^D E \, dD + \frac{T}{4\pi} \int_0^E \frac{\partial D}{\partial T} dE \qquad (33.7)$$

in which both integrations are carried out keeping the temperature fixed. The first integral obviously represents the isothermal electrical charging work, so that the second must be the quantity of heat $Q$ necessary to keep the temperature constant in the charging process:

$$Q = \frac{T}{4\pi} \int_0^E \frac{\partial D}{\partial T} dE = T \int_0^E \frac{\partial P}{\partial T} dE \qquad (33.8)$$

the latter because $D = E + 4\pi P$. The necessary heat here vanishes in carrying out the process isothermally if $P$ depends upon $E$, but not on $T$. Since in general, however, $P$ diminishes with increasing temperature, heat is mostly liberated in isothermal charging or polarization. (Postive heat liberation.)

If, in particular, $D = \varepsilon E$ and $P = \chi E$, then $\varepsilon = 1 + 4\pi\chi$, and thus from (33.7) we have

$$u_{th}(T, E) - u_{th}(T, 0) = \frac{\varepsilon}{8\pi} E^2 + \frac{1}{2} E^2 T \frac{d\chi}{dT}$$

For a whole series of substances, in particular for dipolar gases, $\chi$ is inversely proportional to the temperature. In this case $T d\chi/dT = -\chi$, so that we have simply

$$u_{th}(T, E) - u_{th}(T, 0) = \frac{1}{8\pi} E^2$$

The development of heat in the polarization of a dielectric also affects its *specific heat*. We can heat a dielectric (at constant volume) either by keeping its polarization $P$ constant, that is without changing its electrical state, or by keeping the field strength $E$ constant, as say in a capacitor kept at constant voltage. The difference of the corresponding specific heats $\gamma_P$ and $\gamma_E$ can be fully stated as soon as the function $P(T, E)$ is known. From (33.5) and (33.6), and adding indices to the quantities to be kept constant in differentiation, we have

$$dQ = T ds = \left\{ \left(\frac{\partial u_{th}}{\partial T}\right)_E - \frac{E}{4\pi} \left(\frac{\partial D}{\partial T}\right)_E \right\} dT + \frac{T}{4\pi} \left(\frac{\partial D}{\partial T}\right)_E dE \qquad (33.9)$$

The expression within the curly brackets is therefore immediately equal to the specific heat at constant field strength:

$$\gamma_E = \left(\frac{\partial u_{th}}{\partial T}\right)_E - \frac{E}{4\pi} \left(\frac{\partial D}{\partial T}\right)_E$$

We obtain the specific heat at constant polarization from this, according to (33.9), by adding the term $\frac{T}{4\pi} \left(\frac{\partial D}{\partial T}\right)_E \left(\frac{\partial E}{\partial T}\right)_P$, since in this case in the heating the field strength has to be changed in order to hold the polarization constant:

$$\gamma_P = \gamma_E + \frac{T}{4\pi} \left(\frac{\partial D}{\partial T}\right)_E \left(\frac{\partial E}{\partial T}\right)_P$$

Because of the relationship

$$0 = \left(\frac{\partial P}{\partial T}\right)_E + \left(\frac{\partial P}{\partial E}\right)_T \left(\frac{\partial E}{\partial T}\right)_P$$

following from $P = P(T, E)$ by $T$-differentiation at constant $P$, we have

$$\gamma_E - \gamma_P = T\left(\frac{\partial P}{\partial T}\right)_E^2 \bigg/ \left(\frac{\partial P}{\partial E}\right)_T \qquad (33.10)$$

From this, for the special case where $P = \chi(T)E$, it follows that

$$\gamma_E - \gamma_P = \frac{T}{\chi}\left(\frac{\partial \chi}{\partial T}\right)^2 E^2$$

Heating at constant field strength therefore requires more energy than heating at constant polarization.

For normally polarizable substances the distinction between $\gamma_E$ and $\gamma_P$ is unimportant. It is however meaningful for anomalous substances such as Rochelle salt and barium titanate mentioned in §26, which, in an electrical fashion, act similarly to, say, iron in magnetism. Analogous thermal effects become important for ferromagnetic substances in the neighbourhood of the Curie point. The same is true for all paramagnetic substances at extremely low temperatures where, by using the cooling effect associated with adiabatic demagnetization, temperatures of about $1/100°K$ can be reached.

## §34. Force effects in the electrostatic field calculated by means of the field energy; several simple examples

In §32 we obtained the field energy from the idea that all work performed in charging a system or in displacing the charges contained in it must find full equivalent representation in the total field energy of the system. In a similar manner the work which we must do in displacing conductors or insulators in an electric field leads to an increase in the field energy. If this energy is known for every arrangement of the system, then, working backwards, we can conclude what individual forces and/or torques are present. Let us illustrate this with three simple examples.

### (1) *The potential balance* (figure 29).

It is well known that the voltage on a parallel-plate capacitor can be determined by measuring with a balance the force with which the two capacitor plates attract one another. This force can be given immediately for a capacitor in a vacuum: since the field which the one

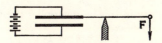

Fig. 29.—Potential balance

plate produces at the location of the other is equal to $\frac{1}{2}E$, where $E$ is the field in the capacitor, we have

$$F = \tfrac{1}{2}eE \tag{34.1}$$

This value can also be easily derived from the energy theorem: the total field energy in the capacitor, amounting to $E^2 S d / 8\pi$, increases in proportion to the plate separation $d$, if at constant charge $e$ (and therefore constant field $E$) the plates are further separated. In the increase of the distance $d$ by the small amount $\delta$, the field energy is therefore increased by $E^2 S \delta / 8\pi = eE\delta/2$. This energy contribution is introduced by the work $F\delta$ which has to be performed in pulling away the plate against the opposing force of attraction. A comparison of the two energy contributions leads immediately to formula (34.1).

(2) *Force on a dielectric in a parallel-plate capacitor.*

We now consider a parallel-plate capacitor between the plates of which a dielectric filling (solid or liquid) has been introduced. The value of the force with which the substance in the capacitor is drawn inward is not immediately obvious. With the help of the energy theorem, however, we can easily calculate it: If the insulator with permittivity $\varepsilon$ already covers an area $a$ of the full capacitor-plate surface $A$, the total field energy is given by

$$U = \frac{1}{8\pi} \int ED \, dv = \frac{1}{8\pi} E^2 [(A-a) + \varepsilon a] \, d$$

In the increase of $a$ by $\delta a$ the field energy is increased by

$$\delta U = \frac{\varepsilon - 1}{8\pi} E^2 \, \delta a \, d$$

provided that during this inserting of the insulator the capacitor voltage $V$ (and therefore $E$) is held constant, as by connecting the capacitor to a battery. Since however the charge

$$e = \frac{1}{4\pi} E[(A-a) + \varepsilon a]$$

also increases, the battery in this case supplies the work

$$\delta W = V \, \delta e = \frac{\varepsilon - 1}{4\pi} E^2 \, \delta a \, d$$

Obviously, only half of this energy contribution is necessary for increasing the field energy. The other half goes into the work $F\delta x$ performed by the field in the displacement of the insulator in the capacitor through the distance $\delta x$ against the external force (e.g. a spring) which holds the insulator in equilibrium. We therefore have

$$F = \frac{\varepsilon - 1}{8\pi} E^2 \frac{\partial a}{\partial x} d$$

We also have, for the special case of a dielectric liquid drawn into a capacitor (height method of determining $\varepsilon$),

$$F = \frac{\varepsilon - 1}{8\pi} E^2 q \qquad (34.2)$$

where $q$ is the top-surface area of the liquid between the capacitor plates.

(3) *Torque on a dielectric ellipsoid in a uniform field.*

As a third example we consider a dielectric ellipsoid suspended at its midpoint so as to be free to rotate in a uniform field $\mathbf{E}_0$. We inquire concerning its equilibrium position in this field.

This question too can be answered by way of the field energy. For this purpose we subtract from the actual field energy $U$ of the system including the ellipsoid, the field energy $U_0$ of the system before introducing the ellipsoid:

$$U - U_0 = \frac{1}{8\pi} \int (\mathbf{E} \cdot \mathbf{D} - \mathbf{E}_0 \cdot \mathbf{D}_0) \, dv \qquad (34.3)$$

We now transform this expression by means of the two relations

$$\int \mathbf{E} \cdot (\mathbf{D} - \mathbf{D}_0) \, dv = 0 \quad \text{and} \quad \int \mathbf{E}_0 \cdot (\mathbf{D} - \mathbf{D}_0) \, dv = 0 \quad (34.4)$$

whose correctness we show for the first equation. With $\mathbf{E} = -\operatorname{grad}\phi$, and applying Gauss's integral theorem to the whole space between any possible metal surfaces at hand (boundary surfaces between different insulators being here imagined as replaced by continuous transitions), we have

$$\int \mathbf{E} \cdot (\mathbf{D} - \mathbf{D}_0) \, dv = -\oiint \phi (D_n - D_{0n}) \, dS + \int \phi (\operatorname{div}\mathbf{D} - \operatorname{div}\mathbf{D}_0) \, dv$$

The second integral on the right vanishes here since in space $\mathbf{D}$ and $\mathbf{D}_0$ both possess the same sources. The first integral also vanishes since, on metal surfaces, $\phi$ is constant and $\oint D_n dS = \oint D_{0n} dS$ holds there if the total charge does not change. With (34.4), (34.3) can be rewritten as

$$U - U_0 = \frac{1}{8\pi} \int (\mathbf{E} \cdot \mathbf{D}_0 - \mathbf{E}_0 \cdot \mathbf{D})\, dv$$

Now the integrand here vanishes at all places outside the ellipsoid because the same interrelationships exist everywhere between $\mathbf{D}$ and $\mathbf{E}$ as exist between $\mathbf{D}_0$ and $\mathbf{E}_0$. Within the ellipsoid, however, we have $\mathbf{D}_0 = \mathbf{E}_0$ and $\mathbf{D} = \varepsilon\mathbf{E}$, so that we have

$$U - U_0 = -\frac{\varepsilon - 1}{8\pi} \int \mathbf{E} \cdot \mathbf{E}_0\, dv \qquad (34.5)$$

where the integral is to extend only over the ellipsoid.

Expression (34.5), valid in the first instance for any arbitrary form of insulator, simplifies in the special case of an ellipsoid because when an ellipsoid is brought into a uniform field $\mathbf{E}_0$, the field $\mathbf{E}$ within it remains uniform, as we saw in §30. Therefore, from (30.3) and (30.4) for $\mathbf{E} = \mathbf{E}_0 + \mathbf{E}'$, we have

$$E_x = \frac{E_{0x}}{1 + \chi A} \qquad E_y = \frac{E_{0y}}{1 + \chi B} \qquad E_z = \frac{E_{0z}}{1 + \chi C} \qquad (34.6)$$

where the coordinate axes coincide with the principal axes of the ellipsoid, and $A$, $B$, $C$ are the depolarization factors given by (30.10). Thus we have

$$U - U_0 = -\frac{(\varepsilon - 1)abc}{6} \left( \frac{E_{0x}^2}{1 + \chi A} + \frac{E_{0y}^2}{1 + \chi B} + \frac{E_{0z}^2}{1 + \chi C} \right) \qquad (34.7)$$

From this formula we see immediately that, since $\varepsilon > 1$, it is energetically more favourable for the ellipsoid to be in the field than outside it. The ellipsoid is therefore pulled into the field. Further, according to (30.10), if $a \geqq b \geqq c$, then we have for the depolarization factor that $A \leqq B \leqq C$, and thus, since $\chi > 0$, energetically we have the most favourable situation when the ellipsoid is oriented with its long axis in the direction of the field. If we would like to know the torque experienced by an ellipsoid whose longest axis makes an angle $\theta$ with the field and whose intermediate axis is held perpendicular to the field, we have, in (34.7), to put $E_{0x} = E_0 \cos\theta$, $E_{0y} = 0$, $E_{0z} = E_0 \sin\theta$, and we shall have for the torque around the intermediate axis (since $\varepsilon - 1 = 4\pi\chi$)

$$T = -\frac{\partial U}{\partial \theta} = -\frac{(\varepsilon-1)^2 abc}{24\pi} \frac{C-A}{(1+\chi A)(1+\chi C)} E_0^2 \sin 2\theta$$

We thus find two equilibrium orientations, one for $\theta = 0$, and the other for $\theta = \pi/2$. The first of these, which is also identical with $\theta = \pi$, is stable; the second is unstable.

If there were a substance with $\varepsilon < 1$, an ellipsoid made of such a substance would be expelled from a field. If, however, the ellipsoid were constrained to remain in the field, at the same time being free to rotate, then, since $\chi < 0$, it would as before orient itself with its longest axis in the direction of the field.

## §35. General calculation of the force on an insulator in an electric field

We now wish to investigate the force on an insulator in an electrostatic field, and we inquire concerning the force $\mathbf{f}\,dv$ exerted upon a volume element $dv$ of this medium. We can think of the *force density* here as being expressible either in the form

$$\mathbf{f}' = (\rho + \rho_P)\mathbf{E} = (\rho - \operatorname{div} \mathbf{P})\mathbf{E} = \frac{1}{4\pi} \mathbf{E} \operatorname{div} \mathbf{E} \qquad (35.1)$$

or in the form

$$\mathbf{f}'' = \rho\mathbf{E} + (\mathbf{P}.\nabla)\mathbf{E} \qquad (35.2)$$

in which we start either from the polarization charges $\rho_P\,dv$ in the volume $dv$, in the sense of (26.2) or (27.2), or we consider the dipole moment $\mathbf{P}\,dv$ of this volume element and use formula (31.6) for the force of the field on this moment.

As two quite different expressions appear equally to be suggested for the force density we see that in the setting up of physical quantities (such as the force density), which are of the second order in the field quantities, we have to proceed very cautiously. For example, let us examine critically the term $(\mathbf{P}.\nabla)\mathbf{E}$ in (35.2). Certainly $\mathbf{P}\,dv = \sum \mathbf{p}_j$, in which here we sum over all dipoles $\mathbf{p}_j$ contained in $dv$. Certainly, also, the dipole force on the element $dv$ is equal to the vector sum of the forces on all dipoles in $dv$. But the force on the individual dipoles is by no means $(\mathbf{p}_j.\nabla)\mathbf{E}$. Rather here, as in §26, we have to deal with the "effective" field strength $\mathbf{F}$, and not with the "average" field strength $\mathbf{E}$. In this connection it has to be observed that in all such product quantities of the second order, the average value of a product of two functions is not in general equal to the product of their individual average values.

In order to obtain the correct value of the macroscopic force density $\mathbf{f}$ we make use of the energy theorem, as in §34. For this we imagine the material in the field to be in some way moved, and we let $\mathbf{s}(x, y, z)$ be the displacement (assumed always to be small) of the material particle located at $(x, y, z)$. Now obviously $\mathbf{s} \cdot \mathbf{f} \, dv$ is the work performed in the movement, by the force density $\mathbf{f}$, on the volume element $dv$. Then, since the work performed by the field on the matter in all space is equal to the decrease in the total field energy, the energy theorem requires that

$$\int \mathbf{s} \cdot \mathbf{f} \, dv = -\delta U = -\delta \left( \frac{1}{4\pi} \int dv \int_0^{\mathbf{D}} \mathbf{E} \cdot d\mathbf{D} \right) \qquad (35.3)$$

If it is possible to bring the right-hand side of this equation into the form $\int \mathbf{s} \cdot \mathbf{g}(\mathbf{E}, \mathbf{D}) \, dv$, $\mathbf{g}$ being a function of the electric field quantities and independent of $\mathbf{s}$, then, by comparing the integrands in (35.3), it can be concluded that $\mathbf{f} = \mathbf{g}(\mathbf{E}, \mathbf{D})$.

This conclusion clearly supposes that the material does not move as a rigid body with constant speed; for in this case we can reach conclusions only with respect to the total force $\int \mathbf{f} \, dv$ (which would of course be interesting), but not concerning the force density. Nor does it suffice for a unique evaluation of $\mathbf{f}$ to limit ourselves to the special case of incompressible media; in this case we should have $\operatorname{div} \mathbf{s} = 0$, and, since $\operatorname{div}(\mathbf{s}\psi) = \mathbf{s} \operatorname{grad} \psi$, the gradient of an arbitrary continuous function $\psi$ could be added to $\mathbf{f}$, without thereby changing the value of the integral $\int \mathbf{s} \cdot \mathbf{f} \, dv$. In order, therefore, to obtain a unique $\mathbf{f}$-value, the vector $\mathbf{s}$, a completely arbitrary space-varying vector, has to be introduced into the calculations and the required variation of density of the material thus taken into account. In order to avoid lengthy calculations we shall limit ourselves in the following to the case of liquid and gaseous matter.

Further, for the sake of brevity we consider only insulators for which we can write $\mathbf{D} = \varepsilon \mathbf{E}$, $\varepsilon$ being a scalar permittivity independent of the field. Since the field in this case is uniquely defined when $\varepsilon(x, y, z)$ and the charge density $\rho(x, y, z)$ are given, and when the net charge on any possible metal bodies is known, we can trace back the change of the field energy $U$ from the motion of insulators to the changes thus required in $\varepsilon$ and $\rho$. In these calculations we shall for simplification imagine the discontinuous transitions of $\varepsilon$ and $\rho$ at the boundary between different insulators as replaced by continuous transitions. We shall speak later of typical boundary-surface forces.

First, since $\mathbf{D} = \varepsilon\mathbf{E}$, we write $U$ in the form

$$U = \frac{1}{8\pi} \int \frac{\mathbf{D}^2}{\varepsilon} \, dv$$

In the motion of the material, $\mathbf{D}$ changes by $\delta\mathbf{D}$, $\rho$ by $\delta\rho$, and $\varepsilon$ by $\delta\varepsilon$; we have therefore

$$\delta U = \frac{1}{4\pi} \int \frac{\mathbf{D}\,\delta\mathbf{D}}{\varepsilon} \, dv - \frac{1}{8\pi} \int \frac{\mathbf{D}^2\,\delta\varepsilon}{\varepsilon^2} \, dv = \frac{1}{4\pi} \int \mathbf{E}\,\delta\mathbf{D} \, dv - \frac{1}{8\pi} \int \mathbf{E}^2\,\delta\varepsilon \, dv$$

Since $\mathbf{E} = -\operatorname{grad}\phi$ and $\operatorname{div}\mathbf{D} = 4\pi\rho$, the integrand of the first integral on the right can be transformed as follows:

$$\mathbf{E}\,\delta\mathbf{D} = -\operatorname{div}(\phi\,\delta\mathbf{D}) + \phi\operatorname{div}\delta\mathbf{D} = -\operatorname{div}(\phi\,\delta\mathbf{D}) + 4\pi\phi\,\delta\rho$$

Since with Gauss's theorem the volume integral over $\operatorname{div}(\phi\,\delta\mathbf{D})$ can be transformed to an integral over the infinitely distant surface and over any possibly existing metal surfaces—and thus vanishes—we obtain for the change $\delta U$ in the field energy

$$\delta U = \int \phi\,\delta\rho \, dv - \frac{1}{8\pi} \int \mathbf{E}^2\,\delta\varepsilon \, dv \tag{35.4}$$

We have now to find the relationship between the change of $\rho$ and $\varepsilon$ and the given displacement $\mathbf{s}$ of the material. In this displacement there passes through a surface element $dS$ fixed in space and with normal direction $\mathbf{n}$, a quantity of electricity $\rho s_n \, dS$, bound together with the matter. Thus for a volume fixed in space we have the relation

$$\int \delta\rho \, dv = -\oiint \rho s_n \, dS = -\int \operatorname{div}(\rho\mathbf{s}) \, dv \tag{35.5}$$

the last because of Gauss's theorem. In working with a small volume element we therefore obtain

$$\delta\rho = -\operatorname{div}(\rho\mathbf{s}) \tag{35.6}$$

An analogous formula holds for the change of the mass density $\sigma$ of the substance:

$$\delta\sigma = -\operatorname{div}(\sigma\mathbf{s}) \tag{35.7}$$

Within an incompressible medium ($\sigma = \text{const}$) this equation would actually lead to $\operatorname{div}\mathbf{s} = 0$.

In the calculation of $\delta\varepsilon$ we limit ourselves to the *assumption*, certainly true for liquids and gases, *that $\varepsilon$ is a unique function of the density $\sigma$*. With (35.7) we then have

$$\delta\varepsilon = \frac{d\varepsilon}{d\sigma}\,\delta\sigma = -\frac{d\varepsilon}{d\sigma}\operatorname{div}(\sigma\mathbf{s})$$

With this and with (35.6) we obtain from (35.4) that

$$\delta U = -\int \phi\operatorname{div}(\rho\mathbf{s})\,dv + \frac{1}{8\pi}\int \mathbf{E}^2\frac{d\varepsilon}{d\sigma}\operatorname{div}(\sigma\mathbf{s})\,dv$$

Now we make use of Gauss's theorem twice, writing

$$\phi\operatorname{div}(\rho\mathbf{s}) = \operatorname{div}(\phi\rho\mathbf{s}) - \rho\mathbf{s}\operatorname{grad}\phi$$

$$\mathbf{E}^2\frac{d\varepsilon}{d\sigma}\operatorname{div}(\sigma\mathbf{s}) = \operatorname{div}\left(\mathbf{E}^2\frac{d\varepsilon}{d\sigma}\sigma\mathbf{s}\right) - \sigma\mathbf{s}\operatorname{grad}\mathbf{E}^2\frac{d\varepsilon}{d\sigma}$$

Thus the integrals over the infinitely distant surface and over the metal surfaces (assumed to be at rest) vanish and there remains, since now in both terms $\mathbf{s}$ comes out as a factor

$$\delta U = -\int \mathbf{s}\cdot\left[\rho\mathbf{E} + \frac{\sigma}{8\pi}\operatorname{grad}\left(\mathbf{E}^2\frac{d\varepsilon}{d\sigma}\right)\right]dv$$

We have thus actually found an expression having the form of the work integral in (35.3), and from it we can immediately obtain the force density $\mathbf{f}$:

$$\mathbf{f} = \rho\mathbf{E} + \frac{\sigma}{8\pi}\operatorname{grad}\left(\mathbf{E}^2\frac{d\varepsilon}{d\sigma}\right) \tag{35.8}$$

Since $\operatorname{grad}\varepsilon = \dfrac{d\varepsilon}{d\sigma}\operatorname{grad}\sigma$, we can also write this in the form

$$\mathbf{f} = \rho\mathbf{E} - \frac{1}{8\pi}\mathbf{E}^2\operatorname{grad}\varepsilon + \frac{1}{8\pi}\operatorname{grad}\left(\mathbf{E}^2\frac{d\varepsilon}{d\sigma}\sigma\right) \tag{35.9}$$

We wish now to examine this expression.

The first part of **f** in (35.8) and (35.9) gives the known force on the true charges.

The second part of **f** in (35.9) is effective where $\varepsilon$ varies in space; in particular, at an insulator-vacuum boundary surface it furnishes a force perpendicular to the surface, urging the latter into the vacuum. Thus it is also responsible for the force calculated in the second example of §34—the force with which a homogeneous dielectric, e.g. a glass plate, is drawn inside a parallel-plate capacitor. In this case the force contributions on the upper and lower sides of the capacitor act in opposition and there remains over only a part from the side face of the plate in the capacitor. Since E here is parallel to the boundary surface and thus has the same value inside as out, the volume integration over the second part of the force in (35.9) in this case gives $\dfrac{\varepsilon - 1}{8\pi}\mathbf{E}^2 q$, where $q$ is the area of the side face. This agrees exactly with the value (34.2).

Finally, the third part of **f** in (35.9), as a gradient, leads to no resulting total force on the dielectric—for, the volume integral formed with it can be transformed to a surface integral over a surface lying external to the dielectric, where $\sigma$ vanishes. It does however become of decisive significance whenever a change of form or volume of the dielectric in an electric field comes into question. We shall occupy ourselves with such questions in §37.

Before this, however, we wish by means of (35.9) to calculate the force exerted on a region of matter of finite size as a consequence of the force density **f**. We shall thereby become acquainted with a remarkable property of the electric force which is characteristic of the Faraday-Maxwell field theory.

## §36. The Maxwell stresses

According to Faraday and Maxwell, there are no action-at-a-distance forces; instead, all force actions are transmitted in a continuous way from one body to another through the electromagnetic field. For an elastic material under stress the concept of a continuous (field-wise) transmission is entirely familiar to us. Now in a similar fashion Faraday regarded the seat of the electromagnetic force action as a peculiar state of stress in the space occupied by the electric or magnetic fields.

Let us imagine a system in electrostatic equilibrium, divided into two parts 1 and 2 by a closed surface $S$. Then, according to the

Faraday concept, the total force which is exerted on 1 by 2 must in some manner or other pass through the surface $S$. It should be entirely immaterial whether a part or even the whole of the surface passes through free space.

The rigorous development of this concept was given by Maxwell. He showed that the total force-effect $\mathbf{F} = \int \mathbf{f} dv$ on part 1 can be represented as surface forces acting on the boundary $S$ of this part. If, as in §14, we designate with $\mathbf{T}_n dS$ a force which shall act on a surface element $dS$ of this boundary, with outward-drawn normal $\mathbf{n}$, then, according to Maxwell, it is possible to find a function $\mathbf{T}_n$ which depends only on $\varepsilon$ and the field quantities at the location of $dS$ such that

$$\mathbf{F} = \int \mathbf{f} dv = \oiint \mathbf{T}_n dS \tag{36.1}$$

The transformation is furnished by Gauss's theorem when it is possible to represent the components of $\mathbf{f}$ in the form of a divergence—for, by (14.4) the $x$-component, for example, of equation (36.1) can be written in the form

$$F_x = \oiint (T_{xx} n_x + T_{yx} n_y + T_{zx} n_z) \, dS$$

$$= \int \left( \frac{\partial T_{xx}}{\partial x} + \frac{\partial T_{yx}}{\partial y} + \frac{\partial T_{zx}}{\partial z} \right) dv \tag{36.2}$$

We accordingly look for quantities $T_{xx}$, $T_{yx}$, $T_{zx}$ such that we have

$$f_x = \frac{\partial T_{xx}}{\partial x} + \frac{\partial T_{yx}}{\partial y} + \frac{\partial T_{zx}}{\partial z} \tag{36.3}$$

Corresponding relationships must also then hold for the $y$- and $z$-components of $\mathbf{f}$.

In order to give expression (35.9) for $f_x$ the form (36.3), we write for the first part of $f_x$

$$\rho E_x = \frac{1}{4\pi} E_x \operatorname{div} \mathbf{D} = \frac{1}{4\pi} \operatorname{div} (E_x \mathbf{D}) - \frac{1}{4\pi} \mathbf{D} \operatorname{grad} E_x \tag{36.4}$$

and for the second part

$$-\frac{1}{8\pi} \mathbf{E}^2 \frac{\partial \varepsilon}{\partial x} = -\frac{1}{8\pi} \frac{\partial}{\partial x} (\varepsilon \mathbf{E}^2) + \frac{\varepsilon}{4\pi} \mathbf{E} \frac{\partial \mathbf{E}}{\partial x} \tag{36.5}$$

Since $\mathbf{E}$ is irrotational, the last term here is identical with $\dfrac{\varepsilon}{4\pi}\mathbf{E}\,\mathrm{grad}\,E_x$,

and thus in the expression for $f_x$ it cancels with the last term of (36.4). For the force density we therefore obtain the desired form

$$f_x = \frac{\partial}{\partial x}\left\{\frac{\varepsilon E_x^2}{4\pi} - \frac{\mathbf{E}^2}{8\pi}\left(\varepsilon - \frac{d\varepsilon}{d\sigma}\sigma\right)\right\} + \frac{\partial}{\partial y}\left\{\frac{\varepsilon E_x E_y}{4\pi}\right\} + \frac{\partial}{\partial z}\left\{\frac{\varepsilon E_x E_z}{4\pi}\right\}$$

and from this the component matrix of the *Maxwell stress tensor*

$$\mathbf{T} \equiv \frac{1}{4\pi}\begin{pmatrix} \varepsilon E_x^2 - \dfrac{\mathbf{E}^2}{2}\left(\varepsilon - \dfrac{d\varepsilon}{d\sigma}\sigma\right) & \varepsilon E_x E_y & \varepsilon E_x E_z \\[2ex] \varepsilon E_x E_y & \varepsilon E_y^2 - \dfrac{\mathbf{E}^2}{2}\left(\varepsilon - \dfrac{d\varepsilon}{d\sigma}\sigma\right) & \varepsilon E_y E_z \\[2ex] \varepsilon E_x E_z & \varepsilon E_y E_z & \varepsilon E_z^2 - \dfrac{\mathbf{E}^2}{2}\left(\varepsilon - \dfrac{d\varepsilon}{d\sigma}\sigma\right) \end{pmatrix} \quad (36.6)$$

For the special case of a surface element *in vacuo* this matrix assumes the simple form

$$\mathbf{T} \equiv \frac{1}{4\pi}\begin{pmatrix} E_x^2 - \tfrac{1}{2}\mathbf{E}^2 & E_x E_y & E_x E_z \\[1ex] E_x E_y & E_y^2 - \tfrac{1}{2}\mathbf{E}^2 & E_y E_z \\[1ex] E_x E_z & E_y E_z & E_z^2 - \tfrac{1}{2}\mathbf{E}^2 \end{pmatrix} \quad (36.7)$$

It is noteworthy that the first-assumed force densities $\mathbf{f}'$ and $\mathbf{f}''$ of equations (35.1) and (35.2) can in like manner be represented as the divergence of stress quantities $\mathbf{T}'$ and $\mathbf{T}''$, and that these tensors for a vacuum agree with the vacuum value $\mathbf{T}$ of the Maxwell stress tensor. If, therefore, we ask for the *total force* on an arbitrary body in an electric field, we can calculate with each of the integral expressions of the chain of equations

$$\mathbf{F} = \int \mathbf{f}\,dv = \int \mathbf{f}'\,dv = \int \mathbf{f}''\,dv = \oint \mathbf{T}_n\,dS \quad (36.8)$$

so long as in the last integral one of the closed surfaces encloses the entire body.

We can elucidate the stress field given by $\mathbf{T}$ in the following way. We imagine a certain surface element, and we lay out the coordinate system so that the $\mathbf{n}$-direction of the surface element coincides with the positive $z$-direction, and the $\mathbf{E}$-vector lies in the $xz$-plane (figure 30). Designating the angle between the surface normal and the field strength by $\theta$, we now have for the components of the field strength

$$E_x = E \sin\theta \qquad E_y = 0 \qquad E_z = E \cos\theta \quad (36.9)$$

and, from (36.7), for the components of the surface force

$$T_{zx} = \frac{E^2}{8\pi}\sin 2\theta \qquad T_{zy} = 0 \qquad T_{zz} = \frac{E^2}{8\pi}\cos 2\theta \quad (36.10)$$

The magnitude of the surface force is therefore equal to $E^2/8\pi$, independent of the orientation of the surface element with respect to the field. The force itself lies in the plane determined by $\mathbf{E}$ and $\mathbf{n}$ and, as shown in figure 30, forms with $\mathbf{E}$ the same angle as $\mathbf{E}$ forms with the

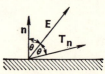

Fig. 30.—The angle between the Maxwell surface force $\mathbf{T_n}$ and the normal $\mathbf{n}$ is bisected by the direction of the field strength $\mathbf{E}$

normal direction $\mathbf{n}$. If, as a special case, $\mathbf{E}$ is perpendicular to the surface element, this also holds for the $\mathbf{T}_n$-vector; we then have a pure *tensile stress*. When, however, $\mathbf{E}$ lies in the plane of the surface element, $\mathbf{T}_n$ is perpendicular to the surface element, but is directed oppositely to the $\mathbf{n}$-vector and therefore acts as a pure *pressure*.

We illustrate these facts of the case by considering the force which two equal point charges exert on one another for the case where the two charges have the same sign (repulsion) or opposite sign (attraction). Let both the charges lie on the $x$-axis, one at $x = +a$ and the other at $x = -a$. As volume part No. 1 in our system we consider the semi-infinite space $x < 0$, bounded by the $yz$-plane and an infinitely distant surface. On this latter surface no stress acts since, at large distances from the origin, the T-components go to zero at least like $1/r^4$. There remains only the force effect transmitted across the plane of symmetry.

*For equal charges* ($e_1 = e_2 = e$), according to figure 31a, the field in the plane of symmetry is everywhere parallel to the surface; there is therefore a pure pressure stress exerted by the right-hand charge on the left-hand semi-infinite space, and therefore on the left-hand charge. For a point at a distance $b$ from the axis, we have $E = 2eb/r^3$, and therefore $|\mathbf{T}_x| = e^2 b^2/2\pi r^6$. Thus, since $dS = 2\pi b\, db$ and $r^2 = a^2 + b^2$

$$F = \int_0^\infty \frac{e^2 b^2}{2\pi(a^2 + b^2)^3}\pi d(b^2) = \frac{e^2}{4a^2}$$

corresponding to the Coulomb repulsion.

*For equal but opposite charges* $(-e_1 = e_2 = e)$ the field lines, according to figure 31$b$, are everywhere perpendicular to the median plane; we therefore have pure tension. Indeed, $E = E_x = 2ea/r^3$, and therefore $|\mathbf{T}_x| = e^2 a^2 / 2\pi r^6$. We therefore again find the Coulomb force

$$F = \int_0^\infty \frac{e^2 a^2}{2\pi(a^2 + b^2)^3} \pi d(b^2) = \frac{e^2}{4a^2}$$

but now, however, as a force of attraction.

Finally, let us give a simple example in which typical surface-force effects can be handled with the help of the Maxwell stresses. In the foregoing work, by way of simplification, we have imagined dis-

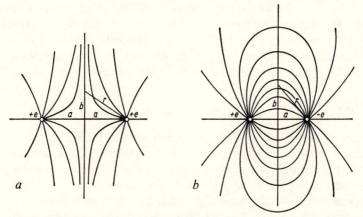

Fig. 31.—Repulsion of point charges of like sign, and attraction of charges of unlike sign, described by the Maxwell stresses transmitted across the plane of symmetry

continuous transitions between different media as having been replaced by continuous transitions. If now, however, we ask somewhat specially about the force action of the charges residing on the boundary surface (on a metal surface, for example), we must then by some means remove this simplifying assumption of a broadened transition. We accordingly ask about the total force on a small region of volume having the shape of a flat cylinder whose base surface and top surface, each of area $dS$, lie respectively below and above the medium boundary or transition zone. The height of the cylinder is small compared to the linear extension of $dS$. Now this force, according to Maxwell, can be calculated from the stresses on the base and top surface of the cylinder, while the contribution of the side surface can be neglected as a small quantity of higher order. Designating the values of the field quantities on the base and upper surface by the subscripts 1 and 2 respectively and choosing the direction of the vector normal to be from 1 to 2, we obtain from (36.2) and (36.6) for the force on the cylinder

$$\int \mathbf{f} \, dv = dS \left\{ \frac{1}{4\pi} \left[ \varepsilon_2 \mathbf{E}_2 (\mathbf{E}_2 \cdot \mathbf{n}) - \varepsilon_1 \mathbf{E}_1 (\mathbf{E}_1 \cdot \mathbf{n}) \right] - \right. $$
$$\left. - \frac{\mathbf{n}}{8\pi} \left[ \mathbf{E}_2^2 \left( \varepsilon - \frac{d\varepsilon}{d\sigma} \sigma \right)_2 - \mathbf{E}_1^2 \left( \varepsilon - \frac{d\varepsilon}{d\sigma} \sigma \right)_1 \right] \right\} \quad (36.11)$$

The factor multiplying $dS$ can therefore be designated as the force on unit area of the boundary surface, in which we can now without difficulty replace the continuous transition by one that is abrupt. Applying equation (36.11) to the simple case of a metal in a vacuum, we have inside the metal that $\mathbf{E}_1 = 0$. We know further that the external field $\mathbf{E}_2 = \mathbf{E}$ is perpendicular to the metal surface; and $\varepsilon_2 = 1$. We then obtain for the force on unit area of the metal surface the value $\frac{1}{4\pi} \mathbf{E}(\mathbf{E} \cdot \mathbf{n}) - \frac{\mathbf{n}}{8\pi} \mathbf{E}^2$, which, since $\mathbf{E} = 4\pi\omega\mathbf{n}$, can also be written in the form $E\omega/2$, in which $\omega$ is the density of the surface charge. The emergence of the factor $\frac{1}{2}$ comes about, as in the first example of §34, from the fact that with respect to the field $E$ existing on the surface, only $\frac{1}{2}E$ is to be regarded as the field contribution of the net charges of the system, while the charged metal surface itself produces on both sides a field $2\pi\omega = \frac{1}{2}E$; this field in fact combines with the field contribution of the net charges outside the metal to give $E$, while cancelling itself inside the metal, but naturally it furnishes no force contribution on the surface.

## §37. Electric force effects in homogeneous liquids and gases

If an electric field be produced within an uncharged dielectric, the resulting electrostatic forces will produce a displacement of the individual parts of the medium with respect to one another. This gives rise to elastic counter-forces. The motion ceases when these two forces hold each other in balance. The phenomenon in which, in the manner indicated, elastic stresses and changes of shape take place in an uncharged insulator is called *electrostriction*.

In liquids and gases in equilibrium there is only one kind of elastic stress, namely the pressure $p$, equal in all directions. If this varies with position, a force $-\mathrm{grad}\,p\,dv$ acts on a volume element $dv$ in consequence of the pressure change. In equilibrium this pressure force must be counterbalanced by the electrostatic force $\mathbf{f}\,dv$ calculated in §35. Thus in an uncharged liquid ($\rho = 0$), on account of (35.8), the condition for equilibrium is

$$\sigma\,\mathrm{grad}\left(\frac{1}{8\pi}\mathbf{E}^2\frac{d\varepsilon}{d\sigma}\right) - \mathrm{grad}\,p = 0 \qquad (37.1)$$

We consider $\sigma$ to be expressed as a function of $p$ through the equation of state of the medium. Equation (37.1) can then be written in the form

$$\mathrm{grad}\left\{\frac{1}{8\pi}\mathbf{E}^2\frac{d\varepsilon}{d\sigma} - \int_{p_0}^{p}\frac{dp}{\sigma}\right\} = 0$$

the expression in brackets having the same value everywhere in the liquid. Comparing two locations designated by 1 and 0, we have

$$\int_{p_0}^{p_1}\frac{dp}{\sigma} = \frac{1}{8\pi}\left[\mathbf{E}_1^2\left(\frac{d\varepsilon}{d\sigma}\right)_1 - \mathbf{E}_0^2\left(\frac{d\varepsilon}{d\sigma}\right)_0\right] \qquad (37.2)$$

If the contributions $E_1$ and $E_0$ of the field strengths are given, then (37.2) furnishes a relationship between the pressures $p_1$ and $p_0$.

We first apply this formula to the case of a *tenuous gas*. We wish to investigate the extent to which the gas pressure in an electric field $\mathbf{E}_1 = \mathbf{E}$ between, say, the plates of a capacitor, differs from the pressure $p_0$ in a field-free space ($\mathbf{E}_0 = 0$). From the equation

$$p = \frac{\sigma RT}{M} = \frac{\sigma kT}{m}$$

($M$ = molecular weight, $m$ = the mass of an individual gas molecule, $R$ = the gas constant, and $k$ = Boltzmann's constant), it follows for the left side of (37.2) that

$$\int_{p_0}^{p_1}\frac{dp}{\sigma} = \frac{kT}{m}\ln\frac{p_1}{p_0}$$

For a gas that is not too dense, we have further (see equation 26.11) that

$$\varepsilon - 1 = 4\pi\chi = 4\pi n\alpha = 4\pi\sigma\alpha/m \qquad (37.3)$$

in which $n = \sigma/m$ is the number of gas molecules per unit volume, and $\alpha$ is the polarizability of a single gas molecule. We therefore obtain

$$\frac{p}{p_0} = \exp\left(\frac{\alpha E^2}{2kT}\right) \tag{37.4}$$

We could, incidentally, have derived this formula in an entirely different way, directly in conjunction with formula (31.9) for the electrical energy of a dipole $\mathbf{p} = \alpha \mathbf{E}$ produced by the field $\mathbf{E}$; since its energy is equal to $-\frac{1}{2}\alpha E^2$, equation (37.4) represents nothing more than the well-known formula for the barometric pressure as a function of height:

$$\frac{p}{p_0} = \exp\left(-\frac{mgh}{kT}\right)$$

in which the potential energy $(mgh)$ of the gas molecules in the gravitational field takes the place of the electrical energy $(-\frac{1}{2}\alpha E^2)$.

Equation (37.4) permits a rough estimate of the order of magnitude of pressure change produced by electrostriction. If gas molecules are conceived as small conducting spheres of radius $a \approx 10^{-8}$ cm (the molecular radius), then from (26.6) we can say approximately that $\alpha = a^3 \approx 10^{-24}$ cm³. In a field of $3 \times 10^7$ V/m, corresponding to $E = 1000$ CGS units, with $k = 1.38 \times 10^{-16}$ erg/deg, and $T = 300°$K, the exponent in (37.4) is approximately equal to $10^{-5}$. It is always therefore a very small effect and great care is required in its measurement.

For liquids, we have to proceed somewhat differently. So long as we remain wholly within the liquid we can employ equation (37.2), in which, to a good approximation, we can assume that $\sigma$ is constant. In this case, therefore, when we again take $\mathbf{E}_1 = \mathbf{E}$ and $\mathbf{E}_0 = 0$, we have

$$p - p_0 = \frac{1}{8\pi} \mathbf{E}^2 \frac{d\varepsilon}{d\sigma} \sigma \tag{37.5}$$

At the boundary surface of the liquid, however, we have to deal with a rapid transition of the density $\sigma$ from its value inside to the value zero outside. Here, therefore, we shall revert to equation (36.11) which represents the electrical force on a thin surface layer, and we shall equate its value per unit area at equilibrium to the pressure difference between outside and inside.

By way of illustration, we examine the arrangement shown in figure 32. Between the plates A and B of the charged capacitor partially immersed in liquid, the latter has the pressure $p$ given by (37.5), when the pressure in its field-free part is $p_0$. For the exposed surface of the liquid in the capacitor, at which $E_1 = E_2 = E$ are perpendicular to $n$, we have from (36.11) the relationship

$$p' - p = \frac{1}{8\pi} E^2 \left( \varepsilon - 1 - \frac{d\varepsilon}{d\sigma} \sigma \right) \tag{37.6}$$

Equilibrium therefore exists between the external pressures $p_0$ and $p'$ on the one hand, and the internal electrical forces on the other, when the equation

$$p' - p_0 = \frac{\varepsilon - 1}{8\pi} E^2 \tag{37.7}$$

holds. This result is in agreement with equation (34.2); our present consideration, however, permits a deeper view into the activity of this force effect, i.e. into the spatial distribution of the force density.

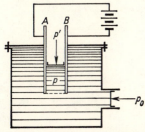

Fig. 32.—For the calculation of the pressure relations in a dielectric liquid in which a charged capacitor is partially immersed

Finally, we consider the force action of the liquid on the capacitor plates. Here the following particular problem faces us: When the charged parallel-plate capacitor is *in vacuo*, one plate of the capacitor experiences from the other a force $\frac{1}{2}\omega E_0$ per square centimetre (with $\omega = e/S = E_0/4\pi$). If now we immerse the capacitor in the liquid while keeping the total charge constant, the force of attraction diminishes (as we can conclude from the energy considerations of §34), being $1/\varepsilon$ of its former value. How is this fact to be understood?

The field strength $E$ in the liquid is indeed $\varepsilon$ times smaller than previously *in vacuo* ($E_0 = \varepsilon E$); at the location of the metal plate, however (where the charges reside) its value has in no way changed. The force exerted on the charges of the plate is therefore, as before, given by $\frac{1}{2}\omega E_0 = E_0^2/8\pi$. The reduction of the force acting on the plate comes into existence through the additional pressure action of the liquid in the field. It follows actually from (36.11), if we designate by 1 the interior of the liquid and by 2 the boundary between the liquid and the metal plate, with $E_1 = E$, $E_2 = E_0$, $\varepsilon_1 = \varepsilon$, $\varepsilon_2 = 1$, that

$$p'' - p = \frac{1}{8\pi}\left(E_0^2 - \varepsilon E^2 - E^2 \frac{d\varepsilon}{d\sigma}\sigma\right) \qquad (37.8)$$

so that from (37.5), the pressure $p''$ on the metal plate differs from the external pressure $p_0$ by

$$p'' - p_0 = \frac{1}{8\pi}(E_0^2 - \varepsilon E^2) \qquad (37.9)$$

The net force acting on the plate is then obtained by subtracting this pressure from the Coulomb force $E_0^2/8\pi$ on the charges. In this way we arrive at the correct value for the net force $\varepsilon E^2/8\pi = E_0^2/8\pi\varepsilon$ as given by the energy principle.

# C

---

Electric
current
and
the
magnetic
field

# CHAPTER C I

## The Laws of the Electric Current

### §38. Current strength and current density

Of the equations which we have used to describe the electrostatic field there are several which will remain valid for more general electromagnetic processes. These are: the relation between the divergence of **D** and the electric charges

$$\operatorname{div} \mathbf{D} = 4\pi\rho \qquad (38.1)$$

the expression for the density of electrical energy, and the relation between the two vectors **D** and **E**. Two relations belong exclusively to electrostatics; they are

$$\mathbf{E} = -\operatorname{grad}\phi \text{ and } \phi = \text{constant in conductors;}$$

in other words **E** is irrotational (vortex-free), and **E** vanishes inside homogeneous conductors of electricity.

We wish now to give up the latter of the above two requirements in that, say, experimentally we join with a metal wire the two plates of a capacitor charged to the potential difference $\phi_1 - \phi_2$. Immediately after completion of this connection the potential in the wire is certainly not constant, since on its ends it possesses the values $\phi_1$ and $\phi_2$. An electric field therefore exists in the wire and the condition for electrostatic equilibrium is no longer fulfilled. Actually the two charges $-e$ and $+e$ of the capacitor tend to neutralize one another through the wire. Equilibrium is not again established until the charges of the capacitor and the field have vanished. During this equalization a current of strength $I$ flows in the wire, i.e. in one second a quantity of electricity $I$ flows through a cross-section of the wire. Corresponding to this definition, the current strength is equal to the time decrease of the capacitor charge:

$$I = -\frac{de}{dt} \qquad (38.2)$$

That during the change of the capacitor charge with time there is something actually going on in the wire can be inferred because a heat development takes place in the wire, and in its neighbourhood a magnetic field exists. The appearance of the magnetic field represents a complication in the process which, in its full generality, we shall handle only in later chapters. So long as the current strength $I$ is constant, the magnetic field which it generates does not change. Hence, up to a certain point, we can deal with the laws of steady currents (i.e. currents which do not vary with time) without taking account of the accompanying magnetic field. In our example we can, to be sure, only approximate to the condition of a steady current—by making the resistance of the wire and the capacitance of the capacitor both as large as possible. The realization of a truly steady current is not possible by purely electrostatic means. This is only to be achieved by artificially stabilizing the potential difference of the capacitor plates by means of appliances such as voltaic cells, storage batteries, or thermocouples—devices which are foreign to electrostatics. We shall return to these devices in later sections.

Along with the current strength $I$ we introduce the current density **g**. It is defined as a vector whose direction is the same as the direction of the transport of the electricity in the conductor and whose magnitude is established by the fact that $g_n dS$ is the quantity of electricity which in one second goes through the surface element toward the side given by the normal vector **n**. We therefore have

$$I = \int g_n \, dS \qquad (38.3)$$

integrated over the cross-section of the conductor.

*Experimentally, electric charge can be neither created nor annihilated.* Therefore, an electric charge can change with time only by the flow of a current. We accordingly have for the *equation of continuity* of charge

$$\frac{d}{dt}\left(\int \rho \, dv\right) = -\oiint g_n \, dS \qquad (38.4)$$

If, using Gauss's theorem, we transform the surface integral into a volume integral, we then pass from the integral form (38.4) of the equation of continuity to its differential form

$$\frac{\partial \rho}{\partial t} = -\operatorname{div} \mathbf{g} \qquad (38.5)$$

which we have already used in the form (35.6) in the derivation of the force density in the electrostatic field from the field energy.

In particular, for a *continuous current* it follows from the equation of continuity that $\operatorname{div}\mathbf{g} = 0$. In this case, moreover, the same current $I$ must flow through every cross-section of the wire. And finally, for branching of the current (see figure 33) we have Kirchhoff's law

$$I = I_1 + I_2$$

Up to now we have considered the motion only of the *true charges*. In addition to these, however, according to §26 *polarization charges* can also exist, being given by $\rho_P = -\operatorname{div}\mathbf{P}$ inside a dielectric, and by

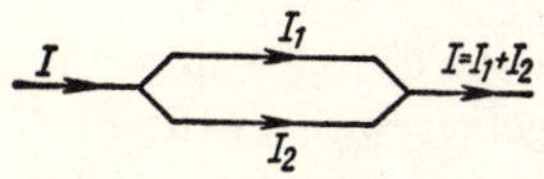

Fig. 33.—Kirchhoff's law for current branching

$\omega_P = P_n$ on the surface. Since $P_n dS$ was defined as the total charge which, in the polarization of the medium, passed through $dS$ in the direction of $\mathbf{n}$, it is suggested that in the steady change of the state of polarization a *polarization current density*

$$\mathbf{g}_P = \frac{\partial \mathbf{P}}{\partial t} \tag{38.6}$$

be introduced, in analogy with the true current density $\mathbf{g}$. Clearly, $\mathbf{g}_P$, together with the charge density $\rho_P$, also satisfies the equation of continuity

$$\frac{\partial \rho_P}{\partial t} = -\operatorname{div}\mathbf{g}_P \tag{38.7}$$

From equation (38.5) we can now obtain a basic result which is important for the further development of the Maxwell theory. If we differentiate (38.1) with respect to the time, we find, in combination with (38.5) that

$$\frac{1}{4\pi}\operatorname{div}\frac{\partial \mathbf{D}}{\partial t} = -\operatorname{div}\mathbf{g}$$

Therefore the quantity

$$\mathbf{c} = \mathbf{g} + \frac{1}{4\pi}\frac{\partial \mathbf{D}}{\partial t} = \mathbf{g} + \mathbf{g}_P + \frac{1}{4\pi}\frac{\partial \mathbf{E}}{\partial t} \tag{38.8}$$

is divergence-free; for this we always have div $\mathbf{c} = 0$ inside, and $c_n$ continuous across boundary surfaces. Through the *displacement current* $\partial\mathbf{D}/\partial t$ the *conduction current* $\mathbf{g}$ is supplemented, giving a source-free total current. A simple illustration of the foregoing is given here by a parallel-plate capacitor short-circuited by a wire: The conduction current $I$ flowing in the wire stops at the plates of the capacitor. Since however in the discharging of the capacitor the electrical displacement $D = 4\pi\omega$ changes ($\omega$ being the charge density), there exists in the space between the capacitor plates a displacement current of total strength

$$\frac{1}{4\pi} \frac{dD}{dt} S = \frac{d\omega}{dt} S = \frac{de}{dt}$$

Thus in the transfer of the charge $e$ from the negative to the positive plate, this current, together with discharge current $I$ flowing in the opposite direction, forms a source-free current circuit.

*Remarks:* In the MKSA system the equation of continuity retains the form (38.5). The formulae for the density of polarization charge and the polarization current are similarly unchanged. Only the expression (38.8) for the total current requires changing:

$$\mathbf{c} = \mathbf{g} + \frac{\partial\mathbf{D}}{\partial t} = \mathbf{g} + \frac{\partial\mathbf{P}}{\partial t} + \varepsilon_0 \frac{\partial\mathbf{E}}{\partial t}$$

## §39. Ohm's law

According to Georg Simon Ohm, in all solid and liquid electrical conductors, the strength $I$ of a steady current passing through a conductor is proportional to the voltage $V$ applied:

$$I = \frac{V}{R} \tag{39.1}$$

The constant of proportionality is called the *resistance* of the conductor; it depends only upon the material and its dimensions.* For example, we have for a cylindrical wire of length $l$ and cross-section $q$

$$R = \frac{1}{\sigma} \frac{l}{q} \tag{39.2}$$

---

* Owing to the rather complicated relationships attending the passage of electric currents through gases, we must postpone the development of these relationships to the third volume. Apparently complicated relations also appear in resistance measurements in solid and liquid conductors if the temperature of the conductor is not carefully held constant. This is due to the strong temperature-dependence of the resistance and to the Joule heating which is always present.

where the quantity $\sigma$ depends only upon the wire material. The quantity $\sigma$ is called the *conductivity* and $1/\sigma$ the *specific resistance* or *resistivity*. The resistivity is equal to the resistance of a cylinder of length 1 cm, and cross-section 1 cm². Applying equation (39.1) to this cylinder, for isotropic substances we obtain as the *differential form of Ohm's law*, in vector formulation,

$$\mathbf{g} = \sigma\mathbf{E} \qquad (39.3)$$

The current density here is proportional to the magnitude of $\mathbf{E}$; it also has the same direction as $\mathbf{E}$.

For *anisotropic* solids (e.g. crystals or elastically stressed materials) the conductivity depends in general upon the direction of the current flow. The conductivity $\sigma$ is then not a scalar but a symmetric tensor, and (39.3) is to be read as a tensor equation in which the vectors $\mathbf{g}$ and $\mathbf{E}$ are in general no longer parallel; rather, we now have for $g_x$, for example

$$g_x = \sigma_{xx}E_x + \sigma_{xy}E_y + \sigma_{xz}E_z$$

with analogous expressions for $g_y$ and $g_z$. In the following, however, we shall limit ourselves to isotropic substances.

From equation (39.3) the integral form (39.1) of Ohm's law can be concluded, that is, for arbitrary arrangements and forms of the conductor—for, owing to the linear relationship between the current density and the field strength, an increase of $\mathbf{E}$ by a factor $F$ leads to a like increase not only in $\mathbf{g}$ but also in $V$ and $I$, whereby the proportionality of these two quantities is maintained.

Thus there remains only the occasional problem of determining the *magnitude* of the resistance. In the most commonly met case, that of stationary currents, we can calculate the $\mathbf{E}$-field from the laws of electrostatics. However, we no longer have $\phi$ constant within the conductors; rather, the field in the conductors has first to be calculated. In the stationary case this field in the conductor is vortex-free; and, because of the equation of continuity (38.5) as well as the assumption of a stationary current ($\partial\rho/\partial t = 0$), we must have that

$$\operatorname{div}\mathbf{g} = \operatorname{div}(\sigma\mathbf{E}) = 0$$

everywhere inside the conductor, while on the boundary surface of the conductor $g_n = \sigma E_n$ is continuous. In particular, on boundary surfaces bordering insulators (e.g. a vacuum) $g_n$ and also therefore $E_n$ must vanish; the field strength on such boundary surfaces must therefore be purely tangential. (Otherwise, owing to the finiteness of the $g_n$ component, electric charges would be brought to the surface until, in consequence of the additional field produced by these charges, the normal component of the total field would vanish. Such surface charges are, for example, the reason why in all current-carrying thin wires, however bent, the E-field in the wire always has the direction of the wire.)

Owing to the great difficulties in the solution of the electrostatic problems encountered, a complete calculation of the resistance is possible for only a few simple forms. We carry out such calculations in three examples:

In *thin wires*, owing to the surface requirement, the E-field has the direction of the wire axis and, since curl$E = 0$ or $E = -\text{grad}\,\phi$, it is always constant over the wire cross-section $q$. The same holds for the current density g, so that we have $I = qg = q\sigma E$. Since, further, for a stationary current, $I$ has the same value for every wire cross-section, we obtain for the line integral over the field strength, taken, say, along the wire axis,

$$V = \int \mathbf{E} \cdot d\mathbf{r} = \int E \, ds = I \int \frac{ds}{q\sigma}$$

so that for the resistance $R$, because of (39.1), we obtain

$$R = \int \frac{ds}{q\sigma} \tag{39.4}$$

For a straight cylindrical wire this formula clearly agrees with (39.2).

In a *coaxial cable* the current-carrying parts consist of two coaxial cylinders, the inner having radius $a_1$, and the outer having an inner radius $a_2$; and, of course, $a_2 > a_1$. Between the cylinders there is a layer of insulation of very small but still always finite conductivity. In order to determine the resistance of this insulation we neglect the ohmic potential drop in the metallic conductors in comparison with that in the insulator which is incomparably larger. In the space between the conductors, then, there is the same field distribution as in a cylindrical capacitor. For the radial field we have

$$E = \frac{V}{r \ln(a_2/a_1)} \qquad V = \int_{a_1}^{a_2} E \, dr$$

and therefore for the current strength through a portion of the cable insulation of length $l$,

$$I = 2\pi r l g = 2\pi r l \sigma E = \frac{2\pi l \sigma V}{\ln(a_2/a_1)}$$

The required resistance is equal to

$$R = \frac{\ln(a_2/a_1)}{2\pi l \sigma}$$

Let a current be led by means of a well-insulated conductor to a metallic sphere of radius $a$, from which sphere the current can flow off through a surrounding medium of infinite extent ("grounding"). In the vicinity of the sphere we have to reckon with the Coulomb field

$$E = \frac{Va}{r^2} \qquad V = \int_a^\infty E \, dr$$

We shall then have

$$I = 4\pi r^2 \sigma E = 4\pi \sigma a V$$

so that in this case we obtain for the resistance to ground

$$R = \frac{1}{4\pi \sigma a}$$

We wish now briefly to occupy ourselves with the question of the *grounds for the validity of Ohm's law*. Obviously the transport of true charges is effected by the motion of atomic charge-carriers (e.g. electrons in metals; ions in electrolytes). In this case the current density is given by

$$\mathbf{g} = \Sigma e_k \mathbf{v}_k \tag{39.5}$$

in which the summation is to be taken over all particles which, in one second, pass through a surface perpendicular to $\mathbf{g}$ having an area of $1 \, \text{cm}^2$. Since these particles move in an electric field they would be continuously accelerated if they did not at the same time (according to Drude) collide with other constituents of matter and thus give up again to these the energy they obtained from the field. This process can be treated formally as if the charge carriers move in a viscous medium having large frictional resistance and thus (according to Stokes) possessing a velocity which is proportional to the effective force. With $\mathbf{v}_k \sim \mathbf{E}$ we have from (39.5) that $\mathbf{g}$ is also proportional to $\mathbf{E}$, and this is just the statement of Ohm's law. As to the calculation of the constant of proportionality met with here, i.e. the electrical conductivity, this is a subject with which we shall be occupied in the third volume.

*Remark:* Ohm's law in the MKSA system retains the form it has in (39.1) or (39.3). In the MKSA system, however, resistance and conductivity are measured in different dimensions from the Gaussian system. While in the latter the total resistance has the dimension of a reciprocal velocity and the specific resistance that of a time, i.e. $1/\sigma$ is to be measured in seconds, the MKSA system leads to a new typical resistance unit, namely, $1\,\mathrm{V/A} = 1$ ohm ($\Omega$). Corresponding to the conversion factors for voltage and current strength there is a CGS resistance unit = $9 \times 10^{11}\Omega$, and therefore the measure numbers of resistances in MKSA units have to be divided by $9 \times 10^{11}$, and the measure numbers of conductivity to be multiplied by $9 \times 10^{11}$ in order to arrive at the corresponding values in the Gaussian system.

## §40. Joule heating

As indicated in §39, carriers of charge in their motion through a conductor continually lose energy, be it because of their collisions with other constituents of matter, or by processes summarized under the heading of frictional resistance. This energy reappears in the form of Joule heat in the current-carrying conductor.

We first elucidate this assertion with the example of the discharging of a capacitor of capacitance $C$ through a resistance $R$. If this discharging does not take place too quickly, it will be determined by the equations

$$ e = CV \qquad -\frac{de}{dt} = I = \frac{V}{R} $$

having the solution

$$ \frac{V}{V_0} = \frac{e}{e_0} = \frac{I}{I_0} = \exp\left(-\frac{t}{RC}\right) $$

Thus voltage, charge, and current decay exponentially, the quicker, in fact, the smaller are $C$ and $R$. As an example, for a capacitor of capacitance $1\mu\mathrm{F} = 9 \times 10^5\,\mathrm{cm}$ discharged through a resistance of $1\Omega$, $RC = 1 \times 10^{-6}\,\mathrm{s}$. Thus in $10^{-6}\,\mathrm{s}$ the capacitor is discharged to $1/e$ of its initial value.

With the decay of the capacitor charge, the electrical field energy of the capacitor diminishes according to the relation

$$ U = \frac{CV^2}{2} = \frac{CV_0^2}{2} \exp\left(-\frac{2t}{RC}\right) $$

Thus in one second the part

$$ -\frac{dU}{dt} = \frac{2U}{RC} = \frac{V^2}{R} = IV = I^2R \tag{40.1} $$

of the field energy disappears by being converted to Joule heat. In the time $dt$ the charge $I\,dt$ passes through the voltage drop $V$ within the conductor, thereby giving up the liberated potential energy $IV\,dt$ as heat energy in the wire.

In an analogous way, in equilibrating processes taking place so slowly that the E-field can at all times be calculated according to the laws of electrostatics, the Joule heat can be obtained from the time rate of change of the electric field energy. From (35.4) we have for matter at rest

$$\frac{dU}{dt} = \int \phi \frac{\partial \rho}{\partial t} \, dv$$

Thus, on account of the equation of continuity (38.5), we obtain by integration by parts

$$-\frac{dU}{dt} = \int \phi \operatorname{div} \mathbf{g} \, dv = -\int \mathbf{g} . \operatorname{grad} \phi \, dv = \int \mathbf{g} . \mathbf{E} \, dv$$

Here, therefore, in unit volume and unit time the energy

$$\mathbf{g} . \mathbf{E} = \frac{\mathbf{g}^2}{\sigma} = \sigma \mathbf{E}^2 \tag{40.2}$$

is transformed from electrical energy into heat.

A further example in the calculation of Joule heat will be considered at the end of the following section. Only in §54, however, will we be able to give the general foundation of Joule's law from the complete set of Maxwell equations. We shall show that of the three equal quantities (40.2) in the example just considered, $\mathbf{g}^2/\sigma$ in particular represents the Joule heat liberated per second in unit volume.

*Remark:* The Joule heat is also given in the MKSA system by $I^2R$ or $\mathbf{g}^2/\sigma$. Now, however, the power is obtained, not in erg/s or erg/cm³s, but in watts (W) or W/m³. Thus, we have $1W = 1 \times 10^7$erg/s.

## §41. Impressed forces. The galvanic chain

In the work of §§39 and 40 we always started with the idea that the motion of charge carriers in conductors, and thus the flow of electric current, was produced solely by the electric field **E**. Now there are however other non-electrical "causes" by which a current can be made to flow through a conductor. We call such a cause an "impressed force". If, formally, we write this as the product $e\mathbf{E}^{(e)}$ of the carrier charge $e$ involved, and an *impressed field strength* $\mathbf{E}^{(e)}$ herewith defined, we can take this field strength into account in Ohm's law, for example, through the expression

$$\mathbf{g} = \sigma(\mathbf{E} + \mathbf{E}^{(e)}) \tag{41.1}$$

In order to make the introduction of the quantity $\mathbf{E}^{(e)}$ better understood we shall consider several examples.

Let there be a variation of concentration within *a dilute aqueous solution of a strong electrolyte* (e.g. HCl). A diffusion process will then set in with the tendency of smoothing out the variation of concentration. Now the electrolyte is practically completely dissociated into $H^+$ and $Cl^-$ ions which diffuse independently of one another. The mobility, and therefore also the diffusion velocity of the $H^+$ ions is, however, much greater than that of the $Cl^-$ ions. The consequence is an electric current in the direction of diminishing concentration, since more $H^+$ ions are set in motion toward the places of lower concentration than $Cl^-$ ions. We recognize the motion of diffusion here as the origin of an impressed force, and therewith of a finite $\mathbf{E}^{(e)}$. This current now produces a positive charging of the dilute portions of the solution and a negative charging at the positions of concentration, and thereby gives rise to an electric field of such direction as to inhibit further diffusion of the $H^+$ ions, and to accelerate the $Cl^-$ ions. Finally, a state of equilibrium is reached in which, by means of the E-field, the difference in the diffusion velocity of the two kinds of ions is exactly cancelled. We then have the state of zero current, designated from (41.1) by $\mathbf{E} + \mathbf{E}^{(e)} = 0$.

Between the vectors $\mathbf{E}$ and $\mathbf{E}^{(e)}$ there exists the following fundamental distinction: $\mathbf{E}^{(e)}$ is present only in the interior of the electrolyte, and indeed only at places where a concentration gradient different from zero exists; in the surrounding vacuum or in dielectrics $\mathbf{E}^{(e)}$ is always zero. By contrast, the E-field, according to the laws of electrostatics, exhibits itself also in the external region; and thus for any arbitrary closed curve we have $\oint \mathbf{E} \cdot d\mathbf{r} = 0$. The correspondingly constructed path-integral $\oint \mathbf{E}^{(e)} \cdot d\mathbf{r}$ can, however, vanish only when, first, $\mathbf{E}^{(e)}$ can be written as the negative gradient of a potential function $\phi^{(e)}$, so that we can speak of an "impressed potential difference"; and second, when the path of integration lies wholly within the electrolyte.

We wish to supplement these considerations with an essentially formal observation. Let $n^+$ be the number of $H^+$ ions and $n^-$ the number of $Cl^-$ ions in $1\,cm^3$, $n^+$ and $n^-$ being functions of position. Further, let $D^+$ and $D^-$ be the respective diffusion constants, $B^+$ and $B^-$ the mobilities, and $+e$ and $-e$ the charges on the two kinds of ions. In general, the diffusion constant $D$ and the mobility $B$ of each kind of particle, of density $n$, are so defined that $-D\,\mathrm{grad}\,n$ gives the number of particles which, per second, go through an area of $1\,cm^2$ perpendicular to the direction of $\mathrm{grad}\,n$, while $BF$ gives the velocity with which a particle under the action of the force $\mathbf{F}$ moves through the liquid. The motion of the $H^+$ ions thus leads to an electric current-density $\mathbf{g}^+$ and that of the $Cl^-$ ions to a density $\mathbf{g}^-$ given by

$$\left.\begin{array}{l} \mathbf{g}^+ = e\{-D^+\,\mathrm{grad}\,n^+ + n^+ B^+ e\mathbf{E}\} \\ \mathbf{g}^- = -e\{-D^-\,\mathrm{grad}\,n^- - n^- B^- e\mathbf{E}\} \end{array}\right\} \qquad (41.2)$$

We therefore have for the total current density

$$\begin{aligned} \mathbf{g} &= \mathbf{g}^+ + \mathbf{g}^- \\ &= -e(D^+\,\mathrm{grad}\,n^+ - D^-\,\mathrm{grad}\,n^-) + e^2(n^+ B^+ + n^- B^-)\mathbf{E} \qquad (41.3) \end{aligned}$$

We can write this relation in the form (41.1) if we introduce the conductivity $\sigma$ (a function of position) given by

$$\sigma = e^2(B^+ n^+ + B^- n^-)$$

and an impressed field strength $\mathbf{E}^{(e)}$ defined by the relation

$$e\mathbf{E}^{(e)} = -\frac{D^+\,\mathrm{grad}\,n^+ - D^-\,\mathrm{grad}\,n^-}{B^+ n^+ + B^- n^-} \qquad (41.4)$$

The *equilibrium state* is reached when $\mathbf{g}^+$ and $\mathbf{g}^-$ both vanish:

$$D^+\,\mathrm{grad}\,n^+ = n^+ B^+ e\mathbf{E} \qquad D^-\,\mathrm{grad}\,n^- = -n^- B^- e\mathbf{E}$$

Since $\mathbf{E} = -\mathrm{grad}\,\phi$, these equations can be immediately integrated giving

$$n^+ = \text{const} \times \exp\left(-\frac{B^+ e\phi}{D^+}\right) \qquad n^- = \text{const} \times \exp\left(+\frac{B^- e\phi}{D^-}\right) \quad (41.5)$$

Since, according to statistical mechanics (see the barometric-height formula) the exponents here must be equal to the negative quotient of the potential energy $\pm e\phi$ of the particle in the $\mathbf{E}$-field by the product $kT$ (with $k = $ Boltzmann's constant $= 1\cdot38 \times 10^{-16}\,erg/deg$, $T = $ absolute temperature), the Einstein relation

$$\frac{D^+}{B^+} = \frac{D^-}{B^-} = kT$$

can be concluded from (41.5). From (41.4) we therefore obtain for the impressed field strength

$$\mathbf{E}^{(e)} = -\frac{kT}{e} \cdot \frac{B^+ \operatorname{grad} n^+ - B^- \operatorname{grad} n^-}{B^+ n^+ + B^- n^-} \tag{41.6}$$

If, as in HCl-electrolytes, $B^+ \gg B^-$, and if everywhere there is approximate electrical neutrality ($|n^+ - n^-| \ll n^+ \approx n$), equation (41.6) simplifies to

$$\mathbf{E}^{(e)} = -\frac{kT}{e} \frac{\operatorname{grad} n}{n} \quad \text{and therefore} \quad \phi^{(e)} = \frac{kT}{e} \ln n + \text{constant}$$

Between two points in the electrolyte with concentrations $n_1$ and $n_2$ we therefore have an *impressed voltage*

$$V^{(e)} = \int_1^2 \mathbf{E}^{(e)} \cdot d\mathbf{r} = \phi_1^{(e)} - \phi_2^{(e)} = \frac{kT}{e} \ln \frac{n_1}{n_2}$$

The order of magnitude of this voltage can be obtained by means of an arbitrarily chosen numerical example. With $\ln(n_1/n_2) = 1$, i.e. $n_1/n_2 = 2\cdot7$, $T = 300°\text{K}$, $e = e_0 = 4\cdot8 \times 10^{-10}\text{CGS}$ units, we find $V^{(e)} = 8\cdot6 \times 10^{-5}\text{CGS}$ units $= 0\cdot026\,\text{V}$.

Another example of the emergence of an impressed force is furnished by the *contact of a metal with an electrolyte*. If, for example, a copper bar is immersed in a dilute solution of copper sulphate, first a small amount of copper in the form of $Cu^{++}$ ions goes into solution. A small electric current thereby flows from the copper into the electrolyte. Here the solution pressure of the copper gives occasion for the existence of an impressed force. The current comes to a standstill when it has produced a negative charging of the copper and a positive charge in the solution such that in the immediate neighbourhood of the copper an electric field directed toward the copper exists. In equilibrium this field strength just compensates the impressed force of the solution pressure. Then, since $\mathbf{E} + \mathbf{E}^{(e)} = 0$ between the interior of the metal (1) and the homogeneous solution (2), we have as a consequence of the impressed voltage an equal and oppositely directed electrical voltage

$$V_{12} = \int_1^2 \mathbf{E} \cdot d\mathbf{r} = -\int_1^2 \mathbf{E}^{(e)} \cdot d\mathbf{r} = -V_{12}^{(e)} \tag{41.7}$$

As a rule the transition region in electrolytes in which $\mathbf{E}^{(e)}$ is sensibly different from zero is so thin that, with some justification, it is spoken of as a potential jump at the metal-electrolyte interface. It is on this kind of phenomenon that the action of the galvanic cell effectively depends.

Similar potential jumps exist at the *place of contact of two different metals*. In this case the difference in the character of the electronic motion in the two metals first produces a current. This is subsequently brought to a standstill upon the creation of a certain potential difference between the metals. A more detailed discussion of this process will be given in the third volume of this work.

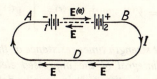

Fig. 34.—Galvanic chain AB, with circuit-closing wire BDA

Finally, we consider here a so-called *voltaic circuit*, i.e. a number of different series-connected conductors (metals and electrolytes) which within themselves and at their boundaries might contain certain impressed voltages (figure 34). The beginning (1) and the end (2) of the circuit shall, however, consist of the same metal. Then, in the case of zero current, there exists between the ends a potential difference which, as in (41.7), is given by the line integral over all impressed field strengths, the path of integration lying everywhere within the circuit. As soon as the two ends of the circuit are brought together in contact (wire BDA), a state of equilibrium is no longer possible—for the line integral $\oint \mathbf{E}^{(e)} \cdot d\mathbf{r}$, taken along the path ABDA, is different from zero, since throughout the homogeneous short-circuiting wire $\mathbf{E}^{(e)} = 0$, while $\oint \mathbf{E} \cdot d\mathbf{r} = 0$ holds because the electrostatic field strength is irrotational. A net electric current must therefore flow.

We consider the case in which a *steady current* is flowing. For this we have everywhere that $\mathrm{div}\,\mathbf{g} = 0$, and from (41.1)

$$\mathrm{div}\,(\sigma \mathbf{E}) = -\mathrm{div}\,(\sigma \mathbf{E}^{(e)})$$

When $\mathbf{E}^{(e)}$ and $\sigma$ are given, this equation in combination with $\mathrm{curl}\,\mathbf{E} = 0$ permits the calculation of $\mathbf{E}$, and therewith of $\mathbf{g}$. The relations for a linear current circuit (see §39) are particularly simple. If $q$ is the cross-sectional area of the conducting circuit and $ds$ a line element in the direction of the current path, then $\mathbf{g} \cdot d\mathbf{r} = I\,ds/q$, when $I$ has the same value at all cross-sections. Integrating (41.1) over the complete closed loop, from A through B and D, and back to A, we obtain

$$I \oint \frac{ds}{\sigma q} = \oint (\mathbf{E} + \mathbf{E}^{(e)}) \cdot d\mathbf{r} = \oint \mathbf{E}^{(e)} \cdot d\mathbf{r} = \int_1^2 \mathbf{E}^{(e)} d\mathbf{r}$$

since the line integral over $\mathbf{E}$ vanishes. Since, according to (39.4) the factor of $I$ is equal to the resistance $R$ of the whole conductor circuit, we obtain finally

$$IR = V^{(e)} \quad \text{with} \quad V^{(e)} = \int_1^2 \mathbf{E}^{(e)} \cdot d\mathbf{r} \tag{41.8}$$

*The product of current strength times resistance of the whole closed circuit is equal to the line integral of the impressed electrical field strength.*

On this basis the quantity $V^{(e)}$ is usually defined (somewhat unfortunately) as the *electromotive force* (e.m.f.) of the circuit. It is identical with the line integral of the open circuit only when the beginning and end members of the circuit are made of the same material. It depends upon the state of the individual elements of the circuit, and it is subject to change in circuits whose e.m.f. depends critically upon diffusion processes, and, even at constant temperature, where there are changes in concentration in the course of time.

We now examine the *heat production* in a circuit in which impressed forces are maintaining a steady current. Since $\mathbf{E} = -\operatorname{grad}\phi$ and $\operatorname{div}\mathbf{g} = 0$, we have, in integrating over the whole region traversed by current,

$$\int \mathbf{g} \cdot \mathbf{E}\,dv = -\int \mathbf{g} \cdot \operatorname{grad}\phi\,dv = +\int (\phi \operatorname{div}\mathbf{g})\,dv = 0$$

and from (41.1), after scalar multiplication by $\mathbf{g}$ and integrating over the whole space, it follows that

$$\int \frac{\mathbf{g}^2}{\sigma}\,dv = \int \mathbf{g} \cdot \mathbf{E}^{(e)}\,dv \tag{41.9}$$

The integral on the left side gives the Joule heat developed per second in the circuit. It is also equal to the work performed by the impressed forces. The energy involved comes ultimately from the process which, in the first instance, is responsible for the appearance of $\mathbf{E}^{(e)}$. In the concentration circuit it is produced at the expense of the free energy which is appropriate to the more concentrated solution as against that which is less concentrated. In a voltaic element it is bound up with the chemical processes of dissolution or separation; in the thermocouple the energy comes from the heat reservoirs by means of which the temperature difference of the junctions is maintained. In all cases the power output $\int g.E^{(e)} dv$ is furnished by energy sources which are foreign to strict electrostatics, the same being true of the Joule heat developed. We therefore have the unique picture of a stationary field (not varying with time) through whose agency, however, energy of a non-electrical nature is changed into another form, namely heat.

## §42. Inertia effects of electrons in metals

A further very instructive example of impressed forces is furnished by investigations in which the *inertia of conduction electrons* is made manifest. The first investigation of this kind is due to K. F. Nichols. He proceeded on the following assumption: When a metal disc is rotated on its axis the conduction electrons, under the influence of the centrifugal force, must move toward the rim of the disc; the rim thus becomes negatively charged, and the middle of the disc positively. Equilibrium takes place when the electric field due to the charges just compensates the effect of the centrifugal force on the electrons. The centrifugal force here acts like an outwardly directed impressed field strength of magnitude

$$E^{(e)} = mr\omega^2/e \qquad (42.1)$$

where $\omega$ is the angular velocity of the disc, and $m$ and $e$ are the mass and charge of the electron. By integrating $E^{(e)}$ over $r$ from the centre of the disc ($r = 0$) to the edge ($r = R$) we find as the impressed voltage (e.m.f.)

$$V^{(e)} = mR^2\omega^2/2e$$

For the purpose of obtaining an idea of the order of magnitude of this effect we let $R = 10\,\text{cm}$, $\omega = 100\,\text{s}^{-1}$, as in Nichols' investigation; with $e = 4\cdot8 \times 10^{-10}$ CGS units, $m = 0\cdot9 \times 10^{-27}$ g, we find $V^{(e)} \approx 3 \times 10^{-10}\,\text{V}$. A static measurement of this effect, owing to its minuteness, is hardly to be considered. Nichols therefore provided brush contacts at the rim of the disc and on the axis, and he connected these to a galvanometer of very low internal resistance (figure 35). A current must therefore flow from the rim of the disc, through the galvanometer, to the middle of the disc, seeking to equalize the negative charge on the edge of the disc and the positive charge at the centre. Nichols could not however establish the flow of the current with certainty. In addition to the expected effect there were thermal

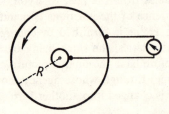

Fig. 35.—Electron centrifuge (after Nichols)

effects at the brush contacts and similar disturbances. These, owing to their irregularity and uncontrollability, did not permit definite establishment of the existence of the expected effect.

By contrast, similar investigations by R. C. Tolman and his co-workers led to a positive result. The first series of investigations was based upon the idea that in the sudden deceleration of a moving

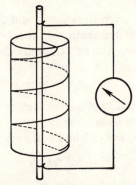

Fig. 36.—Demonstration of the inertia of conduction electrons (after Tolman)

metallic object the electrons, originally moving with the metal, owing to their momentum, continue on a bit farther in the metal until, either through the existence of counter-fields (charges at the boundaries) or because of ohmic resistance which they meet in their motion through the metal lattice, they are brought to rest. This persisting relative motion of the electrons with respect to the metal lattice must manifest itself as a brief surge of current.

For demonstrating this, a coil of wire was employed, the ends being connected through the axis to a ballistic galvanometer (figure 36). In the sudden stopping of the coil, initially rotating about its axis, a measurable deflection of the galvanometer was actually obtained, and so the total quantity of electricity transported in the deceleration process through a cross-section could be measured.

In order to arrive at a quantitative value of this current surge we first consider in a general way a piece of metal moving with velocity $\mathbf{v}(t)$. If we would wish for the free electrons to share at all times in the motion of the metal, we must superimpose an electric field of such strength $\mathbf{E}^{(b)}$ that the electrons undergo the same acceleration as the metal lattice under the mechanically forced motion:

$$e\mathbf{E}^{(b)} = m\frac{d\mathbf{v}}{dt}$$

In Tolman's arrangement, however, no such field exists. The electrons do not therefore take part in the acceleration $d\mathbf{v}/dt$ of the metal ions. This is another way of saying that *relative to the metal* the electrons experience an acceleration $-d\mathbf{v}/dt$. Thus, *they move relative to the metal as if in it an impressed force*

$$e\mathbf{E}^{(e)} = -m\frac{d\mathbf{v}}{dt} \tag{42.2}$$

*were active.* In place of the purely mechanical origin of the relative motion of the electrons we can substitute here a fictitious purely electrical agency in which we introduce an impressed field strength $\mathbf{E}^{(e)}$ defined by the equation above. That the theory of Nichols' experiment is contained in equation (42.2) can, incidentally, be seen by substituting for $d\mathbf{v}/dt$ in the rotating disc its value $-\omega^2\mathbf{r}$, the centripetal acceleration. In this way the relation (42.1) is again obtained.

We can now calculate the current density $\mathbf{g}$ from Ohm's law

$$\mathbf{g} = \sigma(\mathbf{E} + \mathbf{E}^{(e)}) = \sigma\left(\mathbf{E} - \frac{m}{e}\frac{d\mathbf{v}}{dt}\right) \tag{42.3}$$

in which $\mathbf{E}$ is any electric field which might possibly be present in the deceleration procedure. It should be observed that the introduction of the conductivity at this point takes full account of the deceleration of the electrons in their motion relative to the metal, and that $d\mathbf{v}/dt$ is not the acceleration of the electrons, but the acceleration of the metal.

Integrating (42.3) over the complete current circuit, since $\mathbf{E}.d\mathbf{r} = 0$, we obtain as in §41

$$IR = \oint \mathbf{E}^{(e)} \cdot d\mathbf{r} = -\frac{ml}{e}\frac{dv}{dt} \qquad (42.4)$$

Here $l$ is the length of the coiled wire. We need not know the details concerning the time-duration of the deceleration since we do not wish to know the duration of the current, but only the integral in time over the entire duration of the acceleration process, and therefore the total quantity of electricity $\int I dt$ passing through a cross-section during the process. For this, by integrating equation (42.4) with respect to time over the duration of the deceleration process, we obtain the value $mlv_0/eR$, when prior to the deceleration the coil was rotating with a velocity $v_0$ and afterwards was at rest.

Since the above data are known, by ballistic measurement of $\int I dt$ the quotient $e/m$, i.e. the specific charge of the conduction electrons, can be determined. In the investigations of Tolman and Steward a value was obtained which at most differed from the accepted value of $e/m$ by 10 per cent.

In further investigations of Tolman and others, a wooden cylinder covered with copper sheet was made to undergo rotatory oscillations around its axis. Owing to their inertia, the electrons in this case, according to (42.4), carried out a forced oscillation 90° out of phase with respect to the oscillation of the metal, and the alternating current thus produced could be measured with the help of an inductively coupled pick-up coil. Here also the experimental result was in satisfactory agreement with theory.

# CHAPTER C II

## Force Effects in the Magnetic Field

### §43. The magnetic field vectors

Up to the time of Oersted's discovery (1820) of the production of magnetic fields by electric currents, and of Faraday's discovery (1831) of the induction of currents by the time-variation of magnetic fields, the theory of magnetism was a subject of physical investigation entirely independent of electrical theory. It concerned the force effects between permanent magnets and the fields producing these forces. Its special interest was the magnetic field of the earth. There was a broad formal analogy between the laws of the magnetostatic field and the electrostatic field to the extent that comparison was made of

the electric field $\quad\quad$ **E** with the magnetic field $\quad\quad$ **H**

the electric displacement $\quad$ **D** with the magnetic induction **B**

the dielectric polarization **P** with the magnetization $\quad\quad$ **M**

The constitutive equations

$$\mathbf{D} = \mathbf{E} + 4\pi\mathbf{P} \quad \text{and} \quad \mathbf{B} = \mathbf{H} + 4\pi\mathbf{M}$$

then correspond exactly, as do the differential relations

$$\operatorname{curl}\mathbf{E} = 0 \quad \text{and} \quad \operatorname{curl}\mathbf{H} = 0$$

A difference with respect to the divergence equations

$$\operatorname{div}\mathbf{D} = 4\pi\rho \quad \text{and} \quad \operatorname{div}\mathbf{B} = 0$$

does however exist. There are no "magnetically charged" bodies; an iron bar can indeed be polarized, but it can never be magnetically charged.

Now Coulomb's law giving the force between two electrically charged test bodies constitutes the point of departure of electrostatics; it leads to the establishment and realization of the units of charge and of field strength, and it thereby affords a quantitative proof of the basic equations of electrostatics. For the experimental investigation of magnetic fields, we find the permanent magnet a substitute for the non-existent magnetic probe charges. While the electrical polarization **P** of a dielectric (excepting some few substances such as Rochelle salt, barium titanate, etc.) can only be maintained by means of an external electric field, the practically very important substances, iron in particular, under certain conditions possess a permanent magnetic polarization **M** ("magnetization") even without the existence of an external magnetic field. A piece of such a substance (called magnetically "hard") can be so magnetized that for not too strong external fields the magnetization **M** depends only to a very moderate extent upon **H**. For the subsequent considerations of these paragraphs we keep fixed in our minds the extreme case of an ideally hard magnetic bar; i.e. we consider its magnetization or its magnetic moment

$$\mathbf{m} = \int \mathbf{M} \, dv$$

to be a firmly pre-established constant, independent of any external field.

With small bars of such permanently magnetized material ("magnetic needles") we can measure a magnetic field just as well as we can measure an electric field with the help of electrical probe charges. In the choice of units we can be guided, like Gauss, by the formal analogy with the electrostatic field and, exactly as before with the electric probe charge, we can now at once make use of the magnetic needle both as an indicator and as a source of a magnetic field.

As a magnetic dipole the magnetic needle experiences a torque

$$\mathbf{T} = \mathbf{m} \times \mathbf{H}_0 \qquad T = mH_0 \sin\theta$$

in a uniform magnetic field $\mathbf{H}_0$, as for example the local field of the earth. Here $\theta$ is the angle between the magnetic field and the magnetic needle. If, in particular, the needle is suspended above its centre of gravity so as to be free to rotate, and if $\Theta$ is its moment of inertia, then, with respect to its equilibrium position, the needle oscillates in conformity with the equation of motion

$$\Theta \frac{d^2\theta}{dt^2} = -mH_0 \sin\theta$$

For small displacements ($\sin\theta \approx \theta$) we obtain from this as the general solution: $\theta = \theta_0 \cos 2\pi \nu(t - t_0)$, the oscillation frequency being

$$\nu = \frac{1}{2\pi}\sqrt{\frac{mH_0}{\Theta}}$$

Thus, by measuring this frequency, the *product $mH_0$* can be determined.

We now consider the magnetic needle itself as the origin of a magnetic field **H**. As a dipole field, and corresponding to equation (24.9) for the electrical case, this is given by

$$\mathbf{H} = -\frac{\mathbf{m}}{r^3} + \frac{3(\mathbf{m}\cdot\mathbf{r})\mathbf{r}}{r^5}$$

and therefore

$$H_r = \frac{2m}{r^3}\cos\psi \qquad H_\psi = \frac{m}{r^3}\sin\psi$$

when we pass from the position vector **r** to the polar coordinates $(r, \psi)$, the direction of **m** being the polar axis.

We now fix the needle so that it stands perpendicular to the uniform

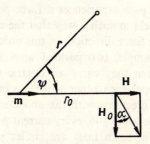

Fig. 37.—For the absolute measurement of **m** and $\mathbf{H}_0$ according to Gauss

field $\mathbf{H}_0$ (see figure 37), and we bring a second (freely suspended) small magnetic needle to a point with polar coordinates $r = r_0$, $\psi = 0$. The second magnetic needle will then set itself in the direction of the resultant of **H** and $\mathbf{H}_0$ at this point, i.e. deflected from the direction of $\mathbf{H}_0$ through the angle $\alpha$, with

$$\tan\alpha \approx \frac{H}{H_0} = \frac{2m}{r_0^3 H_0}$$

From this by measuring $\alpha$ and $r_0$ we now find the *quotient $m/H_0$*.

By combining the two results we can measure in absolute units the earth's field $H_0$ as well as the moment $m$ of our magnetic needle. Moreover, with the help of our needle, thus calibrated, we are now able to measure both the direction and magnitude of any arbitrarily given magnetic field, so long only as the needle is so small that in its immediate vicinity the field can be considered to be uniform. In particular, by this means the set of basic equations of the magnetostatic field can be examined and established at the outset.

*Remark:* In the MKSA system of units the magnetic units become linked with the volt-ampere system through the laws of Oersted and Faraday. We shall have to enter more fully into these questions in subsequent paragraphs. It will then be clear how, and with what constants, the preceding formulae for MKSA units are to be transcribed.

## §44. The force on a current-carrying conductor. The Lorentz force

The electromagnetic-field force of greatest importance in electro-technology is the force which a magnetic field exerts upon a current-carrying conductor. It is this force which drives all electric motors. We wish now to establish a formula for this force.

If we bring a thin straight wire carrying current $I$ into a uniform magnetic field **H**, the wire experiences a force perpendicular to itself and to the magnetic field in such a way that the current direction, the field direction, and the force direction (in this order) form a right-hand system. Thus, for example, two parallel wires in mutual proximity attract each other when they carry an electric current in the same direction, and they repel each other when the current directions are opposite. This is in agreement with the above statement; for, as we shall discuss in more detail in §46, every current produces around itself a magnetic field whose force lines are circles perpendicular to the direction of the current and whose direction relates to that of the current in the sense of the rotation of a right-hand screw (see figure 38).

We arrive at simple and illuminating relationships when we examine, not the wire as a whole with its internally flowing current $I$, but a small volume element $dv$ in which there is the current density **g**. There is then exerted a force **f**$dv$ on this volume element in which the force density

$$\mathbf{f} = \mathbf{g} \times \mathbf{B}/c \qquad (44.1)$$

corresponds in direction to the rule given above. At this point the following is to be noted: First of all in (44.1) we have not taken the magnetic field as the magnetic field strength **H**, but as the magnetic induction **B**. Since in the use of Gaussian units **B** = **H** *in vacuo*, and since in non-ferromagnetic substances the difference between **B** and **H** is always very small compared with **H**, it would be immaterial here whether (44.1) were written with **B** or with **H**. Experiments with ferromagnetics, however, where **B** and **H** may differ by several powers of ten, show clearly that **B** is the quantity to be put into the formula for the force density.

Secondly, in (44.1) we have introduced a constant of proportionality $1/c$. In the MKSA system (see the Remark) this would not be

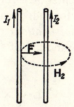

Fig. 38.—Attraction of two wires carrying currents in the same direction

necessary since for that case we can arbitrary set $c = 1$ and thereby, through (44.1) establish the dimensions and unit of measure of induction **B**. For the Gaussian system of units, however, the dimensions and measure unit of **B** are, in line with the discussion in §43, already established. The constant $c$ thus becomes a constant of proportionality to be determined experimentally. That it has the dimensions of a velocity follows because in the Gaussian system of measure **B** has the same dimensions as **E**, **g** × **B** and therefore also **g** . **E** are to be measured in ergs/cm³ s, while the force density has the dimensions of an energy density per unit length. The measured numerical value of $c$, though to be sure not measured directly on the basis of the law (44.1), has the value

$$c = 3 \cdot 0 \times 10^{10} \, \text{cm/s} \qquad (44.2)$$

The constant of proportionality $c$ *is therefore precisely equal to the velocity of light in vacuo*. We may consider this to be the first suggestion of the close connection existing between electromagnetic and optical phenomena.

The force law (44.1) is connected very closely with the force on a moving charged particle in a magnetic field. According to H. A. Lorentz, this particle (with charge $e$ and velocity $\mathbf{v}$) experiences in a **B**-field the force

$$\mathbf{F} = e\mathbf{v} \times \mathbf{B}/c \qquad (44.3)$$

It is called the "Lorentz force". If it is summed over all moving charged particles in unit volume of a conductor, then, since

$$\mathbf{f} = \sum \mathbf{F}_i = \sum e_i \mathbf{v}_i \times \mathbf{B}/c = \mathbf{g} \times \mathbf{B}/c$$

we arrive back at (44.1). While this last law can always be obtained only as an abstraction from the force effect of an extended current-carrying conductor, the Lorentz-force formula (44.3) is immediately experimentally demonstrable for individual charged particles. This is accomplished in *deflection investigations* on electron beams (cathode rays) and on ion beams. The motion of such a particle in an electric field $\mathbf{E}$ and a magnetic field $\mathbf{B}$ is described by the equation of motion*

$$m\frac{d\mathbf{v}}{dt} = e\left(\mathbf{E} + \frac{\mathbf{v}}{c} \times \mathbf{B}\right) \qquad (44.4)$$

We remark first of all that the magnetic field has no influence upon the magnitude of the velocity. In fact, scalar multiplication of (44.4) by $\mathbf{v}$ gives

$$\frac{d}{dt}(\tfrac{1}{2}m\mathbf{v}^2) = e\mathbf{E} \cdot \mathbf{v}$$

A change of the kinetic energy is effected only by the electric field $\mathbf{E}$. If, further, $\mathbf{E}$ comes from the gradient of a static potential ($\mathbf{E} = -\operatorname{grad}\phi$), we obtain for the energy law (also when any arbitrary magnetic field is present) the expression

$$\tfrac{1}{2}m\mathbf{v}^2 + e\phi = \text{constant} \qquad (44.5)$$

We now consider the action of a constant magnetic field alone. We wish to have the field lying in the direction of the positive $z$-axis: $B_x = B_y = 0$, $B_z = B$. In this case, from (44.4), we obtain as components of the equation of motion

$$m\frac{dv_x}{dt} = \frac{eB}{c}v_y \qquad m\frac{dv_y}{dt} = -\frac{eB}{c}v_x \qquad m\frac{dv_z}{dt} = 0 \qquad (44.6)$$

---

* We assume here that the velocity $v$ of the particle is small compared to the velocity of light $c$. Otherwise, owing to the relativistic variation of mass with velocity as $v$ approaches $c$, we must substitute for the term $mdv/dt$ on the left side of (44.4) the expression $md[v/\sqrt{(1-v^2/c^2)}]/dt$. (For this, see §89.)

The third equation states: The component of the motion lying along the direction of **B** is not affected by the field. We need therefore concern ourselves only with the projection of the motion on the $xy$-plane. If for this we introduce the abbreviation that

$$\omega = \frac{eB}{mc} \tag{44.7}$$

the first two equations (44.6) read

$$\frac{dv_x}{dt} = \omega v_y \qquad \frac{dv_y}{dt} = -\omega v_x$$

If we differentiate the first with respect to the time and eliminate $dv_y/dt$ by means of the second, we obtain for $v_x$ the following differential equation of second order:

$$\frac{d^2 v_x}{dt^2} = -\omega^2 v_x$$

We thus find as the most general solution (with the two constants of integration $v_0$ and $t_0$)

$$v_x = v_0 \cos \omega(t-t_0) \qquad v_y = -v_0 \sin \omega(t-t_0)$$

A further integration with respect to the time gives

$$x-x_0 = \frac{v_0}{\omega} \sin \omega(t-t_0) \qquad y-y_0 = \frac{v_0}{\omega} \cos \omega(t-t_0)$$

since $v_x = dx/dt$, $v_y = dy/dt$.

The projection of the trajectory on a plane normal to **B** is therefore a circle centred on the point $\{x_0, y_0\}$, with radius

$$R = \frac{v_0}{\omega} = \frac{mv_0 c}{eB} \tag{44.8}$$

Together with the velocity component in the $z$-direction, we find as the most general motion in a uniform magnetic field a helix with pitch $l = 2\pi v_z/\omega = 2\pi mv_z c/eB$, which is independent of $v_0$. Incidentally, the value (44.8) also follows immediately from the remark that in the circular motion the centrifugal force $mv_0^2/R$ is just compensated by the Lorentz force $ev_0 B/c$.

Equation (44.4) represents the theoretical basis of all investigations concerning the deflection of beams of charged particles in given fields. Conversely, from these investigations and by means (44.4) the deflecting magnetic field **B** can be found. This method of **B**-determination, as well as determining the **B**-field by means of the force on a current-carrying wire in the sense of (44.1), makes us independent of the field determination and the establishment of units as described in the method of §43.

We admittedly obtain in this way not **B** itself but the quantity **B**/$c$. It could therefore be held to be more useful if also in the Gaussian system this quantity **B**/$c$ = **B\***, dimensionally different from E, were defined as the magnetic induction. This is actually done in the electromagnetic CGS system and more particularly in the MKSA system. However, it stands in opposition to reflections based on principle if we wish to hold to those Gaussian-system electrical field quantities proved in atomic physics—for, by introducing **B\*** instead of **B**, not only would the existing symmetry between electrostatics and magnetostatics be disturbed, but also the symmetry of the complete set of Maxwell equations (see §53), not to mention the loss of symmetry that would be suffered in the basic electromagnetic equations of special relativity. In equations (44.1) and (44.3) we wish therefore to retain the factor $c$, bearing in mind, however, that without the previous establishment of the units of **B** by the Gauss method the constant $c$ cannot be determined from the force effect on moving charges or currents.

*Remarks:* As has already been mentioned, the MKSA system does not establish the magnetic units by means of the Gauss method of §43. Rather, for the determination of the induction **B** it makes use of the force effects on currents or moving charges which it expresses in the form

$$\mathbf{f} = \mathbf{g} \times \mathbf{B} \quad \text{or} \quad \mathbf{F} = e\mathbf{v} \times \mathbf{B}$$

Consequently, in the technical measure system $\mathbf{v} \times \mathbf{B}$ is dimensionally equal to the electric field strength, and therefore **B** is to be measured in $Vs/m^2$. Since $1V = 1/300$ of the Gaussian unit of potential, $1Vs/m^2$ is equal to $1/3 \times 10^6$ of the corresponding Gaussian unit. Now, however, to the quantity **B** of the technical measure system there corresponds the quantity **B\*** and not the quantity $\mathbf{B} = c\mathbf{B^*}$ of the Gaussian system of units. Thus, in order to transform the MKSA unit into the corresponding Gaussian CGS unit [1 gauss (G)], the MKSA unit has to be multiplied by the measure number of $c$ in CGS units, i.e. by $3 \times 10^{10}$. Thus $1 Vs/m^2 = 1 \times 10^4 G$, and consequently the numerical value of **B** in $Vs/m^2$ has to be multiplied by $10^4$ in order to obtain its value in "G".

## §45. The Faraday law of induction

In 1831 Michael Faraday made the fundamental discovery that in a closed wire circuit an electric current exists when a magnet in the neighbourhood of the circuit is in motion or when the wire circuit is displaced in a magnetic field. A more detailed investigation yielded the information that in both cases the induced current depends upon the change with time of the flux of induction

$$\Phi = \int B_n \, dS \qquad (45.1)$$

linked by the induced current. (It should be noted that owing to the source-free nature of **B**, the flux of induction $\Phi$ is independent of the precise location of the surface spanning the closed wire circuit.) According to Faraday the induction law in general is

$$IR = -\frac{1}{c}\frac{d\Phi}{dt} = -\frac{1}{c}\frac{d}{dt}\int B_n\, dS \qquad (45.2)$$

*The product of the current and the resistance in the wire circuit is at any moment equal to* $1/c$ *times the time rate of decrease of the flux of induction through any surface bounded by the current circuit.* The current is therefore to be reckoned positive when its direction of circulation is connected with the chosen normal direction **n** of the surface by the right-hand screw rule. In technical parlance the quantity $-\dfrac{1}{c}\dfrac{d\Phi}{dt}$ is sometimes referred to as "magnetic decay".* According to the induction law it is therefore entirely immaterial whether the "decay" comes from a time variation of the magnetic field with the wire circuit at rest, or from the wire circuit moving in a temporally constant but spatially non-uniform magnetic field.

Equation (45.2) furnishes us with a new method, important in practice, *for the measurement of a given magnetic field.* For this we choose a probe coil which is so small that within its boundaries the magnetic field to be measured can be considered to be uniform. This coil is connected by means of well-twisted wires to a ballistic galvanometer. So long as the coil remains at rest in a temporally constant field **B** the galvanometer carries no current. Let the initial flux of magnetic induction through the surface $S$ bounded by the coil be $\Phi_0 = (B_n)_0\, S$. If now we withdraw the coil from the field to a field-free location, there flows in the coil during the motion the current $I$, given by the induction law. The total quantity of electricity indicated by the ballistic galvanometer (for infinitely fast, i.e. ballistic motion of the coil) is then

$$e = \int_0^t I\, dt = -\frac{1}{Rc}(\Phi_t - \Phi_0) = \frac{\Phi_0}{Rc}$$

The galvanometer deflection thus measures directly the component $B_n$ of the magnetic induction that is perpendicular to the coil surface at the place occupied by the coil prior to its motion.

---

* Ger.: "magnetischer Schwund" (Tr.).

The technique can also be carried out in such a way that the coil is left at its initial location. In this case it is rotated 180° around an axis in the plane of $S$ and this causes $B_n$ to change its sign, yielding for $e$ exactly twice the value given above.

If we should be dealing with an induction-law situation which depends only upon the relative motion between the field-producing magnet and the wire circuit, then, in *interpreting* the experiment, a distinction has to be made between (*a*) the case of a moving wire circuit in a static magnetic field and (*b*) that of a wire circuit at rest in a magnetic field which is changing with time.

In the first case, namely that of the *moving circuit*, we can immediately derive the induction law from the action of the *Lorentz force* as explained in §44. Thus, if we move a metal wire, initially devoid of current, with velocity **v** in a magnetic field, the conduction electrons within the wire naturally take part in this motion. They therefore experience the Lorentz force $\mathbf{F} = e\mathbf{v} \times \mathbf{B}/c$ and on this account, for the case where the Lorentz force possesses a component in the direction of the wire, the conduction electrons will move along the wire. We can forthwith look upon this force as a kind of applied force, and we can therefore speak of an "applied" field strength

$$\mathbf{E}^{(\text{ind})} = \mathbf{v} \times \mathbf{B}/c \tag{45.3}$$

which, because of its connection with the induction law, we wish to designate by means of the index "ind". We have thus found the connection with the ideas of §41, equation (41.8) in particular, and for the induced current we can write the expression

$$IR = \oint \mathbf{E}^{(\text{ind})} \cdot d\mathbf{r} = \frac{1}{c}\oint d\mathbf{r} \cdot (\mathbf{v} \times \mathbf{B}) \tag{45.4}$$

We have now only to show that, for the case considered, this expression is in agreement with the induction law (45.2), or that we can write

$$\oint d\mathbf{r} \cdot (\mathbf{v} \times \mathbf{B}) = -\frac{d\Phi}{dt} = -\frac{d}{dt}\int B_n \, dS \tag{45.5}$$

For this purpose we examine figure 39, which shows the location of the wire circuit at time $t$ (index 1) and at time $t + dt$ (index 2). These two positions are connected by the displacement vectors $\mathbf{v}\,dt$ of the individual points of the circuit. The vector product $d\mathbf{r} \times \mathbf{v}\,dt$ has the magnitude of the surface element $dS$ determined by $d\mathbf{r}$ and $\mathbf{v}dt$, and the direction $\mathbf{n}$ of the outward-drawn normal. On account of the distribution law (6.1) for the scalar product, we have

$$dt \oint d\mathbf{r} \cdot (\mathbf{v} \times \mathbf{B}) = \oint (d\mathbf{r} \times \mathbf{v}\, dt) \cdot \mathbf{B} = \int B_n \, dS$$

in which the last-written surface integral is to extend over the external surface of the structure illustrated in figure 39. It therefore represents the flux of induction $\Phi_M$ through this external surface.* Now, however,

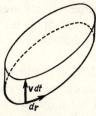

Fig. 39.—Change of flux through a moving surface

owing to the source-free nature of $\mathbf{B}$, the total flux of induction out of the structure must vanish; we must have that $\Phi_M = -(\Phi_2 + \Phi_1)$ or, if the normal directions on both end surfaces of the structure be chosen in the direction $1 \rightarrow 2$, $\Phi_M = -(\Phi_2 - \Phi_1) = -d\Phi$. Equation (45.5) is therefore proved, and *the induction law for moving conductors has been derived from the Lorentz force.*

Not derivable from the foregoing, but much farther reaching is *the induction law for the case of a wire at rest in a time-varying magnetic field.* Here we cannot speak of a current-driving force from the $\mathbf{B}$-field, since a charge at rest in a magnetic field experiences no Lorentz force. There remains here no other possibility of interpretation of the induction law than the assumption that through the change with time of a magnetic field there exists (or is "induced") an electric field that is no longer irrotational, having a non-vanishing circuital voltage $\oint \mathbf{E} \cdot d\mathbf{r}$. We then have for the current impelled by this circuital voltage

$$IR = \oint \mathbf{E} \cdot d\mathbf{r} \tag{45.6}$$

---

* Clearly $\Phi_M$, and therefore the induced current, are different from zero only when $\mathbf{B}$ possesses a component perpendicular to the wire circuit and to its direction of motion; when, that is, "magnetic field lines are cut by the moving wire".

As a comparison of this expression with the induction law shows, we must in this case have that *the electrical circuital voltage is equal to the magnetic decay:*

$$\oint \mathbf{E} \cdot d\mathbf{r} = -\frac{1}{c}\frac{d}{dt}\int B_n \, dS = -\frac{1}{c}\int \left(\frac{\partial \mathbf{B}}{\partial t}\right)_n dS \qquad (45.7)$$

Since the material constant $R$ relating to the matter forming the wire is absent in the above formula, this suggests an extensive and fundamental generalization for all subsequent considerations of equation (45.7), which at first pertained only to the wire circuit. We assume that for the validity of this equation the existence of the wire is entirely immaterial, and that *for any arbitrary closed curve* the circuital voltage is correctly given by (45.7).

In preparation for the later justification of this hypothesis by proving its consequences, we can now by means of it pass immediately from the integral form of the induction law to its differential form. Thus, if equation (45.7) holds for any arbitrarily positioned surface element, by applying Stokes's theorem we obtain

$$\operatorname{curl} \mathbf{E} = -\frac{1}{c}\frac{\partial \mathbf{B}}{\partial t} \qquad (45.8)$$

A time-varying magnetic field therefore induces an electric field which in contrast to electrostatic fields, is no longer irrotational and thus cannot be represented as the gradient of a potential. The equation $\operatorname{curl} \mathbf{E} = 0$, valid in electrostatics, is for the general case to be replaced by equation (45.8).

As an immediate demonstration of the correctness of the assertion that at any place where **B** changes with time an electric field is induced, we examine the operation of the *betatron* (sometimes called the "electron sling"). This apparatus for inductively accelerating electrons to a very high energy consists essentially of a highly-evacuated non-conducting annular chamber ("doughnut") which is placed symmetrically between the pole pieces of an electromagnet (figure 40). Electrons inside the doughnut will always travel on the given circular orbit (radius $r$) without striking the glass walls of the chamber when the magnetic field $B_g$ acting on them produces at all times the necessary centripetal acceleration

$$\frac{mv^2}{r} = |e|\frac{c}{v}B_g \quad \text{i.e.} \quad mv = |e|\frac{r}{c}B_g$$

If now the field between the pole pieces increases, there is induced within the chamber a circuital voltage which is capable of accelerating the electrons. Thus we have

$$\frac{d}{dt}(mv) = eE = -\frac{e}{c}\frac{1}{2\pi r}\frac{d}{dt}\int B_n\,dS$$

since

$$\oint \mathbf{E} \cdot d\mathbf{r} = 2\pi r E$$

Integrating the equation leads to the expression

$$mv = -\frac{e}{c}\frac{1}{2\pi r}\int B_n\,dS$$

At any given moment the electron momentum is therefore equal to the instantaneous flux of induction through the circular orbit multiplied by $e/2\pi rc$, if we have $v = 0$ and $\Phi = 0$ at the beginning of the acceleration process. Thus at all times the orbit-condition equation

$$B_g = \frac{1}{2\pi r^2}\int B_n\,dS$$

must obtain; if therefore the field at lesser radius and almost up to the radius of the vacuum chamber is uniform and of value $B$, then we must have approximately that $B_g = \frac{1}{2}B$.

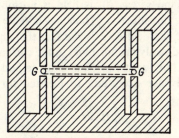

Fig. 40.—Schematic representation of a betatron. G: cross-section through the vacuum chamber ("doughnut")

*Remarks:* In the MKSA system the foregoing formulae have, within the factor $1/c$, the same form throughout. In place of the Gaussian induction $\mathbf{B} = \mathbf{B}^*c$, the corresponding B-quantity of the MKSA system enters here. We designate it by the symbol $\mathbf{B}^*$. In integral form here the induction law therefore reads

$$IR = -\frac{d\Phi}{dt} = -\frac{d}{dt}\int B_n\,dS$$

and in differential form

$$\operatorname{curl}\mathbf{E} = -\frac{\partial \mathbf{B}}{\partial t}$$

# CHAPTER C III

## Magnetic Fields of Currents and Permanent Magnets

### §46. The magnetic field of steady currents. Oersted's law

In accordance with the discovery of Oersted, an electric current is always accompanied by a magnetic field. The magnetic field of a straight wire, infinitely long and carrying a current, consists, as already mentioned at the beginning of §44, of field lines which link the wire in the fashion of circles whose planes are perpendicular to the wire. The direction of **H** in such a circle forms, together with the direction of the current, a right-hand screw. This field is no longer circulation-free; rather, the linking line-integral $\oint \mathbf{H} . d\mathbf{r}$ has a value different from zero.

An analysis of the field reveals the fact that the value of this line-integral is always directly proportional to the current threading the path of integration. Thus, with the Gaussian method of establishing the measure unit, we have according to Oersted for the magnetic field strength

$$\oint \mathbf{H} . d\mathbf{r} = \frac{4\pi}{c} I \tag{46.1}$$

*with the same constant c as in the induction law.*

If, incidentally, it were desired to circumvent or to avoid establishing the magnetic units by Gauss's method, we could start with the force law (44.1) or with the induction law. By this means, however, **B** and $c$ could not be measured separately, but only the quotient $\mathbf{B}/c$. Since in the Gaussian measure system *in vacuo* this quotient is equal to $\mathbf{H}/c$, the left side of equation (46.1), multiplied by $1/c$, would also be directly measurable in CGS units. Moreover, since $I$, with its unit, is fixed, the quantity $4\pi/c^2$ on the right-hand side could then be immediately measured.

As with the induction law, we can go from the general equation (46.1) to a differential law if we assume that (46.1) holds everywhere within a current-carrying conductor. Through a surface element $dS$ the current $g_n dS$ therefore flows and, by applying Stokes's law, we arrive at the equation

$$\operatorname{curl} \mathbf{H} = \frac{4\pi}{c} \mathbf{g} \tag{46.2}$$

which is equivalent to (46.1). The magnetic field therefore possesses circulation at any place where a finite current density exists.

Starting with (46.1) or (46.2) we can determine the magnetic field of a given steady current by a variety of methods: *by direct application of* (46.1); by the method—to be discussed presently—of the *magnetic double layer*; by means of the *Biot-Savart law*; or, finally, by the way involving the *vector potential*.

The first method, *the direct application of* (46.1), leads quickly to the result in those cases where, at the outset, something is known about the distribution of the field either from symmetry or from some other circumstance. As an example, we have the case of a *straight wire* of circular cross-section, of radius $a$. If here as path of integration we choose a circle of radius $r$, centred on the wire axis, we have

$$2\pi r H = \frac{4\pi}{c} I \qquad \text{therefore} \quad H = \frac{2I}{rc} \quad \text{outside the wire } (r > a) \quad (46.3)$$

$$2\pi r H = \frac{4\pi}{c} \frac{r^2}{a^2} I \qquad \text{therefore} \quad H = \frac{2rI}{a^2 c} \quad \text{inside the wire } (r < a) \quad (46.4)$$

For a very long *straight current-carrying solenoid* it is known that a field exists essentially only in its interior and that in that region the field is parallel to the solenoid axis. We therefore choose as the path of integration a small rectangle, two opposite sides of which are parallel to the axis of the coil and 1 cm long, one of them lying inside the coil and the other outside. For this path, so long as we are sufficiently far from the ends, $\oint \mathbf{H} \cdot d\mathbf{r}$ is simply equal to the field $H$ within the solenoid. If the coil has $n$ turns per centimetre, the current flowing through our rectangle is $nI$. We therefore find for the constant solenoid field

$$H = \frac{4\pi}{c} nI \qquad (46.5)$$

This result is still practically correct when the wire is coiled around a ring (toroid), provided the diameter of the toroid (its thickness) is small compared to its large diameter (see figure 41, where the path of integration is shown on the upper left).

*The method of the magnetic double layer* (magnetic shell) is based on the following consideration: If we examine the field of a closed filamentary-current path, we find that for every integration path not linked by the current $\oint \mathbf{H} . d\mathbf{r} = 0$, while for every path which links the current once, we have that $\oint \mathbf{H} . d\mathbf{r} = \pm 4\pi I/c$. In the sense of §11, $\mathbf{H}$ outside the wire therefore represents a vortex-free vector field with a constant circulation $Z = 4\pi I/c$ when going round the wire.

We can therefore derive the magnetic field of this current-carrying wire from a scalar potential $\phi$ which, however, in contrast to the electrostatic and magnetostatic potential, is no longer single-valued, but changes by the amount $\pm 4\pi I/c$ for each complete circuit round the wire. In order to make the potential single-valued we imagine an

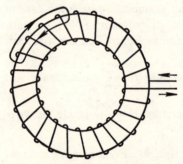

Fig. 41.—For calculating the magnetic field through a toroidal coil

arbitrary surface placed upon the current path so that the latter bounds this surface. We now require that upon passing through this surface the potential shall undergo a discontinuity in that (depending upon the direction of the passage through) the potential makes a jump of $\pm 4\pi I/c$.

Now however such a potential relationship points (as can easily be made clear by the example of a parallel-plate capacitor with small plate-spacing) to a homogeneous double-layer lying in the surface and having a moment $\tau = I/c$. By "double layer" we understand a structure consisting of two juxtaposed continuous surfaces (separation $d$) wherein the upper carries a surface charge of density $+\omega$, and the lower a surface charge of density $-\omega$. We designate the moment $\tau$ of the double layer by the product $\omega d$. A surface element $dS$ of the double layer, having normal $\mathbf{n}$, accordingly acts like a dipole with dipole moment

$$d\mathbf{m} = \mathbf{n}\tau \, dS = \frac{\mathbf{n}I \, dS}{c} \qquad (46.6)$$

Corresponding to (24.7) we then have for the potential of the entire double layer

$$\phi(\mathbf{r}) = \int \frac{d\mathbf{m} \cdot (\mathbf{r} - \mathbf{r}')}{|\mathbf{r} - \mathbf{r}'|^3} = \frac{I}{c} \int \frac{\mathbf{n} \cdot (\mathbf{r} - \mathbf{r}')}{|\mathbf{r} - \mathbf{r}'|^3} \, dS' \qquad (46.7)$$

We shall discuss this formula thoroughly in §47.

As our first application of the foregoing, let us calculate the potential and field on the axis of a current-carrying circular ring of radius $a$. We choose the plane of the ring as the surface of the double layer and also for the $xy$-plane, the origin being at the centre of the circle. Thus from (46.7) it follows for a point on the $z$-axis that

$$\phi = \frac{Iz}{c} \int \frac{2\pi r' \, dr'}{(z^2 + r'^2)^{3/2}} = \frac{2\pi I}{c} \left[ \frac{-z}{(z^2 + r'^2)^{1/2}} \right]_{r'=0}^{r'=a}$$

$$= \frac{2\pi I}{c} \left[ \frac{z}{|z|} - \frac{z}{(z^2 + a^2)^{1/2}} \right]$$

In passing through the point $z = 0$, the quantity $\phi(0, 0, z)$ changes discontinuously, as required, by $4\pi I/c$. By comparison, the field

$$H = H_z = -\frac{d\phi(0, 0, z)}{dz} = \frac{2\pi I}{c} \frac{a^2}{(z^2 + a^2)^{3/2}}$$

remains continuous and for $z = 0$ has the value

$$H(0) = \frac{2\pi I}{ca}$$

This formula forms the basis of measurements made with the tangent galvanometer, the instrument by which the field $H(0)$ given above is compared with the magnetic field of the earth.

In order to obtain an expression for the magnetic field of a ring current from the general potential expression (46.7) we proceed from the relationship

$$\delta\mathbf{r} \cdot \mathbf{H}(\mathbf{r}) = -\delta\mathbf{r} \cdot \text{grad}\, \phi(\mathbf{r}) = -\phi(\mathbf{r} + \delta\mathbf{r}) + \phi(\mathbf{r})$$

For $\phi(\mathbf{r})$ here we imagine the value (46.7) to have been substituted, where the surface integral is to extend over any definite surface bounded by the ring current. We can obtain $\phi(\mathbf{r} + \delta\mathbf{r})$ from $\phi(\mathbf{r})$ either by advancing from $\mathbf{r}$ to $\mathbf{r} + \delta\mathbf{r}$, or by keeping the field point fixed and displacing the entire current ring by $-\delta \mathbf{r}$. In the latter case the surface $S_1$ goes over to the surface $S_2$ parallel to it. Let $S_0$ be the band-shaped surface generated by the displacement of the current ring. Now in $\phi(\mathbf{r})$ we can transform (without altering its value) the surface integral over $S_1$ into another surface integral over the surface $S_0 + S_2$, so that we have

$$\delta \mathbf{r} \cdot \mathbf{H}(\mathbf{r}) = \frac{I}{c} \int_{S_0} \frac{\mathbf{n} \cdot (\mathbf{r} - \mathbf{r}')}{|\mathbf{r} - \mathbf{r}'|^3} dS'$$

Since, obviously, in the displacement of the current ring by $-\delta \mathbf{r}$ the surface $\mathbf{n} dS' = \delta \mathbf{r} \times d\mathbf{r}'$ would result from the ring element $d\mathbf{r}'$, we obtain finally, owing to the arbitrary choice of $\delta \mathbf{r}$ and because of the distributive law for the spar product, that

$$\mathbf{H}(\mathbf{r}) = \frac{I}{c} \oint \frac{d\mathbf{r}' \times (\mathbf{r} - \mathbf{r}')}{|\mathbf{r} - \mathbf{r}'|^3} \tag{46.8}$$

where the integral is to extend over the entire current path.

Equation (46.8) is called the *Biot-Savart law*. It suggests that the total field $\mathbf{H}$ be thought of as decomposed into the field contributions of individual current elements $I d\mathbf{r}'$, even though such isolated elements do not exist physically. The field contribution at the point $\mathbf{r}$ of such a current element is always perpendicular to the direction of the current as well as to the connecting vector $\mathbf{r} - \mathbf{r}'$ extending from the source point to the field point. With increasing distance from the current element the contribution falls off inversely with the square of the distance.

Finally, we consider the *method of the vector potential*—for the general case of an arbitrary stationary distribution of current density $\mathbf{g}$. We have here to determine the field from the two equations

$$\operatorname{curl} \mathbf{H} = \frac{4\pi}{c} \mathbf{g} \qquad \operatorname{div} \mathbf{B} = 0 \tag{46.9}$$

where, for the case of a vacuum (which we assume here), $\mathbf{B} = \mathbf{H}$. For this, however, the example in §12b of the determination of a source-free vector field from its vortices has exactly anticipated us. As in §12b, the second of the equations above can be solved in general (since $\operatorname{div} \operatorname{curl} \mathbf{A} \equiv 0$) by the statement that

$$\mathbf{B} = \operatorname{curl} \mathbf{A} \tag{46.10}$$

in which **A** is an arbitrary, sufficiently continuous, and differentiable vector field. The vector **A** is called the *vector potential* belonging to the magnetic induction vector **B**.

If, since **B** = **H**, we introduce **A** into the first equation (46.9), then, having regard for (11.17), we obtain

$$\text{curl curl } \mathbf{A} \equiv \text{grad div } \mathbf{A} - \nabla^2 \mathbf{A} = \frac{4\pi}{c} \mathbf{g}$$

For simplification we now arbitrarily set

$$\text{div } \mathbf{A} = 0 \qquad (46.11)$$

which is always allowed, since every vector field **A** is fully determined only when *both its vortices and sources* are specified. We therefore obtain for the determining equation

$$\nabla^2 \mathbf{A} = -\frac{4\pi}{c} \mathbf{g} \qquad (46.12)$$

On account of its similarity to Poisson's equation (19.10) of electrostatics (for the scalar potential $\phi$), equation (46.12) corresponding to the $\phi$-representation (19.11) is solved by the integral

$$\mathbf{A}(\mathbf{r}) = \frac{1}{c} \int \frac{\mathbf{g}(\mathbf{r}') \, dv'}{|\mathbf{r} - \mathbf{r}'|} \qquad (46.13)$$

Since $\mathbf{g}(\mathbf{r}') \, dv' = I \, d\mathbf{r}'$, expression (46.13) goes over, for the special case of a linear current distribution, to the expression

$$\mathbf{A}(\mathbf{r}) = \frac{I}{c} \oint \frac{d\mathbf{r}'}{|\mathbf{r} - \mathbf{r}'|} \qquad (46.14)$$

Owing to the equality of **B** and **H** assumed here, and because also

$$\text{curl} \frac{\mathbf{g}(\mathbf{r}')}{|\mathbf{r} - \mathbf{r}'|} = \text{grad} \frac{1}{|\mathbf{r} - \mathbf{r}'|} \times \mathbf{g}(\mathbf{r}') = \mathbf{g}(\mathbf{r}') \times \frac{\mathbf{r} - \mathbf{r}'}{|\mathbf{r} - \mathbf{r}'|^3}$$

we obtain from (46.10) and (46.13) for the magnetic field itself as the generalized form of the Biot-Savart law

$$\mathbf{H}(\mathbf{r}) = \frac{1}{c} \int \frac{\mathbf{g}(\mathbf{r}') \times (\mathbf{r} - \mathbf{r}')}{|\mathbf{r} - \mathbf{r}'|^3} \, dv' \qquad (46.15)$$

while from (46.14), by the same calculations, we arrive at equation (46.8).

*Remark:* In the MKSA system the constant of proportionality in equation (46.1) is arbitrarily set equal to 1; the relation is therefore written in the form

$$\oint \mathbf{H} \cdot d\mathbf{r} = I \tag{46.16}$$

and the unit of magnetic field strength is thereby connected with that of current strength. Accordingly, $\mathbf{H}$ is to be measured in A/m, so that altogether we obtain for the electromagnetic field quantities the following symmetrical unit scheme:

$$\mathbf{E}[\text{V/m}] \qquad \mathbf{D}[\text{As/m}^2] \qquad \varepsilon_0 = \frac{1}{4\pi \times 9 \times 10^9} \frac{\text{As}}{\text{Vm}}$$

$$\mathbf{H}[\text{A/m}] \qquad \mathbf{B}[\text{Vs/m}^2] \qquad \mu_0 = 4\pi \times 10^{-7} \frac{\text{Vs}}{\text{Am}}$$

The last-mentioned quantity $\mu_0$, to which the vacuum permittivity constant $\varepsilon_0$ corresponds, is defined as the constant of proportionality in the constitutive equation

$$\mathbf{B} = \mu_0 \mathbf{H} \tag{46.17}$$

of the magnetic field quantities for the case of a vacuum. It is called the *vacuum permeability* or the *induction constant*. The numerical value of $\mu_0$ follows, according to the definition of the unit of current, the *ampere*, given in §18, from the attraction of two parallel wires carrying equal currents. In fact, the current in each wire is 1A when the force of attraction between the two wires, separated by 1m, is $2 \times 10^{-7} \text{kg/s}^2$ per metre length of wire. Now the magnetic field at distance $r$ from a straight wire, infinitely long, carrying current $I$ is, corresponding to (46.3), given in the MKSA system by $H = I/2\pi r$. The force on the other wire per metre of wire length therefore becomes $F = IB = \mu_0 I^2/2\pi r$. If now we put $F = 2 \times 10^{-7}$ kg/s$^2$, $I = 1$A and $r = 1$m, then, as indicated above, $\mu_0 = 4\pi \times 10^{-7}\text{Vs/Am} = 1 \cdot 256 \times 10^{-6}\text{Vs/Am}$.

The physical quantity $\mathbf{H}$ of the MKSA system, which we wish temporarily to designate by $\mathbf{H}^*$, again does not agree exactly with the corresponding quantity $\mathbf{H}$ of the Gaussian system. Rather, from a comparison of the two defining equations (46.1) and (46.16), it follows that $\mathbf{H} = \dfrac{4\pi}{c}\mathbf{H}^*$. Together with the calculation of induction $\mathbf{B} = c\mathbf{B}^*$ given on p. 166, the constitutive equation (46.17) of the MKSA system, which now reads $\mathbf{B}^* = \mu_0 \mathbf{H}^*$, leads to the corresponding equation

$$\mathbf{B} = \frac{\mu_0 c^2}{4\pi}\mathbf{H}$$

of the Gaussian system. Here, since in empty space $\mathbf{B} = \mathbf{H}$, we have in this system that

$$\mu_0 = \frac{4\pi}{c^2} \approx \frac{4\pi}{9} \times 10^{-20} \frac{\text{s}^2}{\text{cm}^2} \tag{46.18}$$

We can easily convince ourselves by means of the conversion factors for the units V, A and m, of the agreement of this $\mu_0$-value with the above given MKSA system value. (See table, p. 431).

## §47. The ring current as a magnetic dipole

As was shown in §46, the magnetic field of a filamentary ring current of strength $I$ is completely equivalent to the field of a uniform magnetic double layer extending across and bounded by the current ring, the magnetic moment of a surface element $dS$ of this double layer being given by (46.6) as $d\mathbf{m} = \mathbf{n} I \, dS/c$. At distances which are large compared to its linear dimensions, this ring current therefore produces the same magnetic field as a dipole of total moment

$$\mathbf{m} = \frac{I}{c} \int \mathbf{n} \, dS = \frac{I}{c} \mathbf{S} \tag{47.1}$$

where $\mathbf{S}$ denotes a vector whose components are equal to the projections on the coordinate planes of the surface defined by the current ring.

Now in like manner, however, according to Ampère, any other stationary ring current with arbitrary current-density distribution $\mathbf{g}(\mathbf{r})$ acts, at sufficiently great distances, like a magnet of moment

$$\mathbf{m} = \frac{1}{2c} \int \mathbf{r}' \times \mathbf{g}(\mathbf{r}') \, dv' \tag{47.2}$$

(It should be observed that for the case of a filamentary current distribution this formula leads back directly to equation 47.1.)

For proving Ampère's assertion we start with the fact that the magnetic field of a dipole $\mathbf{m}$, namely

$$\mathbf{H} = -\frac{\mathbf{m}}{r^3} + \frac{3(\mathbf{m} \cdot \mathbf{r}) \mathbf{r}}{r^5} \tag{47.3}$$

can be represented in two different ways:

$$\mathbf{H} = -\operatorname{grad} \phi \quad \text{with} \quad \phi = \frac{\mathbf{m} \cdot \mathbf{r}}{r^3} \tag{47.4}$$

$$\mathbf{H} = \operatorname{curl} \mathbf{A} \quad \text{with} \quad \mathbf{A} = \frac{\mathbf{m} \times \mathbf{r}}{r^3} \tag{47.5}$$

We must therefore show that for the case of stationary ring currents and for large distances $(r \gg r')$, expression (46.13) for the vector potential of an arbitrary current distribution can be written in the form $\mathbf{m} \times \mathbf{r}/r^3$, the value of $\mathbf{m}$ being given by (47.2).

Now for large distances we have

$$\frac{1}{|\mathbf{r}-\mathbf{r}'|} = \frac{1}{r} + \frac{\mathbf{r}.\mathbf{r}'}{r^3} + \dots$$

From (46.13) we first therefore obtain

$$\mathbf{A}(\mathbf{r}) = \frac{1}{cr}\int \mathbf{g}(\mathbf{r}')\,dv' + \frac{1}{cr^3}\int (\mathbf{r}.\mathbf{r}')\mathbf{g}(\mathbf{r}')\,dv' + \dots \qquad (47.6)$$

Here, for a *stationary* current distribution, the first integral on the right vanishes. We see this immediately if we multiply the equation of continuity $\operatorname{div}\mathbf{g}(\mathbf{r}') = 0$ by say $x'$ and integrate over all space; it follows then by partial integration owing to the vanishing of the surface integral that

$$0 = \int x'\operatorname{div}\mathbf{g}(\mathbf{r}')\,dv' = -\int g_x\,dv'. \qquad (47.7)$$

by which the above assertion is proved. On multiplying the equation of continuity by $x'^2$ or $x'y'$, however, we obtain in an analogous way

$$\left.\begin{array}{l} 0 = \displaystyle\int x'^2\operatorname{div}\mathbf{g}(\mathbf{r}')\,dv' = -\int 2x'g_x\,dv' \\[2ex] 0 = \displaystyle\int x'y'\operatorname{div}\mathbf{g}(\mathbf{r}')\,dv' = -\int (y'g_x + x'g_y)\,dv' \end{array}\right\} \qquad (47.8)$$

We make use of these relationships for transforming the second integral of (47.6). Neglecting terms in $1/r$ of higher power, we obviously have

$$\mathbf{A}(\mathbf{r}) = \frac{1}{2cr^3}\int \{(\mathbf{r}.\mathbf{r}')\mathbf{g} + (\mathbf{r}.\mathbf{g})\mathbf{r}'\}\,dv' + \frac{1}{2cr^3}\int \{(\mathbf{r}.\mathbf{r}')\mathbf{g} - (\mathbf{r}.\mathbf{g})\mathbf{r}'\}\,dv'$$

Here, on account of (47.8), the first integral vanishes, as may be immediately shown by writing out in components. With the expression (47.2) for the magnetic moment $\mathbf{m}$, the second integral meanwhile can, as asserted, be written in the form

$$\mathbf{A} = \frac{1}{2cr^3}\int (\mathbf{r}' \times \mathbf{g}) \times \mathbf{r}\,dv' = \frac{\mathbf{m} \times \mathbf{r}}{r^3}$$

The equivalence, however, between a stationary ring current and a magnetic dipole goes even further: We inquire concerning the torque

**T** which the ring current experiences in a uniform external magnetic field **B**. Because of the Lorentz force we obviously have that

$$\mathbf{T} = \frac{1}{c} \int \mathbf{r}' \times (\mathbf{g} \times \mathbf{B}) \, dv' \tag{47.9}$$

If, similarly to the above, we write that

$$\mathbf{T} = \frac{1}{2c} \int \{\mathbf{r}' \times (\mathbf{g} \times \mathbf{B}) + \mathbf{g} \times (\mathbf{r}' \times \mathbf{B})\} \, dv' +$$

$$+ \frac{1}{2c} \int \{\mathbf{r}' \times (\mathbf{g} \times \mathbf{B}) - \mathbf{g} \times (\mathbf{r}' \times \mathbf{B})\} \, dv'$$

here also, on account of (47.8), the first integral vanishes, and, having regard for (6.6), we obtain

$$\mathbf{T} = \frac{1}{2c} \int (\mathbf{r}' \times \mathbf{g}) \times \mathbf{B} \, dv' = \mathbf{m} \times \mathbf{B} \tag{47.10}$$

which is just the value of the torque which we also expect for a magnetic dipole.

In a similar manner it can be shown that the ring current in a non-uniform magnetic field **B** experiences the force (also valid for a magnetic dipole)

$$\mathbf{F} = \operatorname{grad}(\mathbf{m} \cdot \mathbf{B}) = (\mathbf{m} \cdot \nabla)\mathbf{B} \tag{47.11}$$

the latter equality holding because curl **B** = 0. For proof, the magnetic field at the location of the ring current is developed in a power series in the components of **r** and the series broken off after the terms of first order.

*Thus, corresponding to the Ampère conception, the magnetization of matter can always be regarded as a consequence of the magnetic moments associated with the orbiting of electrons in the individual building blocks of matter.* With regard to the origin of the magnetic field of a magnetized body, we therefore arrive at two pictures which are identical in their consequence but different in their concept, according as we look upon the elementary magnets as dipoles or as ring currents. We illustrate these two conceptions in the example of a uniformly magnetized circular cylinder (figure 42): Let us imagine the magnetization to be due to *atomic magnetic dipoles*. Thus, within the framework of magnetostatics (see

§43), owing to the absence of true currents g, we must start with the equations

$$\operatorname{curl} \mathbf{H} = 0 \qquad \operatorname{div} \mathbf{H} = \operatorname{div}(\mathbf{B} - 4\pi\mathbf{M}) = -4\pi \operatorname{div} \mathbf{M} \qquad (47.12)$$

The H-field is everywhere irrotational and (for uniform magnetization) has its sources in the form of magnetic charges $\omega_M = M_n$ on the two

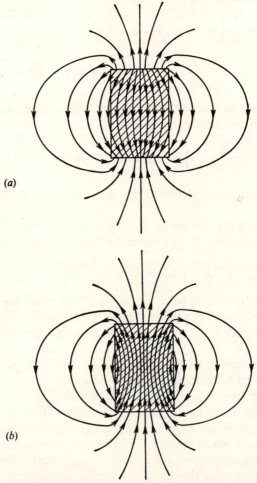

(a)

(b)

Fig. 42.—Field of a permanent magnet: (a) shows the **H**-lines for sources on the end-faces, (b) the **B**-lines with eddies (vortices) on the side-wall

end faces of the cylinder. We thus arrive at the field-line picture shown in figure 42a, there being continuity of the tangential component of **H** on the side of the cylinder.

If, however, we start from the concept of the *molecular current*, then, of necessity, we observe the **B**-field which (in conformity with 47.12) is described by the equations

$$\text{curl}\,\mathbf{B} = \text{curl}\,(\mathbf{H} + 4\pi\mathbf{M}) = 4\pi\,\text{curl}\,\mathbf{M} \qquad \text{div}\,\mathbf{B} = 0 \quad (47.13)$$

The **B**-field is therefore source-free throughout, but it has vortices at the places where curl $\mathbf{M} \neq 0$, i.e. for uniform magnetization, on the side of the cylinder. We have there, in fact, for the discontinuity in the tangential component of **B**

$$(B_t)_{\text{outside}} - (B_t)_{\text{inside}} = -4\pi M \qquad\qquad (47.14)$$

We now therefore arrive at the field-line picture (figure 42b). Since $\mathbf{B} = \mathbf{H}$ in the external regions, this picture at those places agrees with figure 42a. Inside the cylinder, since $\mathbf{B} = \mathbf{H} + 4\pi\mathbf{M}$, the two pictures are different from one another.

If then, as mentioned, the two concepts described lead to the same result, we must make it clear that in looking back on the Ampère interpretation of atomic magnetic moments (which is known through modern atomic physics to be correct), the second concept is in accord with the natural character of magnetism. *The induction* **B**, *and not the field strength* **H**, is the fundamental quantity, and we must consider the vector $\mathbf{H} = \mathbf{B} - 4\pi\mathbf{M}$, as well as its sources appearing in figure 42a, only as quantities for calculation which, exactly as with the electrical displacement **D**, are utilized for more convenient formulation of equations.

This concept, however, now leads to a noteworthy reinterpretation of equation (46.2), namely

$$\text{curl}\,\mathbf{H} = \frac{4\pi}{c}\mathbf{g} \qquad\qquad (47.15)$$

In its place we have now to write

$$\text{curl}\,\mathbf{B} = \frac{4\pi}{c}(\mathbf{g} + \mathbf{g}_M) \quad \text{with} \quad \mathbf{g}_M = c\,\text{curl}\,\mathbf{M} \qquad (47.16)$$

Thus, while the vorticity of **H** is given by the *true current density* **g**, the vorticity of **B** is determined both by the density **g** of the true current and by the density $g_M$ of the *magnetization current*. The latter is every-where source-free and it is different from zero where the magnetization has vorticity—in particular therefore (as in the example of figure 42) on the surface of magnetized bodies.

We give now an intuitive picture of the mechanism of the magnetiza-tion current on the basis of a highly simplified atomic model. Let the individual atoms consist simply of identical plane ring currents of surface $S$ and current-strength $I$. Let their number per cm³ be $n$. We ask what is the total current which passes through a rectangle which, in the $y$- and $z$-directions, has the lengths of side $a$ and $b$ (figure 43). We see

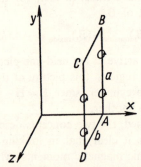

Fig. 43.—For calculating the magnetization current

immediately that only those atoms can contribute to a resulting current $abg_x$ through the rectangle, whose surface $S$ is penetrated by one of the four sides of the rectangle—for only in such atoms does the ring current make but one passage through the surface of the rectangle.

First we calculate the contribution of the atoms encountered along the side AB, of length $a$. We assume that, for all atoms, their projec-tions $S_y$ on the $xz$-plane are of equal magnitude. Then, on the side AB, all atoms are penetrated whose mid-points lie within a cylinder of cross-section $S_y$ and with axis AB. The number of such atoms is $naS_y$, and their contribution to the total current through the rectangle is $InaS_y$. If the $S_y$-values are not all equal, we must then take the average value, introducing the magnetization through

$$In\overline{S} = c\mathbf{M}$$

and obtaining as the contribution of the side AB the value $acM_y$.

In a similar way we find as the contribution of the side CD the value $-acM_y$, in which the value of $M_y$ at the location $z = b$ is naturally to be taken. Thus the two sides of the rectangle that are parallel to the y-axis furnish the quantity

$$ac(M_y)_{z=0} - ac(M_y)_{z=b} \rightarrow -abc\, \partial M_y/\partial z$$

Correspondingly, we obtain from the other two sides of the rectangle

$$-bc(M_z)_{y=0} + bc(M_z)_{y=a} \rightarrow abc\, \partial M_z/\partial y$$

Combining, we therefore have

$$abg_x = abc\left(\frac{\partial M_z}{\partial y} - \frac{\partial M_y}{\partial z}\right)$$

or, in general,

$$\mathbf{g}_M = c\,\text{curl}\,\mathbf{M}$$

in agreement with equation (47.16).

We now wish, however, to supplement this intuitive consideration with a mathematically more rigorous and generally more valid derivation of $\mathbf{g}_M$. We calculate the field of a magnetized body by means of the vector potential $\mathbf{A}$, by the summation of formula (47.5) over all elementary magnets in the matter:

$$\mathbf{A(r)} = \sum_i \frac{\mathbf{m}_i \times (\mathbf{r} - \mathbf{r}_i)}{|\mathbf{r} - \mathbf{r}_i|^3} = \int \frac{\mathbf{M}\,dv' \times (\mathbf{r} - \mathbf{r}')}{|\mathbf{r} - \mathbf{r}'|^3}$$

in which in $\mathbf{M}\,dv'$ we have combined the magnets lying within $dv'$ with the magnetization $\mathbf{M}$ which in general depends upon $\mathbf{r}'$. If now we set $(\mathbf{r} - \mathbf{r}')/|\mathbf{r} - \mathbf{r}'|^3 = \text{grad}'\,1/|\mathbf{r} - \mathbf{r}'|$, in which grad' means differentiation with respect to the components of $\mathbf{r}'$, we can transcribe the above equation to

$$\mathbf{A(r)} = -\int \text{curl}'\frac{\mathbf{M}}{|\mathbf{r} - \mathbf{r}'|}\,dv' + \int \frac{\text{curl}'\,\mathbf{M}}{|\mathbf{r} - \mathbf{r}'|}\,dv'$$

The first integral here can be transformed into a surface integral extending over the surface of the body, or preferably (considering the discontinuous transition on this surface to have been replaced by a continuous one) over a surface outside the body where $\mathbf{M}$ vanishes. In this case only the second integral supplies a finite contribution, and this contribution corresponds exactly, as a comparison with (46.13) shows, to the current density $\mathbf{g}_M$ in (47.16).

*Remark:* In the MKSA system the fundamental equations of magnetism read

$$\operatorname{curl} \mathbf{H} = \mathbf{g} \qquad \operatorname{div} \mathbf{B} = 0 \qquad \mathbf{B} = \mu_0 \mathbf{H} + \mathbf{M}$$

The magnetization current density here is accordingly given by

$$\mathbf{g}_M = \frac{1}{\mu_0} \operatorname{curl} \mathbf{M}$$

For the magnetic moment $\mathbf{M}$ of a circular current we have instead of (47.1) or (47.2)

$$\mathbf{m} = \mu_0 I S \quad \text{or} \quad \mathbf{m} = \frac{\mu_0}{2} \int \mathbf{r} \times \mathbf{g}(\mathbf{r}) \, dv$$

We can easily convince ourselves that the MKSA system quantities $\mathbf{M}$ and $\mathbf{m}$, here introduced, are equal to $4\pi/c$ times the similarly named quantities of the Gaussian system. (See table, p. 431.)

## §48. Magnetization and magnetic susceptibility

If we group all materials according to their magnetic behaviour, we arrive at the following classifications:

(*a*) *The magnetization is proportional to the magnetic field.*

In this case we write

$$\mathbf{M} = \kappa \mathbf{H} \tag{48.1}$$

with the *magnetic susceptibility* $\kappa$, which may be proportional to the temperature, but is independent of the field strength. We then also have

$$\mathbf{B} = \mathbf{H} + 4\pi \mathbf{M} = \mu \mathbf{H} \quad \text{with} \quad \mu = 1 + 4\pi\kappa \tag{48.2}$$

The quantity $\mu$ is called the *magnetic permeability*.

For materials characterized by these simple equations the two following groups merit special mention:

(1) *Diamagnetic bodies.*—For diamagnetic bodies $\mu$ is less than 1, and $\kappa$ is therefore negative and in general independent of the temperature. Indeed $\kappa$ is always very small compared to 1; it has, for example, the values

| | |
|---|---|
| Hydrogen | $\kappa = -0.2 \times 10^{-9}$ |
| Water | $-0.72 \times 10^{-6}$ |
| Gold | $-2.7 \times 10^{-6}$ |
| Bismuth | $-13 \times 10^{-6}$ |

In diamagnetic substances $\mathbf{M}$ is therefore directed oppositely to the field $\mathbf{H}$. *Qualitatively*, diamagnetism can be explained by the concept that in the application of an external magnetic field there are induced in the individual atoms ring currents whose magnetic moment is directed oppositely to the field **H**. The *quantitative* description must, however, be deferred to the appropriate sections of Volumes II and III of this work. Diamagnetism is a general property of matter and as such is present in all substances. However, it is so small that it is almost never observed when the substance in question is, in addition, paramagnetic or ferromagnetic.

(2) *Paramagnetic bodies.*—In paramagnetic bodies $\mu$ is greater than 1, and thus $\kappa$ is positive and, as a rule is inversely proportional to the absolute temperature $T$ (Curie's law), and proportional to the density $\sigma$:

$$\kappa = \frac{C\sigma}{T} \tag{48.3}$$

At room temperature (20° C) we have as examples the following numerical values:

|  |  |
|---|---|
| Oxygen | $\kappa = 0 \cdot 14 \times 10^{-6}$ |
| Platinum | $21 \times 10^{-6}$ |
| Manganese | $54 \times 10^{-6}$ |

The phenomenon of paramagnetism can be represented as follows: Already beforehand, the individual molecules possess a fixed magnetic moment, and when an external field is applied these tiny elementary magnets become partially aligned. The aligning action of the external field is opposed by the disordering tendency of thermal motion. Thus, in conformity with Curie's law, the same external field at lower temperature produces a higher degree of order than at a higher temperature (see the corresponding interpretation of $\chi$ in the electrical case in §26).

The magnetic field in the neighbourhood of diamagnetic or paramagnetic bodies can be calculated in full analogy with the electric field in the neighbourhood of normally polarizable media (see chapter B II, in particular §30). In addition, the formulae for the field energy and for the force effects in the field (chapter B III) can be immediately taken over to the magnetic case. In particular it follows from considerations such as those in §34, example 3, that paramagnetic bodies are attracted into a magnetic field, while diamagnetic bodies are expelled. Nevertheless, in both instances, in accordance with equation (34.8), a freely suspended ellipsoid aligns itself with its longest axis in the field direction.

(*b*) *The magnetization is not proportional to the field strength.*

To this classification belong generally the ferromagnetic materials iron, cobalt, nickel, the Heusler alloys and other alloys, as well as the non-conducting *ferrimagnetic* manganese compounds (e.g. "ferrite"). Just why this substance is magnetic is a very complicated matter and is in large measure dependent upon factors which often appear to be trivial. Because additional details are involved, this subject is deferred to Volume III.

We shall have to content ourselves here with some very sketchy notes on ferromagnets. Their most striking characteristic is that the magnetic moment is substantially greater, often more than a million times greater, than that of other substances, for the same field strength. **M** therefore no longer changes linearly with **H**; moreover, at relatively low and easily produced field strengths a *saturation* is reached. The saturation magnetization $M$ which, even for very strong fields, cannot be markedly exceeded, has for example the following values:

$$
\begin{array}{lll}
\text{Iron} & 4\pi M = & 22{,}000 \text{ G} \\
\text{Nickel} & = & 6{,}000 \text{ G} \\
\text{Cobalt} & = & 18{,}000 \text{ G}
\end{array}
$$

These saturation values are nearly independent of the state or degree of "working" of the material and also of the quantity of minor chemical impurities present. In contrast, however, the "magnetization curve", i.e. the manner by which **M** increases with increasing **H** is in the most intimate way dependent upon the prior treatment of the material. Here again, two different extremes can be distinguished.

(1) *Magnetically soft substances* are those in which **M** is at least a single-valued function of **H**. In typical cases the graph of this function has the shape shown in figure 44. At the outset $M$ increases steeply with $H$; later the climb becomes steadily flatter, and finally (at saturation) becomes practically horizontal. When the curve commences with an almost straight-line rise, we may speak of an "initial susceptibility", which can be defined either as $M/H$ or else as $\partial M/\partial H$. Its value for different kinds of iron lies between 50 and 1000. Completely soft, i.e. absolutely reversible, ferromagnetics are probably not to be found in nature. We can speak only of "softer" or "harder" substances according as their hysteresis loop is less or more broad (see below).

(2) *Magnetically hard substances.*—In these **M** is in general not a single-valued function of **H**; instead the magnetization is largely determined by the field strength to which the test sample was previously exposed. The typical shape of a magnetization curve is shown in figure 45. If on the initially unmagnetized sample an increasing field

$H$ be allowed to act, then $M$ at first follows a curve AB which, qualitatively, is not basically different from the curve (figure 44) of a magnetically soft substance. If now however $H$ be allowed to diminish, then $M$ at first decreases very much slower than its former rate of rise (curve BCDE). For the field $H = 0$ we have a "remanent magnetization" of

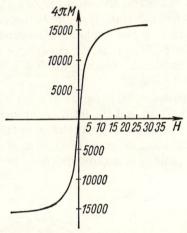

Fig. 44.—Magnetization curve for magnetically soft iron

value AC. Indeed, it is only by applying the "coercive force" AD, in the opposite direction to $H$ that we arrive at a zero value for the magnetization. *Remanence* and *coercive force* constitute a kind of measure for the magnetically hard substance. Then finally, for large negative values of $H$, we arrive again at saturation, E. From there upwards, with appropriate values of $H$, $M$ pursues the curve EFGB, whereby the "hysteresis loop" is closed. If the field $H$ be allowed to vary back and forth from saturation in one direction to saturation in the opposite direction, $M$ in general retraces the same loop.

Quite a different result is however obtained if we go only to a certain place on the curve, say to D′, and then allow $H$ to increase again. For not too great an increase of $H$, an almost straight line is obtained, such as the dashed line D′C′E′ in figure 45. It can also be followed backwards again unchanged. (More accurately stated we have, according to Rayleigh, small ellipses with major axis in the direction of D′C′E′.) If in all further magnetic applications of the material we remain within the limits D′E′, we can speak within those limits of a reversible magnet-ization, and we can characterize the material by means of a "magnetic equation of state"

$$\mathbf{M} = \kappa'\mathbf{H} + \mathbf{M}_0' \qquad (48.4)$$

where $\kappa'$ is the slope of the line D′E′, and $M'_0$ is the interval AC′.

In the absence of currents and other magnets, the field **H** within a permanent magnet has, substantially, a direction opposite to that of the magnetization.  Such a magnet would therefore correspond to the part of the hysteresis curve CD and can, for example, be represented by the point D′ already considered.  However, the field in its interior depends upon the shape of the magnet.

Thus if a magnetized body, in particular a ferromagnet, be brought into an otherwise uniform field $\mathbf{H}_0$, the resulting magnetization **M** and the field **H** existing inside the magnet (plotted as abscissa in figures 44 and 45) both depend upon the geometrical shape of the magnet.  As in the corresponding problem in electrostatics (see §30) we arrive at easily

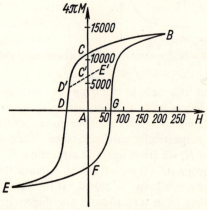

Fig. 45.—Magnetization curve for magnetically hard steel

handled relations only for the case where the body (magnet) has the form of an ellipsoid, or one of its special cases: a cylinder, sphere, or plate. The components of **H** are then connected with those of $\mathbf{H}_0$ and **M**, according to (30.1) and (30.3), through the relationships

$$H_x = H_{0x} - AM_x \quad H_y = H_{0y} - BM_y \quad H_z = H_{0z} - CM_z \qquad (48.5)$$

in which $A$, $B$, $C$ are the demagnetization factors of the ellipsoid, corresponding to equation (30.10). Owing to the large value of $\mathbf{M}$ in comparison with $\mathbf{H_0}$ it is necessary in ferromagnets to be specially mindful of this demagnetization.

All ferromagnetic materials show the property emphasized by the name only so long as their temperature remains under a certain value $\Theta$, which is characteristic of the material in question, and is called the *Curie point*. For iron the Curie point is 774°C, for nickel 372°C, and for cobalt 1131°C. Above the Curie point all ferromagnetic materials show normal paramagnetism, with this difference, however, that in the Curie law (48.3) the absolute temperature $T$ has to be replaced by the interval $T - \Theta$ from the Curie point:

$$\kappa = \frac{C\sigma}{T - \Theta} \qquad (48.6)$$

(the Curie-Weiss law).

To explain this law Pierre Weiss made the assumption that an "inner" field (which is proportional to the existing magnetization $\mathbf{M}$, say $N\mathbf{M}$, but is of non-magnetic origin) acts on the individual elementary magnets in addition to the external applied field $\mathbf{H}$. We call $N$ the *Weiss factor* of the inner field. Its theoretical interpretation was given very much later by Heisenberg with the help of quantum mechanics (see Volume III). The resulting field $\mathbf{H} + N\mathbf{M}$ now, according to Curie's law (48.3), fixes the magnetization

$$\mathbf{M} = \frac{C\sigma}{T}(\mathbf{H} + N\mathbf{M})$$

Solving this equation for $\mathbf{M}$, we have

$$\mathbf{M} = \frac{C\sigma}{T - C\sigma N}\mathbf{H}$$

We find the susceptibility given by equation (48.6) with the value $\Theta = C\sigma N$ for the Curie point.

*Remark:* In the MKSA system the expression (48.1) is written in the form

$$\mathbf{M} = \kappa\mu_0\mathbf{H}$$

while for equation (48.2) we must substitute

$$\mathbf{B} = \mu_0\mathbf{H} + \mathbf{M} = \mu\mu_0\mathbf{H} \text{ with } \mu = 1 + \kappa$$

Thus in the magnetic case, as in the electrical, the susceptibility in the MKSA system is defined as $4\pi$ times the susceptibility in the Gaussian system.

# CHAPTER C IV

## Electrodynamics of Quasi-stationary Currents

### §49. Self-induction and mutual induction

The great practical importance of (preponderantly linear) arrangements of current systems in physics and engineering provides ample justification for our occupying ourselves in a special section with such arrangements. We limit ourselves here to *quasi-stationary currents* and by this we understand alternating currents of such low frequency that the laws of electric and magnetic fields which we have already found can be applied to our problems. Because of the corrections which, for high-frequency alternating currents, have to be applied to the equation $\mathrm{curl}\,\mathbf{H} = 4\pi\mathbf{g}/c$, and the fact that their consequences lead to high-frequency techniques and to the electromagnetic theory of light, we defer a study of such currents to Section D.

The essentially new concept with which we must occupy ourselves in the following work is that of *self-induction and mutual induction*. We consider the flux of induction

$$\Phi_1 = \int B_n \, dS$$

which, in a system of current-carrying conductors, passes through or is linked by the first current circuit. In our system the vector $\mathbf{B}$ is uniquely defined at any point by the currents $I_1, I_2, \ldots, I_n$ in the various conductors. Let us assume further for simplification that the permeability $\mu$ is everywhere independent of $\mathbf{H}$. Thus the contributions of the individual currents to the resultant vector $\mathbf{B}$ are directly proportional to the respective currents. We can therefore write

$$\frac{1}{c}\Phi_1 = L_{11} I_1 + L_{12} I_2 + \ldots + L_{1n} I_n$$

Correspondingly, for each of the $n$ current circuits we have

$$\frac{1}{c}\Phi_k = \sum_{l=1}^{n} L_{kl} I_l \tag{49.1}$$

The $L_{kl}$ are called the *inductance coefficients*. We speak of the self-inductance of the $k$th circuit for the case $k = l$, and of the mutual inductance of the two circuits $k$ and $l$, when $k \neq l$.

## (a) Mutual inductance of two linear current circuits

The $L_{kl}$ with $k \neq l$ obviously depend only on the relative positions of the current circuits $k$ and $l$. If in the entire region occupied by the magnetic field we can say to a good approximation that $\mu \approx 1$, it is possible to give a general formula for the mutual inductance $L_{12}$ of the two circuits 1 and 2. $L_{12} I_2$ is indeed defined as $1/c$ times the flux of induction which is sent through circuit 1 by the current $I_2$ flowing in circuit 2. Denoting by $d\mathbf{r}_1$ a line element of circuit 1 and by $d\mathbf{r}_2$ a line element of circuit 2, we have (since $\mathbf{B} = \text{curl}\,\mathbf{A}$) that

$$L_{12} I_2 = \frac{1}{c} \int (B_2)_n \, dS_1 = \frac{1}{c} \oint \mathbf{A}_2 \cdot d\mathbf{r}_1$$

in which $\mathbf{A}_2$ is the vector potential at the place denoted by $d\mathbf{r}_1$, due to current 2. For this, however, we have previously found the value

$$\mathbf{A}_2(\mathbf{r}_1) = \frac{I_2}{c} \oint \frac{d\mathbf{r}_2}{|\mathbf{r}_1 - \mathbf{r}_2|}$$

Thus we obtain that

$$L_{12} = \frac{1}{c^2} \oint \oint \frac{d\mathbf{r}_1 \cdot d\mathbf{r}_2}{|\mathbf{r}_1 - \mathbf{r}_2|} \tag{49.2}$$

To calculate the mutual inductance we have therefore to take the scalar product of every line element of the one circuit by every line element of the other; to divide by the distance between the two elements; and to integrate over both circuits. Thus the symmetry condition $L_{12} = L_{21}$ is obviously always fulfilled.

As an application we treat the case, important in practice, of *two parallel coaxial circular current loops*. Let $a_1$ and $a_2$ be their respective radii, and let $h$ be the distance between their planes. For calculation, we fix in our minds two line elements $d\mathbf{r}_1$ and $d\mathbf{r}_2$ inclined to one another at an angle $\theta$. Their separation is

$$|\mathbf{r}_1 - \mathbf{r}_2| = \sqrt{(h^2 + a_1^2 + a_2^2 - 2a_1 a_2 \cos\theta)}$$

Further, introducing the azimuth $\psi_1$ of $\mathbf{r}_1$, so that $|d\mathbf{r}_1| = a_1 d\psi_1$, and analogously $\psi_2$ with $|d\mathbf{r}_2| = a_2 d\psi_2$, and noting that $\psi_2 = \psi_1 + \theta$, we obtain

$$L_{12} = \frac{1}{c^2} \int_0^{2\pi} \int_0^{2\pi} \frac{a_1 a_2 \cos\theta \, d\psi_1 \, d\psi_2}{\sqrt{(h^2 + a_1^2 + a_2^2 - 2a_1 a_2 \cos\theta)}}$$

$$= \frac{2\pi}{c^2} \int_0^{2\pi} \frac{a_1 a_2 \cos\theta \, d\theta}{\sqrt{(h^2 + a_1^2 + a_2^2 - 2a_1 a_2 \cos\theta)}}$$

the latter after substituting the integration with respect to $\theta$ for the $\psi_2$-integration, and after carrying out the $\psi_1$-integration. In order to work out the remaining integral, given in tables of *complete elliptic integrals* of the first and second kind,

$$K(k) = \int_0^{\pi/2} \frac{d\phi}{\sqrt{(1 - k^2 \sin^2 \phi)}}$$

and
$$E(k) = \int_0^{\pi/2} d\phi \sqrt{(1 - k^2 \sin^2 \phi)} \qquad (49.3)$$

we set $\theta = 2\phi + \pi$, and introduce for brevity

$$k^2 = \frac{4a_1 a_2}{h^2 + (a_1 + a_2)^2} \qquad 1 - k^2 = \frac{h^2 + (a_1 - a_2)^2}{h^2 + (a_1 + a_2)^2} \qquad (49.4)$$

We have, then, after some rewriting that

$$L_{12} = \frac{4\pi k}{c^2} \sqrt{(a_1 a_2)} \int_0^{\pi/2} \frac{-\cos 2\phi \, d\phi}{\sqrt{(1 - k^2 \sin^2 \phi)}}$$

$$= \frac{4\pi}{c^2} \sqrt{(a_1 a_2)} \left\{ \left( \frac{2}{k} - k \right) K - \frac{2}{k} E \right\} \qquad (49.5)$$

We consider further the two extreme cases $k \ll 1$ and $k \approx 1$. In the first case we must consider that the integral in (49.5) vanishes when, there, we set $k = 0$. If, however, we develop the denominator of the integrand in powers of $k^2$ and break off after the term of the first power, we find

$$L_{12} = \frac{\pi^2 k^3}{4c^2} \sqrt{(a_1 a_2)} \qquad (49.6)$$

as the value for the mutual inductance of two widely separated coaxial circular loops ($k \approx 2(a_1 a_2)^{\frac{1}{2}}/h$).

If, however, $k \approx 1$, i.e. $a_1 \approx a_2 \approx a$, $h \ll a$, the integrand in (49.5) then has a very sharp maximum at $\phi = \frac{1}{2}\pi$ and for $k = 1$ would itself become infinite in a non-integrable fashion. For calculating the integral we first substitute $\psi = \frac{1}{2}\pi - \phi$ for $\phi$. We then break up the $\psi$-integral into two component integrals

$$\int_0^{\pi/2} \frac{-\cos 2\phi \, d\phi}{\sqrt{(1 - k^2 \sin^2 \phi)}} = \int_0^{\varepsilon} \frac{\cos 2\psi \, d\psi}{\sqrt{(1 - k^2 \cos^2 \psi)}} + \int_{\varepsilon}^{\pi/2} \frac{\cos 2\psi \, d\psi}{\sqrt{(1 - k^2 \cos^2 \psi)}}$$

with $\varepsilon$ so chosen that $1 - k^2 \ll \varepsilon^2 \ll 1$. In the second integral we can then substitute 1 for $k$, and thus we obtain for the value of this integral

$$\int_{\varepsilon}^{\pi/2} \frac{\cos 2\psi \, d\psi}{\sin \psi} = \left[ \ln \tan \tfrac{1}{2}\psi + 2\cos \psi \right]_{\varepsilon}^{\pi/2} \approx -\ln \tfrac{1}{2}\varepsilon - 2$$

In the denominator of the first integral we set $\cos^2 \psi \approx 1 - \psi^2$; we neglect the $\psi$-dependence in the numerator, obtaining

$$\int_0^{\varepsilon} \frac{d\psi}{\sqrt{(1 - k^2 + k^2 \psi^2)}} = \frac{1}{k} \ln \frac{k\varepsilon + \sqrt{(1 - k^2 + k^2 \varepsilon^2)}}{\sqrt{(1 - k^2)}} \approx \ln \frac{2\varepsilon}{\sqrt{(1 - k^2)}}$$

in which we have set $k\varepsilon \approx \varepsilon$ and we have taken into account that $1 - k^2 \ll \varepsilon^2$. We therefore find altogether ($\varepsilon$ drops out) that

$$L_{12} = \frac{4\pi a}{c^2} \left\{ \ln \frac{4}{\sqrt{(1 - k^2)}} - 2 \right\} = \frac{4\pi a}{c^2} \left\{ \ln \frac{8a}{b} - 2 \right\} \qquad (49.7)$$

in which $b = \sqrt{[h^2 + (a_1 - a_2)^2]}$ is the shortest distance between the two circular loops.

### (b) Self-inductance

We apply the above result in the calculation of the *self-inductance of a circular wire ring*. The radius of the circle is $a$, and the radius of the wire $r_0$, it being assumed that $r_0 \ll a$. In this case we are no longer permitted to assume the wire to be filamentary (diameter vanishingly small) since with such a filament the magnetic field, and therewith the total flux of induction $\Phi$ would become infinite. If, however, we calculate with a finite wire thickness, there appears at first to be an indeterminacy with regard to where within the wire we must place the boundary curve for the calculation of the flux of induction.

We get around this difficulty by picturing the wire, of cross-sectional area $q = \pi r_0^2$, as subdivided parallel to its length into many very thin current-carrying threads of cross-sectional area $dS_1, dS_2, \ldots$ We then obtain the self-inductance $L$ of the whole wire as the mutual inductance $L_{ik}$ between the individual threads of the bundle of which the wire is considered to be composed.

We now assume at first that *the current strength I is uniformly distributed over the whole cross-section of the wire*, and that therefore in a thread of cross-section $dS$, a current $IdS/q$ flows. We then have for the flux of induction $\Phi_i$ through the $i$th thread

$$\frac{1}{c}\Phi_i = \frac{I}{q}\sum_k L_{ik}\, dS_k \rightarrow \frac{I}{q}\int L_{ik}\, dS_k \qquad (49.8)$$

in which we have substituted a surface integral over the whole cross-section of the wire for the summation over all remaining threads. For the induction law we now need however, not the flux of induction through a particular thread, but the average value of the fluxes through all threads

$$\Phi = \frac{1}{q}\sum \Phi_i\, dS_i \rightarrow \frac{1}{q}\int \Phi_i\, dS_i \qquad (49.9)$$

for, by applying the induction law in the form of (45.2), say, we have for the current induced in the entire wire

$$I = \sum_i I_i = -\frac{1}{c}\sum_i \frac{1}{R_i}\frac{d\Phi_i}{dt} = -\frac{1}{qRc}\sum_i \frac{d\Phi_i}{dt}\, dS_i = -\frac{1}{Rc}\frac{d\Phi}{dt}$$

in which $R_i = qR/dS_i$ means the resistance of the $i$th thread. If, now, corresponding to the definition of self-inductance, we set $\Phi = cLI$, we find from (49.8) and (49.9) the expression for $L$:

$$L = \frac{1}{q^2}\iint L_{12}\, dS_1\, dS_2 \qquad (49.10)$$

Thus for our circular loop, with equation (49.7) for the mutual inductance of two almost equal coaxial current loops ($a_1 \approx a_2 \approx a$), and with the smallest distance of mutual separation $b_{12} = b$, we have that

$$L = \frac{4\pi a}{c^2}\overline{\left\{\ln\frac{8a}{b}-2\right\}} = \frac{4\pi a}{c^2}\{\ln 8a - \overline{\ln b} - 2\}$$

which, in the sense (49.10), is to be averaged over all $b$-values. In order to carry out this averaging we introduce into the cross-sectional plane the position vectors $\mathbf{r}_1$ and $\mathbf{r}_2$ of the two surface elements $dS_1$ and $dS_2$, reckoned from the centre of the circle, and thus we have to evaluate the quantity

$$\overline{\ln b} = \frac{1}{r_0^4 \pi^2} \iint dS_1 \, dS_2 \ln b$$

$$= \frac{2}{r_0^4 \pi} \int_0^{r_0} r_1 \, dr_1 \int_0^{r_0} r_2 \, dr_2 \int_0^{2\pi} d\phi \ln \sqrt{(r_1^2 + r_2^2 - 2r_1 r_2 \cos \phi)} \quad (49.11)$$

Now for $r_1 > r_2$, since

$$\ln(1-x) = -\sum_{n=1}^{\infty} \frac{x^n}{n}$$

we have the following series development

$$\ln \sqrt{(r_1^2 + r_2^2 - 2r_1 r_2 \cos \phi)} = \ln r_1 + \frac{1}{2} \left\{ \ln \left( 1 - \frac{r_2}{r_1} e^{i\phi} \right) + \ln \left( 1 - \frac{r_2}{r_1} e^{-i\phi} \right) \right\}$$

$$= \ln r_1 - \sum_{n=1}^{\infty} \left( \frac{r_2}{r_1} \right)^n \frac{\cos n\phi}{n}$$

If this development, or that corresponding to $r_1 < r_2$, be put into (49.11), the contribution of the summation over $n$ vanishes because

$$\int_0^{2\pi} \cos n\phi \, d\phi = 0$$

Then, by splitting the $r_2$-integral into the two component integrals with $r_2 < r_1$ and $r_2 > r_1$, we have

$$\overline{\ln b} = \frac{4}{r_0^4} \int_0^{r_0} r_1 \, dr_1 \left\{ \int_0^{r_1} r_2 \, dr_2 \ln r_1 + \int_{r_1}^{r_0} r_2 \, dr_2 \ln r_2 \right\}$$

$$= \frac{1}{r_0^2} \int_0^{r_0} r_1 \, dr_1 \left\{ 2 \ln r_0 - \left( 1 - \frac{r_1^2}{r_0^2} \right) \right\}$$

and after carrying out the remaining integration

$$\overline{\ln b} = \ln r_0 - \tfrac{1}{4}$$

Thus we find for the self-inductance of the wire ring

$$L = \frac{4\pi a}{c^2} \left\{ \ln \frac{8a}{r_0} - \frac{7}{4} \right\} \quad (49.12)$$

As an example, for a wire ring made of non-magnetic material, in air, with $a = 5\,$cm and $r = 0 \cdot 05\,$cm, we have

$$L = 3 \cdot 44 \times 10^{-19}\,\mathrm{s}^2/\mathrm{cm} = 3 \cdot 10 \times 10^{-7}\,\mathrm{henry\ (H)}$$

We introduced the assumption above that the current is uniformly distributed over the cross-section of the wire. Actually, however, for alternating currents the current density tends to increase toward the outer boundary of the wire, and this the more so the higher the frequency of the alternating current and the greater the permeability of the wire material. For high-frequency alternating currents, and also in the case of soft iron wires ($\mu \gg 1$) even when the frequency is low, the current confines itself to a thin layer on the conductor surface. We shall have a fuller discussion of this phenomenon, known as the *skin effect*, in §60. We shall then see that the skin effect is hardly noticeable for non-magnetic wires (and thus our foregoing calculation of $L$ is justified), so long as the frequency of the alternating current lies below the value $c^2/4\pi r_0^2\sigma$ (with $r_0 = $ wire radius, and $\sigma = $ conductivity of wire material). For our wire, having $r_0 = 0 \cdot 05\,$cm, this frequency (if we calculate with the conductivity of copper, namely $\sigma = 5 \cdot 14 \times 10^{17}\,\mathrm{s}^{-1}$) comes out to be $5 \cdot 6 \times 10^4\,\mathrm{s}^{-1}$.

A much simpler calculation than that for the circular loop is that of the self-inductance $L$ of a long coil (solenoid) of cross-section $a$ and length $l$. Let $n$ be the number of turns per cm, and let $\mu$ be the permeability of the core of the coil. Inside the coil $H = 4\pi n I/c$; thus the flux of induction through one turn of the coil is $4\pi\mu n I a/c$, and that through the whole coil is therefore $\Phi = 4\pi\mu n^2 I l a/c$. For the self-inductance we therefore obtain

$$L = \frac{4\pi\mu n^2 l a}{c^2} \qquad (49.13)$$

*Remark:* Since in both the Gaussian unit system and the MKSA system the self-inductance is defined by the equation

$$V^{(\mathrm{ind})} = -L\,dI/dt$$

in order to obtain an $L$-value in Vs/A, i.e. in henrys (H), the $L$-value in the Gaussian system has to be multiplied by $300 \times (3 \times 10^9) = 9 \times 10^{11}$.

## §50. Circuit with resistance and self-inductance. The vector diagram

We consider a system of $n$ closed circuits, the ohmic resistance of the $k$th circuit being $R_k$, the applied e.m.f. being $V_k^{(e)}$, and the flux of induction $\Phi_k$. The behaviour of the current $I_k$ is in general given by the relation

$$I_k R_k = V_k^{(e)} - \frac{1}{c} \frac{d\Phi_k}{dt}$$

If the special conditions of §49 are satisfied, so that in particular the permeability $\mu$ is everywhere a constant of the material, independent of **H**, then $\Phi_k$ is a linear function of the current strength, or, from (49.1),

$$\frac{1}{c} \Phi_k = \sum_{l=1}^{n} L_{kl} I_l$$

with the induction constants $L_{kl}$, the circuits being assumed to be at rest. The induction law therefore gives the $n$ equations

$$I_k R_k + \sum_{l=1}^{n} L_{kl} \frac{dI_l}{dt} = V_k^{(e)} \tag{50.1}$$

These are $n$ equations for the $n$ derivatives $dI_k/dt$. If at time $t$ the values $I_k$ and $V_k^{(e)}$ are given, then, through (50.1), the values of the $I_k$ at time $t+dt$ are determined. The whole "run" of the currents with time is therefore established by (50.1), provided the applied e.m.f.s $V_k^{(e)}$ are known functions of the time.

We wish to discuss this in greater detail using several examples.

*Circuit without applied e.m.f.*—We have here only one equation this being with $V^{(e)} = 0$:

$$IR + L \frac{dI}{dt} = 0$$

Its general solution is

$$I = I_0 e^{-tR/L} \tag{50.2}$$

The current existing at time zero decays away exponentially in a manner such that after the lapse of a time $L/R$ (the time-constant of the circuit) it has fallen to the $e$th part of its initial value.

*Circuit with periodic applied voltage* $V^{(e)} = V_0 \cos \omega t$. (Here $V_0 =$ amplitude of the periodic applied voltage; $\omega =$ the circular frequency of same.) We have then that

$$IR + L \frac{dI}{dt} = V_0 \cos \omega t \tag{50.3}$$

We can find a particular solution of this non-homogeneous equation by putting

$$I = I_0 \cos(\omega t - \phi) \tag{50.4}$$

Equation (50.3) then becomes

$$\{I_0(R \cos \phi + \omega L \sin \phi) - V_0\} \cos \omega t + I_0(R \sin \phi - \omega L \cos \phi) \sin \omega t = 0$$

This equation must be satisfied for all values of $t$; the coefficients of $\cos \omega t$ and $\sin \omega t$ must therefore vanish separately. We thus obtain two equations for $I_0$ and $\phi$ with the solution

$$I_0 = \frac{V_0}{\sqrt{(R^2 + \omega^2 L^2)}} \quad \text{and} \quad \tan \phi = \frac{\omega L}{R} \tag{50.5}$$

The quantity $\sqrt{(R^2 + \omega^2 L^2)}$ is called the *impedance* of our circuit, and $\phi$ is called the *phase angle*.

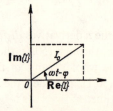

Fig. 46.—Complex representation of the strength of an alternating current

The calculation of the unknown quantities in alternating-current work becomes shorter and more readily visualized if we make use of *complex quantities* and a *graphical representation*. If in place of the real expression (50.4) for the current we substitute the "complex" current*

$$I = I_0 e^{j(\omega t - \phi)} \tag{50.6}$$

we can represent (figure 46) this quantity in the complex plane as a vector with the components

$$\text{Re}\{I\} = I_0 \cos(\omega t - \phi) \quad \text{Im}\{I\} = I_0 \sin(\omega t - \phi) \tag{50.7}$$

The vector thus has length $I_0$ and makes an angle $\omega t - \phi$ with the real axis. In the course of one period the end-point of this vector describes

---

* Here we use $j$ for $\sqrt{(-1)}$ as is the custom with electrical engineers. Later on we shall feel free to use $i$ for the same quantity, as this is the symbol more familiar to physicists.

a circle around the origin such that at any moment the projection OA of this vector on the real axis gives the real value of the current.

For calculating with such complex vectors we have the following rules: Addition of two quantities $Ae^{j\alpha}$ and $Be^{j\beta}$ consists in a geometrical addition of the corresponding vectors (by the parallelogram rule). Multiplication by the imaginary unit $j = e^{j\pi/2}$ means a rotation of 90° in the positive (i.e. counterclockwise) sense. Differentiation with respect to $t$ for processes having circular frequency $\omega$ is equivalent to multiplication by $j\omega$.

With (50.6), our alternating-current equation (50.3) takes the form

$$RI + j\omega LI = V^{(e)}$$

which connects the complex vectors $I$ and $V^{(e)}$ (figure 47). For an arbitrary value of $I$ let $IR$ equal OA at a certain time. The vector $LdI/dt$ is then perpendicular to it (line OB). $V^{(e)}$ is obtained from OA and OB by geometric addition (line OC). If this whole figure is

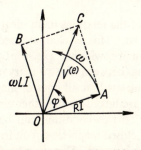

Fig. 47.—Vector diagram for a circuit with self-inductance and resistance

imagined to rotate as a rigid structure around the origin, with the angular frequency $\omega$, then at any moment the projections of OA and OC on the real axis (or on the imaginary axis) give the values of current and voltage. The values given in equation (50.5) for the impedance and phase angle can obviously be directly read from figure 47. The (in general complex) ratio $V^{(e)}/I$ is called the *apparent resistance*, **R**. Thus in the foregoing case $\mathbf{R} = R + j\omega L$. The impedance is therefore the magnitude of **R**. The real part of **R** is called the *effective resistance* and the imaginary part the *reactance*.

As a further example we consider two coupled circuits in one of which there is a periodic e.m.f. $V^{(e)}$—the case of a *transformer* (figure 48). In this case equation (50.1) reads

$$I_1 R_1 + L_{11}\frac{dI_1}{dt} + L_{12}\frac{dI_2}{dt} = V^{(e)} \qquad I_2 R_2 + L_{21}\frac{dI_1}{dt} + L_{22}\frac{dI_2}{dt} = 0$$

For an alternating applied e.m.f., $V^{(e)} = V_0 e^{j\omega t}$, these equations can be completely solved by a method analogous to the one just used, thus: We put $I_1 = I_{10} e^{j(\omega t - \phi_1)}$ and $I_2 = I_{20} e^{j(\omega t - \phi_2)}$. We limit ourselves

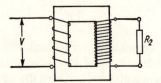

Fig. 48.—Schematic diagram of a transformer

to a *direct* discussion of the relations in an ideal transformer connected to a purely ohmic load. This is characterized by vanishingly small resistance in the primary winding ($R_1 \ll \omega L_{11}$) and ideal "tight" (leakage-free) coupling between the primary and secondary circuit. This latter condition means that all lines of magnetic induction linked by one of the circuits are linked by the other circuit as well. To a high degree of approximation this condition can be realized by arranging the two windings on a closed magnetic-circuit iron core in which almost all the lines of induction travel. Denoting by $\Phi$ the flux of induction in this core, and by $n_1$ and $n_2$ the total number of turns on the primary and secondary windings, then $n_1 \Phi$ and $n_2 \Phi$ are the fluxes of induction through each of the two windings. For $R_1 = 0$ we therefore have

$$\frac{n_1}{c}\frac{d\Phi}{dt} = V^{(e)} \qquad I_2 R_2 + \frac{n_2}{c}\frac{d\Phi}{dt} = 0$$

From these equations we have, first,

$$I_2 R_2 = -\frac{n_2}{n_1} V^{(e)}$$

for the current flowing in the secondary circuit. A most important point is that, as we see from the first equation, the flux $\Phi$ is determined by the primary voltage $V^{(e)}$ alone (independent of the load):

$$\frac{j\omega n_1}{c}\Phi = V^{(e)}$$

Moreover, the magnetic field strength, and therefore also the flux in the iron core, are at any moment proportional to the sum $n_1 I_1 + n_2 I_2$, thus

$$n_1 I_1 + n_2 I_2 = k\Phi$$

where $k$ is a (real) constant of the transformer. We therefore have the vector diagram for the ideal transformer shown in figure 49. There is a vector $V^{(e)}$ for the primary voltage, the vector $k\Phi$ lagging 90° behind this, and the vector $n_2 I_2$ with a further lag of 90°. The vector $n_1 I_1$

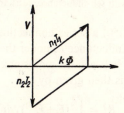

Fig. 49.—Vector diagram of an ideal transformer

for the primary current is then determined as the vector difference $k\Phi - n_2 I_2$. For decreasing $R$—therefore increasing $I_2$ and thus increasing load on the transformer—$n_1 I_1$ rotates in the positive sense away from $k\Phi$, in such a way, however, that the projection of $n_1 I_1$ in the direction of $k\Phi$ remains always equal to $k\Phi$. While in this case the effective resistance in the primary circuit therefore increases, the reactance is independent of $R_2$.

## §51. Circuit with resistance, self-inductance and capacitance

When a capacitor is included in an alternating-current circuit, the current ceases to be source-free, i.e. it loses its solenoidal character, since the plates of the capacitor act as sources or sinks for the current. Then, however, the equation $\operatorname{curl}\mathbf{H} = 4\pi\mathbf{g}/c$, which was set up for steady currents and was used in §49 for the calculation of inductances, also commences to lose its validity since its divergence leads to $\operatorname{div} \mathbf{g} = 0$. We shall be occupied with a thorough-going discussion of the necessary corrections to this equation in §53. So long, however, as the separation of the capacitor plates is small we can neglect this correction. As we shall soon see, in the application of the induction law we shall be obliged to reckon with an uncertainty, but one of no practical importance.

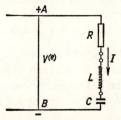

Fig. 50.—Circuit with self-inductance, resistance, and capacitance

We consider (figure 50) a series connection of ohmic resistance $R$, capacitance $C$ and self-inductance $L$, under the influence of an alternating voltage $V^{(e)}$ applied between the mains points A and B. The current $I$ flowing in the circuit is then given by

$$RI = \int_1^2 (\mathbf{E} + \mathbf{E}^{(e)}) \cdot d\mathbf{r}$$

where the line integral is to extend from capacitor plate 1, over BARL, to plate 2. Now we obviously have

$$\int_1^2 \mathbf{E}^{(e)} \cdot d\mathbf{r} = V^{(e)}$$

we have, further, that

$$\int_1^2 \mathbf{E} \cdot d\mathbf{r} = \phi_1 - \phi_2 + \oint \mathbf{E}^{(\mathrm{ind})} \cdot d\mathbf{r} = V_{12} - L\frac{dI}{dt}$$

in which $\phi_1 - \phi_2 = V_{12}$ is the electrostatic potential on the capacitor. Since for the induction term, however, the path of integration must be a closed one, the contribution to the integral from the portion of path between capacitor plates on the right-hand side has to be subtracted. We thus obtain, altogether

$$RI + L\frac{dI}{dt} - V_{12} = V^{(e)} \qquad (51.1)$$

Furthermore, the current $I$ is equal to the time-rate of change of the charge on the capacitor. If $C$ is the capacitance of the capacitor, we then have that

$$I = -C\frac{dV_{12}}{dt} \tag{51.2}$$

Thus, for an alternating current of period $\omega$

$$RI + j\omega LI - \frac{j}{\omega C}I = V^{(e)} \tag{51.3}$$

Thus we have the connection between $V^{(e)}$ and $I$ illustrated in the vector diagram (figure 51).

The apparent resistance of our arrangement is given by the vector $\mathbf{R} = R + j(\omega L - 1/\omega c)$, and the impedance by $\sqrt{[R^2 + (\omega L - 1/\omega C)^2]}$. The latter, for $\omega^2 = 1/LC$, has a minimum which for small values of $R$ is very sharply defined. Thus if the applied voltage $V^{(e)}$ consists of a combination (superposition) of all possible different periods, then, in the main, the current will contain only those periods which are close to this fundamental frequency $1/\sqrt{(LC)}$. We have before us here

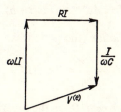

Fig. 51.—Vector diagram for circuit of fig. 50

the case of resonance between the period of the applied voltage and the period of the characteristic oscillations of the circuit, presently to be discussed.

If we connect the points A and B of the terminal leads (figure 50) by means of a heavy wire (short-circuit), $V^{(e)}$ becomes zero, and our equations (51.1) and (51.2) give, after elimination of $V_{12}$,

$$L\frac{d^2I}{dt^2} + R\frac{dI}{dt} + \frac{1}{C}I = 0 \tag{51.4}$$

The general solution of this equation (with the two constants of integration $c_1$ and $c_2$) is

$$I = c_1 e^{k_1 t} + c_2 e^{k_2 t} \quad \text{with} \quad k_{1,2} = -\frac{R}{2L} \pm \sqrt{\left[\left(\frac{R}{2L}\right)^2 - \frac{1}{LC}\right]}$$

We find, therefore, a periodic discharge when $R/2L < 1/\sqrt{(LC)}$, and an aperiodic discharge for $R/2L > 1/\sqrt{(LC)}$. Designating for brevity

$$\delta = \frac{R}{2L} \quad \text{and} \quad \omega_0 = \frac{1}{\sqrt{(LC)}} \tag{51.5}$$

then, for the periodic case,

$$I = A e^{-\delta t} \sin\left[(\omega_0^2 - \delta^2)^{1/2} t + b\right]$$

while for the aperiodic discharge

$$I = A \exp - \left[\delta + \sqrt{(\delta^2 - \omega_0^2)}\right]t + B \exp - \left[\delta - \sqrt{(\delta^2 - \omega_0^2)}\right]t$$

In practical applications, weakly damped oscillations are of special importance. These take place when $\delta$ is so small compared to $\omega_0$ that $\delta^2$ can be neglected in comparison with $\omega_0^2$. We then have

$$I = A e^{-\delta t} \sin(\omega_0 t + b)$$

The *time for one oscillation* (the period) is

$$T = \frac{2\pi}{\omega_0} = 2\pi \sqrt{(LC)}$$

while the *logarithmic decrement D*, as the logarithm of the ratio of the amplitudes of two consecutive oscillations, is

$$D = \log \frac{e^{-\delta t}}{e^{-\delta(t+T)}} = \delta T = \frac{2\pi\delta}{\omega_0} = \pi R \sqrt{\frac{C}{L}} \tag{51.6}$$

$1/D$ gives the number of oscillations in the duration of which the amplitude falls to $1/e$ of its original value.

As a numerical example we consider a Leyden jar of radius 5 cm, wall thickness 0·2 cm and height 20 cm. From the formula for the parallel-plate capacitor $C = \varepsilon S/4\pi d$ we obtain with $\varepsilon = 5$ (glass) and $S = 2\pi \times 5 \times 20 + \pi \times 25)\,\text{cm}^2$ altogether $C = 1400\,\text{cm} = 1·5 \times 10^{-9}\,\text{F}$. As the short-circuiting conductor we take a copper wire bent simply into a circle of the dimensions used in §49 and thus having the self-inductance found there, namely $L = 3·4 \times 10^{-19}\,\text{s}^2/\text{cm} =$

$3 \cdot 1 \times 10^{-7}$H. For the resistance $R$ of the wire we first take the ordinary d.c. resistance $R = 6 \cdot 8 \times 10^{-3} \Omega$. With these values, from (51.5) and (51.6), we get

$$v_0 = \frac{\omega_0}{2\pi} = 7 \cdot 2 \times 10^6 \mathrm{s}^{-1} \quad \text{and} \quad \delta = \frac{R}{2L} = 1 \cdot 1 \times 10^4 \mathrm{s}^{-1}$$

This frequency would correspond to a wavelength $\lambda = c/v_0 = 42$m. The number of oscillations for a damping to $1/e$ is $v_0/\delta = 650$. If the oscillations are excited by a discharge across a spark-gap, we should naturally expect decidedly stronger damping on account of the increase in $R$ due to the added resistance of the spark-gap.

Even without a spark-gap, however, an essentially higher value of $R$ is to be expected on account of the skin effect (already mentioned in §49), in which for such rapid vibrations the current is displaced toward the surface of the wire. In cases of this type we cannot use the whole cross-section of the wire for calculating $R$, but only the cross-section of that layer near the surface which is occupied by the current, its thickness being a function of the frequency $\omega_0$ (see §60).

## §52. The energy theorem for a system of linear currents

In order to complete our observations on the electrodynamics of quasi-stationary currents, we wish now to occupy ourselves with a consideration of the energy relations in such systems. We start with the basic equations valid for a system of linear currents. Corresponding to (50.1) and (51.1), we write the equations in the form

$$V_k^{(e)} = R_k I_k + \frac{1}{c} \frac{d\Phi_k}{dt} - V_k \quad \text{with} \quad I_k = -C_k \frac{dV_k}{dt} \qquad (52.1)$$

Since $V_k^{(e)} I_k$ is the power from the applied voltage (e.m.f.) in the $k$th circuit acting upon the current flowing in this circuit, we obtain the total energy input per second of all applied e.m.f.s in the system by multiplying the first equation (52.1) by $I_k$ and summing over all circuits. There then emerges as the first term on the right-hand side the total Joule heat, $\sum\limits_{k=1}^{n} R_k I_k^2$ produced per second in all circuits. Carrying this quantity over to the left side, we obtain as the *energy per second* supplied to the current system by the current sources, *less the Joule heat*,

$$W \equiv \sum_{k=1}^{n} V_k^{(e)} I_k - \sum_{k=1}^{n} R_k I_k^2 = \frac{1}{c} \sum_{k=1}^{n} I_k \frac{d\Phi_k}{dt} - \sum_{k=1}^{n} I_k V_k \qquad (52.2)$$

The last term on the right is known to us from electrostatics:

$$-\sum_{k=1}^{n} I_k V_k = \sum_{k=1}^{n} C_k V_k \frac{dV_k}{dt} = \frac{dU_{el}}{dt} \quad \text{with} \quad U_{el} = \tfrac{1}{2} \sum_{k=1}^{n} C_k V_k^2 \qquad (52.3)$$

The quantity $U_{el}$ represents the *electrical field energy* which at times is stored in the electrical capacitors of the system. A part of $W$ is therefore used for increasing this energy. There is thus the suggestion that the remainder of $W$ comes out as an increase in the *magnetic field energy* $U_m$ and, in addition, if need be, as *mechanical work* $W_{mech}$ which the field performs in a displacement of current-carrying wires (as, for example, in the electric motor):

$$\frac{1}{c}\sum_{k=1}^{n} I_k \frac{d\Phi_k}{dt} = \frac{dU_m}{dt} + W_{mech} \tag{52.4}$$

The energy theorem (52.2) then immediately assumes the illuminating form

$$W = \frac{d}{dt}\{U_{el} + U_m\} + W_{mech}$$

We must now find out the meaning of $U_m$ and $W_{mech}$.

We first imagine the current-carrying wires of the system *held fixed*. Then no mechanical work is performed ($W_{mech} = 0$). In addition, the inductances $L_{kl}$ of the system remain constant, and from (49.1), namely

$$\frac{1}{c}\Phi_k = \sum_{l=1}^{n} L_{kl} I_l \tag{52.5}$$

as well as from (52.4) it follows in this case that

$$\frac{1}{c}\sum_{k} I_k \frac{d\Phi_k}{dt} = \sum_{k,l=1}^{n} L_{kl} I_k \frac{dI_l}{dt} = \frac{d}{dt}\left\{\tfrac{1}{2}\sum_{k,l=1}^{n} L_{kl} I_k I_l\right\}$$

and therefore, for the quantity $U_m$, we have the expression

$$U_m = \tfrac{1}{2}\sum_{k,l=1}^{n} L_{kl} I_k I_l \tag{52.6}$$

We shall see in §55 that this quantity actually represents the magnetic field energy. It is perhaps sufficient to refer here to the example of a straight coil (solenoid) for which, since $H = 4\pi nI/c$, and with the value (49.13) for its self-inductance, we find the expression

$$U_m = \frac{L}{2}I^2 = \frac{2\pi n^2 \mu lq}{c^2}I^2 = \frac{\mu H^2}{8\pi} lq \tag{52.7}$$

Since $lq$ = volume of coil, equation (52.7) agrees with the expression that would be expected by analogy with the electrostatic field energy.

Let us also mention here the energy relations in an oscillation cycle of the type considered in §51. Let the resistance be small giving small damping. From (52.2), with $W \approx 0$, $W_{mech} = 0$, the electromagnetic energy oscillates to and fro between the energy content of the capacitor and that of the inductor (the coil). With $I = I_0 \cos \omega t$, we have from (52.3) and (52.6), account being taken of the second equation in (52.1), that

$$U_{el} = \frac{I_0^2}{2\omega_0^2 C} \sin^2 \omega_0 t \qquad U_m = \frac{L I_0^2}{2} \cos^2 \omega_0 t$$

and thus, since $\omega_0^2 = 1/LC$, we actually have here that $U_{el} + U_m$ is constant.

We now consider the case in which the individual current circuits, or (non-ferromagnetic) material parts of them, are in some way capable of motion with respect to one another. Let the momentary location of the movable elements of the system be designated by certain parameters $a_1, a_2, \ldots, a_s$. If, for example, a part of a wire is capable of displacement parallel to the $x$-axis, then $a_1$ can simply be the $x$-coordinate of a certain point on this portion of wire. We now define the force $F_r$, belonging to the parameter $a_r$, by establishing that $F_r da_r$ shall be the work expended by the system for a change $da_r$ of $a_r$. If $a_r$ is a length, then $F_r$ is a force in the ordinary sense; if, however, $a_r$ is an angle, then $F_r$ becomes a torque. For one of the changes of length in the system, described by the functions of time $a_1(t), a_2(t), \ldots, a_s(t)$, work

$$W_{mech} = \sum_{r=1}^{s} F_r \frac{da_r}{dt} \qquad (52.8)$$

is expended.

For determining the $F_r$, we draw upon the energy theorem (52.4). We introduce the value (52.5) for the $\Phi_k$ and (52.6) for $U_m$, taking into consideration that the $L_{kl}$ are functions of the $a_r$ and thus also depend upon the time. We therefore find that

$$W_{mech} = \frac{1}{c} \sum_{k=1}^{n} I_k \frac{d\Phi_k}{dt} - \frac{dU_m}{dt} = \frac{1}{2} \sum_{k,l=1}^{n} \frac{dL_{kl}}{dt} I_k I_l \quad \text{with} \quad \frac{dL_{kl}}{dt} = \sum_{r=1}^{s} \frac{\partial L_{kl}}{\partial a_r} \frac{da_r}{dt}$$

As a comparison with (52.8) shows, we have also that

$$F_r = \frac{1}{2} \sum_{k,l=1}^{n} \frac{\partial L_{kl}}{\partial a_r} I_k I_l = \frac{\partial U_m}{\partial a_r} \qquad (52.9)$$

The algebraic sign in equation (52.9) merits special attention. When, in this connection, in ordinary mechanics the potential energy is given as a function of the position coordinates, the forces are found, as we know, as the partial derivative of the negative of the energy (or the negative of the potential) with respect to the corresponding coordinate. Hence, according to (52.9), the negative of the magnetic energy plays the role of the potential. While in mechanics the forces act in such directions that the potential energy is diminished by their action ("work done at the expense of potential energy"), our electrodynamic forces behave in the opposite way: they act in such directions that the energy of the field *increases*. A particularly clear example of this behaviour is shown if the currents are maintained at constant strength during the motion, say by suitable changes in the applied e.m.f.s (batteries switched on or off). In that case $dI_k/dt = 0$ for all $k$, and we obtain simply

$$W_{\text{mech}} = \sum_{r=1}^{s} \frac{\partial U_m}{\partial a_r} \frac{da_r}{dt} = \left( \frac{dU_m}{dt} \right)_{I=const}$$

In this case the energy of the field increases by exactly the amount of the work done. Thus the double energy gain of amount $2\,W_{\text{mech}}$ per second is balanced by the work performed by the applied e.m.f.s which maintain the constancy of the currents.

On account of the directional sense which we have established for mechanical forces, it follows from (52.9) that every current-carrying wire tends to move so as to link the greatest possible flux of magnetic induction. Thus, if the first circuit can be displaced parallel to, say, the $x$-axis while all other circuits remain at rest, and if we describe this displacement by the parameter $a_1$, then, from (52.9), we have for the $x$-component of the force acting on the first circuit,

$$F_1 = \tfrac{1}{2} \sum_{l=1}^{n} \frac{\partial L_{1l}}{\partial a_1} I_1 I_l + \tfrac{1}{2} \sum_{k=1}^{n} \frac{\partial L_{k1}}{\partial a_1} I_k I_1 = \frac{I_1}{c} \frac{\partial \Phi_1}{\partial a_1} \quad (52.10)$$

since here only those $L_{kl}$ depend upon $a_1$ for which either $k$ or $l$ is equal to 1.

The force on the first circuit "in the direction" of the coordinate $a_1$ is therefore, to within the factor $I_1/c$, equal to the space rate of increase (in the direction $a_1$) of the flux of induction $\Phi_1$, the current in the circuit being held constant.

In a similar manner we can calculate the *force on a current-carrying conductor* (mentioned already in §44). With the help, say, of brush contacts, let a portion $\delta r$ of a conductor be freely movable in the direction of the unit-vector **s**. Let its displacement in this direction be given by $a$**s**, $a$ being a variable parameter. In the motion of $\delta r$ through the distance $a$**s**, the surface $a$**s** $\times \delta$**r** is swept out. Thus the flux of induction increases by $a$(**s** $\times \delta$**r**) **. B**. Equation (52.10) therefore gives for the **s**-component of the force on the piece $\delta$**r**

$$\delta F = \frac{I}{c}(\mathbf{s} \times \delta \mathbf{r}) \cdot \mathbf{B} = \frac{I}{c}\mathbf{s} \cdot (\delta \mathbf{r} \times \mathbf{B})$$

and thus, for the force itself, in agreement with (44.1),

$$\delta \mathbf{F} = \frac{I}{c} \delta \mathbf{r} \times \mathbf{B}$$

In this derivation of the forces on current-carrying wires—a derivation employing the energy theorem—we have made essential use of the induction law for the case of movable (or deformable) current loops in a magnetic field. Conversely, on another occasion (in §44), we directly concluded this form of the induction law from the Lorentz force. Thus, in order to avoid reasoning in a circle, we must make it clear that, *as an independent and self-sufficient experimental law, we must choose either the force law* (44.1) *or the induction law for movable (or deformable) current loops as the more fundamental; the other law is then derivable from it.*

In conclusion, let us make a remark concerning calculations. Total-energy expressions in (52.2) are of second order in the quantities determining the electromagnetic field. Thus, for relations concerning alternating currents, and particularly in looking back on the application of complex vectors described in §50, special consideration is necessary.

The time-average of an alternating current $I = I_0 \cos \omega t$ is, of course, zero $(\bar{I} = 0)$, while the average value of the square of the current $(\overline{I^2})$ is obviously equal to $\frac{1}{2}I_0^2$. In this case, then, in the energy loss due to Joule heating, $R\overline{I^2} = \frac{1}{2}RI_0^2$. Thus, in engineering discussions, instead of the current amplitude $I_0$, the "effective" current* $I_{eff} = I_0/\sqrt{2}$ is often given; with it the Joule heating is equal to $RI_{eff}^2$.

If in a circuit such as in figure 47 we have, in addition to an applied voltage $V^{(e)} = V_0 \cos \omega t$, a current $I = I_0 \cos (\omega t - \phi)$ lagging behind the

---

* Also called the "r.m.s. current".

voltage by an amount equal to the phase angle $\phi$, the power supplied by the seat of e.m.f. becomes

$$\overline{V^{(e)}I} = V_0 I_0 \overline{\cos \omega t \cos (\omega t - \phi)} = \tfrac{1}{2} V_0 I_0 \cos \phi$$

the latter because

$$\cos (\omega t - \phi) = \cos \omega t \cos \phi + \sin \omega t \sin \phi$$

For the a.c. circuit of figure 47 having self-inductance and resistance, this power from the voltage source must, of course, be equal to the dissipation, per second, of Joule heat. Mathematically this follows by multiplying equation (50.3) by $I$ and averaging over a period. In this case the inductive term $L\overline{I dI/dt}$ vanishes for alternating current. In a.c. engineering parlance the quantity $\overline{V^{(e)}I} = \tfrac{1}{2} V_0 I_0 \cos \phi$ is called the *effective power*, while the quantity $\overline{V^{(e)} dI/d(\omega t)} = \tfrac{1}{2} V_0 I_0 \sin \phi$ is called the *virtual power* in agreement with the splitting (mentioned in §50) of the complex resistance $\mathbf{R} = R + j\omega L$ into an effective resistance $R$ and a reactance $\omega L$.

If the current is not given as a real function of the time, but rather, in the sense of a vector diagram, as a complex quantity, say $I = I_0 e^{j\omega t}$, and correspondingly for the voltage $V = V_0 e^{j\omega t}$, the foregoing considerations can immediately be applied to these complex representations. We have only to take into consideration that the real quantities for current and voltage are given by

$$\mathrm{Re}\,\{I\} = \tfrac{1}{2}(I + I^*) \qquad \mathrm{Re}\,\{V\} = \tfrac{1}{2}(V + V^*)$$

in which the stars means the transfer over to complex-conjugate quantities. We thus have

$$\overline{(\mathrm{Re}\,\{I\})^2} = \tfrac{1}{4}\overline{(I^2 + 2II^* + I^{*2})} = \tfrac{1}{2}II^* = \tfrac{1}{2}I_0 I_0^*$$

since, in their time average, $I^2$ and $I^{*2}$ vanish on account of their $t$-dependence. Analogously, we have

$$\overline{\mathrm{Re}\,\{V\}\,\mathrm{Re}\,\{I\}} = \tfrac{1}{4}\overline{(VI + VI^* + V^*I + V^*I^*)} = \tfrac{1}{4}(V_0 I_0^* + V_0^* I_0)$$

Thus in this method of calculating the power it is not necessary to determine $\cos \phi$ separately; rather, the phase displacement between current and voltage is already contained in the complex amplitudes $V_0$ and $I_0$.

# D

---

The
general
fundamental
equations
of
the
electromagnetic
field

# CHAPTER D I

## Maxwell Theory for Stationary Media

### §53. Completing the Maxwell equations

We wish now to assemble the Maxwell equations in their final form, for the case first of media at rest. As was often mentioned in Chapter C IV, the law of Oersted,

$$\operatorname{curl} \mathbf{H} = \frac{4\pi}{c} \mathbf{g} \tag{53.1}$$

for the magnetic field of a steady-current distribution, needs to be essentially and decisively extended in order to cover the case where the current filaments are not closed but terminate, for example, on the plates of a capacitor. At such places the divergence of $\mathbf{g}$ is not equal to zero, whereas the left side of (53.1) is always source-free ($\operatorname{div} \operatorname{curl} \equiv 0$). If, therefore, an equation of general validity is desired, either an entirely new relationship has to be sought, or else the right side of (53.1) has to be made source-free by the introduction of an additional vector. Maxwell chose the latter way, already intimated in §38, in which for the current density $\mathbf{g}$ in (53.1) he substituted the vector

$$\mathbf{c} = \mathbf{g} + \frac{1}{4\pi} \frac{\partial \mathbf{D}}{\partial t}$$

as in equation (38.8). This vector is always source-free since

$$\operatorname{div} \mathbf{c} = \operatorname{div} \mathbf{g} + \frac{1}{4\pi} \frac{\partial \operatorname{div} \mathbf{D}}{\partial t} = \operatorname{div} \mathbf{g} + \frac{\partial \rho}{\partial t} = 0$$

*Introducing the "displacement current"* $\dfrac{1}{4\pi} \dfrac{\partial \mathbf{D}}{\partial t}$ *into the fundamental electromagnetic equations forms the nucleus of Maxwell's contribution to electrical theory.* This is intrinsically the only place, but also the decisive place, at which, in its basic assertions, Maxwell's theory differs from the old theory of action-at-a-distance.

Thus, in place of equation (53.1), Maxwell substituted the relationship

$$\operatorname{curl} \mathbf{H} = \frac{4\pi}{c}\mathbf{g} + \frac{1}{c}\frac{\partial \mathbf{D}}{\partial t} \tag{53.2}$$

which we can write in the form

$$\operatorname{curl} \mathbf{B} = \frac{4\pi}{c}\left(\mathbf{g} + \frac{\partial \mathbf{P}}{\partial t} + c\operatorname{curl} \mathbf{M}\right) + \frac{1}{c}\frac{\partial \mathbf{E}}{\partial t} \tag{53.3}$$

which is a more meaningful form from the standpoint of electron theory. The curl of **B** is thus determined not only by the density of the true current **g** and of the magnetization current $\mathbf{g}_M = c\operatorname{curl}\mathbf{M}$, but also by the polarization current $\mathbf{g}_P = \partial \mathbf{P}/\partial t$; and in addition the term $\frac{1}{c}\frac{\partial \mathbf{E}}{\partial t}$ enters. While the emergence of the polarization current on the the right side of (53.3) is immediately understandable and can be readily proved experimentally, the term involving the change of **E** with time appears as something new and unexpected. At first it appears to be required only for purely mathematical reasons. Owing to its smallness it cannot be immediately verified, as, for example, in a discharging capacitor. However, it finds its brilliant and irrefutable confirmation in the propagation of electromagnetic waves (see chapter D II); for, as we shall see, this only becomes possible through the additional Maxwell term which, in contrast to the other terms on the right-hand side of (53.2) or (53.3), can be different from zero even *in vacuo*.

Along with the augmented equation (53.2) we now take the induction law (45.8) as well as the two statements concerning the sources of **D** and **B**, and thus we obtain the four fundamental equations, of remarkably symmetrical structure:

$$\text{(I)}\ \operatorname{curl} \mathbf{H} = \frac{1}{c}\frac{\partial \mathbf{D}}{\partial t} + \frac{4\pi\mathbf{g}}{c} \qquad \text{(II)}\ \operatorname{div} \mathbf{D} = 4\pi\rho$$

$$\text{(III)}\ \operatorname{curl} \mathbf{E} = -\frac{1}{c}\frac{\partial \mathbf{B}}{\partial t} \qquad \text{(IV)}\ \operatorname{div} \mathbf{B} = 0$$

*as the final form of the Maxwell equations for media at rest.* We notice that these four equations are not entirely independent of one another. Thus, by taking the divergence, it follows from III that the time derivative of div **B** must vanish; equation IV, along with III, is satisfied for all times if IV is satisfied in all space only at one particular moment Correspondingly, it follows from I by taking the divergence and making use of the equation of continuity of electric charge, that the time-derivative of II is satisfied. Conversely, in consideration of the discussion at the beginning of these paragraphs, it is obvious that by eliminating **D** from I and II we arrive at the equation of continuity.

The equations I to IV, however, become a complete system in which the temporal course of an electromagnetic phenomenon can be uniquely calculated only when three other equations are introduced. These three equations, called *constitutive equations*, relate the three vectors **D**, **B**, and **g** to the two field-strengths **E** and **H**. The simplest (and up to now exclusively considered) form of these equations, valid for isotropic, normally polarizable and magnetizable media, read

$$\text{(V)} \quad \mathbf{D} = \varepsilon\mathbf{E} \qquad \text{(VI)} \quad \mathbf{B} = \mu\mathbf{H} \qquad \text{(VII)} \quad \mathbf{g} = \sigma(\mathbf{E} + \mathbf{E}^{(e)})$$

These three equations, V to VII, are required in view of the special properties displayed by matter in the presence of a field. They are by no means rigorous and general, except *in vacuo*, in which case we have exactly that $\varepsilon = \mu = 1$, $\sigma = 0$. In particular, they fail completely for ferromagnetic and ferroelectric substances with their non-linear constitutive equations between **B** and **H** or between **D** and **E**. But the relationships V to VII are also fallible even for a linear dependence of the field quantities. This is the case when the fields change very rapidly with the time, as with light waves, and also when we attempt to describe after-effect phenomena in polarization and magnetization. In what way, in these and similar cases, the constitutive equations require modification we shall learn in §58. We shall also attempt there to obtain from electron theory laws governing the frequency dependence of the permittivity and the electrical conductivity.

*Remark:* In the MKSA system the Maxwell equations possess the simpler form:

$$\text{(I)} \quad \operatorname{curl}\mathbf{H} = \frac{\partial\mathbf{D}}{\partial t} + \mathbf{g} \qquad \text{(II)} \quad \operatorname{div}\mathbf{D} = \rho$$

$$\text{(III)} \quad \operatorname{curl}\mathbf{E} = -\frac{\partial\mathbf{B}}{\partial t} \qquad \text{(IV)} \quad \operatorname{div}\mathbf{B} = 0$$

Along with these we have for normal isotropic media the following constitutive equations:

$$\text{(V)} \quad \mathbf{D} = \varepsilon\varepsilon_0\mathbf{E}, \qquad \text{(VI)} \quad \mathbf{B} = \mu\mu_0\mathbf{H}, \qquad \text{(VII)} \quad \mathbf{g} = \sigma(\mathbf{E} + \mathbf{E}^{(e)})$$

## §54. The energy theorem in Maxwell's theory. The Poynting vector

From the system of Maxwell equations I to IV it is possible to derive a very important expression which we shall recognize as the *energy principle in the electromagnetic field*. If we multiply equation I by **E**, and equation III by $-\mathbf{H}$ and add the equations obtained, we have first

$$-\mathbf{H} \cdot \operatorname{curl} \mathbf{E} + \mathbf{E} \cdot \operatorname{curl} \mathbf{H} = \frac{4\pi}{c} \mathbf{g} \cdot \mathbf{E} + \frac{1}{c} \left\{ \mathbf{E} \cdot \frac{\partial \mathbf{D}}{\partial t} + \mathbf{H} \cdot \frac{\partial \mathbf{B}}{\partial t} \right\}$$

From the identity

$$\mathbf{H} \cdot \operatorname{curl} \mathbf{E} - \mathbf{E} \cdot \operatorname{curl} \mathbf{H} = \operatorname{div}(\mathbf{E} \times \mathbf{H})$$

after multiplying by $c/4\pi$, we obtain the equation

$$-\frac{1}{4\pi} \left\{ \mathbf{E} \cdot \frac{\partial \mathbf{D}}{\partial t} + \mathbf{H} \cdot \frac{\partial \mathbf{B}}{\partial t} \right\} = \mathbf{g} \cdot \mathbf{E} + \frac{c}{4\pi} \operatorname{div}(\mathbf{E} \times \mathbf{H}) \qquad (54.1)$$

Since this equation derives solely from the rigorously valid field equations I and III, we must look upon it as an exact expression for bodies at rest. For brevity we write it in the form

$$-\frac{\partial}{\partial t} \{u_{el} + u_m\} = \psi + \operatorname{div} \mathbf{S} \qquad (54.2)$$

in which we introduce* the well-known energy density of the electric field (from (32.6)),

$$u_{el} = \frac{1}{4\pi} \int \mathbf{E} \cdot d\mathbf{D} \rightarrow u_{el} = \frac{1}{8\pi} \mathbf{E} \cdot \mathbf{D} \quad \text{for} \quad \mathbf{D} = \varepsilon \mathbf{E} \qquad (54.3)$$

and the correspondingly constructed energy density of the magnetic field (see §52)

$$u_m = \frac{1}{4\pi} \int \mathbf{H} \cdot d\mathbf{B} \rightarrow u_m = \frac{1}{8\pi} \mathbf{H} \cdot \mathbf{B} \quad \text{for} \quad \mathbf{B} = \mu \mathbf{H} \qquad (54.4)$$

Thus (54.2) affords an insight as to the whereabouts of the energy involved in any decrease with time of the total energy density of the electromagnetic field. The first term on the right-hand side

$$\psi = \mathbf{g} \cdot \mathbf{E} \qquad (54.5)$$

---

* Accurately regarded, as we have seen in §33, $\partial u_{el}/\partial t$ concerns a change in the density, not of the ordinary internal energy, but of the free energy, since isothermal activity has been assumed. For the purpose of this chapter, however, this distinction can be ignored.

obviously represents the work from the field expended upon the electric current density. For the case where VII is valid, and on account of the fact that

$$\mathbf{g} \cdot \mathbf{E} = \frac{\mathbf{g}^2}{\sigma} - \mathbf{g} \cdot \mathbf{E}^{(e)} \qquad (54.6)$$

the quantity of energy injected into the current stream in this manner is recovered (a) in the Joule heat produced per cubic centimetre and per second, and (b) in the work expended against the applied forces. We call the two quantities taken together the *thermochemical power* of the field. Entering in addition, as a further cause of the energy decrease of the field, is the divergence of the vector

$$\mathbf{S} = \frac{c}{4\pi} \mathbf{E} \times \mathbf{H} \qquad (54.7)$$

Thus the law of conservation of energy requires that, in the cubic centimetre considered, there shall be liberated per second a certain amount of energy which proceeds outward through the surface $A$ of this volume. We see this particularly clearly if we integrate (54.2) over an arbitrary volume and then apply Gauss's theorem:

$$-\frac{d}{dt} \int \{u_{el} + u_m\} \, dv = \int \psi \, dv + \oint S_n \, dA \qquad (54.8)$$

Accordingly, $S_n \, dA$ gives the energy which passes per second through the surface element $dA$ in the direction of its normal $\mathbf{n}$. We call $\mathbf{S}$ the *Poynting radiation vector*. We shall have to deal more fully with this vector in the theory of electromagnetic waves. The Poynting vector is thus the immediate expression of the possibility of an energy transfer in the form of radiation (also through empty space), and therefore of the possibility of an electromagnetic interpretation of optical phenomena. On account of its importance we shall now endeavour to become acquainted with its characteristics by means of an elementary example. (See also §61.)

We consider therefore a portion of a straight cylindrical wire (figure 52) of radius $a$, carrying a steady current of strength $I$. A quantity of Joule heat $EI$ is produced per second and per unit length in this wire, where $E$ is the voltage drop per unit length along the wire. We ask now how the electrical energy converted into heat actually arrived at its place of conversion. That it was transported there by means of the conduction electrons appears from the outset to be extremely unlikely—especially since the electrons, despite their large number, move through the wire (even for the strongest current densities) with only a relatively small average velocity (a few centimetres per second).

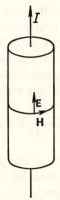

Fig. 52.—Orientation of the E and H vectors on the surface
of a current-carrying conductor

We obtain the answer to our question from (54.8) by considering the Poynting vector **S**. Since **E** lies parallel to the wire axis and **H** circles round the wire, **S** is directed against the surface of the wire. Its value at the surface of the wire, using (46.3), is therefore

$$S = \frac{c}{4\pi} EH = \frac{c}{4\pi} E \frac{2I}{ac} = \frac{EI}{2\pi a}$$

The total energy flow into a 1-cm length of wire, namely $2\pi a S$, is thus quantitatively equal to the heating $EI$ developed in this length of wire.

Thus, at the place in the current system at which electrical energy is required, this energy flows out of the field and into the wire, and the density of this energy flow is given by the Poynting vector **S**. It can in like manner be shown that at the place where an applied force is active, there is a corresponding flow of energy out of the current system and into the field.

It should also be mentioned that the above definition of the Poynting vector is not a mandatory one. Since this vector has been introduced only by way of its divergence, the curl of an arbitrary vector can be added to it without altering the physical facts of the case. On various grounds, however, the definition of **S** given in (54.7) is especially convenient, particularly for the electromagnetic theory of light.

*Remark:* In the MKSA system the relationship (54.1) reads

$$-\left\{\mathbf{E}\,\frac{\partial \mathbf{D}}{\partial t} + \mathbf{H}\,\frac{\partial \mathbf{B}}{\partial t}\right\} = \mathbf{g}\,.\,\mathbf{E} + \text{div}\,(\mathbf{E} \times \mathbf{H})$$

Thus for the Poynting vector, instead of (54.7), we have the expression

$$\mathbf{S} = \mathbf{E} \times \mathbf{H}$$

This leads to the already familiar quantities for the energy density, namely

$$u_{el} = \int \mathbf{E}\,.\,d\mathbf{D} \to u_{el} = \tfrac{1}{2}\mathbf{E}\,.\,\mathbf{D} \quad \text{for} \quad \mathbf{D} = \varepsilon\varepsilon_0 \mathbf{E}$$

$$u_m = \int \mathbf{H}\,.\,d\mathbf{B} \to u_m = \tfrac{1}{2}\mathbf{H}\,.\,\mathbf{B} \quad \text{for} \quad \mathbf{B} = \mu\mu_0 \mathbf{H}$$

## §55. Magnetic field energy. Forces in the magnetic field

We wish here to investigate in a more detailed way the magnetic field energy and the forces derivable from it. According to (54.4), this energy is given by

$$U_m = \int u_m\,dv \quad \text{with} \quad u_m = \frac{1}{4\pi}\int \mathbf{H}\,.\,d\mathbf{B} \qquad (55.1)$$

We have derived this equation in §54 in a general way from the Maxwell equations. We now go back to this derivation for a specially simple physical arrangement, as follows:

A bar of length $l$ and cross-sectional area $q$ has been bent into a circular ring. Let a conducting wire of resistance $R$ be wound uniformly at $n$ turns per centimetre on this ring. With a battery supplying an applied voltage $V^{(e)}$, a current $I$ flows in the wire. The battery then, in the time $dt$, performs work $dW = IV^{(e)}\,dt$. Now, first, from the induction law we have that

$$IR = V^{(e)} - \frac{qln}{c}\frac{dB}{dt}$$

and, second, the magnetic field within the ring coil is determined by $I$ alone:

$$H = \frac{4\pi}{c}nI$$

Thus the work performed is equal to

$$dW = I^2R\,dt + \frac{qln}{c}I\,dB = I^2R\,dt + \frac{ql}{4\pi}H\,dB$$

Since $ql$ is equal to the volume of the bar, we find, along with the Joule heat, as equivalent to the power from the battery, an energy contribution per unit volume of $H\,dB/4\pi$ is supplied to the field in agreement with (55.1).

Further, we have to show that the energy value (52.6) derived in §52 for a system of linear currents is in agreement with (55.1). The latter, in the absence of ferromagnets (since $\mathbf{B} = \mu\mathbf{H}$) leads to

$$U_m = \frac{1}{8\pi}\int \mathbf{H}\cdot\mathbf{B}\,dv \qquad (55.2)$$

Since $\mathbf{B}$ has no sources, by introducing the vector potential $\mathbf{A}$ the induction $\mathbf{B}$ can be replaced by $\mathrm{curl}\,\mathbf{A}$, giving

$$\mathbf{H}\cdot\mathbf{B} = \mathbf{H}\cdot\mathrm{curl}\,\mathbf{A} = \mathbf{A}\cdot\mathrm{curl}\,\mathbf{H} + \mathrm{div}\,(\mathbf{A}\times\mathbf{H})$$

In integrating over all space the surface integral originating in the divergence vanishes so that, for the case of quasi-stationary currents (i.e. with $\mathrm{curl}\,\mathbf{H} = 4\pi\mathbf{g}/c$) we obtain

$$U_m = \frac{1}{2c}\int \mathbf{g}\cdot\mathbf{A}\,dv \qquad (55.3)$$

Now, for a linear current of strength $I$ and cross-section $q$, if $d\mathbf{r}$ is a line element of the conductor, $\mathbf{g}\,dv = I\,d\mathbf{r}$. Since $I$ has the same value at all places on a current-carrying wire, we have for $n$ circuits

$$U_m = \frac{1}{2c}\sum_{k=1}^{n} I_k \oint_k \mathbf{A}\cdot d\mathbf{r} = \frac{1}{2c}\sum_{k=1}^{n} I_k \int_k B_n\,dS = \frac{1}{2c}\sum_{k=1}^{n} I_k \Phi_k$$

Taking account of (52.5) we have thus found equation (52.6) for the magnetic field energy of a system of quasi-stationary currents.

From the expression for the electrical field energy we have calculated in §§34 and 35 the forces to be expected to act on material media, by equating the decrease $-\delta U_{el}$ of this energy for an arbitrary displacement $\mathbf{s}$ of the medium, to the work $\int \mathbf{f}\cdot\mathbf{s}\,dv$ performed by the field in this displacement. We were able there to determine a unique value for the force density $\mathbf{f}$ when $\mathbf{s}$ was actually taken as the most general possible displacement. By restricting ourselves to incompressible media, or even to rigid media, we would have been able to evaluate **f** only to within the gradient of a completely arbitrary function, or to calculate the total force $\mathbf{F} = \int \mathbf{f} dv$ on the body displaced as a whole.

In an analogous way we could attempt here to calculate the force density obtaining at every point of matter in a magnetic field from the magnetic field energy (55.1). For non-ferromagnetic and non-conducting substances the considerations of §35 can actually be taken over and lead here to the force density

$$\mathbf{f} = -\frac{1}{8\pi} \mathbf{H}^2 \operatorname{grad} \mu + \frac{1}{8\pi} \operatorname{grad}\left(\mathbf{H}^2 \frac{d\mu}{d\sigma} \sigma\right) \qquad (55.4)$$

where we regard the permeability as a single-valued function of the density $\sigma$. The term $\rho \mathbf{E}$ coming from the electric field finds no counterpart here owing to the non-existence of true magnetic charges.*

In carrying out corresponding arguments for ferromagnetic bodies or electrical conductors, certain characteristic difficulties appear. In the first case these difficulties originate in that in a displacement, which we might designate by **s**, in a ferromagnet, the magnetization curve will in general vary in a non-calculable way from place to place and from one element of the material to another. Thus, for example, the phenomenon of magnetostriction can be quantitatively calculated only in very special cases (such as in an ellipsoidal body). We can expect only general statements for the total force **F** on ferromagnets, in so far as we displace the body as a whole and inquire concerning the concomitant change in the field energy.

For the case of conducting media a characteristic difficulty presents itself through the application of the law of induction. In the displacement of a metal in a magnetic field a current is always induced. According to the arguments of §45, however, this current has its origin not, say, in the appearance of an induced field, but in the action of the Lorentz force. This force, being foreign to the Maxwell equations, is not found in them. Now, we might propose to derive it from the Maxwell equations by means of energy considerations of the foregoing nature. In §52 we were in fact able to carry out such a derivation for linear current systems. But this was only possible because we made use of the induction law in the technical form which is the only form

---

* With equation (55.4), the considerations of §37 can now be adapted to the case of force effects on diamagnetic or paramagnetic non-conducting liquids and gases.

permitting the change of the flux of induction to be taken into account through the motion of the current path. With the Maxwell equations for media at rest, assembled in §53, we would in principle not have been able to handle this problem. For the same reason we are not now able with these equations to calculate the force on spatially extended conductors in a magnetic field by displacing the conductor and then considering the change of field energy accompanying this displacement. This would only be possible through a knowledge of the Maxwell equations for moving media (see §70).

## §56. The momentum theorem in Maxwell's theory. The momentum density of the radiation field

In spite of the difficulties described in §55 we can calculate the force on conducting and magnetizable bodies in a given field if, instead of the energy theorem whose calculation of force requires a displacement, we make use of the momentum theorem.

For an easier understanding of what follows, we first consider the special case where in all space, besides the electromagnetic field, there exists only a moving charge. The total force on this charge is given from the force density

$$\mathbf{f} = \rho\mathbf{E} + \frac{\mathbf{g}}{c} \times \mathbf{B} \tag{56.1}$$

by volume integration. For carrying out this integration it is convenient to transform equation (56.1).

Upon eliminating $\rho$ and $\mathbf{g}$ by means of the Maxwell equations, we find

$$\mathbf{f} = \frac{1}{4\pi}\left\{\mathbf{E}\,\mathrm{div}\,\mathbf{D} + \mathrm{curl}\,\mathbf{H} \times \mathbf{B} - \frac{1}{c}\frac{\partial\mathbf{D}}{\partial t} \times \mathbf{B}\right\}$$

Adding, for reasons of symmetry, the relationships

$$0 = \frac{1}{4\pi}\left\{-\mathbf{D} \times \mathrm{curl}\,\mathbf{E} - \frac{1}{c}\mathbf{D} \times \frac{\partial\mathbf{B}}{\partial t}\right\} \quad \text{and} \quad 0 = \frac{1}{4\pi}\mathbf{H}\,\mathrm{div}\,\mathbf{B}$$

we arrive finally at the equation

$$\mathbf{f} = \frac{1}{4\pi}\{\mathbf{E}\,\mathrm{div}\,\mathbf{D} - \mathbf{D} \times \mathrm{curl}\,\mathbf{E} + \mathbf{H}\,\mathrm{div}\,\mathbf{B} - \mathbf{B} \times \mathrm{curl}\,\mathbf{H}\} - \frac{1}{4\pi c}\frac{\partial}{\partial t}(\mathbf{D} \times \mathbf{B})$$

$$\tag{56.2}$$

Since for our special case $\mathbf{D} = \mathbf{E}$ and $\mathbf{B} = \mathbf{H}$, we can write, say, the $x$-component in the form

$$f_x = \frac{\partial T_{xx}}{\partial x} + \frac{\partial T_{yx}}{\partial y} + \frac{\partial T_{zx}}{\partial z} - \frac{1}{4\pi c}\frac{\partial}{\partial t}(\mathbf{E} \times \mathbf{H})_x \qquad (56.3)$$

in which for brevity we have placed

$$\left.\begin{array}{l} T_{xx} = \dfrac{1}{4\pi}\{E_x^2 + H_x^2 - \tfrac{1}{2}(\mathbf{E}^2 + \mathbf{H}^2)\}, \dots, \dots, \\[2ex] T_{xy} = \dfrac{1}{4\pi}\{E_x E_y + H_x H_y\} \qquad , \dots, \dots \end{array}\right\} \qquad (56.4)$$

If now we find the total force on the charge by integrating over a bounded region, we have then that

$$F_x = \int f_x\, dv = \oint T_{nx}\, dS - \frac{d}{dt}\left(\frac{1}{4\pi c}\int (\mathbf{E} \times \mathbf{H})_x\, dv\right) \qquad (56.5)$$

As is shown by comparison with the considerations in §36, we can look upon this equation, excepting the last term, as if the force on the charge were calculated from *Maxwell stresses* which act upon a surface enclosing the charge. In fact, for the case of a vacuum, these stresses agree exactly with their electrical counterparts, the stresses of equation (36.7).

The appearance of the term with the time derivative is, however, something essentially new. We can understand this as soon as we remind ourselves that the calculated force leads to a change with time of the mechanical momentum $\mathbf{J}_M$:

$$\mathbf{F} = \frac{d}{dt}\mathbf{J}_M \qquad (56.6)$$

Writing, now, the term in question on the left side of equation (56.5), we then obtain, for the case of a vacuum, the *momentum theorem of Maxwell's theory* in the form

$$\frac{d}{dt}(\mathbf{J}_M + \mathbf{J}_S) = \oint \mathbf{T}_n\, dS \qquad (56.7)$$

Thus to the electromagnetic field, as to the particle, we can ascribe a momentum

$$\mathbf{J}_S = \int \mathbf{j}_S\, dv \quad \text{with} \quad \mathbf{j}_S = \frac{\mathbf{E} \times \mathbf{H}}{4\pi c} = \frac{\mathbf{S}}{c^2} \qquad (56.8)$$

such that *the time rate of change of the sum of the particle momentum and the field momentum is equal to the total force which would be exerted by the Maxwell stresses on the region considered.*

The appearance of a field momentum becomes understandable when we consider that it is possible for electromagnetic radiation to convey not only energy but also momentum from a source of radiation to an absorber, and that this conveyed momentum is measurable as "radiation pressure" on the absorber.

In the calculation of this radiation pressure for the stationary case (as a time-average) the second member on the right in (56.5) drops out and, for a surface-element with normal in the positive $x$-direction, we have that

$$p_{rad} = -T_{xx} = -\frac{1}{8\pi}\{E^2 + H^2 - 2(E_x^2 + H_x^2)\} \qquad (56.9)$$

For a light-wave at normal incidence we have that $E_x = H_x = 0$, as will be shown in the following section. We then have here that

$$p_{rad} = u, \quad \text{with} \quad u = u_{el} + u_m \qquad (56.9a)$$

For isotropically incident radiation (uniform from all directions, as for example in Hohlraum radiation), we have for the average that $\overline{E_x^2} = \overline{E^2}/3$, $\overline{H_x^2} = \overline{H^2}/3$. In this case then

$$p_{rad} = u/3 \qquad (56.9b)$$

We now go on to the general case of an arbitrarily shaped body in a vacuum. We assume with Maxwell that the momentum theorem in the form (56.7) holds also in this case, i.e. when the surface integral extends over the surface of the body and the stress components (56.4) are those valid for a vacuum. There is now the problem of representing the total force **F** on the body, and the field momentum in the body. Going back to formula (56.2), we put in the f-value (56.1). We now integrate its $x$-component over the volume of the body, and we imagine the discontinuous transition on the surface replaced by a continuous one. Thus we find that

$$\int\left\{\rho E + \frac{g}{c}\times B\right\}_x dv - \frac{1}{8\pi}\int\left\{E\cdot\frac{\partial D}{\partial x} - D\cdot\frac{\partial E}{\partial x} + H\cdot\frac{\partial B}{\partial x} - B\cdot\frac{\partial H}{\partial x}\right\}dv +$$

$$+\frac{1}{4\pi c}\frac{d}{dt}\int(D\times B)_x\,dv = \oint T_{nx}\,dS \qquad (56.10)$$

(For this, from the expression on the right side of (56.2), let the part consisting of the second integral be split off; the remainder can be transformed into a surface integral and gives then exactly $T_{nx}$, since there $D = E$ and $B = H$.)

If now the above-mentioned question concerning the splitting off of the left side of (56.10), in the sense of (56.7), cannot be uniquely answered, then the form (56.10) suggests defining the quantity

$$\mathbf{j}_S = \frac{\mathbf{D} \times \mathbf{B}}{4\pi c} \qquad (56.11)$$

and correspondingly, regarding the first line of (56.10) as the total force on the body. This interpretation is supported by the fact that for normally polarizable and magnetizable bodies, i.e. for $\mathbf{D} = \varepsilon\mathbf{E}$, $\mathbf{B} = \mu\mathbf{H}$, $\mathbf{F}$ assumes the expected form

$$\mathbf{F} = \int \left\{ \rho\mathbf{E} + \frac{\mathbf{g}}{c} \times \mathbf{B} - \frac{1}{8\pi}(\mathbf{E}^2 \operatorname{grad} \varepsilon + \mathbf{H}^2 \operatorname{grad} \mu) \right\} dv \quad (56.12)$$

This now justifies our assuming that also in the general case, i.e. for a medium with arbitrary field dependence of the polarization $\mathbf{P}$ and magnetization $\mathbf{M}$, the force is given by the first line in equation (56.10).

The "acceleration theorem" in Maxwell's theory 227

of the left side of (56.10), in the sense of (56.7), cannot be uniquely answered, then the form (56.10) suggests defining the quantity

$$D \times H$$

and correspondingly regarding the first line of (56.10) as the total force on the body. This interpretation is not very direct; for normally polarizable and magnetizable bodies, i.e. for $D = \varepsilon E$, $B = \mu H$, $P$ assumes the expected form

This now justifies our ass...

# CHAPTER D II

## Electromagnetic Waves

### §57. Electromagnetic waves in a vacuum

We wish in this chapter to consider electromagnetic fields which change rapidly both in space and time—in particular, electromagnetic waves. For this purpose we therefore start out with the differential equations of the electromagnetic field (§53).

We first restrict ourselves to the case of wave propagation *in vacuo*. In this case $D = E$, $B = H$; moreover $g = 0$, $\rho = 0$, so that we work with the Maxwell equations in the form

$$\left.\begin{array}{ll} \operatorname{curl} H = \dfrac{1}{c}\dfrac{\partial E}{\partial t} & \operatorname{div} E = 0 \\[2mm] \operatorname{curl} E = -\dfrac{1}{c}\dfrac{\partial H}{\partial t} & \operatorname{div} H = 0 \end{array}\right\} \tag{57.1}$$

From this system of equations we can easily eliminate one of the two vectors E or H. If, for example, we take the curl of the first equation, and then make use of the third equation, we arrive at the relationship

$$\operatorname{curl}\operatorname{curl} H = \frac{1}{c}\frac{\partial \operatorname{curl} E}{\partial t} = -\frac{1}{c^2}\frac{\partial^2 H}{\partial t^2}$$

From the calculation rule (11.17) it follows, since $\operatorname{div} H = 0$, that

$$\nabla^2 H = \frac{1}{c^2}\frac{\partial^2 H}{\partial t^2} \tag{57.2}$$

If, in like manner, we eliminate H by forming the curl of both sides of the induction law we find correspondingly

$$\nabla^2 E = \frac{1}{c^2}\frac{\partial^2 E}{\partial t^2} \tag{57.3}$$

Both vectors E and H thus satisfy the same differential equation. It is called the *wave equation*.

We now seek *particular solutions* of the field equations (57.1) or the wave equations (57.2) and (57.3), and we ask specially for solutions which correspond to *plane homogeneous wave trains*. We describe a wave train as plane when it is possible to place a family of parallel planes so that along each one of these planes the magnetic field strength does not change; each plane is called a wave plane, and their normal direction is called the wave normal. We wish to place the $x$-axis in the wave normal so that the wave planes lie parallel to the $yz$-plane. Since **E** and **H** have to be constant along the wave planes, all partial derivatives with respect to $y$ and $z$ vanish. The $x$-components of the two curl equations and the two divergence equations then read

$$\frac{\partial E_x}{\partial t} = 0 \qquad \frac{\partial H_x}{\partial t} = 0 \qquad \frac{\partial E_x}{\partial x} = 0 \qquad \frac{\partial H_x}{\partial x} = 0$$

Thus the longitudinal components $E_x$ and $H_x$ are constant in both space and time. If they should be different from zero, it could only be a case of a static field superimposed upon the wave process. Since such a field has no influence upon the wave propagation, and hence is of no interest to us, we say that

$$E_x = 0 \qquad H_x = 0 \tag{57.4}$$

The remaining components of the curl equations then read

$$\left.\begin{array}{ll} -\dfrac{\partial H_z}{\partial x} = \dfrac{1}{c}\dfrac{\partial E_y}{\partial t} & \dfrac{\partial H_y}{\partial x} = \dfrac{1}{c}\dfrac{\partial E_z}{\partial t} \\[2ex] \dfrac{\partial E_y}{\partial x} = -\dfrac{1}{c}\dfrac{\partial H_z}{\partial t} & -\dfrac{\partial E_z}{\partial x} = -\dfrac{1}{c}\dfrac{\partial H_y}{\partial t} \end{array}\right\} \tag{57.5}$$

Connected through these four equations are, on the one hand, the components $E_y$ and $H_z$, and on the other $E_z$ and $H_y$. It thus suffices for us to consider the first pair of equations by themselves; the corresponding solutions for the other pair result from a 90° rotation of the coordinate system around the $x$-axis. By eliminating $H_z$ and $E_y$ (respectively), we are led from the first pair of equations to the relations

$$\frac{\partial^2 E_y}{\partial x^2} = \frac{1}{c^2}\frac{\partial^2 E_y}{\partial t^2} \qquad \frac{\partial^2 H_z}{\partial x^2} = \frac{1}{c^2}\frac{\partial^2 H_z}{\partial t^2}$$

These also follow directly from (57.3) and (57.2), respectively, if there, corresponding to the assumption of homogeneous plane waves in the

$x$-direction, the derivatives with respect to $y$ and $z$ are deleted. These partial differential equations are known from the theory of the vibrating string. Using (57.5) we write their general solution in the form

$$E_y = f(x-ct)+g(x+ct) \qquad H_z = f(x-ct)-g(x+ct) \quad (57.6)$$

in which $f$ and $g$ are arbitrary functions of their arguments $x-ct$ and $x+ct$. Obviously, they represent waves which travel in the direction of the positive and negative $x$-axis respectively.

We limit ourselves in the sequel to those partial waves which are given by the function $f(x-ct)$. The form of the function is given by the wave curve at time $t=0$. This wave is displaced with velocity $c$ in the direction of the positive $x$-axis undistorted and independently of the wave form. When using Gaussian units we had introduced the quantity $c$ as a constant of proportionality both in the induction law and in Oersted's law concerning the magnetic field of currents, and we found its numerical value to be equal to the velocity of light. We therefore see that in empty space the velocity of plane electromagnetic waves is equal to the velocity of light $c = 3 \times 10^{10}$ cm/s.

Equality of velocity of propagation is not the only feature shared in common by light waves and electromagnetic waves. Electromagnetic waves, like light waves, oscillate *transversely*. We find, in fact, that neither **E** nor **H** possesses a periodically changing longitudinal component. Both vectors are perpendicular to the wave normals. In empty space, therefore, electromagnetic waves and light waves display a completely corresponding relationship.

These consequences of his field equations were what led Maxwell to the formulation of the electromagnetic theory of light. The electromagnetic theory of light looks upon light rays and heat rays as electromagnetic waves. It is superior to the old mechanical theory of light in that it permits calculation of the velocity of propagation from purely electrical measurements and in that from the outset it allows only *transverse* plane light waves. It was only with difficulty that the old theory, which looked upon light as a wave motion in an elastic medium, could account for the absence of longitudinal light waves. The electromagnetic theory of light excludes "longitudinal light" at the outset.

If light is actually an electromagnetic process, then all optical properties of matter must be fully determined by its electrical constants. It follows, in fact, from the Maxwell theory that the index of refraction, in essence, goes together with the permittivity, and the absorptivity with the electrical conductivity. In the following paragraphs we shall be thoroughly occupied with this fact, which explains the basic distinction between the optical properties of insulators and metals and is to be regarded as *one of the strongest supports for the electromagnetic theory of light*.

As a preparation for this we wish once again to consider the case of plane waves *in vacuo*. Since, in the presence of matter, it will always turn out to be convenient to calculate with wave processes that are purely harmonic in time, or with superpositions of such waves, we wish here in the case of a vacuum to concern ourselves with such waves. According to Fourier, every wave train of the form $f(x-ct)$ can actually be decomposed into an integral over such waves:

$$E_y = f(x-ct) = \int_{-\infty}^{+\infty} A(\omega)\, e^{i\omega(t-x/c)}\, d\omega \qquad (57.7)$$

In this the coefficient $A(\omega)$ is obtained from $f(x-ct)$ by a Fourier transformation:

$$A(\omega) = \frac{1}{2\pi} \int_{-\infty}^{+\infty} f(x-ct)\, e^{-i\omega(t-x/c)}\, dt$$

$$= \frac{1}{2\pi} \int_{-\infty}^{+\infty} f(-c\tau)\, e^{-i\omega\tau}\, d\tau \qquad (57.8)$$

In order for $E_y$ to be real, we must obviously have that

$$A(\omega) = A^*(-\omega) \qquad (57.9)$$

Then, since $E_y$ is real, in going over to the complex conjugate in the integrand in (57.7) we have the relation

$$\int_{-\infty}^{+\infty} A(\omega)\, e^{i\omega(t-x/c)}\, d\omega = \int_{-\infty}^{+\infty} A^*(\omega)\, e^{-i\omega(t-x/c)}\, d\omega$$

which immediately leads to (57.9) when in the second integral $-\omega$ is everywhere substituted for $\omega$.

We now consider one particular Fourier component from the development (57.7). At once, however, using the complex formulation of chapter C IV, we put it into the more general form of a wave travelling in an arbitrary direction

$$\mathbf{E} = \mathbf{E}_0 \exp i(\omega t - \mathbf{k}.\mathbf{r}) \qquad \mathbf{H} = \mathbf{H}_0 \exp i(\omega t - \mathbf{k}.\mathbf{r}) \quad (57.10)$$

Actually **E** and **H** are constant on all planes perpendicular to the vector **k**. As a comparison with (57.7) shows, we must have that

$$k = \frac{\omega}{c} = \frac{2\pi v}{c} = \frac{2\pi}{\lambda} \qquad (57.11)$$

where $v$ is the frequency and $\lambda$ is the wave length of the undulatory process.

In order to increase our confidence in the complex formulation (57.10) of the general plane electromagnetic wave, we wish to derive from the Maxwell equations (57.1) the relations holding between **E** and **H** on the one hand, and between **k** and $\omega$ on the other. For this purpose we establish the fact that differentiation of the functions (57.10) with respect to time leads to these functions being multiplied by the factor $i\omega$. Correspondingly, differentiation with respect to $x$ means multiplication by $-ik_x$. With this rule

$$\frac{\partial}{\partial t} \to i\omega \qquad \nabla \to -i\mathbf{k} \qquad (57.12)$$

by putting the functions (57.10) into the Maxwell equations (57.1) and by cancelling $i$'s, we obtain the algebraic relations

$$\left.\begin{array}{ll} -\mathbf{k} \times \mathbf{H} = \dfrac{\omega}{c}\mathbf{E} & \mathbf{k}.\mathbf{E} = 0 \\[2mm] -\mathbf{k} \times \mathbf{E} = -\dfrac{\omega}{c}\mathbf{H} & \mathbf{k}.\mathbf{H} = 0 \end{array}\right\} \qquad (57.13)$$

From these it first of all follows that both **E** and **H** are perpendicular to **k**, and the waves are therefore transverse. But **E** and **H** are also always perpendicular to one another, so that, in fact, *the directions of* **k**, **E**, *and* **H** (*in that order*) *can be chosen as coordinate directions of a right-hand Cartesian system*. Corresponding to this, we specially chose the $x$-direction above to be in the direction of **k**, **E** to be parallel to **y**, and **H** parallel to **z**. Finally, from (57.13), in which we pass over from the vector equations to the equations for the magnitudes, we find not only that $E = H$, but we also again find the relationship (57.11).

Returning once again to the real general representation (57.6), with the direction of propagation parallel to the positive $x$-axis ($f$-wave), we inquire concerning the energy density $u$ in the wave, and also concerning the Poynting vector **S** valid for it. For the first we have

$$u = \frac{1}{8\pi}(\mathbf{E}^2 + \mathbf{H}^2) = \frac{1}{4\pi}f^2$$

for the latter, since only the $x$-component of the Poynting vector has a finite value,

$$S = S_x = \frac{c}{4\pi}E_y H_z = \frac{c}{4\pi}f^2$$

Thus for every electromagnetic wave travelling *in vacuo*

$$S = cu \qquad (57.14)$$

i.e. in one second just that quantity of energy passes through $1\,\text{cm}^2$ of the $yz$-plane which is found in a cylinder of cross-sectional area 1 and of length $c \times 1$. Correspondingly, according to (56.8), the momentum $j_s.c = u = S/c$ goes through this square centimetre.

*Remark:* In the MKSA system the vacuum equations written with **E** and **H** have the form

$$\left. \begin{array}{ll} \text{curl } \mathbf{H} = \varepsilon_0 \dfrac{\partial \mathbf{E}}{\partial t} & \text{div } \mathbf{E} = 0 \\[2mm] \text{curl } \mathbf{E} = -\mu_0 \dfrac{\partial \mathbf{H}}{\partial t} & \text{div } \mathbf{H} = 0 \end{array} \right\}$$

Thus, as can be easily shown, there enters here in place, say, of the special $f$-solution in (57.6), the solution

$$E_y = \frac{1}{\sqrt{\varepsilon_0}}f\left(x - \frac{t}{\sqrt{(\varepsilon_0\mu_0)}}\right) \qquad H_z = \frac{1}{\sqrt{\mu_0}}f\left(x - \frac{t}{\sqrt{(\varepsilon_0\mu_0)}}\right)$$

which corresponds to a wave travelling with velocity

$$\frac{1}{\sqrt{(\varepsilon_0\mu_0)}} = \sqrt{(4\pi \times 9 \times 10^9)}\sqrt{\frac{10^7}{4\pi}}\sqrt{\left(\frac{\text{Vm}}{\text{As}}\cdot\frac{\text{Am}}{\text{Vs}}\right)} = 3 \times 10^8\,\text{m/s}$$

in the $x$-direction.

We see here in an especially striking way how, out of the quantities $\varepsilon_0$ and $\mu_0$ which are obtainable from purely electrical or magnetic measurements, the equality of the velocity of light to the propagation velocity of electromagnetic waves results.

## §58. Plane waves in stationary homogeneous media

We would like now to investigate the propagation of electromagnetic waves in homogeneous media at rest, restricting ourselves to normally polarizable and magnetizable substances without applied forces, having linear relationships between the field quantities $\mathbf{D}$ and $\mathbf{E}$, $\mathbf{B}$ and $\mathbf{H}$, $\mathbf{g}$ and $\mathbf{E}$. Further, we at first assume that throughout $\rho = 0$; for the case where $\rho \neq 0$, see p. 239. Taking now the constitutive equations V to VII of §53, namely,

$$\mathbf{D} = \varepsilon \mathbf{E} \qquad \mathbf{B} = \mu \mathbf{H} \qquad \mathbf{g} = \sigma \mathbf{E} \qquad (58.1)$$

and the material constants $\varepsilon$, $\mu$, $\sigma$ known from electrostatics, magnetostatics, and the theory of steady currents, we arrive at the equations

$$\left. \begin{array}{ll} \operatorname{curl} \mathbf{H} = \dfrac{\varepsilon}{c} \dfrac{\partial \mathbf{E}}{\partial t} + \dfrac{4\pi\sigma}{c} \mathbf{E} & \operatorname{div} \mathbf{E} = 0 \\[2mm] \operatorname{curl} \mathbf{E} = -\dfrac{\mu}{c} \dfrac{\partial \mathbf{H}}{\partial t} & \operatorname{div} \mathbf{H} = 0 \end{array} \right\} \qquad (58.2)$$

which will constitute our point of departure. From these equations there follow by elimination of $\mathbf{H}$ and $\mathbf{E}$, respectively,

$$\nabla^2 \mathbf{E} = \frac{\varepsilon\mu}{c^2} \frac{\partial^2 \mathbf{E}}{\partial t^2} + \frac{4\pi\mu\sigma}{c^2} \frac{\partial \mathbf{E}}{\partial t} \qquad \nabla^2 \mathbf{H} = \frac{\varepsilon\mu}{c^2} \frac{\partial^2 \mathbf{H}}{\partial t^2} + \frac{4\pi\mu\sigma}{c^2} \frac{\partial \mathbf{H}}{\partial t} \qquad (58.3)$$

which now enter in place of the wave equations (57.2) and (57.3). They are known as "equations of telegraphy" from the role they play in the theory of the transmission of electromagnetic waves along wires.

For the special case of *insulators* ($\sigma = 0$), equation (58.3) reduces to the ordinary wave equation; but now, however, we have as wave velocity no longer the vacuum velocity of light $c$, but the velocity

$$c' = \frac{c}{\sqrt{(\varepsilon\mu)}}$$

Thus, from the rules of optics for the *refractive index*, we have the "Maxwell relation"

$$n = c/c' = \sqrt{(\varepsilon\mu)} \qquad (58.4)$$

which, because $\mu \approx 1$, can usually be written in the form

$$n = \sqrt{\varepsilon} \qquad (58.4a)$$

Thus for insulators *the permittivity should be approximately equal to the square of the optical refractive index*. For long waves (radio-frequency and slow infra-red oscillations) this relationship is actually confirmed by experiment. In the visible region of the spectrum it is also fairly well satisfied for some substances, for example, $H_2$, $CO_2$, $N_2$, and $O_2$.

But for many other substances it fails, when, as a rule, the substances show infra-red selective absorption. With water the failure is especially marked; the static permittivity $\varepsilon$ is 81, but the refractive index ($n$) in the visible region is only 1·33.

The failure of the Maxwell relation (58.4) when calculating with the static $\varepsilon$-value obviously has its basis in the dependence of the polarization of the material on the electrical field strength which is quite different for the extremely high frequency of light waves from what it is for the static case—for, in the rapid oscillations of light waves, the existence of polarization must be treated as a dynamic process because of the inertia of the electrons. In such considerations the frequency of the incident waves plays a decisive role. We shall develop this dynamic theory of the permittivity, and a corresponding theory of electrical conductivity, in the second half of this section. Experiment has now shown that Maxwell's theory of light, with this addition, furnishes a completely concordant description of the optical properties of most material media.

As we shall see shortly, the electron theory leads to a dependence upon frequency of the constants $\varepsilon$ and $\sigma$, hitherto defined through the constitutive equations (58.1). (In the following, the dynamic theory of the permeability $\mu$ will remain beyond our consideration.) Since $\varepsilon \equiv \varepsilon(\omega)$, $\sigma \equiv \sigma(\omega)$, equations (58.1) become meaningless unless we restrict ourselves to special time-dependent Fourier components. If, as in §57, we set

$$\mathbf{D}(t) = \int_{-\infty}^{+\infty} \mathbf{D}_\omega e^{i\omega t}\, d\omega \qquad \mathbf{E}(t) = \int_{-\infty}^{+\infty} \mathbf{E}_\omega e^{i\omega t}\, d\omega \qquad \mathbf{g}(t) = \int_{-\infty}^{+\infty} \mathbf{g}_\omega e^{i\omega t}\, d\omega$$

then, instead of the constitutive equations (58.1), we must write

$$\mathbf{D}_\omega = \varepsilon(\omega)\mathbf{E}_\omega \qquad \mathbf{g}_\omega = \sigma(\omega)\mathbf{E}_\omega \tag{58.5}$$

Taking account of the inversion

$$\mathbf{E}_\omega = \frac{1}{2\pi} \int_{-\infty}^{+\infty} \mathbf{E}(\tau) e^{-i\omega\tau}\, d\tau$$

Instead of the simple relationships (58.1), we arrive now at the integral relationships

$$\left. \begin{aligned} \mathbf{D}(t) &= \frac{1}{2\pi} \int_{-\infty}^{+\infty} \varepsilon(\omega)\, d\omega \int_{-\infty}^{+\infty} \mathbf{E}(\tau) e^{i\omega(t-\tau)}\, d\tau \\ \mathbf{g}(t) &= \frac{1}{2\pi} \int_{-\infty}^{+\infty} \sigma(\omega)\, d\omega \int_{-\infty}^{+\infty} \mathbf{E}(\tau) e^{i\omega(t-\tau)}\, d\tau \end{aligned} \right\} \tag{58.6}$$

We shall later evaluate these on the basis of the formulae derived for $\varepsilon(\omega)$ and $\sigma(\omega)$.

For the time being it suffices for us to establish that, owing to the dependence of $\varepsilon$ and $\sigma$ upon frequency, *equations* (58.2) *are no longer of general validity, but are valid now only for the temporal Fourier components* of **E** and **H**. In this case, however, we can substitute multiplication by $i\omega$ for the time derivative (in the manner of (57.12)), so that in place of the first equation (58.2), for example, we have the relationship

$$\operatorname{curl} \mathbf{H}_\omega = \left(\frac{i\omega\varepsilon}{c} + \frac{4\pi\sigma}{c}\right)\mathbf{E}_\omega = \frac{i\omega}{c}\left(\varepsilon + \frac{4\pi\sigma}{i\omega}\right)\mathbf{E}_\omega$$

Permittivity and conductivity appear here, and therefore also in the following, always in the combination

$$\varepsilon + \frac{4\pi\sigma}{i\omega} = \eta \qquad (58.7)$$

We wish however to go at once a bit further in that, instead of calculating with the time-dependent Fourier components, we work immediately with the space-and-time-dependent Fourier components, i.e. we write **E** and **H** in the special form of plane waves, as in (57.10):

$$\mathbf{E} = \mathbf{E}_0 \exp i(\omega t - \mathbf{k} \cdot \mathbf{r}) \qquad \mathbf{H} = \mathbf{H}_0 \exp i(\omega t - \mathbf{k} \cdot \mathbf{r}) \qquad (58.8)$$

We then find from the Maxwell equations, using the calculation rule (57.12) and dividing by $i$, that instead of (57.13) we have

$$\left.\begin{array}{ll} -\mathbf{k} \times \mathbf{H} = \dfrac{\omega}{c}\eta\mathbf{E} & \mathbf{k} \cdot \mathbf{E} = 0 \\[2ex] -\mathbf{k} \times \mathbf{E} = -\dfrac{\omega}{c}\mu\mathbf{H} & \mathbf{k} \cdot \mathbf{H} = 0 \end{array}\right\} \qquad (58.9)$$

First, from these relationships it again follows that **k**, **E**, and **H** stand pairwise perpendicular to one another. They can therefore be used as base vectors of a right-hand system. Further, since $\eta$ is now in general complex, there will exist between **E** and **H** a complex interdependence, i.e. a *phase displacement*, in contrast to plane waves *in vacuo*, in which **E** and **H** always oscillate in phase with one another.

Further, for real $\omega$, both $\eta$ and **k** become complex; thus, by eliminating **E** and **H** from equations (58.9), as by vectorial multiplication of the first equation by **k** and by making use of the third and fourth equations, it follows that

$$k^2 = \frac{\omega^2}{c^2}\eta\mu \quad \text{and therefore} \quad k = \frac{\omega}{c}\sqrt{(\eta\mu)} \qquad (58.10)$$

For insulators ($\sigma = 0$) and slow oscillation processes ($\varepsilon \approx$ static permittivity) the speed of travel of the wave thus becomes

$$c' = \frac{\omega}{k} = \frac{c}{\sqrt{(\varepsilon\mu)}} = \frac{c}{n}$$

with the (here real) value of the index of refraction $n$ given by (58.4). For complex $\eta$, however, the general index of refraction $c/c' \equiv p$ becomes complex. We write it in the form

$$p = \sqrt{(\eta\mu)} = n - i\kappa \tag{58.11}$$

$n$ and $\kappa$ being real and positive. (That $\kappa$ is always positive follows for real $\varepsilon$ and $\sigma$ from (58.7), but in the following this will also turn out to be so for complex $\varepsilon$ and $\sigma$.) In order to see the significance of the complex index of refraction $p$ in (58.11) we place the $x$-direction in (58.8) in the direction of the **k**-vector. Then, on account of (58.10) and (58.11)

$$\exp i(\omega t - \mathbf{k} \cdot \mathbf{r}) = \exp i\omega(t - px/c) = \exp i\omega(t - nx/c)\exp(-\omega\kappa x/c)$$

Thus in the direction of its travel the wave decays away exponentially; in fact, in the space of a wavelength ($\omega n x/c = 2\pi$) by the factor $e^{-2\pi\kappa/n}$. The quantity $\kappa$ is therefore called the *extinction coefficient*.

For further discussion we must first investigate the frequency dependence of $\varepsilon$, and more especially that of $\sigma$. In order to distinguish them from the static value of $\varepsilon$, which we wish here temporarily to denote by $\varepsilon_0 \equiv \varepsilon(\omega = 0)$, and from the d.c. conductivity $\sigma_0 \equiv \sigma(\omega = 0)$, we speak here of the dynamic values of $\varepsilon$ and $\sigma$, having in mind their dependence upon $\omega$. We begin with a consideration of $\sigma$.

1. *The dynamic value of the conductivity.* Since, with increasing frequency of the alternating electromagnetic field, the conduction electrons, owing to their inertia, follow the field with increasing difficulty, we expect a steady decrease of the conductivity with increasing $\omega$. For a quantitative pursuit of this consideration we take up the idea indicated in §39 concerning the frictional resistance experienced by electrons in their motion through a conductor, and we investigate the electron motion in the changing field on the basis of Newton's equations of motion. For the $k$th electron the equation reads

$$m\frac{d\mathbf{v}_k}{dt} = e\mathbf{E} - \xi\mathbf{v}_k$$

The constant $\xi$ in the Stokes friction term can be related to the d.c. conductivity $\sigma_0$. If we sum this equation of motion over all $N$ electrons in unit volume, then, since $\sum_k e v_k = g$, we obtain as first approximation (as regards the left-hand side)

$$m \frac{\partial g}{\partial t} = Ne^2 E - \xi g$$

In the stationary case ($\partial g / \partial t = 0$) it follows from this that $g = Ne^2 E / \xi$, so we must have that $Ne^2 / \xi = \sigma_0$. We thus find as an extension of Ohm's law

$$\frac{m}{Ne^2} \frac{\partial g}{\partial t} + \frac{g}{\sigma_0} = E \quad \text{or} \quad \frac{\partial g}{\partial t} + \frac{g}{\tau} = \frac{Ne^2}{m} E \qquad (58.12)$$

when we introduce a new constant $\tau$ of the dimension of a time where

$$\tau = \frac{\sigma_0 m}{Ne^2} = \frac{m}{\xi} \quad \text{and hence*} \quad \sigma_0 = \frac{Ne^2 \tau}{m} \qquad (58.13)$$

(For example, for copper we have $Ne^2 / m\sigma_0 = 1/\tau = 4\cdot1 \times 10^{13} \text{s}^{-1}$, $\tau = 2\cdot4 \times 10^{-14} \text{s}$.)

If, specially, $E$ and therefore $g$ have the time-dependence $e^{i\omega t}$, then from (58.12), we have

$$(i\omega\tau + 1)g = \sigma_0 E \quad \text{i.e.} \quad g = \sigma E$$

---

* For some metals (superconductors), upon cooling below a certain temperature, the d.c. resistance suddenly vanishes completely. If we wished to describe phenomenologically this hitherto entirely unexplained effect, we might attempt to assume that $\sigma$, and therefore $\tau$, in the above formulae were infinitely large for temperatures below the point of discontinuity. In this case, however, we could not start with the usual form of Ohm's law, $g = \sigma_0 E$, for from this we would have that $E \rightarrow 0$. Rather, the extended Ohm relation (58.12) would have to be taken. According to this, for superconductors, instead of Ohm's law $g = \sigma_0 E$, the equation

$$\frac{\partial g}{\partial t} = \frac{Ne^2}{m} E = \Lambda E \quad \text{with} \quad \Lambda = \frac{Ne^2}{m} \qquad (58.14a)$$

must hold. If we take the curl of this equation, then, for homogeneous material, because of the induction law we find the relation

$$\frac{\partial \, \text{curl} \, g}{\partial t} = \Lambda \, \text{curl} \, E = -\frac{\Lambda}{c} \frac{\partial B}{\partial t}$$

which, in the case of a purely harmonic alternating current is identical with the relationship

$$\text{curl} \, g = -\frac{\Lambda}{c} B \qquad (58.14b)$$

In the phenomenological theory of superconductivity this relationship is assumed to be generally valid. It is called the *London equation*.
Further details of the phenomenon of superconductivity are postponed to Volume III.

the dynamic conductivity being

$$\sigma = \frac{\sigma_0}{1 + i\omega\tau} \qquad (58.15)$$

So long as $\omega \ll 1/\tau$, $\sigma$ remains nearly real and equal to $\sigma_0$. As $\omega$ increases, $\sigma$ develops a negative imaginary part in increasing measure, and, for $\omega \gg 1/\tau$, becomes nearly purely imaginary, corresponding to the fact that now it is not $\mathbf{g}$ but $\partial\mathbf{g}/\partial t$ that has the same phase as $\mathbf{E}$.

If the dynamic value of $\sigma$, equation (58.14), is put into the relationship (58.7) for the effective permittivity $\eta$, then

$$\eta = \varepsilon + \frac{4\pi\sigma_0}{i\omega(1 + i\omega\tau)} = \varepsilon\left(1 + \frac{\omega_p^2 \tau}{i\omega(1 + i\omega\tau)}\right) \qquad (58.16)$$

with the abbreviation

$$\omega_p = \sqrt{\frac{4\pi\sigma_0}{\varepsilon\tau}} = \sqrt{\frac{4\pi Ne^2}{\varepsilon m}} \qquad (58.17)$$

This frequency quantity $\omega_p$ is called the *plasma frequency*.* (Its value for copper is $1\cdot6 \times 10^{16}\,\text{s}^{-1}$.)

In order to make the concept of the plasma frequency clearer, we shall investigate the dispersal of a local accumulation of charge within a conductor, i.e. the case, hitherto excluded, where $\rho \neq 0$. For this purpose we form the divergence of the extended Ohm's law (58.12) and in this we consider the quantities $\sigma_0$ and $\tau$ in first approximation as being constant in space and time. Since

$$\text{div } \mathbf{g} = -\frac{\partial\rho}{\partial t} \quad \text{and} \quad \text{div } \mathbf{E} = \frac{4\pi\rho}{\varepsilon}$$

we obtain for the charge density $\rho$ the relationship

$$\frac{\partial^2\rho}{\partial t^2} + \frac{1}{\tau}\frac{\partial\rho}{\partial t} + \omega_p^2\rho = 0 \qquad (58.18)$$

This is the equation of a damped oscillation with damping constant $1/\tau$, and with the eigenfrequency $\omega_p$ defined through equation (58.15).

2. *The dynamic value of the permittivity.* The inertia of electrons affects not only conductivity but also the electrical behaviour of atoms. This leads, in the latter case, to a frequency dependence of the atomic

---

* A plasma is described in physics as a system containing a large number of mobile statistically assembled charge bearers. In gas discharge tubes as well as in the uppermost layer of the earth's atmosphere (the "ionosphere") there are gas plasmas consisting of negative and positive ions as well as free electrons. In metals we have to deal with pure electron plasmas existing within the metallic lattice of the rigidly bound ions.

polarizability $\alpha$, and thus to a similar dependence of the permittivity $\varepsilon$. Although the dynamic processes in the atom can be completely understood only within the framework of the quantum theory (see the corresponding considerations in Volume II), it is possible, according to Thomson, to give something of a description of the optical behaviour by means of the concept in which the atomic electrons are elastically bound to the atomic nucleus. Without giving an exact account of the details of this atomic model, we assume that each individual atomic electron satisfies the equation of motion

$$m\left(\frac{d^2\mathbf{r}}{dt^2} + \gamma\,\frac{d\mathbf{r}}{dt} + \omega_0^2\mathbf{r}\right) = e\mathbf{F} \qquad (58.19)$$

We have here written the restoring force on the electron in the form $-m\omega_0^2\mathbf{r}$, and, further, we have introduced a damping force $-m\gamma\,d\mathbf{r}/dt$ (with $\gamma \ll \omega_0$) proportional to the velocity and oppositely directed. We have brought both of these forces over to the left side of the equation. The electrical force $e\mathbf{F}$ remains on the right-hand side. Here $\mathbf{F}$, as defined in §26, is the effective field strength at the location of the atom. Now the atomic dipole has the moment $\mathbf{p} = e\mathbf{r}$; thus from (58.19) we obtain for this the determining equation

$$\frac{d^2\mathbf{p}}{dt^2} + \gamma\,\frac{d\mathbf{p}}{dt} + \omega_0^2\mathbf{p} = \frac{e^2}{m}\,\mathbf{F} \qquad (58.20)$$

From this, for the polarizability $\alpha$ defined by $\mathbf{p} = \alpha\mathbf{F}$, it follows immediately for the case of a static field that

$$\alpha = \alpha_0 = \frac{e^2}{m\omega_0^2} \qquad (58.21)$$

In an alternating field, in which both $\mathbf{p}$ and $\mathbf{F}$ have the time dependence $e^{i\omega t}$, we obtain from (58.20) for the dynamic polarizability

$$\alpha = \frac{e^2/m}{\omega_0^2 - \omega^2 + i\omega\gamma} \qquad (58.22)$$

Figure 53 shows the behaviour of the real and imaginary parts of $\alpha$ in their dependence upon $\omega$. Thus $\alpha$ possesses a resonance at $\omega = \omega_0$. It changes very rapidly within the small frequency region $\omega_0 - \gamma < \omega < \omega_0 + \gamma$ and therewith goes through very pronounced extreme values in both real and imaginary parts. Away from the resonance $\alpha$ is approximately real. It is positive for $\omega < \omega_0$ and negative for $\omega > \omega_0$. In the latter case the atomic electron oscillates, as in any other resonance process, in opposite phase to the applied force. From

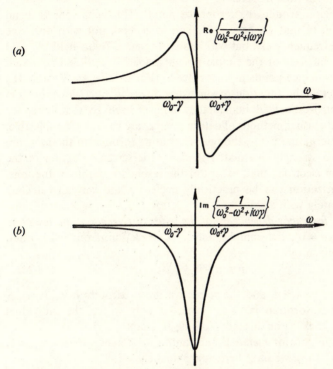

(a)

(b)

Fig. 53.—Shape of (a) the real part, and (b) the imaginary part of the polarizability in the neighbourhood of a resonance

the polarizability (58.22) we find for not-too-dense gases (with $N$ atoms per cm$^3$)

$$\varepsilon = 1 + 4\pi N\alpha = 1 + \frac{4\pi Ne^2/m}{\omega_0^2 - \omega^2 + i\omega\gamma} \qquad (58.23)$$

For dense gases and liquids, another correction term would appear in the denominator because of the difference between **F** and **E** (see 26.8), but we shall not take these complications into account here. Since the real and imaginary parts of $\varepsilon$ have an analogous behaviour to those of

$\alpha$ (figure 53), $\varepsilon$ is real and positive except in a small region around the resonance point $\omega_0$.

Thus in *insulators* the index of refraction increases with increasing frequency ("normal dispersion") and behaves "anomalously" only in a small region around the resonance point. This behaviour is in no way altered in that we have to reckon in general, not with only one resonance frequency $\omega_0$, but with many resonance frequencies $\omega_j$, and that then, in place of the simple formulae (58.24) and (58.25), a sum over all resonance points $\omega_j$ is involved. (For details, see Volume II.)

For metals the situation is somewhat different, since in conductors every electrostatic field immediately adjusts itself so that here, for example, the usual methods of determining $\varepsilon_0$ are basically inapplicable. Electric alternating fields, though penetrating through to the ions, are however so strongly attenuated there by the screening action of the conduction electrons that, except at the resonance points of the ions, their polarization can be practically neglected and we can calculate approximately with $\varepsilon = 1$.

3. *The frequency dependence of the index of refraction.* In insulators (in which $\sigma = 0$), the complex index of refraction, from (58.11) and (58.7), is given by

$$p \equiv n - i\kappa = \sqrt{(\varepsilon\mu)} \qquad (58.24)$$

but in general $\kappa \ll n$, and the refractive index becomes $n \approx \sqrt{(\varepsilon\mu)}$. A significant absorption is to be expected only within the individual resonance lines. (For this, see Volume II, chapter EII.)

The situation for *metals* is substantially different. With $\varepsilon \approx \mu \approx 1$, and with (58.15), we have here approximately

$$p \equiv n - i\kappa = \sqrt{\left[1 + \frac{4\pi\sigma_0}{i\omega(1 + i\omega\tau)}\right]} = \sqrt{\left[1 + \frac{\omega_p^2\tau}{i\omega(1 + i\omega\tau)}\right]} \qquad (58.25)$$

So long as $\omega \ll 1/\tau \ll \omega_p$, i.e. for sufficiently long waves, $i\omega\tau$ in the denominator can be neglected in comparison with 1, and then the 1 preceding the conduction term can be dropped. We then have, approximately,

$$n \approx \kappa \approx \sqrt{\frac{2\pi\sigma_0}{\omega}} = \sqrt{\frac{\omega_p^2\tau}{2\omega}} \qquad (58.26)$$

Thus in the space of one wavelength an electromagnetic wave becomes attenuated by the factor $e^{-2\pi} \approx 0.028$. The penetration depth $d$ is designated as the distance in which the wave has become attenuated by the factor $1/e$. From (58.26) it is

$$d = \frac{c}{\omega\kappa} \approx \frac{c}{\sqrt{(2\pi\sigma_0\,\omega)}} \qquad (58.27)$$

For copper, for example, $(\sigma_0 = 5\cdot 4 \times 10^{17}\ \mathrm{s}^{-1})$ for the vacuum wavelengths

| $\lambda_0 =$ | 1 cm | 1 m | 100 m | 10 km |
|---|---|---|---|---|
| $d =$ | $0\cdot 37\ \mu$ | $3\cdot 7\ \mu$ | $37\ \mu$ | $0\cdot 37\ \mathrm{mm}$ |

These $d$-values also provide something of an estimate of the metal thickness necessary for screening or shielding against the respective wavelengths.

If, however, $\omega \gg 1/\tau$, then, neglecting the 1 in the denominator in comparison with the term $i\omega\tau$, we obtain the relationship

$$p \equiv n - i\kappa \approx \sqrt{\left(1 - \frac{\omega_p^2}{\omega^2}\right)} \qquad (58.28)$$

We have also in this region, so long as $\omega \ll \omega_p$, to reckon with a strong absorption

$$\kappa \approx \frac{\omega_p}{\omega}$$

which now leads to a frequency-independent penetration depth $d \approx c/\omega_p$. (For copper this is about $0\cdot 02\,\mu$.) If, however, $\omega$ increases beyond $\omega_p$, the refractive index becomes practically real. In this region the metal becomes almost transparent.[*] The reason for this is that for the case $\omega > \omega_p$, the displacement current becomes significantly stronger than the conduction current; for such high frequencies, therefore, the conductor behaves like an insulator.

Since, in the frequency dependence of $\varepsilon$ and $\omega$, the connections between D and E and between g and E are now apparent, let us in conclusion evaluate for illustration formulae (58.6), with the expressions (58.23) for $\varepsilon(\omega)$ and (58.15) for $\sigma(\omega)$. We wish to begin with the g-formula, as the simpler. First, by substituting $\sigma(\omega)$ in (58.6) and by exchanging the order of integration, we have

$$\mathbf{g}(t) = \frac{Ne^2}{2\pi i m} \int_{-\infty}^{+\infty} \mathbf{E}(\tau)\,d\tau \int_{-\infty}^{+\infty} \frac{\exp[i\omega(t-\tau)]\,d\omega}{\omega - iNe^2/m\sigma_0}$$

(The change in order of integration is possible since $\mathbf{E}(t)$ can always be so chosen that it is different from zero only in a finite time-interval.) We now evaluate the

---

[*] The same considerations are applicable to the gas plasma in the ionosphere. Here, owing to lower electron density, $\omega_p$ becomes considerably smaller (about 10 to 100 Mc/s), so that the inception of transparency at $\omega = \omega_p$ can be conveniently followed even in the high-frequency region of the radio spectrum.

$\omega$-integral by complex means. Obviously, in the complex $\omega$-plane the integral has only one pole at $\omega = iNe^2/m\sigma_0$. If the path of integration, originally lying along the real axis, be displaced downward into the negative-imaginary half-plane, we encounter no pole; thus, for the case $t - \tau < 0$, we can go with the path of integration to negative-imaginary infinity where, because of the exponential function, the integrand itself so strongly vanishes that the entire integral has the value 0. Thus we need extend the $\tau$-integration only from $-\infty$ to $t$. For positive $t - \tau$ we find it advantageous to take the path of integration upward to positive-imaginary infinity, where again the integral contribution vanishes. This time, however, the integration path becomes involved with the pole $\omega = iNe^2/m\sigma_0$, so that the integral has to be taken around this pole. This gives $2\pi i$ times the residue, or $2\pi i$ times the value of the exponential function at this place. Thus we find in all

$$\mathbf{g}(t) = \frac{Ne^2}{m} \int_{-\infty}^{t} \mathbf{E}(\tau) \exp\left[ -\frac{Ne^2}{m\sigma_0}(t-\tau) \right] d\tau \qquad (58.29)$$

[For proof we can put this equation into (58.12), bearing the upper limit of integration in mind in the differentiation.]

Thus in our dynamic theory, in place of the simple Ohm's law $\mathbf{g} = \sigma_0\mathbf{E}$, an integral interrelationship exists in which the g-value at time $t$ is conditioned by the value of the field strength for all times $\tau < t$. Our calculational treatment leads automatically to an after-effect of $\mathbf{E}$ in the future, but not to a reaction into the past. From (58.29), for $\mathbf{E}$ constant in time, in carrying out the $\tau$-integration we find again the old Ohm's law $\mathbf{g} = \sigma_0\mathbf{E}$. For $\mathbf{E} = \mathbf{E}_0 e^{i\omega t}$ we find the dynamic Ohm's law $\mathbf{g} = \sigma\mathbf{E}$ with the value of $\sigma$ given by (58.15).

In an analogous way, by putting (58.23) into the first equation (58.6), there being now two poles: $\omega = +\frac{1}{2}i\gamma \pm \sqrt{(\omega_0^2 - \frac{1}{4}\gamma^2)}$, we find the somewhat complicated relationship

$$\mathbf{D} - \mathbf{E} = 4\pi\mathbf{P}$$
$$= \frac{4\pi Ne^2}{m\sqrt{(\omega_0^2 - \frac{1}{4}\gamma^2)}} \int_{-\infty}^{t} \mathbf{E}(\tau) \exp\left[ -\frac{1}{2}\gamma(t-\tau) \right] \sin(\omega_0^2 - \frac{1}{4}\gamma^2)^{1/2}(t-\tau) d\tau \quad (58.30)$$

Here also for the polarization $\mathbf{P}$ we automatically obtain an after-effect with an average after-duration $2/\gamma$, in which $\gamma$ represents the damping constant of the atomic oscillation process.

That here, and in the more generalized Ohm's law, we obtain only an effect in the future but not a similar one in the past connects mathematically with the fact that the integrands appearing have poles only in the positive-imaginary half-plane. Physically this means that for the time-dependence $e^{i\omega t}$ assumed above, the complex index of refraction $p$ can be written in the form $n - i\kappa$, with $\kappa > 0$ and that the waves are therefore always attenuated in the direction of their travel.

## §59. The reflection of electromagnetic waves at boundary surfaces

We now consider a plane electromagnetic wave which, having travelled *in vacuo*, enters perpendicularly into a surface (chosen as the $yz$-plane) of a non-ferromagnetic medium. Experimentally, this wave is split at the surface into (*a*) a reflected wave which goes *in vacuo* in the direction of the negative $x$-axis, and into (*b*) a penetrating wave going in the medium in the positive $x$-direction.

In conformity with (58.11), the medium can be described by its complex index of refraction $p = n - i\kappa$. In a vacuum $p = 1$. We note here that, corresponding to the Maxwell equations, at the boundary surfaces under all circumstances, continuity of the tangential component of $\mathbf{E}$ and of the normal component of $\mathbf{B}$ is required, while $D_n$ and $H_t$ are only continuous if, as in the foregoing case, no surface charges or surface currents appear.

Having regard for equations (58.8) and (58.9), we satisfy the Maxwell equations with the following statement:

Incident wave    $(x < 0)$:

$$E_y^{(e)} = a \exp i\omega(t - x/c) \qquad H_z^{(e)} = a \exp i\omega(t - x/c)$$

Reflected wave    $(x < 0)$:

$$E_y^{(r)} = -a' \exp i\omega(t + x/c) \qquad H_z^{(r)} = a' \exp i\omega(t + x/c)$$

Penetrating wave $(x > 0)$:

$$E_y = a'' \exp i\omega(t - px/c) \qquad H_z = \frac{p}{\mu} a'' \exp i\omega(t - px/c)$$

Herewith, on account of (58.10), we would set $k = \omega/c$ in a vacuum, and $k = p\omega/c$ in the medium. For fulfilling the boundary conditions at $x$ the initially undetermined amplitudes $a$, $a'$, $a''$ must now satisfy two relations which follow from the

continuity of $E_t$:

$$E_y^{(e)} + E_y^{(r)} = E_y \qquad a - a' = a''$$

continuity of $H_t$:

$$H_z^{(e)} + H_z^{(r)} = H_z \qquad a + a' = a'' p/\mu$$

Solving for the amplitudes of the reflected wave and the penetrating wave, we find

$$a' = a\frac{p - \mu}{p + \mu} \qquad a'' = a\frac{2\mu}{p + \mu} \tag{59.1}$$

The ratio of the reflected intensity to the incident intensity is called the *reflectivity* $R$ of the material. Since both the reflected wave and the incident wave travel *in vacuo*, we can set the ratio of the two

Poynting vectors equal to the ratio of the square of the absolute magnitude of $a'$ and $a$:

$$R = \frac{a'a'^*}{aa^*} = \frac{(p-\mu)(p^*-\mu)}{(p+\mu)(p^*+\mu)} = \frac{(n-\mu)^2+\kappa^2}{(n+\mu)^2+\kappa^2} \qquad (59.2)$$

For a transparent medium ($\kappa = 0$) we have the well-known relationship

$$R = \frac{(n-\mu)^2}{(n+\mu)^2} \approx \frac{(n-1)^2}{(n+1)^2} \qquad (59.3)$$

With increasing absorption the reflectivity also increases. For metals, where the frequency of the incident radiation is not too high, i.e. in the region of validity of equation (58.14), with $n = \kappa \gg 1$, $\mu \approx 1$, the reflectivity becomes equal approximately to

$$R = \frac{2n^2-2n+1}{2n^2+2n+1} \approx 1 - \frac{2}{n} = 1 - \sqrt{\frac{2\omega}{\pi\sigma_0}} \qquad (59.4)$$

This reflection formula was examined experimentally by Hagen and Rubens for different metals (e.g. silver, copper, nickel, bismuth). For long-wave infra-red, up to about $\lambda = 25\mu$, quantitative agreement with the formula was good. The optical properties of metals up to this wavelength region are therefore determined by the d.c. conductivity $\sigma_0$. For shorter wavelengths the observed reflectivity becomes substantially smaller than that calculated from (59.4), corresponding to our considerations in §58 concerning the dynamic conductivity. Owing to their inertia, the electrons are no longer able to follow exactly the rapidly changing field. This inertia effect is more pronounced in electrolytes (e.g. $H_2SO_4$ in water) which display superior conductivity statically but which at the same time are completely transparent. Since here the carriers of the current consist of ions whose mass is more than a thousand times that of an electron, it is understandable that, with respect to the electric field of light waves, electrolytes behave like insulators.

Returning again to the above formula, we first establish that, on account of (59.1), for real $p$, i.e. $\kappa = 0$, the E-vector of the reflected wave at the boundary surface is directed oppositely to the incident wave, and that in this case, therefore, the E-vector undergoes in the reflection a *phase jump* of 180°. This consideration continues to fall approximately within the region of validity of equation (59.4) even for reflection from metals, for here $a' \approx a(1 - 2\mu/p)$, $a'' \approx 2a\mu/p$. The fields of the two waves in the external region therefore largely compensate one another at the boundary surface, so that in the metal only a relatively weak E-field penetrates. (For very good conductors $\mu/p$ goes to zero.) In contrast to this, the two H-vectors of the waves in the external region both have the same direction and are of approximately equal magnitude so that a strong magnetic wave penetrates into the metal. It is however very quickly weakened by the magnetic field of the volume current from the E-wave—in a distance, in fact, of about the magnitude of the depth of penetration $d$ considered in §58.

*Reflection at oblique incidence* can be handled in a manner similar to that used above in calculating the reflection for normal incidence. We have then to reckon with the three phase functions

$$\exp\left[i(\omega t - \mathbf{k} \cdot \mathbf{r})\right] \quad \exp\left[i(\omega t - \mathbf{k}' \cdot \mathbf{r})\right] \quad \text{and} \quad \exp\left[i(\omega t - \mathbf{k}'' \cdot \mathbf{r})\right]$$

in which $\mathbf{k}$, $\mathbf{k}'$, and $\mathbf{k}''$ are the propagation vectors of the incident, reflected, and transmitted waves. In order for the continuity conditions to be fulfilled not only for all times, but also for all points of the boundary plane, the three vectors $\mathbf{k}$, $\mathbf{k}'$, and $\mathbf{k}''$ must lie in a plane, and must have equally large tangential components. If we designate the "angle of incidence" between $\mathbf{k}$ and the surface normal by $\alpha$, the "angle of reflection" by $\alpha'$, and the "angle of emergence" by $\beta$, we must have that

$$k \sin \alpha = k' \sin \alpha' = k'' \sin \beta \tag{59.5}$$

Since $k = k'$ we thus find not only the reflection law $\alpha = \alpha'$, but also, since $k'' = kp$, the law of refraction

$$\sin \alpha = p \sin \beta \tag{59.6}$$

which, for light entering transparent media, assumes the usual form

$$\sin \alpha = n \sin \beta \tag{59.7}$$

(For absorbing media $\beta$, through $p$, becomes complex, and thus in this case it loses its clear and simple meaning.)

We now state without proof the reflectivity formulae for the two cases of polarization: E parallel to the plane of incidence, and E perpendicular to it:

$$R_{||} = \left|\frac{p \cos \alpha - \mu \cos \beta}{p \cos \alpha + \mu \cos \beta}\right|^2 \qquad R_\perp = \left|\frac{p \cos \beta - \mu \cos \alpha}{p \cos \beta + \mu \cos \alpha}\right|^2 \tag{59.8}$$

For the special case of reflection of light from transparent non-magnetic media ($p = n$, $\mu = 1$), and when we set $n = \sin \alpha/\sin \beta$, these formulae go over to the *Fresnel formulae*

$$R_{||} = \left(\frac{\tan (\alpha - \beta)}{\tan (\alpha + \beta)}\right)^2 \qquad R_{\perp} = \left(\frac{\sin (\alpha - \beta)}{\sin (\alpha + \beta)}\right)^2 \qquad (59.9)$$

With increasing $\alpha$, $R_{\perp}$ increases monotonically from the $R$-value given by equation (59.3), (where $\alpha = 0$), to the value 1 (where $\alpha = 90°$). $R_{||}$, however, for $\alpha$ increasing, at first falls to the value 0, reaching this value for $\alpha + \beta = 90°$, i.e. at $\tan \alpha = n$ (defining the so-called *Brewster polarization angle*), and it then increases monotonically to the value 1. We close the discussion here, referring the reader interested in other optical phenomena at boundary surfaces (e.g. total reflection) to textbooks on optics.

## §60. Current displacement or the skin effect

We consider a straight metal wire with circular cross-section, having radius $r_0$. Let this wire carry an alternating current of frequency $\omega$. On its surface we then have an electric field in the direction of the wire axis, and a magnetic field perpendicular to it but also parallel to the surface. This gives qualitatively the same relations as we have considered above (§58 and §59) for a light wave meeting a metal surface at normal incidence and penetrating within. For the penetration of such a light wave, in the case of a plane metal surface, and for not too high frequency, we have calculated in §58 a depth $d$, being the distance in which the wave amplitude is attenuated to the $e$th part; that is, according to (58.27),

$$d = \frac{c}{\sqrt{(2\pi\sigma\omega\mu)}} \qquad (60.1)$$

We expect similar results for the cylindrical wire surface. Although in any particular case the calculation here is somewhat more complicated than for the plane surface, we can already foresee the following two extreme cases:

1. $r_0 \ll d$: The alternating field, upon penetrating to the wire axis, is not appreciably attenuated; the current density remains almost uniformly distributed over the cross-section of the wire.

2. $r_0 \gg d$: The alternating field is completely attenuated away before having penetrated (from outside in) an appreciable fraction of the wire radius. The current density is therefore sensibly different from zero only in a thin surface layer ("skin"), while the whole inside of the wire is practically current-free.

In order to follow quantitatively the penetration of the alternating field into the wire, we go back to the telegraphic equation (58.3). For our case only the component $E_z = E$, parallel to the $z$-axis, is different from zero, and equation (58.3) then has the form

$$\nabla^2 E = \frac{\varepsilon\mu}{c^2}\frac{\partial^2 E}{\partial t^2} + \frac{4\pi\sigma\mu}{c^2}\frac{\partial E}{\partial t} \qquad (60.2)$$

We limit ourselves further to frequencies so low that the first term on the right side of (60.2) (the term coming from the displacement current) can be neglected in comparison with the second term, which comes from the conduction current. We thus arrive formally at the same differential equation as that upon which the entire *theory of heat conduction* is constructed. Finally, on symmetry grounds we assume that $E$ depends only on the distance $r$ from the wire axis, and in addition we express its time-dependence in the usual form $e^{i\omega t}$.

It follows then from (60.2), with the abbreviation (60.1), that

$$\frac{1}{r}\frac{\partial}{\partial r}\left(r\frac{\partial E}{\partial r}\right) = \frac{4\pi\sigma\mu\omega i}{c^2}E = \frac{2i}{d^2}E \qquad (60.3)$$

This equation is rigorously solved by the Bessel function of zero order with complex argument $(2i)^{1/2}r/d = (1+i)r/d$. We wish here, however, to content ourselves with finding directly approximate solutions for the two extreme cases of interest to us, namely $r_0 \ll d$ and $r_0 \gg d$. For *weak skin effect* it is convenient to develop $E$ in a series of even powers of $r/d$. As can be readily verified by substituting in (60.3) and comparing coefficients, we obtain

$$E = E_0\left\{1 + \frac{ir^2}{2d^2} + \ldots + \frac{1}{(n!)^2}\left(\frac{ir^2}{2d^2}\right)^n + \ldots\right\} \qquad (60.4)$$

For *strong skin effect* all processes take place so close to the wire surface $r \approx r_0$ that the $r$ explicitly appearing in (60.3) can be regarded as being constant, and can be replaced by $r_0$, i.e. the wire surface can be considered as plane. The partial solution from the wire boundary inward then reads

$$E \approx E_0\, e^{(1+i)r/d} \qquad (60.5)$$

We consider further the total current $I = \sigma\int E\,dq$ through the wire. Substituting here the value of $E$ from the differential equation (60.3), we can carry out the *r*-integration with $dq = 2\pi r\,dr$, and so obtain in general

$$I = 2\pi\sigma \int_0^{r_0} E r\,dr = \frac{\pi\sigma d^2 r_0}{i}\left(\frac{\partial E}{\partial r}\right)_{r=r_0} \qquad (60.6)$$

For direct current, i.e. with no skin effect, we would have found for this

$$I_0 = \pi r_0^2 \sigma E_0$$

with the field strength value $E_0$ constant over the wire cross-section. The d.c. resistance per centimetre of wire length is therefore equal to

$$R_0 = \frac{E_0}{I_0} = \frac{1}{\pi r_0^2 \sigma} \qquad (60.7)$$

Correspondingly, we define the complex resistance of the wire per centimetre for alternating current by

$$\mathbf{R} \equiv R + j\omega L_i = \frac{E(r_0)}{I} = \frac{j}{\pi\sigma d^2 r_0}\left(\frac{E}{\partial E/\partial r}\right)_{r=r_0} \qquad (60.8)$$

since, because of the skin effect, it only makes sense to relate the current strength to the field which has access to the wire surface from the outside. $R$ is then the effective resistance per unit of length of the current-conducting regions in the wire; and $L_i(\omega) - L_i(0)$ means the *supplemental inductance* per unit length due to the skin effect to be added to the normal self-inductance $L$ already considered in §49. Instead of going by way of equation (60.8) we could also calculate $R$ and $L_i$ from the idea of inhomogeneous current distribution ($\mathbf{g} = \sigma\mathbf{E}$), in particular, by a corresponding modification of equation (49.10); we would then, however, come to the same results as are given by equation (60.8).

In the case of *weak skin effect*, if to simplify the formulae we introduce the abbreviation

$$z = \frac{r_0}{2d} = \frac{r_0\sqrt{(2\pi\sigma\omega\mu)}}{2c} \qquad (60.9)$$

it follows from (60.8) and (60.4) in retaining the first four series terms that

$$R + j\omega L_i = \frac{j}{\pi\sigma d^2}\,\frac{1 + 2jz^2 - z^4 - 2jz^6/9 + \ldots}{4jz^2 - 4z^4 - 4jz^6/3 + 2z^8/9 + \ldots}$$

$$= \frac{1}{\pi r_0^2 \sigma}(1 + jz^2 + \tfrac{1}{3}z^4 - \tfrac{1}{6}jz^6 + \ldots)$$

Here also we find an increased resistance per unit length

$$R = R_0(1 + \tfrac{1}{3}z^4 + \ldots) = R_0\left(1 + \frac{r_0^4}{48d^4} + \ldots\right)$$

and a self-inductance contribution per centimetre

$$L_i = \frac{\mu}{2c^2}(1 - \tfrac{1}{6}z^4 + \ldots) = \frac{\mu}{2c^2}\left(1 - \frac{r_0^4}{96d^4} + \ldots\right)$$

As an example, let us consider a circular wire loop of loop radius $a$, and therefore of circumference $2\pi a$. To be added to its self-inductance (49.12) calculated in §49 for quasi-stationary currents, we now have the expression

$$2\pi a\{L_i(\omega, \mu) - L_i(0, 1)\} = \frac{\pi a}{c^2}\left((\mu - 1) - \frac{\mu r_0^4}{96d^4} + \ldots\right)$$

since in §49 we specially calculated with $\mu = 1$. For soft-iron wires with, say, $\mu = 500$ the $\omega$-influence upon the total inductance becomes considerable even for small $z$-values.

For *strong skin effect*, with (60.8), (60.5), and (60.9) we obtain simply

$$R + j\omega L_i = \frac{j}{\pi\sigma d^2 r_0}\frac{d}{1 + j} = R_0(1 + j)\frac{r_0}{2d} = R_0(1 + j)z$$

Here, therefore, the ratio of the a.c. resistance to the d.c. resistance is $z:1 = r_0^2\pi : 2r_0\pi d$, i.e. it goes inversely as the cross-sections of the current-carrying layers. We therefore obtain the dependence of the quantity $R/R_0$ upon $z$ given in figure 54. With the $d$-values for copper

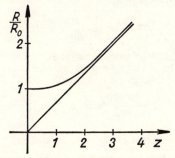

Fig. 54.—Skin effect. Increase in the ohmic resistance $R$ with respect to the quantity $z$ given in equation (60.9)

calculated in §58, we have the following table for the $z$-values in their dependence upon the given wire radii, wave length, and frequency:

Values of $z$ from equation (60.9) for copper wire of radius 1 cm,
0·1 cm, and 0·01 cm, for different wavelengths

| $r =$ | 1 | 0·1 | 0·01 | $\dfrac{\omega}{2\pi} = \dfrac{c}{\lambda}$ |
|---|---|---|---|---|
| $\lambda = 6 \times 10^8$ cm | 0·53 | 0·053 | 0·0053 | 50/s |
| $\lambda = 6 \times 10^6$ cm | 5·3 | 0·53 | 0·053 | 5000/s |
| $\lambda = 6 \times 10^4$ cm | 53 | 5·3 | 0·53 | $5 \times 10^5$/s |
| $\lambda = 6 \times 10^2$ cm | 530 | 53 | 5·3 | $5 \times 10^7$/s |
| $\lambda = 6$ cm | 5300 | 530 | 53 | $5 \times 10^9$/s |

The so-called induction heating of cylindrical bars is an applied practical phenomenon related to the skin effect. This consists in placing the bar to be heated in a longitudinal alternating magnetic field, usually of high frequency. This field produces in the bar an electric field whose force lines are circles concentric with the bar axis. The Joule heating of the ring current thus induced produces the required temperature rise. We have here, therefore, the same electromagnetic situation as would be produced by a linearly polarized wave incident normally upon the surface. Now, however, the **H**-vector oscillates parallel to the wire axis. As opposed to the skin effect, the electric and magnetic field strengths now have their roles exchanged; in particular, equation (60.3) for the penetration of the magnetic field **H** into the bar to be heated ceases to be valid.

*Remark:* In the MKSA system, in which the telegraphic equation (60.2) has the form

$$\nabla^2 \mathbf{E} = \varepsilon \varepsilon_0 \mu \mu_0 \frac{\partial^2 \mathbf{E}}{\partial t^2} + \sigma \mu \mu_0 \frac{\partial \mathbf{E}}{\partial t}$$

the decisive equation (60.3) for calculating the skin effect remains the same as in the Gaussian unit system if the depth of penetration $d$ and the auxiliary quantity $z$ be defined, instead of through (60.1) and (60.9), by

$$d = \sqrt{\left(\frac{2}{\mu \mu_0 \sigma \omega}\right)} \quad \text{and} \quad z = \frac{r_0}{2} \sqrt{\left(\frac{\mu \mu_0 \sigma \omega}{2}\right)}$$

## §61. Electromagnetic waves along ideal conductors

We consider two long straight cylindrical conductors parallel to one another, e.g. two parallel wires (a regular parallel transmission line), or a cable consisting of a central conductor and an insulating sheath, the whole immersed in sea-water. In the latter case the wire constitutes one of the two conductors, and the sea-water the other. Let the resistance of the conductor at first be set equal to zero, so that an ohmic voltage drop along the conductor does not take place. Let there be between the conductors a homogeneous insulator with the material constants $\varepsilon$ and $\mu$.

We now seek the field in the insulator as a special solution of the Maxwell equations, being the solution for which the electric field as well as the magnetic field is everywhere perpendicular to the wire axis. If we choose the latter as the $z$-axis, then we shall have everywhere that $E_z = 0$, $H_z = 0$. Following then from the two divergence equations, as well as from the $z$-components of the curl equations, we have

$$\left. \begin{array}{ll} \dfrac{\partial E_x}{\partial x} + \dfrac{\partial E_y}{\partial y} = 0 & \dfrac{\partial H_x}{\partial x} + \dfrac{\partial H_y}{\partial y} = 0 \\[2mm] \dfrac{\partial E_y}{\partial x} - \dfrac{\partial E_x}{\partial y} = 0 & \dfrac{\partial H_y}{\partial x} - \dfrac{\partial H_x}{\partial y} = 0 \end{array} \right\} \tag{61.1}$$

while the other components of the curl equation then read

$$\left. \begin{array}{ll} -\dfrac{\partial H_y}{\partial z} = \dfrac{\varepsilon}{c} \dfrac{\partial E_x}{\partial t} & -\dfrac{\partial E_y}{\partial z} = -\dfrac{\mu}{c} \dfrac{\partial H_x}{\partial t} \\[2mm] \dfrac{\partial H_x}{\partial z} = \dfrac{\varepsilon}{c} \dfrac{\partial E_y}{\partial t} & \dfrac{\partial E_x}{\partial z} = -\dfrac{\mu}{c} \dfrac{\partial H_y}{\partial t} \end{array} \right\} \tag{61.2}$$

Now these equations are in general solved through the lemma

$$\left. \begin{array}{ll} E_x = -\dfrac{\partial^2 \Psi}{\partial x\, \partial z} & E_y = -\dfrac{\partial^2 \Psi}{\partial y\, \partial z} \\[2mm] H_x = -\dfrac{\varepsilon}{c} \dfrac{\partial^2 \Psi}{\partial y\, \partial t} & H_y = \dfrac{\varepsilon}{c} \dfrac{\partial^2 \Psi}{\partial x\, \partial t} \end{array} \right\} \tag{61.3}$$

if the as yet arbitrary function $\Psi \equiv \Psi(x,y,z,t)$ satisfies the equations

$$\frac{\partial^2 \Psi}{\partial x^2} + \frac{\partial^2 \Psi}{\partial y^2} = 0 \tag{61.4}$$

and

$$\frac{\partial^2 \Psi}{\partial z^2} = \frac{\varepsilon\mu}{c^2} \frac{\partial^2 \Psi}{\partial t^2} \tag{61.5}$$

Obviously $\Psi$, according to (61.5), represents a wave process which propagates itself in a dielectric with the well-known wave velocity

$$w = \frac{c}{\sqrt{(\varepsilon\mu)}} \qquad (61.6)$$

in the direction of the positive or negative $z$-axis, that is, in either direction along the transmission line. With two completely arbitrary functions $f$ and $g$ we can therefore write

$$\Psi(x, y, z, t) = \psi(x, y) \cdot \{f(z - wt) + g(z + wt)\} \qquad (61.7)$$

Then, on account of (61.4), the function $\psi$ which is independent of these propagation phenomena must satisfy the two-dimensional Laplace equation

$$\frac{\partial^2\psi}{\partial x^2} + \frac{\partial^2\psi}{\partial y^2} = 0 \qquad (61.8)$$

It must, in addition, satisfy the boundary conditions of the electrostatic potential on the conductor surface.

The field **E**, **H**, to be derived from (61.3) with (61.7), differs from that of a plane wave in a homogeneous medium only in that in the wave plane ($xy$-plane) both the field direction and the field intensity are functions of position. Since the intensity is essentially different from zero only in the close neighbourhood of the wire, we have the picture of a wave on, and guided by, the double conductor.

The description of the field for any specific case is given by the form of the function $\psi$, that is, by the solution of equation (61.8). Restricting ourselves for the present to the wave travelling in the positive $z$-direction and writing

$$\frac{df(z - wt)}{dz} = F(z - wt)$$

we have from (61.3) with (61.6) that

$$E_x = \sqrt{\left(\frac{\mu}{\varepsilon}\right)} H_y = -\frac{\partial\psi}{\partial x} F \qquad E_y = -\sqrt{\left(\frac{\mu}{\varepsilon}\right)} H_x = -\frac{\partial\psi}{\partial y} F \qquad (61.9)$$

In this case also, then, as for the plane light wave, **E** and **H** are always perpendicular to each other, and we always have, as there, $\varepsilon E^2 = \mu H^2$.

Applying the electric-flux theorem to a 1-cm-high cylindrical portion of the surface of one of the conductors, we obtain that

$$\varepsilon \oint E_n \, dS \rightarrow \varepsilon \oint E_n \, ds = 4\pi e = -\varepsilon F \oint \frac{\partial \psi}{\partial n} \, ds$$

in which $e$ is the charge on the conductor per centimetre of length and in which the line integral goes once around the conductor on its surface. Correspondingly, for the line integral of the magnetic field strength along the same curve we have

$$\oint \mathbf{H}. \, d\mathbf{r} = \oint H_t \, ds = \frac{4\pi I}{c}$$

for the term with the displacement current on the right-hand side vanishes since $E_z = 0$. Since now on account of (61.9) we obviously have that $\varepsilon^{1/2} E_n = \mu^{1/2} H_t$, it follows from the two foregoing equations that

$$e = I \frac{\sqrt{(\varepsilon\mu)}}{c} = \frac{I}{w} = -\frac{\varepsilon F}{4\pi} \oint \frac{\partial \psi}{\partial n} \, ds \qquad (61.10)$$

If, further, we form the line integral of $\mathbf{E}$ from the first to the second conductor, we obtain the voltage between the conductors:

$$V = \int_1^2 \mathbf{E}. \, d\mathbf{r} = (\psi_1 - \psi_2) F$$

The same line integral over $\mathbf{B}$, ($\mathbf{B} = \mu\mathbf{H}$, and $\mathbf{B}$ is perpendicular to $\mathbf{E}$) gives the flux of induction $\Phi$, per centimetre, through the space between the two conductors. Thus, again because of the proportionality between $\mathbf{E}$ and $\mathbf{H}$, we have

$$V = \frac{\Phi}{\sqrt{(\varepsilon\mu)}} = (\psi_1 - \psi_2) F \qquad (61.11)$$

Now the capacitance $C$ and the external inductance $L_a$ of the arrangement, per centimetre, are defined by

$$C = \frac{e}{V} \qquad L_a = \frac{\Phi}{Ic}$$

Thus, on account of (60.10) and (60.11) we have that

$$CL_a = \frac{e}{Ic} \frac{\Phi}{V} = \frac{\varepsilon\mu}{c^2} = \frac{1}{w^2} \quad \text{with} \quad C = -\frac{\varepsilon}{4\pi(\psi_1 - \psi_2)} \oint \frac{\partial \psi}{\partial n} \, ds \qquad (61.12)$$

*The product: capacitance times external self-inductance per unit length is therefore equal to the reciprocal of the square of the velocity of light in the surrounding medium.* (The designation "external self-inductance" and the subscript $a$ shall mean here that for calculating the total inductance $L$, the contribution $L_i$ of the magnetic field within the conductor, as found in §60, would also be taken into consideration.)

The Poynting vector in our arrangement is everywhere parallel to the $z$-axis, and thus, according to (61.9), we have

$$S_z = \frac{c}{4\pi}(E_x H_y - E_y H_x) = \frac{c}{4\pi}\sqrt{\left(\frac{\varepsilon}{\mu}\right)}F^2\left\{\left(\frac{\partial\psi}{\partial x}\right)^2 + \left(\frac{\partial\psi}{\partial y}\right)^2\right\}$$

From (61.8), Green's theorem gives for the total energy flow through the $xy$-plane as the contribution of the conductor surfaces (using $A$ here for surface area).

$$\int S_z \, dA = -\frac{c}{4\pi}\sqrt{\left(\frac{\varepsilon}{\mu}\right)}F^2\left\{\oint_1 \psi\frac{\partial\psi}{\partial n}ds + \oint_2 \psi\frac{\partial\psi}{\partial n}ds\right\}$$

Since $\psi$ on the two conductor surfaces is constant, we obtain with (61.10) and (61.11)

$$\int S_z \, dA = IV$$

that is, just the Joule heat which is obtained when an ohmic resistance of value $R = V/I$ is annexed to the (resistance-free) parallel line. This result was immediately to be expected from the considerations of §54 with respect to the Poynting vector.

Let us mention, parenthetically, the values of $C$ found in two simple but explicit cases:

1. *The parallel transmission line:* If the distance $d$ between the two conductors is large in comparison with their radius $b$, then, for an arbitrary field point we have that $\psi = k \ln (r_2/r_1)$, where $k$ is a constant independent of voltage, and $r_1$ and $r_2$ represent the distance of the field points from the two wire axes. On account of (61.9), this formula can be obtained as the electrostatic potential of two charged thin wires, or as the vector potential of two current-carrying wires (see figure 55). Thus we have $\psi_1 = -\psi_2 \approx k \ln (d/b)$, and $\oint_1(\partial\psi/\partial n)ds = -2\pi k$, so that on account of (61.2) we obtain

$$C = \frac{\varepsilon}{4 \ln (d/b)}$$

2. *The coaxial cable:* (Central conductor of radius $a$; conducting outer sheath of radius $b > a$). In the insulation between the two conductors we have that

$\psi \sim \ln (r/a)$, being the basic solution of (61.8) in plane polar coordinates. It therefore follows from (61.12) that

$$C = \frac{\varepsilon}{2 \ln (b/a)}$$

By making use of the concepts of capacitance and self-inductance of our parallel transmission line we can also obtain the fundamental equations for waves along wires as follows: The charge on an element

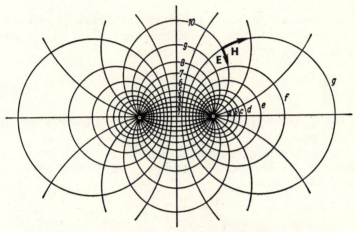

Fig. 55.—Field lines between two parallel wires. If the wires carry currents in opposite directions, then the circles $a, b, \ldots g$ give the positions of constant vector potential, and, conjointly, the magnetic field-lines. If however, the wires are electrostatically charged, the same lines give the equipotentials. The electric field-lines are then given by the circles 1, 2, . . . 10

$dz$ of one of the conductors can only change if the current entering the element differs in amount from the current leaving. The connection between $e$ and $I$ gives the *equation of continuity*

$$\frac{\partial e}{\partial t} = -\frac{\partial I}{\partial z} \tag{61.13}$$

which, since $F \equiv F(z - wt)$, is in agreement with the relationship (61.10).

Next, we may apply the induction law to a strip, of breadth $dz$, lying between the two conductors (the area $abcd$ of figure 56). The flux of induction (divided by $c$) through this strip is, according to the above definition of $L_a$, equal to $L_a l\,dz$, while for the line integral over the field strength we have

$$\oint_{abcd} \mathbf{E} \cdot d\mathbf{r} = V_{ab} - V_{cd} = \frac{\partial V}{\partial z}\,dz$$

and therefore

$$\frac{\partial V}{\partial z} = -L_a \frac{\partial I}{\partial t} \qquad (61.14)$$

Now, however, from (61.13) and (61.14), by elimination of $e$ and $V$ by means of the relation $e = CV$, and since $CL_a = 1/w^2$, we obtain immediately the wave equation (61.15) for $I$

$$\frac{\partial^2 I}{\partial z^2} = \frac{1}{w^2}\frac{\partial^2 I}{\partial t^2} \qquad (61.15)$$

and also analogously for every other electrotechnical quantity associated with our arrangement.

Of the many possible applications of waves along wires we shall discuss only the following: A parallel pair extends from $z = 0$ to $z = l$.

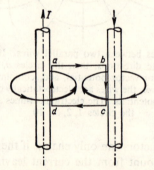

Fig. 56.—For applying the induction law to the parallel twin-pair

Let an alternating-current source $V = V_0 e^{jwt}$ be connected at its beginning ($z = 0$), and let an ohmic resistance be connected at its end ($z = l$). The general solution of equations (61.13) to (61.15) here is obviously

$$\left.\begin{array}{l} V = V_1 e^{j\omega(t-z/w)} + V_2 e^{j\omega(t+z/w)} \\ I = Cw\{V_1 e^{j\omega(t-z/w)} - V_2 e^{j\omega(t+z/w)}\} \end{array}\right\} \qquad (61.16)$$

in which $V_1$ and $V_2$ are two constants of integration to be determined from the boundary conditions. From the boundary condition at $z = 0$ it follows that $V_1 + V_2 = V_0$, and from the condition at $z = l$, it follows that

$$V_1 e^{-j\omega l/w} + V_2 e^{j\omega l/w} = RCw \{V_1 e^{-j\omega l/w} - V_2 e^{j\omega l/w}\} \quad (61.17)$$

By inserting into (61.16) the values of $V_1$ and $V_2$, we obtain for each $R$-value explicit expressions for voltage and current. We discuss here, however, only two special cases:

1. $R = \infty$. The circuit is therefore open at its end. The expression in parentheses in (61.17) must vanish. With the values thus obtaining for $V_1$ and $V_2$ we have from (61.16), introducing the wavelength $\lambda = 2\pi w/\omega$,

$$V = V_0 e^{j\omega t} \frac{\cos 2\pi(l-z)/\lambda}{\cos 2\pi l/\lambda} \qquad I = jV_0 Cw e^{j\omega t} \frac{\sin 2\pi(l-z)/\lambda}{\cos 2\pi l/\lambda}$$

$V$ and $I$ are therefore displaced everywhere by 90°. At points $z = l$, $l - \lambda/2, \ldots$ we have $I = 0$, and at the points $z = l - \lambda/4, l - 3\lambda/4, \ldots$ we have $V = 0$. We thus have *standing waves* with nodes of $I$ and maxima of $V$ at the ends of the line and at points $l - m\lambda/2$, where $m$ is a whole number.

2. If $R = 1/Cw = \sqrt{(L/C)}$, we immediately obtain from (61.17) that $V_2 = 0$. The reflected wave thus disappears and the terminating resistance completely absorbs the waves incident upon it. Thus, from (61.16), $V$ and $I/Cw$ have everywhere equal magnitude and equal phase. We designate as the *wave impedance* **Z** that complex resistance with which a finite transmission line must be terminated so that with respect to incident waves the terminated line behaves like a line of infinite length. For the case here considered,

$$\mathbf{Z} = \sqrt{(L/C)} = 1/Cw = Lw$$

## §62. Waves along wires of finite resistance

For practical telegraphy there is the fundamental question: What modifications are required in the wave propagation on ideal conductors, as described in the foregoing paragraphs, in order to take into account the ever-present ohmic resistance of the conducting wires? It is clear from the outset that the Joule heat which would be developed in the wire would have as a consequence a damping of the waves. In addition to this—as we shall see—both damping and wave velocity will depend upon the frequency, producing a distortion of the signals (of speech, for example) being carried on the conductor.

For a preliminary orientation we consider an infinitely long parallel-pair transmission line at the beginning of which ($z = 0$), a source of alternating current is connected, commencing at a certain moment. If the conductors are without resistance, the power output of the current source has as its sole function the building up of the field between the two conductors; for, in this case, since

$$V = V_0 e^{j\omega(t-z/w)}, \ I = I_0 e^{j\omega(t-z/w)} \text{ with } V_0 = I_0 \mathbf{Z} = I_0 \sqrt{(L/C)} \quad (62.1)$$

we have for the time average of the power, corresponding to the closing remarks in §52, that

$$\overline{\text{Re}\,\{V\}\,\text{Re}\,\{I\}} = \tfrac{1}{2}|I_0|^2\,\text{Re}\,\{\mathbf{Z}\} = \tfrac{1}{2}|I_0|^2\,Lw \qquad (62.2)$$

The interpretation of this is obvious: In one second the wave front advances a distance $w$ centimetres. The quantity $\tfrac{1}{2}|I_0|^2 L$ is the field energy per centimetre of conductor. Equation (62.2) therefore represents the field energy arriving per second.

If now the conductor possesses a finite resistance $R$ per centimetre, then a part of the power output is converted into Joule heat. The farther the wave front advances, the greater becomes the fraction of the input power converted into heat, and the less the energy available for further build-up of the field.

For subjecting these ideas to actual calculation we must take into account the finite ohmic resistance of the transmission line. We do this already in the formulation of the basic equations, specifically in equation (61.14), as follows: We calculate the voltage around the rectangle *abcd* (figure 56). This accounts for the appearance along the paths *bc* and *da* of an ohmic voltage drop $RI$, in total. This quantity is added to the term $\partial V/\partial z$. Then, instead of equations (61.13) and (61.14) for describing the wave propagation, we have the two equations

$$C\frac{\partial V}{\partial t} + \frac{\partial I}{\partial z} = 0 \qquad L\frac{\partial I}{\partial t} + RI + \frac{\partial V}{\partial z} = 0 \qquad (62.3)$$

These equations are solved by the forms

$$I = I_0 e^{j(\omega t - \gamma z)} \text{ and } V = \mathbf{Z}I_0 e^{j(\omega t - \gamma z)} \qquad (62.4)$$

if $\qquad \omega C\mathbf{Z} - \gamma = 0 \text{ and } \omega L - jR - \gamma \mathbf{Z} = 0$

From this, for the propagation constant $\gamma = \alpha - j\beta$, we have

$$\gamma = \sqrt{[\omega C(\omega L - jR)]} \text{ with } \left.\frac{\alpha^2}{\beta^2}\right\} = \tfrac{1}{2}\omega C\left\{\sqrt{[(\omega L)^2 + R^2]} \pm \omega L\right\} \quad (62.5a)$$

and for the wave impedance $\mathbf{Z} = Z - jZ'$

$$\mathbf{Z} = \sqrt{\left(\frac{\omega L - jR}{\omega C}\right)} = \frac{\gamma}{\omega C} \text{ with } Z = \frac{\alpha}{\omega C}, \quad Z' = \frac{\beta}{\omega C} \quad (62.5b)$$

We therefore find damped waves, with the damping factor $e^{-\beta z}$; after travelling a distance $1/\beta$, the amplitudes of current and voltage have fallen to $1/e$ of their original values.

The value of $\beta$ found here could also have been immediately obtained from $Z = \text{Re}\{\mathbf{Z}\}$ by the fact that the power (62.2) introduced at $z = 0$ must be recovered in the power transformed into Joule heat in the wire.

$$\tfrac{1}{2}R\int_0^\infty |I|^2\,dz = \tfrac{1}{2}R|I_0|^2\int_0^\infty e^{-2\beta z}\,dz = \frac{R}{4\beta}|I_0|^2$$

We must therefore have that $Z = R/2\beta$, which, since $\alpha\beta = \tfrac{1}{2}\omega CR$ according to (62.5), is actually the case.

We speak of *weak damping* when $R \ll \omega L$, and of *strong damping* for $R \gg \omega L$. For the first extreme case we obtain for $\beta$ from (62.5a) by development, the frequency-independent value $\beta = \sqrt{(R^2C/4L)}$, and for the second the frequency-dependent value $\beta = \sqrt{(\omega CR/2)}$.

We wish, from the practical point of view, to inform ourselves of the order of magnitude of the foregoing quantities. As normal frequency for telephone service we choose the technically customary value $\omega = 5000\,\text{s}^{-1}$. For the (dimensionless) capacitance per centimetre we say, corresponding to the value $C = \varepsilon/4 \ln(d/b)$ for the parallel twin-pair, that $C = 0.2$; for the resistance per centimetre, about $3.4 \times 10^{-4}\Omega/\text{cm} = 4 \times 10^{-16}\text{s}/\text{cm}^2$, corresponding to a twin pair consisting of copper wires of $1\,\text{mm}^2$ cross-section. From this, for a parallel twin-pair in air, since $LC = 1/w^2 = 1/c^2$, we have the value $L = 6 \times 10^{-21}\text{s}^2/\text{cm}^2$. Thus, with $R/\omega L = 13$, we have the case of strong damping.

For the $\beta$-value of our system we find $\beta = \sqrt{(\omega CR/2)} = 4.5 \times 10^{-7}\text{cm}^{-1}$, corresponding to a range $1/\beta = 22\,\text{km}$. (This is the distance at which the amplitudes have decreased to $1/e$ of their original values. By making full use of modern amplification techniques, however, an appreciably greater damping can be tolerated. Thus the long-distance cable we have calculated becomes usable over a distance equal to an appreciable multiple of the range value found above.)

With regard to technical applications we must observe that in practice, besides the ohmic resistance $R$, a quantity $G$ makes itself felt as a disturbance. It is due to the imperfection of cable insulation and is called the "leakance". Thus $GV$ means the current which flows, per centimetre of cable length, across the insulation from one conductor to the other when a potential difference $V$ exists between the conductors. The leakance has as a consequence the decay with time of the charge $CV$. Equations (62.3) thus assume the following form (as most often used in practice):

$$C\frac{\partial V}{\partial t}+GV+\frac{\partial I}{\partial z}=0 \qquad L\frac{\partial I}{\partial t}+RI+\frac{\partial V}{\partial z}=0$$

With the solution form (62.4) we now obtain instead of (62.5):

$$\gamma = \sqrt{[(\omega C-jG)(\omega L-jR)]} \qquad (62.6)$$

Decomposing $\gamma$ into $\alpha-i\beta$ gives us

$$\left.\begin{array}{c}\alpha^2\\\beta^2\end{array}\right\} = \tfrac{1}{2}\{\sqrt{[(\omega^2 LC-RG)^2+\omega^2(CR+LG)^2]}\pm(\omega^2 LC-RG)\}$$

In the case of light damping, these expressions permit in practice an instructive transformation. If we consider $R$ and $G$ as small quantities and neglect higher powers of them than the second, then

$$\alpha^2 = \omega^2 LC+\tfrac{1}{4}\left(R\sqrt{\frac{C}{L}}-G\sqrt{\frac{L}{C}}\right)^2 \qquad \beta = \tfrac{1}{2}\left(R\sqrt{\frac{C}{L}}+G\sqrt{\frac{L}{C}}\right) \quad (62.7)$$

Thus we see that although we can increase $L$ to diminish damping, a limit to this is set by the appearance of leakance. $L$ being variable, we reach a minimum value of $\beta$ if the two terms of the sum, $R\sqrt{(C/L)}$ and $G\sqrt{(L/C)}$, become equal. This state of least damping brings a further advantage. According to the first equation we then have that $\alpha/\omega = \sqrt{(LC)}$, and this means that the wave velocity is independent of the frequency. Thus, for the condition of minimum damping, the transmission line is distortion-free.

## §63. Waves in hollow conductors

Inside an electrically conducting tube or pipe purely transverse waves of the type considered in §61 are not possible. Here, however, we have waves with longitudinal field components. These waves are of great importance in ultra-high-frequency techniques.* We distinguish an $H$-type wave and an $E$-type wave according as a magnetic or an electric longitudinal component is present.

---

* Such waves do indeed exist in the arrangements treated in §61, but they are much less common than the purely transverse waves.

In order to obtain some idea of the properties of these waves, we shall consider the case of a hollow conductor of rectangular cross-section having perfectly conducting walls. Let the $z$-coordinate lie parallel to the tube axis, and let the inside of the tube extend in the $xy$-plane over the region $0 < x < a$, $0 < y < b$. We look for waves (of frequency $\omega$), which travel in the $z$-direction with the wave number $k = 2\pi/\lambda$. Thus all field vectors shall be represented in the form $f(x, y)e^{i(\omega t - kz)}$. They must satisfy both the Maxwell equations and the boundary conditions (vanishing of the tangential component of **E** as well as the normal component of **H**):

$$
\begin{aligned}
E_y = E_z = H_x = 0 \quad \text{for } x = 0 \quad \text{and} \quad x = a \\
E_x = E_z = H_y = 0 \quad \text{for } y = 0 \quad \text{and} \quad y = b
\end{aligned}
\tag{63.1}
$$

We therefore begin tentatively with $n$ and $m$ whole numbers as follows:

$$
\left.
\begin{aligned}
E_x &= \alpha \cos \frac{n\pi x}{a} \sin \frac{m\pi y}{b} e^{i(\omega t - kz)} & H_x &= \alpha' \sin \frac{n\pi x}{a} \cos \frac{m\pi y}{b} e^{i(\omega t - kz)} \\[2mm]
E_y &= \beta \sin \frac{n\pi x}{a} \cos \frac{m\pi y}{b} e^{i(\omega t - kz)} & H_y &= \beta' \cos \frac{n\pi x}{a} \sin \frac{m\pi y}{b} e^{i(\omega t - kz)} \\[2mm]
E_z &= \gamma \sin \frac{n\pi x}{a} \sin \frac{m\pi y}{b} e^{i(\omega t - kz)} & H_z &= \gamma' \cos \frac{n\pi x}{a} \cos \frac{m\pi y}{b} e^{i(\omega t - kz)}
\end{aligned}
\right\}
\tag{63.2}
$$

Since each of these components must satisfy the wave equation *in vacuo*, we arrive at the relationship

$$
\left(\frac{n\pi}{a}\right)^2 + \left(\frac{m\pi}{b}\right)^2 + k^2 = \frac{\omega^2}{c^2}
\tag{63.3}
$$

Finally, the Maxwell equations lead to relationships between the coefficients $\alpha, \ldots, \gamma'$; with the two new arbitrary constants $\delta$ and $\delta'$ we find after a brief calculation

$$
\left.
\begin{aligned}
\alpha &= \frac{n\pi}{a} k\delta - \frac{m\pi}{b} \frac{\omega}{c} \delta' & \alpha' &= -\frac{m\pi}{b} \frac{\omega}{c} \delta - \frac{n\pi}{a} k\delta' \\[2mm]
\beta &= \frac{m\pi}{b} k\delta + \frac{n\pi}{a} \frac{\omega}{c} \delta' & \beta' &= \frac{n\pi}{a} \frac{\omega}{c} \delta - \frac{m\pi}{b} k\delta' \\[2mm]
\gamma &= i\left(\frac{\omega^2}{c^2} - k^2\right)\delta & \gamma' &= i\left(\frac{\omega^2}{c^2} - k^2\right)\delta'
\end{aligned}
\right\}
\tag{63.4}
$$

Herewith, for the particular case $\delta = 0$, $\delta' \neq 0$ we obtain the H-type wave in a conducting pipe (with $E_z = 0$), and for $\delta' = 0$, $\delta \neq 0$ the E-type (with $H_z = 0$).

Equation (63.3) dominates the most important aspects of hollow-conductor phenomena. It yields a real value of $k$ only when $\omega$ is greater than the "critical" frequency

$$\omega_c = c \sqrt{\left[\left(\frac{n\pi}{a}\right)^2 + \left(\frac{m\pi}{b}\right)^2\right]} \qquad (63.5)$$

when, that is, the wavelength *in vacuo* is shorter than the "critical" wavelength

$$\lambda_c = \frac{2\pi c}{\omega_c} = \frac{2}{\sqrt{[(n/a)^2 + (m/b)^2]}}$$

If we ask what is the lowest frequency which our pipe can pass, there appears an essential difference between the two wave types. In the H-type ($\delta = 0$) we obtain (when $a \geqq b$) the smallest value of $\omega_c$ with $n = 1$, $m = 0$ (fundamental wave). For this wave, not only $E_z$ vanishes, but also $E_x$ and $H_y$. For the E-type ($\delta' = 0$), however, according to (63.2) and (63.4), there exists a solution different from zero only when both $n$ and $m$ are greater than zero. Thus the smallest value of $\omega_c$ obtains for $n = 1$, $m = 1$. That the multiplicity of solutions of the H-type is greater than that of the E-type is rendered plausible by the fact that for $H_z$ no restricting boundary conditions exist, while $E_z$ must be zero on the tube walls.

Upon introducing the critical frequency $\omega_c$, we have from (63.3)

$$k = \frac{1}{c} \sqrt{(\omega^2 - \omega_c^2)} \qquad (63.6)$$

Thus for the *phase velocity* $w$ we have

$$w = \frac{\omega}{k} = \frac{c}{\sqrt{[1 - (\omega_c/\omega)^2]}} \qquad (63.7)$$

It is greater than the velocity of light $c$ and approaches $c$ in the limit $\omega \to \infty$. By contrast, the *group velocity* $v$ is always smaller than $c$. Indeed, we have always that $vw = c^2$.

To substantiate this assertion we consider a wave group with a frequency spectrum given by $f(\omega)$. Let $f(\omega)$ be different from zero only in the immediate neighbourhood of $\omega_0$. The dependence of each field quantity is then given by the factor

$$\phi(z, t) = \int f(\omega) \exp i[\omega t - k(\omega)z] \, d\omega$$

We set $\omega = \omega_0 + \mu$, and we develop $k(\omega) = k(\omega_0) + \mu(dk/d\omega)_{\omega_0}$. Since $f(\omega)$ is different from zero only for small $\mu$, we can break off the development, to a first approximation, with the linear term in $\mu$. We then have

$$\phi(z, t) = \exp i[\omega_0 t - k(\omega_0)z] \int f(\omega_0 + \mu) \exp i\mu[t - (dk/d\omega)_{\omega_0} z] \, d\mu$$

Here $\phi(z, t)$ is represented as a plane wave with an amplitude depending upon $t - z(dk/d\omega)_{\omega_0}$. It describes a wave group which moves with velocity $v = (d\omega/dk)_{\omega_0}$. From (63.6) it follows directly that

$$v = c \sqrt{\left[1 - \left(\frac{\omega_c}{\omega}\right)^2\right]} = \frac{c^2}{w} \tag{63.8}$$

Any wave process *in vacuo* can be described by means of a superposition of plane waves. This applies with the greatest simplicity to the waves we have represented in (63.2), by substituting exponential functions for the trigonometric functions there, through the relationships

$$\cos \phi = \tfrac{1}{2}(e^{i\phi} + e^{-i\phi}) \qquad \sin \phi = \frac{1}{2i}(e^{i\phi} - e^{-i\phi})$$

After carrying out all multiplications, **E** and **H** stand as sums of pure plane waves. We restrict ourselves to carrying out such a decomposition on the fundamental wave of the $H$-type, and we discuss this briefly. We have for it

$$E_y = \beta \sin \frac{\pi x}{a} \exp i(\omega t - kz)$$

$$= \frac{\beta}{2i} \left\{ \exp i\left(\omega t + \frac{\pi}{a}x - kz\right) - \exp i\left(\omega t - \frac{\pi}{a}x - kz\right) \right\} \tag{63.9}$$

It therefore appears as the superposition of two waves whose wave normals lie in the $xy$-plane. The angle $\varepsilon$ between them and the $x$-direction is given by

$$\cos \varepsilon = \mp \frac{\pi}{a} \bigg/ \left[\left(\frac{\pi}{a}\right)^2 + k^2\right]^{\frac{1}{2}}$$

i.e., on account of (63.5), by $\cos\varepsilon = -\omega_c/\omega$ for the first of the waves appearing in (63.9), and by $\cos\varepsilon = +\omega_c/\omega$ for the second. The entire field $E_y$ can therefore be imagined as being produced by the continual reflection of a plane wave incident at angle $\varepsilon$ upon the plane surfaces $x = 0$ and $x = a$. According to equations (63.7) and (63.9), the velocity of propagation $w$ of this wave in the $z$-direction is

$$w = \frac{c}{\sin\varepsilon}$$

It is equal to the velocity with which the wave plane cuts the plane $x = 0$.

## CHAPTER D III

### The Electromagnetic Field of a Given Distribution of Charge and Current

#### §64. The field of a uniformly moving charged particle

We consider the field of a charge moving *in vacuo* with *constant velocity* **v**, described by a charge density $\rho$ which has values different from zero only in a very small region. This field is determined by

$$\left.\begin{array}{ll} \operatorname{curl}\mathbf{H} = \dfrac{1}{c}\dfrac{\partial\mathbf{E}}{\partial t} + \dfrac{4\pi}{c}\rho\mathbf{v} & \operatorname{div}\mathbf{E} = 4\pi\rho \\[3mm] \operatorname{curl}\mathbf{E} = -\dfrac{1}{c}\dfrac{\partial\mathbf{H}}{\partial t} & \operatorname{div}\mathbf{H} = 0 \end{array}\right\} \tag{64.1}$$

with the solution of which we shall now be occupied.

First of all we can establish without any calculation, merely on symmetry grounds, that **E** always lies in the plane containing the field point and the particle trajectory, while **H** is perpendicular to that plane. Moreover, the field comprising **E** and **H** is rotationally symmetric around this trajectory.

Now further, we show that whenever we know the **E**-field, the **H**-field can be immediately determined from the expression

$$\oint\mathbf{H}.d\mathbf{r} = \frac{1}{c}\frac{d}{dt}\int E_n\,dS \tag{64.2}$$

this being identical with the first equation in (64.1). As boundary curve, we choose a circle of radius $a$ perpendicular to the direction of motion and centred on the particle trajectory. In this case $\oint\mathbf{H}.d\mathbf{r} = 2\pi aH$. For calculating the change with time of the flux of **E** through any surface $S$ bordered by the chosen boundary curve we observe that, on account of the motion of the charge, the total flux passing through surface $S$ at the moment $t+dt$ is equal to the flux linked at moment $t$ by the surface $S'$ obtained from $S$ by a rigid displacement of $S$ through a distance $-\mathbf{v}dt$. Hence $d(\int E_n\,dS)$ is equal to the difference of the flux of **E** through $S'$ and $S$, both evaluated at time $t$. Since the vector field of **E** is source-free at all points outside the charge itself, this difference of the fluxes through $S'$ and $S$ is equal to the flux through the cylindrical band connecting the two surfaces, that is, equal to $2\pi a v\, dt\, E_n$, in which $E_n$ now means the component normal to the surface of this band. Putting this into equation (64.2) we find that $H = (v/c)E_n$, or, in view of the directional characteristics of **E** and **H** established above, we have the vector relationship

$$\mathbf{H} = (1/c)\, \mathbf{v} \times \mathbf{E} \qquad (64.3)$$

We wish now to convince ourselves on the basis of the Maxwell equations that this expression for the magnetic field of a charge distribution moving with constant velocity is in fact *rigorously true*. For this purpose we determine the sources and vortices of the magnetic field given by equation (64.3). First we have, **v** being constant, that

$$\operatorname{div}\mathbf{H} = \operatorname{div}\left(\frac{\mathbf{v}}{c}\times\mathbf{E}\right) = -\frac{\mathbf{v}}{c}\cdot\operatorname{curl}\mathbf{E}$$

The last expression vanishes because from the induction law,

$$\operatorname{curl}\mathbf{E} = -\frac{1}{c}\frac{\partial\mathbf{H}}{\partial t} = -\frac{\mathbf{v}}{c^2}\times\frac{\partial\mathbf{E}}{\partial t}$$

we have that $\operatorname{curl}\mathbf{E}$ is perpendicular to **v**. Further we have

$$\operatorname{curl}\mathbf{H} = \operatorname{curl}\left(\frac{\mathbf{v}}{c}\times\mathbf{E}\right) = -\left(\frac{\mathbf{v}}{c}\cdot\nabla\right)\mathbf{E} + \frac{\mathbf{v}}{c}\operatorname{div}\mathbf{E} \qquad (64.4)$$

Here the last expression on the right agrees with the current term of the first Maxwell equation because $\operatorname{div}\mathbf{E} = 4\pi\rho$. Also, that the first two expressions are equal follows because in the special case here considered, the whole field distribution moves unaltered with the particle. For each field component the relation

$$f(x, y, z, t) = f(x + v_x\, dt, y + v_y\, dt, z + v_z\, dt, t + dt)$$

therefore holds. From this it follows that

$$\frac{\partial f}{\partial t} = -\left(v_x\frac{\partial f}{\partial x} + v_y\frac{\partial f}{\partial y} + v_z\frac{\partial f}{\partial z}\right) = -\mathbf{v}\cdot\operatorname{grad}f \qquad (64.5)$$

Thus in our case $\qquad \partial\mathbf{E}/\partial t = -(\mathbf{v}\cdot\nabla)\mathbf{E}$

by which the agreement of equation (64.4) with the first Maxwell equation is proved. When, therefore, we know the electric field of the moving charge, the accompanying magnetic field is given by equation (64.3).

For *sufficiently small velocities* we may safely assume that the E-field of the moving charge does not differ appreciably from the field of a charge at rest, and that therefore we may write

$$\mathbf{E} = e \frac{\mathbf{r}}{r^3} \qquad \mathbf{H} = e \frac{\mathbf{v}}{c} \times \frac{\mathbf{r}}{r^3} \tag{64.6}$$

where $\mathbf{r}$ stands for the vector from the instantaneous location of the particle to the field point. Obviously the magnetic field here is equal to the field which, according to the Biot-Savart law, is produced by a current element $e\mathbf{v}$.

For finding the field for *arbitrarily large velocities* it is expedient to go from the field quantities to the *electromagnetic potentials* $\mathbf{A}$ *and* $\phi$. For this purpose we start with the last of equations (64.1) and satisfy it by the statement that

$$\mathbf{H} = \operatorname{curl} \mathbf{A} \tag{64.7}$$

The induction law then requires that $\mathbf{E} + (1/c)\partial\mathbf{A}/\partial t$ be irrotational. We must therefore require that this quantity be the gradient of a scalar:

$$\mathbf{E} = -\frac{1}{c}\frac{\partial \mathbf{A}}{\partial t} - \operatorname{grad} \phi \tag{64.8}$$

Now we can specify the sources of $\mathbf{A}$ as we like, for a vector is uniquely established only when both its sources and vortices are defined. Following H. A. Lorentz, we make the specification of the sources of $\mathbf{A}$ through the requirement (the so-called "Lorentz convention") that*

$$\operatorname{div} \mathbf{A} + \frac{1}{c}\frac{\partial \phi}{\partial t} = 0 \tag{64.9}$$

---

* The vector and scalar potentials, which describe the magnetic and electric fields through equations (64.7) and (64.8) are not unique. Starting with a given choice for $A$ and $\phi$, the same fields may be obtained from the alternative potentials

$$A' = A + \nabla\psi \tag{64.7a}$$
$$\phi' = \phi - \partial\psi/\partial t \tag{64.8a}$$

where $\psi$ is an arbitrary scalar function. Since it is the fields which are the observable quantities there is no physical basis for choosing between $A$ and $A'$ and the choice may be made on the basis of convenience.

The transformation (64.7a) is called a *gauge transformation*; the equations of Maxwell may be said to be invariant with respect to gauge transformations. The choice of gauge corresponding to equation (64.9) is sometimes referred to as the "Lorentz gauge". Another useful gauge choice, which leads to $\nabla \cdot A = 0$, is discussed in § 46.

When we put equations (64.7) and (64.8) into the first two equations (64.1), we obtain, after some short calculations using (64.9), the two *potential equations*

$$\nabla^2 \mathbf{A} - \frac{1}{c^2}\frac{\partial^2 \mathbf{A}}{\partial t^2} = -\frac{4\pi}{c}\rho\mathbf{v} \qquad \nabla^2 \phi - \frac{1}{c^2}\frac{\partial^2 \phi}{\partial t^2} = -4\pi\rho \qquad (64.10)$$

We notice that the equation for $\mathbf{A}$ differs from that for $\phi$ only in the constant factor $\mathbf{v}/c$ on the right-hand side. When, therefore, we have solved the equation for $\phi$, we immediately find for $\mathbf{A}$ the solution

$$\mathbf{A} = \frac{\mathbf{v}}{c}\phi \qquad (64.11)$$

That this solution also satisfies the requirements of the Lorentz convention (64.9) follows from the relationship (64.5) with $f \equiv \phi$. Moreover, from the two defining equations (64.7) and (64.8), and having regard to (64.5), we obtain

$$\mathbf{E} = -\operatorname{grad}\phi + \frac{\mathbf{v}}{c}\left(\frac{\mathbf{v}}{c}.\operatorname{grad}\phi\right) \qquad \mathbf{H} = -\frac{\mathbf{v}}{c}\times\operatorname{grad}\phi \qquad (64.12)$$

Obviously the relation (64.3) is satisfied by these equations.

In order to solve our problem we have only to integrate the second equation (64.10) for the scalar potential $\phi$. On transforming the time-derivative to a space-derivative by means of (64.5), and taking the $x$-axis along the particle path, we arrive at the relationship

$$\left(1-\frac{v^2}{c^2}\right)\frac{\partial^2 \phi}{\partial x^2}+\frac{\partial^2 \phi}{\partial y^2}+\frac{\partial^2 \phi}{\partial z^2} = -4\pi\rho \qquad (64.13)$$

This obviously differs from the usual potential equation for a charged particle at rest only by the presence of the constant numerical factor $1-v^2/c^2$ multiplying the derivative in $x$. We can therefore formally carry out the solution of our problem as if it were one of simple electrostatics by means of the transformation

$$x = x'\sqrt{(1-\beta^2)} \text{ with } \beta = \frac{v}{c} \qquad (64.14)$$

with $y$ and $z$ unchanged. We have then to solve the equation

$$\frac{\partial^2 \phi}{\partial x'^2}+\frac{\partial^2 \phi}{\partial y^2}+\frac{\partial^2 y}{\partial z^2} = -4\pi\rho(x'\sqrt{(1-\beta^2)},y,z)$$

and, from (19.11) we find directly that

$$\phi(x', y, z) = \iiint \frac{\rho\left[\xi'\sqrt{(1-\beta^2)}, \eta, \zeta\right] d\xi'\, d\eta\, d\zeta}{\sqrt{\left[(x'-\xi')^2 + (y-\eta)^2 + (z-\zeta)^2\right]}}$$

On going back then from $x', \xi'$ to $x, \xi$ we have

$$\phi(x, y, z) = \iiint \frac{\rho(\xi, \eta, \zeta)\, d\xi\, d\eta\, d\zeta}{\sqrt{\{(x-\xi)^2 + (1-\beta^2)\left[(y-\eta)^2 + (z-\zeta)^2\right]\}}} \qquad (64.15)$$

Particularizing, we now look for the solution at the moment for which the charge is located at the origin of the coordinate system, and we restrict ourselves to the case where the charge is a point. We can then set $\xi = \eta = \zeta = 0$ in the denominator and carry out the integration in the numerator, obtaining

$$\phi(x, y, z) = \frac{e}{\sqrt{\{x^2 + (1-\beta^2)\left[y^2 + z^2\right]\}}} \qquad (64.16)$$

If, however, we free ourselves of the special choice of coordinate system by introducing the field-point vector $\mathbf{r} \equiv \{x, y, z\}$ and the velocity $\mathbf{v} \equiv \{v, 0, 0\}$, we obtain

$$\phi(\mathbf{r}) = \frac{e}{\sqrt{\left[r^2(1-\beta^2) + (\mathbf{r} \cdot \mathbf{v}/c)^2\right]}} = \frac{e}{\sqrt{\left[r^2 - (\mathbf{r} \times \mathbf{v}/c)^2\right]}} \qquad (64.17)$$

From this it can be seen with the help of some short calculating, if necessary working with components and using (64.16), that, according to (64.12), the accompanying field is rigorously given by

$$\mathbf{E} = \frac{e\mathbf{r}(1 - v^2/c^2)}{\left[r^2 - (\mathbf{r} \times \mathbf{v}/c)^2\right]^{3/2}} \qquad \mathbf{H} = \frac{\mathbf{v}}{c} \times \mathbf{E} \qquad (64.18)$$

This rigorous solution obviously goes over to the approximate solution (64.6) for small velocities, i.e. where $v \ll c$. It agrees with the approximate solution as regards the field direction. Only with respect to the magnitude of the field strength is there a difference. While the approximate solution $E = e/r^2$ is independent of direction, the magnitude of $\mathbf{E}$ in equation (64.18) displays an increasingly strong dependence upon direction as the velocity increases. In particular,

$$E = \frac{e}{r^2} \cdot (1 - \beta^2) \text{ for } \mathbf{r} \| \mathbf{v} \quad \text{and} \quad E = \frac{e}{r^2} \cdot \frac{1}{\sqrt{(1-\beta^2)}} \text{ for } \mathbf{r} \perp \mathbf{v}$$

As the speed of the particle approaches the speed of light, the whole field concentrates itself more and more into a flattish disc ("pancake") oriented perpendicular to the particle trajectory, in which, in the limiting case, $v = c$, the field strength becomes infinite everywhere.

Of particular interest now is the *convection force* **F** exerted by the moving point charge on a co-moving charge $e'$. In this case, by means of equation (64.12), we find from the Lorentz force that

$$\mathbf{F} = e'\left(\mathbf{E} + \frac{\mathbf{v}}{c} \times \mathbf{H}\right) = -e'(1 - \beta^2)\,\text{grad}\,\phi$$

In contrast to the field of **E**, the field of this force can be written as the gradient of a scalar potential function:

$$\mathbf{F} = -\text{grad}\,e'\psi \quad \text{with} \quad \psi = (1 - \beta^2)\phi \tag{64.19}$$

The function $\psi$, first introduced by Heaviside, is designated the *convection potential*. It is constant on the flattened ellipsoids of revolution

$$x^2 + (1 - \beta^2)(y^2 + z^2) = \text{constant}$$

These Heaviside ellipsoids can be considered as having evolved from a family of concentric spheres subject to a deformation in the direction of the $x$-axis in the ratio $1 : \sqrt{(1 - \beta^2)}$. The intuitive significance of the

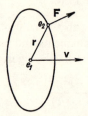

Fig. 57.—Mutual repulsion force **F** of two equal point charges moving with the same velocity **v**

ellipsoids $\psi = \text{constant}$ consists in that the force exerted upon a co-moving charge, that force being the gradient of $\psi$, is always directed normally to the ellipsoid's surface.

We can draw a significant conclusion from this fact. We consider two point charges $e_1$ and $e_2$ (figure 57) which are attached to the ends of a bar of length $r$. So long as the bar with the attached charges is at rest, the force exerted by $e_1$ on $e_2$ has the direction of the line joining

$e_1$ and $e_2$. When, however, the bar with the charges moves with velocity $\mathbf{v}$, this force stands normal to the ellipsoids we have mentioned and it is thus, in general, no longer in the direction along the axis of the bar. The bar, because of this, experiences a torque which, for the case of charges of like sign, has the tendency to rotate it into an orientation parallel to the direction of motion. For charges of opposite sign the bar seeks an orientation perpendicular to the direction of motion. With the help of equation (64.19) we obtain for this torque

$$\mathbf{T} = -\mathbf{r} \times \operatorname{grad} \frac{e_1 e_2 (1-\beta^2)}{\sqrt{[r^2-(\mathbf{r} \times \mathbf{v}/c)^2]}} \qquad (64.20)$$

Upon restricting ourselves to terms of order $\beta^2$, we find by developing the denominator, that

$$\mathbf{T} = -\mathbf{r} \times \operatorname{grad} e_1 e_2 (1-\beta^2) \left( \frac{1}{r} + \frac{(\mathbf{r} \times \mathbf{v}/c)^2}{2r^3} + \ldots \right) \approx \frac{e_1 e_2 (\mathbf{r}.\mathbf{v})(\mathbf{r} \times \mathbf{v})}{r^3 c^2}$$

i.e. the absolute magnitude of $T$ has the value

$$T = \frac{e_1 e_2 v^2}{2rc^2} \sin 2\theta \qquad (64.21)$$

in which we call $\theta$ the angle between $\mathbf{r}$ and $\mathbf{v}$. We obtain the same result, incidentally, from the approximate solution represented by equations (64.6). In this consideration $\mathbf{E}$ is always in the direction of $\mathbf{r}$ and therefore contributes nothing to the torque. With $\mathbf{H}$ as given by equation (64.6), the torque becomes $\mathbf{T} = e_2 \mathbf{r} \times \left( \dfrac{\mathbf{v}}{c} \times \mathbf{H} \right)$.

The fundamental significance of this result is that with it we apparently have the possibility of establishing the existence of an *absolute motion* of the earth, or perhaps of the whole solar system. A charged capacitor suspended on a torsion fibre, for example, must seek an orientation transverse to its direction of motion. From the very painstaking experiments of Trouton and Noble, however, we know that *such a torque does not exist*—agreeing with the basic postulate of relativity theory. A second torque must therefore exist which just compensates the torque found above. As we shall show in the section on the Theory of Relativity, this second torque has its origin in the mechanical stresses which must exist within the bar joining the two charges. Only when this second force is taken into account is the negative result of the Trouton-Noble experiment fully explained.

We have so far tacitly assumed that the velocity $v$ of the particles is less than the velocity of light $c$. This is fully in keeping with the theory of relativity, according to which the velocity of all bodies never exceeds a certain upper limit, the velocity of light.

It is also known that the above-mentioned solution, formulae (64.16) for example, is valid only for $v < c$, because for the case $v > c$ the denominator of (64.16) becomes imaginary for $x^2 < (\beta^2 - 1)(y^2 + z^2)$. The mathematical basis for the failure of the equations is that equation (64.16) is, for this case as an example, no longer an elliptical differential equation having the character of the potential equation, but a hyperbolic differential equation having the character of the wave equation. When duly integrated, the equation leads to a form of wave propagation which is fully analogous to that of an acoustic wave in a medium through which a small body moves with supersonic velocity ("Mach waves"). This phenomenon has been longest known in acoustics. In optics it was first observed by P.A. Cerenkov in 1934, not, to be sure, by the motion of a particle with velocity greater than that of light—such a speed, as already mentioned, is ruled out on fundamental grounds by relativity—but by the motion through matter of fast charged particles whenever $v > c/\sqrt{(\varepsilon\mu)}$. The difficulty in treating the frequency-dependence of $\varepsilon$, however, prevents our going further into this question here.

*Remark:* In the MKSA system of units, the equations (64.1), with which we began our work, can be written in the form

$$\operatorname{curl} \mathbf{H} = \varepsilon_0 \frac{\partial \mathbf{E}}{\partial t} + \rho\mathbf{v} \qquad \operatorname{div} \mathbf{E} = \frac{\rho}{\varepsilon_0}$$

$$\operatorname{curl} \mathbf{E} = -\mu_0 \frac{\partial \mathbf{H}}{\partial t} \qquad \operatorname{div} \mathbf{H} = 0$$

Use of the potential equations permits us to write the following expressions, corresponding to equations (64.7) to (64.9), (with $\varepsilon_0\mu_0 = 1/c^2$):

$$\mathbf{E} = -\frac{\partial \mathbf{A}}{\partial t} - \operatorname{grad} \phi \qquad \mathbf{H} = \frac{1}{\mu_0}\operatorname{curl} \mathbf{A} \qquad \operatorname{div} \mathbf{A} + \varepsilon_0\mu_0 \frac{\partial \phi}{\partial t} = 0$$

The potential equations are then

$$\nabla^2 \mathbf{A} - \frac{1}{c^2}\frac{\partial^2 \mathbf{A}}{\partial t^2} = -\mu_0\rho\mathbf{v} \qquad \nabla^2 \phi - \frac{1}{c^2}\frac{\partial^2 \phi}{\partial t^2} = -\frac{\rho}{\varepsilon_0}$$

In particular, for a charged particle moving with constant velocity, we have in place of equation (64.11):

$$\mathbf{A} = \varepsilon_0\mu_0\mathbf{v}\phi$$

Finally, equations (64.18) are to be replaced by

$$E = \frac{e\mathbf{r}(1 - v^2/c^2)}{4\pi\varepsilon_0[r^2 - (\mathbf{r}\times\mathbf{v}/c)^2]^{3/2}} \qquad \mathbf{H} = \varepsilon_0\mathbf{v}\times\mathbf{E}$$

## §65. Energy and momentum for a uniformly moving charged particle

In §§54 and 56 we have seen that every electromagnetic field possesses not only energy but also momentum. In particular, we may consider the energy law for a moving particle. According to equations (54.8) and (54.5), where the power of the field $\int \mathbf{g}.\mathbf{E}\,dV = e\mathbf{v}.\mathbf{E}$ has been equated to the time-derivative of the kinetic energy $U_k$ of the particle, the energy law reads

$$\frac{d}{dt}(U_k + U_{el} + U_m) = -\oint S_n\,dA \qquad (65.1)$$

in which the integral on the right-hand side represents the total energy radiated per second over the area $A$. Analogously, we have for the momentum law from equation (56.7):

$$\frac{d}{dt}(\mathbf{J}_M + \mathbf{J}_S) = \oint \mathbf{T}_n\,dA \qquad (65.2)$$

in which $\mathbf{J}_M$ and $\mathbf{J}_S$ stand for the total momentum of the charge and its field, and the integral on the right-hand side represents the effect of the Maxwell stresses on a closed surface $A$ surrounding the charge at a great distance. From equation (56.8), the momentum of the field is equal to

$$\mathbf{J}_S = \frac{1}{4\pi c}\int \mathbf{E} \times \mathbf{H}\,dV = \frac{1}{c^2}\int \mathbf{S}\,dV \qquad (65.3)$$

We now calculate this field momentum for the uniformly moving particle considered in §64. From equation (64.3), we have

$$\mathbf{J}_S = \frac{1}{4\pi c^2}\int \mathbf{E} \times (\mathbf{v} \times \mathbf{E})\,dV$$

Here, on symmetry grounds, the momentum components perpendicular to the direction of motion vanish and we obtain for the component in the $\mathbf{v}$-direction ($= x$-direction)

$$(\mathbf{J}_S)_x = J_S = \frac{v}{4\pi c^2}\int \{E_y^2 + E_z^2\}\,dV \qquad (65.4)$$

Correspondingly, from equation (64.3), the total field energy is

$$U = U_{el} + U_m = \frac{1}{8\pi}\int \{E_x^2 + (1+\beta^2)(E_y^2 + E_z^2)\}\,dV \qquad (65.5)$$

The momentum of the particle field is thus proportional to the particle's velocity, so long as we may neglect the dependence of $\mathbf{E}$ upon velocity, as given in (64.18); so long, that is, as $v \ll c$. For this case, and for a spherically symmetric charge distribution

$$\int E_x^2 \, dV = \int E_y^2 \, dV = \int E_z^2 \, dV = \tfrac{1}{3} \int E^2 \, dV$$

And so

$$\mathbf{J}_S = \frac{v}{6\pi c^2} \int E^2 \, dV \qquad U = U_0 = \frac{1}{8\pi} \int E^2 \, dV$$

or, combined,

$$\mathbf{J}_S = \frac{4}{3} \frac{U_0}{c^2} \mathbf{v} \tag{65.6}$$

For a spherical charge of radius $R$ we have for the electrostatic field energy $U_o = fe^2/R$, in which now the dimensionless factor $f$ depends on the nature of the charge distribution in the body. It equals $\tfrac{1}{2}$ for a sphere with the charge confined to its surface, and is $\tfrac{3}{5}$ for a homogeneous volume distribution of charge.

If we conceive our charged particle as possessing no inertial mass, that is, $U_k = 0$ and $\mathbf{J}_M = 0$, we must ascribe to the acceleration of the particle a reactionary force effect so as to furnish the increase with increasing velocity of the growing field momentum $\mathbf{J}_S$ and the accompanying increasing field energy $U$. *Under the influence of an externally applied force a particle without inertial mass would behave, according to* (65.6), *as if it possessed a mass of magnitude* $4U_o/3c^2$. The smaller the radius $R$ of the particle, the greater is this "electromagnetic mass". Through appropriate choice of $R$, therefore, we can account for any observed mass of a charged particle as electromagnetic mass. It must be emphasized, however, that at present there is at hand no definite proof that the mass of either the electron or the proton is purely electromagnetic in character.*

The existence of an electromagnetic inertial force for a charged particle can easily be rendered plausible in the following manner: Let us imagine an electron moving with constant velocity. Moving with it is a magnetic field whose circular lines of force are centred on the electron's line of motion. During this motion there is no net force acting on the electron. If, now, we should wish suddenly to decelerate the electron, this would naturally require that the magnetic field should decrease

---

* It has become customary to define a "radius of the electron" $R_e$ according to the relation

$$\frac{e_0^2}{R_e c^2} = m \quad \text{i.e.} \quad R_e = \frac{e_0^2}{mc^2}$$

Numerical values for the electron charge and mass give $R_e = 2 \cdot 8 \times 10^{-13}$ cm.

during such deceleration. According to the induction law, however, a collapsing magnetic field gives rise to an [electric field. It can be easily shown that the electric field so produced gives at the location of the electron, a force on it of such direction that during the deceleration the electron is urged to accelerate. The inertial force felt during the deceleration is identical with the electrical force $eE$, where $E$ is the field strength given by the induction law, and is only brought into being by a change in the electron's velocity.

We wish now to consider the case of higher charge velocity. The velocity dependence of the momentum of the field merits our particular attention. Now the manner in which this field momentum varies with velocity is closely connected with the particular assumption which we make about how the electric charge is distributed in the moving body. The most obvious assumption, that discussed rather fully by Max Abraham, is that the charged body is a rigid sphere having a given fixed spherically symmetric charge distribution. By contrast, A. H. Bucherer and H. A. Lorentz introduced the idea that the electron, corresponding to the convection potential considered in §64, possessed the form of an ellipsoid of revolution flattened in the direction of its motion, the ratio of its axes being $1:\sqrt{(1-\beta^2)}$. In particular, it was assumed that the electron actually contracted in the direction of motion in the proportions just mentioned, and therefore that

$$\rho(x, y, z) = \frac{1}{\sqrt{(1-\beta^2)}} \rho_0\left(\frac{x}{\sqrt{(1-\beta^2)}}, y, z\right) \qquad (65.7)$$

where the total charge of the moving body is equal to the charge on the body at rest:

$$\iiint \rho(x, y, z)\, dx\, dy\, dz = \iiint \rho_0\left(\frac{x}{\sqrt{(1-\beta^2)}}, y, z\right)\frac{dx}{\sqrt{(1-\beta^2)}}\, dy\, dz = e$$

We wish to defer until later the consideration of this *contraction hypothesis*. We shall become thoroughly acquainted with its general basis in the chapter on the theory of relativity. According to relativity, all scales in motion at any given speed shorten themselves in the direction of motion in the same ratio.

Introducing equation (65.7) for the charge density into the potential equation (64.13), we obtain for an arbitrary (and not necessarily rotationally symmetric) charge distribution the following equation

$$\phi(x, y, z) = \frac{1}{\sqrt{(1-\beta^2)}} \phi_0\left(\frac{x}{\sqrt{(1-\beta^2)}}, y, z\right) \qquad (65.8)$$

as a connection between the scalar potential $\phi$ of the moving charge and the static potential of the same charge at rest.

From this, and with (64.12) we obtain for the component $E_y$ needed in (65.4) the expression

$$E_y(x, y, z) = -\frac{\partial \phi}{\partial y} = \frac{1}{\sqrt{(1-\beta^2)}} E_{0y}\left(\frac{x}{\sqrt{(1-\beta^2)}}, y, z\right)$$

An analogous formula holds for $E_z$. Thus, from (65.4), we have

$$J_s = \frac{v}{4\pi c^2 (1-\beta^2)} \iiint \left\{ E_{0y}^2\left(\frac{x}{\sqrt{(1-\beta^2)}}, y, z\right) + \right.$$

$$\left. + E_{0z}^2\left(\frac{x}{\sqrt{(1-\beta^2)}}, y, z\right) \right\} dx \, dy \, dz$$

Substituting $x$ for $x/\sqrt{(1-\beta^2)}$, we have

$$J_s = \frac{v}{4\pi c^2 \sqrt{(1-\beta^2)}} \iiint \left\{ E_{0y}^2(x, y, z) + E_{0z}^2(x, y, z) \right\} dx \, dy \, dz$$

Considering the field of the charge at rest as spherically symmetric, we find as above

$$\mathbf{J}_S = \frac{4U_0}{3c^2} \frac{\mathbf{v}}{\sqrt{(1-\beta^2)}} \tag{65.9}$$

If we call the constant factor $4U_0/3c^2$ the *rest mass m*, we can write the momentum of the field in the form

$$\mathbf{J}_S = \frac{m\mathbf{v}}{\sqrt{(1-v^2/c^2)}} \tag{65.10}$$

Now investigations of fast-moving charged particles (see Volume II, §3) have shown that their momentum, defined by the expression $d\mathbf{J}/dt = \mathbf{F}$, increases with velocity exactly as given by equation (65.10). Prior to the advent of the theory of relativity this observation held special interest because it was hoped that with its help it would be possible to decide whether the mass of an electron was purely electromagnetic, or whether the electron, besides having electromagnetic mass, carried with it a mass in the ordinary sense. It was often thought that only the electromagnetic part of the mass increased with velocity, that the ordinary mass (mass in the sense of mechanics) could be regarded as a constant of the body considered. With respect to this question, however, the above-mentioned investigations have now lost much of their interest because of the theory of relativity. According to relativity, formula (65.10), which we have here obtained for the electromagnetic part of the mass, is actually quite generally valid for every kind of mass. Experimental distinction between different kinds of mass by investigations such as these seems therefore not to be possible.

## §66. The electromagnetic potential of an arbitrary distribution of charge and current

We come now to a consideration of the field in empty space of an arbitrary distribution of electric charge and/or current. For this we consider the quantities $\rho$ and $\rho \mathbf{v} = \mathbf{g}$ as being arbitrarily given functions of the space and time coordinates.

To integrate the Maxwell equations we use (as in §64) the relations

$$\mathbf{H} = \operatorname{curl} \mathbf{A} \qquad \mathbf{E} = -\frac{1}{c}\frac{\partial \mathbf{A}}{\partial t} - \operatorname{grad} \phi \qquad (66.1)$$

to go over to the potentials $\mathbf{A}$ and $\phi$, and we require* that these satisfy the following auxiliary condition ("Lorentz convention"):

$$\operatorname{div} \mathbf{A} + \frac{1}{c}\frac{\partial \phi}{\partial t} = 0 \qquad (66.2)$$

Thus we obtain, as in §64, the potential equations

$$\nabla^2 \mathbf{A} - \frac{1}{c^2}\frac{\partial^2 \mathbf{A}}{\partial t^2} = -\frac{4\pi}{c}\mathbf{g} \qquad \nabla^2 \phi - \frac{1}{c^2}\frac{\partial^2 \phi}{\partial t^2} = -4\pi\rho \qquad (66.3)$$

For the case of fields that are constant in time these equations immediately pass over to the already fully discussed equations of electrostatics and magnetic fields of steady currents. The outward expansion of the field with time is considered here by bringing in the term with the time derivative.

---

* Instead of the Lorentz convention (66.2), the so-called "Coulomb convention", $\operatorname{div} \mathbf{A} = 0$ is sometimes employed. Then, to be sure, the equation for the scalar potential is as simple as the Poisson equation $\nabla^2 \phi = -4\pi\rho$ of electrostatics, but the equation for the vector potential becomes more complicated than the wave equation.

The solutions of equations (66.3) can be brought into a form very similar to that for static fields. These are

$$\left.\begin{array}{l} \mathbf{A}(x, y, z, t) = \dfrac{1}{c} \displaystyle\iiint \dfrac{\mathbf{g}(\xi, \eta, \zeta, t - r/c)}{r} \, d\xi \, d\eta \, d\zeta \\[4mm] \phi(x, y, z, t) = \displaystyle\iiint \dfrac{\rho(\xi, \eta, \zeta, t - r/c)}{r} \, d\xi \, d\eta \, d\zeta \end{array}\right\} \qquad (66.4)$$

as we shall forthwith verify.* In these expressions $r$, the distance between the source point and the field point, is given by

$$r = \sqrt{[(x - \xi)^2 + (y - \eta)^2 + (z - \zeta)^2]} \qquad (66.5)$$

In these formulae the contribution of the volume element $d\xi \, d\eta \, d\zeta$ to the potentials at the field point $(x, y, z)$ differs from that for the static case only in that for $\mathbf{g}$ and $\rho$ their present values should not be introduced, but rather those values which prevailed within $d\xi \, d\eta \, d\zeta$ at a moment that is earlier by a length of time $r/c$, the signal travel time. The contribution of a source point to the potential arrives at the field point only after the finite time interval $r/c$. For this reason the quantities $\mathbf{A}$ and $\phi$ given by equations (66.4) are called *retarded potentials*.

It should be pointed out that the physical content of solutions (66.4) is not identical with that of equations (66.3). While in the differential equations (66.3) the *sign of the time* is in no way distinguished, i.e. the equations are not altered by an exchange of past with future, solutions (66.4) do make an essential distinction between past and future. Purely mathematically, solutions of (66.4) would also be possible in which values of $\mathbf{g}$ and $\rho$ at the source point were chosen for later time $t + r/c$, giving the so-called "advanced potentials". Such solutions would, however, be contradictory to our basic conceptions, since we consider the charges and currents to be *origins* of the potentials. By excluding these advanced solutions we actually go somewhat beyond the content of differential equations (66.3).

We must now be convinced that equations (66.4) are really solutions of equations (66.3). It suffices to carry out this proof for the scalar potential. To this end we divide the region of integration into two parts $V_1$ and $V_2$, in which $V_1$ stands for a very small volume containing

---

* The solutions (66.4) are particular integrals of the inhomogeneous equations (66.3). For the purpose of adapting to given initial and boundary conditions, we can add to these solutions integrals of the homogeneous potential equations, i.e. the wave equations for $\mathbf{A}$ and $\phi$.

the source point, while $V_2$ is the whole surrounding region. Correspondingly, we make the separation

$$\phi = \phi_1 + \phi_2 = \int_{V_1} \frac{\rho(\xi,\eta,\zeta,t-r/c)}{r} dV + \int_{V_2} \frac{\rho(\xi,\eta,\zeta,t-r/c)}{r} dV$$

Now in the integration over the very small region $V_1$ the retardation obviously plays no part. In the first integration, therefore, we can simply substitute $\rho(\xi,\eta,\zeta,t)$ for $\rho(\xi,\eta,\zeta,t-r/c)$. But then this differs in no way from the static case. We thus obtain $\nabla^2\phi_1 = -4\pi\rho(x,y,z,t)$, while $\phi_1$ itself, and also $\partial^2\phi_1/\partial t^2$ because of the smallness of $V_1$, are negligibly small. Thus the part $\phi_1$ by itself satisfies the potential equation (66.3) for $\phi$.

Moreover, according to §13, the expression

$$\nabla^2 f(r) = \frac{1}{r}\frac{\partial^2 [rf(r)]}{\partial r^2}$$

holds generally for a quantity depending on $r$ alone. We therefore have

$$\nabla^2\phi_2 = \iiint_{V_2} \frac{1}{r}\frac{\partial^2}{\partial r^2}\rho\left(\xi,\eta,\zeta,t-\frac{r}{c}\right)d\xi\,d\eta\,d\zeta$$

Further, for any function $g$ of $t-\dfrac{r}{c}$,

$$\frac{\partial^2 g}{\partial r^2} = \frac{1}{c^2}\frac{\partial^2 g}{\partial t^2}$$

Thus, since

$$\nabla^2\phi_2 = \frac{1}{c^2}\frac{\partial^2}{\partial t^2}\iiint_{V_2}\frac{1}{r}\rho\left(\xi,\eta,\zeta,t-\frac{r}{c}\right)d\xi\,d\eta\,d\zeta = \frac{1}{c^2}\frac{\partial^2\phi}{\partial t^2}$$

$\phi_2$ satisfies the homogeneous wave equation and can therefore be added to the solution $\phi_1$ of the inhomogeneous equation, so that $\phi = \phi_1 + \phi_2$ does in fact satisfy the potential equation (66.3).

We wish now to transform formulae (66.4) for $\phi$ and $\mathbf{A}$ so that they can be applied to charged particles of very small spatial extent. It turns out here (rather troublesomely) that we have to take the integrands for the individual volume elements of the particle at different times $t-r/c$. Thus, for example, the integral $\iiint\rho(\xi,\eta,\zeta,t-r/c)\,d\xi\,d\eta\,d\zeta$ which comes from (66.4) by leaving out the denominator $r$, is not, as in the static case, the total charge of the particle; it would only be this if the integrand had always to be taken for the same time.

The latter can, however, be attained by means of an artifice, i.e. by introducing the so-called one-dimensional $\delta$-function. This function $\delta(u)$ is defined so that $\delta(u) = 0$ for $u \neq 0$; but $\delta(0) \neq 0$, such that $\int_{-\infty}^{+\infty} \delta(u) \, du = 1$. For every regular continuous and differentiable function $f(u)$ it thus follows that

$$f(u)\,\delta(u-a) = f(a)\,\delta(u-a) \quad \text{and} \quad \int_{-\infty}^{\infty} f(u)\,\delta(u-a)\,du = f(a)$$

and for the derivative $\delta'(u)$ there results from partial integration,

$$\int_{-\infty}^{\infty} f(u)\,\delta'(u-a)\,du = -f'(a)$$

With this $\delta$-function the potentials in (66.4) can be written in the form

$$
\mathbf{A}(x, y, z, t) = \frac{1}{c} \iiiint \frac{\mathbf{g}(\xi, \eta, \zeta, \tau)}{r} \delta\left(\tau - t + \frac{r}{c}\right) d\xi \, d\eta \, d\zeta \, d\tau \left.\vphantom{\iiiint}\right\}
$$
$$
\phi(x, y, z, t) = \iiiint \frac{\rho(\xi, \eta, \zeta, \tau)}{r} \delta\left(\tau - t + \frac{r}{c}\right) d\xi \, d\eta \, d\zeta \, d\tau \qquad (66.6)
$$

Now, however, $\rho(\xi,\eta,\zeta,\tau)\,d\xi\,d\eta\,d\zeta = de$ is equal to the charge in the volume element $d\xi\,d\eta\,d\zeta$ at time $\tau$, and likewise, $\mathbf{g}(\xi,\eta,\zeta,\tau)\,d\xi\,d\eta\,d\zeta = \mathbf{v}(\tau)\,de$, $\mathbf{v}$ being the particle velocity. Thus for particles of small dimensions we can immediately carry out the integration over the charge elements, obtaining

$$
\mathbf{A} = \frac{e}{c}\int \frac{\mathbf{v}(\tau)\,d\tau}{r} \delta\left(\tau - t + \frac{r}{c}\right) \qquad \phi = e\int \frac{d\tau}{r} \delta\left(\tau - t + \frac{r}{c}\right) \quad (66.7)
$$

Through this transformation $\mathbf{r}$, as the distance between the particle and the field point, has become a function $[r = r(\tau)]$ of the time. We can now evaluate the integral. Introducing instead of $\tau$ a new integration variable $u = \tau - t + r/c$, where

$$
\frac{du}{d\tau} = 1 + \frac{1}{c}\frac{dr}{d\tau} = 1 - \frac{\mathbf{r} \cdot \mathbf{v}}{rc} \qquad (66.8)
$$

we find, because of the character of the $\delta$-function, that

$$
\mathbf{A} = \frac{e}{c}\left[\frac{\mathbf{v}}{r - (\mathbf{r} \cdot \mathbf{v})/c}\right]_{\tau = t - r/c} \qquad \phi = \frac{e}{[r - (\mathbf{r} \cdot \mathbf{v})/c]_{\tau = t - r/c}} \quad (66.9)
$$

This form of the potentials was first given by A. Liénard and E. Wiechert.

*Remark:* From the potential formulae (66.4) to (66.9) in the Gaussian system, the corresponding formulae in the MKSA system are obtained by multiplication of the $\phi$-expressions by $1/4\pi\varepsilon_0$, and the A-expressions by $\mu_0 c/4\pi$.

## §67. The Hertz solution for the oscillating dipole

For the first example we apply formula (66.9) to a case first investigated by Hertz, that of an *oscillating electric dipole* having a dipole moment $\mathbf{p} = \mathbf{p}(t)$. Let us picture the dipole as a charge pair $\pm e$ with the time-varying dipole separation $\mathbf{s}(t)$. We think perhaps of Thomson's atom model with its elastically bound electron. We can then write immediately

$$e\mathbf{v} = e\dot{\mathbf{s}} = \dot{\mathbf{p}} \quad \left(\text{with } \dot{\mathbf{p}}(t) \equiv \frac{d\mathbf{p}(t)}{dt}\right) \tag{67.1}$$

We further assume that the oscillation amplitude of the dipole is small compared to the wavelength $\lambda$ of the emitted radiation, and is also small compared to the distance $r$ at which we investigate the field of the emitter. Next, then, in formula (66.9) for the vector potential, we can set the denominator equal simply to $r$, since for a harmonic oscillator $v/c \approx \omega s/c = 2\pi s/\lambda$, which we assume to be small compared to unity. Further, we can also neglect the variation of $r$ due to the motion of the dipole and thus we can calculate with a fixed $r$, as the distance of the field point from the dipole. We therefore obtain for $\mathbf{A}$ the relation

$$\mathbf{A} = \frac{\dot{\mathbf{p}}(t-r/c)}{cr} \tag{67.2}$$

In the calculation of $\phi$ from (66.9) we have to be somewhat more careful with what we neglect, otherwise we shall obtain $\phi = 0$, on account of the electrical neutrality of the dipole as a whole. In this case we would have to carry out a development in series, breaking off after the first non-vanishing term. We reach our goal more simply, however, if we introduce formula (67.2) for $\mathbf{A}$ into equation (66.2). After carrying out the time integration, and omitting for simplicity here and in what follows the argument $t-r/c$, which is everywhere the same, we find that

$$\phi = -\operatorname{div}\frac{\mathbf{p}(t-r/c)}{r} = \frac{\dot{\mathbf{p}}\mathbf{r}}{cr^2} + \frac{\mathbf{p}\mathbf{r}}{r^3} \tag{67.3}$$

By direct calculation, incidentally, we can be convinced that this expression for $\phi$ satisfies the potential equation (66.3).

In order to see the physical content of this equation we specially require of **p** that

$$\mathbf{p} = \mathbf{p}_0 \sin \omega \left( t - \frac{r}{c} \right) \quad \text{with} \quad \omega = 2\pi v = \frac{2\pi c}{\lambda} \qquad (67.4)$$

It follows then from (67.3) that

$$\phi = \frac{\mathbf{p}_0 \mathbf{r}}{cr^2} \omega \cos \omega \left( t - \frac{r}{c} \right) + \frac{\mathbf{p}_0 \mathbf{r}}{r^3} \sin \omega \left( t - \frac{r}{c} \right)$$

If now we choose $r \ll \lambda$ we can first of all neglect the $r$-term in the argument of the angle function, and second we can completely eliminate the cosine term since for the latter the amplitude is smaller by a factor $\omega r / c = 2\pi r / \lambda$ than the sine term. In the vicinity of the emitter then, $\phi$ becomes approximately equal to the electrostatic potential of the dipole $\mathbf{p}(t)$ prevailing at time $t$. For greater distances ($r \gg \lambda$), however, the electrostatic dipole term in the $\phi$-formula diminishes with respect to the first term whose value falls off much more slowly, i.e. only as $1/r$, corresponding to an expanding spherical wave. We therefore speak here (for $r \gg \lambda$) of the *wave zone*, in contrast to the *near zone* (for $r \ll \lambda$).

We come now to the calculation of the field intensities. Introducing (67.2) and (67.3) into equations (66.1), we find for arbitrary values of $r$:

$$\left.\begin{array}{l}
\mathbf{E} = -\dfrac{\ddot{\mathbf{p}}}{c^2 r} + \dfrac{(\ddot{\mathbf{p}}\mathbf{r})\mathbf{r}}{c^2 r^3} - \dfrac{\dot{\mathbf{p}}}{cr^2} + \dfrac{3(\dot{\mathbf{p}}\mathbf{r})\mathbf{r}}{cr^4} - \dfrac{\mathbf{p}}{r^3} + \dfrac{3(\mathbf{p}\mathbf{r})\mathbf{r}}{r^5} \\[3mm]
\mathbf{H} = \dfrac{\ddot{\mathbf{p}} \times \mathbf{r}}{c^2 r^2} + \dfrac{\dot{\mathbf{p}} \times \mathbf{r}}{cr^3}
\end{array}\right\} \qquad (67.5)$$

We shall discuss this field for the two extreme cases $r \ll \lambda$ and $r \gg \lambda$.

In the *near zone* the terms with the highest power of $r$ in the denominator predominate; therefore

$$\mathbf{E} = -\frac{\mathbf{p}}{r^3} + \frac{3(\mathbf{p}\mathbf{r})\mathbf{r}}{r^5} \qquad \mathbf{H} = \frac{\dot{\mathbf{p}} \times \mathbf{r}}{cr^3} \qquad (67.6)$$

This is, however (and from what was said above we should not have expected otherwise), the electrostatic field of a dipole of moment $\mathbf{p}$, and the steady magnetic field of a current element $\int \mathbf{g} \, dV = d\mathbf{p}/dt$.

In the *wave zone* only the terms with the lowest power of $r$ in the denominator remain significant, thus

$$\mathbf{E} = -\frac{\ddot{\mathbf{p}}}{c^2 r} + \frac{(\ddot{\mathbf{p}}\mathbf{r})\mathbf{r}}{c^2 r^3} \qquad \mathbf{H} = \frac{\ddot{\mathbf{p}} \times \mathbf{r}}{c^2 r^2} \qquad (67.7)$$

We obviously have here

$$\mathbf{E} = \mathbf{H} \times \frac{\mathbf{r}}{r} \quad \text{and} \quad \mathbf{H} = -\mathbf{E} \times \frac{\mathbf{r}}{r} \qquad (67.8)$$

Thus *in the wave zone the two field intensities have the same value, and they are perpendicular to one another and to the direction of propagation* $\mathbf{r}$. But this is just the picture of a spherical wave radiating out from the "sender" in the direction $\mathbf{r}$.

If, in particular, the dipole oscillates in a fixed direction, as for example along the axis of a space coordinate system $r, \theta, \psi$, then *the $\mathbf{E}$-vector oscillates in a meridian, and the $\mathbf{H}$-vector in a parallel of latitude.* The field amplitude decreases from the equator to the pole; in the direction of oscillation of the dipole ($\theta = 0$) there is no radiation. The Poynting vector

$$\mathbf{S} = \frac{c}{4\pi}\mathbf{E} \times \mathbf{H} = \frac{(\ddot{\mathbf{p}} \times \mathbf{r})^2}{4\pi c^3 r^4}\frac{\mathbf{r}}{r} = \frac{\ddot{\mathbf{p}}^2 \sin^2\theta}{4\pi c^3 r^2}\frac{\mathbf{r}}{r} \qquad (67.9)$$

is in the direction of the radius and falls off as $1/r^2$, in agreement with the inverse square law. We find for the total radiation $S$ through a spherical surface of radius $r$

$$S = \int S_r r^2 \, d\Omega = \frac{\ddot{\mathbf{p}}^2}{4\pi c^3}\int \sin^2\theta \, d\Omega = \frac{2}{3c^3}\left\{\ddot{\mathbf{p}}\left(t - \frac{r}{c}\right)\right\}^2$$

For the special case of a purely harmonic oscillating dipole (67.4) we have for the average over one oscillation period

$$\bar{S} = \frac{2\omega^4\overline{p^2}}{3c^3} = \frac{\omega^4 p_0^2}{3c^3} \qquad (67.10)$$

It is noteworthy that for the existence of the spherical wave (67.7), the retardation in the potential formulae is decisive; that is, the spherical wave is immediately obtained if, throughout, in differentiating with respect to $r$, only the $r$ in the combination $t - r/c$ is taken into account, and the $r$ in the denominator is considered as a constant.

Conversely, the field strengths in the near zone are immediately obtained by neglecting the retardation.

*Remark:* For going over to the MKSA system the expressions for **E** above have to be multiplied by $1/4\pi\varepsilon_0$, and those for **H** by $c/4\pi = 1/4\pi\sqrt{(\varepsilon_0\mu_0)}$. Then for the wave zone, instead of (67.8), we have

$$\mathbf{E} = \sqrt{\left(\frac{\mu_0}{\varepsilon_0}\right)}\,\mathbf{H} \times \frac{\mathbf{r}}{r} \qquad \mathbf{H} = -\sqrt{\left(\frac{\varepsilon_0}{\mu_0}\right)}\,\mathbf{E} \times \frac{\mathbf{r}}{r} \text{ and thus } \varepsilon_0\mathbf{E}^2 = \mu_0\mathbf{H}^2$$

Because of the difference of form of the Poynting vector in the Gaussian and the MKSA system, the E and S formulae for the latter case are obtained from formulae (67.9) and (67.10) by multiplication by $1/4\pi\varepsilon_0$, so that we obtain, for example, for the total radiation of a harmonically oscillating electric dipole, averaged over a period,

$$\overline{S} = \frac{\omega^4\overline{p^2}}{6\pi\varepsilon_0 c^3} = \frac{\mu_0(\varepsilon_0\mu_0)^{1/2}\omega^4\overline{p^2}}{6\pi}$$

## §68. The radiation of electromagnetic waves by an emitter

The remark at the end of the foregoing paragraphs permits us now to calculate the radiation of an arbitrarily specified oscillating distribution of charge and current. For this we assume as above that the dimensions of the emitter are small compared with the wave length of the emitted radiation. Furthermore we restrict our interest to the field in the region of the wave zone, where $r \gg \lambda$.

From these assumptions there follows an essential simplification of the potential formulae (66.4) as well as the calculations connected with them: First, in space differentiation in the wave zone we need differentiate only the retardation term $t - r/c$, but not the $r$-factors otherwise occurring in the integrands. Second, in these factors we can always disregard the finite extension of the emitter. Thus, for example, we substitute $r$ for the distance of the field point from the origin of coordinates lying within the emitter. In order to make this clear we designate in the following the location of the field point with the position vector $\mathbf{r}$, and the location of the source point with the position vector $\mathbf{r}'$. The distance from the source point to the field point, which we have hitherto designated by $r$, is now $|\mathbf{r}-\mathbf{r}'|$. Carrying out all these modifications and simplifications, we obtain from (66.4) for the wave zone

$$\mathbf{A} \to \frac{1}{cr}\int \mathbf{g}\left(\mathbf{r}', t - \frac{|\mathbf{r}-\mathbf{r}'|}{c}\right)dV' \qquad \phi \to \frac{1}{r}\int \rho\left(\mathbf{r}', t - \frac{|\mathbf{r}-\mathbf{r}'|}{c}\right)dV' \quad (68.1)$$

We find **H** by taking the curl of the vector potential:

$$\mathbf{H} \to -\frac{1}{c^2 r}\int \frac{\mathbf{r}-\mathbf{r}'}{|\mathbf{r}-\mathbf{r}'|} \times \frac{\partial \mathbf{g}\left(\mathbf{r}', t-\dfrac{|\mathbf{r}-\mathbf{r}'|}{c}\right)}{\partial t}\,dV'$$

$$\to -\frac{\mathbf{r}}{c^2 r^2} \times \frac{\partial}{\partial t}\int \mathbf{g}\left(\mathbf{r}', t-\frac{|\mathbf{r}-\mathbf{r}'|}{c}\right)dV' \qquad (68.2)$$

We could calculate **E** in an analogous way, making use of the equation of continuity of electric charge. However, we find **E** more simply from the first Maxwell equation; with (68.2) we arrive at the expression

$$\frac{\partial \mathbf{E}}{\partial t} = c\,\mathrm{curl}\,\mathbf{H} = \frac{\mathbf{r}}{c^2 r^3} \times \left[\mathbf{r} \times \frac{\partial^2}{\partial t^2}\int \mathbf{g}\left(\mathbf{r}', t-\frac{|\mathbf{r}-\mathbf{r}'|}{c}\right)\right]dV' \qquad (68.3)$$

which with one integration immediately gives the electric field strength. Introducing for brevity the quantity **q**, where

$$\dot{\mathbf{q}} \equiv \frac{\partial \mathbf{q}}{\partial t} = \int \mathbf{g}\left(\mathbf{r}', t-\frac{|\mathbf{r}-\mathbf{r}'|}{c}\right)dV' \qquad (68.4)$$

we can connect the two expressions for **E** and **H** in a form analogous to (67.7), namely

$$\mathbf{E} = \frac{(\ddot{\mathbf{q}} \times \mathbf{r}) \times \mathbf{r}}{c^2 r^3} \qquad \mathbf{H} = \frac{\ddot{\mathbf{q}} \times \mathbf{r}}{c^2 r^2} \qquad (68.5)$$

For an arbitrary emitter we therefore obtain in the wave zone the same picture of an expanding spherical wave as for an oscillating electric dipole. As we shall see later, however, **q** possesses an additional directional dependence because of the **r**-vector contained in the retardation.

The vector **q** is thus decisive in all radiation problems. On account of our above-mentioned assumption ($r' \ll \lambda \ll r$) we can also write **q** in another way wherein we develop the retardation term in powers of $r'/r$. We have, obviously,

$$|\mathbf{r}-\mathbf{r}'| = \sqrt{(r^2 - 2\mathbf{r}\mathbf{r}' + r'^2)} = r - \frac{\mathbf{r}\mathbf{r}'}{r} + \frac{(\mathbf{r} \times \mathbf{r}')^2}{2r^3} - \dots$$

In the wave zone, however, we need consider only the first two terms in this development since, as $r \to \infty$, the third and higher terms vanish like $1/r$, or else more rapidly. Account must in any case be taken of the second term, since it remains finite for $r \to \infty$; its contribution, however (since $r' \ll \lambda$) is small—of the order of magnitude $\omega r'/c = 2\pi r'/\lambda$. This is recognized if for **g** a pure harmonic time-dependence of the form $\cos \omega t$ is assumed. Consequently for (68.4) we can now write

$$\dot{\mathbf{q}} = \int \mathbf{g}\left(\mathbf{r}', t - \frac{r}{c} + \frac{\mathbf{rr'}}{rc}\right) dV' \qquad (68.6)$$

We now consider several special types of emitter.

If the emitter consists of an oscillating electric dipole, we come back with our present formulae to the case dealt with in §67—for, on account of (67.1), we have then to set $\int \mathbf{g} dV = e\mathbf{v} = \partial \mathbf{p}/\partial t$, and thus, upon neglecting the small correction in the retardation term, we find as the first non-vanishing approximation that $\mathbf{q} \equiv \mathbf{p}(t - r/c)$.

If the emitter consists of two small metal bodies which are connected through a spark gap or by a wire, we have to set $\mathbf{g} dV' = I d\mathbf{r}'$ and, upon again neglecting the correction in the retardation, we obtain

$$\dot{\mathbf{q}} = I\left(t - \frac{r}{c}\right) \int_1^2 d\mathbf{r}' = \mathbf{l}I\left(t - \frac{r}{c}\right) \qquad (68.7)$$

in which the vector **l** extends from the beginning point to the end point of the spark gap or the antenna. If we are dealing with a sinusoidal alternating current, the time-averaged total radiation corresponding to (67.10) is equal to

$$\bar{S} = \frac{2\omega^2 l^2}{3c^3} \overline{I^2} = \frac{8\pi^2 l^2}{3c\lambda^2} \overline{I^2} \qquad (68.8)$$

For *wireless telegraphy* this formula gives the radiation of an antenna of length $l$ when $\lambda \gg l$, the effective antenna current being $I_{\text{eff}} = (\overline{I^2})^{1/2}$. Since in this case, in addition to the radiated power, the oscillation generator has to furnish energy $R\overline{I^2}$ of Joule heating developed in the antenna, where $R$ is the ohmic resistance (considering skin effect), the factor multiplying $\overline{I^2}$ in (68.8) is called the *radiation resistance $R_s$* of the antenna. This is given by

$$R_s = \frac{8\pi^2 l^2}{3c \lambda^2} = 790 \frac{l^2}{\lambda^2} \Omega \quad \text{(since } 80\pi^2 = 790\text{)} \qquad (68.9)$$

A special consideration is furnished by the case (figure 58) in which the two metal bodies with separation $l$ might, for example, be the plates of a capacitor connected to the ends of an almost closed circular wire loop, $l$ being very small. Here the **q**-value given by (68.7) would result in vanishingly small radiation. The arrangement in figure 58 considered however as a ring current $I$ enclosing an area $S$ is just the representation of a *magnetic dipole* with dipole moment $m = IS/c$. Let us observe the radiation of this magnetic dipole at a large distance, with $l$ infinitely small.

Now, so far as (68.7) is concerned, this radiation is negligible because in assuming that $r' \ll \lambda$ in the derivation of (68.7), we have considered

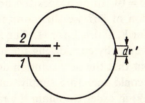

Fig. 58.—For the radiation of a closed oscillating circuit

the dimensions of the emitter as infinitely small compared with the wavelength. In order to obtain a non-vanishing result in this case, we must take into account the correction due to the retardation, at least to the first approximation. Developing the integrands of (68.6) in powers of $\mathbf{r} \cdot \mathbf{r}'/rc$, and again setting $g\,dV' = I\,d\mathbf{r}'$, we obtain

$$\dot{\mathbf{q}} = \oint I\left(t - \frac{r}{c} + \frac{\mathbf{r}\mathbf{r}'}{rc}\right)d\mathbf{r}' = \oint \left\{ I\left(t - \frac{r}{c}\right) + \dot{I}\left(t - \frac{r}{c}\right)\frac{\mathbf{r}\mathbf{r}'}{rc} + \dots \right\}d\mathbf{r}'$$

The first term of the integrand gives no finite contribution because $\oint d\mathbf{r}' = 0$, but the second term gives

$$\mathbf{q} = \dot{I}\left(t - \frac{r}{c}\right)\oint \frac{(\mathbf{r}\mathbf{r}')\,d\mathbf{r}'}{rc}$$

For the $\mathbf{r}'$-integrals occurring here we have

$$\oint x'\,dx' = 0 \qquad \oint x'\,dy' = -\oint y'\,dx' = S_z \qquad \text{etc.}$$

where $S_z$ means the projection of the surface $S$ bounded by the circular current on the $xy$-plane. Condensed, we obtain as in §47, through the introduction of the magnetic moment $\mathbf{m}(t)$ of our circular current,

$$\mathbf{q} = -\frac{\mathbf{r}}{r} \times \mathbf{m}\left(t - \frac{r}{c}\right) \qquad (68.10)$$

and thus for the Poynting vector of the radiation of a magnetic dipole, in place of (67.9), we have

$$\mathbf{S} = \frac{(\ddot{\mathbf{q}} \times \mathbf{r})^2}{4\pi c^3 r^4} \frac{\mathbf{r}}{r} = \frac{\ddot{m}^2 \sin^2 \theta}{4\pi c^3 r^2} \frac{\mathbf{r}}{r}$$

Here again, therefore, we find the $\sin^2 \theta$ directional dependence.

We could have found this result more simply by means of the following observation: If a field $\mathbf{E}, \mathbf{H}$ satisfies the Maxwell equations *in vacuo* (with $\mathbf{g} = 0$, $\rho = 0$), so also does the field $\mathbf{E'} = -\mathbf{H}$, $\mathbf{H'} = \mathbf{E}$. From the Hertz solution of §67 we therefore obtain a new solution when we substitute $\mathbf{H}$ for $\mathbf{E}$, and $-\mathbf{E}$ for $\mathbf{H}$. The Poynting vector is not thereby changed. However, by considering the near zone corresponding to (67.7) it turns out that the new solution describes the field of a magnetic dipole. We could therefore simply substitute the vector $\mathbf{m}$ for the vector $\mathbf{p}$ in (67.9) in order to obtain the radiation from a magnetic dipole.

Corresponding to (67.10), we thus find for the time-averaged total radiation of a magnetic dipole

$$\bar{S} = \frac{2\omega^4 \overline{m^2}}{3c^3} = \frac{32\pi^4 S^2}{3c\lambda^4} \overline{I^2} \tag{68.11}$$

As a comparison with (68.8) shows, there radiates from a current-carrying circular ring (loop) of area $S$ as much energy as would radiate from a straight antenna of length $l = 2\pi S/\lambda$ when carrying a like current. We can now also state the domain of validity of equation (68.8) for our circular loop: This formula holds so long as the plate separation $l$ is substantially greater than $2\pi S/\lambda$. If, however, $l$ is much smaller than this expression, (68.8) is completely invalid and the circular loop radiates as a magnetic dipole.

Finally, we investigate the radiation of an *electric quadrupole*. We consider the special case of a stretched-out quadrupole of the type sketched in figure 24b. We can, however, picture it as two in-line rod antennae carrying oppositely phased alternating currents. With the z-direction lying along the direction of the antenna, the vector $\mathbf{q}$ defined by (68.6) possesses only a z-component. For calculating $\mathbf{q}$ we must, as with the magnetic dipole, develop the correction term in the retardation, obtaining in the first non-vanishing approximation

$$q = q_z = \int g\left(\mathbf{r'}, t - \frac{r}{c}\right) \frac{\mathbf{r} \cdot \mathbf{r'}}{rc} \, dV' \tag{68.12}$$

We now particularize to the case of figure 24b, in which we assume that the quadrupole consists of a central charge $-2e$ at rest, and two oppositely phased oscillating charges $e$ with $z' = +\zeta(t)$ and $z' = -\zeta(t)$. These charges give to (68.12) the current contributions $\int g \, dV' = +e\dot{\zeta}$ for $z' = +\zeta$, and $\int g \, dV' = -e\dot{\zeta}$ for $z' = -\zeta$; altogether

$$q = \frac{2z}{rc} e\zeta\dot{\zeta}$$

On the other hand the quadrupole moment $Q$ of this (rotationally symmetric) charge distribution is, according to (24.17), equal to $2e\zeta^2$, so that we finally obtain

$$\mathbf{q} = [0, 0, q] \quad \text{with} \quad q = \frac{\dot{Q}}{2c}\frac{z}{r} = \frac{\dot{Q}}{2c}\cos\theta \qquad (68.13)$$

where again we designate by $\theta$ the angle between the direction of the quadrupole and the direction of the radiation.

Corresponding to (67.9) the radiation intensity of our array is given by[*]

$$\mathbf{S} = \frac{[\dddot{Q}(t - r/c)]^2}{16\pi c^5} \frac{\sin^2\theta\cos^2\theta}{r^2} \frac{\mathbf{r}}{r} \qquad (68.14)$$

This radiation intensity displays a directional distribution different from the dipole radiation in that not only is there no radiation along the direction of the dipole, but there is also none in the perpendicular plane. The latter fact is obviously the result of interference of the two oppositely oriented dipoles making up the quadrupole. The mean total radiated energy of our quadrupole for harmonic oscillations is then equal to

$$\overline{S} = \frac{\omega^6 \overline{Q^2}}{30c^5} = \frac{\omega^6 Q_0^2}{60c^5}$$

This differs from the mean total radiation of a simple dipole of maximum moment $p_0$ by the factor $\omega^2 Q_0^2/20c^2 p_0^2 = \pi^2 Q_0^2/5\lambda^2 p_0^2$. Within orders of magnitude, it is also smaller by the factor (dipole separation/wavelength)$^2$ than the total radiation of a simple dipole. The same, incidentally, also applies to other oscillating quadrupoles. As an example, the dipole separation of an atom is of the order of magnitude of $10^{-8}$ cm, and the wave length of its spectral lines about one thousand times greater. In atoms, therefore, the quadrupole radiation is about one million times weaker than the dipole radiation, and consequently it is usually ignored. It is observable, however, when for one reason or another the dipole radiation cannot take place ("forbidden" spectral lines).

*Remark:* In the MKSA system, in place of formula (68.8) for the radiation of a rod antenna of length $l$, we have the formula

$$\overline{S} = \frac{2\pi}{3} \sqrt{\frac{\mu_0}{\varepsilon_0}} \left(\frac{l}{\lambda}\right)^2 \overline{I^2}$$

Thus the radiation resistance of the antenna, corresponding to (68.9), is given by

$$R_S = \frac{2\pi}{3} \sqrt{\frac{\mu_0}{\varepsilon_0}} \left(\frac{l}{\lambda}\right)^2 = 790 \left(\frac{l}{\lambda}\right)^2 \Omega$$

---

[*] For the special quadrupole of figure 24*b* we could also have obtained this result by imagining the quadrupole as decomposed into two mutually displaced dipoles oscillating in opposite phase, their fields being individually calculated according to (67.7), and then appropriately combined.

We arrive at the case of an oscillating ring current of magnetic moment $m = \mu_0 IS$ if throughout we substitute $(\mu_0/\varepsilon_0)^{1/2}\mathbf{H}$ for $\mathbf{E}$, and $-(\varepsilon_0/\mu_0)^{1/2}\mathbf{E}$ for $\mathbf{H}$. The radiation of a magnetic dipole is then calculated from the same formula as that for an electric dipole (according to §67) in which $m$ is written there instead of $p$, and $\varepsilon_0$ and $\mu_0$ exchanged; thus

$$\bar{S} = \frac{\varepsilon_0 \sqrt{(\varepsilon_0 \mu_0)}}{6\pi} \omega^4 \overline{m^2} = \frac{8\pi^3}{3} \sqrt{\frac{\mu_0}{\varepsilon_0}} \frac{S^2}{\lambda^4} \overline{I^2}$$

## §69. The field of an arbitrarily moving point charge

We now inquire into the field of a charged particle moving with an arbitrary time-varying velocity $\mathbf{v}(\tau)$. For calculation it is best to start with the potential expressions (66.7), thus

$$\mathbf{A} = \frac{e}{c} \int_{-\infty}^{+\infty} \frac{\mathbf{v}}{r} \delta\left(\tau - t + \frac{r}{c}\right) d\tau \qquad \phi = e \int_{-\infty}^{+\infty} \frac{1}{r} \delta\left(\tau - t + \frac{r}{c}\right) d\tau \quad (69.1)$$

Here $r$ is the distance of the field point from the particle location at time $\tau$. By substituting these expressions in equations (66.1) we obtain for the field strengths

$$\mathbf{E} = -\operatorname{grad} \phi - \frac{1}{c} \frac{\partial \mathbf{A}}{\partial t}$$

$$= e \int \left\{ \frac{\mathbf{r}}{r^3} \delta\left(\tau - t + \frac{r}{c}\right) - \frac{1}{cr}\left(\frac{\mathbf{r}}{r} - \frac{\mathbf{v}}{c}\right) \delta'\left(\tau - t + \frac{r}{c}\right) \right\} d\tau$$

$$\mathbf{H} = \operatorname{curl} \mathbf{A}$$

$$= \frac{e}{c} \int \left\{ \frac{\mathbf{v} \times \mathbf{r}}{r^3} \delta\left(\tau - t + \frac{r}{c}\right) - \frac{\mathbf{v} \times \mathbf{r}}{cr^2} \delta'\left(\tau - t + \frac{r}{c}\right) \right\} d\tau$$

Here, since

$$\frac{d}{d\tau} \delta\left(\tau - t + \frac{r}{c}\right) = \left(1 - \frac{\mathbf{r}\mathbf{v}}{rc}\right) \delta'\left(\tau - t + \frac{r}{c}\right)$$

we can eliminate the derivative of the $\delta$-function with respect to its argument by partial integration with respect to $\tau$, and thus we find that

$$\mathbf{E} = e \int \left\{ \frac{\mathbf{r}}{r^3} + \frac{d}{d\tau}\left(\frac{\mathbf{r}/r - \mathbf{v}/c}{rc - \mathbf{r}\mathbf{v}}\right) \right\} \delta\left(\tau - t + \frac{r}{c}\right) d\tau$$

$$\mathbf{H} = \frac{e}{c} \int \left\{ \frac{\mathbf{v} \times \mathbf{r}}{r^3} + \frac{d}{d\tau}\left(\frac{\mathbf{v} \times \mathbf{r}/r}{rc - \mathbf{r}\mathbf{v}}\right) \right\} \delta\left(\tau - t + \frac{r}{c}\right) d\tau$$

Carrying out the $\tau$-integration now by means of the auxiliary variable $u = \tau - t + r/c$, corresponding to the passage from (66.7) to (66.9), we obtain, after a brief rewriting, the final expressions

$$\left.\begin{aligned}
\mathbf{E} &= e\left\{\frac{(\mathbf{r} - r\mathbf{v}/c)(1 - v^2/c^2)}{(r - \mathbf{r}\mathbf{v}/c)^3} + \frac{\mathbf{r} \times [(\mathbf{r} - r\mathbf{v}/c) \times \dot{\mathbf{v}}]}{c^2(r - \mathbf{r}\mathbf{v}/c)^3}\right\}_{\tau = t - r/c} \\
\mathbf{H} &= e\left\{\frac{(\mathbf{v} \times \mathbf{r})(1 - v^2/c^2)}{c(r - \mathbf{r}\mathbf{v}/c)^3} + \frac{\mathbf{r} \times \{\mathbf{r} \times [(\mathbf{r} - r\mathbf{v}/c) \times \dot{\mathbf{v}}]\}}{c^2 r(r - \mathbf{r}\mathbf{v}/c)^3}\right\}_{\tau = t - r/c}
\end{aligned}\right\} \quad (69.2)$$

Clearly we have rigorously that

$$\mathbf{H} = \frac{\mathbf{r}}{r} \times \mathbf{E} \qquad (69.3)$$

i.e. the magnetic field is perpendicular to the electric field and to the vector $\mathbf{r}$ extending from the particle location at time $\tau = t - r/c$ to the field point.

For the discussion of formulae (69.2) we consider separately the two terms in each expression. Both first terms, the parts $\mathbf{E}_1$ and $\mathbf{H}_1$ not

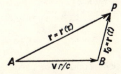

Fig. 59.—The connection between the retarded particle location A, and the simultaneous location B for uniform particle motion

containing the acceleration $\dot{\mathbf{v}}$, give a field which for large distances falls off as $1/r^2$, and therefore has the character of a static field. Since we obtain this partial field $\mathbf{E}_1$ and $\mathbf{H}_1$ from the rigorous solutions (69.2) valid for all $\mathbf{v}(t)$, it must agree with the field (64.18) of a uniformly moving charged particle. The formal difference between the expressions comes from the different meanings of the vectors $\mathbf{r}$ here and in §64. There $\mathbf{r} = \mathbf{r}(t)$ was set equal to the vector from the instantaneous particle location B to the field point P, while here by $\mathbf{r} = \mathbf{r}(\tau) = \mathbf{r}(t - r/c)$ we understand the vector from the particle location A at time $\tau$ to the field point P (see figure 59). Now for the case of constant velocity we obviously have

$$\mathbf{r}(t) = \mathbf{r}(\tau) - \frac{r(\tau)}{c}\mathbf{v}$$

and this is just the denominator in $E_1$. If we write $r(t) = r_0$, but $r(\tau) = r$ as heretofore in these paragraphs, we find from $r_0 = r - r v/c$ for the denominator in (64.18) that

$$\sqrt{\left[ r_0^2 - \left( r_0 \times \frac{v}{c} \right)^2 \right]} = r - \frac{rv}{c}$$

that is, exactly the denominator in equations (69.2). Thus the field parts $E_1$ and $H_1$ actually represent the part of the field moving along with the particle.

By contrast, the second parts $E_2$ and $H_2$ in (69.2) which are proportional to $\dot{v}$ have the character, going outward, of a $1/r$ decreasing wave field in which the three vectors $r$, $E_2$, and $H_2$ are, pairwise, mutually perpendicular.

For the special case of a particle at rest at time $\tau$ this wave field at distance $r$ at the time $\tau + r/c$ is given by

$$\mathbf{E}_2 = \frac{e}{c^2 r} \left( \dot{v} \times \frac{r}{r} \right) \times \frac{r}{r} \qquad \mathbf{H}_2 = \frac{e}{c^2 r} \dot{v} \times \frac{r}{r} \qquad (69.4)$$

Through a spherical surface of radius $r$ around the particle there flows, as in the dipole field, the total quantity of radiation

$$\mathbf{S} = \frac{2 e^2 \dot{v}^2}{3 c^3} \qquad (69.5)$$

This is equal to the decrease per second of the particle and field energy within the spherical surface. Owing to symmetry, the total momentum of the particle and field within the sphere does not change.

If during the time of emission the particle is moving with velocity $v$, the calculation of the radiated energy going through the spherical surface during the time interval $d\tau$ becomes more difficult—for, during this time the emitting particle has moved away from the centre of the sphere a distance $v d\tau$, and the radiation field therefore becomes asymmetric with respect to the stationary sphere. We skip over the tedious calculations and simply state the result:

Owing to the motion of the particle there enters for the value (69.5) of the radiated energy passing through a stationary sphere, a factor $(1 - \beta^2 \sin^2 \gamma)/(1 - \beta^2)^3$, where $\beta = v/c$ and $\gamma$ is the angle between the vectors $v$ and $\dot{v}$. At the same time the total energy found within the sphere changes by $- v \mathbf{S}/c^2$. (In §85 we shall find these relationships again and we shall prove them in a completely different context.)

In Volume II we shall consider more thoroughly the consequences of the above results, and particularly the question of radiation reaction. We mention here, however, that equation (69.5) and its generalization for $v \neq 0$ is of importance for the understanding of the *spectrum of Bremsstrahlung X-rays*. The latter result from the impact of electrons of high velocity, e.g. cathode rays, against a solid body. As is shown in the spectral investigation of the emitted radiation, X-rays consist of two markedly different components. Of these, one component consists of a *line spectrum* and depends essentially upon the *nature of the anticathode*. It is emitted by atoms of the solid body which have been "excited" by impacts with the cathode rays. The other radiation component consists of a continuous spectrum. It is to be understood as radiation of the type just considered, emitted by electrons during their deceleration upon penetrating into the material of the anticathode. According to A. Sommerfeld the Bremsstrahlung spectrum can be approximately arrived at in calculation by working with the acceleration (retardation) which the individual electrons experience in their approximately hyperbolic paths in the immediate neighbourhood of the atomic nuclei of the anticathode material, i.e. just where $\dot{v}^2$ possesses its maximum value and thus where the emission of radiation is greatest. The spectral distribution is then obtained from the Fourier analysis of the deceleration process and its radiation field. The results so obtained for low frequencies agree qualitatively rather well with observations. For the higher frequencies, however, there are certain characteristic differences which appear, and these can only be understood from the standpoint of quantum theory. Meriting first mention here is the fact that the X-ray Bremsstrahlung spectrum does not, corresponding to the result of the calculation just intimated, extend to infinitely high frequencies, but is cut off at a sharply determined upper frequency limit depending upon the energy of the cathode rays (acceleration voltage of the X-ray tube).

The formulae we have mentioned are also of importance for the radiation coming from charged particles in accelerator apparatus (betatron, cyclotron, synchrotron, etc.), in which the particles are constrained to pursue a circular path by means of a magnetic guide field. In these machines, for not too high velocities, the particles experience a continuous acceleration $\dot{\mathbf{v}} = \dfrac{e}{mc} \mathbf{v} \times \mathbf{B}$ and, since $\mathbf{v}$ is at right angles to $\mathbf{B}$, the amount of energy they radiate per second is given by

$$S = \frac{2e^4}{3m^2c^3}\left(\frac{\mathbf{v}}{c} \times \mathbf{B}\right)^2 = \frac{4e^4}{3m^3c^5}\, WB^2 \tag{69.6}$$

where $W = \tfrac{1}{2}mv^2$ is the kinetic energy of the particle. (For greater particle energy a factor $1 + W/2mc^2$ multiplies this expression.) In (circular) acceleration apparatus for electrons (small $m$) this energy loss becomes serious for electron energies of about $1 \times 10^9$ eV. For further details of such radiation emitted from "luminous electrons", its spectral distribution (the intensity maximum for $W \sim 10^8$ eV lies in the visible, and for $10^9$ eV is in the region of soft X-rays), and its directional distribution (for $v \approx c$ the radiation, because of the denominator in (69.2), is concentrated in a region of very small angle around the direction of $\mathbf{v}$), see for example D. Iwanenko and A. Sokolow, *Klassische Feldtheorie* (1953).

## CHAPTER D IV

### The Field Equations in Slowly Moving Non-magnetic Media

#### §70. Derivation of the field equations

Heretofore we have concerned ourselves with the field equations only for media at rest. We wish now, using the electron theory as a basis, to derive these equations for media in motion. Only later, however, in the section on the *Theory of Relativity*, will we become acquainted with what, in a certain sense, is the final solution of the problem. For the case where the (constant) velocity of the medium *is small compared to the velocity of light*, the equations obtained by relativity theory are the same as those which we wish now to derive.

For our case the two Maxwell equations

$$\text{curl } \mathbf{E} = -\frac{1}{c}\frac{\partial \mathbf{B}}{\partial t} \quad \text{and} \quad \text{div } \mathbf{B} = 0 \qquad (70.1)$$

(in which the charge and current do not appear) remain unchanged since they cannot, particularly for non-magnetizable media, be influenced by the motion of matter.

It is otherwise, however, with the behaviour of the $\mathbf{E}$ and $\mathbf{B}$ fields themselves where there is moving matter (velocity of motion $\mathbf{v}$). According to the considerations of §44, the individual charges in the moving body experience the Lorentz force

$$\mathbf{F} = e\left(\mathbf{E} + \frac{\mathbf{v}}{c} \times \mathbf{B}\right) \qquad (70.2)$$

If, for example, *an observer moving along with the matter* measures the force on a co-moving unit charge and associates with this force a field strength $\mathbf{E}^*$, then this apparent field strength must be given by

$$\mathbf{E}^* = \mathbf{E} + \frac{\mathbf{v}}{c} \times \mathbf{B} \qquad (70.3)$$

The special meaning of $\mathbf{E}^*$ is that in matter of conductivity $\sigma$, $\mathbf{E}^*$ gives rise to a *current density* $\mathbf{g}$ in accordance with the expression

$$\mathbf{g} = \sigma\mathbf{E}^* = \sigma\left(\mathbf{E} + \frac{\mathbf{v}}{c} \times \mathbf{B}\right) \qquad (70.4)$$

Similarly, the *polarization* $\mathbf{P}$ in moving matter is determined by the quantity $\mathbf{E}^*$, thus:

$$\mathbf{P} = \frac{\varepsilon-1}{4\pi}\mathbf{E}^* = \frac{\varepsilon-1}{4\pi}\left(\mathbf{E} + \frac{\mathbf{v}}{c} \times \mathbf{B}\right) \qquad (70.5)$$

We turn now to the other two Maxwell equations, namely those in which the charge and current appear. In formulating these equations we make use of a previously obtained result of electron theory, namely that the primary field quantities are $\mathbf{E}$ and $\mathbf{B}$, while the quantities $\mathbf{D}$ and $\mathbf{H}$ represent derived quantities. In what follows we shall therefore, as in §§27 and 48, write the two remaining Maxwell equations with $\mathbf{E}$ and $\mathbf{B}$, being careful however to take *all* contributions to the charges and currents into account.

As for charges, there come into consideration in moving matter, along with the given true charges (density $\rho$), the polarization charges (density $\rho_P$) for which $\rho_P = -\operatorname{div}\mathbf{P}$, as in §26. As before, therefore, we have for the electric flux theorem

$$\operatorname{div}\mathbf{E} = 4\pi(\rho - \operatorname{div}\mathbf{P}) \qquad (70.6)$$

To help us we may therefore now define a dielectric displacement $\mathbf{D}$ by the equation

$$\mathbf{D} = \mathbf{E} + 4\pi\mathbf{P} = \varepsilon\mathbf{E} + (\varepsilon-1)\frac{\mathbf{v}}{c} \times \mathbf{B} \qquad (70.7)$$

the latter expression coming from (70.5), and we have as before

$$\operatorname{div}\mathbf{D} = 4\pi\rho \qquad (70.8)$$

We consider now the last field equation, the relationships of which are somewhat more complicated. According to this equation $\operatorname{curl}\mathbf{B}$ is determined by the current density. In the case of matter at rest this current density, disregarding the Maxwell term $\partial\mathbf{E}/\partial t$, was given by the conduction current $\mathbf{g}$, and the polarization current $\mathbf{g}_P = \partial\mathbf{P}/\partial t$. In moving matter, however, there enter additional current contributions coming from all those charges which, in consequence of the motion, would be carried through an (imaginary) surface element fixed in space. Accordingly, we divide the current into two parts: *the current relative to the moving matter* (a current which would be recognized by an observer moving with the material), and a *convection current* from all the charges in the moving matter. The latter is obviously equal to $(\rho - \text{div}\,\mathbf{P})\mathbf{v}$. The current relative to the moving matter consists of the conduction current $\mathbf{g}$ and the polarization current. The latter is no longer, however, equal to $\partial \mathbf{P}/\partial t$; rather, the quantity of electricity which, in consequence of the change with time of the polarization, passes through an (imaginary) co-moving surface element, is given by

$$\left(\frac{\partial \mathbf{P}}{\partial t}\right)^* = \frac{\partial \mathbf{P}}{\partial t} + (\mathbf{v}\nabla)\mathbf{P} = \frac{\partial \mathbf{P}}{\partial t} + \text{curl}\,(\mathbf{P}\times\mathbf{v}) + \mathbf{v}\,\text{div}\,\mathbf{P} \quad (70.9)$$

Obviously, the last term here cancels with the polarization part of the convection current. Altogether, then, we obtain for the curl of the vector $\mathbf{B}$ the equation

$$\text{curl}\,\mathbf{B} = \frac{1}{c}\frac{\partial \mathbf{E}}{\partial t} + \frac{4\pi}{c}\left[\mathbf{g} + \rho\mathbf{v} + \frac{\partial \mathbf{P}}{\partial t} + \text{curl}\,(\mathbf{P}\times\mathbf{v})\right] \quad (70.10)$$

Thus, along with (70.1) and (70.6), we have found, *for the case of small velocities, the complete set of Maxwell equations for moving non-magnetizable media.*

The only equation of this system which differs from the corresponding equation for media at rest is expression (70.10). That in its given form it is in mathematical agreement with the Maxwell term $\partial \mathbf{E}/\partial t$ can be seen by taking the divergence. This leads to the expression

$$0 = \frac{1}{c}\frac{\partial}{\partial t}\,\text{div}\,(\mathbf{E} + 4\pi\mathbf{P}) + \frac{4\pi}{c}\,\text{div}\,(\mathbf{g} + \rho\mathbf{v})$$

which, on account of (70.6), is actually fulfilled if the equation of continuity of electric charge holds in the form

$$\frac{\partial \rho}{\partial t} + \text{div}\,(\mathbf{g} + \rho\mathbf{v}) = 0$$

For moving media, however, this equation is identical with the immediately obvious expression

$$\left(\frac{\partial \rho}{\partial t}\right)^* + \text{div}\,\mathbf{g} = 0 \quad (70.11)$$

in which the time derivative provided with a star means, as in (70.9), the time derivative as evaluated by an observer moving with the material.

In the paragraphs following we shall discuss several experiments which throw light on the expression for the current in (70.10). Already here, however, we make special reference to the last current term in (70.10), which has the same form as the magnetization current $g_M = c\,\mathrm{curl}\,\mathbf{M}$ defined by (47.16). Thus it comes about that a moving electrically polarized medium, which for a co-moving observer appears to be unmagnetized, carries, for an observer at rest, a magnetization

$$\mathbf{M} = \mathbf{P} \times \frac{\mathbf{v}}{c} \qquad (70.12)$$

It therefore appears reasonable to define a magnetic field strength which is different from $\mathbf{B}$ by the expression

$$\mathbf{H} = \mathbf{B} - 4\pi\mathbf{M} = \mathbf{B} - 4\pi\mathbf{P} \times \frac{\mathbf{v}}{c} \qquad (70.13)$$

In introducing this quantity $\mathbf{H}$ along with the quantity $\mathbf{D}$ into equation (70.10), we obtain the relation

$$\mathrm{curl}\,\mathbf{H} = \frac{1}{c}\frac{\partial \mathbf{D}}{\partial t} + \frac{4\pi}{c}(\mathbf{g} + \rho\mathbf{v}) \qquad (70.14)$$

which differs from the first Maxwell equation for media at rest only in the term for the convection current of the true charges.

## §71. Experimental confirmation of the basic equations

In the following we wish to examine several experimental arrangements which go back in particular to the investigations of H. A. Rowland, W. C. Röntgen, and A. Eichenwald, these investigations having had the aim of demonstrating directly by experiment the various kinds of currents which, as we have found in the foregoing paragraphs, can be produced by bodies in motion. The total current responsible for the production of a magnetic field is given, according to (70.10), by the expression

$$\mathbf{g} + (\rho - \mathrm{div}\,\mathbf{P})\mathbf{v} + \left(\frac{\partial \mathbf{P}}{\partial t}\right)^{*} \equiv \mathbf{g} + \rho\mathbf{v} + \frac{\partial \mathbf{P}}{\partial t} + \mathrm{curl}\,(\mathbf{P} \times \mathbf{v}) \qquad (71.1)$$

The first group of experiments relates to processes in which all time derivatives vanish, i.e. stationary relations obtain both for the co-moving and the stationary observers. As effective current there then remains only $g + (\rho - \operatorname{div} \mathbf{P})\mathbf{v}$. Accordingly, it must be possible to produce the same magnetic field either by a *conduction current,* or by *the motion of a charged body,* or by *the motion of a polarized body.*

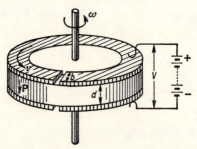

Fig. 60.—Schematic diagram of the investigations of Röntgen and of Eichenwald

Let us for our purposes examine the experimental arrangement shown in figure 60. A circular hard rubber disc of thickness $d$ is mounted on bearings so as to be free to rotate about a vertical axis. Attached rigidly to it above and below are metal rings of radial width $b$. Each metal ring has been cut right through at one place by a narrow radial slit. The two metal rings are connected to the terminals of a voltage source and thus become charged to a potential difference $V$. The apparatus may now be set in motion, there being the possibility of having the metal rings rotate too (Rowland), or remain at rest (Röntgen). In any case, we have the following charges with which to reckon:

True surface charges on the metal plates:  $\omega = \varepsilon V/4\pi d$

Surface charges on the hard rubber disc:  $\omega_P = -(\varepsilon - 1)V/4\pi d$

The designation given relates to the side of the disc connected to the positive terminal of the voltage source. Both designations are of course reversed for the opposite side of the disc. If both disc and rings be placed in rotation, the upper metal ring carries the convection current $vb\omega$, and the adjacent surface of the hard rubber carries the current $vb\omega_P$. In all, then, we have a current $vb(\omega + \omega_P) = vb\,V/4\pi d$.

The existence of this current can be demonstrated by bringing a small magnetized needle into the vicinity of the rotating disc and observing its deflection. We can then check the theory by sending a current through the stationary rings as shown in figure 61. The current, whose value is read on the ammeter shown, is adjusted to give the same deflection on the magnetic needle as was previously obtained with the rotating disc. According to the theory, then, this current must be given by the expression $I = vbV/4\pi d$. It is especially to be noted that with co-moving metal rings *the effect is completely independent of the permittivity of the hard rubber disc*: thus we obtain the same deflection when there is no dielectric between the rotating rings. This result, confirmed by the many noteworthy high-precision experiments of Eichenwald, appears to us now, from the standpoint of electron theory, as rather obvious.

If, in the apparatus described, the *metal plates* be held fixed and only the disc between them allowed to rotate, then there is present as effective current only the displacement of the surface charge of the capacitor, so that in this case the motion of the dielectric is, from the

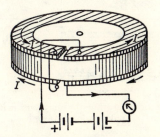

Fig. 61.—Measurement of the current which produces the same magnetic field as the apparatus of figure 60

standpoint of its producing a magnetic effect, equivalent to a current $I = -vb(\varepsilon-1)V/4\pi d$. This result too was confirmed by Eichenwald.

We have so far explained the experiments of Eichenwald in terms of the left-hand side of equation (71.1). In the case last considered (disc alone in motion) only the term $-\mathbf{v}\,\mathrm{div}\,\mathbf{P}$ enters into consideration, this leading by integration over a very flat cylinder of unit cross-section whose end surfaces lie one in air and the other in hard rubber, to a surface current density $v\omega_P \equiv vP$. The same result could, of course, be obtained by the term $\mathrm{curl}(\mathbf{P} \times \mathbf{v})$. In our disc the vector $\mathbf{P} \times \mathbf{v}$ is directed radially, i.e. in the arrangement shown in figure 60 it is directed away from the axis of rotation. In the hard rubber $\mathbf{P} \times \mathbf{v}$ is constant while outside it is equal to zero. It therefore produces a surface current density on the surface of the hard rubber of magnitude $vP$. We come to the same result if, corresponding to equation (70.12), we ascribe to the hard rubber a (radial) magnetization.

By means of a modification of the previously described apparatus the *polarization current* $(\partial \mathbf{P}/\partial t)^*$ may be demonstrated in an interesting

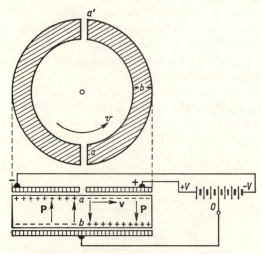

Fig. 62.—Method of measuring the magnetic field produced by a polarization current

fashion. Figure 62 shows the scheme of the system in plan (above), and elevation (below). As before, there is a rotating hard-rubber disc. It moves between two stationary metal rings, the upper of which is made up of two semicircular halves. The lower ring is grounded while the two upper halves are connected to equal but opposite sources of electrical potential, $+V$ and $-V$. A delicately pivoted magnetized needle free to rotate about a vertical axis is located above the disc in the neighbourhood of the rotation axis. In this arrangement there flows, as before, the convection current $-\mathbf{v} \operatorname{div} \mathbf{P}$, only with the difference that now this current in the right and left halves flows in *opposite* directions. With respect to their effect on the magnetic needle, then, the two currents cancel one another. Besides these currents, however, there is also a *polarization current* which flows in a direction *parallel* to the axis of rotation, i.e. at $a$ on the disc from above downward, while at $a'$ from below upward. We obtain the magnitude of the total current flowing from $a$ to $b$ from (71.1) by integrating over a horizontal rectangle located where the current is flowing, and having a size sufficient to enclose the flow-lines of all currents. If the $x$-axis has the direction of $\mathbf{v}$, the $y$-axis the direction of $\mathbf{P}$, and if $v_x = v$, and $P_y = P$, then we have

$$I = b \int \left| \left( \frac{\partial \mathbf{P}}{\partial t} \right)^* \right| dx = bv \int \frac{\partial P}{\partial x} dx$$

$$= 2bvP = \frac{2bv(\varepsilon - 1)V}{4\pi d}$$

The same result may be obtained, incidentally, by examining the current paths in figure 62. During rotation of the hard rubber disc there flows *from* right and *from* left toward $a$, at the *upper* surface, the current $bvP$, while at the *lower* surface at $b$, *to* right and *to* left, a like current flows away. The current path becomes closed by virtue of the vertically directed currents (as from $a$ to $b$), such as we calculated above.

A series of experiments carried out by H. A. Wilson showed directly that within a substance moving in a magnetic field the Lorentz force $e\mathbf{v} \times \mathbf{B}/c$ is actually exerted on all the charges of which the substance is constituted; that therefore a co-moving observer would find an electric field strength $\mathbf{E}^* = \mathbf{v} \times \mathbf{B}/c$. Wilson's arrangement consists of a hollow cylinder, of wall thickness $b$, which is rotated about its own axis while at the same time being immersed in an axially directed magnetic field. A radially directed electric field of strength $\mathbf{E}^*$ ensues within the cylinder. For the case of a *conducting* cylinder, this field produces *a net migration of the conduction electrons* to (say) the outer surface of the cylinder until the opposing field produced by the surface charges is just able to hold the Lorentz force in equilibrium. Thus there exists between outer and inner surfaces a *potential difference* of magnitude $bE^*$. If a resistor of value $R$ be connected by brushes to the inner and outer surfaces of the cylinder, a current flows in the resulting closed circuit, its magnitude being $I = bE^*/R = bvB/cR$. If the cylinder is made of an *insulating material*, the Lorentz force produces within it a *radially directed dielectric polarization* $\mathbf{P} = (\varepsilon - 1)\mathbf{E}^*/4\pi$. We can show this experimentally in the following way: The rotor (a dielectric) is provided at its inner and outer surfaces with snugly fitting cylindrical metal coverings. The two metal cylinders are then connected to the terminals of a suitable electrometer. On rotating the dielectric rotor in an axially directed magnetic field a charging of the electrometer is obtained, corresponding to the polarization just calculated.

## §72. Fizeau's investigation

Along with the purely electromagnetic experiments on moving bodies with which we have become acquainted in the foregoing paragraphs, we consider now a purely *optical* investigation related to the propagation

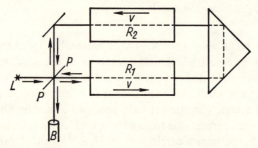

Fig. 63.—Measurement of the Fizeau entrainment coefficient

of light in moving media. According to the theory the light must, as we shall see later, travel *faster in the direction of motion of the medium* than in the opposite direction, i.e. the light is, so to speak, *carried along* with the moving medium.

The first experimental investigation of the magnitude of this transportation was carried out by Fizeau. His experimental arrangement is shown in figure 63. Light from the source L, divided into two parts by the half-silvered mirror PP, pursues paths in opposite directions around the circuit as shown in the figure. The two beams are again united in the half-silvered mirror and arrive at an interference apparatus B for observation. Along their way the two light beams pass through two tubes $R_1$ and $R_2$ through which water flows in opposite directions. One light beam goes through both tubes in the same direction as the water flow; the other beam in the opposite direction. This has the result that the first beam requires less time than the second for the journey from the half-silvered mirror through the tube system and back again. Thus the two beams arrive at the interference apparatus with a difference of phase. From the position in the apparatus of the interference fringes obtained, we can determine this time difference and thereby also the amount of speed increase.

We wish now to obtain a clear picture of the expected result on the basis of the formulae developed in §70. The Maxwell equations for the electromagnetic field in an uncharged non-magnetic insulating medium ($\rho = 0$, $\mathbf{g} = 0$, $\mu = 1$) moving with velocity $\mathbf{v}$ are:

$$\left. \begin{array}{lll} \text{curl } \mathbf{H} = \dfrac{1}{c} \dfrac{\partial \mathbf{D}}{\partial t} & \text{div } \mathbf{D} = 0 & \mathbf{D} = \mathbf{E} + 4\pi \mathbf{P} \\[3mm] \text{curl } \mathbf{E} = -\dfrac{1}{c} \dfrac{\partial \mathbf{B}}{\partial t} & \text{div } \mathbf{B} = 0 & \mathbf{H} = \mathbf{B} - 4\pi \mathbf{P} \times \dfrac{\mathbf{v}}{c} \end{array} \right\} \quad (72.1)$$

According to (70.5), $\mathbf{P}$ is given by

$$\mathbf{P} = \frac{\varepsilon - 1}{4\pi} \mathbf{E}^* = \frac{\varepsilon - 1}{4\pi} \left( \mathbf{E} + \frac{\mathbf{v}}{c} \times \mathbf{B} \right) \quad (72.2)$$

Owing to the dependence of $\varepsilon$ on frequency, care is required in applying the foregoing relation to a process involving light propagation. If we consider the propagation of a plane wave, as described in (57.10), given by the expressions

$$\mathbf{E} = \mathbf{E}_0 \exp i(\omega t - \mathbf{k} \cdot \mathbf{r}) \qquad \mathbf{B} = \mathbf{B}_0 \exp i(\omega t - \mathbf{k} \cdot \mathbf{r}) \quad (72.3)$$

there then acts at the location of an atom moving with the matter, the location of the atom being designated by $\mathbf{r} = \mathbf{r}_0 + \mathbf{v}t$, the apparent field intensity

$$E^* = E_0^* \exp i(\omega t - \mathbf{k} \cdot \mathbf{r}_0 - \mathbf{k} \cdot \mathbf{v}t)$$

The field's apparent frequency at the location of this atom is given by

$$\omega^* = \omega - \mathbf{k}\mathbf{v} \quad (72.4)$$

We have thus to put into (72.2) the permittivity appropriate to this "Doppler-shifted" frequency $\omega^*$; thus we have in the following that

$$\varepsilon = \varepsilon(\omega^*) \quad (72.5)$$

We return now to equations (72.1). From these we obtain, with the unit phase factor from (72.3) [compare (57.13) or (58.9)]

$$
\left.
\begin{aligned}
-\mathbf{k} \times \mathbf{H} = \frac{\omega}{c} \mathbf{D} \qquad \mathbf{k}\,\mathbf{D} = 0 \\[2mm]
-\mathbf{k} \times \mathbf{E} = -\frac{\omega}{c} \mathbf{B} \qquad \mathbf{k}\,\mathbf{B} = 0
\end{aligned}
\right\}
\qquad (72.6)
$$

From these equations we see that $\mathbf{D}$ and $\mathbf{B}$ are perpendicular to $\mathbf{k}$, but that in general (and in contrast to the case of media at rest), this is no longer true for the electric vector $\mathbf{E}$. We have instead

$$
0 = \mathbf{k}\,\mathbf{D} = \mathbf{k}\,\mathbf{E} + 4\pi \mathbf{k}\,\mathbf{P} = \varepsilon \mathbf{k}\,\mathbf{E} + (\varepsilon - 1)\mathbf{k}\!\left(\frac{\mathbf{v}}{c} \times \mathbf{B}\right)
$$

Accordingly, $\mathbf{k}.\mathbf{E}$ is always different from zero whenever the three vectors $\mathbf{k}$, $\mathbf{v}$, and $\mathbf{B}$ do not lie in a plane. Analogously, we have for the magnetic field strength $\mathbf{H} = \mathbf{B} - 4\pi \mathbf{P} \times \mathbf{v}/c$. Here then, as a consequence, the Poynting vector does not in general have the direction of $\mathbf{k}$.

For our further calculations it will be expedient to express $\mathbf{E}$ and $\mathbf{H}$ in equations (72.6) in terms of $\mathbf{D}$ and $\mathbf{B}$. We have obviously, on account of (72.2),

$$
\mathbf{E} = \frac{\mathbf{D}}{\varepsilon} - \frac{\varepsilon - 1}{\varepsilon}\,\frac{\mathbf{v}}{c} \times \mathbf{B} \qquad \mathbf{P} = \frac{\varepsilon - 1}{4\pi\varepsilon}\left(\mathbf{D} + \frac{\mathbf{v}}{c} \times \mathbf{B}\right) \qquad (72.7)
$$

Limiting ourselves to the case of *small velocities* so that we may neglect terms of the order $(v/c)^2$ compared to 1, and remembering that $\mathbf{k}.\mathbf{D} = 0$ and $\mathbf{k}.\mathbf{B} = 0$, we obtain from (72.6) the relationships

$$
-\mathbf{k} \times \mathbf{B} = \left(\frac{\omega}{c} - \frac{\varepsilon - 1}{\varepsilon}\,\frac{\mathbf{k}\mathbf{v}}{c}\right)\mathbf{D} \qquad -\frac{1}{\varepsilon}\mathbf{k} \times \mathbf{D} = -\left(\frac{\omega}{c} - \frac{\varepsilon - 1}{\varepsilon}\,\frac{\mathbf{k}\mathbf{v}}{c}\right)\mathbf{B}
$$

Making use of the familiar vector formula

$$
\mathbf{A} \times (\mathbf{B} \times \mathbf{C}) = \mathbf{B}(\mathbf{A}.\mathbf{C}) - \mathbf{C}(\mathbf{A}.\mathbf{B})
$$

and eliminating $\mathbf{B}$ and $\mathbf{D}$ respectively by substitution, we arrive at the expression

$$
\left(\frac{\omega}{c} - \frac{\varepsilon - 1}{\varepsilon}\,\frac{\mathbf{k}\mathbf{v}}{c}\right)^2 = \frac{k^2}{\varepsilon}
$$

Upon introducing in addition the angle $\alpha$ between the direction of propagation of the waves and the direction of motion of the material, we find for the wave velocity

$$w = \frac{\omega}{k} = \frac{c}{\sqrt{\varepsilon}} + \frac{\varepsilon - 1}{\varepsilon} v \cos \alpha \qquad (72.8)$$

Further, we make use of the refractive index formula $\sqrt{\varepsilon} = n$, taking for $n$ the value appropriate to the frequency $\omega^*$ as given by (72.4). In the first approximation we have for $\omega^*$

$$\omega^* = \omega \left( 1 - \frac{v \cos \alpha}{w} \right) \approx \omega \left( 1 - \frac{nv}{c} \cos \alpha \right)$$

thus we have, also approximately,

$$\sqrt{[\varepsilon(\omega^*)]} = n(\omega^*) = n(\omega) - \frac{dn(\omega)}{d\omega} \cdot \frac{nv\omega}{c} \cos \alpha$$

From (72.8) then we obtain as the *general law for the propagation of light in slowly moving media* (with $\omega_0 = c/n$)

$$w = w_0 + \left( 1 - \frac{1}{n^2} + \frac{\omega}{n} \frac{dn}{d\omega} \right) v \cos \alpha \qquad (72.9)$$

We might picture this result as if the light, in its propagation in the direction of the motion, were "blown" along. It is astonishing, however, that this entrainment of the light by the moving matter is not a complete speeding up but, by contrast with the case of the propagation of sound in a moving liquid, is only partial. The factor

$$\eta = 1 - \frac{1}{n^2} + \frac{\omega}{n} \frac{dn}{d\omega} \qquad (72.10)$$

characterizing the degree of speeding up is designated the *Fizeau entrainment coefficient*. In Fizeau's apparatus one light beam went through the two tubes (total length $2l$) in the direction of flow of the water, while the other beam went through in the opposite direction. The difference in travel time is thus

$$\Delta t = \frac{2l}{w_0 - \eta v} - \frac{2l}{w_0 + \eta v} \approx \frac{4l\eta v}{w_0^2}$$

For the case of water Fizeau was able to make an interferometric determination of $\Delta t$, thus quantitatively confirming the entrainment coefficient formula (72.10).

## §73. The Michelson experiment

In the foregoing paragraphs we have seen that the electron theory readily explains the phenomenon of the partial entrainment of light waves in moving media, as first shown by Fizeau. It is also possible to describe the propagation of light in a moving dispersive medium in the following manner: The complete propagation consists of the superposition of an *incoming vacuum wave* upon *spherical waves* emitted by individual dipoles. In order to explain the Fizeau result we have to assume that the *vacuum wave* is *not in general influenced* by the motion of the medium. The whole effect comes about solely because the superimposed spherical waves are now emitted by *moving* dipoles. This combination of the plane waves in the ether at rest and the spherical waves emitted by the *moving* dipoles gives, in fact, the entrainment coefficient observed by Fizeau. For a long time the result of the Fizeau experiment was regarded as an experimental demonstration of the existence of a *stationary ether* which did not take part in the motion of matter.

On the basis of the foregoing conception it appeared necessarily then to follow that *the value of the velocity of light as found by an observer in motion relative to the ether would depend on the observer's direction of motion*. If $c$ is the velocity of light relative to the stationary ether, and $v$ that of the observer, then the observer would find a velocity of light of value $c-v$ or $c+v$, according as his motion was with or against that of the light. If these relationships of motion for a particular observer were unknown to him, he could inform himself of them by experiment by allowing a visual light signal to be emitted in all directions from a point, and then measuring the time of arrival of this signal at various points of a sphere centred on the point of emission. If actually he were in motion with respect to the ether, then, by virtue of the ether "wind", the light signal would have to be borne along, in the sense that it would reach the surface of the sphere first at a point in the direction contrary to his motion, and last at a point in the same direction as his motion.

This experiment is exactly that carried out by Michelson for the purpose of establishing the direction of the earth's motion relative to the hypothetical stationary ether. The speed of the earth in its orbit amounts to about 30 km/s. Since, to start with, we do not know the direction of the ether's motion with respect to us, it would be conceivable that at a certain place in the earth's orbit the ether itself would move with just the earth's velocity, the earth thus being at rest there with respect to the ether. It should accordingly be possible half a year later to observe an ether wind of double the earth's velocity.

Michelson's apparatus, in plan view, is sketched in figure 64. The whole arrangement is firmly mounted on a massive base which floats

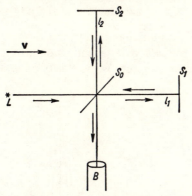

Fig. 64.—Arrangement of Michelson's experiment

in mercury and can thus, without disturbance, be rotated about a vertical axis through any desired angle. Light from the source L is partly reflected and partly transmitted by the half-silvered mirror $S_0$. The reflected part, travelling toward and being reflected from the mirror $S_2$, passes through $S_0$, and goes on to B for observation. The other part is reflected by $S_1$ and $S_0$, and in B is brought into interference with the first-mentioned part. By suitable means *interference fringes* are observed in B. Any change of either of the two optical paths $l_1$ or $l_2$ is made evident by a displacement of the interference fringes.

A displacement equal to the breadth of one fringe results, for example, if one of the two mirrors, $S_1$ or $S_2$, be displaced a distance equal to one-half a light wavelength. This is the principle by which *Michelson* linked the length of the *standard metre bar* with the wavelength of the red line of cadmium. A different application consists in placing an optically refractive medium in one of the two light paths. The observation of the fringe displacement then allows a very accurate measurement of the *refractive index* of the medium introduced.

In our case the apparatus was to be used for observing an effect on the propagation of light produced by the orbital motion of the earth. The arrow **v** shows the direction in which the apparatus might move with respect to the ether. We must calculate the travelling times $t_1$ and $t_2$ of the two rays into which the light, initially incident on $S_0$, is split. Let $t_1$ be the time for the light ray to travel from $S_0$ to $S_1$ and back again to $S_0$; analogously for $t_2$. Since the portion of path $l_1$ lies in the direction of the velocity of motion, we obtain for $t_1$

$$t_1 = \frac{l_1}{c-v} + \frac{l_1}{c+v} = 2\frac{l_1}{c}\frac{1}{1-v^2/c^2}$$

for, on account of the motion of the apparatus with respect to the ether, the light, *relative* to the apparatus, moves on its outward journey with velocity $c-v$, and on its return with velocity $c+v$. For calculating the

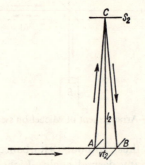

Fig. 65.—For calculating the travel time in Michelson's experiment

travelling time of the *second* ray, we must consider the path actually traversed by it (figure 65). Let A be the location of the half-silvered mirror at time 0, and let B be its location at time $t_2$, the moment when the light beam reflected from $S_2$ returns to $S_0$. The part AB is then equal to $vt_2$. The whole path traversed by the light ray is obviously given by $L = \sqrt{(4l_2^2 + v^2 t_2^2)}$; it must be equal to $ct_2$. Thus we find for the time $t_2$,

$$t_2 = \frac{2l_2}{c}\frac{1}{\sqrt{(1-v^2/c^2)}}$$

The difference in paths leads to a difference in times which is accessible to observation. This time difference is given by

$$t_2 - t_1 = \frac{2}{c}\left[\frac{l_2}{\sqrt{(1-v^2/c^2)}} - \frac{l_1}{(1-v^2/c^2)}\right] \tag{73.1}$$

If now the entire apparatus be rotated through an angle of 90°, the two light paths $l_1$ and $l_2$ exchange their relationships to the direction of motion through the ether. After this rotation, then, we obtain a difference of travel times of value

$$t_2' - t_1' = \frac{2}{c}\left[\frac{l_2}{1-v^2/c^2} - \frac{l_1}{\sqrt{(1-v^2/c^2)}}\right] \qquad (73.2)$$

During the rotation of the apparatus it must be possible, of course, for the resulting displacement of the fringes to be observed. If $\theta$ is the difference in the two travel-time differences as given above, then for $\theta$ we have by subtraction

$$\theta = \frac{2(l_2+l_1)}{c}\left[\frac{1}{1-v^2/c^2} - \frac{1}{\sqrt{(1-v^2/c^2)}}\right] \qquad (73.3)$$

or, for small velocities ($v^2 \ll c^2$),

$$\theta = \frac{(l_1+l_2)}{c}\frac{v^2}{c^2} \qquad (73.4)$$

Let us consider the question of what *order of magnitude* this fringe displacement might have. For this purpose we must make a comparison of our calculated time $\theta$ with the oscillation period time $\tau = \lambda/c$ of the light. A change $\tau$ in the travel-time difference must produce a displacement of the interference fringes equal to at least something like a fringe width. With the value $v = 3 \times 10^6$ cm/s for the earth's velocity, and a light wavelength $\lambda = 5 \times 10^{-5}$ cm, we would want a total length of light path $l_1 + l_2 = 50$ m. A light-path length of this order can in fact be obtained by letting the light beams traverse the paths $l_1$ and $l_2$ not merely twice, but, by means of multiple reflections, a large number of times.

In a repetition of the experiment carried out by *Joos*, the sensitivity of this arrangement was extended to the point where an ether wind of velocity as small as 1·5 km/s could have been observed. Actually, however, *within the accuracy of the measurement, no displacement on rotating the apparatus through 90° was observed*. Repetitions of the experiment at different times of the year always gave the same negative result.

These investigations show that *the velocity of light does not depend upon the state of motion of the observer*. The interpretation of this result on the basis of the electron theory runs into very serious difficulties. On the one hand, for the interpretation of the earlier experiments, the assumption of an ether at rest seems hardly avoidable; on the other hand, however, every experimental attempt to reveal the presence of this resting ether has ended with a negative result.

### §74. Search for an explanation of the negative result of the Michelson experiment

In order to explain the negative result of the Michelson experiment, there was proposed a whole series of hypotheses, all of which were eventually found to be untenable. We wish to discuss the most important of these.

The first explanation consists in the assumption that *the ether at the earth's surface is carried along by the earth*, adhering (so to speak) to the earth like the earth's atmosphere. This explanation became very improbable in the light of Fizeau's experiment on light propagation in media in motion, this experiment requiring that the ether be *not* carried along, or at most only partially carried along, by a moving body. Still, we might think that the degree to which the ether was carried along depended on the mass of the moving body, and that the large mass of the earth could therefore carry the ether, while the small mass of the liquid in the Fizeau apparatus was not able to accomplish this. Highly synthetic and not very satisfying, these assumptions, as well as the general idea of a convected ether, were completely refuted by the phenomenon of *aberration*. Aberration consists in the fact that a fixed star as observed by a telescope on the earth is not to be seen in exactly the direction in which it is really located, but the telescope must be tilted through a very small angle which depends on the component of the earth's velocity perpendicular to the line connecting the earth with the fixed star under observation.

As an example, let us consider a light ray arriving perpendicular to the moving surface of the earth. Upon entering the stratified earth-carried ether, the wave fronts of such a light ray would always remain parallel to the earth's surface and the light ray would also arrive for observation as a plane wave which continued to move in a direction perpendicular to the earth's surface. Contrary to an erroneous idea held by Stokes, we would not be able to observe any aberration in the case of an earth-carried ether.

Another attempted explanation was suggested by Ritz. He sought to explain the negative result of the Michelson experiment by assuming that the velocity of light depends *on the state of motion of the emitting light source* such that to the velocity of light *in vacuo* must be vectorially added the velocity of the light source itself. We can, in fact, be readily convinced that on this assumption the Michelson experiment would indeed be explained. Against this Ritz hypothesis, however, it is first of all to be objected that the hypothesis is completely untenable from the *theoretical* standpoint of a field concept which describes the motion of light by a differential equation, because it cannot be understood how the velocity of propagation of light from a source located at any point of space should be related to the condition of the light source. But the hypothesis can be dismissed on purely *experimental* grounds, and in a rather drastic way, by observations on *double stars*. For a double star consisting of a central member and another member orbiting round the first, the light from the orbiting member must, according to the Ritz hypothesis, arrive at the earth with different velocities for different places in the star's orbit. Owing to the great distance which has to be covered between star and earth, small differences in velocity would lead to large differences in travel time, so that in some cases we would perceive the rotating member at several different places in its orbit simultaneously. As was pointed out particularly by de Sitter, however, there exists not even the suggestion of such an effect. The Ritz hypothesis, moreover, is refuted by a negative result of the Michelson experiment carried out, not with an earth-bound light source, but with the light of a star. This experiment was actually carried out by Tomaschek.

The third attempted explanation, to be regarded as an immediate precursor to the Theory of Relativity, consists of the *Lorentz contraction hypothesis*, enunciated independently also by Fitzgerald. This hypothesis was mentioned in §65. It asserts that for a body moving with velocity $v$ all dimensions lying in the direction of motion become shortened by the factor $\sqrt{(1 - v^2/c^2)}$, while the lateral dimensions remain unchanged. Since, however, all scales laid upon the moving body for establishing the existence of the contraction are in like manner themselves shortened, the contraction effect would never be revealed to a co-moving observer.

We wish now to convince ourselves that by means of the contraction hypothesis the negative result of the Michelson experiment can actually be explained. For this we must introduce the contraction hypothesis into equations (73.1) and (73.2) of the foregoing paragraph. In the time-difference expression (73.1) the length lying in the direction of motion is $l_1$; in (73.2) it is $l_2$. If for these two lengths the contraction by the factor $\sqrt{(1-v^2/c^2)}$ be introduced, the two travel times in question become equal, and the same value of the time difference

$$t_2 - t_1 = t_2' - t_1' = \frac{2}{c} \frac{l_2 - l_1}{\sqrt{(1-v^2/c^2)}}$$

is obtained for both. It follows from this that for a rotation of the apparatus, a change in the difference of travel times is no longer to be expected.

Although the contraction hypothesis is to be regarded as an immediate forerunner of the Theory of Relativity, it should be emphasized that the concept, when standing by itself, contradicts the fundamental Principle of Relativity. Thus, if the moving observer compares the scale moving with him with a scale at rest, he can obviously confirm that his own scale is actually shortened. In principle, then, we should have available a means of experimentally determining the state of absolute rest simply by observing which of various more or less rapidly moving unit-length scales possessed the greatest length.

# E

---

The
theory
of
relativity

# CHAPTER E I

## The Physical Basis of Relativity Theory and Mathematical Aids

### §75. Revision of the space-time concept

Let us review the experimental facts given in the preceding section: *Fizeau's entrainment experiment* as well as the phenomenon of *aberration* have been most simply and easily explained by the assumption of a *light-ether which is absolutely at rest*, i.e. by the assumption that we can define a preferred coordinate system in which the velocity of light in all directions has the same value $c$. After the failure of all those experiments which attempted to reveal this preferred coordinate system, especially the experiments of Michelson and of Trouton and Noble (§64), special assumptions had to be introduced to explain how it comes about that the existence of this fixed ether cannot be observed. We thus have the unsatisfactory situation of a structure of hypotheses wherein a second hypothesis has been introduced only to explain that nothing can be observed about the stationary ether—which was just the thing introduced by the first hypothesis.

In this situation the discovery of relativity by Einstein provided the crucial step. Einstein started directly from the negative result of the Michelson experiment and from it conjectured that perhaps it is *basically impossible* to distinguish which of two systems moving with constant velocity with respect to one another is at rest, and which is in motion. He therefore asked: How must the laws of nature be formulated in order that it shall be impossible in principle to make a distinction between two systems which find themselves in relative motion with respect to one another?

Such a relativity principle had already been known for a long time in classical physics. Let us derive the equations of motion of mass points whose force interactions depend on a potential which is a function of the distance only.

$$m_1 \ddot{x}_1 = -\frac{\partial \Phi}{\partial x_1}, \dots, \qquad \Phi = \Phi(r_{12}, r_{13}, \dots)$$

We now observe this motion from a coordinate system moving at a velocity **v** with respect to the first system. Then, between the coordinates of the mass points in the first and in the second coordinate systems, there exist obviously the equations

$$\mathbf{r}'_i = \mathbf{r}_i - \mathbf{v}t \qquad (75.1)$$

and also

$$r'_{12} = r_{12} \ldots$$

We can immediately translate the equations of motion into the new coordinate system and obtain as a law of motion

$$m_1 \frac{d^2 x'_1}{dt^2} = -\frac{\partial \Phi'}{\partial x'_1} \ldots \qquad \Phi' = \Phi(r'_{12}, r'_{13}, \ldots)$$

which, apart from the primes attached to the coordinates, completely agrees with the first law of motion. The transformation (75.1) which connects the coordinates of one system with those of another is called a *Galilean transformation* and the fact just found is expressed by saying: *The Newtonian equations of motion of classical physics are invariant with respect to a Galilean transformation.* From the standpoint of pure mechanics, two coordinate systems connected by the transformation (75.1) are equivalent. Naturally they are no longer equivalent as soon as we include *electromagnetic processes*, since a light signal which expands in the unprimed coordinate system in two opposite directions with the same velocity $c$ would have in the primed coordinate system, the velocity $c + v$ or $c - v$, depending on the direction in which the light expansion was observed.

According to Einstein the whole difficulty is that heretofore in physics the concepts of space and time have been naïvely accepted as something absolute. Especially has *time* been considered as flowing uniformly, and it has been thereby assumed that we can properly speak of an *absolute simultaneity* of two events taking place at separated places.

It makes sense for the physicist to speak of the place and time of an event only when the measures for space and time are introduced as a result of well-defined procedures which can be carried out at any time. The aids which enable such measurements to be carried out are *scales* and *clocks*, whose existence we expressly assume. In order to describe an event in a chosen coordinate system we have to establish at which place and at which time the event takes place. This task is carried out when we make a mark at each point of the space indicating to us its space coordinates, and when in addition there is at this place a clock from which we can read the moment at which the event at this location takes place. We wish to make it clear that for the measure numbers of the location and of the time of a point event, only readings of the place mark and the clock which coincide with the point event are used.

We consider now how to make the *place marks* in our coordinate system and how to set the *clocks* which have been brought to the various places. We fix the place marks in the usual way by repeated scaling with a unit scale whose existence we have assumed. For instance on the $x$-axis we obtain the mark $x = 6$ by scaling with the unit scale six times from the starting point. We now place a clock at each such marked place. We have naturally to assume that all clocks are running at the same rate, meaning that their speed does not depend on their location. Our main task now consists in *setting the clocks* so that they all show the same time. In this procedure we may not synchronize the clocks at the zero point of the coordinate system and then bring them to the several previously marked positions. We cannot know *a priori* whether the speed of the clock suffers any temporary change during the transport from one place to another. If we would wish to avoid introducing a new hypothesis we have instead first to bring the clocks to their places, and *then* to set them to the time of the clock at the zero point.

We may adopt the following procedure: At the moment when the clock at the origin indicates zero, we emit from the origin a light signal in the direction of the clock requiring setting, this clock being at distance $r$. An observer at the position of this clock has the instruction to adjust his clock to the time $t = r/c$, which is the time at which the light signal passes. We imagine now that all clocks in the coordinate system are adjusted in this way. *Only after this is done* may we define *the concept of simultaneity* in our system: Two events occurring at two different places of the system have to be called *simultaneous* when the clocks at the respective places show the same time. This procedure for adjusting the clocks forms, as the *definition of simultaneity*, the crux of the special theory of relativity.

We recognize that with this definition the result of Michelson's experiment becomes a principle—for we have made *the direct basis of the time concept* the fact that light expands equally in all directions with the same velocity $c$.

We wish to consider how measurements are to be carried out on objects which are moving relative to the system of marks and clocks described above. In order to measure the length of a moving scale, for example, we have at the same time to observe the location of its initial point and its end point. We know for this that there is no sense at all in speaking of the length of a moving scale if *beforehand* the concept of simultaneity has not been clearly defined. We have to proceed with this length measurement in the following way: Each observer in our coordinate system has to be instructed to note the time at which the end of the moving scale or the beginning has passed at his particular place. From all data so obtained we have to select for the determination of the length those two observers, one of whom saw the beginning, and the other the end of the scale passing by at the same time. The time duration of a process occurring on the moving body would be measured in a similar way. Naturally also in this case the relevant time data are from those two clocks in whose immediate vicinity the event started and ended respectively.

Up to this point we have spoken only of a single coordinate system, and we have thought about how to make measurements in it in an unambiguous way. Exactly the same procedure can be followed for any other coordinate system which is moving with constant speed with respect to the first. In the next paragraph we shall deduce how the location and time data of these two coordinate systems turn out to be related. It is *a priori* not at all obvious that *the concept of simultaneity is the same for both systems.* We shall see that two events which are simultaneous in one system are not necessarily simultaneous in another.

### §76. The Lorentz transformation

We consider two coordinate systems moving with respect to one another, and we assume that in each of these systems there are scales and clocks which are located and adjusted according to the procedures of the foregoing paragraph. Each real point event, as for example the arrival of a light flash on a screen or the passage through zero of the pointer of an instrument, will then be measured in both systems. If we distinguish by primes the data of the second observer from those of the first, then the first observer will ascribe to the event the coordinates $x, y, z, t$; and the second the coordinates $x', y', z', t'$. When, in the following, we speak of the *transformation* from one coordinate system to another, we understand this to be the analytical relationship between the coordinates $x', y', z', t'$ and $x, y, z, t$. As a point event we might consider the pointer position $t$ of the clock at point $x, y, z$ in the first coordinate system. For this case the transformation relations have to tell us: (*a*) at which place $(x', y', z')$ of the second coordinate system this event has been observed, and (*b*) what was the pointer position $t'$ of the clock located at that place.

To find the relationship between the primed and unprimed coordinate systems we make the following assumptions:

(1). *The relationship shall be linear.* This is necessary in order not to give to the origin of a coordinate system or to some other point a physical preference over all other points.

(2). The two systems are moving with respect to one another with a constant speed of translation, i.e. each point $x', y', z'$, of the second system is moving with respect to the first system with the velocity **v**. Conversely, each point $x, y, z$ of the first system is moving with respect to the second with velocity $-\mathbf{v}$. This stipulation characterizes the *special* theory of relativity. (Only in the general theory of relativity, with which this book is not concerned, are related *accelerated* systems considered.)

(3). *A measurement of the velocity of light in both systems and in any direction will yield the value c.* This requirement is the starting point of all our considerations; it is already contained in the definitions of the times $t$ and $t'$ and in Einstein's prescription for setting clocks.

(4). *It shall not be possible by any physical measurement to establish any fundamental difference between the two systems.* This requirement contains forthwith the whole programme of the theory of relativity. Extending beyond the measurements of the velocity of light, it requires validity in the totality of all physical phenomena. (We shall need this requirement for our treatment of the special process, the Lorentz contraction.)

We now deduce the transformation formulae for the case where the velocity **v**, mentioned in (2) above, has the same direction as the positive $x$-axis, and where the $x$-axis continuously coincides with the $x'$-axis. Further, the origin of coordinates of the one system at time $t = 0$ has to coincide with the origin of the other system at time $t' = 0$. The $x$- and $x'$-axes coincide only if for $y = 0$, $z = 0$ it always follows that $y' = 0$, and $z' = 0$. The transformation formulae for $y$ and $z$ have then to be of the form

$$y' = \alpha y + \beta z \qquad z' = \gamma y + \delta z \qquad (76.1)$$

We wish also to exclude spatial distortions as non-essential, and we therefore require that the $xy$-plane ($z = 0$) pass over to the $x'y'$-plane ($z' = 0$), etc. From the equal status of the $y$- and $z$-directions, it follows that

$$y' = \varepsilon y \qquad z' = \varepsilon z$$

The factor $\varepsilon$ has the meaning that a unit scale in the first system, lying in the $y$-direction and observed from the second system, has the length $\varepsilon$. The second observer would thus observe a dilation by a factor $\varepsilon$, while inversely, the first observer would measure the length $1/\varepsilon$ in the second system on a scale at rest. If these reciprocally established length changes were different from one another, an objective difference between the two systems would exist; but this is excluded by assumption (4). We must therefore have that $\varepsilon = 1/\varepsilon$, or $\varepsilon = 1$. Thus,

$$y' = y \qquad z' = z \tag{76.2}$$

There now remain the transformation equations for $x$ and $t$. According to our assumption, the point $x' = 0$ moves with velocity $v$ along the positive $x$-axis, i.e. the statement $x' = 0$ must be identical with $x = vt$. Correspondingly, the statement $x = 0$ is equivalent to $x' = -vt'$. The transformation has therefore to have the form

$$x' = \gamma'(x - vt) \qquad x = \gamma(x' + vt') \tag{76.3}$$

where the quantities $\gamma$ and $\gamma'$ have still to be determined. First of all we convince ourselves that in accordance with requirement (4), $\gamma$ and $\gamma'$ must be equal. For this purpose we measure from the second system the length of a scale of length $l$ at rest in the first system. We then bring the same scale to the velocity $v$ of the second system and measure its length from the first system. We require that we get the same result both times. In the first measurement, and in the first system, the beginning and end of the scale are given by $x = 0$ and $x = l$. In the second system these two points at the time $t' = 0$ have, according to the second equation of (76.3), the coordinates $x' = 0$ and $x' = l/\gamma$. Measured from the second system, therefore, the scale has the length $l/\gamma$. It appears *shortened* by the factor $1/\gamma$. Now in the second measurement the scale has to be at rest in the second system. The beginning and end are therefore given by $x' = 0$ and $x' = l$. Measured from the first system, these points at the time $t = 0$, and according to the first equation of (76.2), are located at $x = 0$ and $x = l/\gamma'$. Now the scale appears reduced by the factor $1/\gamma'$. Thus requirement (4) demands in fact that $\gamma = \gamma'$, as asserted.

Now for the determination of the value of $\gamma$ there remains the decisive use of requirement (3) on the constancy of the velocity of light: Let a light signal be emitted at the origin of coordinates and at the time $t = t' = 0$, such that the signal makes a momentary flash on a screen located somewhere on the $x$-axis. This "point event" of the flash is described by *one* of the observers by quoting $x$ and $t$; by the *other* by quoting $x'$ and $t'$; and thus we must have $x = ct$ and $x' = ct'$, with the same value of $c$ for both observers. Putting these values in (76.3), we have that

$$ct' = \gamma t(c - v) \qquad ct = \gamma t'(c + v)$$

Through multiplication of both equations it follows that

$$\gamma = \frac{1}{\sqrt{(1 - v^2/c^2)}} \tag{76.4}$$

In order to find all the desired transformation equations we have still to calculate $t'$ as a function of $x$ and $t$, with the aid of (76.3):

$$t' = \gamma\left[t + \frac{x}{v}\left(\frac{1}{\gamma^2} - 1\right)\right]$$

Thus, with the value (76.4) for $\gamma$, we obtain

$$x' = \frac{x - vt}{\sqrt{(1 - v^2/c^2)}}, \quad y' = y, \quad z' = z, \quad t' = \frac{t - (v/c^2)x}{\sqrt{(1 - v^2/c^2)}} \tag{76.5}$$

On solving these equations for the unprimed quantities the following equations (76.6) are obtained, differing from (76.5) (as must be) only by the sign of $v$:

$$x = \frac{x' + vt'}{\sqrt{(1 - v^2/c^2)}}, \quad y = y', \quad z = z', \quad t = \frac{t' + (v/c^2)x'}{\sqrt{(1 - v^2/c^2)}} \tag{76.6}$$

The relation given by equations (76.5) and (76.6) between the coordinates of the two systems is called the *Lorentz transformation*. It has to replace the Galilean transformation (75.1). It is seen immediately that in the limiting case $c = \infty$ the transformations are identical. In particular, in this case according to (76.5), $t = t'$; i.e. we obtain an absolute simultaneity. As has been clearly shown by Einstein, the usual conception of simultaneity contains the assumption that for the observation of the simultaneity of two events separated in space, we have available *signals of infinite propagation velocity.*

The use of the velocity of light, $c$, for the definition of simultaneity already implicitly contains the proposition that a signal velocity greater than $c$ is in principle impossible. If there could be an action which propagates itself with a velocity $v > c$, we could find a coordinate system for which the action progresses into the *past*.

We have determined our transformation in such a manner that the laws of light expansion are independent of a constant relative translation velocity. This is also recognized in the following way: A spherical wave starting from the origin at time $t = 0$ has, at time $t$, reached the spherical surface given by the equation

$$x^2 + y^2 + z^2 - c^2 t^2 = 0$$

The arrival of the light wave on this spherical surface, marked, say, by the flashing of white screens, has to be described from the second system by

$$x'^2 + y'^2 + z'^2 - c^2 t'^2 = 0$$

By means of (76.5) it is easy to be convinced that as a consequence of the Lorentz transformation we always have that

$$x^2 + y^2 + z^2 - c^2 t^2 = x'^2 + y'^2 + z'^2 - c^2 t'^2 \qquad (76.7)$$

Later on we must make this circumstance the starting point of a deeper conception. In *differential* form a similar result is obtained—namely, the motion of light is always a solution of the differential equation

$$\frac{\partial^2 \phi}{\partial x^2} + \frac{\partial^2 \phi}{\partial y^2} + \frac{\partial^2 \phi}{\partial z^2} - \frac{1}{c^2} \frac{\partial^2 \phi}{\partial t^2} = 0$$

In order to transform this equation to the equation for a coordinate system moving with velocity $v$, the arguments $\phi(x,y,z,t)$ have to be replaced by arguments given in (76.6). We have then, for example:

$$\frac{\partial \phi}{\partial x'} = \frac{\partial \phi}{\partial x} \frac{1}{\sqrt{(1 - v^2/c^2)}} + \frac{\partial \phi}{\partial t} \frac{v/c^2}{\sqrt{(1 - v^2/c^2)}}$$

$$\frac{\partial \phi}{\partial t'} = \frac{\partial \phi}{\partial x} \frac{v}{\sqrt{(1 - v^2/c^2)}} + \frac{\partial \phi}{\partial t} \frac{1}{\sqrt{(1 - v^2/c^2)}}$$

By computing the second derivatives in an analogous way we find that

$$\frac{\partial^2\phi}{\partial x^2}+\frac{\partial^2\phi}{\partial y^2}+\frac{\partial^2\phi}{\partial z^2}-\frac{1}{c^2}\frac{\partial^2\phi}{\partial t^2}=\frac{\partial^2\phi}{\partial x'^2}+\frac{\partial^2\phi}{\partial y'^2}+\frac{\partial^2\phi}{\partial z'^2}-\frac{1}{c^2}\frac{\partial^2\phi}{\partial t'^2} \qquad (76.8)$$

*The differential expression characteristic of light propagation is invariant with respect to the Lorentz transformation.*

## §77. Consequences of the Lorentz transformation

(a) *Scales and clocks.* We have already seen in the foregoing paragraphs that a scale which, in the state of rest, has the length $l$ seems for an observer moving relative to it to have the smaller length $l\sqrt{(1-v^2/c^2)}$. This effect is called the *Lorentz contraction*. As we saw in §74, the contraction hypothesis introduced by Lorentz did not satisfy the postulate of relativity.

In a similar way, from the "unprimed system" we can check the running rate of a clock at rest in the "primed system". According to equations (76.6), at location $x'$ and times $t_1' = t'$ and $t_2' = t'+1$, there correspond the times

$$t_1 = \frac{x'v/c^2+t'}{\sqrt{(1-v^2/c^2)}} \quad \text{and} \quad t_2 = \frac{x'v/c^2+t'+1}{\sqrt{(1-v^2/c^2)}}$$

Seen from the first system the clock in the second system has carried out a rotation of its pointer hand from $t'$ to $t'+1$ in the time

$$t_2-t_1 = \frac{1}{\sqrt{(1-v^2/c^2)}}$$

that is, *the moving clock runs slower than the resting clock by the factor* $1/\sqrt{(1-v^2/c^2)}$.

An important quantity often met in applications is the *proper time* of a moving body. We define this time as the indication $\tau$ on a clock moving along with the observed body. The connection between this proper time and the time $t$ as observed in another chosen coordinate system may be derived immediately on the basis of the foregoing procedures. If the clock in the second system registers, not the time interval 1, but the interval $d\tau$, the observer reads the interval $dt$, with

$$dt = \frac{d\tau}{\sqrt{(1-v^2/c^2)}} \qquad (77.1)$$

*The observer with the resting clock finds a larger value for the time interval than that corresponding to the proper time, i.e. than the time interval indicated by the clock moving with the body.*

This time dilation is especially impressive in fast-moving atomic particles which, within themselves, carry a "clock" keeping the proper time. This is the case, for example, with radioactive nuclei which, on the average after a certain time $\tau_0$, go over to the product body by the emission of a charged particle. The average lifetime $\tau_0$ is defined as that time in which the number of still-undecayed nuclei has become reduced to $1/e$ of the original number. If such substances are observed in a system with respect to which they move with velocity $v$, the lifetime undergoes a lengthening by the factor $1/\sqrt{(1 - v^2/c^2)}$, corresponding to (77.1). In the radioactive particles of cosmic radiation, in particular in $\mu$-mesons which are produced in the upper regions of the earth's atmosphere by collisions of energetic particles coming from cosmic space, this dilation factor ($v$ being almost equal to $c$) can reach values approaching $10^4$. The consequence is that these $\mu$-mesons, for which $\tau \approx 2 \times 10^{-6}$s, reach the earth's surface in enormous numbers before their decay and are observed—this in spite of their relatively long travel time (approximately $3 \times 10^{-5}$s) through about 10 km thickness of homogeneous atmosphere.

(*b*) *Geometrical representation of the Lorentz transformation.* These and similar consequences of equations (76.5) become more clear when, as *Minkowski* first showed, the contents of the equations are visualized in a geometrical fashion. We leave out of consideration the *y*- and

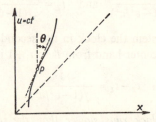

Fig. 66.—World line of a point in the *xu*-plane. Its velocity is $c \tan \theta$

*z*-axes since these are not altered by the transformation (76.5) when the motion of both systems is along the *x*-axis. We represent all possible point events in the first system by a space-time diagram in which an abscissa denotes the location *x*, and an ordinate denotes $u = ct$, the time multiplied by *c*. The motion of a material point appears in our scheme (figure 66) as a curve (*world line* of the point), the tangent of which forms with the time axis the angle $\theta = \tan^{-1}(dx/du) = \tan^{-1}(v/c)$, *v* being the momentary velocity of the point. Since we exclude from consideration any velocity faster than light, the angle of inclination of these curves with the time axis has always to be smaller than 45°. The motion diagram for a light wave is a straight line inclined at 45° with respect to the axis.

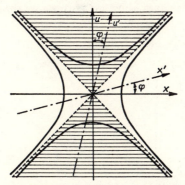

Fig. 67.—Transfer from system $(x, u)$ to the system $(x', u')$. The hyperbolas intersect the respective axes at unit values of the variables

In addition to the first (unprimed) system (I), we consider (figure 67) a second system (II) moving with respect to the first with velocity $v$ along the $x$-axis. With the abbreviations $u = ct$, $u' = ct'$, $\beta = v/c$, the equations to be discussed, namely (76.5) and (76.6), read:

$$x' = \frac{x - \beta u}{\sqrt{(1 - \beta^2)}}, \qquad u' = \frac{-\beta x + u}{\sqrt{(1 - \beta^2)}}$$

$$x = \frac{x' + \beta u'}{\sqrt{(1 - \beta^2)}}, \qquad u = \frac{\beta x' + u'}{\sqrt{(1 - \beta^2)}} \qquad (77.2)$$

In our scheme there corresponds to system II (itself in motion with respect to I) a coordinate system whose axes we find in the following way: By definition, the points $(x = 0, u = 0)$ and $(x' = 0, u' = 0)$ fall together. The point $x' = 0$ is moving with respect to I with velocity $v$; its world line is therefore a straight line through 0 forming with the time axis the angle $\phi = \tan^{-1}\beta$. Since $x' = 0$ for this line, it is identical with the time axis of the primed system. By setting $u' = 0$ in (77.2) we obtain the space axis, whose equation thus reads $ct = u = \beta x$; this is a straight line which forms the same angle $\phi = \tan^{-1}\beta$ with the $x$-axis.

From this presentation we recognize with special emphasis the relative character of simultaneity. All point events on the $x'$-axis appear to the second observer as *simultaneous*, while for the first observer they are *progressive*. For the first observer the point event $A'$ (figure 68) occurs for him later by a time interval $u/c = (1/c) \cdot AA'$ than the event 0.

For a complete representation of the Lorentz transformation we still need the units on the axes. For this purpose we draw in figure 67 the two rectangular hyperbolas

$$x^2 - u^2 = 1 \quad \text{and} \quad u^2 - x^2 = 1 \tag{77.3}$$

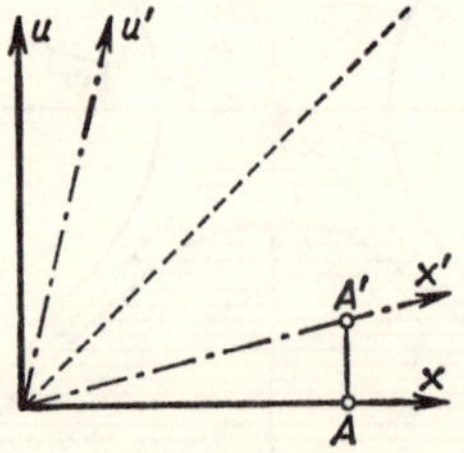

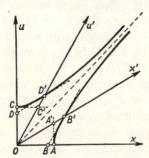

Fig. 68.—For the relativity of simultaneity

They cut the $u$- and $x$-axes of the unprimed system at the points $u = 1$, $x = 0$, and $x = 1$, $u = 0$. That they cut the axes of the primed system in the points $(u' = 1, x' = 0)$ and $(u' = 0, x' = 1)$ follows immediately from relation (76.7) with $u = ct$ and $u' = ct'$, to give

$$x^2 - u^2 = x'^2 - u'^2 = \pm 1$$

With the aid of figure 69, the phenomenon of *reciprocal scale-short-ening* can be described in the following way: Let OA be a unit scale at rest in the first system. The world lines of its end points are ODC and AA'. For an observer at rest in the second system, the corresponding and simultaneous position of the scale $(u' = 0)$ gives the location of its beginning and end points at the world points O and A' respectively.

Fig. 69.—For comparing units of scales or clocks in mutual motion

The scale, therefore, is shorter for him than his unit length OB'. Conversely, the beginning and end of a unit length OB' at rest in the second system with world lines OC'D' and BB' appears for the first observer at the time $u = 0$ at the world points O and B. It again appears shorter than the unit scale OA of the resting system.

The comparison of clocks takes place in a completely analogous manner. A clock at rest in the second system moves on the world line OC'D'. At the world point D' ($u' = 1$) it has completed one revolution; but already before this (at the world point C') the clock of the first system with which it is in space coincidence has completed a revolution ($u = 1$). The moving clock therefore runs slower than the clock at rest. A clock at rest (at $x = 0$) in the first system has completed a revolution at world point C, while already at D a clock located at the same place in the second system has completed a revolution.

In this way we can be fully and intuitively convinced that the assertion of the reciprocal shortening of scales and the retardation of clocks contains absolutely nothing paradoxical as soon as we have renounced the notion of absolute simultaneity.

(c) *The Einstein addition theorem for velocities.* We wish now, with the aid of the Lorentz transformation, to derive the *Einstein theorem for the addition of two velocities.* According to the old kinematics this consisted simply in the vectorial addition of the velocities: If **v** is the velocity of, say, a ship (i.e. a coordinate system) with respect to a certain coordinate system, and if a mass point moves with velocity **u'** on the ship, then the mass point moves with velocity **u** = **v** + **u'** relative to an observer at rest.

In relativity theory this relationship is substantially more complicated. We examine further the two coordinate systems connected by equations (76.5) and we assume that a mass point moves relative to the primed system with the velocity $u'$ in the $x'y'$-plane so that its trajectory makes an angle $\theta'$ with the $x'$-axis. Its equation of motion, in the primed system, is then

$$x' = u't' \cos \theta' \qquad y' = u't' \sin \theta' \qquad z' = 0 \qquad (77.4)$$

We wish now to describe the same world line from the point of view of the *first* system, i.e. we wish to determine two quantities $u$ and $\theta$ in such a manner that the equations

$$x = ut \cos \theta \qquad y = ut \sin \theta \qquad z = 0 \qquad (77.5)$$

shall, upon carrying out the transformation (76.6), be identical with (77.4). To this end we substitute for the space coordinates in the transformation equations (76.6), the expressions (77.4) and (77.5), and so obtain the three relations

$$ut \cos \theta = \frac{u' \cos \theta' + v}{\sqrt{(1 - \beta^2)}} t' \qquad ut \sin \theta = u't' \sin \theta'$$

$$t = \frac{1 + (u'v/c^2) \cos \theta'}{\sqrt{(1 - \beta^2)}} t'$$

After division of the first two equations by the third, it follows that

$$u \cos \theta = \frac{u' \cos \theta' + v}{1 + (u'v/c^2) \cos \theta'} \qquad u \sin \theta = \frac{u' \sin \theta' \sqrt{(1 - \beta^2)}}{1 + (u'v/c^2) \cos \theta'} \quad (77.6)$$

and hence for the *magnitude and direction* of the desired velocity we have

$$\left.\begin{aligned} u^2 &= \frac{u'^2 + v^2 + 2u'v \cos \theta' - (u'^2 v^2/c^2) \sin^2 \theta'}{[1 + (u'v/c^2) \cos \theta']^2} \\ \tan \theta &= \frac{u' \sin \theta' \sqrt{(1 - \beta^2)}}{u' \cos \theta' + v} \end{aligned}\right\} \quad (77.7)$$

If, in particular, $u'$ lies in the direction of $v$, then we have that $\theta' = 0$ and also $\theta = 0$, and it follows that

$$u = \frac{u' + v}{1 + u'v/c^2} \quad (77.8)$$

The resulting velocity is always smaller than the sum $u' + v$ of the two added velocities. In particular it follows from (77.8) that by the addition of two velocities, both being smaller than $c$, it is *impossible* to exceed the velocity of light. In the special case where $u' = c$ it always follows that $u = c$ no matter how large a value we may choose for $v$. This result is really self-evident inasmuch as the transformation formulae (76.5) were formulated on just the assumption that an event, if moving in one system with the velocity $c$, will do likewise in any other system.

An especially instructive application of equation (77.8) due to M. von Laue is to be had in the following interpretation of the *Fizeau entrainment experiment*: A light signal travels in a medium of refractive index $n$ (disregarding dispersion, and therefore the distinction between group and wave velocities) with velocity $u' = c/n$. If this medium, now, for its part, moves with the velocity $v$, then, according to (77.8), an observer at rest obtains the velocity

$$u = \frac{c/n + v}{1 + v/nc}$$

and, in first approximation (for small $v/c$), we have

$$u = \left(\frac{c}{n} + v\right)\left(1 - \frac{v}{nc}\right) = \frac{c}{n} + v\left(1 - \frac{1}{n^2}\right) \qquad (77.9)$$

in agreement with our electron-theoretic considerations (§72), and with Fizeau's measurements.

## §78. Programme of the special theory of relativity

In the preceding paragraphs we have developed a kinematics, i.e. essentially a description of the behaviour of moving scales and clocks, which satisfies both the principle of the constancy of the velocity of light and the principle of relativity. The characteristic basic equations of our kinematics, equations (76.5), originated expressly out of the demand that through observations of moving scales and clocks, it should be impossible to establish any essential difference between two coordinate systems in mutual relative motion of uniform translation. The assertion of relativity theory now, however, goes much further. It requires that such a distinction shall not in principle be observable; that *all natural laws*, not only that of light propagation, shall be invariant with respect to a *Lorentz transformation*. We have shown above that the equations of motion of classical mechanics are invariant under a Galilean transformation. From the proof of this assertion given there, however, it follows directly that the same equations cannot possibly be invariant under a Lorentz transformation. From this remark we see that it is not possible to abide by the Einstein relativity principle without at the same time making basic modifications in classical mechanics. For the purpose, now, of establishing whether any equation of physics, e.g. the induction law $\mathrm{curl}\,\mathbf{E} + \dfrac{1}{c}\dfrac{\partial \mathbf{H}}{\partial t} = 0$, satisfies the requirements of the principle of relativity, we must establish the transformation laws for $\mathbf{E}$ and $\mathbf{H}$; that is, to find out what fields $\mathbf{E}'$ and $\mathbf{H}'$ a moving observer would see. Then it has also to be established that these quantities satisfy the equation $\operatorname{curl}' \mathbf{E}' + \dfrac{1}{c} \dfrac{\partial \mathbf{H}'}{\partial t'} = 0$. Einstein set out in this way, and he found that the Maxwell equations of electrodynamics do indeed satisfy the principle of relativity. This method, often quite laborious in any specific case, can be replaced by a mathematical procedure indicated by Minkowski. Minkowski's method permits formulation of every natural law in a manner such that its invariance under a Lorentz transformation is immediately guaranteed. We shall become acquainted with the method in the next two sections. In a certain sense the procedure has its example in the familiar vector calculus of three-dimensional space. This calculus originated from the wish not to have the coordinate system (whose role is completely unimportant) appearing in equations (as, for example, in the Newtonian equations $m\dot{\mathbf{v}} = \mathbf{F}$). The introduction of a particular coordinate system cannot, however, be avoided when we ask for an actual *numerical* relationship between $\dot{\mathbf{v}}$ and $\mathbf{F}$. In that case we can in general describe the required relationship through the three equations $m\dot{v}_i = F_i$ ($i = 1$, 2, 3) for the three components in some suitably chosen coordinate system. These three equations we shall call equations between the vectors $\dot{\mathbf{v}}$ and $\mathbf{F}$ provided, after a rotation of the coordinate system, the equation $m\dot{v}_i' = F_i'$ ($i = 1, 2, 3$) between $\dot{v}_i'$ and $F_i'$ are valid in the rotated coordinate system. On both sides of a physically meaningful equation there must be only quantities of like transformation character, such as vector = vector, tensor = tensor, etc.

It will be readily illustrated by an *example* in three dimensions how this particular requirement facilitates the interpretation of physical laws. Let it be required to find the stress components $p_{xx}, p_{xy}, \ldots$ in an arbitrarily moving viscous compressible liquid. We shall assume that the stresses depend linearly on the first derivatives $\left( \dfrac{\partial v_x}{\partial x}, \dfrac{\partial v_x}{\partial y}, \ldots \right)$ of the velocity components $v_x, v_y, v_z$ with respect to the coordinates. (For the indices $x, y, z$, we substitute the indices 1, 2, 3.) Now it is known that the $p_{ik}$ form a symmetric tensor. The quantity $\delta_{ik}$ always exists as a symmetric tensor, where $\delta_{ik} = 1$ for $i = k$ and $\delta_{ik} = 0$ for $i \neq k$. Moreover from the quantities $\dfrac{\partial v_i}{\partial x_k}$ the symmetric tensor $\dfrac{\partial v_i}{\partial x_k} + \dfrac{\partial v_k}{\partial x_i}$ may be constructed (§14.15). Thus we next obtain

$$p_{ik} = R\delta_{ik} + \mu \left( \frac{\partial v_i}{\partial x_k} + \frac{\partial v_k}{\partial x_i} \right)$$

The factors $R$ and $\mu$ must be scalar-invariant. The only invariant which can be linearly derived from the tensor $\dfrac{\partial v_i}{\partial x_k}$ is its *spur* $\dfrac{\partial v_1}{\partial x_1} + \dfrac{\partial v_2}{\partial x_2} + \dfrac{\partial v_3}{\partial x_3}$. Thus we obtain as the most general expression for the required stresses

$$p_{ik} = \delta_{ik} \left\{ p + \lambda \left( \frac{\partial v_1}{\partial x_1} + \frac{\partial v_2}{\partial x_2} + \frac{\partial v_3}{\partial x_3} \right) \right\} + \mu \left( \frac{\partial v_i}{\partial x_k} + \frac{\partial v_k}{\partial x_i} \right)$$

with the three scalar numbers $p$ (hydrostatic pressure), $\mu$ (viscosity), and $\lambda$ (volume viscosity).

The Minkowski treatment of relativity theory consists now in an imaginary extension of the three-dimensional vector calculus in which the requirement of invariance against a rotation in three dimensions shall, in four dimensions, represent a corresponding invariance against a Lorentz transformation. As we shall see in the following paragraphs, this latter can be described as a "rotation" of the quantities $x, y, z, ict$ in four-dimensional space.

## §79. The general Lorentz group

Within the context of the programme developed in the foregoing paragraphs we now begin the development of the mathematical aids to relativity theory; that is, we set ourselves the task of finding the most general transformation which, in passing from one reference system to another, connects these equally favoured systems. Let the coordinates in one reference system be $x_1 = x$, $x_2 = y$, $x_3 = z$, $x_4 = ict$; let those in another equivalent system be $x_1{}' = x'$, $x_2{}' = y'$, $x_3{}' = z'$, $x_4{}' = ict'$. Now it will involve no essential restriction if we allow the zero points of both systems, i.e. the points $x_i = 0$, and $x_i{}' = 0$, to coincide; for this only means a special choice of the origin of the space and time zero-points. Two reference systems are now to be considered as equivalent if the following two conditions are fulfilled:

(1). *Every uniform straight line*, i.e. force-free trajectory of a mass point, shall go over to another such. This means geometrically that every straight line in four-dimensional $x_i$-space will carry over to a straight line in the $x_i{}'$-system.

(2). *The quadratic form*

$$\sum_{i=1}^{4} x_i^2 = x^2 + y^2 + z^2 - c^2 t^2$$

shall remain invariant:

$$\sum_{i=1}^{4} x_i^2 = \sum_{i=1}^{4} x_i'^2 \tag{79.1}$$

We base this second requirement on the fact that the velocity of light must have the same value $c$ in both coordinate systems; and out of this the further requirement immediately follows that $\sum x_i^2 = 0$ must go over to $\sum x_i'^2 = 0$, this being what we generalize in condition (79.1).

We collect all transformations fulfilling (1) and (2) in the *general Lorentz group*, for we shall soon show that these transformations form a group.

On account of requirement (1) (and the condition that $x_i = 0$ must conform with $x_i' = 0$) the Lorentz transformations must be linear transformations of the form

$$x_i' = \sum_{k=1}^{4} A_{ik} x_k \qquad (79.2)$$

Every Lorentz transformation is therefore uniquely determined by a matrix $A$:

$$A = \begin{pmatrix} A_{11} & A_{12} & A_{13} & A_{14} \\ A_{21} & A_{22} & A_{23} & A_{24} \\ A_{31} & A_{32} & A_{33} & A_{34} \\ A_{41} & A_{42} & A_{43} & A_{44} \end{pmatrix} \qquad (79.3)$$

As in three-dimensional space the position vector is given by its components $x_1 = x$, $x_2 = y$, $x_3 = z$, and as any other system of three components $k_1 = k_x$, $k_2 = k_y$, $k_3 = k_z$ is similarly called a vector if the $k_i$ transform in a space rotation of the coordinate system in the same way as the $x_i$, so then we shall give the name four-vector, $\mathbf{a} \equiv \{a_i\}$, to any system of four components $a_1$, $a_2$, $a_3$, $a_4$, which, like $x_1$, $x_2$, $x_3$, $x_4$, transforms according to (79.2). It is seen immediately that the sum of two four-vectors $\mathbf{a} + \mathbf{b} \equiv \{a_i + b_i\}$ is again a four-vector; and so also is the manifold $\alpha \mathbf{a} \equiv \{\alpha a_i\}$. The vector $\mathbf{x}$ is therefore the four-vector with components $x_i$; in short, then, we can write the transformation formulae (79.2) in the form

$$\mathbf{x}' = A\mathbf{x} \qquad (79.4)$$

That $x_1$, $x_2$, $x_3$ are real and $x_4$ is pure imaginary, and that the same must hold for the $x_i'$, follows immediately from (79.2), since this (and only this) is the case whenever $A_{ik}$ (with $i$, $k = 1, 2, 3$) and $A_{44}$ are real, and $A_{i4}$, $A_{4i}$ (with $i = 1, 2, 3$) are pure imaginary.

We consider further the so-called *product* of two transformations. This is defined as follows: If, through a transformation $B$ given by the equation $\mathbf{x}'' = B\mathbf{x}'$, we go from reference system $\mathbf{x}'$ to another equally valid system $\mathbf{x}''$, then by substitution of (79.4) in this equation there follows a linear transformation, namely $\mathbf{x}'' = B\mathbf{x}' = B(A\mathbf{x})$, for which we write $\mathbf{x}'' = (BA)\mathbf{x}$. The product of the two matrices $B$ and $A$ is defined in this way. By explicit calculation we find immediately for the elements $(BA)_{ik}$ of $BA$ the form

$$(BA)_{ik} = \sum_l B_{il} A_{lk} \tag{79.5}$$

The product of two matrices is thus obtained by pairing the rows of the first factor with the columns of the second.* Let it be emphasized that in general $AB \neq BA$; i.e. we usually arrive at different results according as the transformation $A$ and then the transformation $B$ are carried out, or first $B$ and then $A$. That the associative law $C(BA) = (CB)A$ holds follows easily by writing out in components by means of (79.5).

For the identity transformation $E$ which leaves everything unchanged, $\mathbf{x}' = E\mathbf{x} = \mathbf{x}$. Its matrix $E$, the unit matrix, has the elements $E_{ik} = \delta_{ik}$ with $\delta_{ii} = 1$, and $\delta_{ik} = 0$ for $i \neq k$. Obviously $EA = AE = A$.

For two four-vectors, $\mathbf{a}$, $\mathbf{b}$ we define the *inner product* by

$$(\mathbf{a},\mathbf{b}) = \sum_i a_i b_i = (\mathbf{b},\mathbf{a}) \tag{79.6}$$

Requirement (2), which we impose on the Lorentz transformations, thus has the same meaning as the requirement that, for any four-vector, $\mathbf{a}^2 = (\mathbf{a},\mathbf{a})$ be Lorentz invariant. That $(\mathbf{a},\mathbf{b})$ is also Lorentz-invariant follows from $(\mathbf{a}+\mathbf{b})^2 = \mathbf{a}^2 + 2(\mathbf{a},\mathbf{b}) + \mathbf{b}^2$ on account of the Lorentz invariance of $\mathbf{a}^2$, $\mathbf{b}^2$, and $(\mathbf{a}+\mathbf{b})^2$.

Associated with a matrix $A$ we define the transposed matrix $\tilde{A}$ by the relationship $(A\mathbf{a},\mathbf{b}) = (\mathbf{a},\tilde{A}\mathbf{b})$. By explicit calculation it follows easily that for the elements, $\tilde{A}_{ik} = A_{ki}$; consequently $\tilde{A}$ derives from $A$ by interchanging rows and columns. Since $(AB\mathbf{a},\mathbf{b}) = (B\mathbf{a},\tilde{A}\mathbf{b}) = (\mathbf{a},\tilde{B}\tilde{A}\mathbf{b})$, and on the other hand $(AB\mathbf{a},\mathbf{b}) = [\mathbf{a},(AB)^{\sim}\mathbf{b}]$, it follows that $(AB)^{\sim} = \tilde{B}\tilde{A}$.

Since now $(\mathbf{a},\mathbf{b})$ is Lorentz-invariant, we must have that $(\mathbf{a}',\mathbf{b}') = (A\mathbf{a},A\mathbf{b}) = (\mathbf{a}\tilde{A},A\mathbf{b}) = (\mathbf{a},\mathbf{b})$. For arbitrary four-vectors, however, this is the case if and only if

$$\tilde{A}A = E \; (= \text{unit matrix}) \tag{79.7}$$

---

* It is easily seen that the elements of the matrix $BA$ possess the same reality conditions as those for the matrices $A$ and $B$. If $A$ as well as $B$ leave the direction of the time axis unchanged, so that $A_{44} > 0$ and $B_{44} > 0$, it follows that $(BA)_{44} > 0$, in so far as $A$ and $B$ actually represent Lorentz transformations, i.e. satisfy condition (79.7).

Thus, we have found the condition following from the second requirement on p. 333, namely that *A represents a Lorentz transformation.* Explicitly written, this equation reads

$$\sum_i A_{ik} A_{il} = \delta_{kl} \qquad \sum_i A_{ki} A_{li} = \delta_{kl} \qquad (79.7a)$$

The identity transformation $E$ is obviously also a Lorentz transformation. If, further, $A$ and $B$ are two Lorentz transformations, so also is their product $BA$; for this case

$$(BA)^\sim(BA) = \tilde{A}\tilde{B}BA = \tilde{A}EA = \tilde{A}A = E$$

If for the determinant of the matrix $A$ we write $|A|$, it follows that $|\tilde{A}| = |A|$. For a Lorentz transformation we must therefore have that $|\tilde{A}A| = |\tilde{A}||A| = |A^2| = |E| = 1$, i.e. $|A| = \pm 1$.* Since $|A| \neq 0$, we can solve the linear equations (79.2) for the $x_k$ obtaining with the matrix $A^{-1}$:

$$\mathbf{x} = A^{-1}\mathbf{x}' \qquad (79.8)$$

On putting (79.4) into (79.8), and inversely (79.8) into (79.4), it follows that

$$A^{-1}A = AA^{-1} = E \qquad (79.9)$$

Thus $A^{-1}$ is called the matrix inverse to $A$, or the inverse transformation. Multiplying (79.9) by $\tilde{A}$ leads immediately to the relation $A^{-1} = \tilde{A}$.

For the Lorentz transformations $A, B, C, \ldots$ we therefore have that

(1) $AB$ is a Lorentz transformation;
(2) so also is $A(BC) = (AB)C$;
(3) the identity transformation $E$, with $EA = A$, and
(4) the inverse transformation $A^{-1}$, with $A^{-1}A = E$;

and all these transformations form a group, the general Lorentz group.

We now consider all Lorentz transformations for which the time, and therefore $x_4$, does not change. These form a group, or rather a sub-group, of the general Lorentz group. Since $x'_4 = x_4$, we have that $x_1^2 + x_2^2 + x_3^2 = x^2 + y^2 + z^2$ is an invariant, and the group here con-

---

* As in three-dimensional space where $|A| = +1$ means a rotation while $|A| = -1$ leads from a right-hand to a left-hand system and thus consists of a rotation along with a reflection, so in four-dimensional space, where the Lorentz transformations are rotations, it has to be required of these transformations that $|A| = +1$.

sidered is the well-known group of space rotations of a coordinate system. The relevant matrices have the form

$$B = \begin{pmatrix} B_{11} & B_{12} & B_{13} & 0 \\ B_{21} & B_{22} & B_{23} & 0 \\ B_{31} & B_{32} & B_{33} & 0 \\ 0 & 0 & 0 & 1 \end{pmatrix} \qquad (79.10)$$

In §76 we have already become acquainted with a special Lorentz transformation which left $x_2 = y$ and $x_3 = z$ unchanged. From (76.5) we can write its matrix

$$L_\beta = \begin{pmatrix} \dfrac{1}{\sqrt{(1-\beta^2)}} & 0 & 0 & \dfrac{i\beta}{\sqrt{(1-\beta^2)}} \\ 0 & 1 & 0 & 0 \\ 0 & 0 & 1 & 0 \\ -\dfrac{i\beta}{\sqrt{(1-\beta^2)}} & 0 & 0 & \dfrac{1}{\sqrt{(1-\beta^2)}} \end{pmatrix} \qquad (79.11)$$

We now maintain that by successively carrying out: (1) a space-like rotation $B_1$, according to (79.10), (2) a spatial transformation $L_\beta$ according to (79.11), and (3) a further space-like rotation $B_2$, we can produce any arbitrary Lorentz transformation

$$A = B_2 L_\beta B_1 \qquad (79.12)$$

If this is correct, then for many special problems we can restrict ourselves (as we have already done) to the special Lorentz transformation $L_\beta$.

We demonstrate (79.12) without calculation, i.e. by purely geometric considerations. Let $x_i'$ be the coordinates transformed by $A$. The $x_4'$-axis, i.e. $x_1' = x_2' = x_3' = 0$ gives, in the unprimed system, a straight line (explicitly, $x_i = A_{4i} x_4'$; $i = 1, 2, 3, 4$) whose space projection in the unprimed system (explicitly, $x_i = A_{4i} x_4'$, with $i = 1, 2, 3$) is also known. Now there is a space-like rotation $B_1$ which rotates the $x_1$-axis in the direction of this projection, i.e. in the direction of the velocity of the primed system relative to the unprimed. Let the system originating from $B_1$ be designated by $x_i''$. The $x_4'$-axis therefore lies in the $x_1'' x_4''$ coordinate plane, and we could, as was fully described in §77(*b*), go over, by transformation of $x_1''$, $x_4''$ alone, i.e. by an $L_\beta$ given by (79.11), to coordinates $x_i'''$ such that the $x_4'''$-axis coincides with the

$x'_4$-axis. The system $x'''_i$ can therefore differ from system $x'_i$ only by a space-like rotation $B_2$.

## §80.  Four-vectors and four-tensors

As, in three-dimensional space, vector and tensor calculus is the most fitting expedient for writing equations in a form that is invariant against spatial rotations, so also a corresponding four-dimensional vector and tensor calculus is useful for expressing physical equations in a Lorentz-invariant form. Moreover, the character of every physical quantity is immediately seen (i.e. according as it is a scalar, vector, or tensor) as it transforms in changing the reference system.

In the foregoing paragraphs we have defined the concept of the four-vector $\mathbf{a} \equiv \{a_i\}$, and also the square of its length $\mathbf{a}^2 = \sum_{i=1}^{4} a_i^2$. Since the fourth component of a four-vector is always imaginary, $\mathbf{a}^2$ can be larger, equal to, or smaller than zero. We call the vector $\mathbf{a}$ "space-like" for the case where $\mathbf{a}^2 > 0$, and "time-like" for $\mathbf{a}^2 < 0$. These definitions are Lorentz-invariant since $\mathbf{a}^2$ is Lorentz-invariant and its sign cannot therefore change in transformations. Such invariance is specially to be noted in regard to the "distance vector" $\mathbf{x} - \mathbf{y}$. It follows immediately that there is no reference system in which the time-like separated points $x_i$ and $y_i$ (for which therefore $l^2 = \sum (x_i - y_i)^2 < 0$) appear as simultaneous; for then the fourth component of the connecting vector would be equal to zero, and thus for the separation we should have $l^2 > 0$, in contradiction to the assumption that $l^2 < 0$. It likewise follows that two space-like separated points (for which therefore $l^2 > 0$) cannot in any reference system correspond to two temporally consecutive events at the same location. The region comprising those points $x_i$ whose separation from $y_i$ is space-like, is separated from the region comprising those points $x_i$ whose separation from $y_i$ is time-like, by the cone $\sum_{i=1}^{4} (x_i - y_i)^2 = 0$ whose apex lies at the point $y_i$. In figure 67 the regions whose points are time-like with respect to the origin are shaded, while the unshaded portion gives the points which are space-like with respect to the origin. There are always, however, transformations for which two points having space-like separation come to lie on the $x_1$-axis, and there are also transformations for which two points having time-like separation come to lie on the $x_4$-axis.

Since $x_i$ represents* a four-vector, $dx_i$ is also a four-vector. From equation (77.1) it follows for the proper time that

$$d\tau^2 = dt^2 \left(1 - \frac{v^2}{c^2}\right) = dt^2 - \frac{dx^2 + dy^2 + dz^2}{c^2} = -\frac{1}{c^2} \sum_{i=1}^{4} dx_i^2 \qquad (80.1)$$

consequently $d\tau$ is a scalar. If now the world line of a material point is given as a function of the proper time,

$$x_i = x_i(\tau) \qquad (80.2)$$

we define the *four-velocity* as the four-vector

$$u_i = \frac{dx_i}{d\tau} \qquad (80.3)$$

Its connection with the normal velocity follows from the chain rule e.g. $u_1 = v_x . dt/d\tau$. From (80.1) we therefore have

$$\left. \begin{array}{ll} u_1 = \dfrac{v_x}{\sqrt{(1-\beta^2)}}, & u_2 = \dfrac{v_y}{\sqrt{(1-\beta^2)}} \\[2ex] u_3 = \dfrac{v_z}{\sqrt{(1-\beta^2)}}, & u_4 = \dfrac{ic}{\sqrt{(1-\beta^2)}} \end{array} \right\} \qquad (80.4)$$

The simplest calculation of $\mathbf{u}^2 = \sum u_i^2$ is made in the system in which the mass point is at rest (i.e. $u_1 = u_2 = u_3 = 0$; $u_4 = ic$). We thus obtain $\mathbf{u}^2 = -c^2$. Since $\mathbf{u}^2$ is a scalar, this holds in all reference systems. $u_i$ is therefore a time-like vector.

The four-acceleration is defined by

$$b_i = \frac{du_i}{d\tau} \qquad (80.5)$$

and is therefore a four-vector. We have, for example,

$$\left. \begin{array}{l} b_1 = \dfrac{d}{dt}\left(\dfrac{v_x}{\sqrt{(1-\beta^2)}}\right)\dfrac{dt}{d\tau} = \dfrac{\dot{v}_x}{1-\beta^2} + \dfrac{v_x(\mathbf{v}\dot{\mathbf{v}})}{c^2(1-\beta^2)^2} \\[3ex] b_4 = \dfrac{d}{dt}\left(\dfrac{ic}{\sqrt{(1-\beta^2)}}\right)\dfrac{dt}{d\tau} = \dfrac{i}{c}\dfrac{(\mathbf{v}\dot{\mathbf{v}})}{(1-\beta^2)^2} \end{array} \right\} \qquad (80.6)$$

---

* In the following we shall often characterize four-vectors and four-tensors by their components writing, for example, simply $x_i$ instead of $\mathbf{x} \equiv \{x_i\}$.

Therefore, in the rest system of the mass point, $b_1 = \dot{v}_x, \ldots, b_4 = 0$, and thus $\mathbf{b}^2 = \sum b_i^2 > 0$, i.e. $b_i$ is a space-like vector. From the equation $\sum u_i^2 = -c^2$ there follows by differentiation with respect to $\tau$ the relationship $(\mathbf{u}, \mathbf{b}) = \sum u_i b_i = 0$.

The invariance of (79.6) is characteristic of a four-vector, for we have the theorem: If a linear form

$$\sum_{i=1}^{4} a_i v_i \tag{80.7}$$

of an arbitrary vector $v_i$ is Lorentz-invariant, $a_i$ is a four-vector. This is at once clear if we consider that through (80.7) (since $v_i$ is an arbitrary vector) the transformation of the quantities $a_i$ is uniquely established and on the other hand (80.7) is invariant if $a_i$ transforms like a four-vector.

If a bilinear form with two arbitrary vectors $v_i$, $w_i$

$$\sum_{i, k=1}^{4} T_{ik} v_i w_k \tag{80.8}$$

is Lorentz-invariant, $T_{ik}$ is called a tensor (second rank). From this it follows that with the Lorentz transformation (79.2) we have

$$T'_{ik} = \sum_{l, m} A_{il} A_{km} T_{lm} \tag{80.9}$$

as is very easily shown with the help of (79.9). A special tensor is $T_{ik} = a_i b_k$, the "tensor" product of the two vectors $a_i$ and $b_k$; for

$$\sum_{i, k} T_{ik} v_i w_k = \sum_{i, k} a_i b_k v_i w_k = \left(\sum_i a_i v_i\right)\left(\sum_k b_k w_k\right) = (\mathbf{a}, \mathbf{v})(\mathbf{b}, \mathbf{w})$$

is, as the product of the two scalars $(\mathbf{a}, \mathbf{v})$ and $(\mathbf{b}, \mathbf{w})$, invariant. Thus a tensor has, in general, sixteen components. The sum of two tensors $T_{ik}$ and $S_{ik}$ is again a tensor $T_{ik} + S_{ik}$; for if $\sum T_{ik} v_i w_k$ and $\sum S_{ik} v_i w_k$ are invariant, so also is $\sum T_{ik} v_i w_k + \sum S_{ik} v_i w_k = \sum (T_{ik} + S_{ik}) v_i w_k$. In addition it follows very easily that $\alpha T_{ik}$ is a tensor if $\alpha$ is a scalar. Furthermore $\sum_k T_{ik} a_k$ is a vector, and $\sum_i T_{ii}$ and $\sum_{i, k} T_{ik} T_{ik}$ are scalars.

If $T_{ik} = T_{ki}$, $T_{ik}$ is called a symmetric tensor. It has only ten different components. If $T_{ik} = -T_{ki}$, then $T_{ik}$ is called an anti-symmetric tensor or also a "six-vector", since it has only six essentially different components. The symmetry properties just mentioned are invariant, as is easily concluded from (80.9).

Every tensor $T_{ik}$ can be written, as in three dimensions (§15), as the sum of a symmetric and an anti-symmetric part:

$$T_{ik} = \tfrac{1}{2}(T_{ik} + T_{ki}) + \tfrac{1}{2}(T_{ik} - T_{ki}) \qquad (80.10)$$

$$\underbrace{\phantom{\tfrac{1}{2}(T_{ik} + T_{ki})}}_{\substack{\text{symmetric}\\\text{part}}} \quad \underbrace{\phantom{\tfrac{1}{2}(T_{ik} - T_{ki})}}_{\substack{\text{anti-symmetric}\\\text{part}}}$$

The "vector product" $a_i b_k - a_k b_i$ is a special anti-symmetric tensor.

After this short sketch of tensor algebra, we must briefly consider tensor analysis. If $\phi(x_i)$ is a scalar function of $x_i$, then

$$d\phi = \sum_{i=1}^{4} \frac{\partial \phi}{\partial x_i} dx_i \qquad (80.11)$$

is Lorentz-invariant. Since however $dx_i$ is a four-vector, then, according to the theorem proved above, $\partial \phi / \partial x_i$ is a four-vector, the gradient of $\phi$. From this it follows that the operator $\partial / \partial x_i$ transforms like a vector. Thus

$$\sum_{i=1}^{4} \frac{\partial a_i}{\partial x_i} \qquad (80.12)$$

is a scalar, the *four-dimensional divergence*. Likewise, then [compare (76.8)]

$$\Box = \sum_{i=1}^{4} \frac{\partial^2}{\partial x_i^2} = \frac{\partial^2}{\partial x^2} + \frac{\partial^2}{\partial y^2} + \frac{\partial^2}{\partial z^2} - \frac{1}{c^2} \frac{\partial^2}{\partial t^2} = \nabla^2 - \frac{1}{c^2} \frac{\partial^2}{\partial t^2} \qquad (80.13)$$

must be an invariant differential operator—a result which takes us back to our original point of departure since we have required that wave propagation according to (80.13) shall be invariant.

Out of the vector $a_i$ and the operator $\partial / \partial x_i$ we can construct as vector product the anti-symmetric tensor

$$F_{ik} = \frac{\partial a_k}{\partial x_i} - \frac{\partial a_i}{\partial x_k} \qquad (80.14)$$

the *four-dimensional* curl of the vector $a_i$. For a tensor $T_{ik}$,

$$\sum_{k=1}^{4} \frac{\partial T_{ik}}{\partial x_k} \qquad (80.15)$$

is a vector, the *divergence* of the tensor $T_{ik}$.

In conclusion we wish to give the transformation formulae of a four-vector and a four-tensor for the special transformations (79.11). For a four-vector $a_i$ we have

$$a_1' = \frac{a_1 + i\beta a_4}{\sqrt{(1-\beta^2)}}, \quad a_2' = a_2, \quad a_3' = a_3, \quad a_4' = \frac{a_4 - i\beta a_1}{\sqrt{(1-\beta^2)}} \tag{80.16}$$

For a symmetric tensor $T_{ik} = T_{ki}$, and we have:

$$
\left.
\begin{aligned}
&T_{11}' = \frac{T_{11} + 2i\beta T_{14} - \beta^2 T_{44}}{1-\beta^2} \qquad T_{44}' = \frac{T_{44} - 2i\beta T_{14} - \beta^2 T_{11}}{1-\beta^2} \\[2mm]
&T_{22}' = T_{22} \qquad\qquad T_{33}' = T_{33} \qquad\qquad T_{23}' = T_{23} \\[2mm]
&T_{14}' = \frac{T_{14} + i\beta(T_{44} - T_{11}) + \beta^2 T_{41}}{1-\beta^2} \qquad T_{12}' = \frac{T_{12} + i\beta T_{42}}{\sqrt{(1-\beta^2)}} \\[2mm]
&T_{13}' = \frac{T_{13} + i\beta T_{43}}{\sqrt{(1-\beta^2)}} \quad T_{24}' = \frac{T_{24} - i\beta T_{21}}{\sqrt{(1-\beta^2)}} \quad T_{34}' = \frac{T_{34} - i\beta T_{31}}{\sqrt{(1-\beta^2)}}
\end{aligned}
\right\} \tag{80.17}
$$

For an anti-symmetric tensor $T_{ik} = -T_{ki}$, and we have:

$$
\left.
\begin{aligned}
&T_{12}' = \frac{T_{12} - i\beta T_{24}}{\sqrt{(1-\beta^2)}} \quad T_{13}' = \frac{T_{13} - i\beta T_{34}}{\sqrt{(1-\beta^2)}} \quad T_{23}' = T_{23} \\[2mm]
&T_{14}' = T_{14} \qquad\qquad T_{24}' = \frac{T_{24} + i\beta T_{12}}{\sqrt{(1-\beta^2)}} \quad T_{34}' = \frac{T_{34} + i\beta T_{13}}{\sqrt{(1-\beta^2)}}
\end{aligned}
\right\} \tag{80.18}
$$

# CHAPTER E II

## The Relativistic Electrodynamics of Empty Space

### §81. The field equations

We shall occupy ourselves in this section with a *relativistic* development of electrodynamics in accordance with the programme outlined in §78. Mathematically speaking this means a rewriting of the fundamental equations of electrodynamics in *four-dimensional* vector form. This, as we shall see immediately, can be easily carried out since these equations are already Lorentz-invariant, proof being in just the possibility of a four-dimensional presentation of electrodynamics.

This rewriting of the basic equations brings with it two advantages: *First*, purely formally and in contrast to the three-dimensional formalism, they possess a highly simplified presentation remarkable for its symmetry. *Second*, and this is of particular importance to us, there is physically a series of special connections between apparently separated entities of the Maxwell theory, as between electric and magnetic field strength, between energy flow, Poynting vector, Maxwell stress tensor, etc.; this being of great value for a deeper understanding of electromagnetic processes.

For simplicity we begin by rewriting the *potential equations*

$$\nabla^2 \mathbf{A} - \frac{1}{c^2}\frac{\partial^2 \mathbf{A}}{\partial t^2} = -\frac{4\pi}{c}\rho\mathbf{v} \qquad \nabla^2 \phi - \frac{1}{c^2}\frac{\partial^2 \phi}{\partial t^2} = -4\pi\rho \quad (81.1)$$

in addition to which we have the Lorentz convention or gauge (see footnote on page 269)

$$\operatorname{div}\mathbf{A} + \frac{1}{c}\frac{\partial \phi}{\partial t} = 0 \qquad\qquad (81.2)$$

Combination of (81.1) with (81.2) yields the *equation of continuity* of electric charge:

$$\operatorname{div}(\rho\mathbf{v}) + \frac{\partial \rho}{\partial t} = 0 \qquad\qquad (81.3)$$

Now in (81.1), left, we have differential expressions having their origin in the application of the differential operator $\nabla^2 - \dfrac{1}{c^2}\dfrac{\partial^2}{\partial t^2}$ to a component of $\mathbf{A}$, or to $\phi$. Corresponding to (80.13), we designate this operator by $\square$. From the vector standpoint it is a scalar, just as is the three-dimensional Laplace operator in three-dimensional space.

Of the four similarly formed equations (81.1), three are comprehended in an ordinary vector equation. Let us then assume that, by an extension of this to a four-equation, $\phi$ would play the role of the fourth component (additional to the three space-like components) of the four-vector potential $\mathbf{A}$. This idea is also suggested by equation (81.2) which, on introducing the coordinates $(x_1, x_2, x_3, x_4)$ in place of $(x, y, z, t)$, takes the form

$$\frac{\partial A_x}{\partial x_1} + \frac{\partial A_y}{\partial x_2} + \frac{\partial A_z}{\partial x_3} + \frac{\partial i\phi}{\partial x_4} = 0$$

If, in this equation, we regard $A_x$, $A_y$, $A_z$ as the three space-like components $\Phi_1$, $\Phi_2$, $\Phi_3$ of a *four-potential* whose fourth component is $\Phi_4 = i\phi$, then (81.2) takes the invariant form

$$\sum_{\nu=1}^{4} \frac{\partial \Phi_\nu}{\partial x_\nu} = 0 \qquad (81.4)$$

Correspondingly, we can also conceive of the vector $\rho\mathbf{v}$ and the scalar $ic\rho$ as the space-like and the time-like components respectively of a *four-current* $s_\nu = (\rho v_x, \rho v_y, \rho v_z, ic\rho)$ which, through (81.3), is expressible as

$$\sum_{\nu=1}^{4} \frac{\partial s_\nu}{\partial x_\nu} = 0 \qquad (81.5)$$

The equations (81.1) then assume the four-dimensional vector form

$$\square \Phi_\nu = -\frac{4\pi s_\nu}{c} \qquad (81.6)$$

With this four-dimensional interpretation of the potential and the current density, it is readily seen how these quantities change upon going over to another coordinate system. We wish especially to know how the current vector $s_\nu$ changes on going from the charge's rest system (index 0) to another system in which the charge moves in the

$x$-direction with velocity $v$. For this we have the transformation formulae

$$s_1 = \frac{s_1^0 - i\beta s_4^0}{\sqrt{(1-\beta^2)}} \qquad s_2 = s_2^0 \qquad s_3 = s_3^0 \qquad s_4 = \frac{s_4^0 + i\beta s_1^0}{\sqrt{(1-\beta^2)}}$$

with
$$s_1^0 = s_2^0 = s_3^0 = 0 \qquad s_4^0 = i\rho_0 c$$

We therefore obtain

$$\left. \begin{array}{ll} s_1 = \rho v_x = \dfrac{\rho_0 v}{\sqrt{(1-\beta^2)}} & s_2 = \rho v_y = 0 \\[3ex] s_3 = \rho v_z = 0 & s_4 = i\rho c = \dfrac{i\rho_0 c}{\sqrt{(1-\beta^2)}} \end{array} \right\} \quad (81.7)$$

The equations state that the charge in the new system is moving with velocity $v$ (as we should expect), and further that in this system the charge density has the value

$$\rho = \frac{\rho_0}{\sqrt{(1-\beta^2)}} \tag{81.8}$$

The charge density therefore increases in making a transfer from the system in which the charge is at rest to a system with respect to which the charge moves. By contrast, however, we see in the following that *the quantity of charge de in a given element of volume is the same for both systems*—for we have

$$de = \rho \, dV = \frac{\rho_0}{\sqrt{(1-\beta^2)}} dV^0 \sqrt{(1-\beta^2)} = \rho_0 \, dV^0 = de^0 \tag{81.9}$$

In regard to $dV$, the step from the second to the third member in the above equation comes from the Lorentz contraction of the length dimension in the direction of the motion. It can therefore be said that the moving charge density appears larger because, owing to the Lorentz contraction, the same charge is conveyed in a smaller volume. It should be noted in passing that the charge on an electron has the same value in all coordinate systems. For this compare (65.7).

The form of the equations (81.7) suggests that for the case of an arbitrary direction of the velocity these equations generalize to

$$s_1 = \rho v_x = \frac{\rho_0 v_x}{\sqrt{(1-\beta^2)}} \qquad s_2 = \rho v_y = \frac{\rho_0 v_y}{\sqrt{(1-\beta^2)}}$$

$$s_3 = \rho v_z = \frac{\rho_0 v_z}{\sqrt{(1-\beta^2)}} \qquad s_4 = i\rho c = \frac{i\rho_0 c}{\sqrt{(1-\beta^2)}}$$

The quantities multiplying the *rest charge density* $\rho_0$ are, however, just the components of the four-velocity $u_v$, so that the four-current can also be written as

$$s_v = \rho_0 u_v \qquad (81.10)$$

(This equation is a true vector equation since, obviously, $\rho_0$ is a scalar, and $s_v$ and $u_v$ are four-vectors.)

We pass on now from the potentials **A** and $\phi$ to the field quantities **E** and **H**. These are related to the potentials through the equations

$$\mathbf{H} = \text{curl}\,\mathbf{A} \qquad \mathbf{E} = -\text{grad}\,\phi - \frac{1}{c}\frac{\partial \mathbf{A}}{\partial t} \qquad (81.11)$$

Introducing the components $\Phi_v$ of the four-potential and the four-coordinates $x_v$, we form the following relationships:

$$H_x = \frac{\partial \Phi_3}{\partial x_2} - \frac{\partial \Phi_2}{\partial x_3} \qquad E_x = i\left(\frac{\partial \Phi_4}{\partial x_1} - \frac{\partial \Phi_1}{\partial x_4}\right)$$

$$H_y = \frac{\partial \Phi_1}{\partial x_3} - \frac{\partial \Phi_3}{\partial x_1} \qquad E_y = i\left(\frac{\partial \Phi_4}{\partial x_2} - \frac{\partial \Phi_2}{\partial x_4}\right)$$

$$H_z = \frac{\partial \Phi_2}{\partial x_1} - \frac{\partial \Phi_1}{\partial x_2} \qquad E_z = i\left(\frac{\partial \Phi_4}{\partial x_3} - \frac{\partial \Phi_3}{\partial x_4}\right)$$

The structure of these assembled equations suggests that a skew-symmetric tensor be defined by the following equation:

$$F_{v\mu} = \frac{\partial \Phi_\mu}{\partial x_v} - \frac{\partial \Phi_v}{\partial x_\mu} \qquad (81.12)$$

Its components then correlate with the field quantities **E** and **H** in the following way:

$$F_{v\mu} = \begin{pmatrix} 0 & H_z & -H_y & -iE_x \\ -H_z & 0 & H_x & -iE_y \\ H_y & -H_x & 0 & -iE_z \\ iE_x & iE_y & iE_z & 0 \end{pmatrix} \qquad (81.13)$$

We thus obtain the result that in four-dimensional formalism the electromagnetic field is no longer to be described by two vectors, but by a *skew-symmetric tensor of the second rank*. The electric and magnetic field strengths are therefore no longer independent of one another, but are unified in a tensor. In a Lorentz transformation the components of **E** and **H** do not transform as such, but rather are mixed. We write the transformation relations for going from an unprimed system to a primed system moving in the *x*-direction; from (80.18) we have

$$
\begin{aligned}
H'_x &= H_x & E'_x &= E_x \\
H'_y &= \frac{1}{\sqrt{(1-\beta^2)}}(H_y + \beta E_z) & E'_y &= \frac{1}{\sqrt{(1-\beta^2)}}(E_y - \beta H_z) \\
H'_z &= \frac{1}{\sqrt{(1-\beta^2)}}(H_z - \beta E_y) & E'_z &= \frac{1}{\sqrt{(1-\beta^2)}}(E_z + \beta H_y)
\end{aligned}
\right\}
\quad (81.14)
$$

If the primed system signifies the rest system of an electron, then in this system the force $e\mathbf{E}' = e\mathbf{E}^0$ acts on the electron. From (81.14) it is seen that this leads directly to the expression for the Lorentz force $e(\mathbf{E} + \mathbf{v}/c \times \mathbf{H})$, when terms of relativistic magnitude, i.e. where $\beta^2$ is small compared to 1, are neglected. We therefore see that whereas in the Maxwell theory the Lorentz force (from which the force on a current-carrying conductor in a magnetic field, as well as the induction law for moving current coils can be derived) represents a foreign element and has to be evaluated as an independent experimental theorem, *in relativity theory it can be derived as an immediate consequence of the transformation formulae.*

On the basis of the transformation formulae (81.14) we can be easily satisfied that expressions **E.H** and $\mathbf{E}^2 - \mathbf{H}^2$ are both invariants; that therefore their value is unchanged in going over to the primed system. We shall make use of these facts later.

We wish now to write down the Maxwell equations themselves in the four-dimensional formalism and subsequently to verify their correctness. They read:

$$
\sum_{\mu=1}^{4} \frac{\partial F_{\nu\mu}}{\partial x_\mu} = \frac{4\pi s_\nu}{c} \qquad \frac{\partial F_{\nu\mu}}{\partial x_\lambda} + \frac{\partial F_{\mu\lambda}}{\partial x_\nu} + \frac{\partial F_{\lambda\nu}}{\partial x_\mu} = 0 \qquad (81.15)
$$

The first equation system comprises four equations ($\nu = 1, 2, 3, 4$) and is identical with the equations

$$
\begin{aligned}
\operatorname{curl} \mathbf{H} &= \frac{1}{c}\frac{\partial \mathbf{E}}{\partial t} + \frac{4\pi\rho\mathbf{v}}{c} & \text{for} \quad \nu &= 1, 2, 3 \\
\operatorname{div} \mathbf{E} &= 4\pi\rho & \text{for} \quad \nu &= 4
\end{aligned}
\right\}
\quad (81.16)
$$

which are immediately obtained by introducing (81.13) and (81.7) into (81.15). In the second system of equations $\nu$, $\mu$, and $\lambda$ can, of themselves and independently of one another, assume all values between 1 and 4. It can be seen, however, that owing to the skew-symmetry of $F_{\nu\mu}$, only in those equations for which $\nu$, $\mu$, and $\lambda$ are different from one another does the left side not vanish identically. For $(\nu, \mu, \lambda)$ we therefore obtain equations in accordance with the following trios of values: $(2,3,4)$, $(3,4,1)$, $(4,1,2)$, and $(1,2,3)$. On introducing (81.13) we thus obtain

$$\operatorname{curl} \mathbf{E} = -\frac{1}{c}\frac{\partial \mathbf{H}}{\partial t} \qquad \operatorname{div} \mathbf{H} = 0 \qquad (81.17)$$

In the four-dimensional form the Maxwell equations therefore assume an especially simple form which is required, on the one hand by the equality of status of space and time coordinates, and on the other hand by a similar sort of equality between $\mathbf{E}$ and $\mathbf{H}$. Of course in determining the field from a given charge distribution, we would in general first determine the potentials, and then, through (81.12) and (81.13), calculate the field intensities.

In our derivation of the four-dimensional Maxwell equations it is self-evident that these equations do not contradict the *potential equations* (81.6) and (81.4). We wish, nevertheless, to give here a simple proof that the potential equations are derivable from the field equations. If, as in (81.12), the $F_{\nu\mu}$ are written down as the four-dimensional curl of a four-vector, the second system of equations is identically satisfied. Introducing (81.12) in the first gives

$$\frac{4\pi s_\nu}{c} = \sum_\mu \frac{\partial}{\partial x_\mu}\left(\frac{\partial \Phi_\mu}{\partial x_\nu} - \frac{\partial \Phi_\nu}{\partial x_\mu}\right) = \frac{\partial}{\partial x_\nu}\sum_\mu \frac{\partial \Phi_\mu}{\partial x_\mu} - \sum_\mu \frac{\partial^2 \Phi_\nu}{\partial x_\mu{}^2}$$

The second term on the right side is equal to $-\Box\,\Phi_\nu$; the first term vanishes when we impose the Lorentz convention (81.4) as an additional requirement. This is permitted because through (81.12) the $\Phi_\nu$ are not uniquely determined, but rather only to within a gradient $\partial\Psi/\partial x_\nu$, in which $\Psi$ can be a completely arbitrary function of the $x_\nu$. Through the additional conditions (81.4) a condition equation for $\Psi$ is given. If, that is, the $\Phi_\nu$ do not satisfy equations (81.4), then $\Psi$ is to be determined so that

$$\sum_\nu \frac{\partial}{\partial x_\nu}\left(\Phi_\nu + \frac{\partial \Psi}{\partial x_\nu}\right) = 0 \quad \text{or} \quad \Box\Psi = -\sum \frac{\partial \Phi_\nu}{\partial x_\nu}$$

Then for $\Phi_\nu{}^* = \Phi_\nu + \dfrac{\partial \Psi}{\partial x_\nu}$ equation (81.4) is satisfied. We therefore obtain the result

$$\Box\Phi_\nu = -\frac{4\pi s_\nu}{c}$$

in *agreement* with (81.6).

### §82. The force density

For the force acting on the electric charge contained in unit volume, the electron theory of Lorentz states that

$$\mathbf{f} = \rho\left(\mathbf{E} + \frac{\mathbf{v}}{c} \times \mathbf{H}\right) \tag{82.1}$$

$\mathbf{f}$ is a force density and therefore has the dimensions of *force per unit volume*. If for $\rho$, $\rho\mathbf{v}$, $\mathbf{E}$, and $\mathbf{H}$ the corresponding components of the four-current $s_\nu$ and of the field tensor $F_{\nu\mu}$ are introduced, then (82.1) can also be written in the form

$$
\left.
\begin{aligned}
f_x &= \frac{1}{c}( \qquad\quad s_2 F_{12} + s_3 F_{13} + s_4 F_{14}) \\
f_y &= \frac{1}{c}(s_1 F_{21} \qquad\quad + s_3 F_{23} + s_4 F_{24}) \\
f_z &= \frac{1}{c}(s_1 F_{31} + s_2 F_{32} \qquad\quad + s_4 F_{34})
\end{aligned}
\right\} \tag{82.1a}
$$

as can be readily understood on the basis of (81.13). The symmetrical structure of (82.1a) suggests that a four-vector

$$f_\nu = \frac{1}{c} \sum_{\mu=1}^{4} F_{\nu\mu} s_\mu \tag{82.2}$$

be defined, whose first three components accord with the expressions on the right side of the equation system (82.1a); for we have $F_{\nu\nu} = 0$. The point here is the following: From the vector character of the four-current and the tensor character of the field intensities $F_{\nu\mu}$, it follows that the components of $\mathbf{f}$ can be expressed as space-like components of a four-vector $f_\nu$. We are therefore permitted to identify $f_x, f_y, f_z$ with these components of $f_\nu$; and, by adjoining a fourth component $f_4$, to make up a four-vector. From (82.2), the meaning of $f_4$ is seen to be

$$f_4 = \frac{1}{c}(s_1 F_{41} + s_2 F_{42} + s_3 F_{43}) = \frac{i\rho\mathbf{v}\mathbf{E}}{c} = \frac{i\mathbf{v}\mathbf{f}}{c} \tag{82.3}$$

It is pure imaginary, as indeed it must be, and within the factor $i/c$ it represents the work done per unit volume and per unit time by the force density $\mathbf{f}$.

In accordance with (82.3), the fourth component $f_4$ of the four-force vanishes in the rest system of the charge; no work is done on non-moving charges. This result may also be obtained directly from (82.2); for in the rest system only the fourth component of the four-current $s_\nu$ is different from zero. Thus in the rest system we have $f^0_\nu = F^0_{\nu 4} s^0_4 / c$, the fourth component $f^0_4$ vanishing because $F_{44} = 0$. We have in general that

$$\sum_{\nu=1}^{4} f_\nu s_\nu \equiv 0 \qquad (82.4)$$

for the left side is an invariant whose value we can find through the use of an arbitrarily chosen coordinate system. If we choose the rest system, then, in this system only $s^0_4$ is different from zero; $f^0_4$, however is equal to zero. We can also verify the relationship (82.4) without any special choice of coordinate system, for from (82.2) it follows that

$$\sum_{\nu=1}^{4} f_\nu s_\nu = \frac{1}{c} \sum_\nu \sum_\mu F_{\nu\mu} s_\nu s_\mu$$

On the right we have the inner product of a skew-symmetric tensor $F_{\nu\mu}$ times the symmetric tensor $s_\nu s_\mu$. An exchange of the summation indices $\nu$ and $\mu$ would therefore require a change of sign of the expression; on the other hand, through such a change its value would not be altered. It must therefore be equal to zero.

Up to now we have been concerned only with the force density and we have established that it can be understood to be the space-like part of a four-vector $f_\nu$. We now ask for the *total force* which will be exerted on a particular charge distribution, or on a particular volume of space containing charge. By definition, and through integration over the volume concerned, we obtain from the force density

$$\mathbf{F} = \int \mathbf{f} \, dV \qquad (82.5)$$

Thus, for example, the total force (Lorentz force) exerted on an electron by a field having a sufficiently slow variation with time is

$$\mathbf{F} = \int \rho \left( \mathbf{E} + \frac{\mathbf{v}}{c} \times \mathbf{H} \right) dV = e \left( \mathbf{E} + \frac{\mathbf{v}}{c} \times \mathbf{H} \right) \qquad (82.5a)$$

$\mathbf{F}$ cannot be described as a four-vector, for while $\mathbf{F}$ is indeed the space-like portion of a vector, the volume element $dV$ is not an invariant (as, of course, it should not be); rather, it varies in the transfer from one system to another by virtue of the Lorentz contraction of its length. It is therefore a question of some interest how the force (82.5) varies in going from one system to another. We shall limit ourselves to the special case in which one of the two systems involved is the *rest system* of the charge. In order to obtain the transformation formula for the force, we start with the known transformation formula for the force density, which, because $f_4^0 = 0$, has the form:

$$\left.\begin{aligned} f_x &\equiv f_1 = \frac{f_1^0}{\sqrt{(1-\beta^2)}} \equiv \frac{f_x^0}{\sqrt{(1-\beta^2)}} \\ f_y &\equiv f_2 = f_2^0 \equiv f_y^0 \\ f_z &\equiv f_3 = f_3^0 \equiv f_z^0 \end{aligned}\right\} \tag{82.6}$$

We integrate this formula on right and left over the volume, bearing in mind that

$$dV = dV^0 \sqrt{(1-\beta^2)} \tag{82.7}$$

From (82.5) we then obtain the force **F** as measured in a coordinate system with respect to which the charge moves with the velocity **v** (the force being $\mathbf{F}^0$ in the rest system):

$$F_x = F_x^0 \qquad F_y = F_y^0 \sqrt{(1-\beta^2)} \qquad F_z = F_z^0 \sqrt{(1-\beta^2)} \tag{82.8}$$

We shall meet these equations again in the relativistic treatment of mechanics.

We again briefly consider equation (82.8) in the example of the Lorentz force on a moving electron. This is given by equation (82.5a). In the electron's rest system the only force acting is

$$\mathbf{F}^0 = e\mathbf{E}^0 \tag{82.5b}$$

Now we know from §81 how the electric field changes in going to another system; we found there that

$$E_x^0 = E_x$$

$$E_y^0 = \frac{1}{\sqrt{(1-\beta^2)}}\left(E_y - \beta H_z\right)$$

$$E_z^0 = \frac{1}{\sqrt{(1-\beta^2)}}\left(E_z + \beta H_y\right)$$

On multiplying these equations by $e$ and introducing the force components (82.5a) and (82.5b), we immediately obtain (82.8).

The same result is obtained if we consider two charged particles of like sign mutually at rest in an otherwise field-free space. Although in their rest system only an electrostatic repulsion force $\mathbf{F}^0$ exists, in the system with respect to which the particles move with velocity $v$ in the $x$-direction there enters, besides the electro-static repulsion, a magnetic attractive force which leads to a diminution of $F_y$ and $F_z$ in the relations given by (82.8).

## §83. The energy-momentum tensor of the electromagnetic field

We now wish to summarize the topics already treated in chapter D I concerning the interrelationships between *energy density*, *momentum density*, and *ponderomotive forces*, which, using the four-dimensional approach, we shall give a Lorentz-invariant form. With the rewriting in four-dimensional form the question of the transformation of the above-mentioned quantities from one system to another in motion relative to it, is immediately settled. The transformation formulae thus derived will, in the application of relativity theory to matters of mechanics, prove to be most significant.

We start by rewriting the force expression

$$f_v = \frac{1}{c} \sum_{\mu=1}^{4} F_{v\mu} s_\mu \tag{83.1}$$

and we wish to show that it can be expressed in the form

$$f_v = \sum_{\mu=1}^{4} \frac{\partial T_{v\mu}}{\partial x_\mu} \tag{83.2}$$

namely, as the four-dimensional divergence of a four-tensor of the second rank. For $s_\mu$ in (83.1) we introduce its value given in the first of equations (81.15), obtaining for $f_v$

$$f_v = \frac{1}{4\pi} \sum_{\mu=1}^{4} \sum_{\lambda=1}^{4} F_{v\mu} \frac{\partial F_{\mu\lambda}}{\partial x_\lambda}$$

For this we can write

$$f_v = \frac{1}{4\pi} \sum_{\mu=1}^{4} \sum_{\lambda=1}^{4} \frac{\partial (F_{v\mu} F_{\mu\lambda})}{\partial x_\lambda} - \frac{1}{4\pi} \sum_{\mu=1}^{4} \sum_{\lambda=1}^{4} F_{\mu\lambda} \frac{\partial F_{v\mu}}{\partial x_\lambda}$$

In view of the skew-symmetry of the field tensor, and through an exchange of the summation indices, we transform the second term, obtaining

$$\sum_{\mu=1}^{4} \sum_{\lambda=1}^{4} F_{\mu\lambda} \frac{\partial F_{v\mu}}{\partial x_\lambda} = \frac{1}{2} \sum_{\mu=1}^{4} \sum_{\lambda=1}^{4} \left( F_{\mu\lambda} \frac{\partial F_{v\mu}}{\partial x_\lambda} + F_{\lambda\mu} \frac{\partial F_{v\lambda}}{\partial x_\mu} \right)$$

$$= \frac{1}{2} \sum_{\mu=1}^{4} \sum_{\lambda=1}^{4} F_{\mu\lambda} \left( \frac{\partial F_{v\mu}}{\partial x_\lambda} + \frac{\partial F_{\lambda v}}{\partial x_\mu} \right)$$

The expression in parentheses on the right is, from the second of equations (81.15), equal to $-\partial F_{\mu\lambda}/\partial x_\nu$, so that this series of equations can be further written

$$= -\frac{1}{2} \sum_{\mu=1}^{4} \sum_{\lambda=1}^{4} F_{\mu\lambda} \frac{\partial F_{\mu\lambda}}{\partial x_\nu} = -\frac{1}{4} \frac{\partial}{\partial x_\nu} \sum_{\mu=1}^{4} \sum_{\lambda=1}^{4} (F_{\mu\lambda})^2$$

We therefore have that

$$f_\nu = \frac{1}{4\pi} \sum_{\lambda=1}^{4} \frac{\partial}{\partial x_\lambda} \left( \sum_{\mu=1}^{4} F_{\nu\mu} F_{\mu\lambda} \right) + \frac{1}{16\pi} \frac{\partial}{\partial x_\nu} \sum_{\mu=1}^{4} \sum_{\lambda=1}^{4} (F_{\mu\lambda})^2 \quad (83.3)$$

This expression agrees with (83.2) if, following an exchange of indices, we set

$$T_{\nu\mu} = \frac{1}{4\pi} \sum_{\kappa=1}^{4} F_{\nu\kappa} F_{\kappa\mu} + \frac{1}{16\pi} \delta_{\nu\mu} \sum_{\kappa=1}^{4} \sum_{\lambda=1}^{4} (F_{\kappa\lambda})^2 \quad (83.4)$$

where $\delta_{\nu\mu}$ is 0 or 1 according as $\nu \neq \mu$, or $\nu = \mu$.

According to (83.4) the tensor $T_{\nu\mu}$ is symmetrically constructed. In keeping with general usage we call it the *energy-momentum tensor*. For the purpose of further exploring its meaning, we next write its components on the basis of (81.13) as expressions in **E** and **H**. In the second term of (83.4) only the diagonal elements contribute, thus

$$\frac{1}{16\pi} \sum_{\kappa=1}^{4} \sum_{\lambda=1}^{4} F_{\kappa\lambda}^2 = \frac{1}{8\pi} (\mathbf{H}^2 - \mathbf{E}^2)$$

The other part of (83.4), namely

$$\frac{1}{4\pi} \sum_{\kappa=1}^{4} F_{\nu\kappa} F_{\kappa\mu}$$

has three terms in the diagonal elements. Thus, for example,

for $\nu = \mu = 1$: $\qquad \frac{1}{4\pi}(-H_y^2 - H_z^2 + E_x^2)$

for $\nu = \mu = 4$: $\qquad \frac{1}{4\pi}(E_x^2 + E_y^2 + E_z^2) = \frac{1}{4\pi}\mathbf{E}^2$

The non-diagonal elements have only two terms. For the purely space-like members, e.g. for $\nu = 1$, $\mu = 2$, they have the form

$$\frac{1}{4\pi}(H_x H_y + E_x E_y)$$

The mixed elements, such as where $v = 1$, $\mu = 4$, take the form

$$\frac{i}{4\pi}(E_z H_y - E_y H_z)$$

If, for brevity, we write

$$T_{xx} = \frac{1}{4\pi}[E_x^2 + H_x^2 - \tfrac{1}{2}(\mathbf{E}^2 + \mathbf{H}^2)], \quad T_{xy} = T_{yx} = \frac{1}{4\pi}(E_x E_y + H_x H_y)$$

$$T_{yy} = \frac{1}{4\pi}[E_y^2 + H_y^2 - \tfrac{1}{2}(\mathbf{E}^2 + \mathbf{H}^2)], \quad T_{yz} = T_{zy} = \frac{1}{4\pi}(E_y E_z + H_y H_z) \quad (83.5a)$$

$$T_{zz} = \frac{1}{4\pi}[E_z^2 + H_z^2 - \tfrac{1}{2}(\mathbf{E}^2 + \mathbf{H}^2)], \quad T_{zx} = T_{xz} = \frac{1}{4\pi}(E_z E_x + H_z H_x)$$

further, $$\mathbf{S} = \frac{c}{4\pi}[\mathbf{E} \times \mathbf{H}] \qquad \mathbf{j} = \frac{1}{4\pi c}[\mathbf{E} \times \mathbf{H}] \qquad (83.5b)$$

and $$u = \frac{1}{8\pi}(\mathbf{E}^2 + \mathbf{H}^2) \qquad (83.5c)$$

then the energy-momentum tensor reads

$$(T_{v\mu}) = \begin{pmatrix} T_{xx} & T_{xy} & T_{xz} & -icj_x \\ T_{yx} & T_{yy} & T_{yz} & -icj_y \\ T_{zx} & T_{zy} & T_{zz} & -icj_z \\ -\dfrac{i}{c}S_x & -\dfrac{i}{c}S_y & -\dfrac{i}{c}S_z & u \end{pmatrix} \qquad (83.6)$$

We wish to justify the symbolism introduced in (83.5)-(83.6) for the various components of the tensor $T_{v\mu}$, revealing their *physical* significance in (83.2). For this we integrate (83.2) over a fixed volume $V$ in space (with $dV = dx_1 dx_2 dx_3$), and obtain for the time-like components

$$\int f_4 \, dV = \int \left( \frac{\partial T_{41}}{\partial x} + \frac{\partial T_{42}}{\partial y} + \frac{\partial T_{43}}{\partial z} \right) dV + \frac{1}{ic} \frac{\partial}{\partial t} \int T_{44} \, dV$$

Multiplying this equation by $-ic$ we get $-icf_4 = (\mathbf{f} \cdot \mathbf{v})$, and from (82.3) we obtain the power density, the left side being the work performed by the field on the charges contained within the volume $V$. The whole equation must therefore be identical with the energy law (54.8). Consequently we can conclude that

$$T_{44} = u \qquad icT_{41} = S_x \qquad icT_{42} = S_y \qquad icT_{43} = S_z \qquad (83.7)$$

wherein $u$ is the *energy density* and $\mathbf{S}$ is the vector representing energy flow. Our last equation therefore actually goes over to the well-known form

$$-\int (\mathbf{fv})\, dV - \int S_n\, dA = \frac{d}{dt} \int u\, dV \qquad (83.8)$$

and thus explains the designations used in the last row of (83.6) where $A$ is now used for area.

We now examine the first of the four equations (83.2). Integrating over a space-like volume $V$ gives

$$\int f_1\, dV = \int \left( \frac{\partial T_{11}}{\partial x} + \frac{\partial T_{12}}{\partial y} + \frac{\partial T_{13}}{\partial z} \right) dV + \frac{1}{ic} \frac{\partial}{\partial t} \int T_{14}\, dV$$

On the left here, since $f_1 = f_x$, we have the $x$-component of the resulting force $\mathbf{F}$ which acts on all charges contained within $V$. The complete equation must therefore have the same significance as the *momentum theorem*, equation (56.5). Since, in accordance with the rules of classical mechanics, the force $\mathbf{F}$ gives an increase of the momentum $\mathbf{J}_M$ of the inert masses bound up with the charges, we shall write the left side of the equation (pending later justification of this interdependence in relativistic mechanics) in the form

$$\mathbf{F} = \frac{d\mathbf{J}_M}{dt} \qquad (83.9)$$

The first term on the right side permits transforming to a surface integral

$$\oint (T_{11} \cos \langle \mathbf{n}, \mathbf{x} \rangle + T_{12} \cos \langle \mathbf{n}, \mathbf{y} \rangle + T_{13} \cos \langle \mathbf{n}, \mathbf{z} \rangle)\, dA$$

and can be interpreted as the $x$-component of all *surface forces* acting on the surface of $V$. The last term is to be interpreted as a *momentum density* connected with the electromagnetic field:

$$j_x = -\frac{1}{ic} T_{14} = \frac{i}{c} T_{14} \qquad (83.10)$$

In all, therefore, and in agreement with (56.7), the first three equations (83.2) give us

$$\frac{d}{dt} \left\{ \mathbf{J}_M + \int \mathbf{j}\, dV \right\} = \oint \mathbf{T}_n\, dA \qquad (83.11)$$

in which the vector $\mathbf{T}_n$ is constructed from the *unit-normal vector* $\mathbf{n}$ and the components of the *stress tensor* (83.5a) thus:

$$(\mathbf{T}_n)_x = T_{xx} \cos \langle \mathbf{n}, \mathbf{x} \rangle + T_{xy} \cos \langle \mathbf{n}, \mathbf{y} \rangle + T_{xz} \cos \langle \mathbf{n}, \mathbf{z} \rangle \quad (83.12)$$

Our justification for the symbolism introduced in (83.6) has not yet made any reference to the special connection given in (83.5) between the components of $T_{\nu\mu}$ and the electromagnetic field quantities. It would hold for any force density which can be written in the form (83.2). A property of particular importance in our special tensor (83.4) resides in its symmetry. From this there follows from (83.7) and (83.10) the fundamental relationship

$$\mathbf{S} = \mathbf{j}c^2 \quad (83.13)$$

between *energy flux* and *momentum density*, though to be sure only for the electromagnetic field *in vacuo*. Later we shall be able to interpret this same equation in a much more general way, and then *all* mechanical forces and momenta will be included in our formulae.

So long as no charges are present, the force density in equation (83.2) equals zero. As an example, let us consider a light ray travelling *in vacuo*, having been emitted at an earlier time by some light source. In the entire region of such a wave train the energy momentum tensor satisfies the equation

$$\sum_{\mu=1}^{4} \frac{\partial T_{\nu\mu}}{\partial x_\mu} = 0 \quad (83.14)$$

We wish here to append a proof of the following theorem:
*If a four-dimensional tensor $T_{\nu\mu}$ differs from zero only in a finite region of space, and if in this region, within the meaning of equation (83.14), it is source-free, then the integral over all space,*

$$\int T_{4\mu} dV \quad (83.15)$$

*has the character of a four-vector.*

For the special case of our hitherto considered electromagnetic tensor $T_{\nu\mu}$, this means that the total momentum and the total energy of a closed wave train

$$\left. \begin{array}{l} J_x = \dfrac{i}{c} \displaystyle\int T_{14} dV \\[2ex] J_y = \dfrac{i}{c} \displaystyle\int T_{24} dV \\[2ex] J_z = \dfrac{i}{c} \displaystyle\int T_{34} dV \\[2ex] \dfrac{i}{c} U = \dfrac{i}{c} \displaystyle\int T_{44} dV \end{array} \right\} \quad (83.15a)$$

transform in going over to a moving coordinate system like the four components of a vector.

For the proof we introduce the following lemma: Let $A_\nu$ be a four-vector different from zero only in a finite region of space, its divergence vanishing everywhere. From the fact that $A_\nu$ is source-free it follows next, with the help of Gauss's theorem, that

$$\int \sum_\nu \frac{\partial A_\nu}{\partial x_\nu} dx_1 dx_2 dx_3 dx_4 = \int A_\nu dS = 0 \qquad (83.16)$$

where the integration is to extend over the surface of any region of four-dimensional space. Accordingly, we make the special choice of a region which is bounded by the two regions $x_4 = $ constant, and $x_4' = $ constant, i.e. two simultaneous cross-sections of our radiation-filled world tube, the simultaneity being understood to be at one time in the sense of the first observer, and at the other time in the sense of the second observer.

For the space region thus chosen, Gauss's theorem has the form

$$\int A_4 dx_1 dx_2 dx_3 = \int A_4' dx_1' dx_2' dx_3' \qquad (83.17)$$

*The integral extending over all space and over the time-like components of the source-free four-vector $A_\nu$ is invariant against a Lorentz transformation.*

With this lemma our question is settled in the following manner: Let $p_\nu$ be an arbitrary four-vector which, however, is constant in space and time. With this we construct the vector

$$\sum_{\nu=1}^{4} p_\nu T_{\mu\nu}$$

which now satisfies all conditions of our lemma if $T_{\mu\nu}$ satisfies equation (83.14). We can therefore conclude that the quantity

$$\sum_{\nu=1}^{4} p_\nu \int T_{4\nu} dV = \sum_{\nu=1}^{4} p_\nu' \int T_{4\nu}' dV'$$

is an *invariant*. Since, however, $p_\nu$ is an arbitrary four-vector, the *four-character* of the quantity $\int T_{4\nu} dV$ is demonstrated.

## §84. The plane light-wave

In this section we shall apply the results of relativistic electrodynamics to the case of a *plane light-wave*. In so doing we shall answer a whole series of questions in optics (Doppler effect, aberration, reflection at a moving mirror). Naturally these problems can be handled without recourse to relativity theory, i.e. solely on the basis of Maxwell's equations which, as we have already seen, satisfy the relativity postulate. The use of relativity methods, however, brings an essential simplification. It permits the laws of the Doppler effect, etc., to be immediately derived merely by applying the transformation formulae (connecting one system to another that moves with respect to it) to the quantities characteristic of light waves.

We begin with the description of a linearly polarized wave, of frequency $v$ and wave length $\lambda$ (in the unprimed system), moving in the direction $\mathbf{n} \equiv (n_x, n_y, n_z)$. The wave is described by the equations

$$\mathbf{E} = \mathbf{E}_0 \, e^{i\Phi} \qquad \mathbf{H} = \mathbf{H}_0 \, e^{i\Phi} \tag{84.1}$$

with the abbreviation

$$\Phi = 2\pi v \left( t - \frac{x n_x + y n_y + z n_z}{c} \right) \tag{84.2}$$

for the "phase" of the motion of the light. $\mathbf{E}_0$ and $\mathbf{H}_0$ are two constant vectors which are perpendicular to one another and to $\mathbf{n}$, and of equal magnitude. From the standpoint of the primed system the waves are described by equations of the form

$$\mathbf{E}' = \mathbf{E}'_0 \, e^{i\Phi'} \qquad \mathbf{H}' = \mathbf{H}'_0 \, e^{i\Phi'} \tag{84.3}$$

in which $\Phi'$ is given by the expression analogous to (84.2), namely

$$\Phi' = 2\pi v' \left( t' - \frac{x' n'_x + y' n'_y + z' n'_z}{c} \right) \tag{84.4}$$

We consider next the *transformation of the phase*. This must obviously be an invariant of the Lorentz transformation—for, the statement that at a world point the phase is equal to zero (or an integral multiple of $2\pi$) is certainly an objective statement, quite independent of the coordinate system. We have therefore that

$$\Phi = \Phi' \tag{84.5}$$

or, if (84.2) and (84.4) are introduced and if, on the basis of the Lorentz transformation (76.5), the unprimed coordinates are substituted for the primed, we have

$$v \left( t - \frac{x n_x + y n_y + z n_z}{c} \right)$$
$$= v' \left( \frac{t - vx/c^2}{\sqrt{(1-\beta^2)}} - \frac{1}{c} \left[ \frac{x - vt}{\sqrt{(1-\beta^2)}} \, n'_x + y n'_y + z n'_z \right] \right)$$

Since this equation must hold as an identity in $x,y,z,t$, the following transformation formulae for *frequency* and *direction* of a plane wave are immediately obtained:

$$v = v'\frac{1+\beta n'_x}{\sqrt{(1-\beta^2)}} \qquad vn_x = v'\frac{\beta+n'_x}{\sqrt{(1-\beta^2)}} \left.\begin{array}{c} \\ \\ \\ \\ \end{array}\right\} \qquad (84.6)$$
$$vn_y = v'n'_y \qquad vn_z = v'n'_z$$

The expressions (84.6) include, first of all, the phenomenon well known as the *Doppler effect*: the change in colour of light emitted from a moving source. Thus, a light source at rest in the primed system emits in all directions, i.e. for all **n**′, the colour $v'$. Then, for an observer moving toward the light source with velocity $v$ parallel to the $x$-axis, the colour depends upon the observation direction **n**, or more accurately stated, upon the cosine $n_x$ of the angle between the direction of observation and the direction of motion. This interrelationship is most simply obtained from the first equation (84.6) by exchanging the unprimed quantities for the primed, and at the same time substituting $-\beta$ for $\beta$ (*relativity* of the two systems), giving

$$v = v'\frac{\sqrt{(1-\beta^2)}}{1-\beta n_x} \qquad (84.7)$$

This equation describes the Doppler effect for an *arbitrary* direction of motion. If, as a special case, the light source moves directly toward an observer, or directly away, then we obtain the *longitudinal* Doppler effect, familiar also from *elementary* physics. From (84.7), with $n_x = n'_x = \pm 1$, we have

$$v = v'\frac{\sqrt{(1-\beta^2)}}{1\mp\beta} \approx v'(1\pm\beta+\ldots) \qquad (84.7a)$$

A different expression is obtained for the case where the motion of the light source is perpendicular to the direction of observation ($n_x = 0$):

$$v = v'\sqrt{(1-\beta^2)} \qquad (84.7b)$$

Since this change of colour in the transverse Doppler effect is, in contrast to the longitudinal effect, *proportional to $\beta^2$*, its smallness has so far prevented experimental verification.

The assertions contained in the last three equations (84.6) concerning the change of direction of a light ray in the transfer to a moving coordinate system underlie the phenomenon of *aberration* in observing the light coming to us from a fixed star. Let a fixed star, for practical purposes at infinity, be at rest in the primed coordinate system and let it emit a light wave in the direction $n'_x = 0$, $n'_y = 1$, $n'_z = 0$. In the primed system the light wave is perpendicular to the direction of motion, in our case perpendicular to the direction of the earth's trajectory. For an observer moving with velocity $v$ in the negative $x'$-direction, but at rest on the earth, we have from (84.6) for the direction of the light ray:

$$\left.\begin{array}{l} n_x = \dfrac{\beta + n'_x}{1 + \beta n'_x} = \beta \\[3mm] n_y = \dfrac{n'_y \sqrt{(1-\beta^2)}}{1 + \beta n'_x} = \sqrt{(1-\beta^2)} \\[3mm] n_z = \dfrac{n'_z \sqrt{(1-\beta^2)}}{1 + \beta n'_x} = 0 \end{array}\right\} \qquad (84.8)$$

For this observer the wave front therefore appears inclined at an angle $\alpha$, that is

$$\tan \alpha = \frac{n_x}{n_y} = \frac{\beta}{\sqrt{(1-\beta^2)}} \quad \text{or} \quad \sin \alpha = \beta \qquad (84.9)$$

This tilting of the wave front is an immediate consequence of the Einstein definition of simultaneity. In the primed system the phase planes, i.e. the planes upon which the phase $\Phi'$ has a certain constant value at a certain time $t'$, are parallel to the $x'z'$-plane. Pictured as something objective and independent of the coordinate system, these planes are no longer considered as phase planes from the standpoint of the unprimed system. This is because on those planes, in the unprimed system, the phase $\Phi$ does not have the same value at all points at constant time—for, two events (in our case an event occurring when the phase acquires a given value at a certain point in space) which in one system take place at the same time, are found in another system to take place at different times.

Finally, we consider *the reflection of a plane light-wave by a moving mirror*, a physically important phenomenon upon which, among other derivations, the thermodynamic derivation of the Wien displacement law in radiation theory rests. The phenomenon follows in a simple way from a double application of the transformation expressions (84.6) if we consider the ordinary laws of reflection to hold in our own reference system. We imagine the plane of the mirror to be perpendicular to the $x$-axis. With respect to the unprimed system the mirror has velocity $v$ in the positive $x$-direction. A plane light-wave, of frequency $v_0$, whose propagation direction makes an angle $\theta_0$ with the $x$-axis, is incident upon this mirror. After some brief calculating we obtain for the frequency $v$ of the reflected wave and for the angle between its propagation direction and the $x$-axis, the relations

$$\left.\begin{aligned}
v &= v_0 \frac{1 - 2\beta \cos \theta_0 + \beta^2}{1 - \beta^2} \approx v_0 (1 - 2\beta \cos \theta_0) \\[2mm]
\cos \theta &= -\frac{\cos \theta_0 - 2\beta + \beta^2 \cos \theta_0}{1 - 2\beta \cos \theta_0 + \beta^2} \approx -\cos \theta_0 + 2\beta \sin^2 \theta_0 \\[2mm]
\sin \theta &= \frac{\sin \theta_0 (1 - \beta^2)}{1 - 2\beta \cos \theta_0 + \beta^2} \approx \sin \theta_0 + 2\beta \sin \theta_0 \cos \theta_0
\end{aligned}\right\} \quad (84.10)$$

(including their non-relativistic approximations for the case always realized in practice, namely where $\beta^2 \ll 1$).

With regard to the transformation properties of a light wave, we have so far made use only of the invariance of the phase against a Lorentz transformation. We wish now to investigate how the *amplitude of a light wave* changes in going from one coordinate system to another. We first of all establish, then, that on account of the relativistic invariance of the Maxwell equations in all coordinate systems, the **E** and **H** vectors of a light wave always remain at right angles to one another, and that if **E** and **H** are equal in one system they are equal in all others. Mathematically, this means that, as we have already established in §81, the expressions (**E.H**) and $\mathbf{E}^2 - \mathbf{H}^2$ are invariants against a Lorentz transformation; for plane waves both expressions have the value zero.

We shall now derive the transformation of the separate amplitudes for the special case of a *plane wave*, where both the wave normal and the electric vector lie in the $xy$-plane. If we designate the amplitude by $A$ and the direction cosines of the wave normals by $n_x$ and $n_y$, then the field strengths (84.1) of the wave are given by

$$H_z = A e^{i\Phi} \qquad E_x = -A n_y e^{i\Phi} \qquad E_y = A n_x e^{i\Phi} \quad (84.11)$$

Corresponding expressions hold for the primed system. Now, from §81, the following transformation relations hold between the unprimed and the primed field quantities:

$$H_z = \frac{H'_z + \beta E'_y}{\sqrt{(1-\beta^2)}} \qquad E_y = \frac{E'_y + \beta H'_z}{\sqrt{(1-\beta^2)}} \qquad E_x = E'_x \quad (84.12)$$

and out of these we obtain the following relations:

$$A = A'\,\frac{1+\beta n'_x}{\sqrt{(1-\beta^2)}} \qquad n_x A = A'\,\frac{n'_x + \beta}{\sqrt{(1-\beta^2)}} \qquad n_y A = A'n'_y \quad (84.13)$$

By dividing these equations by (84.6), identical transformation formulae for the direction of the wave normals are, on the one hand, obtained:

$$n_x = \frac{n'_x + \beta}{1 + \beta n'_x} \qquad n_y = \frac{n'_y\sqrt{(1-\beta^2)}}{1 + \beta n'_x} \quad (84.14)$$

On the other hand a comparison of the first equation (84.13) with the first equation (84.6) shows that the amplitude $A$ transforms like the frequency $\nu$:

$$\frac{A}{\nu} = \frac{A'}{\nu'} \quad (84.15)$$

We proceed now to a consideration of the *total energy of a wave train* in a given finite *volume moving with the wave*. This is

$$U = \frac{1}{8\pi}\int (\mathbf{E}^2 + \mathbf{H}^2)\,dV = \frac{1}{4\pi}\int \mathbf{E}^2\,dV \quad (84.16)$$

We wish to investigate how great is the energy $U'$ of the same wave train when observed from the *primed* system. We have already learned how the *amplitude* transforms. We have now to obtain the transformation of (moving) *volume*. Here we meet with the difficulty that the observed volume moves with the velocity of light, so that we cannot ascribe to it any rest volume. In order to get around this difficulty we consider a certain region of rest volume $V_0$ in motion with respect to the primed coordinate system with velocity $u'$, this velocity making an angle $\theta'$ with the $x'$-axis. Seen from the primed system the volume is

$$V' = V_0\sqrt{\left(1 - \frac{u'^2}{c^2}\right)} \quad (84.17)$$

It moves with respect to the unprimed system with a velocity $u$, given by the addition theorem of velocities, and for which in (77.7) we have found the value

$$u^2 = \frac{u'^2 + 2u'v\cos\theta' + v^2 - (u'^2v^2/c^2)\sin^2\theta'}{[1+(u'v/c^2)\cos\theta']^2}$$

In the unprimed system the region under consideration has the volume

$$V = V_0\sqrt{\left(1-\frac{u^2}{c^2}\right)} = V_0\frac{\sqrt{(1-u'^2/c^2)}\sqrt{(1-v^2/c^2)}}{1+(u'v/c^2)\cos\theta'} \quad (84.18)$$

Comparison with (84.17) shows that $V$ is connected with $V'$ through the relation

$$V = V'\frac{\sqrt{(1-\beta^2)}}{1+(u'/c)\beta\cos\theta'}$$

Without further trouble we can pass over in this equation to the limit $u' \to c$, obtaining the following *transformation relation for the volume moving with the velocity of light*:

$$V = V'\frac{\sqrt{(1-\beta^2)}}{1+\beta\cos\theta'} = V'\frac{\sqrt{(1-\beta^2)}}{1+\beta n'_x} \quad (84.19)$$

Comparing this expression with the frequency equation (84.6) we see that $V$ transforms like $1/v$, and that therefore $Vv$ is an invariant. Consequently from (84.16) we know that $U$ must transform like $v$, and so we obtain the relation

$$\frac{U}{v} = \frac{U'}{v'} \quad (84.20)$$

i.e. *energy values of a wave train as measured by different observers compare in the same way that frequencies are compared.*

This result is of special interest in the *quantum theory of radiation*, according to which light behaves in energy interactions with matter as if it consisted of discrete light quanta of energy $hv$, $h$ being Planck's constant. If we imagine the wave train to consist of $Z$ such quanta, then its energy would be $U = Zhv$. The invariance of the quantity $U/v$ requires that the product $Zh$ be also an invariant. $Z$ is of course invariant since it is an ordinary number. It therefore follows from our result (84.20) that *Planck's constant* is an invariant against a Lorentz transformation.

Along with the total energy we consider the *total momentum* of a plane light-wave, the magnitude and direction of which are given by the expression

$$\mathbf{J} = \frac{1}{4\pi c}\int \mathbf{E} \times \mathbf{H}\, dV = \frac{\mathbf{n}}{4\pi c}\int \mathbf{E}^2\, dV = \frac{\mathbf{n}U}{c} \qquad (84.21)$$

Designating the invariant $U/cv$ by $C$, we have now

$$C = \frac{1}{4\pi vc}\int \mathbf{E}^2\, dV \qquad (84.22)$$

and we can write the momentum components and the energy in the form

$$\mathbf{J} = \{J_1, J_2, J_3\} = C v \mathbf{n} \qquad J_4 = i\frac{U}{c} = Civ \qquad (84.23)$$

In accordance with the concluding considerations of the last section, this means that the total momentum and total energy can be united in a four-vector, since the quadruplet

$$(v n_x, v n_y, v n_z, iv) = (v_1, v_2, v_3, v_4) \qquad (84.24)$$

is to be looked upon as comprising the components of a four-vector. This follows not only from the transformation formulae (84.6) for these components, but we could have concluded this also from a consideration of the phase (84.2) which, by applying the symbolism of four-vectors, can be written

$$\Phi = -\frac{2\pi}{c}\sum_{i=1}^{4} v_i x_i \qquad (84.25)$$

Since the phase is an invariant, the four components $v_i$ must represent the four components of a four-vector. This vector, incidentally, has the special property that its magnitude $\sum_i v_i^2$ vanishes. The same then naturally also holds for the momentum-energy vector (84.23), and this requirement is again identical with the relation $J = U/c$.

## §85. The radiation field of a moving electron

As a special application of the foregoing discussion regarding the energy content and momentum content of wave trains, we examine the radiation of an electron in arbitrary motion. We have already investigated this radiation in §69 where, for the case in which the electron was at rest at the time of emission, we found for the change with time of its energy and its momentum

$$\frac{dU}{dt} = -S = -\frac{2e^2\dot{\mathbf{v}}_0^2}{3c^3} \qquad \frac{d\mathbf{J}}{dt} = 0 \qquad (85.1)$$

From these relationships we now wish to calculate the energy loss and momentum loss for an electron moving with velocity **v**. We shall succeed in doing this by writing in the four-dimensional form formulae (85.1), valid in the rest system of the electron, and then transforming these formulae to a moving coordinate system.

In order to facilitate the transfer to four-vectors (having introduced the proper time $\tau$ of the electron), we write (85.1) in the form

$$\frac{dU^0}{d\tau} = -\frac{2e^2\dot{\mathbf{v}}_0^2}{3c^3} \qquad \frac{d\mathbf{J}^0}{d\tau} = 0 \qquad (85.2)$$

We establish $-dU^0$ and $-d\mathbf{J}^0$ as the energy and momentum of the part of the radiation field emitted during the time interval $d\tau$. This part of the field radiates independently of the rest of the field and does not interfere with the radiation emitted previously or subsequently. We may therefore regard it as a closed wave train. For such, however, according to §83, the energy and momentum form a four-vector which for our case we shall designate by $-d\mathbf{J}_v$. We can then combine the two equations (85.2) into a four-dimensional vector equation

$$\frac{d\mathbf{J}_v}{d\tau} = -\frac{2e^2\dot{\mathbf{v}}_0^2}{3c^5}u_v \qquad (85.3)$$

which, in the rest system ($u_1^0 = u_2^0 = u_3^0 = 0$; $u_4^0 = ic$; $J_4^0 = iU^0/c$) goes over to equation system (85.2).

We have now to express the rest-system acceleration $\dot{\mathbf{v}}_0$ in terms of the acceleration $\dot{\mathbf{v}}$ in an arbitrary system. For this we recall the procedures of §80. There we recognized the acceleration as a space-like vector, so that by using (80.6) we can write

$$\dot{\mathbf{v}}_0^2 = \sum_{v=1}^{4} b_v^2 = \frac{\dot{\mathbf{v}}^2 - (\dot{\mathbf{v}} \times \mathbf{v}/c)^2}{(1-\beta^2)^3} \qquad (85.4)$$

Putting this expression into (85.3), introducing the time $t$ in place of $\tau$ (with $d\tau = dt\sqrt{(1-\beta^2)}$), we obtain in the usual form the two expressions

$$\frac{dU}{dt} = -\frac{2e^3}{3c^3}\frac{\dot{\mathbf{v}}^2 - (\dot{\mathbf{v}} \times \mathbf{v}/c)^2}{(1-\beta^2)^3} \qquad \frac{d\mathbf{J}}{dt} = -\frac{2e^2\mathbf{v}}{3c^5}\frac{\dot{\mathbf{v}}^2 - (\dot{\mathbf{v}} \times \mathbf{v}/c)^2}{(1-\beta^2)^3} \qquad (85.5)$$

in agreement with the corrections given in §69 to formulae (85.1) for the case of a finite electron velocity.

We wish to consider briefly, and to formulate with relativistic invariance, the auxiliary conditions which have made the differentiations in §69 so detailed, namely the prescription required by retardation that for the field at a place $\mathbf{r}$ and time $t$, the electron's location and condition of its velocity at time $t' = t - |\mathbf{r} - \mathbf{r}'|/c$ are the definitive data. This prescription is identical with the requirement

$$\sum R_\nu{}^2 = (\mathbf{r} - \mathbf{r}')^2 - c^2(t - t')^2 = 0 \qquad (85.6)$$

in which $R_\nu$ means the coordinates of the four-dimensional distance vector from the source point $(x', y', z', ict')$ to the field point $(x, y, z, ict)$.

Had we attempted to solve the problem treated in §69—the calculation of the field of an arbitrarily moving electron—by using four-dimensional formalism in addition to the formulae of §81, and had we thus in particular proceeded from the potential equation (81.6), we should have met with considerable mathematical difficulty; the key to the solution in the three-dimensional case was the application of *Green's theorem* using the basic solution $f = 1/r$ of Laplace's equation $\nabla^2 f = 0$. The four-dimensional counterpart of $f = 1/r$ is $F = 1/R^2$, the solution of $\Box F = 0$, in which $\mathbf{R} \equiv \{R_\nu\}$ again means the four-dimensional distance vector. In using Green's theorem in four dimensions, however (in contrast to the three-dimensional case) there enter infinitely large contributions from the surface integrals—for the potentials do not vanish for $t \to \infty$ on account of the constancy of total charge in space, which influences a field even after an infinitely long time. A further difficulty is that the basic solution $1/R^2$ in four dimensions is not only infinitely great at the field point (like the solution $1/r$ in three dimensions) but is also infinitely great on the three-dimensional "conical surface" $R^2 \equiv r^2 - c^2t^2 = 0$. For four-dimensional potential equations recourse must be had to other methods of integration, specifically, by going back to solution procedures already discussed in §66 and §69.

# CHAPTER E III

## The Relativistic Electrodynamics of Material Bodies

### §86. The field equations

Proceeding from the conceptions of the electron theory, we have derived the field equations in chapter DIV for slowly moving bodies. Even in the special case of non-magnetizable bodies treated there a detailed consideration of the motions of the electrons in matter was necessary. Now it is a very noteworthy achievement of relativity theory that, as Minkowski first showed, the field equations, *solely* from the requirement of their invariance under Lorentz transformations, can be written down for *moving* bodies as soon as they are known for the *rest* case, and this without having recourse to any additional atomic hypotheses.

We know the field equations for the rest case. They are identical with the *usual Maxwell equations*. In the rest system, which we designate by the superscript $^0$, they are:

$$\operatorname{curl} \mathbf{H}^0 = \frac{1}{c}\frac{\partial \mathbf{D}^0}{\partial t} + \frac{4\pi}{c}\mathbf{g}^0 \qquad \operatorname{div} \mathbf{D}^0 = 4\pi\rho^0$$

$$\operatorname{curl} \mathbf{E}^0 = -\frac{1}{c}\frac{\partial \mathbf{B}^0}{\partial t} \qquad \operatorname{div} \mathbf{B}^0 = 0$$

(86.1)

In addition there enter the constitutive equations

$$\mathbf{D}^0 = \varepsilon\mathbf{E}^0 \qquad \mathbf{B}^0 = \mu\mathbf{H}^0 \qquad (86.2)$$

and

$$\mathbf{g}^0 = \sigma\mathbf{E}^0 \qquad (86.3)$$

if we limit ourselves to bodies whose properties are uniquely described by the three constants relating to material media, namely $\sigma$ (electrical conductivity), $\varepsilon$ (permittivity), and $\mu$ (magnetic permeability).

In setting up the field equations for an arbitrary coordinate system we, like Minkowski, allow ourselves to be guided by analogy with the corresponding vacuum equations (§81) to which in the absence of matter (i.e. for $\varepsilon = \mu = 1$; $\sigma = 0$) our desired field equations must revert. We have seen in §81 that for the case of a vacuum the field equations can be stated in four-dimensional form if we combine the field quantities **E** and **H** in a certain way in a skew-symmetric tensor. In our case, however, there result not two but four field vectors with altogether twelve components, and there is the obvious suggestion that instead of introducing only *one* field tensor, two should be introduced. Now in those equations (86.1) in which the charge enters, only **H** and **D** appear, and these would be combined in such a tensor. The other two quantities **E** and **B** appearing in the remaining equations (86.1) would form a second tensor. Accordingly, we define two tensors $F_{\nu\mu}$ and $H_{\nu\mu}$ as follows:

$$F_{\nu\mu} = \begin{pmatrix} 0 & B_z & -B_y & -iE_x \\ -B_z & 0 & B_x & -iE_y \\ B_y & -B_x & 0 & -iE_z \\ iE_x & iE_y & iE_z & 0 \end{pmatrix} \quad H_{\nu\mu} = \begin{pmatrix} 0 & H_z & -H_y & -iD_x \\ -H_z & 0 & H_x & -iD_y \\ H_y & -H_x & 0 & -iD_z \\ iD_x & iD_y & iD_z & 0 \end{pmatrix} \quad (86.4)$$

It is easily established that for the rest system the two equation systems

$$\sum_{\mu=1}^{4} \frac{\partial H_{\nu\mu}}{\partial x_\mu} = \frac{4\pi s_\nu}{c} \qquad \frac{\partial F_{\nu\mu}}{\partial x_\lambda} + \frac{\partial F_{\mu\lambda}}{\partial x_\nu} + \frac{\partial F_{\lambda\nu}}{\partial x_\mu} = 0 \qquad (86.5)$$

agree with equations (86.1) if, in this system, the current vector has the components $s_\nu^0$, with

$$s_\nu^0 = (g_x^0, g_y^0, g_z^0, ic\rho^{\,0}) \qquad (86.6)$$

In the four-dimensional form the relationships (86.5) thus represent an equation system which, in the rest coordinate system, is identical with the Maxwell equations (86.1). In accordance with the basic principles of relativity theory, however, equations (86.5) must be valid in *all* coordinate systems; thus they represent the desired *equations for moving bodies*. *In vacuo* they agree with the equation system (81.15) since in this case, as indeed follows from the Maxwell equations and as will be fully shown later, **E** becomes equal to **D**, **B** becomes equal to **H**, and thus the two tensors $F_{\nu\mu}$ and $H_{\nu\mu}$ become identical.

Through (86.4), therefore, for all coordinate systems, we have defined field vectors **E**, **B**, **H**, **D**, their connection with the corresponding vectors in the rest system being determined by the tensor properties of $F_{\nu\mu}$ and $H_{\nu\mu}$. Thus we have the equations:

$$E_x = E_x^0 \qquad\qquad B_x = B_x^0$$

$$E_y = \frac{1}{\sqrt{(1-\beta^2)}}(E_y^0 + \beta B_z^0) \qquad B_y = \frac{1}{\sqrt{(1-\beta^2)}}(B_y^0 - \beta E_z^0)$$

$$E_z = \frac{1}{\sqrt{(1-\beta^2)}}(E_z^0 - \beta B_y^0) \qquad B_z = \frac{1}{\sqrt{(1-\beta^2)}}(B_z^0 + \beta E_y^0)$$

$$D_x = D_x^0 \qquad\qquad H_x = H_x^0 \qquad\qquad (86.7)$$

$$D_y = \frac{1}{\sqrt{(1-\beta^2)}}(D_y^0 + \beta H_z^0) \qquad H_y = \frac{1}{\sqrt{(1-\beta^2)}}(H_y^0 - \beta D_z^0)$$

$$D_z = \frac{1}{\sqrt{(1-\beta^2)}}(D_z^0 - \beta H_y^0) \qquad H_z = \frac{1}{\sqrt{(1-\beta^2)}}(H_z^0 + \beta D_y^0)$$

The electric field strength **E** is not determined by the electric and magnetic field strengths in the rest system of the body, but from the electric field strength and the magnetic *induction* in this system, fully corresponding to the expression for the Lorentz force which we have often used.

When terms of order $\beta^2$ are neglected, the transformation equations (86.7) assume the simple form:

$$\mathbf{E} = \mathbf{E}^0 - \frac{\mathbf{v}}{c} \times \mathbf{B}^0 \qquad \mathbf{B} = \mathbf{B}^0 + \frac{\mathbf{v}}{c} \times \mathbf{E}^0$$

$$\mathbf{D} = \mathbf{D}^0 - \frac{\mathbf{v}}{c} \times \mathbf{H}^0 \qquad \mathbf{H} = \mathbf{H}^0 + \frac{\mathbf{v}}{c} \times \mathbf{D}^0 \qquad (86.8)$$

The *force density* $f_i$ for a moving charge element derives from the three-dimensional force density for charges at rest, thus

$$\mathbf{f}^0 = \rho^0 \mathbf{E}^0$$

In four-dimensional formalism this equation is

$$f_\nu = \frac{1}{c} \sum_{\mu=1}^{4} F_{\nu\mu} s_\mu \qquad (86.9)$$

According to (86.4) and (86.6) the vector of the three-dimensional force density is therefore

$$\mathbf{f} = \rho\mathbf{E} + \frac{\mathbf{g}}{c} \times \mathbf{B}$$

in agreement with classical electrodynamics.

The *four-current* $s_\nu$ requires special consideration. From (86.6), its three space-like components in the rest system constitute the *conduction current* $\mathbf{g}^0$, while its time-like component is proportional to the *charge density* $\rho^0$. For an arbitrary coordinate system with respect to which the body moves in the $x$-direction with velocity $v$, we expect, in addition to the conduction current, the appearance of a *convection current* occasioned by the motion of the charge, whose density is $\rho^0$. We therefore make a distinction between the *conduction part* $(s_\nu)_L$ and the *convection part* $(s_\nu)_K$ of the current, the sum of the two being the total current $s_\nu$. If we follow the individual parts of these two currents in the transfer from the rest system to a moving system we obtain the following scheme, based on the Lorentz transformation:

|  | Convection Current | | Conduction Current | |
|---|---|---|---|---|
|  | Rest System | Moving System | Rest System | Moving System |
| $s_1$ | $0$ | $\dfrac{\rho^0 v}{\sqrt{(1-\beta^2)}} = (g_x)_K$ | $g_x{}^0$ | $\dfrac{g_x{}^0}{\sqrt{(1-\beta^2)}} = (g_x)_L$ |
| $s_2$ | $0$ | $0$ | $g_y{}^0$ | $g_y{}^0 = (g_y)_L$ |
| $s_3$ | $0$ | $0$ | $g_z{}^0$ | $g_z{}^0 = (g_z)_L$ |
| $s_4$ | $ic\rho^0$ | $\dfrac{ic\rho^0}{\sqrt{(1-\beta^2)}} = ic\rho_K$ | $0$ | $\dfrac{i\beta g_x{}^0}{\sqrt{(1-\beta^2)}} = ic\rho_L$ |

(86.10)

With respect to the convection current there is nothing new to mention; it agrees with the current considered in §70. In regard to the conduction current, however, there is the surprising appearance of a *time-like* component which means that *every current-carrying conductor appearing to a co-moving observer to be uncharged* ($\rho^0 = 0$), *carries an electrical charge of density*

$$\rho_L = \frac{\beta g_x^0}{c\sqrt{(1-\beta^2)}} = \frac{\mathbf{v}\mathbf{g}_L}{c^2} \qquad (86.11)$$

This result is an immediate consequence of the Einstein definition of simultaneity and is clearly understandable from it. Let us consider a metal rod at rest, through which an electric current flows in the longitudinal direction. There are in this rod positive ions at rest, and electrons moving in a direction opposite to the current. Let us draw (figure 70) the world lines of the electrons and ions in the $x^0ct^0$-plane. Thus for the ions we obtain the dashed lines which are parallel to the $ct^0$-axis. For the electrons, however, we have the slanted solid lines. On account of the rod's electrical neutrality, there naturally issue from a finite section of the rod equal numbers (on the average) of world lines

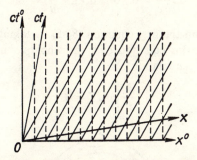

Fig. 70.—Conceptual interpretation of the charge density according to von Laue, making use of the world lines for the positive ions (dashed lines) and the negative electrons (solid lines)

of each kind. We now consider this world-line picture from a moving coordinate system with axes $Ox$ and $Oct$. We see that in a given section of the $x$-axis there are no longer equal numbers of electrons and ions to be found. In the case illustrated in the figure only ten electrons are associated with something like eleven ions, so that in this region the rod appears to possess a net positive charge.

In §81, in the context of electron theory, we stated the rule that the *total charge* $\int \rho \, dV$ in a given region *is invariant under a Lorentz transformation*. This rule yielded the fact that in going over from the rest system to a moving system the charge density becomes greater in the ratio $1:\sqrt{(1-\beta^2)}$, the volume however becoming smaller in the proportion $\sqrt{(1-\beta^2)}:1$ because of the Lorentz contraction. This rule naturally holds for an *electric circuit*, provided we speak of the total charge of the complete closed circuit. It does not, however, hold for a *part* of a current-carrying wire, since the differing simultaneity in systems in relative motion also leads to differing statements about the amount of charge contained in a given part of the wire. That the total charge of a closed system is, however, a relativistic invariant and is thus constant in time, follows immediately from electron theory— for the *charge* of single ions and electrons *has the same value* $\pm e$ in *all systems*; in other words the *number* of ions and electrons in a closed system does not vary. Therefore the total charge in all reference systems must be the same, and consequently constant in time. The last follows, incidentally, from the equation of continuity of charge:

$$\sum_i \frac{\partial s_i}{\partial x_i} = 0 \quad \text{or} \quad \text{div } \mathbf{g} + \frac{\partial}{\partial t}\rho = 0 \tag{86.12}$$

which follows from the first equation (86.5) by taking the divergence. By integration of this equation over the whole volume of the closed system we obtain the relation

$$\frac{d}{dt}\int \rho \, dV = 0 \tag{86.13}$$

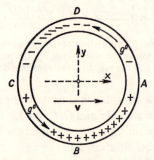

Fig. 71.—Charge density in a current-carrying metal ring in motion

A system which, for a co-moving observer, is *uncharged* ($\rho^0 = 0$) carries no net charge for any other observer; for *every* closed system the volume integral over the charge density of the conduction current vanishes,

$$\int \rho_L \, dV = 0 \tag{86.14}$$

even when there is a net charge in the rest system.

By way of illustration we consider a metal ring (figure 71) lying in the $xy$-plane and carrying a current $I^0$. When this ring moves in the direction of the positive $x$-axis with the velocity $\mathbf{v}$, then, according to the results just obtained, it acquires a charge density which is net positive on the half-circle ABC, and negative throughout CDA, but such that the total charge (upon which the current density depends) is equal to zero. The ring possesses an *electric moment* whose direction is perpendicular to the velocity and to the axis of the ring current. We shall meet this phenomenon again later, finding that with every *moving magnetic dipole* $\mathbf{m}$ (of which, as regards the magnetic field produced at large distances, our circular ring can be considered an example) there is necessarily associated an *electric dipole* given by the expression

$$\mathbf{p} = \frac{\mathbf{v}}{c} \times \mathbf{m} \tag{86.15}$$

The foregoing treatment is incomplete in that we have yet to consider what form the *constitutive equations* (86.2) and (86.3) take in any arbitrary system. They can of course be obtained by introducing the transformation formulae (86.7) in the constitutive equations for the rest system. We wish however to choose the way, already used several times in these paragraphs, of transcribing in four-dimensional formalism the equations valid in the rest system.

We begin with the transcribing of *Ohm's law* and we must find an equation between the conduction current $(s_v)_L$ and the field quantities which, for the rest system, agrees with (86.3). We assert that the equation

$$(s_v)_L = \frac{\sigma}{c} \sum_\mu F_{v\mu} u_\mu \tag{86.16}$$

where $u_\mu$ is the four-velocity of the matter, satisfies this requirement: it is a *bona fide* four-dimensional equation and, as may be easily shown, it goes over in the rest system to equation (86.3). In ordinary vector formulation we have for the first three components

$$\mathbf{g}_L = \frac{\sigma}{\sqrt{(1-\beta^2)}} \left( \mathbf{E} + \frac{\mathbf{v}}{c} \times \mathbf{B} \right) \tag{86.16a}$$

while from the fourth equation it follows that

$$ic\rho_L = \frac{i\sigma}{\sqrt{(1-\beta^2)}} \mathbf{E} \frac{\mathbf{v}}{c} = i\mathbf{g}_L \frac{\mathbf{v}}{c} \tag{86.16b}$$

This is the equation already found in (86.10) for the charge density of the conduction current.

Further, it is easily shown that the two *constitutive equations*

$$\left.\begin{array}{c} \sum_{\mu} H_{\nu\mu} u_{\mu} = \varepsilon \sum_{\mu} F_{\nu\mu} u_{\mu} \\[2mm] F_{\nu\mu} u_{\lambda} + F_{\mu\lambda} u_{\nu} + F_{\lambda\nu} u_{\mu} = \mu(H_{\nu\mu} u_{\lambda} + H_{\mu\lambda} u_{\nu} + H_{\lambda\nu} u_{\mu}) \end{array}\right\} \quad (86.17)$$

represent the correct four-dimensional equations for the two relations (86.2). Through (86.17), the vectors **E**, **H**, **D**, **B** for any arbitrary system are connected with one another in a proportionately more complicated fashion. In ordinary vector formulation they read

$$\mathbf{D} + \frac{\mathbf{v}}{c} \times \mathbf{H} = \varepsilon\left(\mathbf{E} + \frac{\mathbf{v}}{c} \times \mathbf{B}\right), \quad \mathbf{B} - \frac{\mathbf{v}}{c} \times \mathbf{E} = \mu\left(\mathbf{H} - \frac{\mathbf{v}}{c} \times \mathbf{D}\right) \quad (86.18)$$

Solving for the two vectors **D** and **B** we obtain

$$\mathbf{D} = \frac{1}{1-\varepsilon\mu\beta^2}\left\{\varepsilon\mathbf{E}(1-\beta^2) + (\varepsilon\mu-1)\left[\frac{\mathbf{v}}{c} \times \mathbf{H} - \varepsilon\frac{\mathbf{v}}{c}\left(\frac{\mathbf{v}}{c}\mathbf{E}\right)\right]\right\}$$

$$\mathbf{B} = \frac{1}{1-\varepsilon\mu\beta^2}\left\{\mu\mathbf{H}(1-\beta^2) - (\varepsilon\mu-1)\left[\frac{\mathbf{v}}{c} \times \mathbf{E} - \mu\frac{\mathbf{v}}{c}\left(\frac{\mathbf{v}}{c}\mathbf{H}\right)\right]\right\} \quad (86.19)$$

as can be easily shown either by direct calculation or by substituting (86.19) in (86.18). We know from (86.19) that as stated earlier, *for a vacuum* (where $\varepsilon = \mu = 1$), **D** becomes equal to **E**, and **B** equal to **H**, so that in this case the two field tensors $F_{\nu\mu}$ and $H_{\nu\mu}$ are identical.

## §87. The moments tensor

The means by which the results of the foregoing paragraphs have been obtained from the Maxwell equations for media at rest have been purely formal: the sole basis was the requirement of relativistic invariance. There was no mention made of the *dielectric polarization* **P** or *the magnetization* **M**, both being indispensable concepts in any atomistic representation. We now wish to examine these quantities from the standpoint of relativity theory and to investigate the result obtained to the extent that it provides an intuitive interpretation within the framework of *electron theory*. For this purpose we define the *moments tensor* $M_{\nu\mu}$ by the equation

$$F_{\nu\mu} = H_{\nu\mu} + 4\pi M_{\nu\mu} \quad (87.1)$$

If for the components of this new tensor the designation given by the following scheme be introduced,

$$M_{\nu\mu} = \begin{pmatrix} 0 & M_z & -M_y & iP_x \\ -M_z & 0 & M_x & iP_y \\ M_y & -M_x & 0 & iP_z \\ -iP_x & -iP_y & -iP_z & 0 \end{pmatrix} \qquad (87.2)$$

we see immediately that (87.1) is identical with the equations

$$\mathbf{B} = \mathbf{H} + 4\pi\mathbf{M}, \qquad \mathbf{E} = \mathbf{D} - 4\pi\mathbf{P} \qquad (87.3)$$

of the Maxwell theory.

The four-dimensional presentation at once yields the general transformation formulae for the polarization and the magnetization. We designate these quantities, as measured by an observer moving with the matter, by $\mathbf{P}^0$ and $\mathbf{M}^0$. Then, according to (80.18), another observer with respect to whom the matter moves in the x-direction with velocity $v$, will obtain for the polarization and the magnetization the values:

$$P_x = P_x^0 \qquad\qquad M_x = M_x^0$$

$$P_y = \frac{1}{\sqrt{(1-\beta^2)}}(P_y^0 - \beta M_z^0) \qquad M_y = \frac{1}{\sqrt{(1-\beta^2)}}(M_y^0 + \beta P_z^0) \quad (87.4)$$

$$P_z = \frac{1}{\sqrt{(1-\beta^2)}}(P_z^0 + \beta M_y^0) \qquad M_z = \frac{1}{\sqrt{(1-\beta^2)}}(M_z^0 - \beta P_y^0)$$

These formulae provide an interesting connection between the two three-dimensional vectors under consideration. A body which, in its own rest system, is electrically polarized but *not magnetized*, appears to a moving observer to be not only polarized but *magnetized* as well. Conversely, a body which appears in the rest system to be only *magnetized* (such as a magnetized iron rod) will, owing to its motion, appear to a moving observer to carry an *electric moment*.

For a closer look into these matters we examine first the case of a body which *in its rest system is electrically polarized*, but *not magnetized*. Our earlier consideration (§70) showed that this polarization $\mathbf{P}$, through motion, implies a contribution $\partial\mathbf{P}/\partial t + \mathrm{curl}(\mathbf{P} \times \mathbf{v})$ to the total current; thus the first three Maxwell equations (for slowly moving bodies) can be written in the form

$$\mathrm{curl}\left(\mathbf{B} - 4\pi\mathbf{P} \times \frac{\mathbf{v}}{c}\right) = \frac{1}{c}\frac{\partial}{\partial t}(\mathbf{E} + 4\pi\mathbf{P}) + \frac{4\pi}{c}(\mathbf{g} + \rho\mathbf{v}) \qquad (87.5)$$

We arrive at this same result with the help of our newly obtained transformation formulae (87.4) if in them we let $\mathbf{M}^0 = 0$ and if in addition we restrict ourselves to small velocities (i.e. where $\beta^2 \ll 1$):

$$\mathbf{P} = \mathbf{P}^0 \qquad \mathbf{M} = \mathbf{P}^0 \times \frac{\mathbf{v}}{c} \tag{87.6}$$

Putting this result into the Maxwell equation (87.5), and taking into account the relationship between $\mathbf{B}$ and $\mathbf{H}$ given by equation (87.3), we obtain the simple expression

$$\operatorname{curl} \mathbf{H} = \frac{1}{c} \frac{\partial \mathbf{D}}{\partial t} + \frac{4\pi}{c} (\mathbf{g} + \rho \mathbf{v})$$

being the first field equation for slowly moving unmagnetized bodies. This expression is in complete agreement with our earlier considerations (§70) based on electron theory. It should be noticed that even in non-magnetizable bodies the vectors $\mathbf{B}$ and $\mathbf{H}$ are no longer identical; rather they differ by the magnetization $\mathbf{P} \times \mathbf{v}/c$ which itself is different from zero.

We consider now a body which in its rest system has *no electric moment* but has a *magnetic moment*—as, for example, a moving permanent magnet. In this case formulae (87.4) furnish a new result, one not expected from classical electron theory. If here again we restrict ourselves to terms of the first order in $v/c$ we find from (87.4) that a *moving permanent magnet* is accompanied by *an electric moment* given by

$$\mathbf{P} = \frac{\mathbf{v}}{c} \times \mathbf{M}^0 \tag{87.7}$$

We wish to render plausible the existence of this effect by considering a single atom possessing a magnetization electron which moves in a circular orbit. Let the trajectory of this electron be given in the rest system of the atom by the equations

$$x' = r' \cos \omega'(t' - t_0) \qquad y' = r' \sin \omega'(t' - t_0) \tag{87.8}$$

The orbit, lying in the $x'y'$-plane, has radius $r'$; the circular frequency of the electron is $\omega'$. The world line of the electron is pictured in figure 72, which, however, gives only the projection of the world line on the $x'ct'$-plane. The $y'$-axis has to be imagined as being perpendicular to the plane of the paper. The points $0', 1', 2', \ldots$ given in figure 72 are the places where the electron goes through the $x'ct'$-plane, i.e. places where $y' = 0$. $y'$ itself is therefore positive from $1'$ to $2'$; it is negative from $0'$ to $1'$, and from $2'$ to $3'$, etc. The projections of these points on the $ct'$-axis are naturally equidistant, for the electron remains just as long in the region of positive $y'$ as in that of negative $y'$. The *time average of $y'$* is equal to *zero*.

If now we pass over to a system with respect to which the atom moves with velocity $v$ in the $x$-direction, then, according to Minkowski, this entails a newly oriented motion diagram, the new axes being rotated with respect to the old through an angle $\phi = \tan^{-1}\beta$. The $y$-coordinates are not however changed; the points $0'$, $1'$, $2'$, ... are also for the new system the points where the electron passes through the $xct$-plane. The time values for these transits are, however, altered, as can be

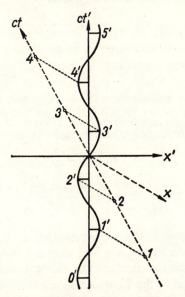

Fig. 72.—Observation of circular motion (87.8) in two coordinate systems in motion with respect to one another

seen from the projection of these points on the $t$-axis. The electron takes longer, for example, to go from 2 to 3 than from 1 to 2. Over a complete cycle, therefore, the *time average* $\bar{y} = \dfrac{1}{T}\displaystyle\int_0^T y\,dt$ comes out *negative*. This result of a time average different from zero is just another way of saying that an electric moment exists—for the nucleus always lies in the $xct$-plane ($y = 0$), while the electron on the average lies in the plane given by $y = \bar{y}$. The magnitude $p$ of the dipole is $p = -e_0\bar{y}$; the direction for the case considered is that of the $y$-axis, i.e. perpendicular to the direction of motion.

The resulting electric moment is easily calculated. For this we need merely apply the Lorentz transformation to the equations of motion (87.8); for the relationship between $t$ and $t'$ we then obtain

$$t = \frac{t' + \dfrac{vx'}{c^2}}{\sqrt{(1-\beta^2)}} = \frac{t' + \dfrac{vr'}{c^2}\cos\omega' t'}{\sqrt{(1-\beta^2)}}$$

and therefore

$$\frac{dt}{dt'} = \frac{1 - \dfrac{\omega' v r'}{c^2} \sin \omega' t'}{\sqrt{(1 - \beta^2)}}$$

Since $y = y' = r' \sin \omega' t'$ and $T = T'/\sqrt{(1 - \beta^2)}$, with $T' = 2\pi/\omega'$, the time-average of $y$ in the new coordinate system is equal to

$$\bar{y} = \frac{1}{T} \int_0^{T'} y \frac{dt}{dt'} dt' = -\frac{\omega' v r'^2}{2c^2}$$

On multiplying by the electric charge, we obtain for the *electric moment of the moving atom* the value

$$p_y = -e_0 \bar{y} = \frac{e_0 v r'^2 \omega'}{2c^2}$$

The magnetic moment associated with the circular motion (87.8) is given by the simple relationship $m = IS/c$, where $S$ is the area of the electron orbit and $I$ is the quantity of charge which flows per second across any cross-section cutting through the electron orbit. Accordingly, the magnetic moment of the atom is

$$m = m_z = -\frac{e_0 r'^2 \omega'}{2c}$$

The electric moment is then connected with the magnetic moment through the relationship

$$p_y = -m_z \frac{v}{c} \qquad \mathbf{p} = \frac{\mathbf{v}}{c} \times \mathbf{m} \qquad (87.9)$$

If, again, we compare the two symmetrical results (87.6) and (87.7) of relativistic reasoning, we see that although (87.6) is immediately understandable from the standpoint of classical electron theory, (87.7) very definitely requires the Einstein concept of time. As can be seen at once from figure 72, adherence to the concept of absolute simultaneity would never lead to an electric moment such as that required by equation (87.7). This stands as an exact parallel to the fact that the magnetic field of a moving polarized medium can be directly calculated from the Maxwell equations by a careful consideration of the transport of charge, while for calculating the effective E-field for motion with respect to a magnet we need the concept of the Lorentz force which is foreign to the Maxwell theory but is derivable by relativity theory.

## §88. Unipolar induction

In §71 we learned about Eichenwald's investigation. This provided us with the proof that the motion of a polarized dielectric produces a magnetic field. The magnetization

$$\mathbf{M} = \mathbf{P}^0 \times \frac{\mathbf{v}}{c} \qquad (88.1)$$

comes from the dielectric which in the rest system is unmagnetized. While this effect can be readily understood on the basis of *ordinary electron* theory, there is the *inverse effect*, namely the electrical polarization of a moving magnet, given by the equation

$$\mathbf{P} = \frac{\mathbf{v}}{c} \times \mathbf{M}^0 \qquad (88.2)$$

which is a direct consequence of relativity theory. An experimental demonstration of the reality of this second effect has apparently not been reported in the literature. From this we might suppose that the effect has escaped observation on account of its feebleness. Actually,

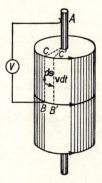

Fig. 73.—Schematic diagram of the unipolar machine

however, the effect corresponding to (88.2) has been known in the technical field for a long time under the name *unipolar induction*. The polarization (88.2) is not a phenomenon to be detected only by apparatus of very high sensitivity; rather it permits the production of currents measured in thousands of amperes. From the usual descriptions in the technical literature, however, it is seldom clear that unipolar induction is basically a relativistic effect falling within the compass of equation (88.2).

In technology the unipolar machine often takes the form of a cylindrical iron body which rotates on its axis and is magnetized parallel to it (figure 73). By means of two brush contacts, A (on the axis), and B (on the equator), a current can be taken whose e.m.f. is customarily calculated according to the following prescription: The induction law

$$\oint \mathbf{E} \, d\mathbf{s} = -\frac{1}{c}\frac{d}{dt}\int B_n \, dS \qquad (88.3)$$

is applied to the material integration path ACBVA. In the time $dt$ this goes over to the path AC′B′VA, wherein the angle CAC′ is equal to $\omega dt$ ($\omega$ is the angular velocity of rotation). The increase of flux $\int B_n \, dS$ in time $dt$ is just that flux which passes through the sector of the iron surface ACBB′C′A. The e.m.f. is then equal to

$$V^{(e)} = -\frac{1}{c}\int_A^B (\mathbf{v} \times d\mathbf{s})\mathbf{B} = -\frac{\omega}{c}\int_A^B r \,|\,\mathbf{B} \times d\mathbf{s}\,| \qquad (88.4)$$

The integration path from A to B lies in the meridian plane containing both these points; $r$ is the distance of the path element $d\mathbf{s}$ from the axis of rotation. Within the meridian plane the path of integration can be arbitrarily chosen. This is because of the source-free nature of $\mathbf{B}$ as well as the fact that the direction of $\mathbf{B}$ is always in the meridian plane.

The same value for the potential between A and B, incidentally, can also be obtained by a consideration of the free electrons in metal. These electrons take part in the motion $\mathbf{v}$ of the rotating metal body, and the Lorentz force $e\mathbf{v} \times \mathbf{B}/c$ acts on them. Equilibrium can exist, however, only when throughout the volume of the metal this force is compensated by an electric field $\mathbf{E} = -\mathbf{v} \times \mathbf{B}/c$. Between the points A and B, a potential difference must therefore exist, given by

$$V^{(e)} = \phi_B - \phi_A = -\int_A^B \mathbf{E} \, d\mathbf{s} = \frac{1}{c}\int_A^B (\mathbf{v} \times \mathbf{B}) \, d\mathbf{s} = -\frac{1}{c}\int_A^B (\mathbf{v} \times d\mathbf{s}) \, \mathbf{B}$$

in agreement with the result obtained above.

In what follows this unquestionably correct procedure for calculating the electrical action will be carried through, not for the case of a rotating magnet, but, in the interest of having a broader perspective, for an iron bar in motion of *simple translation*. Our aim will be primarily to understand how the electric field comes into being. We consider a very long iron bar, of rectangular cross-section (figure 74). Its long axis lies in the $x$-direction. Let it be magnetized in the $y$-direction so that its upper surface is a north pole and its lower surface a south pole. Whenever this bar moves in the $x$-direction with velocity

$v$, it can be used as a unipolar machine. Between the two brush contacts A and B there is an e.m.f. produced, of magnitude

$$V^{(e)} = -\frac{1}{c}\int_A^B (\mathbf{v} \times d\mathbf{s})\mathbf{B} = -\frac{v}{c}\int_A^B B_n\,ds$$

Instead of this we may ask what electric field there is in the neighbourhood of the moving bar. For the set-up we are considering, a particularly simple form of the relativity principle gives the following

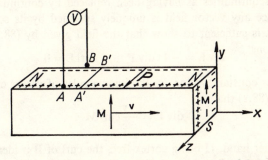

Fig. 74.—Unipolar induction in an arrangement involving translational motion. The electrical polarization of a moving magnet

answer: On a charge $e$ at rest, the motion of the bar produces the same force as when the bar is at rest and the charge moves with velocity $-\mathbf{v}$. In this latter case, however, the force is equal to $-e\mathbf{v} \times \mathbf{B}/c$. Thus the moving bar will give occasion for the existence of an electrical field strength

$$\mathbf{E} = -\frac{\mathbf{v}}{c} \times \mathbf{B} \qquad (88.5)$$

We say that it must come to the same thing whether the charges move against the force lines or the force lines move against the charges. Thus we imagine for the latter case that the force lines of moving magnets move with them as if they were rigidly attached needle spines. This description is not at present reconcilable, however, with any reasonable field theory—for the field **B** which the long moving bar produces where the charge is located is *constant in time* throughout. By a measurement of **B** in the neighbourhood of the charge it is completely impossible to decide whether the bar is stationary or moving. While the existence of this field (88.5) is not questioned, its origin in this description remains quite problematical. Actually this field is of a *purely electrostatic nature*; *it proceeds from the electrical polarization* of the magnet, as required by (88.2). In the illustration (figure 74) this polarization is so directed that the front face of the bar is positively charged, and the rear face negative.

We now convince ourselves that the field produced by these charges is to be directly described by (88.5). To simplify the proof we consider **M** and **P** to be continuous functions of position; also, we consider any surface discontinuities as having been replaced by continuous transitions. Since any vector field is uniquely specified by its sources and vortices, it is sufficient to show that the field given by (88.5) satisfies the equations

$$\text{div}\,\mathbf{E} = -4\pi\,\text{div}\,\mathbf{P} \qquad \text{curl}\,\mathbf{E} = 0 \qquad (88.6)$$

Since **v** is a constant vector, by the known rules of vector calculus we find from (88.2) that

$$\text{div}\,\mathbf{P} = -\frac{\mathbf{v}}{c}\,\text{curl}\,\mathbf{M}$$

On the other hand, **H** being vortex-free, the curl of **B** is identical with the curl of $4\pi\mathbf{M}$. Consequently, according to (88.5),

$$\text{div}\,\mathbf{E} = \frac{\mathbf{v}}{c}\,\text{curl}\,\mathbf{B} = 4\pi\frac{\mathbf{v}}{c}\,\text{curl}\,\mathbf{M}$$

Thus the first of equations (88.6) is proved. To prove the second equation of (88.6) we make use of the generally valid vector equation (**v** being constant):

$$-\text{curl}\left(\frac{\mathbf{v}}{c} \times \mathbf{B}\right) = -\frac{\mathbf{v}}{c}\,\text{div}\,\mathbf{B} + \frac{1}{c}(\mathbf{v}\,\text{grad})\,\mathbf{B}$$

Here div **B** is always zero. The second term vanishes for the case of an infinitely long bar because in this case **B** cannot depend on $x$. Equation (88.5) is therefore really identical with the electrostatic field described by (88.6).

Equation (88.5), incidentally, is also correct when, in place of a long bar, we have an *arbitrarily oriented permanent magnet* moving with velocity **v**. Then, however, the E-field is no longer vortex-free; rather, according to the last equation, we have

$$\text{curl}\,\mathbf{E} = \frac{1}{c}(\mathbf{v}\,\text{grad})\,\mathbf{B}$$

Since for a co-moving observer, however, the total derivative of **B**, namely $\partial\mathbf{B}/\partial t + (\mathbf{v}\cdot\text{grad})\mathbf{B} = 0$, we have that $\text{curl}\,\mathbf{E} = -\dfrac{1}{c}\dfrac{\partial\mathbf{B}}{\partial t}$ as required by the induction law. The field (88.5) naturally also exists within the iron. In this region the effect on the co-moving conduction electrons is compensated by the Lorentz force $e\mathbf{v}\times\mathbf{B}/c$.

The existence of an electric field produced by a moving magnet, although known for a long time in electrotechnology, is actually only to be understood by the *relativistic* formula (88.2). In turn, as we have seen in the foregoing paragraphs, this phenomenon can be regarded as a direct consequence of the *Einstein definition of simultaneity*.

# CHAPTER E IV

## Relativistic Mechanics

### §89. The mechanics of mass points

In §75 we showed that the basic equations of Newtonian mechanics are invariant with respect to a *Galilean transformation*. An essential characteristic of this transformation is that when two velocities are to be added they are simply added vectorially, i.e. in the following way:

If an observer sees a given mass point moving with a velocity $\mathbf{u}'$, while at the same time he himself is moving with velocity $\mathbf{v}$ as seen by a second observer, then the second observer would assign the value $\mathbf{u} = \mathbf{v} + \mathbf{u}'$ to the velocity of the mass point. As opposed to this, however, the foregoing chapters on Einstein's principle of relativity together with the results of Michelson's investigation have led us to make the claim that *all laws of nature must be invariant with respect to the Lorentz transformation,* so that in particular the Einstein addition theorem for velocities must possess universal validity. We can no longer therefore look upon the old Newtonian equations of motion as rigorously true natural laws. Rather we must seek to modify these laws so that they too satisfy the requirements of the principle of relativity. In this proposed modification we can allow ourselves to be guided by the observation that *for the extreme case of very small velocities, the new equations must pass over to the old Newtonian equations.* Actually, all discrepancies predicted for classical electron theory by relativity theory are of the order of $(v/c)^2$.

We begin by examining *the motion of a mass point in a given force field.* The Newtonian equations of motion read

$$m \frac{d\mathbf{v}}{dt} = \mathbf{F} \tag{89.1}$$

From these, in the well-known way, there follows the energy law

$$\frac{d}{dt} (\tfrac{1}{2} m \mathbf{v}^2) = \mathbf{F} \mathbf{v} \tag{89.2}$$

which states that the time rate of change of the kinetic energy of a mass point is equal to the work performed in unit time. We wish to consider how we may endow these equations with relativistic invariance. There are two different starting points from which we can proceed. We can *either* rewrite in relativistic form the equation (89.1) relating to the mass point, *or* we can proceed from the basic equations of electrodynamics using the concept of the force density introduced at the outset. In §82 we have fully discussed the transformation law relating to the electrodynamic force density. An immediate carrying over of this transformation law to the force density of mechanics might at first sight appear suspect. Actually, however, the principle of relativity requires that the transformation law be the same for all forces, regardless of origin—for, the statement that different forces acting on a mass point are in equilibrium, has an objective content which in its nature is independent of any actual reference system. If there were different transformation laws for different kinds of forces, then it could happen that observers in motion with respect to one another would arrive at contradictory conclusions regarding the existence of the equilibrium.

We now follow the first course, namely that of the *direct relativistic rewriting of the equations of motion* (89.1). For this purpose we substitute for the velocity **v** the *four-velocity* $u_\nu$ with components

$$(u_1, u_2, u_3) = \frac{\mathbf{v}}{\sqrt{(1-\beta^2)}} \qquad u_4 = \frac{ic}{\sqrt{(1-\beta^2)}} \qquad (89.3)$$

In addition, in place of the time element $dt$ which depends on a particular coordinate system, we introduce the invariant proper time $d\tau$. We characterize the inertia of the mass point by an associated *invariant mass* which we designate by $m_0$. We then obtain the four-dimensional relationship for the equations of motion, as first given by Minkowski:

$$m_0 \frac{du_\nu}{d\tau} = F_\nu \quad (\nu = 1, 2, 3, 4) \qquad (89.4)$$

in which the four-vector $F_\nu$ appearing on the right side is usually called the *Minkowski force vector*. In order to grasp the physical content of our modified equation of motion, and, in particular, to facilitate its comparison with (89.1), we resubstitute $dt\sqrt{(1-\beta^2)}$ for the proper time $d\tau$, and for the $u_\nu$ we substitute their values given by (89.3). We thereby obtain for the first three components of relations (89.4) the vector equation

$$m_0 \frac{d}{dt} \frac{\mathbf{v}}{\sqrt{(1-\beta^2)}} = \mathbf{F} \tag{89.5}$$

if we assume the connection between the *first three components of the Minkowski force* $F_\nu$ and the three-dimensional force vector $\mathbf{F}$ to be given by

$$(F_1, F_2, F_3) = \frac{\mathbf{F}}{\sqrt{(1-\beta^2)}} \tag{89.6}$$

In order to find out the meaning of the *fourth component of* $F_\nu$, we multiply the equation of motion (89.4) by $u_\nu$ and sum over $\nu$ from 1 to 4. The left member is then zero because of the identity $\sum_\nu u_\nu^2 = -c^2$; therefore

$$\sum_\nu F_\nu u_\nu = 0 \quad \text{or} \quad \frac{\mathbf{F}\mathbf{v}}{1-\beta^2} + \frac{F_4 ic}{\sqrt{(1-\beta^2)}} = 0$$

The fourth component of the Minkowski force therefore has the meaning

$$F_4 = \frac{i}{c} \frac{\mathbf{F}\mathbf{v}}{\sqrt{(1-\beta^2)}} \tag{89.7}$$

Consequently, the fourth component of the equation of motion is

$$\frac{d}{dt} \frac{m_0 c^2}{\sqrt{(1-\beta^2)}} = \mathbf{F}\mathbf{v} \tag{89.8}$$

The right side of this equation is identical with the right side of the *energy equation* (89.2) *of classical mechanics*. We are therefore led to regard the energy of our mass point as

$$E = \frac{m_0 c^2}{\sqrt{(1-v^2/c^2)}} \tag{89.9}$$

i.e. the time derivative of this quantity is equal to the power produced by the applied force. If we develop this energy $E$ in a power series in increasing powers of $v/c = \beta$, we obtain

$$E = m_0 c^2 + \tfrac{1}{2} m_0 v^2 + \dots \tag{89.9a}$$

We shall later designate the first term $(m_0 c^2)$ of this development as the *rest energy* of the mass point. It is constant, irrespective of the motion, and for the present it has no particular significance in the kinematics of our point. The second term is identical with the ordinary *kinetic energy* of classical mechanics. In the case of small velocities our energy is in fact, except for an additive term, equal to the kinetic energy in the old sense. As we shall see later, a rather profound meaning attaches to this additive constant. We can recognize this in a more formal way in that by multiplication of the four-velocity $u_\nu$ by the scalar rest mass $m_0$ we can derive a new four-vector

$$J_\nu = m_0 u_\nu \qquad (89.10)$$

whose components have the meaning

$$(J_1, J_2, J_3) = \mathbf{J} = \frac{m_0 \mathbf{v}}{\sqrt{(1-\beta^2)}} \qquad J_4 = \frac{i}{c}E = \frac{im_0 c}{\sqrt{(1-\beta^2)}} \qquad (89.11)$$

The first three components of $J_\nu$ are therefore identical with the *mechanical momentum vector* $\mathbf{J}$; the time-like component, on the other hand, is identical with the quantity $iE/c$. The invariant magnitude of this vector is given by

$$-\sum_{\nu=1}^{4} J_\nu^2 = \frac{E^2}{c^2} - J^2 = m_0^2 c^2 \qquad (89.11a)$$

If we wish to approach the dynamics of a mass point by way of the concept of the *force density*, we are obliged to treat the mass point as if it were a *continuum*, and to ascribe to this continuum a scalar function of position $\mu_0$ which we shall call the *rest density*. Since on the basis of the results of electrodynamics the force density $f_\nu$ has the form of a four-vector, we expect as equations of motion

$$\mu_0 \frac{du_\nu}{d\tau} = f_\nu \qquad (89.12)$$

which refer to a particular mass element of the continuum. If now the four-velocity $u_\nu$ is constant everywhere within this continuum, the equations of motion for the entire body can be derived from (89.12) by integrating over the volume. We therefore multiply (89.12) by the scalar element of rest volume $dV_0 = dV/\sqrt{(1-\beta^2)}$ and, with the abbreviation $m_0 = \int \mu_0 dV_0$, obtain as the *equation of motion of the mass point*

$$m_0 \frac{du_\nu}{d\tau} = \frac{\int f_\nu dV}{\sqrt{(1-\beta^2)}} \qquad (89.13)$$

If now the resulting three-dimensional force is introduced through

$$F_x = \int f_1 \, dV \qquad F_y = \int f_2 \, dV \qquad F_z = \int f_3 \, dV \quad (89.13a)$$

then, taking account of (89.6), the Minkowski equations of motion (89.4) are again obtained.

The most apparent new result of relativistic mechanics consists in *the increase of the inertial mass with velocity*; that is, the new equations (89.11) for the momentum can also be written in the familiar form $\mathbf{J} = m\mathbf{v}$, if only it be added that *the mass m itself depends upon the velocity*:

$$m = \frac{m_0}{\sqrt{(1 - v^2/c^2)}} \qquad (89.14)$$

We notice that for *small* velocities $m$ is equal to the rest mass; on the other hand, when $v$ approaches the *velocity of light*, the value of $m$ becomes *infinitely large*. We have in fact already become acquainted with such a relationship for the mass by considering the electromagnetic field accompanying a moving electron. Actually this increase of mass has been experimentally observed for *very fast cathode rays* and, more particularly, for *β-rays from radioactive nuclei*. (We recall our work near the end of §65 concerning the variation of mass with velocity in connection with the deflection investigations of Kaufmann.) Prior to the relativity theory this behaviour of the electron was looked upon as evidence for the electromagnetic character of its mass. Now, however, we see that this dependence upon velocity is quite general for any inertial mass, whether its origin be of an electromagnetic nature or not.

## §90. The inertia of energy

Through equation (89.11) we have become acquainted with the four-vector $J_v$ which provides an invariant connection between the energy and momentum of a single mass point. In a coordinate system in which the space-like momentum $\mathbf{J}$ is equal to zero, we have

$$J_1^0 = J_2^0 = J_3^0 = 0 \qquad J_4^0 = \frac{iE_0}{c}$$

$E_0$ being the *rest energy*. If we observe the mass point from a coordinate system with respect to which it moves with velocity $\mathbf{v}$ in the direction of the positive $x$-axis, it follows from the Lorentz transformation of the four-vector that

$$J_1 = \frac{-i\beta J_4^0}{\sqrt{(1-\beta^2)}} \qquad J_4 = \frac{J_4^0}{\sqrt{(1-\beta^2)}}$$

or, in three-dimensional formalism:

$$\mathbf{J} = \frac{\mathbf{v}}{\sqrt{(1-\beta^2)}} \frac{E_0}{c^2} \qquad E = \frac{E_0}{\sqrt{(1-\beta^2)}} \qquad (90.1)$$

In this form our results are capable of a large and far-reaching generalization with which we shall become acquainted in these paragraphs by means of several simple examples. This generalization consists in the assertion that *every closed system for which the rest energy is $E_0$ possesses an inertial mass of magnitude*

$$m_0 = \frac{E_0}{c^2} \qquad (90.2)$$

The rest energy is then defined as the total energy in a coordinate system in which the resulting momentum $\mathbf{J}$ is equal to zero. In §84 the truth of the assertion (90.2) was demonstrated for the special case of a *pure electromagnetic field* (wave train of finite length). By means of two *mechanical examples* we shall now see that even to heat energy there belongs a mass inertia within the meaning of equation (90.2).

As the first example we consider *a system of mass points capable of mutual elastic collision*, rather like the situation which we imagine for the familiar ideal gas of kinetic theory. If we designate the individual rest masses by $m_1^0, m_2^0, \ldots$, and their velocities by $\mathbf{v}_1, \mathbf{v}_2, \ldots$, then, for elastic collisions we have the two conservation laws of classical mechanics for momentum and energy

$$\mathbf{J} = \sum_s m_s^0 \mathbf{v}_s = \text{const.} \qquad E = \sum_s \tfrac{1}{2} m_s^0 \mathbf{v}_s^2 = \text{const.}$$

In *relativity theory* these two conservation laws are combined in the single theorem that *the four-vector for the total energy and total momentum*

$$J_v = \sum_s m_s^0 u_{s,v} \qquad (90.3)$$

*is constant in time.* In three-dimensional formulation the conservation laws are

$$\mathbf{J} = \sum_s \frac{m_s^0 \mathbf{v}_s}{\sqrt{(1-\mathbf{v}_s^2/c^2)}} \qquad E = \sum_s \frac{m_s^0 c^2}{\sqrt{(1-\mathbf{v}_s^2/c^2)}}$$

We observe our gas from the special coordinate system in which, as a whole, the gas is at rest, i.e. in the system for which its mechanical momentum is equal to zero. In this coordinate system, when the $s$th gas molecule has the velocity $\mathbf{v}_s^0$, the energy possessed by it is

$$E_0 = \frac{c}{i} I_4^0 = \sum_s \frac{m_s^0 c^2}{\sqrt{[1 - (\mathbf{v}_s^0)^2/c^2]}} \tag{90.4}$$

If, now, we consider the gas from a system with respect to which the gas moves with the velocity $\mathbf{v}$, it follows from the known transformation formulae for the momentum that

$$\mathbf{J} = \frac{\mathbf{v}E_0}{c^2\sqrt{(1 - \mathbf{v}^2/c^2)}} = \frac{\mathbf{v}M_0}{\sqrt{(1 - \mathbf{v}^2/c^2)}} \tag{90.5}$$

if there is ascribed to the whole gas a rest mass of magnitude

$$M_0 = \frac{E_0}{c^2} = \sum_s \frac{m_s^0}{\sqrt{[1 - (\mathbf{v}_s^0)^2/c^2]}} \tag{90.6}$$

Although we consider the gas now, not as a mechanical system, but, from the standpoint of macroscopic heat theory as an extended body with a definite heat content, this does not alter our consideration in the slightest. In equation (90.6) we recognize, however, that the rest mass of the whole gas comprises not only the rest mass of the individual molecules, but, in addition, its total kinetic energy, which is macroscopically synonymous with the heat content.

We wish now to establish this important result in a somewhat different way, namely with the help of the Einstein *addition theorem for velocities*. This will bring us to the same result without our having to make use of the energy concept. We employ the same symbolism as before, but for simplicity we say that the gas in the second system moves with velocity $v$ in the $x$-direction. We now apply the addition theorem for velocities to find from the velocities in the "rest system" the velocities $\mathbf{v}_s$ in the moving system. For this, in §77 we have derived the relations

$$v_{s,x} = \frac{v_{s,x}^0 + v}{1 + \dfrac{v_{s,x}^0 v}{c^2}} \qquad v_{s,y} = \frac{v_{s,y}^0 \sqrt{\left(1 - \dfrac{v^2}{c^2}\right)}}{1 + \dfrac{v_{s,x}^0 v}{c^2}} \tag{90.7}$$

from which, as may be easily shown, it follows that

$$\sqrt{\left(1 - \frac{v_s^2}{c^2}\right)} = \frac{\sqrt{\left(1 - \dfrac{(v_s^0)^2}{c^2}\right)} \sqrt{\left(1 - \dfrac{v^2}{c^2}\right)}}{1 + \dfrac{v_{s,x}^0 v}{c^2}} \tag{90.8}$$

The total momentum for an arbitrary system is then given by

$$J = \sum_s \frac{m_s^0 \mathbf{v}_s}{\sqrt{\left(1 - \frac{v_s^2}{c^2}\right)}} \tag{90.9}$$

Upon introducing the values (90.7) and (90.8), and bearing in mind (90.6), we obtain

$$
\left.
\begin{aligned}
J_x &= \sum_s m_s^0 \frac{v_{s,x}^0 + v}{\sqrt{\left(1 - \frac{(v_s^0)^2}{c^2}\right)}\sqrt{\left(1 - \frac{v^2}{c^2}\right)}} = \frac{1}{\sqrt{\left(1 - \frac{v^2}{c^2}\right)}}(J_x^0 + M_0 v) \\
&= \frac{M_0 v}{\sqrt{\left(1 - \frac{v^2}{c^2}\right)}} \\
J_y &= \sum_s m_s^0 \frac{v_{s,y}^0}{\sqrt{\left(1 - \frac{(v_s^0)^2}{c^2}\right)}} = J_y^0 = 0
\end{aligned}
\right\} \tag{90.10}
$$

just as in the momentum expression of (90.5). In this derivation the energy was not mentioned; instead, we have derived the rest mass of our gas solely from the relativistic theorem for the addition of velocities.

We justified equation (90.2) for the special case in which the thermal energy consists only of the kinetic energy of the free gas atoms. *The gas as a whole actually becomes heavier when there is an increase in the average energy of the individual atoms.*

We wish now to convince ourselves, by the example of *inelastic collisions*, that this inertial property of heat energy is wholly independent of any molecular concept concerning the mechanism of heat motion. We investigate the special case in which two material spheres collide inelastically in such a manner that after the collision they remain united with one another. From the standpoint of *Newtonian mechanics* this occurrence can be described by the law of *conservation of momentum*. If $m_1$ and $m_2$ are the masses of the colliding spheres and $\mathbf{v}_1$ and $\mathbf{v}_2$ their velocities prior to the collision, and if $\mathbf{v}$ is their velocity after the collision (i.e. when united), then the momentum conservation law gives

$$m_1 \mathbf{v}_1 + m_2 \mathbf{v}_2 = (m_1 + m_2)\mathbf{v} \tag{90.11}$$

The energy law of classical mechanics (for an elastic collision) is clearly not applicable here. Kinetic energy is lost in the collision; it is converted into heat. The magnitude $W$ of the kinetic energy converted into heat is equal to the difference between the kinetic energy before the collision and that afterwards. According to (90.11) then,

$$2W = m_1 \mathbf{v}_1^2 + m_2 \mathbf{v}_2^2 - (m_1 + m_2) \mathbf{v}^2 = \frac{m_1 m_2}{m_1 + m_2} (\mathbf{v}_1 - \mathbf{v}_2)^2 \qquad (90.12)$$

We could perhaps be tempted to rewrite the Newtonian conservation law (90.11) by introducing the four-velocities $u_v^{(1)}$, $u_v^{(2)}$, etc. as well as the rest masses $m_1^0$, $m_2^0$, in the form

$$m_1^0 u_v^{(1)} + m_2^0 u_v^{(2)} = (m_1^0 + m_2^0) u_v$$

This formulation, while indeed relativistically invariant, is nevertheless absurd as we may easily convince ourselves by examining the fourth component of this equation. The four components of the velocity $u_v$ are over-determined in these equations, since between them the relationship $\sum u_v^2 = -c^2$ must always hold.

We can give a four-dimensional formulation of the momentum law (90.11) that is free from contradictions only when we take into account that *a difference can exist between the rest mass of the united bodies after the collision and the sum of the rest masses before the collision.* Then our four-dimensional formulation is

$$m_1^0 u_v^{(1)} + m_2^0 u_v^{(2)} = M^0 u_v \quad (v = 1, 2, 3, 4) \qquad (90.13)$$

Here, as opposed to (90.11), we have obtained an essentially new result. While the three equations (90.11) are to be regarded as the three components of the new velocity $\mathbf{v}$, there emerges from (90.13) a fourth equation for the rest mass of the united spheres after their collision. In three-dimensional formulation this equation is

$$\frac{m_1^0}{\sqrt{(1 - \mathbf{v}_1^2/c^2)}} + \frac{m_2^0}{\sqrt{(1 - \mathbf{v}_2^2/c^2)}} = \frac{M^0}{\sqrt{(1 - \mathbf{v}^2/c^2)}} \qquad (90.14)$$

Considered from the standpoint of a coordinate system in which the resulting velocity $\mathbf{v}$ is zero, equation (90.14) says the following:

The mass $M^0$ of the inelastically collided and united spheres comprises not only the two rest masses $m_1^0$ and $m_2^0$, but also a third mass $m_3 = W/c^2$, where $W$ represents that part of the kinetic energy of the two colliding bodies that is converted into heat. It can appear, for example, as a rotational energy of the two bodies going round their common centre of gravity following their collision. In this case, too, there must be an increase of the inertial mass.

As a further example of the inertia of energy we consider a body (rest mass $M^0$) which, during a finite time interval emits a quantity of energy $E^0$ in the form of *electromagnetic radiation* (i.e. light or radiant heat). This radiation, moreover, is to come out of the emitting body in such a way that the resultant momentum of the emitted energy is equal to zero. The radiation could, for example, come out in the form of a spherical wave, or, for flat bodies, in the form of two plane wave trains travelling in opposite directions. It is clear that the body, at rest at the outset, remains at rest after the emission of the radiation.

Let us observe this phenomenon from a coordinate system with respect to which the body moves with velocity **v**. In this coordinate system, according to the transformation formula for a four-vector, there is associated with this radiation a total *momentum* whose value is

$$\mathbf{J} = \frac{E^0 \mathbf{v}}{c^2 \sqrt{(1-\beta^2)}} \tag{90.15}$$

During the emission of the radiation our body has provided this momentum, without changing its velocity. *This is only possible, however, by the body changing its rest mass.* If we designate the rest mass of the body after the emission of the light wave by $M'^0$, the momentum law requires that

or that

$$\frac{M^0 \mathbf{v}}{\sqrt{(1-\beta^2)}} = \frac{M'^0 \mathbf{v}}{\sqrt{(1-\beta^2)}} + \frac{E^0 \mathbf{v}}{c^2 \sqrt{(1-\beta^2)}} \tag{90.16}$$

$$M'^0 = M^0 - \frac{E^0}{c^2}$$

The radiating body has actually had taken from it a quantity of mass equal to $1/c^2$ times the energy which it emitted. So far as our prime equation (90.2) is concerned, however, it does not matter whether the energy prior to the emission was stored in the form of heat, electrical field energy, or mechanical energy. This example is of special interest because with its help Einstein for the first time derived the law of the *inertia of energy as a general law of nature.*

The inertia of energy is especially evident and lends itself to experimental measurement in nuclear reactions, as for example in radioactive decay. In these reactions the sum $M_1$ of the masses of the participating particles before the process does not in general agree with the sum $M_2$ after the process. The difference $(M_1 - M_2)c^2$ is the liberated energy or, better, the difference of the total kinetic energies after and before the process. In a similar way the *mass defects* in atomic weights, i.e. the difference between the mass of the whole nucleus and its individual constituents, traces back to the binding energy of the constituents.

## §91. Mechanical stresses

(*a*) *Energy-momentum balance of the electron.* We continue with the further development of relativistic mechanics. We shall have to be content here with the handling of two concrete problems embodying important questions of relativistic dynamics. The first of these is the question first raised in §65, of the *energy-momentum balance of the electron.* The second concerns the *negative result of the Trouton-Noble experiment.*

We proceed from the result obtained in §65, according to which the total momentum of the electromagnetic field of an electron moving with the constant velocity **v** is given by

$$\mathbf{J} = \frac{4U_0}{3c^2} \frac{\mathbf{v}}{\sqrt{(1-\beta^2)}} \tag{91.1}$$

where $U_0$ is the total energy of the field of an electron at rest. Now this result contains a difficulty in the matter of seeking *a purely electromagnetic interpretation for the mass of an electron*—for, obviously, the theory yields a value for the electromagnetic momentum that is 4/3 larger than that we obtain if, according to the Einstein principle of inertia of energy, a rest mass of $U_0/c^2$ is ascribed to the electron and we then calculate as if the electron were a mass point having that mass.

Before we turn to a clarification of the puzzling factor 4/3, we wish to derive the result (91.1) by using the mathematical aids to relativity theory. For this purpose we go back to §83 where we have derived the *energy-momentum tensor $T_{\nu\mu}$ for the electromagnetic field.* The nine purely space-like components represent the components of the Maxwell stress tensor; $T_{14}, T_{24}, T_{34}$ mean (within the factor $-ic$) the components of the momentum density; $T_{41}, T_{42}, T_{43}$ mean (within the factor $-i/c$) the components of the Poynting energy-flow vector; and finally, $T_{44}$ is the energy density of the electromagnetic field. This tensor, when employed for the special case of *the field of an electron in its rest system*, gives vanishing values for the space-time components because $\mathbf{H} = 0$.

Accordingly, the electrostatic field of an electron at rest possesses no momentum and produces no energy flow.

If now $T_{\nu\mu}$ is the energy-momentum tensor of the electron in a system with respect to which the electron moves with velocity $v$ in the $x$-direction, then the components $J_x$ of the total momentum, and the total energy $U$ of the field, are respectively given by

$$J_x = \frac{i}{c}\int T_{14}\, dV \qquad U = \int T_{44}\, dV \qquad (91.2)$$

The components $J_y$ and $J_z$ vanish because of symmetry.

By (80.17) the two stress components $T_{14}$ and $T_{44}$ are derived from the tensor $T^0_{\nu\mu}$ for the electron at rest, giving

$$T_{14} = \frac{-i\beta(T^0_{44}-T^0_{11})}{1-\beta^2} \qquad T_{44} = \frac{T^0_{44}-\beta^2 T^0_{11}}{1-\beta^2}$$

since, for the electron at rest, the two mixed components $T^0_{14}$ and $T^0_{41}$ vanish. Since $dV = dV_0 \sqrt{(1-\beta^2)}$, we can carry out the integration over space in the rest system, obtaining

$$\left.\begin{aligned} J_x &= \frac{v}{c^2\sqrt{(1-\beta^2)}}\int (T^0_{44}-T^0_{11})\, dV_0 \\ U &= \frac{1}{\sqrt{(1-\beta^2)}}\int (T^0_{44}-\beta^2 T^0_{11})\, dV_0 \end{aligned}\right\} \qquad (91.3)$$

The volume integral over $T^0_{44}$ is the total field energy $U_0$ in the rest system. On the other hand, according to §83,

$$\int T^0_{11}\, dV_0 = \frac{1}{4\pi}\int \{(E^0_x)^2 - \tfrac{1}{2}(\mathbf{E}^0)^2\}\, dV_0 = -\tfrac{1}{3}U_0$$

since, by symmetry,

$$\int (E^0_x)^2\, dV_0 = \tfrac{1}{3}\int (\mathbf{E}^0)^2\, dV_0$$

We therefore obtain for $\mathbf{J}$ exactly the result (91.1), while for $U$ we find

$$U = \frac{U_0}{\sqrt{(1-\beta^2)}}(1+\tfrac{1}{3}\beta^2) \qquad (91.4)$$

The emergence of the factor 4/3 in (91.1) indicates to us that the dynamic inner structure of an electron is not to be understood solely on the basis of its field. This factor enters the derivation because in the expressions (91.3), in addition to the term $T_{44}^0$ to which the rest energy of the field corresponds, there emerges a term $T_{11}^0$ having its origin in the Maxwell stresses. In the dynamics of the electron, other forces than just those of the electromagnetic field must be active. In all attempts based on classical physics to construct a *model of the electron* this is recognized. Since charge elements of like sign always repel, such a model can only be stable when there are other forces involved which hold the electrical repulsive forces in equilibrium. Although at this time it is impossible for us to say whether these forces are the results of mechanical stresses or are of other origin, we can at least say with certainty that: under the Lorentz transformation these forces must behave like pure electromagnetic forces, so that when they maintain equilibrium in a rest system, they will do likewise with respect to a moving system.

In addition to the electromagnetic tensor $T_{\nu\mu}$, we introduce a second tensor $P_{\nu\mu}$ of mechanical or other origin. Its structure shall be the same as $T_{\nu\mu}$, i.e. symmetrical, and possessing components $P_{\nu 4}^0$ and $P_{4\nu}^0$ (with $\nu \neq 4$) which vanish in the rest-system. It is to be added to $T_{\nu\mu}$ to complete the dynamics of the electron. The electron at rest must remain in equilibrium under the simultaneous actions of $T_{\nu\mu}^0$ and $P_{\nu\mu}^0$. For the force density coming from these tensors then, we must have in the rest system, corresponding to (83.2), that

$$f_\nu^0 = \sum_{\mu=1}^{4} \frac{\partial}{\partial x_\mu}(T_{\nu\mu}^0 + P_{\nu\mu}^0) = 0 \quad \text{(for } \nu = 1, 2, 3) \qquad (91.5)$$

From this the important relation follows that

$$\int (T_{\nu\mu}^0 + P_{\nu\mu}^0)\, dV^0 = 0 \quad \text{(for } \nu = 1, 2, 3) \qquad (91.6)$$

For proof, we observe that for the rest system $T_{\nu4}{}^0 = P_{\nu4}{}^0 = 0$, for $\nu = 1, 2, 3$. According to (91.5) then, for example, the three-dimensional vector with components $T_{1\mu}{}^0 + P_{1\mu}{}^0$, with $\mu = 1, 2, 3$, is source-free. With the help of Gauss's theorem it is easily shown that the volume integral over all components of a source-free vector is equal to zero if, as $r$ approaches infinity, the vector vanishes more rapidly than $1/r^2$.

With this the energy-momentum balance of the electron now falls into place; that is, if throughout (91.3) we write $T^0_{\nu\mu} + P^0_{\nu\mu}$ instead of $T^0_{\nu\mu}$, then on account of (91.6),

$$\mathbf{J} = \frac{\mathbf{v}}{c^2 \sqrt{(1-\beta^2)}} \int (T^0_{44} + P^0_{44}) \, dV^0$$

$$U = \frac{1}{\sqrt{(1-\beta^2)}} \int (T^0_{44} + P^0_{44}) \, dV^0$$

which can be written in the form

$$\left. \begin{array}{cc} \mathbf{J} = \dfrac{m_0 \mathbf{v}}{\sqrt{(1-\beta^2)}} & U = \dfrac{m_0 c^2}{\sqrt{(1-\beta^2)}} \\[2ex] \text{with} \qquad m_0 c^2 = \displaystyle\int (T^0_{44} + P^0_{44}) \, dV^0 \end{array} \right\} \tag{91.7}$$

in complete agreement with Einstein's theorem of the inertia of energy.

We incline in general to the assumption that the rest energy of an electron is of a purely electromagnetic nature, that therefore $P^0_{44} = 0$. This does not of course alter the validity of result (91.7)—for it was not the introduction of the mechanical rest-density $P^0_{44}/c^2$ that was essential to our considerations, but the introduction of stresses holding the electrical forces in equilibrium.

(b) *The Trouton-Noble experiment.* We come now to a consideration of Trouton and Noble's investigation. As is well known, this investigation consisted in an attempt to establish the existence of an absolute velocity by means of a rotatable (delicately suspended) oriented charged capacitor. According to classical theory, on a charged capacitor in translational motion there should be a resulting torque to which the capacitor would have to respond by means of a rotation. Even, however, with the most painstaking execution of the experiment, not the barest suggestion of such a rotation could be found. From the standpoint of classical physics the negative result of this investigation was just as puzzling as the negative result of the Michelson experiment. In relativity theory, however, the result is understandable: if the capacitor does not turn for an observer relative to whom it is at rest, it cannot turn for an observer relative to whom it moves. We shall not, however, content ourselves with this summary statement. We shall seek to understand the result by means of a model.

As a *simplified model* of the Trouton-Noble capacitor we choose two small charged bodies which are held at a constant distance apart $l_0$ by a rigid connection (bar). If the two bodies carry equal charges $e$ of the same sign, they repel one another in the rest system with the force $e^2/l_0^2$; if the charges are equal but opposite, they attract one another with this same amount of force. In order to simplify the argument, we shall assume the first case (equal charges of the same sign). If both charges move with velocity $v$ in the direction of the

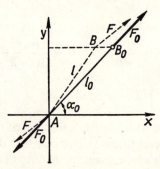

Fig. 75.—For the investigation of Trouton and Noble

$x$-axis, then the forces between the two charges *no longer act along the line joining them*. We have already obtained this result in §64 from a rigorous solution of the field equations. As we saw there, a *torque* exists which, for the case where the charges attract one another, has the tendency to set the connecting bar transverse to the direction of motion. In the opposite case (repelling charges) the effect is to turn the bar so that it tends to align itself with the direction of motion.

We wish now to derive this result from the standpoint of relativity theory. This must lead to the same result as the Maxwell equations, since the latter satisfy the principle of relativity. We hold fixed the charged bodies A and $B_0$ shown (figure 75). These charges are kept apart the fixed distance $l_0$ and repel one another in the direction of the bar with the force $F_0$. The bar itself makes with the $x$-axis the angle $\alpha_0$, and its projections on the coordinate axes are given by:

$$l_x^0 = l_0 \cos \alpha_0 \qquad l_y^0 = l_0 \sin \alpha_0$$

Acting at its ends are the forces

$$F_x^0 = \mp F_0 \cos \alpha_0 \qquad F_y^0 = \mp F_0 \sin \alpha_0 \qquad \text{with} \quad F_0 = \frac{e^2}{l_0^2}$$

in which the upper signs refer to A, and the lower to $B_0$. If now the entire assembly moves with velocity $v$ in the $x$-direction, then, on account of the Lorentz contraction, the new values for the projections of the bar are

$$l_x = l_x^0 \sqrt{(1-\beta^2)} = l_0 \sqrt{(1-\beta^2)} \cos \alpha_0 \qquad l_y = l_y^0 = l_0 \sin \alpha_0$$

On the other hand, from the transformation laws for a force we have

$$F_x = F_x^0 = \mp F_0 \cos \alpha_0 \qquad F_y = F_y^0 \sqrt{(1-\beta^2)} = \mp F_0 \sqrt{(1-\beta^2)} \sin \alpha_0$$

For the moving observer the forces no longer lie along the line joining the charges: they form a smaller angle with the $x$-axis than that of the bar. There results a torque around the $z$-axis which is given by

$$T = l \times F \qquad T_z = l_x F_y - l_y F_x = -l_0 F_0 \beta^2 \cos \alpha_0 \sin \alpha_0 \quad (91.8)$$

This torque obviously agrees with that found in §64 from the field vectors of point charges.

Thus, from the standpoint of classical physics, there results from the electrical forces a torque for which there is no compensating moment. Experimentally, however, there is no concomitant time change of angular momentum since no rotation was observed. In *relativity theory*, on the other hand, the situation is essentially this: We must include in our considerations here the *mechanical forces* which hold the electrical forces in equilibrium. Without these mechanical forces the system would not be stable. If, however, we include these forces, then the negative result of the Trouton-Noble experiment is understandable—for, in the rest system the electrical and mechanical forces are in equilibrium; and since, further, both kinds of force transform in the same way in going over to a new system, the forces likewise exactly cancel one another in the new system. Or, in other words: *The torque due to the electrical forces is exactly compensated by the equal and opposite torque due to the mechanical forces.* It is just this circumstance that classical mechanics cannot explain, namely that mechanical forces possess a torque in a moving system.

In conclusion, therefore, it is instructive to consider the following problem: Let there be given a bar at rest, of length $l_0$, and upon each end of this bar let equal and opposite tensile stresses $F_0$ act, in the direction of the long axis. In the bar therefore there exists a tension of magnitude $F_0/a_0$, where $a_0$ stands for the bar's cross-sectional area.

We wish now to consider the stress from the point of view of a reference system with respect to which the bar moves with velocity $v$ in the $x$-direction. We take the case in which the bar moves in the direction of its length, this having been chosen as the $x$-axis. That in this case an additional momentum density must exist can be seen as follows: Let A and B be the two ends of the bar which are moving with velocity $v$ in the $x$-direction. The force $F_0$ acts at B and "pulls" it, i.e. it performs on B, per second, work $F_0 v$. A like amount of work is performed at point A against the "restraining action" of the force $F_0$. Power $F_0 v$ must therefore flow constantly from B to A. This requires a component $S_x = F_0 v/a_0$ of the density of energy flow. Now owing to the symmetry of the momentum-energy tensor there is associated with every energy flow S a momentum density $\mathbf{j} = \mathbf{S}/c^2$. In the bar, therefore, there must exist an additional momentum density

$$J_x = -\frac{l_0 a_0}{c^2} S_x = -\frac{v l_0 F_0}{c^2} \qquad (91.9)$$

We can also obtain this result by a consideration of the momentum that will be transmitted to, and removed from, the bar during a "switching-on" of the tensile forces. We say that in the rest system before the time $t^0 = 0$ the bar was stress-free. At the moment $t^0 = 0$ the tension forces are turned on (as for example by

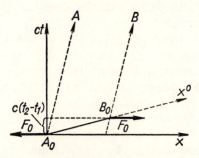

Fig. 76.—World lines of the end-points of a bar which, at time $t^0 = 0$, is placed under tension

attaching at this moment, at right and left on the bar, springs under tension). During this turning-on process the bar naturally remains at rest. (We assume the bar to be so hard that we can neglect any effect of elastic dilatation.) Upon observing this turning-on process from a moving system, an entirely different picture results. In figure 76 the world lines of the two end-points of the bar are portrayed in the new coordinate system. $A_0$ and $B_0$ are the world points at which the two tension forces commence their action. (In the new system these forces retain the magnitude $F_0$ because force components in the direction of motion are not altered in a Lorentz transformation.) As seen from figure 76, in the new system the two forces do not both commence action at the same moment; rather, as follows from the usual formulae for the Lorentz transformation, there elapses the finite time interval $t_2 - t_1 = v l_0 / c^2 \sqrt{(1 - \beta^2)}$, during which only the applied force on the left-hand end-face is active. This unilateral force takes from the bar the momentum

$$J_x = -(t_2 - t_1)F_0 = -\frac{v l_0 F_0}{c^2 \sqrt{(1 - \beta^2)}} \qquad (91.10)$$

during this time interval without thereby altering its velocity. Clearly this value agrees in first approximation with that of (91.9).

For the case where the bar makes an angle $\alpha_0$ with the direction of its motion (= the $x$-axis) it possesses an additional momentum due to the tension forces acting at its ends. By reasoning as above, we find that this momentum has the components

$$J_x = -\frac{vl_0 \cos \alpha_0}{c^2\sqrt{(1-\beta^2)}} F_0 \cos \alpha_0 \qquad J_y = -\frac{vl_0 \cos \alpha_0}{c^2\sqrt{(1-\beta^2)}} F_0 \sin \alpha_0 \sqrt{(1-\beta^2)}$$

From the general expression for angular momentum $\mathbf{L}$ we have for the bar, owing to the constancy of the momentum density, that

$$\mathbf{L} = \int \mathbf{r} \times \mathbf{j}\, dV = \mathbf{r}_s \times \mathbf{J}$$

in which $\mathbf{r}_s$ is the position vector of the centre of gravity of the bar. Since $d\mathbf{r}_s/dt = \mathbf{v}$, the angular momentum changes with time. We have then for the $z$-component (normal to the plane determined by the bar and the direction of motion, this being the only component different from zero)

$$\frac{dL_z}{dt} = (\mathbf{v} \times \mathbf{J})_z = -vJ_y = -\frac{v^2 l_0 F_0}{c^2} \cos \alpha_0 \sin \alpha_0 \qquad (91.11)$$

In order to produce this continuous change of angular momentum, a continuous torque $T_z = dL_z/dt$ must act. Now, as we have seen in (91.8), such a torque will actually be produced on the bar by the tension forces which, in the rest system, act in the direction of the bar axis and have the value $F_0$.

The negative result of the Trouton-Noble experiment can consequently be interpreted thus: The torque calculated from electrodynamics produces just the rate of change of angular momentum given by (91.11), this being connected with the energy flow from one capacitor plate to the other and with the consequent additional momentum perpendicular to the direction of motion.

**F**

---

Exercise
problems
and
solutions

# CHAPTER F I

## Exercises

### Part A. Vector and tensor calculus

1. Vector **A** has the component $A_x = 2$; it forms an angle of 60° with the positive $y$-axis; in addition, its projection on the $yz$-plane forms an angle of 60° with the positive $z$-axis. Determine: the length of **A**; the magnitude of its components; the angle between **A** and the positive $x$-axis; the angle between **A** and the positive $z$-axis.

2. Determine the angle between any two space diagonals of a die (cube).

3. The unit vectors **x**, **y**, **z**, of a Cartesian coordinate system form three edges of a cube of unit volume. The three surface diagonals from the origin determine a regular tetrahedron. What is the angle between any pair of these surface diagonals? What is the area of each face of the tetrahedron, and what is its volume?

4. Let **A** be a vector extending from the origin in some direction, and let $\mathbf{a}_0$ be the corresponding unit vector. Further, let **r** be the vector drawn from the origin to the variable point P. Show that $\mathbf{a}_0 . \mathbf{r} = A$ is the equation of the plane which passes through the end point of **A** and is perpendicular to it.

5. The perpendicular extending from the origin to a certain plane is given by the vector $\mathbf{A} = (1, 2, 2)$. A line having direction cosines $(1/\sqrt{2}, 1/\sqrt{2}, 0)$ extends from the origin. Find the coordinates of the point at which this line intersects the plane.

6. The vector $(1, 1, 2)$ is drawn from the origin. Find the vector extending perpendicularly from the point having coordinates $(1, 2, 3)$ to the line of the given vector.

7. The end points of the vectors **a**, **b**, **c**, each extending from the origin, determine a plane. What is the distance of this plane from the origin? (Observe that the volume of a tetrahedron formed by the three edges of a parallelopiped is equal to 1/6 the volume of the parallelopiped.)

8. A rigid body rotates at 300 revolutions per minute around an axis which points into the first octant, making an angle of 60° with both the $x$- and $y$-axes. What is the velocity of a point $\mathbf{r} = (0, 0, 2)$ cm?

9. A body rotates around a fixed axis; the rotation vector is $\mathbf{w} = (-50, +80, +100)/s$. The velocity **v** of the point $\mathbf{r} = (4, 5, 6)$ has components $v_x = 20$ cm/s, and $v_y = 30$ cm/s. Find the shortest distance of the axis of rotation from the origin.

10. Let a unit vector **t** be associated with every point of a space curve having the direction of the velocity in the traversal of the curve in a given sense (tangent vector). What is the geometrical meaning of $d\mathbf{t}/ds$, where $s$ is the arc length of the curve from a fixed point?

11. Let **r** be the position vector, and let **a** be a constant vector. Calculate the gradient of the scalar product **a** . **r**, and the divergence and curl of the vectors **r** and **a** $\times$ **r**.

12. The points of a plane rotate around a fixed point with angular velocity $\omega(r)$ which depends upon their distance $r$ from the pivot point. How must $\omega(r)$ be constituted so that the velocity field is irrotational (vortex-free)?

13. A central force in space is given by $\mathbf{v} = \mathbf{r}f(r)$. Determine $f(r)$ so that $\mathbf{v}$ is source-free and irrotational.

14. Show that the same vector field is described by $\mathbf{v} = -\operatorname{grad}\phi$ with $\phi = \mathbf{p}.\mathbf{r}/r^3$ as is described by $\mathbf{v} = \operatorname{curl}\mathbf{A}$ with $\mathbf{A} = \mathbf{p} \times \mathbf{r}/r^3$. Calculate div $\mathbf{v}$ and curl $\mathbf{v}$ for this field.

15. Let a vector field be given by $\mathbf{v} = c\mathbf{r}/r$. Find (a) the line integral $\displaystyle\int_{r_1}^{r_2} \mathbf{v} . d\mathbf{r}$ for an arbitrary path, (b) the surface integral $\oint v_n\, dS$ for a spherical surface of radius $a$ centred on the origin, (c) this surface integral for a spherical surface of radius $a$ whose centre lies on the $x$-axis at a distance $b$ from the origin.

16. For the vector field $\mathbf{v} = \mathbf{w} \times \mathbf{r}$ with constant $\mathbf{w}$, calculate the line integral $\oint \mathbf{v}.d\mathbf{r}$, first for a circle of radius $a$ around the axis of rotation, and then (using Stokes's theorem) for an arbitrary closed curve.

17. Find div $\mathbf{v}$ and $\nabla^2\phi$ in parabolic coordinates $\xi$, $\eta$, $\alpha$, defined by $x = \xi\eta\cos\alpha$, $y = \xi\eta\sin\alpha$, $z = \frac{1}{2}(\xi^2 - \eta^2)$.

18. A symmetrical tensor is given by $T_{ik} = \partial^2\phi/\partial x_i\,\partial x_k$ with $\phi = (\mathbf{a}.\mathbf{r})(\mathbf{b}.\mathbf{r})$ in which $\mathbf{a}$ and $\mathbf{b}$ are two arbitrary constant vectors. Determine its eigen values $\lambda^{\mathrm{I}}$, $\lambda^{\mathrm{II}}$, $\lambda^{\mathrm{III}}$ and the unit vectors $\mathbf{n}^{\mathrm{I}}$, $\mathbf{n}^{\mathrm{II}}$, $\mathbf{n}^{\mathrm{III}}$ in the directions of its principal axes.

## Part B. The electrostatic field

1. With what force would two charges, each of 1 coulomb, but of opposite sign, attract one another at a distance of 1 km?

2. A quantity $e$ of positive electric charge is uniformly distributed as a volume charge within a sphere of radius $a$. Inside this charge cloud there is a negative point charge $-e$. What is the force acting on the point charge as a function of the distance of this charge from the centre of the sphere?

3. A point charge of 50 electrostatic units is located 1 cm in front of (a) a metallic plane, (b) a glass block (dielectric half-space) of permittivity $\varepsilon = 7$. What force acts on the charge?

4. What is the force between a metallic sphere of radius $R$ carrying a charge $E$, and a small body of charge $e$, distant $r$ from the centre of the sphere? Is it possible under these circumstances for the sphere and the small body to attract each other even if $E$ and $e$ have the same sign?

5. What is the maximum charge that can be conveyed to a sphere of diameter 10 cm if the breakdown voltage of air is 20,000 V/cm?

6. With what radius of curvature must the edges of a conductor carrying a potential of 10,000 V be provided if the breakdown voltage of air is 20,000 V/cm? (Set the potential gradient at the surface of the rounded edge equal approximately to that of a sphere.)

7. Let the field strength on the surface of a rounded edge of radius $r$ and length $l\,(\gg r)$ be equal approximately to the field strength at the middle of an ellipsoid of revolution of length $l$, with semi-minor axis $r$. Show that the field strength on the edge is less than on the ellipsoid surface of the same radius of curvature.

8. What is the surface density of electric charge on the surface of the earth at a place where the potential gradient is 250 V/m? How great is the force here on 1 m$^2$ of the earth's surface?

9. A potential of 100 V is applied between a tungsten wire 0·05 mm in diameter and a coaxial anode 1 cm away. What is the value of the field strength on the surface of the wire?

10. A soap bubble hanging from a blowing-pipe is collapsing because of the hole in the blow-pipe and the action of the surface tension, whose value is 50 ergs/cm$^2$. Is it possible to check completely the collapse by charging the bubble electrically? If so, at what radius does the soap bubble remain in equilibrium when the charge is so great that the electric field strength at the bubble surface equals the breakdown strength in air of 20,000 V/cm?

11. A uniform field of strength E exists in an extended dielectric having permittivity $\varepsilon$. What is the field strength in a cavity when this has the form (a) of a very long thin cylinder parallel to the field lines, (b) of a thin plate perpendicular to the field lines, (c) of a sphere?

12. (a) With what force (per cm$^2$) do the plates of a parallel-plate capacitor attract each other when the voltage is 1000 V, and the plate separation is 1 mm?
    (b) What is the force when, after charging to 1000 V, the capacitor is disconnected from the battery and is then immersed in petroleum ($\varepsilon = 2$)?
    (c) What is the force when the capacitor is first filled with petroleum and then charged to 1000 V?

13. Two series-connected capacitors of capacitance 0·5 $\mu$F and 0·2 $\mu$F are connected to 220 V d.c. mains. What is the charge on each capacitor, and what is the voltage across each?

14. Two initially uncharged capacitors of capacitance 1 $\mu$F and 10 $\mu$F are joined in series and connected to a battery whose terminals have potentials of $+100$ V and $-100$ V with respect to the earth. The wire joining the two capacitors is now grounded. What quantity of electricity flows through the grounding wire in the grounding process?

15. Calculate the work necessary to bring a dipole p from a point at which the field strength is E into a field-free region (at infinity). Let the angle between p and E be $\alpha$. In particular, how much work is needed to bring a freely rotatable dipole to infinity?

16. How much force is experienced by (a) a dipole of moment $p$, (b) a stretched quadrupole of moment Q and distance apart $r$, due to a point charge $e$ when this point charge is on the axis of the dipole or quadrupole?

17. Each of two equal dipoles of moment $p$ is suspended so as to be free to rotate about its centre of mass. The dipoles are separated from one another by a distance $a$. How will they orient themselves under the action of their mutual forces? It should be observed that their mutual interaction energy in the position of stable equilibrium possesses an absolute minimum.

18. Let the potential $\phi(x, y)$ for two infinitely long metal cylinders be found such that on the cylinders $\phi_1$ and $\phi_2$ have constant values and the equation $\nabla^2\phi = 0$ is satisfied everywhere in the space between the cylinders. How is the capacitance of the cylinder obtained?

19. Through the function $w = u + iv = f(z)$ of the complex variable $z = x + iy$, let there be coordinated with every point of the $xy$-plane a value $u = u(x, y)$ and a value $v = v(x, y)$. Show then that $u$ and $v$ satisfy the Laplace equations

$\nabla^2 u = 0$ and $\nabla^2 v = 0$. If 1 and 2 are two curves on which, for example, $u$ takes on the values $u_1$ and $u_2$, then $u$ is the solution of the potential problem for the field that would be produced by two cylindrical metal bodies having peripheral outline curves 1 and 2 when these have the voltage $u_1 - u_2$ between them.

20. The complex function $\zeta = \xi + i\eta = F(z) = F(x + iy)$ coordinates every point $(x, y)$ of the $z$-plane with a point $(\xi, \eta)$ of the $\zeta$-plane. Let there be given in the $\zeta$-plane an arrangement B of conductors (better stated, cross-sections B of cylindrical conductors), and let the potential problem here be solved, i.e. let a function $\phi(\xi, \eta)$ be found which is constant on the lines B, and which satisfies Laplace's equation. Let $\phi$ be the real part of the function $f(\zeta) = \phi + i\psi$. Further, in the $z$-plane, let an arrangement A be given for which the solution of the potential problem is sought. Show that if a function $\zeta = F(z)$ be given which maps the $z$-plane on the $\zeta$-plane such that figure A goes over to figure B, then the real part of $f(F[z])$ is the desired potential function for arrangement A. In what relationship does the capacitance of two conductors in A stand to the capacitance of those corresponding in B?

21. Determine the capacitance (per cm) of a double conductor, i.e. of two parallel cylinders of equal radius $r$ and axis-separation $d$ ($> 2r$), wherein a cylindrical capacitor is carried over, with the help of the function $\zeta = c^2/z$ and by suitable choice of the zero point of the representation, into the double conductor. The capacitance of the cylindrical capacitor is $1/[2\ln(b/a)]$ per cm, where $a$ and $b$ are the radii of the inner and outer cylinders. (In this representation all circles not passing through the zero point go over again into circles, while circles through the zero point go over into straight lines.)

## Part C. The electric current and the magnetic field

1. What is the length of the tungsten filament wire in an electric light bulb if it consumes 50 W on 220 V, and if the diameter of the wire is $25\mu$? The specific resistance of the tungsten is $5.3 \times 10^{-6}\,\Omega\text{cm}$ at 18°C and it rises approximately in proportion to the absolute temperature. Consider the operating temperature of the incandescent filament to be 2500°K. How much greater is the current at the moment of switching on (wire temperature 18°C) than in steady operation?

2. A copper conductor of 1 mm² cross-section is protected by a safety fuse of silver wire whose diameter is 0·2 mm. Calculate for a short-circuit current of 20 A approximately how long (neglecting heat conduction) it takes to melt the fuse wire, and how warm the copper wire is at this moment. (For silver the specific heat is 0·055 cal/gm degC, and the specific resistance is $1.6 \times 10^{-6}\,\Omega\text{cm}$. For copper the corresponding values are 0·091 cal/gm degC, and $1.7 \times 10^{-6}\,\Omega\text{cm}$. The melting-point of silver is 961°C.)

3. Connected to the same 220 V mains are: 6 electric light bulbs each 220 V, 50 W; a light bulb for 8 V and 6 A with the appropriate series protective resistance; an electric motor rated 1/6 hp at 220 V and having 75 per cent efficiency. What is the resistance of the entire load? What is the total current and power? What resistance has the series protective resistor for the lamp?

4. An electric heating-pot rated 220 V and 3 A brings 1 litre of water from 18°C to boiling in 11 minutes. What is the efficiency, i.e. what percentage of the energy supplied is actually used in warming the water?

5. We have $n$ accumulators (storage cells) at our disposal, each having internal resistance $R_i$ and output voltage $V^{(e)}$. The accumulators are grouped in sets consisting of $k$ series-connected accumulators each. The $n/k$ sets are themselves connected in parallel. How great must $k$ be in order to obtain, for a load-resistance $R$, the greatest power output, and how great is this power?

6. Let the capacitance $C$ be known for an arrangement consisting of two metal bodies of arbitrary shape. The whole space between the bodies is now filled with a substance of specific resistance $\rho$. Let $\rho$ be much greater than the specific resistance of the metal so that the voltage-drop in the metal electrodes can be neglected. What, now, is the resistance $R$ of the arrangement for the passage of the current from one metal body to the other? In particular, calculate the internal resistance of a Daniell cell consisting of coaxial copper and zinc cylinders with radii $a$ and $b$ and height $h$. Calculate by considering the arrangement as a cylindrical capacitor, and designate by $\rho$ the specific resistance of the acid solution of copper sulphate.

7. A parallel-plate capacitor is filled with a substance of permittivity $\varepsilon$ and conductivity $\sigma$. It is first connected to the terminals of a battery. At time $t = 0$ the connection is broken so that the capacitor is gradually discharged. How long is the relaxation time, i.e. the time at which the charge (or the voltage) of the capacitor has dropped to $1/e$ of the original value?

8. A metal sphere of radius $a$ is surrounded by a spherical medium of radius $b$, having material constants $\varepsilon$ and $\sigma$. At time $t = 0$ a quantity of charge $Q$ is uniformly distributed over the surface of the sphere. Calculate the Joule heat involved in the subsequent flow of this charge, and show that it is equal to the decrease in the electrostatic energy consequent upon the dispersal of the charge.

9. To a good approximation the magnetic field of the earth can be considered to be the field of a magnetic dipole located at the centre of the earth.
   (a) How large is the moment $m$ of this dipole if at magnetic latitude 45° we take the average value of the horizontal component of $H$ to be 0·23 Oe?
   (b) Give the inclination $i$ as a function of the magnetic latitude $\beta$.

10. Let an iron sphere of radius $a = 5$ cm be uniformly magnetized to saturation ($4\pi M = 22{,}000$ G). How great is its dipole moment? What values have $B$ and $H$ inside the sphere? What is the value of the surface divergence of $\mathbf{M}$ and the corresponding surface current density in A/cm?

11. Let there be given two small magnets $m_1$ and $m_2$; let the radius vector from the first to the second be $\mathbf{r}$. (a) What is the mutual energy of the dipoles? (b) How great a force acts between them? (c) How do the dipoles orient themselves when they are suspended so as to be free to rotate, and what is the force then acting between them?

12. Calculate the magnetic field on the axis of a straight solenoid (length $l$, radius $a$, total number of turns $n$), carrying a current $I$. First find the field on the axis of a single ring-shaped current circuit using the Biot-Savart law; then sum up or integrate this field over all $n$ current rings. Discuss the result for the case where $l \gg a$.

13. An iron ring of diameter $d = 20$ cm and cross-sectional area $a = 10$ cm² is wound evenly with $n = 600$ turns of wire. What is the value of the flux of induction $\Phi = \int B_n dS$ in the ring assuming $\mu = 500$ and a current of 1 A?

14. The iron ring in the previous example contains an air-gap $\delta$ cm wide, but not so wide that any account need in this case be taken of possible flux-fringing

effects in the air beyond the air-gap. How does the flux of induction depend on $\delta$? Calculate this flux, the field energy in the iron, the field energy in the air-gap, the total field energy, and the self-inductance for the three $\delta$-values 0·1, 1, and 5 mm.

15. With how much force (per cm) do the two wires of a parallel pair separated by 30 cm repel one another when each carries a current of 50 A?

16. A "string galvanometer" consists of a thin vertical current-carrying wire under tension in a uniform horizontally directed magnetic field. The deflection of the middle portion of the wire perpendicular to the magnetic force lines is observed with a microscope. How much is the deflection for a current of 1 mA, a magnetic field strength of 500 Oe, a wire length of 5 cm, and a tension $Z = 0·2$ g-weight? The shape of the current-carrying wire is a parabola with the equation $y = px^2/2Z$, where $p$ is the transverse "loading" (force) per unit length.

17. A ring 20 cm in diameter of copper of 1 mm² cross-sectional area rotates in the earth's magnetic field with a speed of 300 revolutions per minute around its vertical diameter. How does the strength of the current in the ring depend upon the angle between the normal direction of the ring and the direction north? How much Joule heat is liberated per second? How great a torque is necessary? How great is the magnetic field strength produced at the centre of the ring? Through what angle would a magnetic needle located here be deflected?

18. An air-core choke-coil of 0·3 H self-inductance and 20 Ω resistance is connected to an alternating voltage of 220 $V_{eff}$ with a frequency of 50 c/s. What quantity of heat is developed per minute in the coil?

19. A resistor of 10 Ω, a coil with self-inductance of 0·5 H, and a capacitor of 0·5 μF are connected in series to a sinusoidal alternating voltage of 220 $V_{eff}$ and 50 c/s. What is the effective value of the current flowing, what is its phase displacement with respect to the voltage, the effective power, and the reactive power?

20. In the "star" (Y) connection of a three-phase system, alternating currents of equal frequency and amplitude, but with phase displaced 120° with respect to one another, flow in the three branches which meet at the centre of the star. (*a*) Show that at all times the sum of the currents flowing to the centre point vanishes. (*b*) Three coils are symmetrically disposed around a circle, their axes all passing through the centre point. They carry currents supplied by a three-phase system. Each coil produces a magnetic field at the centre point in the direction of the coil axis. What direction has the resulting field strength at the centre point?

21. A choke coil with $L = 1$ H and $R = 1$ Ω is connected at time $t = 0$ to a battery with constant voltage $V$. What is the behaviour of the current with respect to the time? How long does it take for the current to reach 99·9 per cent of its eventual steady value?

22. A straight wire of length and direction s moves with velocity v in a magnetic field **B**. The ends of the wire are in sliding contact with a fixed conductor which, with the moving wire, forms a closed circuit. What is the value of the induced e.m.f.?

23. The two rails of a railroad track are insulated from one another and also from the earth. They are connected together through a millivoltmeter. What voltage

will be indicated when a train travelling at 100 km/h goes over the rails? The rail separation is 1435 mm, and the vertical component of the earth's magnetic field is 0·15 G.

## Part D. The fundamental equations of the electromagnetic field

1. The energy of solar radiation passing through a $1 \text{cm}^2$ area at the earth and perpendicular to the direction of the radiation is 2·2 cal per min (the "solar constant"). Calculate the r.m.s. average of the electric field strength (in V/cm), and the magnetic field strength (in G) in sunlight.

2. How many watts of power must an electric lamp radiate in order to produce at 1 m the brightness of sunlight?

3. A plane light-wave is incident at an angle $\alpha$ upon the plane boundary surface of a metal. Does the depth of penetration show an appreciable dependence on the angle of incidence?

4. Let a plane light-wave travelling in an insulator approach at angle $\alpha$ a plane surface bounding air. From the law of refraction, determine the limiting angle $\alpha_l$ for total reflection. For $\alpha > \alpha_l$ discuss the behaviour of the wave in air.

5. A plane wave falls at normal incidence upon the boundary surface of an extended conductor. What is the force (calculated from the momentum transfer) exerted on the conductor by the radiation? How great is the radiation pressure of the sun on the earth's surface (neglecting reflection and assuming normal incidence throughout)?

6. In the interior of an absorbing body the E-vector of a light wave produces a current density g upon which the Lorentz force $\mathbf{F} = \mathbf{g} \times \mathbf{B}/c$ acts. Using Maxwell's equations, show that the radiation pressure in problem 5 is identical with the integral $p = \displaystyle\int_0^\infty F_x \, dx$.

7. A plane light-wave described by the vector potential $\mathbf{A} = (0, A(x - ct), 0)$ falls upon an electron initially at rest. Here $A = (x - ct)$ is an arbitrary function of its argument, vanishing for $t \to -\infty$. Show that the wave causes the electron to experience a force in the $x$-direction.

8. If a suitable liquid (e.g. nitrobenzene) is brought into a strong electric field $\mathbf{E}_0$, the dipoles in the liquid acquire a partial orientation in the field direction (say the $z$-direction). The permittivity then becomes direction-dependent and can be described by means of a symmetric tensor with $\varepsilon_{xx} = \varepsilon_{yy} < \varepsilon_{zz}$, $\varepsilon_{xy} = \varepsilon_{xz} = \varepsilon_{yz} = 0$. Determine the propagation of a plane wave in a medium made anisotropic in the manner indicated.

9. If a substance be brought into a strong uniform magnetic field $\mathbf{H}_0$, then, owing to the Lorentz force, all atomic electrons acquire an additional rotational component which, because the electrons are negatively charged, is in the positive sense around the field direction. Show in the example of the Thomson atom model how the polarizability of an atom in an alternating electric field is altered by such a field $\mathbf{H}_0$.

10. For the situation described in problem 9, investigate the propagation of plane waves, in particular for the case in which the wave normal lies in the direction of $\mathbf{H}_0$. (The H-field in the light wave gives a negligibly small contribution to the Lorentz force.)

## Part E.  Relativity theory

1. Give the vector equations of the Lorentz transformation for the case in which the primed coordinate system moves with velocity **v** with respect to the unprimed.

2. If we pass from a first system ($S_1$) by means of a Lorentz transformation with $v_1$ to a second system ($S_2$), and then, joining up with this second system, pass by means of a transformation with $v_2$ to a third system ($S_3$), we cannot in general go from $S_1$ to $S_3$ by means of a Lorentz transformation (say with $v_3 = v_1 + v_2$). Rather, for this direct transfer we need in general, besides this Lorentz transformation, an additional rotation perpendicular to $v_1$ and $v_2$. Calculate this additional rotation (Thomas precession) for the special case of $v_1 = -v$, $v_2 = v + \delta v$, with $|\delta v| \ll v$.

3. As from the components of a three-dimensional tensor, so similarly from the components of a tensor of four dimensions it is possible to derive certain expressions that are invariant against general coordinate transformation. Thus it is shown directly in §81 that $E.H$ and $E^2 - H^2$ are invariants of the field tensor (81.13). To which invariants of a general tensor do these structures correspond? Are there still other tensor invariants?

4. A particle falls in a constant force-field $F$. How is the law of free fall modified by the variation of mass with velocity?

5. Derive the trajectory of a particle in a force-field $F$ for the case in which the initial velocity $v_0$ is perpendicular to the force.

6. How, in consequence of the variation of mass with velocity, does a charged particle move in a uniform magnetic field?

7. A particle of rest mass $m$ and initial velocity $v_0 = c\beta_0$ collides with a particle of mass $M$, initially at rest. What is the maximum energy that can be transferred in this elastic impact?

8. At what angles $\theta$ and $\eta$ with respect to the initial velocity do the particles $m$ and $M$ leave the point of impact in the collision described in problem 7 when the energy $T$ is transferred?

9. Let the collision of two particles of mass $m_1$ and $m_2$ be observed in the "centre-of-mass system", i.e. the system in which the total momentum of the two particles vanishes. In this system the collision appears as a change in the direction of the velocity of the two particles by the same angle $\Theta$ without changing the magnitudes $u_1$ and $u_2$ of the velocities. By means of a Lorentz transformation, calculate the collision process in the "laboratory system", in which before the collision the second particle is at rest, while the first has the velocity $v_0$. Calculate the dependence on the angle of deflection $\Theta$ in the centre-of-mass system and the energy transferred in the laboratory system.

# CHAPTER F II

## Solutions

### A. Vector and tensor calculus

1. $A = \sqrt{6}$; $\mathbf{A} = \{2, \sqrt{(3/2)}, \sqrt{(1/2)}\}$; $\langle \mathbf{A}, \mathbf{x} \rangle = 54°40'$; $\langle \mathbf{A}, \mathbf{z} \rangle = 16°47'$.

2. If we consider the special case of the unit cube spanned by the unit vectors $\mathbf{x}, \mathbf{y}, \mathbf{z}$, then the space diagonals can be represented by means of the components $(+1, +1, +1)$, $(-1, +1, +1)$, $(+1, -1, +1)$, and $(+1, +1, -1)$. From the scalar product the value $1/\sqrt{3}$ follows for the cosine of the angle, and this gives for the angle itself $70·5°$.

3. The three face diagonals have the components $(1, 1, 0)$, $(1, 0, 1)$, $(0, 1, 1)$. The angle between them is $60°$; the surface of the tetrahedron is $\frac{1}{2}\sqrt{3}$; and the volume of the tetrahedron is $1/3$.

4. The equation represents a plane because it is linear in the coordinates of P. If it be written in the form $A(\mathbf{a}_0 . \mathbf{r} - A) \equiv \mathbf{A}.(\mathbf{r} - \mathbf{A}) = 0$ it follows immediately that the point $\mathbf{r} = \mathbf{A}$ lies in the plane, and that any other vector $\mathbf{r} - \mathbf{A}$ drawn from this point is perpendicular to $\mathbf{A}$.

5. $(3, 3, 0)$.

6. $(\frac{1}{2}, -\frac{1}{2}, 0)$.

7. The area of the triangle formed by the end points of the vectors $\mathbf{a}, \mathbf{b}, \mathbf{c}$ is $S = \frac{1}{2}|(\mathbf{a}-\mathbf{b}) \times (\mathbf{a}-\mathbf{c})| = \frac{1}{2}|\mathbf{a} \times \mathbf{b}+\mathbf{b} \times \mathbf{c}+\mathbf{c} \times \mathbf{a}|$; the volume of the tetrahedron is $V = \frac{1}{6}\mathbf{a}.(\mathbf{b} \times \mathbf{c})$; its height, and therefore the required distance of the plane is $h = 3V/S$.

8. $\mathbf{v} = (31·4, -31·4, 0)\,\text{cm/s}$.

9. $6·3\,\text{cm}$.

10. From $\mathbf{t}^2 = 1$ it follows that $\mathbf{t}.d\mathbf{t}/ds = 0$. The vector $d\mathbf{t}/ds$ points toward the centre of curvature, and its magnitude is equal to the reciprocal of the radius of curvature.

11. $\text{grad}\,(\mathbf{a}.\mathbf{r}) = \mathbf{a}$; $\text{div}\,\mathbf{r} = 3$; $\text{div}\,(\mathbf{a} \times \mathbf{r}) = 0$; $\text{curl}\,\mathbf{r} = 0$; $\text{curl}\,(\mathbf{a} \times \mathbf{r}) = 2\mathbf{a}$.

12. $\omega = k/r^2$.

13. $f = k/r^3$.

14. $\mathbf{v} = -\dfrac{\mathbf{p}}{r^3} + \dfrac{3(\mathbf{p}\cdot\mathbf{r})\mathbf{r}}{r^5}$; $\text{div}\,\mathbf{v} = 0$; $\text{curl}\,\mathbf{v} = 0$.

15. (a) $c(r_2 - r_1)$; (b) $4\pi ca^2$; (c) $\dfrac{4\pi c}{3} 3(a^2 - b^2)$ for $a > b$, and $\dfrac{8\pi ca^3}{3b}$ for $a < b$.

16. (a) $2wa^2\pi$; (b) $2wS_w$, where $S_w$ means the projection of the circumscribed surface on a plane perpendicular to $\mathbf{w}$.

17. For the line element we have $ds^2 = (\xi^2 + \eta^2)(d\xi^2 + d\eta^2) + \xi^2 \eta^2 d\alpha^2$. From this it follows that

$$\operatorname{div} \mathbf{v} = \frac{1}{\sqrt{(\xi^2 + \eta^2)}} \left[ \frac{1}{\xi} \frac{\partial(\xi v_\xi)}{\partial \xi} + \frac{1}{\eta} \frac{\partial(\eta v_\eta)}{\partial \eta} + \frac{\xi v_\xi + \eta v_\eta}{\xi^2 + \eta^2} \right] + \frac{1}{\xi \eta} \frac{\partial v_\alpha}{\partial \alpha}$$

$$\nabla^2 \phi = \frac{1}{\xi^2 + \eta^2} \left[ \frac{1}{\xi} \frac{\partial}{\partial \xi} \left( \xi \frac{\partial \phi}{\partial \xi} \right) + \frac{1}{\eta} \frac{\partial}{\partial \eta} \left( \eta \frac{\partial \phi}{\partial \eta} \right) \right] + \frac{1}{\xi^2 \eta^2} \frac{\partial^2 \phi}{\partial \alpha^2}$$

18. $\lambda^{\mathrm{I}} = (\mathbf{a.b}) - ab$, $\lambda^{\mathrm{II}} = (\mathbf{a.b}) + ab$, $\lambda^{\mathrm{III}} = 0$.

$$\mathbf{n}^{\mathrm{I}} = \frac{ba - ab}{\sqrt{\{2ab[ab - (\mathbf{a.b})]\}}} \qquad \mathbf{n}^{\mathrm{II}} = \frac{ba + ab}{\sqrt{\{2ab[ab + (\mathbf{a.b})]\}}} \qquad \mathbf{n}^{\mathrm{III}} = \frac{\mathbf{a} \times \mathbf{b}}{\sqrt{\{a^2 b^2 - (\mathbf{a.b})^2\}}}$$

## B. The electrostatic field

1. 917 kg-force.

2. $-\dfrac{e^2 r}{a^3}$.

3. (a) 625 dynes;  (b) 470 dynes.

4. Force of attraction $= \dfrac{e^2 R^3 (2r^2 - R^2)}{r^3 (r^2 - R^2)^2} - \dfrac{Ee}{r^2}$.

5. 1667 e.s.u. $= 5.56 \times 10^{-7}$ coulomb.

6. 0.5 cm.

7. If $V$ is the applied voltage, the field strength on the edge is $V/r$, and in the middle of the ellipsoid of revolution it is $V/[r \ln(l/r)]$.

8. $\omega = 6.6 \times 10^{-4}$ e.s.u./cm$^2$, $F = 0.028$ dyne/m$^2$.

9. 6670 V/cm.

10. $r = 11$ mm.

11. (a) E;  (b) $\varepsilon \mathbf{E} = \mathbf{E} + 4\pi \mathbf{P}$;  (c) $\dfrac{3\varepsilon}{1 + 2\varepsilon} \mathbf{E} = \mathbf{E} + \dfrac{4\pi}{1 + 2\varepsilon} \mathbf{P}$.

12. (a) 44.3 dynes/cm$^2$;  (b) 22.2 dynes/cm$^2$;  (c) 88.6 dynes/cm$^2$.

13. $3.15 \times 10^{-5}$ coulomb $= 94{,}500$ e.s.u.;  $V_1 = 63$ V, $V_2 = 157$ V.

14. $9 \times 10^{-4}$ coulomb $= 2.7 \times 10^6$ e.s.u.

15. $\mathbf{p.E} = pE \cos \alpha$. For dipoles that are free to rotate, we must place $\alpha = 0$.

16. (a) $2ep/r^3$;  (b) $3eQ/r^4$; in both cases repulsion if $e$ has the same sign as the charge in the dipole (or quadrupole) nearest to $e$.

17. The dipoles are parallel to each other and have the direction of the line joining them.

18. $C = \dfrac{1}{4\pi |\phi_2 - \phi_1|} \oint \left| \dfrac{\partial \phi}{\partial n} \right| ds$; the integral is to extend over the periphery of the cylinder designated by subscript $-1$.

19. Differentiation of $w = u(x, y) + iv(x, y) = f(x + iy)$ once with respect to $x$, and a second time with respect to $y$ leads to the Cauchy-Riemann differential equations $\partial u / \partial x = \partial v / \partial y$, $\partial u / \partial y = - \partial v / \partial x$. By elimination of $v$ and $u$, we obtain the desired Laplace equations.

20. For the first part of the example it suffices to remark that the real part of $f[F(z)]$ satisfies the Laplace equation and also is constant on the desired line of $A$.—The capacitances of the constructed arrangement and the original arrangement are equal. We show this with respect to the formula for capacitance in the above solution to exercise 18. First, the potential values on the cylindrical surfaces will not be changed by the representation; secondly, on account of the Cauchy relations applied to $f(\zeta) = \phi + i\psi$, we have

$$\oint \frac{\partial \phi}{\partial n}\, ds \equiv \oint \left(\frac{\partial \phi}{\partial \xi}\, d\eta - \frac{\partial \phi}{\partial \eta}\, d\xi\right) = \oint \left(\frac{\partial \psi}{\partial \eta}\, d\eta + \frac{\partial \psi}{\partial \xi}\, d\xi\right) = \oint d\psi$$

and the value of the line integral over the closed path also remains unchanged.

21. In order to find the zero point of the representation (map) which carries the twin parallel conductors over to the cylindrical capacitor, we construct the circle which cuts perpendicularly the circles of the two parallel wires in two points. Each of the two intersection points of this auxiliary circle with the line joining the two conductors can be used as the zero point of the representation $\zeta = F(z)$, for then the auxiliary circle goes over into the line $x = 0$, which must be perpendicular to the representation of the two conductor circles because of the angle-preserving feature of the transformation $\zeta = F(z)$. On symmetry grounds, then, it follows that both these pictures must consist of concentric circles. Their radii then come out to be

$$r_{1,2} = \frac{c^2}{2r}\left[\frac{d}{\sqrt{(d^2 - 4r^2)}} \mp 1\right]$$

and therefore the reciprocal capacitance of the twin conductors per cm is given by

$$\frac{1}{C} = 2 \ln \frac{r_2}{r_1} = 4 \ln \left[\frac{d}{2r} + \sqrt{\left(\frac{d^2}{4r^2} + 1\right)}\right]$$

## C. The electric current and the magnetic field

1. Wire length $\sim 1\,$m; $I_0/I = 8\cdot6$ for $I_0 = 1\cdot95\,$A.

2. $0\cdot35\,$s; during this time the copper wire has warmed up $0\cdot7°$C.

3. $27\cdot2\,\Omega$; $8\cdot1\,$A; $1780\,$W; protective resistor: $35\cdot4\,\Omega$.

4. 79 per cent.

5. $k = \sqrt{(nR/R_i)}$; power $= nV^{(e)2}/4R_i$.

6. $R = \rho/4\pi C$; particular case $R = \rho \ln (b/a)/2\pi h$.

7. $\varepsilon/4\pi\sigma$.

8. $\dfrac{Q^2}{2\varepsilon}\left(\dfrac{1}{a} - \dfrac{1}{b}\right)$.

9. (a) $m = 8\cdot37 \times 10^{25}\,$G; (b) $\tan i = 2 \tan \beta$.

10. $m = 916{,}000\,$G, $H = -7350\,$Oe, $B = 14{,}700\,$G; the surface divergence of $\mathbf{M}$ (surface density of magnetization) is $-M \cos \theta$, the corresponding surface current density is $cM \sin \theta$, with the maximum value $cM = 17{,}500\,$A/cm; $\theta$ here is the polar angle of the sphere with the polar axis in the $\mathbf{M}$-direction.

11. (a) $\dfrac{1}{r^3}\left\{\mathbf{m}_1 . \mathbf{m}_2 - 3\left(\mathbf{m}_1 . \dfrac{\mathbf{r}}{r}\right)\left(\mathbf{m}_2 . \dfrac{\mathbf{r}}{r}\right)\right\}$

   (b) $\dfrac{3}{r^5}\left\{(\mathbf{m}_1 . \mathbf{r})\mathbf{m}_2 + (\mathbf{m}_2 . \mathbf{r})\mathbf{m}_1 + (\mathbf{m}_1 . \mathbf{m}_2)\mathbf{r} - 5\left(\mathbf{m}_1 . \dfrac{\mathbf{r}}{r}\right)\left(\mathbf{m}_2 . \dfrac{\mathbf{r}}{r}\right)\mathbf{r}\right\}$

   (c) Stable equilibrium for $\mathbf{m}_1\|\mathbf{m}_2\|\mathbf{r}$; the attractive force then becomes $-6m_1 m_2/r^4$.

12. $H = \dfrac{2\pi nI}{lc}\left(\dfrac{z+l/2}{\sqrt{[(z+l/2)^2+a^2]}} - \dfrac{z-l/2}{\sqrt{[(z-l/2)^2+a^2]}}\right)$; $z$ here means the distance

   from the plane of symmetry of the coil. In the case $l \gg a$ the expression in parentheses for $|z| < l/2 - a$ becomes practically equal to 2, and thus $H$ becomes equal to the field in an infinitely long coil; while for $|z| > l$, the field $H$ falls off like $z^{-3}$, i.e. like a dipole field.

13. 60,000 maxwells.

14. $\Phi = \dfrac{0\cdot4\pi nIq\mu}{\pi d + (\mu - 1)\delta}$ maxwells.

| Air-gap thickness..............(mm) | 0·1 | 1 | 5 |
|---|---|---|---|
| Flux of induction..........(maxwells) | 55,500 | 33,400 | 12,100 |
| Energy in iron...............(joules) | 0·154 | 0·055 | 0·007 |
| Energy in the air-gap.........(joules) | 0·012 | 0·044 | 0·029 |
| Total energy................(joules) | 0·166 | 0·099 | 0·036 |
| Self-inductance.............(henrys) | 0·332 | 0·198 | 0·072 |

15. 1·67 dynes/cm.

16. 7·8 $\mu$.

17. If $R$ is the wire resistance, $r$ the radius of the ring, and $H$ the horizontal component of the earth's field, then we have

$$I = \dfrac{\omega \pi r^2 H}{cR} \sin \omega t = 2\cdot1 \sin \omega t \text{ mA;}$$

   the heat developed is $5\cdot6 \times 10^{-9}$ cal/s;
   the torque = $6\cdot7 \times 10^{-6}(1 - \cos 2\omega t)$ g-weight cm;
   the field strength perpendicular to the plane of the ring is

$$\dfrac{2\pi^2\omega rH}{c^2R} \sin \omega t = 1\cdot3 \times 10^{-4} \sin \omega t \text{ Oe}$$

   with the component $\dfrac{2\pi^2\omega rH}{c^2R} \sin^2 \omega t$

   perpendicular to the earth's field. The deflection of the magnetic needle at the centre of the ring: $\alpha \approx 1'$, in the direction of rotation of the ring.

18. 1490 cal/min.

19. $I_{eff} = 0\cdot0354$ A; $\phi = 89° 54\cdot5'$ (leading); effective power = $0\cdot00125$ W, reactive power = $0\cdot779$ W.

20. (a) $\sin \omega t + \sin (\omega t + 2\pi/3) + \sin (\omega t + 4\pi/3) = 0$;
    (b) $H = $ constant, $\mathbf{H}$ rotates with the frequency $\omega$.

21. $I = \dfrac{V}{R}(1 - e^{-Rt/L})$. The time required is $6 \cdot 9 \mathrm{s}$.

22. $V = \mathbf{s} \cdot (\mathbf{v} \times \mathbf{B})/c$.

23. $0 \cdot 6 \mathrm{mV}$.

# D. The fundamental equations of the electromagnetic field

1. $7 \cdot 5 \mathrm{V/cm}$; $2 \cdot 5 \times 10^{-2} \mathrm{G}$.

2. $19 \mathrm{kW}$.

3. Since $n \approx k \gg 1$ it follows from $\sin \alpha = (n - ik) \sin \beta$ that $\beta$, as regards its magnitude, is small compared to 1; thus we have that $\beta \approx \sin \alpha/(n - ik)$ and $\cos \beta \approx 1 - \sin^2 \alpha/2(n - ik)^2$. Therefore the depth of penetration varies with $\alpha$ only in terms of higher order.

4. $\sin \alpha_l = 1/n$. For $\alpha > \alpha_l$ the wave in air decays exponentially with a depth of penetration $d = c/\omega \sqrt{(n^2 \sin^2 \alpha - 1)}$, and with a Poynting vector parallel to the boundary surface.

5. The radiation pressure is $\overline{E^2}/4\pi$; for sunlight $5 \times 10^{-5} \mathrm{dyne/cm^2}$.

6. From Maxwell's equations we have for a wave travelling in the $x$-direction and polarized in the $y$-direction, in a non-magnetic medium that

$$F_x = \frac{1}{c} g_y H_z = -\frac{1}{4\pi} \left\{ \frac{1}{2} \frac{\partial H_z^2}{\partial x} + \frac{1}{2} \frac{\partial E_y^2}{\partial x} + \frac{1}{c} \frac{\partial}{\partial t} (E_y H_z) \right\}$$

and thus for the time average $F_x = -\dfrac{1}{8\pi} \dfrac{\partial}{\partial x} (E_y^2 + H_z^2)$, etc.

7. The velocity of the electrons at any chosen time is given by

$$\dot{x} \left( 1 - \frac{\dot{x}}{2c} \right) = \frac{e^2}{m^2 c^3} A^2(x - ct) > 0, \qquad \dot{y} = -\frac{e}{mc} A(x - ct), \qquad \dot{z} = 0.$$

8. The medium becomes doubly refracting (Kerr effect); a wave travelling perpendicularly to the $E_0$-direction becomes split into two components with different velocities of propagation. Thus a linearly polarized wave whose plane of polarization makes an angle with $E_0$, is in general elliptically polarized after traversing the medium.

9. From $m(\ddot{\mathbf{r}} + \omega_0^2 \mathbf{r}) = -e_0 \left( \mathbf{E} + \dfrac{\dot{\mathbf{r}}}{c} \times \mathbf{H}_0 \right)$ with $\mathbf{E} = \mathbf{E}_0\, e^{i\omega t}$, it follows that

$$\mathbf{p} = \frac{e_0^2/m}{(\omega_0^2 - \omega^2) \left\{ 1 - \left[ \dfrac{\omega e_0 H_0}{mc(\omega_0^2 - \omega^2)} \right]^2 \right\}} \times$$

$$\times \left\{ \mathbf{E} - \left[ \frac{\omega e_0}{mc(\omega_0^2 - \omega^2)} \right]^2 (\mathbf{E}.\mathbf{H}_0)\mathbf{H}_0 - \frac{i\omega e_0 (\mathbf{E} \times \mathbf{H}_0)}{mc(\omega_0^2 - \omega^2)} \right\}$$

In the special case of $\mathbf{H}_0 \| z$ we obtain an $\varepsilon$-tensor with $\varepsilon_{xx} = \varepsilon_{yy} \neq \varepsilon_{zz}$; $\varepsilon_{xz} = \varepsilon_{yz} = 0$; $\varepsilon_{xy} = -\varepsilon_{yx}$ is pure imaginary.

10. The medium becomes circularly birefringent (Faraday effect). For the particular case $\mathbf{f} \| \mathbf{H}_0$ we obtain two different directions of propagation for the two circularly polarized waves ($E_y = \pm iE_x$, $H_x = \mp iH_y$).

## E. Relativity theory

1. $\mathbf{r}' = \mathbf{r} - \dfrac{(\mathbf{r}.\mathbf{v})\mathbf{v}}{v^2} + \left(\dfrac{\mathbf{r}.\mathbf{v}}{v^2} - t\right)\dfrac{\mathbf{v}}{\sqrt{(1 - v^2/c^2)}}$;   $t' = \dfrac{t - \dfrac{\mathbf{r}.\mathbf{v}}{c^2}}{\sqrt{(1 - v^2/c^2)}}$

2. The rotation vector $\mathbf{v} = -\dfrac{\mathbf{v} \times \delta\mathbf{v}}{v^2}\left(\dfrac{1}{\sqrt{(1 - v^2/c^2)}} - 1\right) \approx -\dfrac{\mathbf{v} \times \delta\mathbf{v}}{2c^2}$ for $v \ll c$

3. For a skew-symmetric tensor of second rank $F_{\mu\nu}$ there are only the two invariants $\sum_\nu \sum_\mu F_{\mu\nu}^2$ and Det $(F_{\mu\nu})$; in the case of the electromagnetic field-tensor the first is $2(\mathbf{H}^2 - \mathbf{E}^2)$, the second $-(\mathbf{E}.\mathbf{H})^2$. For a symmetric tensor there are in addition other different invariants which, as in §15, can be derived from the principal axis transformation—in particular, the spur of the tensor $\sum_\nu T_{\nu\nu}$.

4. $z = \dfrac{mc^2}{K}\left\{\sqrt{\left[1 + \left(\dfrac{Kt}{mc}\right)^2\right]} - 1\right\}$;   that is, a hyperbola in the $z - t$ diagram with the asymptotic final velocity $c$.

5. $x = l\dfrac{v_0}{c}\ln\left\{\dfrac{ct}{l} + \sqrt{\left[1 + \left(\dfrac{ct}{l}\right)^2\right]}\right\}$     $z = l\left\{\sqrt{\left[1 + \left(\dfrac{ct}{l}\right)^2\right]} - 1\right\}$

   with $\dfrac{1}{l} = \dfrac{K\sqrt{(1 - v_0^2/c^2)}}{mc^2}$     Trajectory: $z = 2l\sinh^2(xc/2lv_0)$.

6. Helix, or in the special case, a circle, which is traversed at the frequency

$$\omega = \dfrac{eH}{mc}\sqrt{\left(1 - \dfrac{v^2}{c^2}\right)}$$

7. Maximum energy transferred $= \dfrac{2Mm^2v_0^2}{\sqrt{(1 - \beta_0^2)}\,[(m^2 + M^2)\sqrt{(1 - \beta_0^2)} + 2mM]}$

   For $\beta_0 \ll 1$ we obtain from this $\dfrac{4Mm}{(m + M)^2}\dfrac{mv_0^2}{2}$.

8. If $p_0$ and $E_0$ are the momentum and energy of the particle of mass $m$ before the collision, then after the collision this particle has the momentum

$$p = \sqrt{[(E_0 - T)^2/c^2 - m^2c^2]}$$

The other particle then has the momentum

$$P = \sqrt{(2MT + T^2/c^2)}.$$

For the angle sought we have

$$\cos \theta = \frac{p_0^2 - (M + E_0/c^2)T}{p_0 p} \qquad \cos \eta = \frac{(M + E_0/c^2)T}{p_0 P}$$

9. The relationships between the velocities $u_1$ and $u_2$ in the centre-of-mass system and initial velocity in the laboratory system are

$$u_1 = m_2 v_0/[m_1\sqrt{(1 - \beta_0^2)} + m_2], \qquad u_2 = m_1 v_0/[m_1 + m_2\sqrt{(1 - \beta_0^2)}]$$

If the Lorentz transformation be carried out on the four-vector of the momentum it follows then from the fourth component, $m_2 T = p^2(1 - \cos \Theta)$, with $p = p_1 = p_2 = $ momentum magnitudes in the centre-of-mass system.

# G

---

List
of
formulae

# CHAPTER G I

## Vector and Tensor Calculus

### 1. Vector algebra

Vectors are indicated as such throughout the book by the use of bold-face type, e.g. **A**, their magnitude by the use of absolute bars or Roman (Latin) letters, e.g. $|\mathbf{A}|$ or $A$. For two vectors **A** and **B**, $\langle \mathbf{A}, \mathbf{B} \rangle$ means the angle between them, $\mathbf{A}.\mathbf{B}$ is their scalar product, and $\mathbf{A} \times \mathbf{B}$ their vector product. The unit vectors in the directions of the coordinates are designated by **x**, **y**, **z**.

Components of a vector:

$$A_n = A \cos \langle \mathbf{A}, \mathbf{n} \rangle$$
$$= A_x \cos \langle \mathbf{x}, \mathbf{n} \rangle + A_y \cos \langle \mathbf{y}, \mathbf{n} \rangle + A_z \cos \langle \mathbf{z}, \mathbf{n} \rangle$$

Scalar product:

$$\mathbf{A}.\mathbf{B} = AB \cos \langle \mathbf{A}, \mathbf{B} \rangle = A_x B_x + A_y B_y + A_z B_z$$

Vector product:

$$\mathbf{A} \times \mathbf{B} = -\mathbf{B} \times \mathbf{A} = \begin{vmatrix} \mathbf{x} & \mathbf{y} & \mathbf{z} \\ A_x & A_y & A_z \\ B_x & B_y & B_z \end{vmatrix} \qquad |\mathbf{A} \times \mathbf{B}| = AB \sin \langle \mathbf{A}, \mathbf{B} \rangle$$

Spar product:

$$\mathbf{A}.(\mathbf{B} \times \mathbf{C}) = \mathbf{B}.(\mathbf{C} \times \mathbf{A}) = \mathbf{C}.(\mathbf{A} \times \mathbf{B}) = \begin{vmatrix} A_x & A_y & A_z \\ B_x & B_y & B_z \\ C_x & C_y & C_z \end{vmatrix}$$

Product rules:

$$\mathbf{A} \times (\mathbf{B} \times \mathbf{C}) = \mathbf{B}(\mathbf{A}.\mathbf{C}) - \mathbf{C}(\mathbf{A}.\mathbf{B})$$

$$\mathbf{A} \times (\mathbf{B} \times \mathbf{C}) + \mathbf{B} \times (\mathbf{C} \times \mathbf{A}) + \mathbf{C} \times (\mathbf{A} \times \mathbf{B}) = 0$$

$$(\mathbf{A} \times \mathbf{B}).(\mathbf{C} \times \mathbf{D}) = (\mathbf{A}.\mathbf{C})(\mathbf{B}.\mathbf{D}) - (\mathbf{A}.\mathbf{D})(\mathbf{B}.\mathbf{C})$$

$$(\mathbf{A} \times \mathbf{B})^2 = A^2 B^2 - (\mathbf{A}.\mathbf{B})^2$$

## 2. Vector analysis

(a) *Differential relationships*:

Del or nabla operator:

$$\nabla \equiv \mathbf{x}\frac{\partial}{\partial x} + \mathbf{y}\frac{\partial}{\partial y} + \mathbf{z}\frac{\partial}{\partial z}$$

Differentiation with respect to the direction of the unit vector $\mathbf{n}$ is often described as

$$\mathbf{n}\cdot\nabla = \frac{\partial}{\partial n}$$

Gradient:

$$\operatorname{grad}\phi \equiv \nabla\phi = \mathbf{x}\frac{\partial\phi}{\partial x} + \mathbf{y}\frac{\partial\phi}{\partial y} + \mathbf{z}\frac{\partial\phi}{\partial z}$$

Divergence:

$$\operatorname{div}\mathbf{A} \equiv \nabla\cdot\mathbf{A} = \frac{\partial A_x}{\partial x} + \frac{\partial A_y}{\partial y} + \frac{\partial A_z}{\partial z}$$

Curl:

$$\operatorname{curl}\mathbf{A} \equiv \nabla\times\mathbf{A}$$
$$= \mathbf{x}\left(\frac{\partial A_z}{\partial y} - \frac{\partial A_y}{\partial z}\right) + \mathbf{y}\left(\frac{\partial A_x}{\partial z} - \frac{\partial A_z}{\partial x}\right) + \mathbf{z}\left(\frac{\partial A_y}{\partial x} - \frac{\partial A_x}{\partial y}\right)$$

Calculation rules:

$$\operatorname{div}\phi\mathbf{A} = \phi\operatorname{div}\mathbf{A} + \mathbf{A}\operatorname{grad}\phi$$

$$\operatorname{curl}\phi\mathbf{A} = \phi\operatorname{curl}\mathbf{A} + \operatorname{grad}\phi\times\mathbf{A}$$

$$\operatorname{div}(\mathbf{A}\times\mathbf{B}) = \mathbf{B}\operatorname{curl}\mathbf{A} - \mathbf{A}\operatorname{curl}\mathbf{B}$$

$$\operatorname{curl}(\mathbf{A}\times\mathbf{B}) = (\mathbf{B}\cdot\nabla)\mathbf{A} - (\mathbf{A}\cdot\nabla)\mathbf{B} + \mathbf{A}\operatorname{div}\mathbf{B} - \mathbf{B}\operatorname{div}\mathbf{A}$$

$$\operatorname{div}\operatorname{grad}\phi \equiv (\nabla\cdot\nabla)\phi \equiv \nabla^2\phi = \frac{\partial^2\phi}{\partial x^2} + \frac{\partial^2\phi}{\partial y^2} + \frac{\partial^2\phi}{\partial z^2}$$

$$\operatorname{curl}\operatorname{grad}\phi = 0$$

$$\operatorname{div}\operatorname{curl}\mathbf{A} = 0$$

$$\operatorname{curl}\operatorname{curl}\mathbf{A} = \operatorname{grad}\operatorname{div}\mathbf{A} - \nabla^2\mathbf{A}$$

For the differential relationships in curvilinear coordinates, see §13.

(*b*) *Integral relationships*:

Symbols:

$$\oint \mathbf{A} \cdot d\mathbf{r}$$

signifies a line integral over a closed curve (circulation).

$$\oiint A_n \, dS$$

signifies a surface integral over a closed surface (total flux).

Surface integrals and volume integrals are often written with only a single integral sign.

Gauss's theorem:

$$\iiint \operatorname{div} \mathbf{A} \, dv = \oiint A_n \, dS$$

Stokes's theorem:

$$\iint (\operatorname{curl} \mathbf{A})_n \, dS = \oint \mathbf{A} \cdot d\mathbf{r}$$

Green's theorem:

$$\iiint (\psi \nabla^2 \phi - \phi \nabla^2 \psi) \, dv = \oiint \left( \psi \frac{\partial \phi}{\partial n} - \phi \frac{\partial \psi}{\partial n} \right) dS$$

## 3. Tensor algebra

If instead of the indices $x$, $y$, $z$, the indices 1, 2, 3 are used, a tensor $\mathbf{T}$ is then described by the nine components $T_{ik}$ with $i, k = 1, 2, 3$. For the tensor components with respect to two directions $\mathbf{n}$ and $\mathbf{m}$ we have

$$T_{nm} = \sum_{i, k=1}^{3} T_{ik} \, n_i \, m_k$$

Symmetric tensor:

$$T_{ik} = T_{ki}$$

Skew-symmetric tensor:

$$T_{ik} = -T_{ki}$$

The eigenvalue equation

$$\sum_k T_{ik} a_k = \lambda a_i \quad \text{for} \quad i = 1, 2, 3$$

gives with

$$\begin{vmatrix} T_{11} - \lambda & T_{12} & T_{13} \\ T_{21} & T_{22} - \lambda & T_{23} \\ T_{31} & T_{32} & T_{33} - \lambda \end{vmatrix} = 0$$

the three eigenvalues, $\lambda^{\mathrm{I}}$, $\lambda^{\mathrm{II}}$, $\lambda^{\mathrm{III}}$ of the tensor and the associated principal-axis directions $\mathbf{a}^{\mathrm{I}}$, $\mathbf{a}^{\mathrm{II}}$, $\mathbf{a}^{\mathrm{III}}$. Referred to these three mutually perpendicular directions as coordinate axes, it follows that

$$T_{11} = \lambda^{\mathrm{I}} \qquad T_{22} = \lambda^{\mathrm{II}} \qquad T_{33} = \lambda^{\mathrm{III}} \qquad T_{ik} = 0 \quad \text{for} \quad i \neq k$$

A tensor with spur zero is designated as the deviator: here the spur, being invariant against rotations of coordinates, is defined by $Sp(\mathbf{T}) = T_{11} + T_{22} + T_{33}$. The associated deviator is obtained from a tensor $\mathbf{T}$ by subtracting one-third of the spur from all components $T_{ii}$.

## CHAPTER G II

### Electrodynamics

The Gaussian CGS system and the Giorgi MKSA system are employed. If a given formula is different in these two systems, then, in the following, both formulae will be given side by side.

### 1. The field equations and the constitutive equations

|  | CGS System | MKSA System |
|---|---|---|

Maxwell equations:

$$\text{curl}\,\mathbf{H} = \frac{1}{c}\frac{\partial \mathbf{D}}{\partial t} + \frac{4\pi}{c}\mathbf{g} \qquad \text{curl}\,\mathbf{H} = \frac{\partial \mathbf{D}}{\partial t} + \mathbf{g}$$

$$\text{curl}\,\mathbf{E} = -\frac{1}{c}\frac{\partial \mathbf{B}}{\partial t} \qquad \text{curl}\,\mathbf{E} = -\frac{\partial \mathbf{B}}{\partial t}$$

$$\text{div}\,\mathbf{D} = 4\pi\rho \qquad \text{div}\,\mathbf{D} = \rho$$

$$\text{div}\,\mathbf{B} = 0 \qquad \text{div}\,\mathbf{B} = 0$$

Constitutive equations:

$$\mathbf{D} = \mathbf{E} + 4\pi\mathbf{P} \qquad \mathbf{D} = \varepsilon_0\mathbf{E} + \mathbf{P}$$

$$\mathbf{B} = \mathbf{H} + 4\pi\mathbf{M} \qquad \mathbf{B} = \mu_0\mathbf{H} + \mathbf{M}$$

Equation of continuity:

$$\text{div}\,\mathbf{g} + \frac{\partial \rho}{\partial t} = 0 \qquad \text{div}\,\mathbf{g} + \frac{\partial \rho}{\partial t} = 0$$

If the quantities $\mathbf{D}$ and $\mathbf{H}$, considered in electron theory to be secondary quantities, are eliminated from the Maxwell equations, we obtain the new equations:

$$\text{curl}\,\mathbf{B} = \frac{1}{c}\frac{\partial \mathbf{E}}{\partial t} + \frac{4\pi}{c}(\mathbf{g} + \mathbf{g}_P + \mathbf{g}_M) \qquad \frac{1}{\mu_0}\text{curl}\,\mathbf{B} = \varepsilon_0\frac{\partial \mathbf{E}}{\partial t} + (\mathbf{g} + \mathbf{g}_P + \mathbf{g}_M)$$

$$\text{div}\,\mathbf{E} = 4\pi(\rho + \rho_P) \qquad \varepsilon_0\,\text{div}\,\mathbf{E} = \rho + \rho_P$$

The meanings here are:

$\rho_P = -\text{div}\,\mathbf{P}$, the density of polarization charge

$\mathbf{g}_P = \dfrac{\partial \mathbf{P}}{\partial t}$, the density of polarization current

$$\mathbf{g}_M = c\,\text{curl}\,\mathbf{M} \quad\Big|\quad \mathbf{g}_M = \frac{1}{\mu_0}\,\text{curl}\,\mathbf{M}, \text{ the density of magnetization current}$$

## 2. The material constants

Normal isotropic substances (i.e. non-ferroelectric non-ferromagnetic substances) are, in their relations to static fields, characterized by special material constants:

| | | |
|---|---|---|
| Electrical conductivity: | $\mathbf{g} = \sigma(\mathbf{E}+\mathbf{E}^{(e)})$ | $\mathbf{g} = \sigma(\mathbf{E}+\mathbf{E}^{(e)})$ |
| Permittivity (dielectric constant) | $\mathbf{D} = \varepsilon\mathbf{E}$ | $\mathbf{D} = \varepsilon\varepsilon_0\,\mathbf{E}$ |
| or electric susceptibility: | $\mathbf{P} = \chi\mathbf{E}$ | $\mathbf{P} = \chi\varepsilon_0\,\mathbf{E}$ |
| | $\varepsilon = 1+4\pi\chi$ | $\varepsilon = 1+\chi$ |
| Permeability | $\mathbf{B} = \mu\mathbf{H}$ | $\mathbf{B} = \mu\mu_0\,\mathbf{H}$ |
| or magnetic susceptibility: | $\mathbf{M} = \kappa\mathbf{H}$ | $\mathbf{M} = \kappa\mu_0\,\mathbf{H}$ |
| | $\mu = 1+4\pi\kappa$ | $\mu = 1+\kappa$ |

In fields changing with time, $\sigma$, $\varepsilon$, and $\mu$ are in general frequency-dependent; the foregoing relationships then only hold between the Fourier components of the field strengths.

## 3. Energy and force expressions

Energy theorem:
$$\frac{\partial u}{\partial t}+\text{div}\,\mathbf{S} = -\mathbf{g}.\mathbf{E} = -\frac{\mathbf{g}^2}{\sigma}+\mathbf{g}.\mathbf{E}^{(e)}$$

| | | |
|---|---|---|
| Energy density of the field: | $du = \dfrac{1}{4\pi}(\mathbf{E}.d\mathbf{D}+\mathbf{H}.d\mathbf{B})$ | $du = \mathbf{E}.d\mathbf{D}+\mathbf{H}.d\mathbf{B}$ |
| specially, for normal substances: | $u = \dfrac{1}{8\pi}(\mathbf{E}.\mathbf{D}+\mathbf{H}.\mathbf{B})$ | $u = \tfrac{1}{2}(\mathbf{E}.\mathbf{D}+\mathbf{H}.\mathbf{B})$ |
| Poynting vector: | $\mathbf{S} = \dfrac{c}{4\pi}\mathbf{E}\times\mathbf{H}$ | $\mathbf{S} = \mathbf{E}\times\mathbf{H}$ |
| Force on moving charge: | $\mathbf{F} = e\left(\mathbf{E}+\dfrac{\mathbf{v}}{c}\times\mathbf{B}\right)$ | $\mathbf{F} = e(\mathbf{E}+\mathbf{v}\times\mathbf{B})$ |
| Force density in matter with $\varepsilon = \mu = 1$: | $\mathbf{f} = \rho\mathbf{E}+\dfrac{\mathbf{g}}{c}\times\mathbf{B}$ | $\mathbf{f} = \rho\mathbf{E}+\mathbf{g}\times\mathbf{B}$ |

Force density in matter with $\mu = 1$, $\sigma = 0$ (conductivity):

$$\mathbf{f} = \rho\mathbf{E}-\frac{1}{8\pi}\mathbf{E}^2\,\text{grad}\,\varepsilon+\frac{1}{8\pi}\,\text{grad}\left(\mathbf{E}^2\,\frac{d\varepsilon}{d\sigma}\,\sigma\right)$$

(The $\sigma$ in this formula is the matter density.)

The force density can always be represented as the divergence of the Maxwell stress tensor.

## 4. Wave propagation

For normal homogeneous uncharged media ($\varepsilon$, $\mu$, $\sigma$ constant in space, $\mathbf{E}^{(e)} = 0$, $\rho = 0$) every field component $F$ satisfies the wave equation

$$\nabla^2 F = \frac{\varepsilon\mu}{c^2}\frac{\partial^2 F}{\partial t^2} + \frac{4\pi\sigma\mu}{c^2}\frac{\partial F}{\partial t} \qquad \nabla^2 F = \mu\mu_0\left(\varepsilon\varepsilon_0\frac{\partial^2 F}{\partial t^2} + \sigma\frac{\partial F}{\partial t}\right)$$

For material constants that are frequency-dependent this equation has meaning only for the individual temporal Fourier components of $F$.

From this, the wave velocity *in vacuo* $\qquad c = 1/\sqrt{(\varepsilon_0\,\mu_0)}$

Wave velocity in insulators $\qquad c/n = c/\sqrt{(\varepsilon\mu)}$

Index of refraction $\qquad n = \sqrt{(\varepsilon\mu)}$

The calculation of fields *in vacuo* is facilitated by going over to the potentials $\mathbf{A}$ and $\phi$ by means of

$$\begin{cases} \mathbf{B} = \operatorname{curl}\mathbf{A} \\[2mm] \mathbf{E} = -\frac{1}{c}\frac{\partial\mathbf{A}}{\partial t} - \operatorname{grad}\phi \end{cases} \qquad \begin{aligned} &\mathbf{B} = \operatorname{curl}\mathbf{A} \\[2mm] &\mathbf{E} = -\frac{\partial\mathbf{A}}{\partial t} - \operatorname{grad}\phi \end{aligned}$$

Potential equations:

$$\begin{cases} \nabla^2\mathbf{A} - \frac{1}{c^2}\frac{\partial^2\mathbf{A}}{\partial t^2} = -\frac{4\pi\mathbf{g}}{c} \\[3mm] \nabla^2\phi - \frac{1}{c^2}\frac{\partial^2\phi}{\partial t^2} = -4\pi\rho \end{cases} \qquad \begin{aligned} &\nabla^2\mathbf{A} - \varepsilon_0\,\mu_0\frac{\partial^2\mathbf{A}}{\partial t^2} = -\mu_0\,\mathbf{g} \\[3mm] &\nabla^2\phi - \varepsilon_0\,\mu_0\frac{\partial^2\phi}{\partial t^2} = -\frac{\rho}{\varepsilon_0} \end{aligned}$$

Lorentz convention $\quad \operatorname{div}\mathbf{A} + \frac{1}{c}\frac{\partial\phi}{\partial t} = 0 \qquad \operatorname{div}\mathbf{A} + \varepsilon_0\,\mu_0\frac{\partial\phi}{\partial t} = 0$

General integral of the $\phi$-equation in the CGS system:

$$\phi(\mathbf{r}, t) = \int \frac{\rho\left(\mathbf{r}', t - \frac{|\mathbf{r} - \mathbf{r}'|}{c}\right) dv'}{|\mathbf{r} - \mathbf{r}'|}$$

## 5. Electrotechnical concepts

Capacitance: $\qquad\qquad C = Q/V \qquad\qquad\qquad\qquad C = Q/V$

Parallel-plate capacitor (plate separation $d$, surface $S$): $\qquad \dfrac{S\varepsilon}{4\pi d} \qquad\qquad\qquad \dfrac{S\varepsilon\varepsilon_0}{d}$

Cylindrical capacitor (radii $r_1 < r_2$; length $l$): $\qquad \dfrac{l\varepsilon}{2\ln(r_2/r_1)} \qquad\qquad \dfrac{2\pi l\varepsilon\varepsilon_0}{\ln(r_2/r_1)}$

Capacitance of a sphere (radius $r$): $\qquad\qquad r\varepsilon \qquad\qquad\qquad 4\pi r\varepsilon\varepsilon_0$

Ellipsoid of rotation (axes $a > b$): $\qquad \dfrac{\sqrt{(a^2-b^2)}\,\varepsilon}{\ln[a+\sqrt{(a^2-b^2)}]-\ln b} \qquad \dfrac{4\pi\sqrt{(a^2-b^2)}\,\varepsilon\varepsilon_0}{\ln[a+\sqrt{(a^2-b^2)}]-\ln b}$

Inductance: $\qquad L = \dfrac{\mu}{c^2}\oiint\dfrac{d\mathbf{r}_1 . d\mathbf{r}_2}{r_{12}} \qquad L = \dfrac{\mu\mu_0}{4\pi}\oiint\dfrac{d\mathbf{r}_1 . d\mathbf{r}_2}{r_{12}}$

Self-inductance of a coil (length $l$, cross-section $a$, $n$ turns): $\qquad \dfrac{4\pi n^2\mu a}{l} \qquad\qquad \dfrac{n^2\mu\mu_0 a}{l}$

Resistance: $\qquad\qquad R = V/I$

Field energy of a capacitor:

$$U_{\mathrm{el}} = \tfrac{1}{2}QV = \tfrac{1}{2}CV^2$$

Field energy of a current system:

$$U_{\mathrm{mag}} = \tfrac{1}{2}\sum\sum L_{jk}I_j I_k$$

Oscillatory circuit with capacitance, self-inductance, and resistance:

Oscillation equation: $\qquad L\dfrac{d^2I}{dt^2} + R\dfrac{dI}{dt} + \dfrac{I}{C} = 0$

Eigen frequency (resonance frequency) of the undamped circuit:

$$\omega_0 = 1/\sqrt{(LC)}$$

Logarithmic decrement: $\qquad D = \pi R\sqrt{(C/L)}$

## 6. Conversion table: MKSA units to Gaussian

In this table, as in certain places in the text, the symbols for quantities in the MKSA system have been temporarily written, for distinction, with a small star. If the symbols in the two systems represent different physical quantities, then in the comparison of units we have an arrow instead of an equal sign. In this case the unit quantity in the MKSA system is not equal to the similarly named quantity in the Gaussian system given on the right, but corresponds to it only in the sense of the relationship given in the second column of the table.

The numerical factor 3 in the table should, more accurately, be replaced by 2·99790, the number associated with the velocity of light.

| Quantity | | Unit conversion | |
|---|---|---|---|
| Quantity of electricity | $Q^* = Q$ | $1 \, \text{Coulomb (C)} = 3 \times 10^9$ | CGS units |
| Electric current | $I^* = I$ | $1 \, \text{Ampere (A)} = 3 \times 10^9$ | CGS units |
| Electric potential | $V^* = V$ | $1 \, \text{Volt (V)} \quad = \dfrac{1}{300}$ | CGS units |
| Electric field | $E^* = E$ | $1 \, \text{V/m} \quad = \dfrac{1}{30{,}000}$ | CGS units |
| Electric displacement | $D^* = \dfrac{1}{4\pi} D$ | $1 \, \text{As/m}^2 \quad \rightarrow 4\pi \times 3 \times 10^5$ | CGS units |
| Dielectric polarization | $P^* = P$ | $1 \, \text{As/m}^2 \quad = 3 \times 10^5$ | CGS units |
| Magnetic field | $H^* = \dfrac{c}{4\pi} H$ | $1 \, \text{A/m} \quad \rightarrow 4\pi \times 10^{-3}$ | Oersted (Oe) |
| Magnetic induction | $B^* = \dfrac{1}{c} B$ | $1 \, \text{Vs/m}^2 \quad \rightarrow 1 \times 10^4$ | Gauss (G) |
| Magnetization | $M^* = \dfrac{4\pi}{c} M$ | $1 \, \text{Vs/m}^2 \quad \rightarrow \dfrac{1}{4\pi} \times 10^4$ | Gauss (G) |
| Magnetic flux | $\Phi^* = \dfrac{1}{c} \Phi$ | $1 \, \text{Weber (Wb)} \rightarrow 1 \times 10^8$ | Maxwell (Mx) |
| Power | $I^* V^* = IV$ | $1 \, \text{Watt (W)} \quad = 1 \times 10^7$ | erg/s |
| Capacitance | $C^* = C$ | $1 \, \text{Farad (F)} \quad = 9 \times 10^{11}$ | cm |
| Inductance | $L^* = L$ | $1 \, \text{Henry (H)} \quad = \frac{1}{9} \times 10^{-11}$ | $\text{s}^2/\text{cm}$ |
| Resistance | $R^* = R$ | $1 \, \text{Ohm} \, (\Omega) \quad = \frac{1}{9} \times 10^{-11}$ | s/cm |
| Conductivity | $\sigma^* = \sigma$ | $1 \, \text{Siemens (S)} = 9 \times 10^9$ | 1/s |

Permittivity of a vacuum:

$$\varepsilon_0^* = \varepsilon_0 = 8{\cdot}86 \times 10^{-12}\,\text{As/Vm} = \frac{1}{4\pi \times 9 \times 10^9}\,\text{As/Vm} = \frac{1}{4\pi}$$

Permeability of a vacuum:

$$\mu_0^* = \mu_0 = 4\pi \times 10^{-7}\,\text{Vs/Am} = \frac{4\pi}{c^2} = \frac{4\pi}{9} \times 10^{-20}\,\text{s}^2/\text{cm}^2$$

# CHAPTER G III

## Relativity Theory

All physical relationships must be invariant against a Lorentz transformation. Such a special transformation is given (with $\beta = v/c$) by

$$x' = \frac{x - vt}{\sqrt{(1-\beta^2)}} \qquad y' = y \qquad z' = z \qquad t' = \frac{t - xv/c^2}{\sqrt{(1-\beta^2)}}$$

and inversely by

$$x = \frac{x' + vt'}{\sqrt{(1-\beta^2)}} \qquad y = y' \qquad z = z' \qquad t = \frac{t' + x'v/c^2}{\sqrt{(1-\beta^2)}}$$

Every Lorentz transformation can be represented as a rotation in the four-dimensional space in which $x_1 = x$, $x_2 = y$, $x_3 = z$, $x_4 = ict$. Relativistic invariance of an expression is guaranteed if it appears as a scalar equation or as a vector equation or tensor equation in four-dimensional space. Four-vectors or four-tensors are:

Position vector: $\quad x_v = \{\mathbf{r}, ict\}$

Four-velocity: $\quad u_v = \left\{ \dfrac{\mathbf{v}}{\sqrt{(1-\beta^2)}}, \ \dfrac{ic}{\sqrt{(1-\beta^2)}} \right\} \qquad \sum u_v^2 = -c^2$

Four-momentum: $J_v = \left\{ \mathbf{p}, \dfrac{iE}{c} \right\} \qquad \dfrac{E^2}{c^2} - p^2 = m_0^2 c^2$

Four-force: $\quad F_v = \left\{ \dfrac{\mathbf{F}}{\sqrt{(1-\beta^2)}}, \ \dfrac{i\mathbf{F}.\mathbf{v}}{c\sqrt{(1-\beta^2)}} \right\} \qquad \sum F_v u_v = 0$

Wave vector: $\quad k_v = \{\mathbf{k}, ik\} \qquad\qquad \mathbf{k} = \dfrac{2\pi}{\lambda}\,\mathbf{n} \qquad\qquad k = \dfrac{\omega}{c}$

Four-current: $\quad g_v = \{\mathbf{g}, ic\rho\}$

Field tensors:

$$F_{\nu\mu} = \begin{pmatrix} 0 & B_z & -B_y & -iE_x \\ -B_z & 0 & B_x & -iE_y \\ B_y & -B_x & 0 & -iE_z \\ iE_x & iE_y & iE_z & 0 \end{pmatrix} \qquad H_{\nu\mu} = \begin{pmatrix} 0 & H_z & -H_y & -iD_x \\ -H_z & 0 & H_x & -iD_y \\ H_y & -H_x & 0 & -iD_z \\ iD_x & iD_y & iD_z & 0 \end{pmatrix}$$

Moments tensor: $M_{\nu\mu} = \dfrac{1}{4\pi}(F_{\nu\mu} - H_{\nu\mu}) = \begin{pmatrix} 0 & M_z & -M_y & iP_x \\ -M_z & 0 & M_x & iP_y \\ M_y & -M_z & 0 & iP_z \\ -iP_x & -iP_y & -iP_z & 0 \end{pmatrix}$

Special transformations from the rest system (index 0—designated by primes above) to the moving system.

Momentum energy:

$$\mathbf{p} = \frac{m_0 \mathbf{v}}{\sqrt{(1-\beta^2)}} \qquad E = \frac{m_0 c^2}{\sqrt{(1-\beta^2)}}$$

Current charge:

$$g_x = \frac{g_{x0} + v\rho_0}{\sqrt{(1-\beta^2)}} \qquad \rho = \frac{\rho_0 + g_{x0}\, v/c^2}{\sqrt{(1-\beta^2)}}$$

Electric field *in vacuo* in the rest system:

$$E_x = E_{x0} \qquad E_{y,\,z} = \frac{(E_{y,\,z})_0}{\sqrt{(1-\beta^2)}} \qquad \mathbf{H} = \frac{\dfrac{\mathbf{v}}{c} \times \mathbf{E}_0}{\sqrt{(1-\beta^2)}}$$

Magnetic field *in vacuo* in the rest system:

$$H_x = H_{x0} \qquad H_{y,\,z} = \frac{(H_{y,\,z})_0}{\sqrt{(1-\beta^2)}} \qquad \mathbf{E} = -\frac{\dfrac{\mathbf{v}}{c} \times \mathbf{H}_0}{\sqrt{(1-\beta^2)}}$$

Electric moment in the rest system:

$$P_x = P_{x0} \qquad P_{y,\,z} = \frac{(P_{y,\,z})_0}{\sqrt{(1-\beta^2)}} \qquad \mathbf{M} = -\frac{\dfrac{\mathbf{v}}{c} \times \mathbf{P}_0}{\sqrt{(1-\beta^2)}}$$

Magnetic moment in the rest system:

$$M_x = M_{x0} \qquad M_{y,\,z} = \frac{(M_{y,\,z})_0}{\sqrt{(1-\beta^2)}} \qquad \mathbf{P} = \frac{\dfrac{\mathbf{v}}{c} \times \mathbf{M}_0}{\sqrt{(1-\beta^2)}}$$

# INDEX

# VOLUME II

# *Quantum Theory of Atoms and Radiation*

Revised by Prof. Günther Leibfried, Technische Hochschule, Aachen
and Dr. Wilhelm Brenig, Technische Hochschule, Munich

TRANSLATED BY IVOR DE TEISSIER

# FOREWORD

Richard Becker intended that the second volume of *Electromagnetic Fields and Interactions* should contain an introduction to the quantum theory of radiation and the electron in its new edition. He considered that the following two provisions were essential for this purpose: firstly, the presentation should be complete in itself, and should include the detailed mathematical basis of quantum mechanics and, in particular, a summary of the theory of Hilbert space. Secondly, this volume should lay the foundations for the treatment of the properties of matter in electromagnetic fields, which is dealt with in the third volume. Becker himself considered the latter point to be of secondary importance; it does, however, determine the scope of the present volume. Even supposing that the quantum mechanical approach were unnecessary for the applications treated in the third volume, a knowledge of the formal basis of the quantum theory and an understanding of its methods are still indispensable for anyone who wishes to have a proper comprehension of the subject.

Richard Becker was unable to complete his *Quantum Theory*: at the time of his death about half this volume was in a more or less completed state. We have endeavoured to continue the presentation in a manner which, we hope, would have satisfied the author. Our task was rendered easier by many fruitful discussions with Becker over a number of years, as a result of which we felt able to convey his intentions and his approach to the subject.

The only knowledge assumed here is a familiarity with the basic principles of classical theoretical physics; we hope, therefore, that this volume will constitute a useful introduction for students. A revision of the classical properties of electrons is followed by a detailed presentation of the principles of the quantum theory. Later chapters are chiefly concerned with the application of the theory to problems involving respectively one and more than one electron. Finally, the principles of quantum field theory are developed to a stage sufficient to permit the treatment of the quantum theory of the Maxwell field and the Dirac field theory of the electron.

A number of exercises is included in each chapter, the solutions of which are given at the end of the book. They should help the reader to grasp the subject matter, since they offer him the opportunity of solving real problems, and enable him to make sure that he understands what he has read. Several important applications are treated in these exercises; we therefore recommend their study, together with their solutions, even for those readers who do not intend to work out the problems in the first instance.

Frau H. Geib and Fräulein F. Albus have been responsible for most of the work involved in the preparation of the manuscript. We have received considerable help from Dr. W. Ludwig in the preparation of the text, and from Dr. G. Süssmann and Dipl.-Ing. K. Fischer, who undertook a critical revision of the manuscript. Our sincere thanks are due to them.

G. LEIBFRIED
W. BRENIG

AACHEN and MUNICH, *September*, 1958

# GERMAN EDITOR'S FOREWORD

Messrs. Leibfried and Brenig were the last colleagues of R. Becker to work in close contact with him. They are therefore most qualified to complete his posthumous half-finished manuscript of the second volume of *Electromagnetic Fields and Interactions*, which he had planned as a work of three volumes. The present volume shows how successful they have been in this task, and I should like to express my warmest thanks to them and to their collaborators.

My own participation in this volume has been limited to some suggestions, and to co-ordination with the other two volumes of this work. We have tried to ensure that this textbook in three volumes should have a uniform character in spite of the multiplicity of collaborators.

A few changes and corrections have been made in the ninth edition.

My special thanks are due to Messrs. B. G. Teubner, publishers, whose co-operation was much appreciated.

F. SAUTER

COLOGNE, 1964

# CONTENTS

*Contents*

Contents

# A

---

The
classical
principles
of
electron
theory

# CHAPTER AI

## Motion of an electron in electric and magnetic fields

### §1. The equation of motion

From the fundamental experiments of P. E. A. Lenard (cathode rays), R. A. Millikan (measurement of the elementary charge), and J. J. Thomson (motion of electrons in electromagnetic fields) it is clear that we may conceive the electron as a charge distribution concentrated in a very small region of space, with total charge* $e$ and mass $m$. Then the force $\mathbf{F}$ acting on an electron moving with velocity $\mathbf{v}$ in an electric field $\mathbf{E}$ and a magnetic field $\mathbf{H}$ is given by †

$$\mathbf{F} = e\mathbf{E} + \frac{e}{c}\mathbf{v} \times \mathbf{H} \tag{1.1}$$

It is assumed that $\mathbf{E}$, $\mathbf{H}$ and $\mathbf{v}$ are practically constant within the region of the charge distribution; $\mathbf{E}$ and $\mathbf{H}$ may then be taken as the fields at the electron's "position". For normal applications the precise charge distribution is unimportant, provided that the fields do not vary appreciably within the region occupied by the electron.

The force $\mathbf{F} = m\,d\mathbf{v}/dt$ in accordance with Newton's law (force = mass × acceleration). Substituting in (1.1), we obtain the equation of motion

$$m\frac{d\mathbf{v}}{dt} = e\mathbf{E} + \frac{e}{c}\mathbf{v} \times \mathbf{H} \tag{1.2}$$

$$m\ddot{\mathbf{r}} = e\mathbf{E} + \frac{e}{c}\dot{\mathbf{r}} \times \mathbf{H} \tag{1.2a}$$

where $\mathbf{r}(t) = [x(t), y(t), z(t)]$ represents the position of the electron. This equation is sufficient in most cases to describe the motion of an electron in a given field $\mathbf{E}$, $\mathbf{H}$.

---

* In this work $e$ always represents the actual electron charge: it therefore has the negative value $e = -e_0$, where $e_0 = 4\cdot8 \times 10^{-10}$ e.s.u. $= 1\cdot6 \times 10^{-19}$ coulomb (C).
† Cf. Vol. I, §§18 and 44. As in Vol. I the Gauss system of measurement is used in all formulae. The force on a charged particle in a vacuum may therefore be expressed by $\mathbf{H}$ instead of $\mathbf{B}$.

It should be mentioned, however, that neither side of equation (1.2) is strictly correct. The left side is incorrect because the mass is no longer constant at very great velocities, but becomes infinitely large as $v$ approaches the velocity of light $c$. Instead of $dm\mathbf{v}/dt$ we should strictly put $\dfrac{d}{dt}\dfrac{m\mathbf{v}}{\sqrt{(1-v^2/c^2)}}$. On the right side of the equation a further term is required to take account of the fact that the accelerating field which acts on an electron moving with non-uniform velocity can no longer be considered constant over the region occupied by the latter. An accelerated electron is a source of electromagnetic radiation, the energy of which must be drawn from the electron's kinetic energy. This "reaction of the electron with itself" will be discussed in §4 (radiation damping). This term plays no part in our immediate applications, however, and we may therefore consider the external fields $\mathbf{E}$ and $\mathbf{H}$ to be constant in the electron region.

We see firstly, from (1.2), that the magnetic field has no influence on the magnitude of the velocity. The scalar product with $\mathbf{v}$ gives

$$\frac{d}{dt}(\tfrac{1}{2}mv^2) = e(\mathbf{E}\cdot\mathbf{v})$$

A change in kinetic energy is produced by the electric field alone. For an electrostatic field $\mathbf{E}$ represented by a potential $\phi$ ($\mathbf{E} = -\operatorname{grad}\phi$)

$$\frac{d}{dt}(\tfrac{1}{2}mv^2) = -e\left(\frac{\partial\phi}{\partial x}\frac{dx}{dt} + \frac{\partial\phi}{\partial y}\frac{dy}{dt} + \frac{\partial\phi}{\partial z}\frac{dz}{dt}\right) = -e\frac{d\phi}{dt}$$

This expression transforms the potential $\phi$, which is a function of position alone, into a time-dependent function taken at the position of the electron. The theorem of the conservation of energy

$$\tfrac{1}{2}mv^2 + e\phi = \text{constant}$$

or

$$\tfrac{1}{2}mv_2^2 - \tfrac{1}{2}mv_1^2 = e(\phi_1 - \phi_2) \tag{1.3}$$

is therefore also valid in the presence of a steady magnetic field.

We now consider the effect of a uniform steady magnetic field alone, the direction of which is taken to be parallel to the positive $z$-axis of the system of coordinates: $\mathbf{H} = (0, 0, H)$. For this case the equations of motion are

$$m\ddot{x} = \frac{e}{c}\dot{y}H \qquad m\ddot{y} = -\frac{e}{c}\dot{x}H \qquad m\ddot{z} = 0 \tag{1.4}$$

The third equation shows that the component of motion in the direction of **H** is unaffected by the field. We therefore only need to consider the projection of the motion on the *xy*-plane (the component of **r** perpendicular to **H**.)

To integrate the first two equations of (1.4) we describe the motion in the complex plane *x-iy* by the introduction of the quantity

$$\zeta = x + iy$$

If the second equation of (1.4) is multiplied by *i* and added to the first, we obtain the following equation for $\zeta$:

$$\ddot{\zeta} = -i\frac{eH}{mc}\dot{\zeta} = -i\omega\dot{\zeta} \qquad (1.4a)$$

where $\omega = \dfrac{eH}{mc}$.

Taking initial values $\zeta_{t=0} = \zeta_0$ and $\dot{\zeta}_{t=0} = \dot{\zeta}_0$, the first integral of (1.4a) is

$$\dot{\zeta} = \dot{\zeta}_0 e^{-i\omega t}$$

and the second,

$$\zeta(t) = \zeta_0 + \int_0^t \dot{\zeta}\, dt = \zeta_0 + \frac{\dot{\zeta}_0}{i\omega}(1 - e^{-i\omega t}) \qquad (1.5)$$

The vector $\zeta(t)$ then appears in the complex plane as the sum of the three complex vectors

$$\zeta_0 \qquad +\frac{\dot{\zeta}_0}{i\omega} \qquad -\frac{\dot{\zeta}_0}{i\omega}e^{-i\omega t}$$

In figure 1, let A be the initial position of the electron ($\overrightarrow{OA} = \zeta_0$), and $\overrightarrow{AA'}$ the direction of its initial velocity. The direction of $\mathbf{R} = \overrightarrow{AB} = \dot{\zeta}_0/i\omega$ is at right angles to $\overrightarrow{AA'}$, measured in a counter-clockwise sense, because $\omega$ is negative for an electron. Finally, $\overrightarrow{BC} = -\mathbf{R}e^{-i\omega t}$. The electron therefore describes an orbit with centre B and radius

$$R = v\left|\frac{mc}{eH}\right| \qquad (1.6)$$

with the constant angular velocity

$$\omega = \frac{eH}{mc} \qquad (1.6a)$$

Equation (1.6) also follows directly from the fact that, for circular motion, the centrifugal force $mv^2/R$ must be balanced by the "Lorentz force" $evH/c$.

Equation (1.6) suggests a valuable method for the production of cathode rays of precisely determined velocity. If a series of small

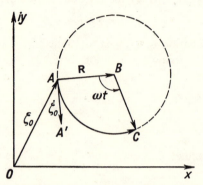

Fig. 1.—Electron track AC in a uniform magnetic field, with initial position $\zeta_0$ and initial velocity $\dot{\zeta}_0$ ($\omega$ is negative as in (1.6a), because $e$ is negative)

apertures is arranged along a circle of radius $R$, perpendicular to a magnetic field $\mathbf{H}$, and a beam of cathode rays (also perpendicular to $\mathbf{H}$) is directed on to the first aperture, the only beam electrons to pass through the apertures will be those whose velocity exactly satisfies the relation

$$v = \left| \frac{eH}{mc} \right| R$$

In a steady electric field, the acceleration is in the direction of the field. For initial values of position $\mathbf{r}_0$ and of velocity $\dot{\mathbf{r}}_0$ at time $t = 0$, the solution of (1.2) is

$$\mathbf{r}(t) = \mathbf{r}_0 + \dot{\mathbf{r}}_0 \, t + \frac{e}{2m} \mathbf{E} t^2 \qquad (1.7)$$

The equation of motion and the electron path are identical with those of a particle in a uniform gravitational field acting in the direction of $\mathbf{E}$. Figure 2 shows the parabolic track of an electron in a steady electric field acting along the $x$-axis $[\mathbf{E} = (E, 0, 0)]$, the initial velocity of which, $\mathbf{v}_0$, is normal to $\mathbf{E}$.

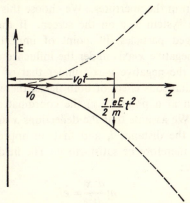

Fig. 2.—In a uniform electric field the track is a parabola

## §2. The charge-mass ratio

The charge-mass ratio $e/m$ can be determined from an investigation of the paths described by electrons acted upon by electric and magnetic fields in accordance with equation (1.2). It must be emphasized that such experiments can never permit the values of the charge or the mass to be separately determined, since only the quotient $e/m$ occurs in the fundamental equation (1.2).

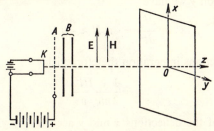

Fig. 3.—Measurement of $e/m$ by simultaneous deflection in uniform electric and magnetic fields

We now consider the application of the equation of motion (1.2) to the case of a cathode-ray beam of velocity $v$ (parallel to the $z$-axis), acted upon simultaneously by steady electric and magnetic fields, the directions of which are parallel to the $x$-axis (figure 3). The beam is laterally defined by suitable apertures, and in the absence of deflecting fields would strike the point O of a screen placed perpendicular to it at a distance $l$ from the apertures. We choose this point as the origin of a coordinate system $x, y$ on the screen. If the beam consists of negatively charged particles, its point of impact will be displaced parallel to the negative $x$-axis under the influence of the electric field, and parallel to the negative $y$-axis under that of the magnetic field. Under the simultaneous effect of both fields, therefore, the beam will strike the screen at a point $(x, y)$ the coordinates of which we will now calculate. We assume that the deflections $x$ and $y$ are very small compared with the distance $l$, and that an approximate method of calculation will therefore be satisfactory. The field **E** acts only along the $x$-coordinate:

$$m \frac{d^2x}{dt^2} = eE$$

The time is reckoned from the instant the electron passes the aperture; therefore $x = 0$ and $\dot{x} = 0$ at $t = 0$, and

$$mx = \tfrac{1}{2}eEt^2$$

The time taken by the electron to travel from the aperture to the screen is $l/v$, so that

$$x = \frac{1}{2}\frac{e}{m}\frac{El^2}{v^2}$$

For the deflection $y$ we have

$$m \frac{d^2y}{dt^2} = \frac{e}{c}v_z H \approx \frac{e}{c} v H$$

For small deflections integration therefore yields

$$y = \frac{1}{2}\frac{e}{mc}\frac{Hl^2}{v}$$

Measurement of the deflections $x$ and $y$ accordingly yields values for $e/m$ and $v$, as follows:

$$\frac{y}{x} = \frac{H}{E}\frac{v}{c} \quad \text{and} \quad \frac{y^2}{x} = \frac{1}{2}\frac{e}{m}\frac{H^2 l^2}{c^2 E}$$

The tracks of particles of the same velocity but of different charge-mass ratios lie on a straight line through the origin of the coordinates. From the experimental point of view, however, the second case is of greater importance: if the beam is composed of particles of one type only, all with the same charge-mass ratio, but with different velocities, the resultant image on a photographic plate (replacing the screen) will be a parabola, known as the Thomson parabola (figure 4).

On performing the above experiment in as many different ways as possible, the theoretically predicted parabolas were always obtained, provided that the velocities of the cathode rays were not too great.

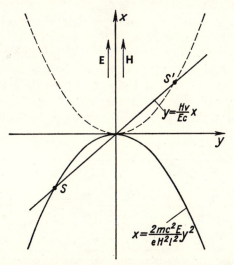

Fig. 4.—Evaluation of results of the deflection experiment. The electrons meet the photographic plate at S. For positively charged particles such as canal rays, the point of impact is given by the intersection S′ of the straight line with the broken parabola

The experiments proved that these rays consist of a single type of particle, negatively charged. Measurements of the important ratio $-e/m$ yielded the value $1.76 \times 10^8\,\text{C/g}$, which is about 1800 times greater than that attributable to the proton $(96,500\,\text{C/g})$.

For very fast cathode rays, the velocities of which approach that of light, experiments yield a curve which departs appreciably from a parabola in the neighbourhood of the vertex (see §3). At this stage the increase in the inertial mass with velocity becomes apparent, an effect which we did not take into account in the above analysis.

Another method of measurement of the charge-mass ratio is based on the fact that, according to equation (1.6*a*), the angular velocity $\omega$

with which an electron describes an orbit in the magnetic field is independent of the radius. The orbital period is

$$\tau = \frac{2\pi}{\omega} = \frac{2\pi mc}{eH}$$

H. Busch has shown that this fact may be used to focus a beam of cathode rays by means of a longitudinal magnetic field. The arrangement is shown diagrammatically in figure 5.

Electrons emitted by an incandescent filament are accelerated along the $z$-axis of the figure by means of a potential difference $V$. Having attained the velocity $v$ (given by $\frac{1}{2}mv^2 = eV$), they pass through an aperture B and impinge on a fluorescent screen S, which is situated perpendicular to the $z$-axis at a distance $l$ from the aperture. In the first instance the beam produces a diffuse spot on the screen, since it has left the hole B with a finite angular aperture. If a uniform magnetic

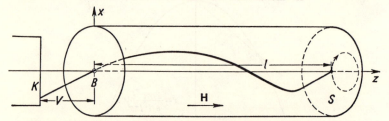

Fig. 5.—Focusing of a divergent cathode-ray beam by a uniform magnetic field parallel to the beam direction. The projection of the electron trajectory on the screen is indicated by the broken circle

field is now switched on, parallel to the beam, all the electron orbits will become helical. One turn of the helix is described in time $\tau$; the pitch of all helices is therefore $v_z\tau$. If the field strength is such that the pitch is exactly equal to the interval $l$, all electrons will have executed precisely one revolution, and will therefore be concentrated on the screen into a spot of the same size as the aperture. The resultant focusing condition is $l = v_z\tau$, or, since

$$v_z \sim v = \sqrt{\frac{2eV}{m}}, \qquad \frac{e}{m} = \frac{8\pi^2 c^2 V}{H^2 l^2}$$

In order to determine the charge-mass ratio, given the potential difference $V$ and the path length $l$, it is therefore only necessary to determine the field strength $H$ at which the spot first becomes sharp. This method has also been developed to a stage of great accuracy.

The fundamental importance of the deflection experiments described above is illustrated by the fact that, in addition to the determination of the charge-mass ratio for slow electrons, they also permit the measurement of the relativistic variation of mass, an analysis of the velocity distribution of electrons occurring in radioactive decay processes, and the determination of the charge-mass ratio for canal rays, or positively charged ions.

Following the above description of procedures for determining the charge-mass ratio of cathode rays, we now give a short account of the methods of determining atomic and molecular masses by deflection experiments in electric and magnetic fields. The direct determination of $e/m$ for individual groups of charged particles (ions) is particularly important, since it afforded the first means of isotope investigation; chemical methods of determining atomic weights only give average values for

Fig. 6.—Parabolic spectrum of methane [O. Eisenhut and R. Conrad, *Z. Elektrochem.* 36, (1930) 654]

large numbers of particles. Using deflection methods, J. J. Thomson was the first to demonstrate the existence of isotopes, by showing that ordinary neon, the chemically determined atomic weight of which is 20·2, consists in fact of a mixture of two isotopes of atomic masses 20·0 and 22·0.

Thomson performed his experiments using the "parabola method" described above, in which the charged particles are subjected to parallel electric and magnetic fields. The points at which the particles meet a plate perpendicular to the beam give rise to parabolas; the points of impact of all particles with the same charge-mass ratio lie along a single parabola, the faster and therefore less easily deflected particles striking nearer the vertex (figure 6).

An essentially different method was employed by Thomson's pupil F. W. Aston, who passed a beam of ions first through an electric field, and then through a magnetic field at right angles, which compensated the deflection produced by the first field. The arrangement of Aston's apparatus is shown diagrammatically in figure 7. A collimated beam of ions is passed between the plates of a capacitor, which deflects the individual particles to an extent dependent on their masses and velocities and so produces a fan-like dispersion of the beam. A narrow pencil of rays is selected from the wide beam by means of the slit B and passed through a magnetic field, which deflects it in the reverse direction. In figure 7 the magnetic field is at right

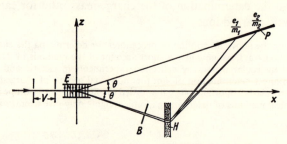

Fig. 7.—Aston's mass spectrograph. Trajectories are shown for two different values of $e/m$

angles to the plane of the paper. If the magnetic field is of the correct strength relative to the electric field, the beam diverging from the slit B is rendered convergent; the rays most strongly deflected by the capacitor also suffer the greatest deflection in the magnetic field, and the tracks of all particles having the same charge-mass ratio converge to a definite point. The essential feature of Aston's method is that the "image points" of rays with different values of $e/m$ lie approximately in a straight line (cf. Exercise 2, p. 16). Hence, if a photographic plate P is placed along this line, a set of sharp images is obtained which correspond to the individual points of convergence of the rays and so enable the separate charge-mass ratios to be determined. A *mass spectrum* is thus obtained of the atoms contained in the primary rays.

A description of a modern mass spectrograph, with detailed references, has been given by H. Ewald, *Z. Naturforsch.* 1 (1946), 131. See also: *Handbuch der Physik*, Vol. XXXIII, p. 546ff., Berlin, 1956.

## §3. Variability of mass at high velocities

We now wish to see how the Thomson parabolas change their shape when allowance is made for the relativistic variation of mass with speed (cf. Vol. I, §89). The momentum **p** of an electron moving with velocity **v** is now

$$\mathbf{p} = \frac{m\mathbf{v}}{\sqrt{(1-\beta^2)}} \quad \text{where} \quad \beta = \frac{v}{c} \tag{3.1}$$

From Newton's basic equation (force = rate of change of momentum) and equation (1.1) we obtain the relativistic equation of motion of the electron

$$\frac{d\mathbf{p}}{dt} = \frac{d}{dt}\frac{m\mathbf{v}}{\sqrt{(1-\beta^2)}} = e\mathbf{E} + \frac{e}{c}\mathbf{v}\times\mathbf{H} \tag{3.2}$$

Hence we have for the equations of motion applicable to Thomson's experiment, where $\mathbf{E} = (E, 0, 0)$ and $\mathbf{H} = (H, 0, 0)$:

$$\frac{d}{dt}\frac{mv_x}{\sqrt{(1-\beta^2)}} = eE \qquad \frac{d}{dt}\frac{mv_y}{\sqrt{(1-\beta^2)}} = \frac{e}{c}Hv_z$$

$$\frac{d}{dt}\frac{mv_z}{\sqrt{(1-\beta^2)}} = -\frac{e}{c}Hv_y \tag{3.3}$$

If the deflections along the $x$- and $y$-axes are very small, i.e. $v_x$ and $v_y$ are always very small compared to $v_z$, then $v = \sqrt{(v_x^2 + v_y^2 + v_z^2)}$ may be taken as practically constant and equal to $v_z$; the rate of change of $v_z$ itself is only a small quantity of higher order, as may easily be seen from the third equation of motion. Then both the other equations may be approximately integrated; the first yields the following expression for the deflection along the $x$-axis:

$$x = \frac{1}{2}\frac{eE}{m}\frac{l^2}{v^2}\sqrt{(1-\beta^2)} \tag{3.3a}$$

where $t = l/v_z = l/v$ is the time interval from the slit to the plate.
    The deflection along the $y$-axis is

$$y = \frac{1}{2}\frac{eH}{mc}\frac{l^2}{v}\sqrt{(1-\beta^2)} \tag{3.3b}$$

If we once more form the quotient

$$\frac{y}{x} = \frac{Hv}{Ec} \tag{3.4}$$

we see that the traces of particles of equal velocity but different charge-mass ratios still lie on a straight line through the origin. If, however, we now investigate the traces made by a single type of particle (i.e. of fixed charge-mass ratio), we find that the results of the two cases differ. If the velocity $v$ is eliminated from both expressions (3.3), we no longer obtain a parabola as in the case for which $\beta \ll 1$, but parabola-like curves of the fourth degree, similar to the solid curve shown in figure 8 (the broken curve is the parabola of figure 4). The fact of chief interest is that these curves do not meet the $y$-axis tangentially, but intersect it at a definite angle $\alpha$. Since the smaller values of $x$ and $y$ correspond to the higher values of $v/c$, the neighbourhood of the origin

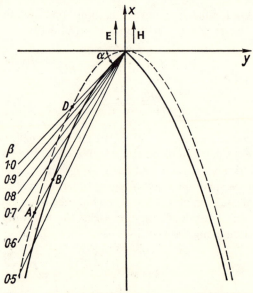

Fig. 8.—Increase of mass with velocity. The experimental arrangement of figure 3 yields the solid curve in place of the Thomson parabola (broken curve)

illustrates the differences between the old and new theories. The non-relativistic theory predicts that only particles of infinite velocity would strike the origin; the present theory, however, indicates that this would be achieved by particles with the limiting velocity of light. From equation (3.4), the angle $\alpha$ is seen to be given by

$$\tan \alpha = \lim_{v \to c} \frac{y}{x} = \frac{H}{E}$$

The effect of the variation of mass may be demonstrated by means of the following construction. In figure 8, the broken curve denotes a parabola calculated from the formula $\dfrac{y^2}{x} = \dfrac{1}{2} \dfrac{e}{m} \dfrac{H^2}{E} \dfrac{l^2}{c^2}$, assuming constant mass; this parabola represents the trace that would be produced by particles of a single type but with different velocities. Further, according to the relation (3.4), the traces of particles of different masses lie on straight lines through the origin; in figure 8, for example, these lines are drawn for $v/c = 0.5$, $0.6$, etc. In order to find a point on the modified curve, for a given velocity, the intersection of the appropriate straight line with the parabola is marked, and the distance along the line to this point is shortened in the ratio $\sqrt{(1 - \beta^2)} : 1$. For example, the point A, corresponding to $\beta = 0.6$, is transformed to the point B by means of the factor $\sqrt{[1 - (0.6)^2]} = 0.8$. In particular, it is evident from this construction that the point of intersection D of the line $\beta = 1$ with the original parabola is transferred to the origin.

The experiments described above indicate that the mass of a rapidly moving electron does in fact increase in the manner predicted by equation (3.1).

## Exercises

### 1. *Rutherford's scattering formula*

Calculate the deflection of an electron with initial velocity $v_0$ by a proton considered to be at rest at the origin of the coordinates. The equation of motion of the electron is $m\ddot{\mathbf{r}} = -\dfrac{e^2}{r^3}\,\mathbf{r}$.

(a) Verify that the following three conservation theorems are valid, by reason of the equation of motion:

1. Energy: $\qquad \dfrac{dE}{dt} = \dfrac{d}{dt}\left(\tfrac{1}{2}m\dot{\mathbf{r}}^2 - \dfrac{e^2}{r}\right) = 0$

2. Angular momentum: $\dfrac{d\mathbf{M}}{dt} = \dfrac{d}{dt}\,m\mathbf{r} \times \dot{\mathbf{r}} = 0$

3. $\qquad\qquad\qquad \dfrac{d\mathbf{e}}{dt} = \dfrac{d}{dt}\left\{\dot{\mathbf{r}} \times \mathbf{M} - e^2\dfrac{\mathbf{r}}{r}\right\} = 0$

(b) The electron's initial velocity is $v_0$, and $v_e$ is its velocity after scattering, at a great distance from the proton. The magnitudes of these velocities are the same, in virtue of the conservation of energy. It follows from the theorem on conservation of angular momentum that the orbit is plane, and normal to $\mathbf{M}$. The deflection angle $\theta$ depends on the impact parameter $b$ (see figure 9). Show the vector $\mathbf{e}$ in a graphical construction, using the initial and final velocity vectors, and the fact that $\mathbf{e}$ is invariable with respect to time. Derive a relation between $\theta$ and $b$ from this construction.

(c) The elementary effective cross-section $dQ$ producing scattering in the element of solid angle $d\Omega$ is defined as follows (cf. figure 9). Imagine a large number of electrons all with the same velocity $v_0$, and of uniform density, travelling towards the proton. A surface $dQ$ is placed across this electron stream, such that all particles passing through $dQ$ are deflected into the element of solid angle $d\Omega$. Calculate the elementary effective cross-section from the relation between $b$ and $\theta$ which was

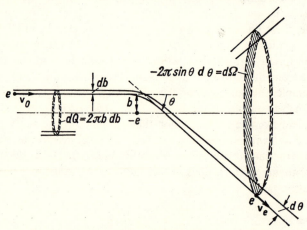

Fig. 9.—Derivation of the Rutherford scattering formula

derived in (*b*) above. Here $dQ = 2\pi b\,db$ and $d\Omega = -2\pi \sin\theta\,d\theta$ (because $d\theta$ is negative). (Since the geometry is cylindrically symmetrical about the direction of incidence, $dQ$ does not depend on the azimuthal angle.)

## 2. *Aston's mass spectrograph*

In Aston's mass spectrograph (figure 10), particles of charge $e$ are first of all accelerated by a potential difference $V$. All those particles with the same kinetic energy $\frac{1}{2}mv^2$ are then deflected through the same angle $\theta$ by the electric field $E$.

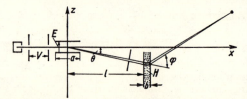

Fig. 10.—The focusing conditions for Aston's mass spectrograph

As a result of the distribution of the particle velocities $v$ about the average value $\bar{v}$, the beam diverges slightly; this deflection is cancelled out by the succeeding deflection through the angle $\phi$, which is produced by the magnetic field $H$ ("velocity focusing"). At what point $(x_f, z_f)$ does focusing occur? What is the position of the registering photographic plate?

# CHAPTER A II

## The classical model of the atom

### §4. Free oscillation of the elastically bound electron

One of the most important objectives of theoretical physics consists in the detailed description of the structure of the atom. How may we construct an atomic model from positive and negative charges so that its properties, when calculated from the fundamental electrodynamic equations, agree with experimentally observed phenomena? Historically, three steps must be distinguished, corresponding to successive approximations to the required objective: they are represented, firstly, by the atomic models due to J. J. Thomson (until 1912), secondly, by the model due to N. Bohr (1912–25), and finally, by the statements of quantum mechanics (W. Heisenberg, E. Schrödinger, since 1925). Each step in turn was so successful in its applications that it was permissible to consider previous conceptions to be definitely superseded. Today the axioms of quantum mechanics are accepted as the point of departure in the systematic representation of atomic phenomena. We shall, however, analyse in some detail the properties of Thomson's model of the atom; our justification is that this model affords a particularly clear and simple means of understanding a number of important electrical and magnetic properties of atoms. We must be prepared for the introduction of corrections of varying degree into the results obtained in this manner when the stricter methods of Bohr's theory and of quantum mechanics are applied. The latter make no attempt to give a clear picture of the atom, but many of their conclusions may be interpreted by postulating that the atom behaves as though it contained elastically bound charges as assumed in Thomson's model. In any case, a knowledge of the properties of this model is essential, because the terms used to interpret the formulae of quantum mechanics have largely been drawn from the concepts developed with the aid of Thomson's simple model.

The existence of (nearly) sharp spectral lines, characteristic of each type of atom, provides the experimental starting-point of modern atomic research. (Spectra of this type are produced, for example, by the rare gases when rendered luminous in a Bunsen burner flame or a discharge tube.) In accordance with the electromagnetic theory of light, any given spectral line must be assumed to originate from an emitter oscillating with a frequency $v$ corresponding to the colour of the line. The simplest emitter of this type would consist of an electron bound to an equilibrium position in such a manner as to permit it to execute harmonic oscillations of just this frequency. Its equation of motion along the axis O$x$ would then be

$$m\ddot{x} + m(2\pi v)^2 x = 0$$

for which the general solution is

$$x = A \cos 2\pi v t + B \sin 2\pi v t$$

In order that the electron should perform oscillations of frequency $v$ it must be bound to the equilibrium position $x = 0$ in such a manner that, when displaced from the latter by a distance $x$, it is drawn back towards it by a restoring force $-m(2\pi v)^2 x$. A model of this sort was proposed by Thomson, as follows.

Let an electron of charge $e$ move within a sphere of radius $a$ which is filled with a uniformly distributed positive charge of total value $-e$. We wish to calculate the force acting on the electron when it is at a distance $r < a$ from the centre of the sphere. The charge outside the sphere of radius $r$ makes no contribution to this force. The charge inside this sphere is $-e(r/a)^3$, and acts on the electron as though it were concentrated at the centre. The force on the electron is therefore

$$\mathbf{F} = -\frac{e^2 \mathbf{r}}{a^3}$$

This gives rise to harmonic oscillations of frequency

$$v = \frac{1}{2\pi} \sqrt{\frac{e^2}{a^3 m}}$$

If we introduce the known values of $e$ and $m$ into the above expression, together with the order of magnitude of the characteristic optical frequencies of free gas atoms ($10^{15} \text{s}^{-1}$), we obtain a value of about $10^{-8}$ cm for the radius of the positive sphere. This is precisely the order of magnitude of atomic radii as determined from the kinetic theory of gases.

This result could be considered to provide support for the Thomson model of the atom. Later, however, Rutherford's experiments on the scattering of $\alpha$-particles were to afford direct proof that the positive charge of atoms is concentrated into a space no greater than $10^{-12}$ to $10^{-13}$ cm; its dimensions are therefore about 100,000 times smaller than is required to account for the optical spectral lines by Thomson's theory. The very result, $a \sim 10^{-8}$ cm, previously considered to be supporting evidence for this theory, now made it quite untenable as a result of Rutherford's experiments. The latter, in fact, constitute the starting-point of Bohr's quantum theory.

Without going further into the details of Thomson's model, let us now consider the motion of an electron acted on by a force of magnitude $-f\mathbf{r}$. From mechanics alone, its equation of motion would be

$$m\ddot{\mathbf{r}} + f\mathbf{r} = 0 \tag{4.1}$$

with the general solution

$$\mathbf{r} = \mathbf{a}\cos\omega t + \mathbf{b}\sin\omega t, \quad \text{where} \quad \omega^2 = \frac{f}{m} \tag{4.2}$$

The constants of integration are the arbitrary vectors $\mathbf{a}$ and $\mathbf{b}$: they may, for instance, serve to specify the initial position and velocity.

The energy $W$ of this mechanical system is

$$W = \tfrac{1}{2}m\dot{\mathbf{r}}^2 + \tfrac{1}{2}f\mathbf{r}^2$$

From (4.2), therefore,

$$W = \tfrac{1}{2}m\omega^2(\mathbf{a}^2 + \mathbf{b}^2) \tag{4.3}$$

Now, since an electron possesses charge as well as mass, it emits electromagnetic radiation in the course of its oscillations, transmitting radiation energy to space at a definite rate as it does so. Equation (4.3) shows the total amount of energy that can be emitted for given initial values of $\mathbf{a}$ and $\mathbf{b}$. It would therefore be a contradiction of the theorem of the conservation of energy if the equation of motion (4.1) were to be taken as strictly correct; on the contrary, we expect the amplitudes $\mathbf{a}$ and $\mathbf{b}$ to decrease with time in proportion to the emitted radiation.

In Vol. I, §67ff. we deduced an expression for the radiation energy $S$ emitted per second by an oscillating electric charge. It is

$$S = \frac{2e^2}{3c^3}\overline{\ddot{\mathbf{r}}^2} \tag{4.4}$$

The conservation theorem requires that this should be equal to the rate of decrease of energy of the oscillating electron:

$$\frac{dW}{dt} = -S = -\frac{2e^2}{3c^3}\ddot{\mathbf{r}}^2 \tag{4.5}$$

Equations (4.4) and (4.5) are only significant if the energy $ST$ emitted during a single period $T = 2\pi/\omega$ is small compared with the existing energy $W$ or, in other words, if the oscillations are lightly damped. With this restriction in mind, we can immediately obtain an expression for the damping from equation (4.5). From equation (4.2)

$$\ddot{\mathbf{r}}^2 = \tfrac{1}{2}\omega^4(\mathbf{a}^2 + \mathbf{b}^2)$$

From (4.4) and (4.3), therefore,

$$S = \frac{e^2\omega^4}{3c^3}(\mathbf{a}^2 + \mathbf{b}^2) = \frac{2e^2\omega^2}{3mc^3}W$$

The expressions for emission and damping are therefore

$$S = \gamma W = -\dot{W}; \quad W = W_0 e^{-\gamma t} \quad \text{where} \quad \gamma = \frac{2e^2\omega^2}{3mc^3} \tag{4.6}$$

The time

$$\tau = \frac{1}{\gamma} = \frac{3mc^3}{2e^2\omega^2} \tag{4.7}$$

is the period required for the energy of the radiating atoms to decrease to a fraction $1/e$ of the initial value. Its approximate values are given for

$$\frac{\omega}{2\pi} = 10^{14}\,\text{s}^{-1}; \quad \tau = 4 \times 10^{-7}\,\text{s}$$

$$\frac{\omega}{2\pi} = 10^{15}\,\text{s}^{-1}; \quad \tau = 4 \times 10^{-9}\,\text{s}$$

The range of application of the above analysis depends on the ratio of the energy radiated in each period to the total energy, $ST/W = \gamma T = T/\tau$. Light damping occurs when $\gamma T$ is much smaller than unity. Now $\gamma T = \frac{4\pi}{3}\frac{e^2\omega}{mc^3} = (8\pi^2/3) \times R_{el}/\lambda$ if we introduce the classical "electron radius"* $R_{el} = e^2/mc^2 = 2\cdot8 \times 10^{-13}$cm and

---

* The classical electron radius is that radius $R_{el}$ which a uniformly distributed charge $e$ must possess, in order that the electrical energy ($\approx e^2/R_{el}$) of this distribution should be equal to the relativistic rest energy $mc^2$ of a stationary particle of mass $m$.

the wavelength $\lambda = 2\pi c/\omega$. Provided, therefore, that the wavelength is large compared with $R_{el}$, the oscillation is very lightly damped. Since $\lambda$ is always much greater than $R_{el}$ for atomic radiation, the application of equations (4.4) and (4.5) is justified.

The corresponding expressions for the attenuation of the amplitudes **a** and **b** are

$$\mathbf{a} = \mathbf{a}_0 \exp -\tfrac{1}{2}\gamma t, \qquad \mathbf{b} = \mathbf{b}_0 \exp -\tfrac{1}{2}\gamma t$$

the logarithmic decrement of which is

$$\delta = \frac{\gamma}{2}\frac{2\pi}{\omega} = \frac{2\pi e^2 \omega}{3mc^3}$$

We do not violate the principle of the conservation of energy if we replace equation (4.2) by

$$\mathbf{r} = (\mathbf{a}_0 \cos \omega t + \mathbf{b}_0 \sin \omega t)\exp -\tfrac{1}{2}\gamma t \qquad (4.8)$$

Naturally, equation (4.8) is not a solution of the original equation of motion (4.1), but is merely the result of a subsequent and therefore unsatisfactory correction to the solution of this equation.

It is natural to go one stage further, and to investigate the possibility of introducing a supplementary force **F** into equation (4.1) in order to avoid violating the principle of conservation of energy. Provided that we confine ourselves to nearly periodic motion we can indeed find a possible expression for such a force. Introducing it into equation (4.1):

$$m\ddot{\mathbf{r}} + f\mathbf{r} = \mathbf{F}$$

which, after multiplication by $\dot{\mathbf{r}}$ may be written as

$$\frac{d}{dt}(\tfrac{1}{2}m\dot{\mathbf{r}}^2 + \tfrac{1}{2}f\mathbf{r}^2) = \mathbf{F} \cdot \dot{\mathbf{r}}$$

On the left-hand side we have the rate of change of energy, which is prescribed by equation (4.5); the force **F** must therefore be such that

$$\mathbf{F} \cdot \dot{\mathbf{r}} = -\frac{2e^2}{3c^3}\ddot{\mathbf{r}}^2 \qquad (4.9)$$

This equation cannot in general be solved by simple methods. Here, however, we shall limit ourselves to satisfying the conservation principle as a time average during nearly periodic motion. Using the identity

$$\frac{d}{dt}(\dot{\mathbf{r}} \cdot \ddot{\mathbf{r}}) = \ddot{\mathbf{r}}^2 + \dot{\mathbf{r}} \cdot \dddot{\mathbf{r}}$$

we may write
$$\mathbf{F} \cdot \dot{\mathbf{r}} = \frac{2e^2}{3c^3} \left[ \dot{\mathbf{r}} \cdot \dddot{\mathbf{r}} - \frac{d}{dt} (\dot{\mathbf{r}} \cdot \ddot{\mathbf{r}}) \right]$$

If we now take the average value of the above expression, say from time $t = 0$ to $t = t_1$, then

$$\overline{\mathbf{F} \cdot \dot{\mathbf{r}}} = \frac{2e^2}{3c^3} \overline{(\dot{\mathbf{r}} \cdot \dddot{\mathbf{r}})} - \frac{2e^2}{3c^3} \frac{[(\dot{\mathbf{r}} \cdot \ddot{\mathbf{r}})_{t=t_1} - (\dot{\mathbf{r}} \cdot \ddot{\mathbf{r}})_{t=0}]}{t_1}$$

If $t_1$ is sufficiently great, the second term is small compared with the first, and may be neglected. We may then satisfy our equation by putting

$$\mathbf{F} = \frac{2e^2}{3c^3} \dddot{\mathbf{r}} \tag{4.10}$$

and thus obtain for the corrected equation of motion of the elastically bound electron

$$m\ddot{\mathbf{r}} + f\mathbf{r} - \frac{2e^3}{3c^3} \dddot{\mathbf{r}} = 0 \tag{4.11}$$

This equation must of course lead to the solution (4.8). If we put

$$\mathbf{r} = \mathbf{A} \, e^{i\omega t}$$

where $\mathbf{A}$ is a complex vector, (4.11) yields the following equation for $\omega$:

$$-\omega^2 + \frac{2e^2}{3mc^3} i\omega^3 + \frac{f}{m} = 0$$

Putting $\sqrt{\dfrac{f}{m}} = \omega_0$ and $\dfrac{2e^2\omega^2}{3mc^3} = \gamma$ the equation becomes

$$-\omega^2 + i\omega\gamma + \omega_0^2 = 0$$

If we limit ourselves to the case for which $\gamma \ll \omega$, the two approximate solutions are

$$\omega_1 = \omega_0 + \tfrac{1}{2}i\gamma, \qquad \omega_2 = -\omega_0 + \tfrac{1}{2}i\gamma$$

We thus obtain for the general solution of (4.11)

$$\mathbf{r} = \mathbf{A} \exp -(\tfrac{1}{2}\gamma - i\omega_0)t + \mathbf{B} \exp -(\tfrac{1}{2}\gamma + i\omega_0)t \tag{4.12}$$

which is identical with (4.8), apart from its complex form. Since $\mathbf{r}$ is real it must equal its complex conjugate $\mathbf{r}^*$; therefore $\mathbf{B} = \mathbf{A}^*$, and there are only six arbitrary constants in (4.12)—three each for the real and the imaginary parts of the vector **A**.

It must be emphasized that the expression (4.10) for the radiation damping is only valid for nearly periodic motion, for which

$$\frac{2e^2}{3c^3}\dddot{\mathbf{r}} \approx -\frac{2e^2\omega_0^2}{3c^3}\dot{\mathbf{r}} \tag{4.10a}$$

Absurd results are obtained if (4.10) is applied to other forms of motion, such as the retardation of a free electron in a constant opposing field. In this case only the second time derivative would be different from zero, and equation (4.11) would therefore predict no radiation damping at all.

The above derivation of the radiation damping is unsatisfactory, because it is not at all clear how the emitted spherical wave influences the electron's motion. In order to gain a closer understanding of the nature of this "self reaction" it is necessary to compute the resultant force on all electron volume elements.

This "self force" $\mathbf{F}_s$ is given by $\mathbf{F}_s = \int \rho \left( \mathbf{E} + \frac{\mathbf{v}}{c} \times \mathbf{H} \right) dV$ where the charge density $\rho$ and the velocity **v** of the moving electron are functions of $x, y, z$, and $t$, and **E** and **H** represent the field strengths produced by the charge and current distributions at the point $x,y,z$, as shown in Vol. I, §66. If we consider the limiting or non-relativistic case, for which $|\mathbf{v}| \ll c$, then $\mathbf{F}_s = \int \rho \mathbf{E} dV$ approximately.

Consider the particular case of an electron moving along the $x$-axis as a rigid spherically symmetrical charge distribution with centre at $x_0(t)$. For the calculation of $\mathbf{F}_s$ and **E**, we can confine ourselves to the $x$-components:

$$E_x = -\frac{\partial \phi}{\partial x} - \frac{1}{c} \dot{A}_x$$

where the retarded potentials $\phi$, $A_x$ are

$$\phi(x,y,z,t) = \int \frac{1}{r} \rho_0 \left[ \xi - x_0 \left( t - \frac{r}{c} \right), \eta, \zeta \right] d\xi \, d\eta \, d\zeta$$

$$A_x(x,y,z,t) = \frac{1}{c} \int \frac{1}{r} \dot{x}_0 \left( t - \frac{r}{c} \right) \rho_0 \left[ \xi - x_0 \left( t - \frac{r}{c} \right), \eta, \zeta \right] d\xi \, d\eta \, d\zeta$$

and 

$$r^2 = (x - \xi)^2 + (y - \eta)^2 + (z - \zeta)^2$$

$\rho_0(\xi, \eta, \zeta)$ is the rigid charge distribution of the electron at rest at the point $\xi, \eta, \zeta = 0$. If $x_0$ and $\rho_0$ can be developed as a power series, we may express

$$\{\rho_0\} = \rho_0 [\xi - x_0(t - r/c), \eta, \zeta]$$

as a power series in terms of $r/c$:

$$\{\rho_0\} = \rho_0 + \frac{r}{c} \frac{\partial \rho_0}{\partial \xi} \dot{x}_0 - \left( \frac{r}{c} \right)^2 \frac{1}{2} \left[ \frac{\partial \rho_0}{\partial \xi} \ddot{x}_0 - \ldots \right] + \left( \frac{r}{c} \right)^3 \frac{1}{6} \left[ \frac{\partial \rho_0}{\partial \xi} \dddot{x}_0 + \ldots \right] + \ldots$$

to the third power of $r/c$, and leaving out terms that may be omitted for reasons of symmetry in the following integration for $F_{sx}$.

We then obtain the following expression for the contribution of $\phi$ to the force $F_{sx}$, after integration by parts to eliminate the derivative $\dfrac{\partial \rho_0}{\partial \xi}$:

$$\frac{1}{2c^2}\ddot{x}_0 \int \rho_0(x,\dots)\rho_0(\xi,\dots)\left[\frac{1}{r}-\frac{(\xi-x)^2}{r^3}\right]d\xi\dots dx\dots$$

$$-\frac{1}{3c^3}\dddot{x}_0\int \rho_0(x,\dots)\rho_0(\xi,\dots)d\xi\dots dx\dots$$

Similarly, the contribution of $A_x$ to the force $F_{sx}$ is

$$-\frac{1}{c^2}\ddot{x}_0\int \rho_0(x,\dots)\rho_0(\xi,\dots)\frac{1}{r}d\xi\dots dx\dots$$

$$+\frac{1}{c^3}\dddot{x}_0\int \rho_0(x,\dots)\rho_0(\xi,\dots)d\xi\dots dx\dots$$

Now
$$U_0=\frac{1}{2}\int\frac{\rho_0(x,\dots)\rho_0(\xi,\dots)}{r}d\xi\dots dx\dots$$

is the electrical rest energy of the electron, and

$$X=\frac{1}{2}\int\frac{\rho_0(x,\dots)\rho_0(\xi,\dots)}{r}\frac{(x-\xi)^2}{r^2}d\xi\dots dx\dots=\frac{1}{3}U_0$$

since $X+Y+Z=U_0$, because $\rho_0$ is spherically symmetrical. The series expansion for the self-force $F_{sx}$ is therefore

$$F_{sx}=-\frac{4}{3}\frac{U_0}{c^2}\ddot{x}_0+\frac{2}{3}\frac{e^2}{c^3}\dddot{x}_0+\dots$$

The first term represents the familiar contribution, $\dfrac{4}{3}\dfrac{U_0}{c^2}$, to the electron mass (cf. Vol. I, §§65 and 91). The second term agrees exactly with expression (4.10) for the self-reactive force of the electron due to emitted radiation. The omitted terms contain higher derivatives of $x_0$.

In considering the useful range of application of the above series, we observe that it represents a development in powers of $R_{el}/c=e^2/mc^2$. Now if $x_0$ is expressed as a Fourier series terminating at $\omega_m$, where $\omega_m$ corresponds to the highest significant frequency, and $\tau=1/\omega_m$, then, since $\left|\dfrac{d^n x_0}{dt^n}\right|\gtrsim\tau^{-n}\,|x_0|$, $F_{sx}$ may be expressed in powers of $R_{el}/c\tau$, interrupted after the term $\dddot{x}_0$, provided that

$$\tau\gg\frac{R_{el}}{c}=\frac{e^2}{mc^3}$$

This provision implies that the point $x_0$ should not have moved significantly during the time required by light to traverse the region occupied by the electron. For instance, if $x_0=ae^{i\omega t}$, we must have

$$\omega\ll\frac{mc^3}{e^2}$$

If this motion is caused by a light wave, then

$$\lambda \gg R_{el} = \frac{e^2}{mc^2}$$

The formula derived for $F_{sx}$ is applicable to radiation in the normal optical region, for which it predicts the damping effect discussed above.

In other cases, however, such as that of the free electron, it is not permissible to solve the equation $m\ddot{x}_0 + \frac{2}{3}\frac{e^2}{c^3}\dddot{x}_0 = 0$ by putting $x = ae^{\alpha t}$, $\alpha = mc^2/e^2$, because the omitted terms of the series expansion are of the same order of magnitude as the others.

Types of motion for which $\tau \gtrsim e^2/mc^3$ can only be treated in the light of a more precise knowledge of the "structure" of the electron, for which the primitive representation by means of a rigid sphere is quite inadequate.

More thorough analyses of the "self force" have been given by P. A. M. Dirac, *Proc. Roy. Soc.* Series A 167 (1938) 148, by F. Bopp, *Z. Naturforsch.* 1 (1946) 53, 237, and by H. Steinwedel, *Fortschr. d. Phys.* I, 1953 and 1954, 7.

## §5. The width of the emitted spectral line

Pure monochromatic light would be equivalent to an infinitely long simple harmonic wave train of form $a\sin 2\pi(vt - x/\lambda)$. Any departure from this form results in the presence of a mixture of different frequencies $v$, i.e. of different colours, as may be seen from spectral analysis. The frequencies contained in a given wave train may be determined by means of Fourier analysis; using this method, the wave train is represented as a linear superposition of infinitely long simple harmonic wave trains. The essence of Fourier's theorem is that any such wave form may be represented in this manner. In particular, if the wave train is approximately harmonic in form and has a definite frequency $v_0$, its deviation from a strictly harmonic form is characterized by a definite widening $\delta v$ of the corresponding spectral line of frequency $v_0$.

In many cases, the amount of broadening, $\delta v$, may be evaluated approximately almost without calculation. Consider, for instance, a wave train of finite length $L$, represented at a given instant (say $t = 0$) by

$$f = a\sin 2\pi\frac{x}{\lambda_0} \quad \text{for } 0 < x < L$$

$$f = 0 \quad \text{for } x < 0 \text{ and } x > L$$

The number of wavelengths contained in the train is $n = L/\lambda_0$, where $n$ is considered to be large compared to 1. In order to represent this train by a set of infinitely long waves, components are required of wavelengths such as to interfere destructively with the wave train $\lambda_0$ (considered here to be infinitely long) outside the interval $L$. For this purpose, constituents with $n+1$ and $n-1$ wavelengths in the interval are necessary; if these waves are in phase with the original wave in the middle of the interval, they are 180° out of phase at the boundaries. Hence, in the Fourier expansion, waves will occur whose wavelength is given by $n \pm 1 = L/\lambda'$. In terms of frequencies:

$$v_0 = \frac{nc}{L} \quad \text{and} \quad v' = \frac{(n \pm 1)c}{L}$$

We then obtain for the order of magnitude of the line width:

$$|\delta v| = \frac{c}{L} = \frac{1}{\tau}$$

where $\tau$ represents the time required by the wave train to pass a stationary observer. Expressed in terms of wavelength, the line width is

$$|\delta \lambda| = \frac{|\delta v|}{c} \lambda^2 = \frac{\lambda^2}{L}$$

The expression for the relative width is particularly simple:

$$\frac{|\delta v|}{v} = \frac{|\delta \lambda|}{\lambda} \approx \frac{1}{n}$$

which is the reciprocal of the number of wavelengths present in the wave train.

Now the wave emitted according to (4.11) by an oscillating electron is not suddenly chopped off but has the form of a damped wave train, the amplitude of which falls to $1/e$ of the original after a time $\tau = 2/\gamma$. We assume, however, that as regards line width it is comparable to a chopped wave of just this time interval. Then

$$\delta v \approx \tfrac{1}{2}\gamma \tag{5.1}$$

and the numerical value of the line width measured in terms of frequency is therefore given directly by the damping factor $\gamma$.

We now wish to verify the above treatment by means of the strict Fourier analysis of the oscillation equation (4.12). Since the field strengths of the wave region are linearly dependent on acceleration, the Fourier analysis of $\ddot{r}$ is also directly valid for the emitted wave train.

For this purpose, let us consider an electron which is initially constrained. If it is released at time $t = 0$, it moves according to equation (4.12), in which the amplitudes $\mathbf{A}$ and $\mathbf{A}^*$ are determined by the initial conditions $\mathbf{r}(0) = \mathbf{r}_0$, $\dot{\mathbf{r}}(0) = 0$. Then

$$\ddot{\mathbf{r}}(t) = \begin{cases} (\tfrac{1}{2}\gamma - i\omega_0)^2 \mathbf{A} \exp -(\tfrac{1}{2}\gamma - i\omega_0)t + \\ \qquad + (\tfrac{1}{2}\gamma + i\omega_0)^2 \mathbf{A}^* \exp -(\tfrac{1}{2}\gamma + i\omega_0)t \text{ for } t > 0 \\ \qquad\qquad\qquad 0 \qquad\qquad\qquad\qquad\quad \text{for } t < 0 \end{cases} \quad (5.2)$$

Confining ourselves initially to the first term of (5.2), we must seek a function $\mathbf{c}(\omega)$ such that

$$\int_{-\infty}^{\infty} \mathbf{c}(\omega) e^{i\omega t} d\omega = \begin{cases} (\tfrac{1}{2}\gamma - i\omega_0)^2 \mathbf{A} \exp -(\tfrac{1}{2}\gamma - i\omega_0)t \text{ for } t > 0 \\ \qquad\qquad 0 \qquad\qquad\qquad\qquad \text{for } t < 0 \end{cases} \quad (5.3)$$

We know from Fourier's theorem that any function $f(t)$ for which $\int_{-\infty}^{\infty} |f(t)| \, dt$ exists may be represented by

$$f(t) = \int_{-\infty}^{\infty} c(\omega) e^{i\omega t} d\omega \quad \text{with} \quad c(\omega) = \frac{1}{2\pi} \int_{-\infty}^{\infty} f(t) e^{-i\omega t} dt$$

From this and from (5.3) it follows that

$$\mathbf{c}(\omega) = \frac{(\tfrac{1}{2}\gamma - i\omega_0)^2}{2\pi} \frac{\mathbf{A}}{\tfrac{1}{2}\gamma - i(\omega_0 - \omega)} \quad (5.4)$$

The same treatment of the second term yields a similar expression with $-\omega_0$ instead of $\omega_0$. We therefore obtain the result

$$\ddot{\mathbf{r}}(t) = \frac{1}{2\pi} \int_{-\infty}^{\infty} \left\{ \frac{(\tfrac{1}{2}\gamma - i\omega_0)^2}{\tfrac{1}{2}\gamma - i(\omega_0 - \omega)} \mathbf{A} + \frac{(\tfrac{1}{2}\gamma + i\omega_0)^2}{\tfrac{1}{2}\gamma + i(\omega_0 + \omega)} \mathbf{A}^* \right\} e^{i\omega t} d\omega \quad (5.5)$$

When studying the distribution of intensity in a spectrogram we must bear in mind that we are not concerned with the relation between intensity and time; experimentally it is always the total intensity that is measured at any point of the spectrum. We are therefore interested in the spectral distribution of the quantity

$$\int_0^{\infty} S \, dt = \frac{2e^2}{3c^3} \int_0^{\infty} \ddot{\mathbf{r}}^2 \, dt$$

such that

$$\int_0^{\infty} S \, dt = \int_0^{\infty} S(\omega) \, d\omega \quad (5.6)$$

Equation (5.5) may be written

$$\ddot{\mathbf{r}} = \int_{-\infty}^{+\infty} \mathbf{g}(\omega)\, e^{i\omega t}\, d\omega$$

$\ddot{\mathbf{r}}$ is a real number, therefore $\mathbf{g}(\omega) = \mathbf{g}^*(-\omega)$. We may therefore also write

$$\ddot{\mathbf{r}} = \int_{-\infty}^{+\infty} \mathbf{g}^*(\omega)\, e^{-i\omega t}\, d\omega$$

It follows that

$$\int_0^\infty \ddot{\mathbf{r}}^2\, dt = \int_{-\infty}^{+\infty} d\omega \mathbf{g}^*(\omega) \int_0^\infty \ddot{\mathbf{r}}(t)\, e^{-i\omega t}\, dt$$

From Fourier's theorem the time integral is $2\pi \mathbf{g}(\omega)$, therefore

$$\int_0^\infty \ddot{\mathbf{r}}^2\, dt = 2\pi \int_{-\infty}^{+\infty} \mathbf{g}(\omega)\mathbf{g}^*(\omega)\, d\omega = 4\pi \int_0^\infty \mathbf{g}(\omega)\mathbf{g}^*(\omega)\, d\omega$$

since the integrand is symmetrical in $\omega$, as may be seen from (5.5). We have now accomplished our analysis and determined the amount of radiation energy arising from the elementary region $\omega, \omega + d\omega$ of the spectrum:

$$S(\omega)\, d\omega = \frac{2e^2}{3c^3} 4\pi \mathbf{g}(\omega)\mathbf{g}^*(\omega)\, d\omega \qquad (5.7)$$

In order to evaluate (5.7) it is necessary to form the product $\mathbf{g}(\omega)\mathbf{g}^*(\omega)$, using (5.5), and this results in a rather long formula. A simpler expression is obtained if we confine ourselves to the neighbourhood of the frequency corresponding to $\omega_0$. This simplification is satisfactory in practice, because almost the whole spectral intensity is distributed in a narrow band of frequencies of approximate width $\gamma$, as we saw in our earlier simplified treatment. We therefore restrict ourselves to frequencies such that $|\omega - \omega_0| \ll \omega_0$. In addition, $\gamma$ is always much smaller than $\omega_0$, as we saw in the last paragraph. On examining the expression for $\mathbf{g}(\omega)$ given by (5.5) we see that in this frequency region the first term is much larger than the second. Since we are only concerned with the value of the vector sum in (5.5) we may restrict ourselves to the first term. We then obtain

$$\mathbf{g}(\omega)\mathbf{g}^*(\omega) = \frac{\omega_0^4}{4\pi^2} \frac{\mathbf{A}\mathbf{A}^*}{(\omega - \omega_0)^2 + \frac{1}{4}\gamma^2} \qquad (5.8)$$

(In the above expression we have neglected the quantity $\frac{1}{4}\gamma^2$ in the numerator, since it is much smaller than $\omega_0^2$.) Inspection shows that equation (5.8) possesses a maximum of value $\omega_0^4 AA^*/\pi^2\gamma^2$ at $\omega = \omega_0$. We also see that the intensity function has decreased to half its maximum value at a frequency separation given by $|\omega - \omega_0| = \frac{1}{2}\gamma$. The "half value" width of the line is therefore

$$\delta\omega = \gamma \quad \text{or} \quad \delta v = \frac{\gamma}{2\pi} \tag{5.9}$$

which agrees roughly with the earlier approximate estimate (5.1). It is now clear that we were justified in restricting our more complete analysis to the immediate neighbourhood of the spectral line.

As a check, we shall calculate the total radiated energy from its spectral distribution, using equations (5.6), (5.7) and (5.8):

$$\int_0^\infty S(t)\,dt = \int_0^\infty S(\omega)\,d\omega = \frac{2e^2}{3c^3}\frac{\omega_0^4}{\pi}AA^*\int_0^\infty \frac{d\omega}{(\omega-\omega_0)^2+\frac{1}{4}\gamma^2}$$

Introducing the variable $x = \dfrac{2(\omega-\omega_0)}{\gamma}$

$$\int_0^\infty S(t)\,dt = \frac{2e^2}{3c^3}\frac{\omega_0^4}{\pi}AA^*\frac{2}{\gamma}\int_{-2\omega_0/\gamma}^\infty \frac{dx}{x^2+1}$$

Since $\gamma \ll \omega_0$, the value of the integral is not appreciably altered if we substitute $-\infty$ for the lower limit. Then $\int_{-\infty}^\infty \dfrac{dx}{x^2+1} = \pi$, and inserting the value for the damping factor $\gamma$ given by (4.6), we have for the total radiated energy

$$\int_0^\infty S(t)\,dt = 2m\omega_0^2 AA^*$$

which is necessarily equal to the total mechanical energy of the elastically bound electron at time $t = 0$:

$$W_0 = (\tfrac{1}{2}m\dot{r}^2+\tfrac{1}{2}fr^2)_{t=0}$$

From the equations of motion (4.11) and (4.12), neglecting $\gamma$ in comparison with $\omega_0$:

$$W_0 = 2m\omega_0^2 AA^*$$

The total initial energy of oscillation therefore reappears as the total energy radiated.

For an understanding of the optical properties of matter it is particularly important to know how an elastically bound electron behaves in an alternating field, and in the field due to incident light waves in particular. We shall restrict ourselves to the consideration of a linearly polarized wave train, the field strength of which at the electron is $\mathbf{E} = \mathbf{E}_0 e^{i\omega t}$. The electron's equation of motion is

$$\ddot{\mathbf{r}} + \gamma\dot{\mathbf{r}} + \omega_0^2 \mathbf{r} = \frac{e}{m}\mathbf{E}_0 e^{i\omega t} \tag{5.10}$$

in which we allow for the loss of energy by radiation through the introduction of the damping term $\gamma\dot{\mathbf{r}}$, where $\gamma$ is the damping factor given by equation (4.6).

The inhomogeneous equation (5.10) may be satisfied by putting

$$\mathbf{r} = \mathbf{r}_0 e^{i\omega t} \tag{5.11}$$

where

$$\mathbf{r}_0 = \frac{e}{m}\frac{\mathbf{E}_0}{\omega_0^2 - \omega^2 + i\omega\gamma} \tag{5.11a}$$

If we write $\mathbf{r}_0 = \mathbf{s}_0 e^{-i\phi}$, where $\mathbf{s}_0$ and $\phi$ are real quantities, we obtain the following result: the motion produced by the field $\mathbf{E} = \mathbf{E}_0 e^{i\omega t}$ may be described by the vector $\mathbf{r} = \mathbf{s}_0 e^{i(\omega t - \phi)}$, of amplitude

$$\mathbf{s}_0 = \frac{e}{m}\frac{\mathbf{E}_0}{\sqrt{[(\omega_0^2 - \omega^2)^2 + \omega^2\gamma^2]}} \tag{5.12a}$$

and phase

$$\phi = \frac{\omega\gamma}{\omega_0^2 - \omega^2} \tag{5.12b}$$

(The general integral of (5.10) is obtained by the addition of an arbitrary solution of the homogeneous equation; since this solution decays with time, after a sufficiently long period the electron is solely influenced by the light wave, and only the forced oscillation remains.)

For the amplitude and phase angle of the forced oscillation we obtain the familiar curves of figure 11. Outside the region of resonance (i.e. when $|\omega - \omega_0| \gg \gamma$), it is in phase with the exciting force at lower frequencies, and at very low frequencies its amplitude approaches the value applicable to steady fields, $\mathbf{r}_0 = eE/m\omega_0^2$. On the other side of the resonance frequency the phase of the oscillation is opposed to the force. As the frequency increases without limit the amplitude tends to zero. In the region of resonance (i.e. when $|\omega - \omega_0| \approx \gamma$), the amplitude rises steeply, attaining a maximum value $\dfrac{e}{m}\dfrac{\mathbf{E}_0}{\omega_0\gamma}$ for $\omega \approx \omega_0$.

The "half-value width" for the square of the amplitude is once more $\delta\omega \approx \gamma$, as in the case of free oscillations.

In consequence of the "frictional term" in equation (5.10), work must be performed by the field on the electron in order to maintain

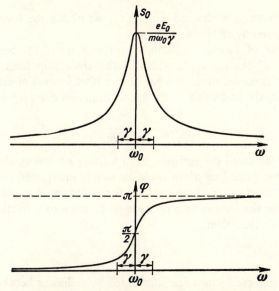

Fig. 11.—Amplitude $s_0$ and phase shift $\phi$ of a forced oscillation in the resonance region $\omega_0$, from equations (5.12)

oscillations. For periodic motion, we may deduce the average value of the power produced by a force acting on the electron from the equation

$$m\ddot{\mathbf{r}} + m\gamma\dot{\mathbf{r}} + m\omega_0^2 \mathbf{r} = \mathbf{F}$$

The power produced by $\mathbf{F}$ at any instant is $\mathbf{F}.\dot{\mathbf{r}}$. If we take the time average of the above expression, the terms $\dot{\mathbf{r}}.\ddot{\mathbf{r}} = \frac{1}{2}\frac{d\dot{\mathbf{r}}^2}{dt}$ and $\mathbf{r}.\dot{\mathbf{r}} = \frac{1}{2}\frac{d\mathbf{r}^2}{dt}$ drop out, and we obtain

$$\overline{\mathbf{F}.\dot{\mathbf{r}}} = m\gamma\overline{\dot{\mathbf{r}}^2} = \tfrac{1}{2}m\gamma\omega^2\mathbf{r}_0^2 \tag{5.13}$$

in view of (5.11). If we are dealing with radiation damping alone, in which the electron loses energy only from emitted radiation, the value deduced above for the average power must correspond exactly to the energy radiated per second. From (4.4) this is

$$S = \frac{2e^2}{3c^3}\overline{\ddot{\mathbf{r}}^2} = \frac{2e^2}{3c^3}\frac{\omega^4 \mathbf{r}_0^2}{2}$$

If this expression is equated with (5.13) we obtain the correct value for $\gamma$, as given by equation (4.6).

The law of forced oscillations expressed by (5.12) corresponds experimentally to the finite width of the absorption lines observed when light is passed through a gas. If no other sources of disturbance are present, the half-value width due to radiation damping alone is

$$\delta\omega_{rad} = \gamma = \frac{2e^2\omega^2}{3mc^3}$$

When we discussed the emission from a freely oscillating electron we saw that the finite line width could be simply interpreted as the reciprocal of the electron's time constant for radiation; if $\tau_{rad}$ is the time required for the oscillation amplitude to decrease to a fraction $1/e$ of its original value, then

$$\gamma = \frac{2}{\tau_{rad}} \tag{5.14}$$

If the free oscillations are interrupted by collisions between atoms we must expect increased broadening of the spectral lines. This is obvious for the case of emission, since, as we have already seen, line width depends essentially upon the length of the individual wave trains and not on the cause of their limitation. If we designate by $\tau$ the average time between collisions, then a line width of order $\delta\omega \approx 1/\tau$ will be produced purely by the collisions. We shall now show that the same factor applies to the process of absorption, by considering the average motion, under the influence of an alternating electric field, of a large number of electrons bound to colliding atoms. Equation (5.10) is still valid for this average motion, provided that the supplementary term

$$g = \frac{2}{\tau} \tag{5.15}$$

is introduced and added to $\gamma$. We may thus describe the average motion of the electron by means of a damping constant $\gamma + g$ when both sources of damping are present.

For proof, we start from the general solution of equation (5.10)

$$\mathbf{r}(t) = \mathbf{r}_0 \, e^{i\omega t} + (\mathbf{A} \, e^{i\omega_0 t} + \mathbf{B} e^{-i\omega_0 t}) \, e^{-\frac{1}{2}\gamma t} \qquad (5.16)$$

which includes the particular integral (5.11) and the two constants of integration $\mathbf{A}$ and $\mathbf{B}$. The vector $\mathbf{r}_0$ is related to the (complex) amplitude vector $\mathbf{E}_0$ by the equation

$$\mathbf{r}_0 = \frac{e}{m} \frac{\mathbf{E}_0}{\omega_0^2 - \omega^2 + i\omega\gamma} \qquad (5.11a)$$

Now assume that the atom under consideration suffered its latest collision at time $t_1$. If we were to know the values of $\mathbf{r}$ and $\dot{\mathbf{r}}$ immediately after collision, the constants $\mathbf{A}$ and $\mathbf{B}$ would be determined. Now we do not know these values for individual atoms; if, however, we consider a large number of such oscillators, all of which have been subjected to collision at time $t_1$, we can assert that immediately after $t_1$ all directions of the displacement $\mathbf{r}$ and the velocity $\dot{\mathbf{r}}$ are equally probable. The average value of $\mathbf{r}$ and $\dot{\mathbf{r}}$ at time $t_1$ will therefore be zero for all these oscillators. The equations determining the constants of integration are therefore, from (5.16),

$$\mathbf{r}_0 \, e^{i\omega t_1} + (\mathbf{A} \, e^{i\omega_0 t_1} + \mathbf{B} e^{-i\omega_0 t_1}) \, e^{-\frac{1}{2}\gamma t_1} = 0$$

$$i\omega\mathbf{r}_0 \, e^{i\omega t_1} + \{(i\omega_0 - \tfrac{1}{2}\gamma)\mathbf{A} \, e^{i\omega_0 t_1} - (i\omega_0 + \tfrac{1}{2}\gamma)\mathbf{B} e^{-i\omega_0 t_1}\} \, e^{-\frac{1}{2}\gamma t_1} = 0$$

If we insert the resultant values for $\mathbf{A}$ and $\mathbf{B}$ into (5.16), and introduce the time elapsed since the last collision, $\theta = t - t_1$, we obtain

$$\mathbf{r}_\theta(t) = \mathbf{r}_0 \, e^{i\omega t} \left\{ 1 - \frac{\omega_0 + \omega - \frac{1}{2}i\gamma}{2\omega_0} \exp\left[i(\omega_0 - \omega) - \tfrac{1}{2}\gamma\right]\theta - \right.$$
$$\left. - \frac{\omega_0 - \omega + \frac{1}{2}i\gamma}{2\omega_0} \exp -\left[i(\omega_0 + \omega) + \tfrac{1}{2}\gamma\right]\theta \right\}$$

If $\tau$ is the average time between collisions and $N$ is the total number of oscillators present in the volume under consideration, then the number of these oscillators whose last collision took place during the time interval $\theta, \theta + d\theta$ is given by

$$\frac{N}{\tau} e^{-\theta/\tau} \, d\theta$$

The mean contribution of each individual oscillator to the resultant displacement is therefore

$$\overline{\mathbf{r}(t)} = \int_{\theta=0}^{\infty} \mathbf{r}_\theta(t)\, e^{-\theta/\tau} \frac{d\theta}{\tau} \tag{5.17}$$

Integration yields

$$\overline{\mathbf{r}(t)} = \mathbf{r}_0' \, e^{i\omega t} \tag{5.18}$$

where

$$\mathbf{r}_0' = \mathbf{r}_0 \frac{\omega_0^2 - (\omega - \tfrac{1}{2}i\gamma)^2}{\omega_0^2 - \left(\omega - \tfrac{1}{2}i\gamma - \dfrac{i}{\tau}\right)^2}$$

If we neglect the quadratic terms in $\gamma$, as before, then from the above expression and (5.11a) we have

$$\mathbf{r}_0' = \frac{e}{m} \frac{\mathbf{E}_0}{\omega_0^2 - \omega^2 + i\omega(\gamma + 2/\tau)}$$

Inspection shows that the original denominator of (5.11a) has been replaced by a new one, which differs from the first only through the addition of the collision damping constant $2/\tau$ to the radiation constant $\gamma$. Our previous statement is now proved.

Apart from the broadening of the spectral lines by radiation and collision damping there is also the Doppler effect, which is caused by the thermal motion of the emitting and absorbing atoms. For details of this broadening effect, see Exercise 1 below.

## Exercises

### 1. *Broadening of spectral lines*

In addition to the effects of radiation and collision damping on line width, there is also the Doppler broadening effect arising from the thermal motion of the radiating atoms. Atoms moving towards the observer exhibit a shift towards the violet end of the spectrum; those moving away, a red shift. Compare the orders of magnitude of the three effects. How may these three effects be separated? (Note that the gas atoms possess a Maxwellian velocity distribution; the fraction of the $N$ atoms, each of mass $M$, with an $x$-component of velocity lying between $v_x$ and $v_x + dv_x$, is

$$Nw(v_x)\, dv_x = N \exp -\frac{Mv_x^2}{2kT}\, dv_x \bigg/ \int_{-\infty}^{\infty} \exp -\frac{Mv_x^2}{2kT}\, dv_x$$

## 2. *Radiation damping and the electron radius*

Let a plane linearly polarized light wave excite oscillations in a free electron. If we imagine the electron to be a sphere of radius $R_{el}$, then according to this conception $R_{el}$ is determined from the fact that the power radiated by the electron is equal to the wave energy incident on this sphere in unit time.

*Hint*: Take a wave $\mathbf{E} = [E(t), 0, 0]$ polarized parallel to the $x$-axis, note the equation of motion $m\ddot{x} = eE$ when the Lorentz force is neglected, and the average value of the energy flow (Poynting vector) $S = c/4\pi E^2$.

# CHAPTER A III

## The Hamiltonian form of the equations of motion

### §6. The Hamiltonian theory in mechanics

The Hamiltonian form of the equations of motion is of proved value as a powerful general method employed in connection with many different physical problems. Two important applications are in statistical mechanics (with which we are only incidentally concerned

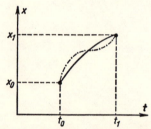

Fig. 12.—Determination of the correct path $x(t)$ (full line). A varied path is shown as a broken line. It must coincide with $x(t)$ at the end points.

here), and in the quantum theory. We shall begin our account of Hamiltonian theory with a mathematical digression on the calculus of variations.

Let us first consider a single function $x(t)$, and a function of the three variables $x$, $\dot{x}$, and $t$, $L(x, \dot{x}, t)$, designated the "Lagrange function". We may choose any function $x(t)$ (figure 12) with fixed initial and final values $x_0$ and $x_1$:

$$x(t_0) = x_0 \quad \text{and} \quad x(t_1) = x_1 \qquad (6.1)$$

Using this function, which is arbitrary except for the above condition, we form the integral

$$G = \int_{t_0}^{t_1} L(x, \dot{x}, t)\, dt \qquad (6.2)$$

Among all the possible functions $x(t)$ that satisfy condition (6.1) we seek the one that makes $G$ an extremum; in doing so we must take account of $\dot{x} = dx/dt$ along all possible paths. We shall call this the "correct function", or in mechanical applications, the "correct path". Thus, let $x(t)$ be the correct function; we now form the varied function

$$x(t) + \alpha\eta(t)$$

where $\alpha$ is a numerical factor, and $\eta(t)$ is an arbitrary function vanishing at the terminal points:

$$\eta(t_0) = \eta(t_1) = 0 \qquad (6.3)$$

$G$ is now a function of $\alpha$:

$$G(\alpha) = \int_{t_0}^{t_1} L(x + \alpha\eta, \dot{x} + \alpha\dot{\eta}, t)\, dt \qquad (6.4)$$

The extremum condition is $(dG/d\alpha)_{\alpha=0} = 0$ for all functions $\eta(t)$ satisfying (6.3). Differentiating under the integral sign we obtain

$$\frac{\partial L}{\partial x}\eta + \frac{\partial L}{\partial \dot{x}}\dot{\eta} = \frac{\partial L}{\partial x}\eta - \left(\frac{d}{dt}\frac{\partial L}{\partial \dot{x}}\right)\eta + \frac{d}{dt}\left(\frac{\partial L}{\partial \dot{x}}\eta\right)$$

Therefore

$$\left(\frac{dG}{d\alpha}\right)_{\alpha=0} = \int_{t_0}^{t_1} \left(\frac{\partial L}{\partial x} - \frac{d}{dt}\frac{\partial L}{\partial \dot{x}}\right)\eta\, dt + \left(\frac{\partial L}{\partial \dot{x}}\eta\right)_{t=t_0} - \left(\frac{\partial L}{\partial \dot{x}}\eta\right)_{t=t_0}$$

The last two terms vanish on account of condition (6.3). In order that $dG/d\alpha$ should vanish for all functions $\eta$, $x(t)$ must satisfy the "Euler equation"

$$\frac{d}{dt}\left(\frac{\partial L}{\partial \dot{x}}\right) = \frac{\partial L}{\partial x} \qquad (6.5)$$

The above treatment may clearly be extended to three functions. If we call these $x_1(t)$, $x_2(t)$, $x_3(t)$, the Lagrange function is $L(x_1, \dot{x}_1, x_2, \dot{x}_2, x_3, \dot{x}_3, t)$. We require that $G$ should be an extremum for

$$G = \int_{t_0}^{t_1} L(x_j, \dot{x}_j, t)\, dt \qquad (6.6)$$

for a variation of $x_1$, $x_2$, $x_3$,

$$x_j + \alpha_j\,\eta_j(t), \text{ where } \eta_j(t_0) = \eta_j(t_1) = 0 \quad (j = 1, 2, 3)$$

The Euler equations are

$$\frac{d}{dt}\left(\frac{\partial L}{\partial \dot{x}_j}\right) = \frac{\partial L}{\partial x_j} \quad (j = 1, 2, 3) \tag{6.7}$$

We now introduce the Hamiltonian function, which is related to the Lagrange function as follows: Let us introduce the "momenta"

$$p_j = \frac{\partial L}{\partial \dot{x}_j}$$

and assume these three equations to be solved for $\dot{x}_1$, $\dot{x}_2$, $\dot{x}_3$. Then the $\dot{x}_j$ appear as functions of the $p_j$ and $x_j$. Using these functions $\dot{x}_j(p_j, x_j)$, we define the Hamiltonian function as

$$\mathscr{H}(p_j, x_j, t) = \sum_{j=1}^{3} p_j \dot{x}_j - L \tag{6.8}$$

From (6.7), it is clear that

$$\dot{x}_j = \frac{\partial \mathscr{H}}{\partial p_j} \qquad \dot{p}_j = -\frac{\partial \mathscr{H}}{\partial x_j} \quad (j = 1, 2, 3) \tag{6.9}$$

If $L$, and hence $\mathscr{H}$, do not depend explicitly on $t$,

$$\frac{d\mathscr{H}}{dt} = 0$$

and from (6.9)

$$\left|\frac{d\mathscr{H}}{dt} = \sum_j \left\{\frac{\partial \mathscr{H}}{\partial p_j}\dot{p}_j + \frac{\partial \mathscr{H}}{\partial x_j}\dot{x}_j\right\} + \frac{\partial \mathscr{H}}{\partial t} = \frac{\partial \mathscr{H}}{\partial t}\right.$$

The value of $\mathscr{H}(p_j, x_j)$ therefore remains constant with respect to time if the "correct" path is chosen for the $p_j(t)$ and $x_j(t)$.

For physical applications $L(x_j, \dot{x}_j, t)$ must be so chosen that the Euler equations do in fact represent the correct equations of motion.

For a particle moving in a field of force of potential energy $V(x, y, z)$, $L$ is equal to the difference between the kinetic and potential energies. In the non-relativistic case,* therefore (writing $x$, $y$, $z$ instead of $x_1$, $x_2$, $x_3$)

$$L = \tfrac{1}{2}m(\dot{x}^2 + \dot{y}^2 + \dot{z}^2) - V(x, y, z) \tag{6.10}$$

---

* Cf. §55 for the relativistic Lagrangian and Hamiltonian functions.

The first of the equations (6.7) then states that

$$\frac{d}{dt} m\dot{x} = -\frac{\partial V}{\partial x}$$

This is the correct Newtonian equation of motion. The momenta are

$$p_x = \frac{\partial L}{\partial \dot{x}} = m\dot{x}, \text{ etc.}$$

From (6.8), the Hamiltonian function is

$$\mathscr{H} = \frac{1}{2m}(p_x^2 + p_y^2 + p_z^2) + V(x, y, z) \tag{6.11}$$

which is the sum of the kinetic and potential energies. In this case, therefore, $\mathscr{H}$ is the total energy, expressed in terms of momenta and coordinates.

Equations (6.6) to (6.9) are clearly also valid for a system of $N$ particles with coordinates $(x_1, x_2, \ldots, x_{3N})$; the index number $j$ now runs from 1 to $3N$. Here too, if the forces between the particles can be represented by means of a potential energy $V(x_1, \ldots, x_{3N})$, $L$ is the difference between the kinetic and potential energies and $\mathscr{H}$ is their sum.

We have seen that the Newtonian equations of motion (6.7) may be considered as Euler equations corresponding to the requirement that the function $G$ should be an extremum for the "correct" paths.* This alternative concept is important, because it enables the equations of motion to be expressed in a form that is invariant with respect to the coordinates. The extremum requirement, known as Hamilton's principle, contains only physical quantities such as kinetic and potential energy, which are independent of the coordinate system. Equations (6.7) are therefore valid for any arbitrary system of coordinates, such as the polar form (cf. Exercise 1, p. 46). It is then only necessary to express $L$ as the difference between the kinetic and potential energies in the coordinate system selected. The same applies to Hamilton's equations. For any arbitrary coordinate system, the "momenta" $p_j = \partial L / \partial \dot{x}_j$ do not in general have the dimensions of a true momentum (e.g. if the coordinate $x_j$ is a dimensionless angular quantity); however, the product of any "momentum" $p_j$ with its associated coordinate $x_j$ always has the dimensions of action (energy $\times$ time). The momentum $p_j$ and coordinate $x_j$ are said to be canonically conjugate.

The Newtonian equations of motion for $N$ particles form a system of $3N$ differential equations of the second order. Therefore, in order to specify a "path" uniquely, we require $6N$ initial conditions, namely the positions $x_j$ and velocities $\dot{x}_j$ at a given time. The motion of the system may then be represented by a curve

---

* $G$ does not always have to be a minimum, always a maximum, or a saddle point. There are cases in which $G$ is a minimum for certain variations, and a maximum for other equally possible path changes. All that is required is that in the first approximation $G$ should not change if a varied path is substituted for the correct path (cf. Exercise 3, p. 47).

in this space of $6N$ dimensions composed of position and velocity coordinates. One point in this space determines the course of the path; for if we define $\dot{x}_j$ as

$$\frac{d}{dt}x_j = \dot{x}_j \qquad (6.12a)$$

then, in conjunction with the Euler equations

$$\frac{d}{dt}\frac{\partial L}{\partial \dot{x}_j} = \frac{\partial L}{\partial x_j} \qquad (6.12b)$$

we have $\partial L/\partial \dot{x}_j = p_j = m\dot{x}_j$ for Cartesian coordinates. Hence, if $x_j(t)$ and $\dot{x}_j(t)$ are given, we can immediately determine $x_j(t+dt)$ and $\dot{x}_j(t+dt)$ from the above equations, from which it follows that the whole path can be constructed. In arbitrary coordinates $p_j$ is no longer proportional to $\dot{x}_j$, but may depend on the other coordinates in a complicated manner. It therefore seems reasonable to introduce the $p_j$ and the $x_j$ as new variables, since their variation with respect to time is directly given by equation (6.12). The path can now be described in a "phase space" of $6N$ dimensions, the coordinates of which are the $p_j$ and $x_j$. The Hamilton equations (6.9) enable the path to be constructed from an initial point in this space, provided that the Hamiltonian function is known. Newton's equations are replaced by (6.12) and (6.9), representing a system of $6N$ equations of the first order; the advantage of this substitution lies in the fact that the symmetrical form of Hamilton's equations makes them more suitable for the treatment of problems in any coordinate system.

*Canonical transformations*

In order to change from one pair of canonically conjugate variables $p$, $x$, to another pair $\bar{p}$, $\bar{x}$, we introduce a function of the old coordinate and the new momentum, $S(x, \bar{p})$, for which

$$p = \frac{\partial S}{\partial x} \quad \text{and} \quad \bar{x} = \frac{\partial S}{\partial \bar{p}} \qquad (6.13)$$

The relation $\bar{p}(p, x)$, $\bar{x}(p, x)$ between the old and the new variables is given implicitly by (6.13). The functional determinant (Jacobian) of the transformation (6.13) is unity. This is most easily demonstrated by expressing the total differential $dS = p \,.\, dx + \bar{x} \,.\, d\bar{p}$ in terms of the new variables:

$$dS = \underbrace{p\frac{\partial x}{\partial \bar{x}}\,d\bar{x}}_{\left(\frac{\partial S}{\partial \bar{x}}\right)_{\bar{p}}} + \underbrace{\left(p\frac{\partial x}{\partial \bar{p}} + \bar{x}\right)d\bar{p}}_{\left(\frac{\partial S}{\partial \bar{p}}\right)_{\bar{x}}}$$

Then the criterion for integrability

$$\frac{\partial}{\partial \bar{p}}\left(p\frac{\partial x}{\partial \bar{x}}\right) = \frac{\partial}{\partial \bar{x}}\left(p\frac{\partial x}{\partial \bar{p}} + \bar{x}\right)$$

yields the functional determinant

$$\frac{\partial p}{\partial \bar{p}}\frac{\partial x}{\partial \bar{x}} - \frac{\partial p}{\partial \bar{x}}\frac{\partial x}{\partial \bar{p}} = \frac{\partial(p, x)}{\partial(\bar{p}, \bar{x})} = 1$$

Hamilton's equations are still valid in terms of the new variables introduced by the transformation (6.13), as we shall show. If we take an arbitrary function $b(p, x)$, its time rate of change is

$$\dot{b} = \dot{p}\frac{\partial b}{\partial p} + \dot{x}\frac{\partial b}{\partial x} = \frac{\partial \mathcal{H}}{\partial p}\frac{\partial b}{\partial x} - \frac{\partial b}{\partial p}\frac{\partial \mathcal{H}}{\partial x} = \frac{\partial(\mathcal{H}, b)}{\partial(p, x)}$$

Since the functional determinant of the transformation is unity, we also have

$$\dot{b} = \frac{\partial(\mathcal{H}, b)}{\partial(p, x)}\frac{\partial(p, x)}{\partial(\bar{p}, \bar{x})} = \frac{\partial(\mathcal{H}, b)}{\partial(\bar{p}, \bar{x})}$$

for any function $b$. If we now put $b$ equal to $\bar{p}$ or to $\bar{x}$, we obtain Hamilton's equations in terms of the new variables and the old Hamiltonian function:

$$-\left(\frac{\partial \mathcal{H}}{\partial \bar{x}}\right)_{\bar{p}} = \dot{\bar{p}} \qquad \left(\frac{\partial \mathcal{H}}{\partial \bar{p}}\right)_{\bar{x}} = \dot{\bar{x}}$$

Equation (6.13) represents the most general form of canonical transformation of canonically conjugate quantities. We may obtain the special case of the "pure coordinate transformation" by putting $S(x, \bar{p}) = g(x)\bar{p}$, from which $\bar{x} = g(x)$ and $\bar{p} = p/g'(x)$. This result may clearly be generalized to several degrees of freedom.

## §7. The electron in a given electromagnetic field

From equation (1.2) we know the electron's equation of motion

$$\frac{d}{dt}m\mathbf{v} = e\mathbf{E} + \frac{e}{c}\mathbf{v} \times \mathbf{H} \tag{7.1}$$

which we may consider to have been experimentally proved in the non-relativistic case. We now seek a Lagrangian function the Euler equation of which will agree with (7.1). For this purpose, let us introduce the potentials $\mathbf{A}$, $\phi$ (cf. Vol. I, §64) from which we may calculate the electric and magnetic fields

$$\mathbf{E} = -\text{grad}\,\phi - \frac{1}{c}\dot{\mathbf{A}}, \qquad \mathbf{H} = \text{curl}\,\mathbf{A}$$

If these expressions for $\mathbf{E}$ and $\mathbf{H}$ are inserted in (7.1), the $x$-component of this equation is

$$\frac{d}{dt}\left(m\dot{x} + \frac{e}{c}A_x\right) = \frac{\partial}{\partial x}\left\{-e\phi + \frac{e}{c}(\dot{x}A_x + \dot{y}A_y + \dot{z}A_z)\right\} \tag{7.2}$$

(For verification, we note that

$$\frac{dA_x}{dt} = \dot{A}_x + \frac{\partial A_x}{\partial x}\dot{x} + \frac{\partial A_x}{\partial y}\dot{y} + \frac{\partial A_x}{\partial z}\dot{z}$$

But (7.2) is the Eulerian form of the equation of motion for which

$$L = \tfrac{1}{2}m(\dot{x}^2 + \dot{y}^2 + \dot{z}^2) - e\phi + \frac{e}{c}(\dot{x}A_x + \dot{y}A_y + \dot{z}A_z) \qquad (7.3)$$

The $x$-component of the momentum is

$$p_x = \frac{\partial L}{\partial \dot{x}} = m\dot{x} + \frac{e}{c}A_x, \quad \text{therefore} \quad \mathbf{p} = m\mathbf{v} + \frac{e}{c}\mathbf{A} \qquad (7.4)$$

At first sight, this appears to be a strange result: the momentum is no longer equal to $m\mathbf{v}$, but contains an additional term $(e/c)\mathbf{A}$, the interpretation of which is not self-evident. The formal Hamiltonian treatment, however, compels us to designate $m\mathbf{v} + (e/c)\mathbf{A}$ as the momentum that is canonically conjugate to $\mathbf{r}$. If we now consider the Hamiltonian function as expressed by (6.8), we first obtain

$$(\mathbf{p} \cdot \dot{\mathbf{r}}) = m\dot{\mathbf{r}}^2 + \frac{e}{c}(\dot{\mathbf{r}} \cdot \mathbf{A})$$

If we use this equation together with (7.3) to form equation (6.8), we have

$$(\mathbf{p} \cdot \dot{\mathbf{r}}) - L = \tfrac{1}{2}m(\dot{x}^2 + \dot{y}^2 + \dot{z}^2) + e\phi$$

This is again the sum of the kinetic and potential energies; the vector potential and hence the magnetic field have disappeared. The above expression, however, is not the Hamiltonian function, which must be obtained by expressing the $\dot{x}_j$ by means of the momenta $p_j$, according to the general rule. If we do this, using (7.4), we finally obtain the (non-relativistic) Hamiltonian function for a charged particle in a field specified by $\phi$ and $\mathbf{A}$:

$$\mathscr{H} = \frac{1}{2m}\left(\mathbf{p} - \frac{e}{c}\mathbf{A}\right)^2 + e\phi \qquad (7.5)$$

We shall frequently have occasion to refer back to this expression.

## §8. Some applications of Hamilton's theory

Hamiltonian theory is of fundamental importance in the transition from classical to quantum mechanics, as we shall see later on. In this section we shall consider some further problems of classical physics in which Hamiltonian theory is of value.

Statistical mechanics and the theory of heat are based in the first instance on the Hamiltonian form of the equations of motion. The fundamental statement of statistical mechanics may be expressed as follows:

"Consider a system of $N$ particles with coordinates $\mathbf{r}_1,\ldots,\mathbf{r}_N$, momenta $\mathbf{p}_1,\ldots,\mathbf{p}_N$, the Hamiltonian function of which, $\mathscr{H}(\mathbf{p}_1,\ldots,\mathbf{r}_N)$, is invariant with respect to time. Assume that it is known only that the energy of the system lies between $E$ and $E+\Delta E$. Then the probability $W\,d\mathbf{p}_1\ldots d\mathbf{r}_N$ of finding the system in an interval $(\mathbf{p}_1,\mathbf{p}_1+d\mathbf{p}_1;\ldots;\mathbf{r}_N,\mathbf{r}_N+d\mathbf{r}_N)$ is given by

$$W\,d\mathbf{p}_1\ldots d\mathbf{r}_N = C\,d\mathbf{p}_1\ldots d\mathbf{r}_N \quad \text{with} \quad E \leqq \mathscr{H} \leqq E+\Delta E \quad (8.1)$$

The constant $C$ is determined by the normalization condition

$$\int\ldots\int_{E\leqq\mathscr{H}\leqq E+\Delta E} W\,d\mathbf{p}_1\ldots d\mathbf{r}_N = 1 \text{ ''}$$

In order that normalization should be possible, the volume of phase space defined by $E \leqq \mathscr{H} \leqq E+\Delta E$ must exist. This may be achieved most simply by imagining the system to be enclosed in a container of volume $V$. Then the function $\mathscr{H}$ contains a share of the potential energy arising between the particles and the wall, which prevents the former from penetrating the latter. The volume $V$ appears as a parameter in the Hamiltonian function. The energy indeterminacy may be arbitrarily small, and is only introduced for convenience in formulation.* The expression (8.1) is to be interpreted as a statistical statement of the course of the system in phase space. In virtue of the principle of conservation of energy, the system lies within the range $E \leqq \mathscr{H} \leqq E+\Delta E$. Equation (8.1) then implies that, for a sufficiently long period of observation, the time during which the phase point $\mathbf{p}_1(t),\ldots,\mathbf{r}_N(t)$ of the system remains within a small element of volume $d\mathbf{p}_1\ldots d\mathbf{r}_N$ lying in the range $E \leqq \mathscr{H} \leqq E+\Delta E$ is directly proportional to the size of this element.

If the system of particles is in contact with a heat source at temperature $T$, the course of the phase point can no longer be described by the Hamiltonian equations alone; the system is now coupled to a statistical ensemble defined solely by its temperature $T$, and because of the energy interchange with the heat source, the phase point path cannot be determined purely on mechanical grounds. In this case (8.1) is replaced by the "canonical distribution"

$$W\,d\mathbf{p}_1\ldots d\mathbf{r}_N = C\,e^{-\mathscr{H}/kT}\,d\mathbf{p}_1\ldots d\mathbf{r}_N \qquad (8.2)$$

* If we wish to avoid the use of $\Delta E$, and to consider $E$ as an exact quantity, we may express $W$ by means of the Dirac $\delta$-function: $W = C\delta(\mathscr{H} - E)$ (cf. Exercise 1, p. 137).

where $k$ is the Boltzmann constant.* Equations (8.1) and (8.2) are independent of the selected coordinate system.

If $f(\mathbf{p}_1,\ldots,\mathbf{r}_N)$ is an observable function of the "phase" $\mathbf{p}_1,\ldots,\mathbf{r}_N$, its average value, $\bar{f}$, is

$$\bar{f} = \frac{\int\ldots\int f W \, d\mathbf{p}_1 \ldots d\mathbf{r}_N}{\int\ldots\int W \, d\mathbf{p}_1 \ldots d\mathbf{r}_N} \tag{8.3}$$

or, with the particular value for $W$ given by (8.2),

$$\bar{f} = \frac{\int\ldots\int f e^{-\mathscr{H}/kT} \, d\mathbf{p}_1 \ldots d\mathbf{r}_N}{\int\ldots\int e^{-\mathscr{H}/kT} \, d\mathbf{p}_1 \ldots d\mathbf{r}_N} \tag{8.4}$$

In the above expressions, $\bar{f}$ may be taken as the time average of $f$ during the period of observation of the system, or as the average of simultaneous observations on a large number of identical systems.

In addition to the phase space volume, $\mathscr{H}$ may also contain as parameters the electric fields $\mathbf{E}$ and $\mathbf{H}$, represented by the potentials $\mathbf{A}(\mathbf{r})$ and $\phi(\mathbf{r})$. Then, generalizing from (7.5):

$$\mathscr{H} = \sum_{\nu=1}^{N} \left\{ \frac{1}{2m_\nu} \left( \mathbf{p}_\nu - \frac{e_\nu}{c} \mathbf{A}(\mathbf{r}_\nu) \right)^2 + e_\nu \, \phi(\mathbf{r}_\nu) \right\} + V(\mathbf{r}_1,\ldots,\mathbf{r}_N) \tag{8.5}$$

The above expression includes the mutual reaction between particles of mass $m_\nu$ and charge $e_\nu$, $V$, and other possible mutual effects such as the reaction with the wall that defines the phase volume.

This definition is of particular importance in the investigation of the electric and magnetic moments of the atom, and of their dependence on temperature in the presence of external electric and magnetic fields. If an atom consists of $N$ charges $e_1,\ldots,e_\nu,\ldots,e_N$ at positions $\mathbf{r}_1,\ldots,\mathbf{r}_\nu,\ldots,\mathbf{r}_N$, its electric and magnetic moments, $\bar{\mathbf{p}}_{el}$, $\bar{\mathbf{p}}_{magn}$, are given by the mean values of

$$\mathbf{p}_{el} = \sum_{\nu=1}^{N} e_\nu \mathbf{r}_\nu \tag{8.6a}$$

and†

$$\mathbf{p}_{magn} = \frac{1}{2c} \sum_{\nu=1}^{N} e_\nu \mathbf{r}_\nu \times \dot{\mathbf{r}}_\nu \tag{8.6b}$$

---

* (8.2) may be deduced from (8.1). For a thorough exposition of statistical mechanics and its bearing on the theory of heat, see R. Becker, *Theorie der Wärme* (Berlin, 1955).

† For (8.6b), cf. Vol. I, §47. Only the mean value of $\mathbf{p}_{magn}$ is significant. The closed orbit of an electron corresponds to a current round a closed circuit, and therefore gives rise to a magnetic moment.

Homogeneous electric and magnetic fields, **E** and **H**, are derived from potentials

$$\phi(\mathbf{r}) = -\mathbf{E}.\mathbf{r} \qquad -\text{grad }\phi = \mathbf{E}$$

$$\mathbf{A}(\mathbf{r}) = \tfrac{1}{2}\mathbf{H} \times \mathbf{r} \qquad \text{curl }\mathbf{A} = \mathbf{H}$$

The potentials appearing in (8.5) may therefore be written

$$\phi = \phi^0 - \mathbf{E}.\mathbf{r} \qquad \mathbf{A} = \tfrac{1}{2}\mathbf{H} \times \mathbf{r} \qquad (8.7)$$

where the potential $\phi^0$ is determined entirely by the mutual electrostatic forces between the individual charges in the atom. We may neglect the effect of the magnetic field due to a single electron on the orbits of the others, since it is very small compared to the electrostatic effect $e_v e_\mu / |\mathbf{r}_v - \mathbf{r}_\mu|$. In any case it could not be taken into account by means of the Hamiltonian function (8.5), since the vector potential **A** would not depend on the coordinates alone, but also on the momenta.

The following simple relationships may now be deduced from equations (8.5) and (8.6), using the values for the potentials given by (8.7):

$$\mathbf{p}_{el} = -\frac{\partial \mathcal{H}}{\partial \mathbf{E}} \qquad (8.8a)$$

$$\mathbf{p}_{magn} = -\frac{\partial \mathcal{H}}{\partial \mathbf{H}} \qquad (8.8b)$$

The derivation of equation (8.8a) is obvious; to prove (8.8b), we differentiate and substitute $m_v \dot{\mathbf{r}}_v$ for $\mathbf{p}_v - e_v \mathbf{A}(\mathbf{r}_v)/c$, in accordance with (7.4).

From (8.4) it follows that, with the thermodynamic " free energy "

$$\bar{\mathbf{p}}_{el} = -\overline{\frac{\partial \mathcal{H}}{\partial \mathbf{E}}} = kT \frac{\partial \ln Z}{\partial \mathbf{E}} = -\frac{\partial F}{\partial \mathbf{E}} \qquad (8.9a)$$

where $Z$ is the "phase integral", defined as

$$Z(\mathbf{E}, \mathbf{H}, T) = \int \ldots \int e^{-\mathcal{H}/kT} \, d\mathbf{p}_1 \ldots d\mathbf{r}_N = e^{-F/kT} \qquad (8.10)$$

Similarly,

$$\bar{\mathbf{p}}_{magn} = -\overline{\frac{\partial \mathcal{H}}{\partial \mathbf{H}}} = kT \frac{\partial \ln Z}{\partial \mathbf{H}} = -\frac{\partial F}{\partial \mathbf{H}} \qquad (8.9b)$$

These relationships may be retained in the quantum theory, in which each system is characterized by its stationary energy value $E_j$. The probability $W_j$ of finding the system in a state corresponding to $E_j$ is

$$W_j = C \exp -\frac{E_j}{kT} \tag{8.11}$$

The energy values depend on the parameters **E** and **H**. Equations (8.9) remain valid in the quantum theory if the phase sum

$$Z(\mathbf{E}, \mathbf{H}, T) = \sum_j \exp -\frac{E_j(\mathbf{E}, \mathbf{H})}{kT} \tag{8.10a}$$

is introduced in place of the phase integral; the summation extends over all possible energy values $E_j$. We shall be particularly concerned with the application of equation (8.9b) in chapter CIII.

In classical physics, however, equation (8.9b) is strictly speaking of no value, since it turns out that $\bar{\mathbf{p}}_{magn}$ always vanishes. This is because the phase integral

$$Z(\mathbf{E}, \mathbf{H}, T) =$$

$$\int \dots \int \exp\left(-\frac{1}{kT}\left\{\left[\sum_\nu \frac{1}{2m_\nu}\left(\mathbf{p}_\nu - \frac{e_\nu}{c}\mathbf{A}(\mathbf{r}_\nu)\right)^2 + e_\nu\,\phi(\mathbf{r}_\nu)\right] + V\right\}\right) d\mathbf{p}_1 \dots d\mathbf{r}_N$$

is independent of **A**, and hence of **H**. Integration over any $\mathbf{p}_\nu$ from $-\infty$ to $+\infty$ always results in a factor $(2\pi m_\nu kT)^{3/2}$ which is independent of $\mathbf{A}(\mathbf{r}_\nu)$. According to classical theory, therefore, no magnetism is present. By contrast the quantum theory is able to explain the magnetic properties of matter both qualitatively and quantitatively. The magnetic moment $\bar{\mathbf{p}}_{magn}$, when formed by means of the phase sum, does not generally vanish. For further discussion the reader is referred to chapter CIII.

## Exercises

1. *The Hamiltonian function in polar coordinates*

A particle is situated in a central field of force, for which the corresponding potential is $U(r)$. Express the kinetic energy in polar coordinates, and hence obtain the Hamiltonian function $\mathscr{H}(p_r, p_\theta, p_\phi, r, \theta, \phi)$ and the equations of motion in polar coordinates.

2. *The Lagrangian function and the conservation of momentum*

The Lagrangian function for $N$ particles is

$$L(\dot{\mathbf{r}}_1, \dots, \dot{\mathbf{r}}_N;\ \mathbf{r}_1, \dots, \mathbf{r}_N) = \sum_{i=1}^N \tfrac{1}{2}m_i\,\dot{\mathbf{r}}_i{}^2 - V(\mathbf{r}_1, \dots, \mathbf{r}_N)$$

If transformations $\mathbf{r}_i \rightarrow \mathbf{r}_i + \mathbf{R}_i$ are introduced that leave $L$ unchanged, then $G = \int_{t_0}^{t_1} L\, dt$ also remains unchanged. If the potential depends only upon the distances between particles, then $L$ is invariant for a "small" common displacement of all points $\mathbf{r}_i \rightarrow \mathbf{r}_i + \mathbf{R}_0$, and for a small common rotation $\mathbf{r}_i \rightarrow \mathbf{r}_i + \mathbf{n} \times \mathbf{r}_i$ ($\mathbf{n}$ is small). If the $\mathbf{R}_i$ are small, then $\mathbf{r}_i + \mathbf{R}_i$ may be looked on as variations of the correct paths. Since $G$ does not change when these variations are introduced, show that the theorems on the conservation of linear and angular momentum may be obtained respectively from the invariance property of $L$ with respect to displacement and rotation. (Note that the correct and the varied paths do not coincide at the limits $t_0$ and $t_1$.)

### 3. *Hamilton's principle of the extremum*

Discuss the behaviour of $G = \int_{t_0}^{t_1} L(x, \dot{x})\, dt$ for arbitrary variations

(a) in the case of a free particle capable of motion along the $x$-axis, for which $L = \tfrac{1}{2} m \dot{x}^2$,

(b) in the case of a linear oscillator, for which $L = \tfrac{1}{2} m(\dot{x}^2 - \omega^2 x^2)$.

If $x(t)$ is the correct path, find $G$ for the varied paths $x(t) + \gamma(t)$, for which $\gamma(t_0) = \gamma(t_1) = 0$. For what variations $\gamma(t)$ is $G$ respectively a maximum and a minimum?

# B

---

The
principles
of
quantum
mechanics

# CHAPTER BI

## The development of the quantum theory

### §9. Planck's radiation formula

The first indication of the inadequacy of the concepts of classical physics appeared in the problem of radiation from an enclosure (black-body radiation). This subject will be treated in some detail in section E; at this stage we shall confine ourselves to the following short discussion.

According to G. R. Kirchhoff's analysis, if an enclosure is at a certain temperature, radiation must exist inside it. If $u$ is the radiation energy density

$$u = \int_0^\infty u_\omega \, d\omega \qquad (9.1)$$

then its spectral distribution function $u_\omega(T)$ must be independent (a) of the nature of the enclosure, (b) of the manner in which it is brought into contact with a heat source at temperature $T$, (c) of the nature of any emitting and absorbing materials present in the enclosure. If the enclosure contains a linear oscillator, such as an elastically bound charged particle oscillating at an angular frequency $\omega$ about its equilibrium position, and if $\varepsilon(T)$ is the mean energy of the oscillator at temperature $T$, then classical physics yields the following result for the equilibrium condition (see § 47):

$$u_\omega(T) = \varepsilon(T) \frac{\omega^2}{\pi^2 c^3} \qquad (9.2)$$

According to the law of equipartition, $\varepsilon(T) = kT$. The energy density $u_\omega(T)$ would therefore be proportional to $\omega^2$, and therefore infinitely large at all temperatures. This catastrophic prediction can only be avoided by drastic assumptions which have no foundation in classical physics. One of these is known as Planck's hypothesis, and assumes that the oscillator cannot take any arbitrary energy value, but only those values that differ by integral multiples of a finite energy $\varepsilon_0$.

In consequence of this assumption the following expression for $\varepsilon$ is obtained in place of the classical value:*

$$\varepsilon = \frac{\varepsilon_0}{\exp(\varepsilon_0/kT)-1} \tag{9.3}$$

It was shown by W. Wien, from very general thermodynamical considerations, that $u_\omega(T)$ must have the form

$$u_\omega(T) = \omega^3 f\left(\frac{\omega}{T}\right)$$

where $f$ is a function only of $\omega/T$. This requirement is satisfied by (9.3) if we put $\varepsilon_0 = \hbar\omega$, where $\hbar$ is a new universal constant.† Then from (9.3)

$$\varepsilon = \bar{E}(T) = \frac{\hbar\omega}{\exp(\hbar\omega/kT)-1} \tag{9.4}$$

If we assume (9.2) to be correct, then

$$u_\omega(T) = \frac{\hbar\omega}{\exp(\hbar\omega/kT)-1}\frac{\omega^2}{\pi^2 c^3} \tag{9.5}$$

This is Planck's radiation formula, which has been shown to give a perfect description of the radiation from a black body.

Two consequences of the greatest importance were deduced from the above considerations by Einstein. These were: the hypothesis of light quanta, and the extension of equation (9.4) to any number of oscillator degrees of freedom. If the oscillator can only gain or lose energy in integral multiples, the radiation field must be able to accept these energy quanta. Einstein expressed this postulate as follows:

"When radiation interacts with matter, it behaves as though it consisted of light quanta of energy $\hbar\omega$ which can only be emitted or absorbed as whole quantities."

---

* In order to derive this formula we start from the partition function $Z$. Writing $\beta = 1/kT$, $Z$ is defined for a linear oscillator by $Z_{cl} = \int_0^\infty e^{-\beta E} dE$ if all values of $E$ are permitted, or by

$Z_{qu} = \sum_{n=0}^{\infty} e^{-\beta n \varepsilon_0}$, if the only values allowed are $E_n = n\varepsilon_0$ ($n = 0, 1, 2, \ldots$). In each case

$$\varepsilon = \bar{E} = -\frac{\partial \ln Z}{\partial \beta}$$

† The experimental value is $\hbar = 1 \cdot 05 \times 10^{-27}$ erg s. $\hbar$ is often called the Dirac constant to distinguish it from Planck's constant $h = 2\pi\hbar = 6 \cdot 63 \times 10^{-27}$ erg s.

The first consequences of this hypothesis concerned the lower (short-wave) limit of X-ray spectra and the photoelectric effect. When an electron possessing energy $e\Phi$ strikes the anode of an X-ray tube, the highest angular frequency $\omega_{max}$ that can occur in the resulting radiation is given by

$$\hbar\omega_{max} = e\Phi$$

In the photoelectric case, if a light quantum strikes a metallic surface and transfers its whole energy $\hbar\omega$ to an electron, the latter can escape from the metal with kinetic energy $\hbar\omega - W$, where $W$ is the work function (i.e. the work done by the electron in escaping). Both statements have been confirmed experimentally and have been used as the basis of precision methods for the determination of $\hbar$.

The extension of (9.4) to the natural modes of vibration of solid bodies forms the basis of the theory of the specific heats of crystals.

In its first stage (until about 1913) the quantum theory might be described as the theory of the linear oscillator, because it only recognized the "permitted" states given by the formula $E_n = n\hbar\omega$.

## §10. The Rutherford-Bohr atomic model

The only "classical" model of the atom that could explain the occurrence of sharp spectral lines is due to J. J. Thomson; it was discussed in §4. Thomson assumed that the positive charge of the nucleus was approximately homogeneously distributed over a sphere of radius of order $10^{-8}$ cm. The electron is then bound to the nucleus by an elastic force and can execute oscillations, the frequency of which lies within the range of visible light.

Thomson's hypothesis became quite untenable as a result of the experiments of Rutherford, who investigated the deflection of $\alpha$-particles by thin metal foils. Rutherford's results, in particular the occasional deflections through more than 90 degrees, forced him to the conclusion that the positive charge of individual metal atoms is concentrated in a space less than $10^{-12}$ cm. The atom accordingly consists of a positively charged nucleus of not more than $10^{-12}$ cm radius, around which the electrons revolve at distances* up to $10^{-8}$ cm.

---

\* It would at first seem possible to attempt to preserve Thomson's model in the light of Rutherford's experiments, simply by interchanging the signs of the charges so that the electron is now regarded as distributed. This assumption leads to the same properties of an "atom"; it corresponds approximately to the results yielded by a quantum mechanical description (cf. §28). Such a model is of comparable value to Thomson's, but is incompatible with the fact that the electron must also be described by a charge distributed over a very small radius ($R_{el} \approx 10^{-13}$ cm).

It was impossible, however, to explain the sharpness of the spectral lines in terms of classical theory and the new model of the atom, as we may see for the simplest case of the hydrogen atom. An electron revolving around the nucleus at a distance of $10^{-8}$ cm must be continually radiating electromagnetic energy. In doing so it loses energy itself, and revolves in spirals round the nucleus which it gradually approaches, and by which it is finally captured. This model is therefore quite unstable, and in addition it cannot possibly radiate at a single well-defined frequency.

To provide a way out of this dilemma two postulates were put forward by N. Bohr:

1. There are definite orbits for the electron, in which it does not radiate. Let the energies of these orbits be

$$E_1, E_2, \ldots, E_n, \ldots \qquad (10.1)$$

2. From time to time, the electron will cross from an orbit $n$ to another orbit $s$, for which $E_s < E_n$. When such a transition occurs, the energy difference $E_n - E_s$ is radiated as a light quantum:

$$\hbar\omega_{ns} = E_n - E_s \qquad (10.2)$$

If the electron is initially in orbit $s$, the same equation gives the frequency at which the transition from $s$ to $n$ can be induced by the absorption of a quantum $\hbar\omega_{ns}$.

In order to describe the optical properties of the atom we need a rule for the evaluation of the terms $E_1, \ldots, E_n, \ldots$. In this connection a hint is provided by the expression $E_n = n\hbar\omega$ for the permitted energy states of the linear oscillator, which we have already encountered. For an oscillator of mass $m$ and angular frequency $2\pi\nu = \omega$, the Hamiltonian function is

$$\frac{1}{2m}p^2 + \tfrac{1}{2}m\omega^2 x^2 = \mathcal{H}(x, p)$$

where the momentum $p$ is canonically conjugate to $x$. Consider the "orbit" in the $xp$-plane (figure 13) for which the energy is $E$. Since

$$\frac{1}{2m}p^2 + \tfrac{1}{2}m\omega^2 x^2 = E \qquad (10.3)$$

it is an ellipse with semi-axes $a = \sqrt{(2mE)}$ and $b = \sqrt{(2E/m\omega^2)}$. The area $\Phi$ of the ellipse is

$$\Phi = \pi ab = 2\pi \frac{E}{\omega}$$

Our postulate $E_n = n\hbar\omega$ therefore states that the only orbits allowed are those for which $\Phi/2\pi = n\hbar$. Now the area is

$$\Phi = \oint p(E, x)\, dx$$

where $p$ is expressed as a function of $x$ and $E$ according to (10.3), and the sign $\oint$ means the integration around a single orbital circuit.

We now have a starting-point for the calculation of the energy values, and one that is capable of considerable generalization. If we

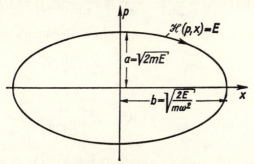

Fig. 13.—The path of a linear oscillator in phase space. The arrow indicates the direction of motion

restrict ourselves at first to one-dimensional motion, with a potential energy $V(x)$ (in place of the particular elastic energy of (10.3)), then (as did Bohr) we assume the following rule for the determination of $E_n$:

$$J(E) = \frac{1}{2\pi} \oint p(E, x)\, dx = n\hbar \tag{10.4}$$

In the above expression $p = \sqrt{\{2m[E - V(x)]\}}$ is the momentum resulting* from the equation $p^2/2m + V(x) = E$.

If the system possesses $f$ degrees of freedom, each orbit is specified by $f-1$ constants $c_1, c_2, \ldots, c_{f-1}$ in addition to the energy. If the coordinates $q_1, \ldots, q_f$ and the canonically conjugate momenta $p_1, \ldots, p_f$

---

* Expression (10.4) is invariant with respect to canonical transformations by means of the function $S$, since, by (6.13), $\bar{p}\,d\bar{x} = p\,dx - d(S - \bar{p}\bar{x})$.

are so chosen that each momentum $p_r$ depends only on the conjugate coordinate $q_r$ and the constants $E, c_1, \ldots, c_{f-1}$, then, as was shown by A. Sommerfeld, equation (10.4) may be generalized to give the quantum conditions

$$J_r = \frac{1}{2\pi} \oint p_r(q_r, E, c_1, \ldots, c_{f-1}) \, dq_r = n_r \hbar \quad (r = 1, 2, \ldots, f) \quad (10.5)$$

required for the determination of the constants $E, c_1, \ldots, c_{f-1}$ as functions of the quantum numbers $n_1, \ldots, n_f$.

We shall describe the procedure for the case of a particle moving within a spherically symmetrical potential $V(r)$. In spherical coordinates $r, \theta, \phi$, the Hamiltonian function is*

$$\mathcal{H}(r, \theta, \phi, p_r, p_\theta, p_\phi) = \frac{1}{2m} \left\{ p_r^2 + \frac{1}{r^2} \left( p_\theta^2 + \frac{1}{\sin^2 \theta} p_\phi^2 \right) \right\} + V(r) \quad (10.6)$$

The variables are already separated: from Hamilton's equations we have

$$\frac{d}{dt} p_\phi = 0, \qquad\qquad p_\phi = c_\phi$$

also

$$\frac{d}{dt} \left( p_\theta^2 + \frac{c_\phi^2}{\sin^2 \theta} \right) = 0, \qquad p_\theta^2 + \frac{c_\phi^2}{\sin^2 \theta} = c_\theta^2$$

and finally

$$\frac{d\mathcal{H}}{dt} = 0, \qquad\qquad \frac{1}{2m} \left( p_r^2 + \frac{c_\theta^2}{r^2} \right) + V(r) = E$$

It is evident that $c_\phi$ is the component of the angular momentum about the z-axis, and that $c_\theta^2$ is the square of the angular momentum.

In particular, if we put the potential $V(r)$ equal to the Coulomb energy $-e^2/r$, we have three phase integrals to evaluate. The results are

$$J_\phi = \frac{1}{2\pi} \oint p_\phi \, d\phi = c_\phi$$

$$J_\theta = \frac{1}{2\pi} \oint \sqrt{\left( c_\theta^2 - \frac{c_\phi^2}{\sin^2 \theta} \right)} \, d\theta = c_\theta - c_\phi \quad (10.7)$$

$$J_r = \frac{1}{2\pi} \oint \sqrt{\left( 2mE + \frac{2me^2}{r} - \frac{c_\theta^2}{r^2} \right)} \, dr = \frac{me^2}{\sqrt{(-2mE)}} - c_\theta$$

---

* Cf. Exercise 1, p. 46.

If these equations are added together, $c_\phi$ and $c_\theta$ are eliminated:

$$J_\phi + J_\theta + J_r = \frac{me^2}{\sqrt{(-2mE)}} \quad \text{or} \quad E = -\frac{1}{2}\frac{me^4}{(J_\phi + J_\theta + J_r)^2}$$

If we now introduce the quantum condition (10.5), the energy $E$ is found to be a function of the three quantum numbers $n_\phi$, $n_\theta$, $n_r$:

$$E = -\frac{1}{2}\frac{me^4}{\hbar^2}\frac{1}{(n_\phi + n_\theta + n_r)^2}$$

Putting $n_\phi + n_\theta + n_r = n$, we have

$$E_n = -2\pi\hbar cR\frac{1}{n^2} \tag{10.8}$$

where $R = \dfrac{1}{4\pi}\dfrac{me^4}{c\hbar^3}$ is known as the Rydberg constant. The only "allowed" energy values are those given by (10.8), with $n = 1,2,3,\dots$ From (10.2) it then follows that the wave numbers (reciprocal wavelengths) of the spectral lines are:

$$\frac{1}{\lambda_{nm}} = \frac{\omega_{nm}}{2\pi c} = \frac{E_n - E_m}{2\pi\hbar c} = R\left(\frac{1}{m^2} - \frac{1}{n^2}\right) \tag{10.9}$$

In particular,

$m = 1$, $n = 2,3,4,\dots$ gives the "Lyman series"

$m = 2$, $n = 3,4,5,\dots$ the "Balmer series"

$m = 3$, $n = 4,5,6,\dots$ the "Paschen series"

of the hydrogen spectrum.

The quantum numbers are often referred to as follows:

$n_\phi + n_\theta + n_r = n$, the principal quantum number,

$n_\phi + n_\theta = l$, the angular-momentum quantum number,

$n_\phi = m$, the magnetic quantum number.

The above treatment of the hydrogen atom must be considered to be a provisional one; a consistent theory can only be obtained by means of quantum mechanics (e.g. the Schrödinger equation). The subject will be discussed in detail in §29, where it will be seen that many of the above results are confirmed.

In the quantum-mechanical treatment of the hydrogen atom, the energy, the angular momentum, and any arbitrary component of the latter are still subject to the quantum conditions, i.e. they can only assume certain discrete values.

## §11. The correspondence principle

Bohr's assumptions of the existence of stationary energy values $E_1, \ldots, E_n, \ldots$ and the emission of light as quanta $\hbar\omega_{ns}$ of frequency

$$\omega_{ns} = \frac{E_n - E_s}{\hbar} \tag{11.1}$$

represent such a fundamental change from the concepts of classical physics that it seems hopeless to try to base them on a model, although this was often attempted during the period of currency of his theory (1913–25).

In spite of this difficulty, an attempt can be made to establish a relationship between the laws of classical physics and those of the

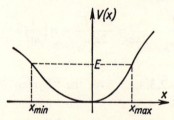

Fig. 14.—Motion of a particle in a field of potential energy $V(x)$. The energy $E$ determines the amplitude of the oscillations

quantum theory. For the purpose of discussion we may take as an example the radiation from a charged particle oscillating parallel to the $x$-axis at a fundamental frequency $\omega = 2\pi/T$ under the influence of potential energy $V(x)$.

If $E$ is the energy of the particle, the maximum and minimum amplitudes are given by $E = V(x_{max}) = V(x_{min})$ (figure 14). The motion can be described by a Fourier series containing the fundamental frequency $\omega$ and integral multiples thereof, $\tau\omega$. (Harmonics of the fundamental frequency are absent only in the case of the linear oscillator, where the potential is a parabolic function.) The oscillation can be expressed in terms of $E$, or in terms of $J$, in view of the relation (10.4). Then

$$x(t) = \sum_{\tau = -\infty}^{\tau = +\infty} x_\tau(J) e^{i\tau\omega t} \quad \text{and} \quad x_{-\tau}(J) = x_\tau^*(J) \qquad (11.2)$$

Let the constant $J = n\hbar$ designate the state of the "atom". On classical theory we should expect the atom to emit spectral lines of angular frequencies $\omega, 2\omega, \ldots, \tau\omega, \ldots$, in virtue of (11.2), and that the intensity of the line $\tau\omega$ would be proportional to the square of the amplitude and therefore to $|x_\tau(J)|^2$, in accordance with Hertz's formula for the radiation from a dipole oscillator.

According to the quantum theory, an atom in state $n$ can change to a lower state $s$, as a result of which it emits one of the lines

$$\omega_{ns} = \frac{E_n - E_s}{\hbar} \qquad (s = 1, 2, \ldots, n-1)$$

It is hopeless to try to establish a simple relationship between the classical and quantum theory predictions regarding the behaviour of an individual atom. The situation becomes more tractable, however, if we consider a gas consisting of a large number (say $N$) of identical independent atoms in a state $J = n\hbar$. According to classical theory, each of the $N$ atoms will emit spectral lines $\omega, 2\omega, \ldots$ as described above. According to quantum theory, on the other hand, each atom is free to change to a lower energy level $E_s$. In general, therefore, all frequencies $\omega_{ns}$ can occur, and the intensity of each line depends on the number of atoms changing from state $n$ to state $s$. This number, however, is proportional to the probability of the transition $n \to s$. It is therefore only when a large number of atoms are considered that the quantum theory predicts a spectrum of many lines, capable of comparison with the "classical" spectrum predicted by (11.2).

The correspondence principle was first enunciated by Bohr. It states that the experimentally realized predictions of the quantum theory must tend asymptotically to those of classical theory, for quantum numbers $n$ increasing without limit and for transitions* $n - s = \Delta n$ for which $\Delta n \ll n$. An example would be a highly excited hydrogen atom, in which the electron revolves slowly round the nucleus at a great distance from it (about $10^{-8}$ cm). In this case the

---

* Another form of the principle states that in the limit as $\hbar \to 0$ the predictions of the quantum theory must coincide with those of the classical theory.

field could certainly be calculated according to the laws of classical electrodynamics. In the limiting case of large quantum numbers, therefore, the spectra of a great number of atoms calculated from the classical and from the quantum theories, should coincide asymptotically as regards all observable particulars, such as the position, intensity, and polarization of the individual spectral lines.

We must now establish a "correspondence" between the classical harmonic frequencies $\tau\omega$ and the Bohr frequencies $\omega_{ns}$ to which they tend at high quantum numbers. This may be achieved by means of a fundamental theorem of classical mechanics.

If the energy $E$ is calculated as a function of $J$ from the phase integral defined in (10.4)

$$J = \frac{1}{2\pi} \oint p(E, x)\, dx \qquad (11.3)$$

the fundamental frequency of the motion is given by

$$\omega = \frac{dE}{dJ} \qquad (11.4)$$

This is a generalization of the equation $E = J\omega$, which is restricted to the case of the linear oscillator. Equation (11.4) may be proved as follows. From the relation $p = \{2m[E - V(x)]\}^{1/2}$ we obtain $\partial p/\partial E = m/p = 1/v$. Therefore

$$\frac{dJ}{dE} = \frac{1}{2\pi} \oint \frac{dx}{v} = \frac{T}{2\pi} = \frac{1}{\omega}$$

since $\oint \dfrac{dx}{v}$ is the time required for one oscillation.

When there are several degrees of freedom, then instead of the one constant, $E$, we have $f$, say: $c_1, c_2, \ldots, c_{f-1}, E$. If we use these to construct the $f$ phase integrals

$$J_j = \frac{1}{2\pi} \oint p_j(x_j, c_1, \ldots, c_{f-1}, E)\, dx_j \qquad (j = 1, 2, \ldots, f)$$

the latter may be used to calculate the constants $c_j$ and hence the energy as functions of the $J_j$, as we did in the case of the hydrogen atom:

$$E = E(J_1, J_2, \ldots, J_f)$$

The variables $x_j$ may then be expressed as Fourier series:

$$x_j(t) = \sum_{\tau_1, \tau_2, \ldots} x_{j\tau_1 \ldots \tau_f}(J_1, \ldots, J_f) \exp i(\omega_1 \tau_1 + \ldots + \omega_f \tau_f) t$$

where

$$\omega_j = \frac{\partial E}{\partial J_j}$$

Using equation (11.4), we may express the frequency $\tau\omega$ of the $\tau$th harmonic as

$$\tau\omega = \tau\frac{dE}{dJ} = \lim_{\varepsilon \to 0}\frac{E(J+\tau\varepsilon)-E(J)}{\varepsilon} \qquad (11.5)$$

since the last expression is the definition of the differential coefficient.

In quantum theory, $J$ can only assume integral values $n\hbar$; in place of the continuous curve $E(J)$ there occur the discrete energy values

$$E_n = E(n\hbar), \qquad E_{n+1} = E\{(n+1)\hbar\}, \ldots$$

In view of the discontinuous nature of the $J$ values, it seems reasonable to substitute a difference quotient for the derivative in (11.5). This means that we no longer pass to the limit $\varepsilon \to 0$, but that we put $\varepsilon = \hbar$. This is in fact the difference between successive values of $J$. From (11.5) we therefore obtain the correspondence*

$$\tau\omega \Leftrightarrow \frac{E_{n+\tau}-E_n}{\hbar} \qquad (11.6)$$

*When the quantum number changes by $\tau$ units, the emitted or absorbed frequency corresponds to the $\tau$th harmonic of the classical motion.*

The recognition of this fact enables us to establish a relationship between each term of the Fourier expansion for the classical orbit in the state $J = n\hbar$, and the corresponding quantum transition:

$$x_\tau(J)\exp i\tau\omega t \Leftrightarrow x_{n+\tau,n}\exp i\frac{E_{n+\tau}-E_n}{\hbar}t \qquad (11.7)$$

This relation may also be written in a slightly modified form:†

$$x_{m-n}(n\hbar)\exp i(m-n)\omega t \Leftrightarrow x_{mn}\exp i\omega_{mn}t \qquad (11.7a)$$

In quantum theory, therefore, although Fourier coefficients associated with electron orbits may not appear, we nevertheless anticipate the existence of the corresponding "transition amplitudes" $x_{mn}$. We shall meet these amplitudes later as the elements of a matrix associated with the $x$-coordinate, and shall often have occasion to employ them.

---

* This correspondence is by no means unequivocal: we should, for instance, be equally justified in using the relation

$$\tau\omega \Leftrightarrow \frac{E_n - E_{n-\tau}}{\hbar}$$

† Here too, we are free to write either $x_{m-n}(n\hbar)$ or $x_{m-n}(m\hbar)$ for the left-hand side. From (11.2) we have $x_{nm} = x^*{}_{mn}$.

The correspondence must be interpreted to mean that at infinitely large values of $n$ the amplitudes $x_{n+\tau,n}$ coincide with the Fourier co-efficients $x_\tau(n\hbar)$ of the classical orbit. It is true that the quantum theory amplitudes have no significance for any single orbit, but the correspondence principle enables us to establish a relationship between the quantities $|x_{mn}|^2$ and the radiation from $N$ atoms in the state $J = n\hbar$.

For this purpose, we make use of classical theory to calculate the energy radiated at frequency $\tau\omega$ by an electron oscillating in accordance with equation (11.2). The motion contributing to the energy at this frequency is

$$x_\tau(t) = x_\tau(J)\, e^{i\tau\omega t} + \{x_\tau(J)\}^*\, e^{-i\tau\omega t}$$

From (4.4) the radiated power is

$$S_{cl} = \frac{2e^2}{3c^3}\,\overline{\ddot{x}^2}$$

Substituting the value given above for $x(t)$:

$$S_{cl} = \frac{4e^2}{3c^3}(\tau\omega)^4\,|x_\tau(J)|^2$$

We assume that a gas consisting of $N$ atoms in state $n$ emits energy at angular frequency $\omega_{mn}$ at the following average rate per atom:

$$S_{qu} = \frac{4e^2}{3c^3}\omega_{mn}^4\,|x_{mn}|^2 \tag{11.8}$$

Also, a single atom emits energy $\hbar\omega_{mn}$ in the course of the transition $m \to n$.

Now if $A_{mn}\delta t$ represents the probability that an atom in state $m$ radiates during the interval $\delta t$ and thereby changes to state $n$ (where $E_n < E_m$), then the number of atoms undergoing this transition will be $NA_{mn}\delta t$, and the total energy emitted will be $\hbar\omega_{mn}NA_{mn}\delta t$. The average emission from a single atom in this interval is therefore

$$S_{qu}\,\delta t = \hbar\omega_{mn}A_{mn}\,\delta t \tag{11.8a}$$

Comparison of the two equations for $S_{qu}$ yields the following expression for the transition probability:

$$A_{mn} = \frac{4e^2\omega_{mn}^3}{3c^3\hbar}\,|x_{mn}|^2 \tag{11.9}$$

So far we have restricted ourselves to motion in one dimension. For the three-dimensional case, we must take account of the additional contributions $y_\tau(J)$ and $z_\tau(J)$, and the corresponding transition amplitudes $y_{mn}$ and $z_{mn}$. The transition probability is then:

$$A_{mn} = \frac{4e^2}{3c^3}\frac{\omega_{mn}^3}{\hbar}(|x_{mn}|^2 + |y_{mn}|^2 + |z_{mn}|^2) \qquad (11.9a)$$

We shall deduce this equation again by strict quantum methods, after having provided general methods for the evaluation of the amplitudes $x_{mn}$.

The foregoing analysis may be used to determine the "matrix elements" $x_{mn}$ of the linear oscillator, for which the classical equation of motion is simple harmonic. Hence from (11.2)

$$x(t) = x_1(J)e^{i\omega t} + x_1^*(J)e^{-i\omega t} \qquad (11.10)$$

In order to evaluate $x_1(J)$ we note, first, that $E = m\omega^2\overline{x^2} = J\omega$ for the oscillator. Also, if we take $x_1(J)$ to be real, then $\overline{x^2} = 2[x_1(J)]^2$ from (11.10). Hence

$$x_1(J) = x_1^*(J) = \sqrt{\frac{J}{2m\omega}}$$

According to the correspondence equation (11.7) we therefore have, for $J = n\hbar$,

$$x_1(n\hbar)\exp i\omega t \Leftrightarrow x_{n,n-1}\exp i\frac{E_n - E_{n-1}}{\hbar}t$$

$$x_1(n\hbar)\exp -i\omega t \Leftrightarrow x_{n-1,n}\exp -i\frac{E_n - E_{n-1}}{\hbar}t$$

The only matrix elements to occur are those for which $|n-m| = 1$. According to (11.9), therefore, this means that the only transitions that take place are those from $n$ to $n-1$ for emission, and from $n$ to $n+1$ in the case of absorption. Also,

$$x_{n,n-1} = x_{n-1,n} = \sqrt{\frac{n\hbar}{2m\omega}} \qquad (11.11)$$

We shall often encounter the above equation.

We shall now calculate the transition probability $A_{n,n-1}$ in order to verify the above analysis. Using (11.9) and (11.11), we have

$$A_{n,n-1} = \frac{2e^2}{3mc^3}n\omega^2$$

If we now put $n\hbar\omega = E$, the energy radiated per second is

$$S = \hbar\omega A_{n,\,n-1} = \frac{2e^2\omega^2}{3mc^3}\,E$$

i.e. the fraction of the energy radiated per second, $\gamma$, is $2e^2\omega^2/3mc^3$. This agrees with the classical value.

An important application of the correspondence equations (11.6) and (11.7) is based on the assumption that only those quantum transitions occur for which a corresponding classical harmonic exists, and that all other transitions are "forbidden". We have already met the simplest example of this assumption in the case of the linear oscillator.

The correspondence principle has been of particular assistance in the determination of the "selection rules" of the atomic spectra. For instance, when we discussed the hydrogen atom in §10 we found that the state of the atom could be described by means of three quantum numbers, $n$, $l$, $m$. This classification may also be employed for many of the more complex atoms, such as those of the alkali metals. The selection rules state that in general, the only transitions to occur are those for which $\Delta l = \pm 1$ and $\Delta m = \pm 1$ or 0. We shall return to this subject later (in §24); at this stage we should merely note that the selection rules become comprehensible if we make the assumption that the corresponding harmonics of classical motion do not exist.

## §12. Deduction of the Heisenberg form of the quantum theory

Heisenberg's approach to quantum mechanics involves the extension of the correspondence principle in a direction to which brief reference was made in §11. The basis of the method is briefly indicated in what follows. We shall again restrict ourselves to a single degree of freedom in order that the crucial aspects of the theory may be easily recognized.

(*a*) *Product formation.*—Consider a system of fundamental angular frequency $\omega$, and assume that two quantities $a(t)$ and $b(t)$ depend on time as follows:

$$a(t) = \sum_\tau a_\tau(J)\,e^{i\tau\omega t} \quad \text{and} \quad b(t) = \sum_\tau b_\tau(J)\,e^{i\tau\omega t} \qquad (12.1)$$

In §11 we showed that a "matrix element" in quantum theory corresponds to a single term of the Fourier expansion, thus:

$$a_\tau(J)\,e^{i\tau\omega t} \Leftrightarrow a_{nm}\,e^{i\omega_{nm}t}$$

where

$$n - m = \tau, \quad J = n\hbar \quad \text{or} \quad m\hbar, \quad \omega_{nm} = (E_n - E_m)/\hbar$$

In order to extend this correspondence to the product $a(t)b(t)$, we shall develop the latter in a Fourier series and then seek the matrix element corresponding to the factor of $e^{i\tau\omega t}$ in this expansion. By (12.1)

$$a(t)\,b(t) = \sum_{\tau} \left\{ \sum_{s} a_s(J)\, b_{\tau-s}(J) \right\} e^{i\tau\omega t} = \sum_{\tau} (ab)_{\tau}\, e^{i\tau\omega t} \quad (12.2)$$

In the above expression we have so arranged the double summation that all terms with the same time factor $e^{i\tau\omega t}$ are collected together in the summation $\sum_s a_s b_{\tau-s} = (ab)_{\tau}$. Similarly, in the case of the quantum mechanical terms

$$a_{nm}\, e^{i\omega_{nm}t} \quad \text{and} \quad b_{nm}\, e^{i\omega_{nm}t}$$

we may seek to associate with the product $ab$ a matrix element $(ab)_{nm}$ that also contains the time factor $e^{i\omega_{nm}t}$. The fact that

$$\omega_{nm} = \omega_{ns} + \omega_{sm}$$

for any value of $s$ necessarily leads to the result

$$(ab)_{nm}\, e^{i\omega_{nm}t} = \sum_{s} a_{ns}\, e^{i\omega_{ns}t} b_{sm}\, e^{i\omega_{sm}t}$$

for the matrix element associated with the product $ab$ and the transition from $E_n$ to $E_m$. Hence we have the following rule for the multiplication of two quantum terms $a_{nm}$ and $b_{nm}$:

$$(ab)_{nm} = \sum_{s} a_{ns}\, b_{sm} \quad (12.3)$$

This rule is the same as the multiplication rule for matrices; in this connection we should note that in general, matrix multiplication does not obey the commutative rule, i.e.

$$(ab)_{mn} \neq (ba)_{mn}$$

(b) *The Poisson bracket.*—If $p$ and $x$ are canonically conjugate variables as defined in Hamiltonian theory, and if $a(p,x)$ and $b(p,x)$ are any two functions of these variables, then the "Poisson bracket" $\{a,b\}$ is defined as

$$\{a, b\} = \frac{\partial a}{\partial p}\frac{\partial b}{\partial x} - \frac{\partial b}{\partial p}\frac{\partial a}{\partial x} = \frac{\partial(a,b)}{\partial(p,x)} \quad (12.4)$$

By the special choice of $a$ and $b$ we obtain

$$\{a, x\} = \frac{\partial a}{\partial p} \qquad \{a, p\} = -\frac{\partial a}{\partial x} \qquad (12.5)$$

$$\{p, x\} = 1 \qquad (12.6)$$

The Hamiltonian equations may also be expressed by Poisson brackets:

$$\{\mathscr{H}, x\} = \frac{\partial \mathscr{H}}{\partial p} = \dot{x} \qquad \{\mathscr{H}, p\} = -\frac{\partial \mathscr{H}}{\partial x} = \dot{p} \qquad (12.7)$$

Hence it follows that for any arbitrary function* $b(p, x)$

$$\{\mathscr{H}, b\} = \dot{b} \qquad (12.7a)$$

According to (12.4) the Poisson bracket can be written as a Jacobian; its value therefore remains unchanged if new variables $\bar{p}$, $\bar{x}$ are introduced by means of the canonical transformation (6.13):

$$\{a, b\} = \frac{\partial(a, b)}{\partial(p, x)} = \frac{\partial(a, b)}{\partial(\bar{p}, \bar{x})} \qquad (12.4a)$$

since the Jacobian of the transformation is unity. For the same reason, equations (12.5,6) remain valid when expressed in the new coordinates.

At this stage it seems natural to introduce $J$ (equation 10.4) as a new variable, because the quantum conditions can be very simply expressed in terms of this quantity. If we put $J = \bar{p}$, and if $w = \bar{x}$ is the coordinate† canonically conjugate to $J$, then

$$\{a, b\} = \frac{\partial(a, b)}{\partial(J, w)} = \frac{\partial a}{\partial J}\frac{\partial b}{\partial w} - \frac{\partial b}{\partial J}\frac{\partial a}{\partial w} \qquad (12.8)$$

The function $S(x, J)$ that produces the required transformation is

$$S(x, J) = \int_0^x p\{E(J), x\}\, dx$$

---

* The function $b$ does not depend explicitly on time; its variation with respect to time is merely the result of the time rate of change of the quantities $p$ and $x$, of which it is a function.

† $J$ and $w$ are termed the action and angle variables. The corresponding transformations may also be performed for periodic motion in problems with several degrees of freedom: cf. the treatment of the hydrogen atom in §10. The Hamiltonian function then depends only on the action variables, and the canonically conjugate angle variables (usually angular quantities) are proportional to the time.

It may easily be shown that this function possesses the properties expressed in (6.13). The canonically conjugate quantity $w$ is obtained by differentiating, taking into account the relation (11.4):

$$w = \frac{\partial S}{\partial J} = \omega \frac{\partial S}{\partial E} = \omega \int_0^x \frac{dx}{v} = \omega t$$

The Hamiltonian function is $\mathcal{H}(J,w) = E(J)$; the Hamiltonian equations $\dot{J} = 0$, $\dot{w} = dE/dJ = \omega(J)$ also give the result $w = \omega t$, as is to be expected. Using this last result, we may write $\frac{\partial b}{\partial w} = \frac{1}{\omega}\frac{\partial b}{\partial t}$; the $\tau$th Fourier component of $\partial b/\partial w$ is therefore

$$\left(\frac{\partial b}{\partial w}\right)_\tau = i\tau b_\tau$$

To find the corresponding expression for $\{a,b\}_\tau$ we put

$$\{a,b\}_\tau = i \sum_{r+s=\tau} \left\{ \left(\frac{\partial a}{\partial J}\right)_r s b_s - r a_r \left(\frac{\partial b}{\partial J}\right)_s \right\}$$

in accordance with (12.2). The derivatives with respect to $J$ are replaced by appropriate difference quotients, so chosen as to eliminate the factors $r$ and $s$:

$$\left(\frac{\partial a}{\partial J}\right)_r \Leftrightarrow \frac{a_r(J) - a_r(J - sh)}{sh} \qquad \left(\frac{\partial b}{\partial J}\right)_s \Leftrightarrow \frac{b_s(J) - b_s(J - rh)}{rh}$$

In the resulting expression

$$\frac{i}{\hbar} \sum_{r+s=\tau} \{a_r(J) b_s(J - rh) - b_s(J) a_r(J - sh)\}$$

let us put $J = \tau h$, and let $r = m - l$, $s = l - n$ in the first term, and $s = m - l$, $r = l - n$ in the second. We then obtain the familiar correspondence

$$\{a,b\}_{m-n} \Leftrightarrow \frac{i}{\hbar}(ab - ba)_{mn} \tag{12.9}$$

The right-hand side of (12.9) may be termed a quantum-mechanical Poisson bracket.

A comparison with the special cases quoted above reveals that, for two canonically conjugate variables $p$ and $x$, the corresponding analogue of (12.6) is

$$(px - xp)_{nm} \equiv [p,x]_{n,m} = \frac{\hbar}{i} \delta_{nm}$$

or
$$px - xp \equiv [p,x] = \frac{\hbar}{i} \mathbf{1} \qquad (12.10)$$

where the symbol $\mathbf{1}$ represents the unit matrix

$$\begin{pmatrix} 1 & 0 & 0 & . & . \\ 0 & 1 & 0 & . & . \\ 0 & 0 & 1 & . & . \\ . & . & . & . & . \end{pmatrix}$$

Again, corresponding to (12.7a), we have the following relation for the rate of change of a quantum-mechanical term:

$$\left(\frac{db}{dt}\right)_{nm} = \frac{i}{\hbar}(\mathcal{H}b - b\mathcal{H})_{nm} \quad \text{and} \quad \frac{db}{dt} = \frac{i}{\hbar}[\mathcal{H},b] \quad (12.11)$$

In order to verify this last relation, we shall restrict ourselves to the case (tacitly assumed in §11) in which $H_{nm}$ is a diagonal matrix, i.e. $H_{nm} = E_n \delta_{nm}$. Then

$$(\mathcal{H}b - b\mathcal{H})_{nm} = \sum_s (\mathcal{H}_{ns} b_{sm} - b_{ns} \mathcal{H}_{sm}) = (E_n - E_m) b_{nm}$$

$$\left(\frac{db}{dt}\right)_{nm} = i\omega_{nm} b_{nm}$$

Hence
$$b_{nm}(t) = b_{nm}(0) e^{i\omega_{nm}t}$$

which is the quantity we started with.

The relations (12.10) and (12.11), which have been derived somewhat intuitively from the correspondence principle, will be seen later to constitute the foundations of the whole quantum theory, in which they will be introduced as "axioms".

(c) *Stationary energy values.*—From the above considerations, Heisenberg, Born, and Jordan deduced the following procedure for the determination of the Bohr stationary energy values in a mechanical system with one degree of freedom and potential energy $V(x)$.

Consider the classical form of the Hamiltonian function

$$\frac{p^2}{2m}+V(x) = \mathscr{H}(p, x)$$

as a matrix equation

$$\left\{\frac{p^2}{2m}+V(x)\right\}_{nm} = \mathscr{H}_{nm} \qquad (12.12)$$

We express the fact that the energy is constant with respect to time by the requirement that the matrix $H_{mn}$ be diagonal, i.e. equal to $E_m\delta_{mn}$. The reason for this postulate is that a term $nm$ of each matrix contains the time factor $e^{i\omega_{nm}t}$, which is only independent of time when $m = n$. We therefore require that the matrix $\mathscr{H}$ should have the form

$$\mathscr{H} = \begin{pmatrix} E_1 & 0 & 0 & . \\ 0 & E_2 & 0 & . \\ 0 & 0 & E_3 & . \\ . & . & . & . \end{pmatrix}$$

where the diagonal terms $E_1, E_2, \ldots$ are initially unknown. We can now formulate the problem precisely: Do matrices $p_{nm}$ and $x_{nm}$ exist, such that firstly

$$(px - xp)_{nm} = \frac{\hbar}{i}\delta_{nm} \qquad (12.13a)$$

and secondly

$$\left\{\frac{p^2}{2m}+V(x)\right\}_{nm} = E_n\delta_{nm} \qquad (12.13b)$$

becomes a diagonal matrix (with initially unknown numbers $E_1, E_2, \ldots$)?

It will be seen later that in the case of potentials for which the question has any significance, these two postulates serve to determine unambiguously the quantities $x_{nm}$, $p_{nm}$, and $E_n$. We shall perform the relevant calculation for the case of the linear oscillator in §15, to which the reader is now referred.

## §13. De Broglie's hypothesis of the wave-like properties of the electron

(a) *De Broglie's hypothesis.*—We have shown how Heisenberg's form of the quantum theory was developed by means of an extension of the correspondence principle. An independent and formally simpler approach was presented by an extremely bold speculation by L. de Broglie, the main features of which we shall now describe.

The approach starts from Einstein's light quantum hypothesis, according to which monochromatic light of angular frequency $\omega$ interacts with matter as though it consisted of discrete light quanta of energy $\hbar\omega$. Without entering too deeply into the difficult subject of "wave corpuscles", we may describe the situation as follows:

A plane wave of angular frequency $\omega$, amplitude vector $\mathbf{A}$, and (vectorial) propagation constant $\mathbf{k}$ is represented by

$$\mathbf{A}\,e^{i(\mathbf{k}.\mathbf{r}-\omega t)} \tag{13.1}$$

The wavelength $\lambda$ is derived directly from the magnitude $k$ of the vector $\mathbf{k}$:

$$k = \frac{2\pi}{\lambda}$$

Also, for propagation in a vacuum

$$\omega = ck$$

The light quantum associated with the wave described by (13.1) possesses an energy $E = \hbar\omega$. Introducing this energy into the phase factor in the wave equation, we may write it as

$$\exp\frac{i}{\hbar}(\hbar\mathbf{k}.\mathbf{r}-Et) \tag{13.2}$$

Since the phase of a light wave (i.e. the exponent $\hbar\mathbf{k}.\mathbf{r}-Et$) must be relativistically invariant, and according to the Special Theory of Relativity (Vol. I, §80) the quantities $\{x,y,z,ict\}$ constitute a four-vector, then the quantities $\{\hbar k_x, \hbar k_y, \hbar k_z, iE/c\}$ must also be the components of such a vector, since the exponent contains the scalar product of both vectors. But the quantity $iE/c$ is the fourth component of the relativistic four-momentum $\{p_x, p_y, p_z, iE/c\}$. If $\hbar\omega$ is interpreted as the energy of the light quantum in (13.1), then the momentum associated with this quantum must be $\mathbf{p} = \hbar\mathbf{k}$. Hence we obtain the following familiar value for the magnitude of the momentum:

$$p = \frac{E}{c} \tag{13.3}$$

The above considerations led to de Broglie's hypothesis, which will now be described:

Since all attempts to interpret light quanta according to classical physics have failed, we must accept the fact that the same entity (in this case a light wave) behaves on some occasions as a wave and on others as a corpuscle, according to the experimental conditions. Now, if the unknown fundamental laws of nature are such as to permit this dualism, it would be unsatisfactory if they only applied in the case of light. On the other hand, if this dualism is really of a more general nature, applying also to an electron, for example, then we may employ the same argument that we have just used for the case of the light wave. This means that if we find that under suitable experimental conditions an electron of momentum $\mathbf{p}$ and energy $E$ behaves like a light wave of angular frequency $\omega = E/\hbar$, then the following relation exists between the momentum and the propagation vector:

$$\mathbf{p} = \hbar\mathbf{k} \tag{13.4}$$

This equation is the well-known relation of de Broglie for momentum and wavelength. Putting $k = 2\pi/\lambda$, we have, from (13.4),

$$\lambda = \frac{2\pi\hbar}{p} = \frac{h}{p} \tag{13.5}$$

For an electron of rest mass $m$, we must replace equation (13.3) by the relation

$$E = c\sqrt{(p^2 + m^2c^2)} \tag{13.6}$$

The $x$-component of the velocity $v_x$, may now be derived:

$$v_x = \frac{\partial E}{\partial p_x} = \frac{cp_x}{\sqrt{(p^2 + m^2c^2)}} \tag{13.7}$$

Hence,

$$p = \frac{mv}{\sqrt{(1 - v^2/c^2)}} \quad \text{and} \quad E = \frac{mc^2}{\sqrt{(1 - v^2/c^2)}}$$

In the energy equation (13.6), two extreme cases are of particular importance:

1. The limiting case $p \gg mc$ leads to the relation for light quanta, $E = cp$, which is also obtained from the same equation if the rest mass $m$ is allowed to become vanishingly small.

2. The non-relativistic case occurs when $p \ll mc$. Equation (13.6) then gives the approximation

$$E = mc^2 + \frac{p^2}{2m} \tag{13.8}$$

to which we shall confine ourselves in the first instance. We may then safely put $E = p^2/2m$ in the expression $\omega = E/\hbar$, since an additive constant is undetectable when the frequencies of matter waves are measured, and is therefore of no significance. If a particle of charge $e$ falls through a potential difference $\Phi$, it acquires kinetic energy $e\Phi$. Using the non-relativistic approximation (13.8), the corresponding momentum is $p = \sqrt{(2me\Phi)}$. De Broglie's relation (13.5) then predicts the following value for the associated wavelength:

$$\lambda = \frac{2\pi\hbar}{\sqrt{(2me\Phi)}}$$

This result may be experimentally verified. If we introduce the known values of $\hbar$, $m$, $e$, and measure $\Phi$ in volts, then

$$\lambda = \sqrt{\left(\frac{150}{\Phi_{(volts)}}\right)} 10^{-8} \, \text{cm} \tag{13.9}$$

This is the wavelength associated with an electron of kinetic energy $e\Phi$.

It is well known that a beam of electrons passing through a crystal gives rise to a diffraction pattern, the position of which can be calculated from the crystal lattice dimensions and the wavelength given by (13.9) in basically the same way as the X-ray diffraction pattern discovered by von Laue. At the time of de Broglie's hypothesis, however, nothing was known of the possibility of such experiments.

(b) *The free electron.*—De Broglie's hypothesis leads directly to the relation (13.4) between momentum and wavelength. The term $\exp\dfrac{i}{\hbar}(\mathbf{p}.\mathbf{r} - Et)$, however, is inadequate to describe a single electron, because its amplitude is unity everywhere in space. In order to obtain a spatially limited expression we make the assumption, essential to all wave theory, that waves may be superposed. Since $E$ is given as a function of $p$, the most general superposition of such waves is

$\psi(x, y, z, t) =$

$$= \frac{1}{(2\pi\hbar)^{3/2}} \iiint\limits_{-\infty}^{\infty} g(p_x, p_y, p_z) \exp\frac{i}{\hbar}(p_x x + p_y y + p_z z - Et) \, dp_x \, dp_y \, dp_z$$

where $g(p_x, p_y, p_z)$ is an arbitrary function. (See figure 15 for the explanation of the factor $1/(2\pi\hbar)^{3/2}$.) In vectorial notation

$$\psi(\mathbf{r}, t) = \frac{1}{(2\pi\hbar)^{3/2}} \int g(\mathbf{p}) \exp\frac{i}{\hbar}(\mathbf{p}\cdot\mathbf{r} - Et)\,d\mathbf{p} \qquad (13.10)$$

We shall restrict ourselves at first to one dimension and discuss the function

$$\psi(x, t) = \frac{1}{(2\pi\hbar)^{1/2}} \int g(p) \exp\frac{i}{\hbar}(px - Et)\,dp \qquad (13.11)$$

We confine ourselves to a narrow frequency band by assuming that the function $g(p)$ is only appreciably different from zero in the neighbourhood of a given momentum $p_0$. We therefore put $g(p) = g(p_0 + q)$, where $g$ is only different from zero for $|q| \ll |p_0|$. Then we can replace $E(p)$ in equation (13.11) by $E(p_0) + (dE/dp)_{p_0} q$, when we obtain

$$\psi(x, t) = \qquad\qquad\qquad\qquad\qquad\qquad\qquad (13.12)$$

$$= \frac{1}{(2\pi\hbar)^{1/2}} \exp\frac{i}{\hbar}[p_0 x - E(p_0)t] \int g(p_0 + q) \exp\left(\frac{i}{\hbar}\left[x - \left(\frac{dE}{dp}\right)_{p_0} t\right]q\right)dq$$

This is a plane wave $\exp i(p_0 x - E_0 t)/\hbar = \exp i(k_0 x - \omega_0 t)$ multiplied by an amplitude factor which is propagated along the $x$-axis with "group velocity" $v_g = (dE/dp)_{p_0}$. According to (13.7), however, $v_g$ is identical with the velocity $v_0$ of a particle with momentum $p_0 = mv_0$.

For the further treatment of the wave packet represented by (13.12), we shall select for $g(p)$ the Gaussian function

$$g(p) = A \exp -\frac{(p - p_0)^2}{4s^2} = A \exp -\left(\frac{q}{2s}\right)^2 \qquad (13.13)$$

where $A$ is an initially unimportant constant. Using this function $g$, the integral in (13.12) can be evaluated by elementary methods. We obtain

$$\psi(x, t) = s\left(\frac{2}{\hbar}\right)^{1/2} A \exp\frac{i}{\hbar}(p_0 x - E_0 t) \exp -\frac{(x - v_0 t)^2}{\hbar^2/s^2} \qquad (13.14)$$

It is important to realize that the expression (13.14) is an approximate solution for small time intervals. The exact determination of the integral in (13.11), using the value of $g$ given by (13.13), leads to the result

$$\psi(x, t) = \frac{s(2/\hbar)^{1/2} A}{\left(1 + \frac{2s^2 it}{m\hbar}\right)^{1/2}} \exp\frac{i}{\hbar}(p_0 x - E_0 t) \exp -\frac{s^2}{\hbar^2}\frac{(x - v_0 t)^2}{1 + \frac{2s^2 it}{m\hbar}} \qquad (13.14a)$$

In particular, the amplitude factor no longer travels as a rigid entity with velocity $v_0$ as in (13.12), but becomes increasingly spread out: in effect, the wave packet breaks up. It is easy to understand the reason for the spreading of the originally compact wave packet: it occurs because $\psi$ is composed of waves possessing different phase velocities, since $E/p$ is not constant, and dispersion therefore arises. The wave packet must therefore become increasingly dispersed with time. The rate of this dispersion increases with the difference between the various phase velocities (cf. §16c).

(*c*) In attempting to explain the physical significance of the wave functions given by (13.10) or (13.11), we are inclined to interpret them in the first instance as "really" describing the spatial distribution of the electron; we might, for instance, consider $e|\psi(x)|^2$ to be proportional to the electron charge density expressed as a function of position. Alternatively, we might say that

$$\frac{|\psi|^2 \, \Delta x \, \Delta y \, \Delta z}{\iiint |\psi|^2 \, dx \, dy \, dz}$$

is the fraction of the electron charge contained in the volume $\Delta x \, \Delta y \, \Delta z$. Or again, we may say that the electron charge is distributed throughout space according to the charge density function $|\psi|^2$. Using this concept, we may interpret (13.10), for instance, as the "classical" equation for an electrically charged fluid in which there is no longer any question of localized electrons. This "wave concept" may be successfully applied to the interpretation of many experiments, and in particular to the diffraction of electron beams.

However, the fundamental difficulty of the quantum theory appears on further analysis of the above simple interpretation of the function $\psi$.

The concept that the electron is really distributed throughout space according to the function $|\psi|^2$ is obviously in complete contrast to the basis of electron theory, according to which only integral multiples of the elementary charge occur in all measurements of this quantity.

We met a somewhat similar difficulty in §11. We saw then that a proviso is necessary if we assume the existence of quantum transition elements $x_{mn}$ which determine the intensities of the spectral lines in a similar manner to the Fourier coefficients $x_\tau(J)$ of classical physics; if we make this assumption, we can make no statement about the individual electron orbits. We must confine ourselves to the statistical statement that the quantities $|x_{mn}|^2$ are proportional to the probabilities of the transitions $m \to n$.

Similarly in the present case: if we wish to represent the electron by means of a wave function $\psi(x,t)$ in spite of the existence of discrete units of electric charge, this is only possible* if we restrict ourselves to a statistical statement. This requires that we should no longer say that the electron is "really" distributed throughout space according to $\psi(x,t)$, but that we should assume that $|\psi(x,t)|^2 \Delta x$ is proportional to the experimental probability of finding the electron in the interval $x, x+\Delta x$. In a precisely similar manner, $|g(p)|^2 \Delta p$ determines the experimental probability of finding a momentum the value of which lies between $p$ and $p+\Delta p$.

This interpretation leads to consequences of great importance. In the first place we perceive that the coordinate and momentum probabilities are not independent:

$$|\psi(x,t)|^2 = \frac{2s^2}{\hbar}|A|^2 \exp - \frac{(x-v_0 t)^2}{2(\hbar/2s)^2}$$

$$|g(p)|^2 = |A|^2 \exp - \frac{(p-p_0)^2}{2s^2} \tag{13.15}$$

Both probabilities are represented by Gaussian functions (figure 15); the possible deviations from the most probable values $v_0 t$ and $p_0$

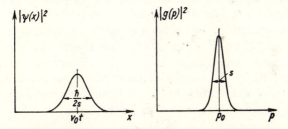

Fig. 15.—Distribution of position and momentum probabilities according to (13.15). The factor $1/\sqrt{(2\pi\hbar)}$ in (13.10) and (13.11) makes the areas under the curves the same for both distributions. [The area is unity for
$$A^2 = 1/s\sqrt{(2\pi)}]$$

are approximately given by the widths $\Delta x = \hbar/2s$ and $\Delta p = s$ of the Gaussian functions. The indeterminacies are reciprocal; if the uncertainty in momentum $\Delta p$ is small, then the uncertainty in position

---

* It might be thought that the electron could be represented by a comparatively concentrated wave packet. This is impossible, however, because such a packet would always spread throughout space in the course of time, as shown by (13.14a).

is large, and vice versa. However, the product $\Delta p \, \Delta x$ of the position and momentum uncertainties is constant:

$$\Delta p \, \Delta x = \tfrac{1}{2}\hbar \tag{13.16}$$

The indeterminacies may be somewhat more precisely formulated by introducing the mean square deviation. The average value of any function of the momentum, $f(p)$, is

$$\bar{f} = \frac{\int f(p)\,|g(p)|^2 dp}{\int |g(p)|^2 dp}$$

The average value of $p$ is $p_0$. The square root of the mean square deviation $\overline{(p-\bar{p})^2}$ is a measure of the dispersion about the mean. If the indeterminacies are defined as $\Delta p = \{\overline{(p-\bar{p})^2}\}^{1/2}$ and $\Delta x = \{\overline{(x-\bar{x})^2}\}^{1/2}$, the required average values may easily be determined from (13.15):

$$\bar{p} = p_0, \quad \overline{(p-\bar{p})^2} = s^2, \quad \bar{x} = v_0 t, \quad \overline{(x-\bar{x})^2} = (\hbar/2s)^2$$

from which (13.16) again follows.

*Thus the simultaneous measurement of the position and momentum of an electron necessarily involves uncertainties which are related in accordance with equation (13.16).*

This may be better understood if we consider how the measurement of the position of an electron might be accomplished in practice. The most direct method would be microscopical observation. For such an observation it would be necessary to illuminate the electron, say with light of wavelength $\lambda$. The position could then be determined with an accuracy $\Delta x$, of order of magnitude $\Delta x \approx \lambda = c/v$. But light of frequency $v$ has a momentum $p = hv/c$, of which an appreciable part must be transferred by the process of measurement to the electron, the momentum of which, after the measurement, is therefore unknown by an extent $\Delta p \approx hv/c$. In the final result, therefore,

$$\Delta p \, \Delta x \approx h$$

The more accurate the measurement of position (i.e. the smaller the value of $\Delta x$), the greater is the uncertainty in the value of the momentum of the electron after the measurement. This simple exposition of the "uncertainty relation" (13.16) and an analogous discussion of other possible methods of measurement were first given by W. Heisenberg.

The interpretation of the wave function $\psi$ as a probability has a further consequence, not specifically related to quantum mechanics, but occurring in classical physics wherever probability theory must be employed.

Our interpretation implies that if we examine the elementary interval $\Delta x$ for the presence of an electron we must expect one of two results: either the electron is completely contained in $\Delta x$, or not at all. If a measurement shows that the electron is contained in $\Delta x$, this statement only has meaning if the electron is "really" in $\Delta x$ immediately after the measurement, as shown by a second measurement of its position immediately after the first. However, since the wave function

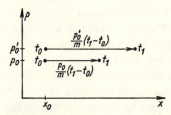

Fig. 16.—Paths in phase space for motion in the absence of external forces

$\psi(x,t)$ determines the probability of occurrence of the electron, it is necessary to postulate that the first measurement caused the $\psi$-function, previously distributed in space, to be concentrated in the elementary interval $\Delta x$.

Under certain circumstances, therefore, the process of measurement has a very drastic effect on the course of the state function designated by $\psi(x,t)$. This effect is called the "reduction of the wave function", by which is meant the reduction of $\psi$ to the region $\Delta x$ prescribed by the measuring equipment. The function is therefore provisionally transformed by the measurement into another function of state.

That there is nothing mysterious about this process may be illustrated by the simple example of a particle in uniform motion in the absence of external forces, according to classical mechanics. In phase space, this motion is represented by a line parallel to the $x$-axis. Figure 16 shows two such paths, both of which start from the same point $x_0$ at time $t_0$, but with different initial values of the momentum $p_0$ and $p_0'$. Each path will be traversed at a different velocity, according to the value of the corresponding momentum. If the initial position and momentum are not known exactly, but can only be measured with certain experimental or at any rate unavoidable errors, $\Delta p$ and $\Delta x$, then the initial conditions are indeterminate and it is impossible to specify the path with complete precision. We can only say that every point within the element of phase space $\Delta p \Delta x$ represents a possible initial point of the path. Every initial point gives rise to a definite path, specified by the equations of motion; in the case of the point itself, however, we can only say that there is a probability $w(p, x) dp dx$ that it lies within the infinitesimal element of phase space $dp dx$. Then at time $t_0$ the value of $w$ is constant within the rectangular area shown in figure 17, and zero outside this area. The probability at a later instant of time is found from the equations of motion; for instance $w(p, x, t_1)$ is constant within the parallelogram marked "$t_1$" in figure 17. The position uncertainty therefore increases with time because of the initial uncertainty in momentum. A new measurement of position at time $t$, with the same inherent error $\Delta x$, might show that the particle lies within the two limits shown as broken lines in the figure. This measurement consequently enables us to reduce the "probability distribution" from the broken-line area to the smaller region bounded by solid lines; the process is analogous to the reduction of the wave function. This sudden change in the probability distribution is not the result of any variations of a physical nature; it is important to realize that the function $w$ merely describes the extent of our knowledge, or of our

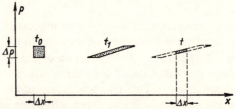

Fig. 17.—The probability distribution at different instants of time, for initial errors $\Delta p$ and $\Delta x$. The distribution is reduced by a measurement of position at time $t$

information on the process. The information gained by a new measurement necessarily leads to a reduction in the area of the probability distribution.

As regards classical physics, it is clear from the above example that by means of a series of measurements we can continuously reduce the volume of phase space in which $w$ is different from zero; in principle, therefore, it should ultimately be possible to specify the physical process with complete precision. In quantum theory the situation is different; in this case a measurement of position with error $\Delta x$ necessarily entails a momentum uncertainty $\Delta p$ according to the relation (13.16). This effect may be crudely represented in the classical model by postulating that the smallest physically significant element of phase space must be of order of magnitude $\hbar$.

There is a further essential difference between quantum and classical mechanics. In the latter case, if the function $w$ is given for a certain definite time, it is known for all later instants of time. In quantum theory, the corresponding situation holds for the "probability amplitude" $\psi$, but not for the probability $w$ itself.

### §14. Schrödinger's wave equation for the electron

(a) *The wave equation for the free electron.*—We shall now derive a differential equation describing the properties of the function which we obtained in the case of the free electron (for motion in one dimension),

$$\psi(x, t) = \frac{1}{(2\pi\hbar)^{1/2}} \int_{-\infty}^{+\infty} g(p) \exp \frac{i}{\hbar} [px - E(p)t] \, dp \qquad (14.1)$$

We shall restrict ourselves to the non-relativistic approximation

$$E(p) = \frac{p^2}{2m} \qquad (14.2)$$

When the operator $-i\hbar \, \partial/\partial x$ acts upon the function $\phi = \exp\frac{i}{\hbar}(px - Et)$, it multiplies the latter by $p$. Similarly, the application of the operator $i\hbar \, \partial/\partial t$ multiplies the function by $E$. Therefore

$$\frac{\hbar}{i} \frac{\partial \phi}{\partial x} = p\phi \qquad -\hbar^2 \frac{\partial^2 \phi}{\partial x^2} = p^2\phi \qquad -\frac{\hbar}{i} \frac{\partial \phi}{\partial t} = E\phi$$

If these operators are applied to (14.1), we see that in virtue of (14.2)

$$-\frac{\hbar^2}{2m} \frac{\partial^2 \psi}{\partial x^2} = -\frac{\hbar}{i} \frac{\partial \psi}{\partial t} \qquad (14.3)$$

The expression (14.3) only contains the first-order derivative with respect to $t$. Therefore, if $\psi(x, t)$ is given for time $t = 0$, the value of this function at all other times is given by (14.3). In fact, a function $\psi(x, 0)$ may always be expressed by the Fourier transform

$$\psi(x, 0) = \frac{1}{(2\pi\hbar)^{1/2}} \int g(p) \exp \left( \frac{i}{\hbar} px \right) dp$$

The function $g(p')$ is obtained by multiplying $\psi(x, 0)$ by $\exp\left(-\frac{i}{\hbar}p'x\right)$ and integrating over all $x$:

$$g(p') = \frac{1}{(2\pi\hbar)^{1/2}} \int \psi(x, 0) \exp -\left(\frac{i}{\hbar}p'x\right) dx$$

If we now put 
$$\psi(x, t) = \frac{1}{(2\pi\hbar)^{1/2}} \int g(p)c(t) \exp\left(\frac{i}{\hbar}px\right) dp$$

as the general solution (with $c(0) = 1$), (14.3) gives the following equation for $c(t)$:

$$\frac{p^2}{2m}c = -\frac{\hbar}{i}\dot{c}, \text{ from which } c(t) = \exp -\frac{i}{\hbar}\frac{p^2}{2m}t$$

Hence we obtain the original formula (14.1) for $\psi(x, t)$.

The same considerations apply to three-dimensional motion. The wave function

$$\phi = \exp\frac{i}{\hbar}(\mathbf{p}.\mathbf{r}-Et) \tag{14.4}$$

possesses the property that

$$\frac{\hbar}{i}\frac{\partial\phi}{\partial x} = p_x\phi, \quad \frac{\hbar}{i}\frac{\partial\phi}{\partial y} = p_y\phi, \quad \frac{\hbar}{i}\frac{\partial\phi}{\partial z} = p_z\phi, \quad -\frac{\hbar}{i}\frac{\partial\phi}{\partial t} = E\phi \tag{14.5}$$

It satisfies the two differential equations

$$-\frac{\hbar^2}{2m}\nabla^2\phi = -\frac{\hbar}{i}\frac{\partial\phi}{\partial t} \tag{14.6}$$

and

$$-\frac{\hbar^2}{2m}\nabla^2\phi = E\phi \tag{14.7}$$

Of these two equations, (14.7) is only valid for those values of $p$ that satisfy the relation $p^2/2m = E$ for the initially assumed energy value. On the other hand the time-dependent equation (14.6) is of much more general application. Its solution is provided by the general expression for a wave packet with any arbitrary function $g(\mathbf{p})$,

$$\psi(\mathbf{r}, t) = \frac{1}{(2\pi\hbar)^{1/2}} \int g(\mathbf{p}) \exp\frac{i}{\hbar}\left(\mathbf{p}.\mathbf{r}-\frac{p^2}{2m}t\right) d\mathbf{p}$$

*The solution of the differential equations* (14.6) *and* (14.7):

When there are no external forces, the classical Hamiltonian function is

$$\frac{p_x^2 + p_y^2 + p_z^2}{2m} = \mathscr{H} \tag{14.8}$$

The quantities **p** and $\mathscr{H}$ occurring in this function are replaced by the operators

$$p_x \rightarrow \frac{\hbar}{i}\frac{\partial}{\partial x}, \ldots, \qquad \mathscr{H} \rightarrow -\frac{\hbar}{i}\frac{\partial}{\partial t} \qquad (14.9)$$

The Hamiltonian function $p^2/2m$ thus becomes the Hamiltonian operator $-(\hbar^2/2m)\nabla^2$. The energy equation (14.8) is transformed into an operator equation, which gives the differential equation (14.6) when applied to a function $\phi(\mathbf{r},t)$:

$$-\frac{\hbar^2}{2m}\nabla^2\phi = -\frac{\hbar}{i}\frac{\partial\phi}{\partial t}$$

On the other hand, if $E$ is again a given quantity, we obtain the particular equation (14.7)

$$-\frac{\hbar^2}{2m}\nabla^2\phi = E\phi$$

### (b) The derivation of the Schrödinger equation

From the above procedure, we can see how the differential equation (14.6) must be modified to take account of a potential energy $V(\mathbf{r})$ in addition to the kinetic energy $p^2/2m$. In this case, the classical form of the Hamiltonian function is

$$\frac{p^2}{2m}+V(\mathbf{r}) = \mathscr{H} \qquad (14.10)$$

where $\mathscr{H} = E$, the total energy, which is constant. We again consider the Hamiltonian function as an operator in accordance with (14.9), and take $V(\mathbf{r})$ to mean an operator signifying "multiply by $V(\mathbf{r})$". We then have the two forms of the Schrödinger equation: the general equation is

$$-\frac{\hbar^2}{2m}\nabla^2\psi+V(\mathbf{r})\psi = -\frac{\hbar}{i}\frac{\partial\psi}{\partial t} \qquad (14.11)$$

If we introduce the trial solution

$$\psi(\mathbf{r},t) = u(\mathbf{r})\exp-\frac{i}{\hbar}Et$$

into (14.11), we obtain the particular equation

$$-\frac{\hbar^2}{2m}\nabla^2 u+V(\mathbf{r})u = Eu \qquad (14.12)$$

In the above equations $\psi$ depends on $\mathbf{r}$ and $t$, but $u$ depends on $\mathbf{r}$ alone.

The solution of the general equation (14.11) expresses $\psi(\mathbf{r}, t)$ as a function of time, for a given initial value $\psi(\mathbf{r}, 0)$. Integration of equation (14.12), on the other hand, yields the particular function $u(\mathbf{r})$ corresponding to a prescribed value of the energy $E$.

If we interpret (14.11) according to the crude wave concept discussed on page 74, then the function

$$\rho = e\psi^*\psi \qquad (14.13)$$

represents a charge density. We shall now investigate the variation of $\rho$ with respect to time:

$$\frac{\partial \rho}{\partial t} = e(\dot{\psi}^*\psi + \psi^*\dot{\psi})$$

Written in terms of the complex conjugate, equation (14.11) becomes

$$-\frac{\hbar^2}{2m}\nabla^2\psi^* + V(\mathbf{r})\psi^* = \frac{\hbar}{i}\dot{\psi}^* \qquad (14.11a)$$

If we now multiply equations (14.11) and (14.11a) respectively by $-e\dfrac{i}{\hbar}\psi^*$ and $e\dfrac{i}{\hbar}\psi$ and add, we obtain

$$\frac{\partial \rho}{\partial t} = -e\frac{\hbar}{2im}(\psi^*\nabla^2\psi - \psi\nabla^2\psi^*)$$

$$= -\operatorname{div}\left\{e\frac{\hbar}{2im}(\psi^*\operatorname{grad}\psi - \psi\operatorname{grad}\psi^*)\right\} \qquad (14.14)$$

This is the equation of continuity* for the electric charge: $\partial\rho/\partial t = -\operatorname{div}\mathbf{j}$. The current density† corresponding to the charge density defined by (14.13) is

$$\mathbf{j} = e\frac{\hbar}{2im}(\psi^*\operatorname{grad}\psi - \psi\operatorname{grad}\psi^*) \qquad (14.15)$$

Since $\psi^*\psi\,d\mathbf{r}$ represents the probability of finding the electron in the volume element $d\mathbf{r}$, we must postulate that

$$\int \psi^*\psi\,d\mathbf{r} = 1$$

---

* The existence of the equation of continuity shows that it is physically justifiable to interpret $e\psi^*\psi$ as a charge density.

† In this volume, the current density is designated by j, in contrast to Vol. I, where it was designated by g. The former notation is the usual one employed in quantum theory.

because the electron must certainly be present somewhere in space. When this condition is fulfilled the function is said to be *normalized*. By (13.10), $g$ is also normalized. Equation (14.14) then states that this normalization is conserved, since from this expression

$$\frac{d}{dt} \int \psi^* \psi \, d\mathbf{r} = 0 \qquad (14.16)$$

if $\mathbf{j}$ vanishes at infinity.

The energy equation (14.12) possesses the fundamental property that no solutions exist for which $\int |u|^2 \, d\mathbf{r} = 1$ for arbitrary values of $E$. Only when $E$ assumes certain definite values $E_1, E_2, \ldots, E_n, \ldots$ do the solutions $u_1, u_2, \ldots, u_n \ldots$ exist. Using the Hamiltonian operator

$$\mathcal{H} = -\frac{\hbar^2}{2m} \nabla^2 + V(\mathbf{r})$$

we may express this situation concisely as follows:

$$\mathcal{H} u_n = E_n u_n \qquad n = 1, 2, 3, \ldots \qquad (14.17)$$

The $E_n$ are called the eigenvalues of the Hamiltonian operator, and the $u_n$ are termed the eigenfunctions belonging to the eigenvalues $E_n$. In this case we say that the eigenvalues are distributed according to a point spectrum. Equation (14.12) may also be soluble for a continuous range of energy values, as in the case of the free electron; we refer to this as a line spectrum. In this case the normalization condition is not fulfilled (cf. §17ff.).

Schrödinger discovered the important fact that the eigenvalues of $\mathcal{H}$ given by (14.17) are identical with Bohr's postulated energy levels, from which the frequencies of the absorbed and emitted spectral lines may be calculated by means of the relation $\omega_{mn} = (E_m - E_n)/\hbar$.

This fact, which at first sight appears most remarkable, may be better understood if the method described in §12 for the calculation of the stationary energy levels is expressed in a somewhat different manner. As originally given, the procedure consisted in finding matrices $p = (p_{mn})$ and $x = (x_{mn})$ that satisfy the commutation relation

$$[p, x] \equiv px - xp = \frac{\hbar}{i}$$

and for which the matrix

$$\mathcal{H} = (\mathcal{H}_{mn}) = \frac{p^2}{2m} + V(x)$$

only possesses diagonal elements, for which $m = n$.

This procedure may be carried out as follows: we first seek any matrices satisfying the commutation relation, and use them to form the matrix $(\mathcal{H}_{mn})$, on which we then perform an orthogonal transformation. As we know from matrix algebra, this is equivalent to finding the eigenvectors $u_n$ of $\mathcal{H}$. The eigenvectors so found satisfy the equation (14.17).

In each case, therefore, we have a problem to solve that involves eigenvalues. The formal agreement between the two procedures is seen to be complete when we note that the differential operators $p = -i\hbar \, \partial/\partial x$ and $x$ satisfy the commutation relation

$$px - xp = \frac{\hbar}{i}$$

If for instance we apply the operator $px - xp$ to a function $f(x)$, we obtain

$$(px - xp)f(x) = \frac{\hbar}{i}\left\{ \frac{\partial}{\partial x}(xf) - x\frac{\partial f}{\partial x} \right\} = \frac{\hbar}{i}f(x)$$

Chapter B II of this book is devoted to a general investigation of these relationships, which at first sight appear so remarkable.

In the next section we shall analyse in some detail the important special case of the linear oscillator, using both the Heisenberg and the Schrödinger methods described in §§ 12 and 14 respectively.

## §15. The Schrödinger and Heisenberg treatments of the linear oscillator

Before proceeding to the abstract formulation of the quantum theory in the next chapter, we shall apply the above procedures to the treatment of a particular example. Our starting-point is the classical Hamiltonian function of a linear harmonic oscillator of angular frequency $\omega$:

$$\frac{p^2}{2m} + \tfrac{1}{2}m\omega^2 x^2 = \mathcal{H} \tag{15.1}$$

In the Schrödinger treatment, $p$ is replaced by $-i\hbar\,\partial/\partial x$. In order to determine the function $\phi(x)$ corresponding to a fixed energy $E$, we must seek solutions of the differential equation

$$-\frac{\hbar^2}{2m}\frac{d^2\phi}{dx^2}+\tfrac{1}{2}m\omega^2 x^2\phi = E\phi \tag{15.2}$$

for which

$$\int_{-\infty}^{+\infty}|\phi(x)|^2\,dx = 1 \tag{15.3}$$

A preliminary qualitative discussion of this equation is relevant. Let us put (15.2) in the form

$$\frac{d^2\phi}{dx^2}\equiv\phi'' = \frac{2m}{\hbar^2}(\tfrac{1}{2}m\omega^2 x^2 - E)\phi$$

and consider this equation as a formula for the graphical construction of the curve $\phi(x)$. We see that the curvature of $\phi(x)$ changes sign both upon a change of sign of $\phi(x)$ itself and when $x$ takes the value $x_g$ defined by $\tfrac{1}{2}m\omega^2 x_g^2 = E$. (The values $\pm x_g$ represent the classical path limits.)

If we begin the construction with $\phi(0) = 1$ and $\phi'(0) = 0$, for instance, then $\phi''$ is negative at first for small $x$, and the curve is concave towards the $x$-axis, like a cosine curve. We now choose $E$ and hence the initial curvature to be so small that the point $x_g$ is reached before $\phi$ can become zero. Then for $x > x_g$, $\phi$ can follow one of three possible courses (figure 18):

1. $E = E_1$ is so small that the consequent change of curvature prevents the curve of $\phi$ from reaching the $x$-axis beyond $x_g$, and it tends instead to $+\infty$ (curve 1).

2. $E = E_2$ is so large that the curve cuts the $x$-axis beyond $x_g$ in spite of the change of curvature. The curvature then changes again and $\phi$ tends to $-\infty$ (curve 2).

3. In between these two cases we must expect to find a limiting case when $E = E_0$, for which the curve of $\phi$ tends asymptotically to the $x$-axis. $E_0$ is an "eigenvalue" of the differential equation (15.2) (curve 3).

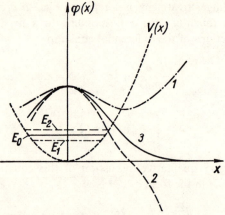

Fig. 18.—Graphical construction for the proper function of a linear oscillator in the lowest energy state. The solid curve is associated with the proper value $E_0$

We shall now calculate one such eigenvalue. Let us try the Gaussian curve

$$\phi = e^{-\frac{1}{2}\alpha^2 x^2}$$

as a solution of (15.2): this choice is prompted by the shape of curve 3 in figure 18. Then

$$\phi' = -\alpha^2 x\, e^{-\frac{1}{2}\alpha^2 x^2} \qquad \phi'' = (-\alpha^2 + \alpha^4 x^2)\phi$$

Substituting for $\phi''$ in (15.2), we have

$$-\frac{\hbar^2}{2m}(-\alpha^2 + \alpha^4 x^2) + \tfrac{1}{2}m\omega^2 x^2 = E$$

In order that the equation should hold for all values of $x$, we must have

$$\alpha^2 = \frac{m\omega}{\hbar} \quad \text{and} \quad E = E_0 = \tfrac{1}{2}\hbar\omega$$

We have now obtained the lowest eigenvalue, and the associated eigenfunction which, when normalized, is

$$\phi = \phi_0 = \frac{1}{(\alpha\sqrt{\pi})^{1/2}}\, e^{-\frac{1}{2}\alpha^2 x^2}$$

We shall meet this solution again in the course of the analysis.

The quantity $\alpha$ has the dimension (length)$^{-1}$, and is defined by the equation

$$\alpha^2 = \frac{m\omega}{\hbar} \tag{15.4}$$

Equation (15.2) may be written in terms of $\alpha$, as follows:

$$\frac{1}{2}\left(\alpha^2 x^2 - \frac{1}{\alpha^2}\frac{d^2}{dx^2}\right)\phi = \frac{E}{\hbar\omega}\phi \tag{15.5}$$

The usual method for the solution of (15.5) requires the use of Hermite polynomials. We shall adopt another course, which will also prove useful for the treatment of the problem by Heisenberg's method.

We introduce the new operators

$$b = \frac{1}{\sqrt{2}}\left(\alpha x + \frac{1}{\alpha}\frac{d}{dx}\right) \qquad b^+ = \frac{1}{\sqrt{2}}\left(\alpha x - \frac{1}{\alpha}\frac{d}{dx}\right) \tag{15.6}$$

which possess the following properties. When the operators are applied in turn to an arbitrary function $x$,

$$b^+ b\phi = \frac{1}{2}\left(\alpha^2 x^2 - \frac{1}{\alpha^2}\frac{d^2}{dx^2}\right)\phi - \frac{1}{2}\phi \tag{15.7}$$

$$bb^+ \phi = \frac{1}{2}\left(\alpha^2 x^2 - \frac{1}{\alpha^2}\frac{d^2}{dx^2}\right)\phi + \frac{1}{2}\phi$$

Hence for any function $\phi$

$$(bb^+ - b^+ b)\phi = \phi$$

and in particular, we may put

$$bb^+ - b^+ b = 1 \tag{15.8}$$

Now, for any two functions $f$ and $g$ that vanish at infinity

$$\int f(x)\frac{dg(x)}{dx}dx = -\int \frac{df(x)}{dx}g(x)dx$$

Hence

$$\int f(x)\,bg(x)\,dx = \int [b^+ f(x)]\,g(x)\,dx \tag{15.9}$$

$b^+$ is said to be the *adjoint operator* to $b$.

If we now put $\lambda = E/\hbar\omega - \frac{1}{2}$ as a temporary abbreviation, the Schrödinger equation assumes the form

$$b^+ b\phi = \lambda\phi \qquad (15.10)$$

If $\phi$ is normalized, then from (15.9) and (15.10)

$$\lambda = \int \phi b^+ b\phi \, dx = \int (b\phi)^2 \, dx \qquad (15.11)$$

In general, therefore, $\lambda$ is positive, and only equal to zero if $b\phi = 0$.

If the operator $b$ is applied to (15.10), then in virtue of (15.8)

$$bb^+ b\phi = (b^+ b + 1)b\phi = \lambda b\phi$$

Therefore

$$b^+ bb\phi = (\lambda - 1)b\phi \qquad (15.12a)$$

Similarly

$$b^+ bb^+ \phi = (\lambda + 1)b^+ \phi \qquad (15.12b)$$

Equation (15.12a) implies that if $\phi$ is an eigenfunction belonging to the eigenvalue $\lambda$, then either $b\phi$ is an eigenfunction belonging to $\lambda - 1$, or $b\phi = 0$. Repeated application of the operator $b$ could yield eigenfunctions belonging to $\lambda - 1, \lambda - 2, \ldots, \lambda - v$, and would therefore finally give an eigenfunction belonging to a negative eigenvalue, which is in contradiction to (15.11). This conflict can only be avoided if the series $\lambda - v$ terminates in such a manner that one function $\phi_0$ occurs in the series $b\phi, b^2\phi, b^3\phi, \ldots$ for which $b\phi_0 = 0$. But $\lambda - v = 0$ for this function $\phi_0$.

There is an element $\phi_0$, therefore, for which $b\phi_0 = 0$ and hence $b^+ b\phi_0 = 0$; the corresponding eigenvalue is $\lambda = 0$. The other eigenvalues of (15.10) are the whole numbers $\lambda = n = 1, 2, \ldots$. The eigenvalues $E_n$ of the oscillator are therefore

$$E_n = (n + \frac{1}{2})\hbar\omega \qquad n = 0, 1, 2, \ldots$$

The eigenfunction $\phi_0$ is a solution of the equation $b\phi_0 = 0$. From (15.6), therefore,

$$\frac{d\phi_0}{dx} = -\alpha^2 x \phi_0$$

from which

$$\phi_0 = C e^{-\frac{1}{2}\alpha^2 x^2}$$

or after normalization,

$$\phi_0 = \frac{1}{(\alpha \sqrt{\pi})^{1/2}} e^{-\frac{1}{2}\alpha^2 x^2} \tag{15.13}$$

as we deduced from our original trial solution.

We shall now make use of equation (15.12$b$), according to which an eigenfunction belonging to $\lambda = n+1$ is obtained as a result of the application of the operator $b^+$ to the eigenfunction belonging to $\lambda = n$. Hence

$$b^+ \phi_n = N\phi_{n+1}$$

where the constant $N$ is determined by the requirement that $\phi_{n+1}$ be normalized if $\phi_n$ is normalized. Hence

$$N^2 = \int (b^+ \phi_n)(b^+ \phi_n)\,dx = \int (bb^+ \phi_n)\phi_n\,dx = n+1$$

since $bb^+ \phi_n = b^+ b\phi_n + \phi_n$ (from 15.8) $= (n+1)\phi_n$. Therefore

$$b^+ \phi_n = (n+1)^{1/2}\phi_{n+1} \quad \text{and} \quad b\phi_n = n^{1/2}\phi_{n-1} \tag{15.14}$$

We can now obtain all the eigenfunctions by the repeated application of the operator $b^+$ to $\phi_0$:

$$\phi_n(x) = \frac{1}{\sqrt{(n!)}}(b^+)^n \phi_0 \tag{15.15}$$

These functions $\phi_n(x)$ are normalized and orthogonal, since

$$\int \phi_n^*(x)\,\phi_m(x)\,dx = \delta_{nm} \tag{15.15a}$$

In order to determine the $\phi_n(x)$ explicitly we make use of the identity

$$b^+ f(x) = \frac{1}{\sqrt{2}}\left(\alpha x - \frac{1}{\alpha}\frac{d}{dx}\right)f(x) = -\frac{1}{\alpha\sqrt{2}} e^{\frac{1}{2}\alpha^2 x^2} \frac{d}{dx}(e^{-\frac{1}{2}\alpha^2 x^2} f)$$

If we now insert in (15.15) the value of $\phi_0$ given by (15.13), we obtain

$$\phi_n(x) = \frac{(-1)^n}{(n!\,2^n \alpha\sqrt{\pi})^{1/2}} e^{\frac{1}{2}\alpha^2 x^2}\left(\frac{1}{\alpha}\frac{d}{dx}\right)^n (e^{-\alpha^2 x^2}) \tag{15.16}$$

The Hermite polynomials are defined by

$$H_n(y) = (-1)^n \exp(-y^2)\left(\frac{d}{dy}\right)^n \exp(-y^2)$$

Putting $\alpha x = y$ and substituting in (15.16), we obtain the following alternative and common form of $\phi_n$:

$$\phi_n(y) = \frac{1}{(n!\,2^n\alpha\sqrt{\pi})^{1/2}} \exp\left(-\tfrac{1}{2}y^2\right) H_n(y)$$

Although (15.16) represents the explicit solution of equation (15.2), it is a most inconvenient formula for purposes of calculation. For instance, if we wish to evaluate the "matrix element"

$$x_{mn} = \int_{-\infty}^{\infty} \phi_m\, x\, \phi_n\, dx$$

this can admittedly be done by the direct use of (15.16). The following method, however, is much more convenient. From (15.6),

$$x = \frac{1}{2^{1/2}\alpha}(b + b^+)$$

From (15.14), therefore,

$$x\phi_n = \frac{1}{2^{1/2}\alpha}\left(n^{1/2}\phi_{n-1} + (n+1)^{1/2}\phi_{n+1}\right)$$

If the above expression is multiplied respectively by $\phi_{n-1}$ and $\phi_{n+1}$ and integrated, then in view of (15.15a) the only matrix elements different from zero are

$$x_{n-1,n} = \frac{1}{\alpha}\sqrt{(\tfrac{1}{2}n)} \qquad x_{n+1,n} = \frac{1}{\alpha}\sqrt{[\tfrac{1}{2}(n+1)]} \qquad (15.17)$$

We shall now describe the treatment of the linear oscillator by Heisenberg's matrix method. We take the equation (15.1) to be a matrix equation, and require the right-hand side to be diagonal, so that

$$\left\{\frac{p^2}{2m} + \tfrac{1}{2}m\omega^2 x^2\right\}_{nm} = E_n\delta_{nm} \qquad (15.18)$$

We then have the following commutative relations between the matrices $p$ and $x$:

$$(px - xp)_{nm} = \frac{\hbar}{i}\delta_{nm} \qquad (15.19)$$

If the quantity $\alpha$, defined as before by (15.4), is introduced into (15.18), then

$$\frac{1}{2}\left(\alpha^2 x^2 + \frac{1}{\hbar^2\alpha^2}p^2\right)_{nm} = \frac{E_n}{\hbar\omega}\delta_{nm} \qquad (15.20)$$

We now consider the matrices

$$b_{nm} = \frac{1}{\sqrt{2}}\left(\alpha x_{nm} + \frac{i}{\hbar\alpha}\,p_{nm}\right) \qquad (b^+)_{nm} = \frac{1}{\sqrt{2}}\left(\alpha x_{nm} - \frac{i}{\hbar\alpha}\,p_{nm}\right) \quad (15.21)$$

Since $x_{mn}^* = x_{mn}$ and $p_{mn} = p_{mn}^*$ (cf. footnote †, p. 61),

$$(b^+)_{nm} = b_{mn}^* \qquad (15.22)$$

As in the case of the operators $b$ and $b^+$, the product of the matrices is

$$b^+ b = \frac{1}{2}\left(\alpha^2 x^2 + \frac{1}{\hbar^2\alpha^2}\,p^2\right) - \frac{1}{2}$$

in virtue of (15.19), and

$$bb^+ - b^+ b = 1 \qquad (15.23)$$

Putting $E_n/\hbar\omega - \frac{1}{2} = \lambda_n$, as before, the matrix equation (15.20) becomes

$$b^+ b = \lambda \qquad (15.24)$$

where $\lambda$ is taken to be a diagonal matrix the elements of which are $\lambda_{nn} = \lambda_n$. For these elements

$$\lambda_n = \sum_s b_{ns}^+ b_{sn} = \sum_s |\,b_{sn}\,|^2 = \sum_s |\,b_{ns}^+\,|^2 \qquad (15.25)$$

If (15.24) is multiplied by $b^+$, we obtain $b^+ b^+ b = b^+\lambda$; from (15.23), $b^+ bb^+ - b^+ = b^+\lambda$. Consequently $\lambda b^+ - b^+\lambda = b^+$. The element $nm$ of the matrix equation (15.24) is

$$b_{nm}^+(\lambda_n - \lambda_m - 1) = 0$$

Hence, either

$$\lambda_n - \lambda_m = 1, \quad \text{or} \quad b_{nm}^+ = 0. \qquad (15.26)$$

If we imagine the $\lambda_n$ to be arranged in order of magnitude, then non-zero matrix elements of $b^+$ exist only for $\lambda_n - \lambda_{n-1} = 1$; for a given $n$, all elements $b_{ns}^+$ are zero except $b_{n,\,n-1}^+$. Equation (15.25) thus reduces to

$$\lambda_n = |\,b_{n,\,n-1}^+\,|^2 \qquad (15.27)$$

If we begin the numbering of the matrix elements with $n = 0, 1, 2, \ldots$ and $m = 0, 1, 2, \ldots$, then it follows from (15.27) that $\lambda_0 = 0$. Hence we again obtain $\lambda_n = n$, where $n = 0, 1, 2, \ldots$, and if we choose the elements $b_{nm}$ to be real, then

$$b_{n,\,n-1}^+ = b_{n-1,\,n} = \sqrt{n} \qquad (15.28)$$

The complete matrices are:

$$b^+ = \begin{array}{c|cccc} {}_n\diagdown{}^m & 0 & 1 & 2 & 3 & . \\ \hline 0 & 0 & 0 & 0 & 0 & . \\ 1 & \sqrt{1} & 0 & 0 & 0 & . \\ 2 & 0 & \sqrt{2} & 0 & 0 & . \\ 3 & 0 & 0 & \sqrt{3} & 0 & . \\ . & . & . & . & . & \end{array} \qquad b = \begin{array}{c|cccc} {}_n\diagdown{}^m & 0 & 1 & 2 & 3 & . \\ \hline 0 & 0 & \sqrt{1} & 0 & 0 & . \\ 1 & 0 & 0 & \sqrt{2} & 0 & . \\ 2 & 0 & 0 & 0 & \sqrt{3} & . \\ 3 & 0 & 0 & 0 & 0 & . \\ . & . & . & . & . & \end{array} \qquad \lambda = \begin{array}{c|cccc} {}_n\diagdown{}^m & 0 & 1 & 2 & 3 & . \\ \hline 0 & 0 & 0 & 0 & 0 & . \\ 1 & 0 & 1 & 0 & 0 & . \\ 2 & 0 & 0 & 2 & 0 & . \\ 3 & 0 & 0 & 0 & 3 & . \\ . & . & . & . & . & \end{array}$$

In addition, we have the matrix $x_{nm}$. From (15.21)

$$x = \frac{1}{\alpha\sqrt{2}}(b + b^+)$$

Hence

$$x_{n,n-1} = x_{n-1,n} = \frac{\sqrt{n}}{\alpha\sqrt{2}} = \sqrt{\frac{n\hbar}{2m\omega}} \qquad (15.29)$$

which agrees with (15.17) and with the value (11.11) previously deduced from the correspondence principle.

## §16. The wave equation of the electron considered as a classical equation

### (a) *The constants of the wave equation*

We saw in §14 that the existence of discrete units of electric charge necessitates a statistical interpretation of the $\psi$-function. On the other hand, if we ignore the corpuscular properties of particles such as electrons, the Schrödinger equation (14.11) may be conceived to be a classical wave equation, which may for instance be used to represent the electron radiation from a cathode as a purely wave-like phenomenon. It is permissible to conjecture that, if diffraction phenomena had been observed when cathode rays were first discovered, this would necessarily have led to the representation of the latter as a wave phenomenon, and hence to equation (14.11). As a classical equation, (14.11) is comparable to Maxwell's equations: like them, it predicts only wave properties and gives no information about the corpuscular nature of the radiation. As in the case of light, the particle-like properties appear only as a result of the quantum theory.

In this case, either the wave or the particle aspect may be chosen, with equal justification. In the domain of classical physics it is necessary to employ either the particle or the wave representation, since these descriptions are mutually exclusive. Each of these classical models leads to the same quantum-mechanical representation; it is a matter of indifference, therefore, whether we "quantize" the classical wave theory or the classical corpuscular theory. The quantum theory is able to describe corpuscular and wave-like properties simultaneously, in so far as the wave representation provides statistical information about the properties of the particles.*

Many phenomena may be most simply expressed in terms of classical wave theory, so that it is important to be fully aware of the classical interpretation of the de Broglie-Schrödinger wave equation, which we shall also attempt to express simultaneously in the language of the quantum theory.

In the first instance we must realize that the Schrödinger equation (14.11) makes no reference to the corpuscular nature of electrons. We may express this otherwise by saying that equation (14.11) contains constants that can be determined from the wave-like properties of the radiation alone. The equation may be written in the form

$$-\frac{\hbar^2}{2m}\nabla^2\psi + e\Phi\psi = i\hbar\dot\psi \tag{16.1}$$

where $\Phi(\mathbf{r})$ represents an electric potential. If we divide by $\hbar^2/2m$ and introduce the new constants

$$\alpha = \frac{2m}{\hbar} \qquad \beta = \frac{2me}{\hbar^2} \tag{16.2}$$

(16.1) assumes the form

$$-\nabla^2\psi + \beta\Phi\psi = i\alpha\dot\psi \tag{16.1a}$$

The wave equation now contains two constants $\alpha$ and $\beta$ which are completely determined by the wave properties of the electron radiation. In the present discussion we shall restrict ourselves to one dimension: $\psi = \psi(x,t)$, $\Phi = \Phi(x)$. The wave equation then takes the form

$$-\psi'' + \beta\Phi\psi = i\alpha\dot\psi \tag{16.3}$$

If the potential $\Phi$ is constant, the solution $\psi = ae^{i(kx-\omega t)}$ gives the following relation between the angular frequency $\omega$ and the wave number $k$, corresponding to the wavelength $\lambda = 2\pi/k$:

$$k^2 + \beta\Phi = \alpha\omega \tag{16.4}$$

---

* When there are several particles, all their coordinates must be included as variables in the $\psi$ function; the wave theory must then apply in a multi-dimensional space.

The group velocity of the radiation is

$$v_g = \frac{d\omega}{dk} = \frac{2}{\alpha}k \tag{16.5}$$

The constant $\alpha$ therefore establishes a connection between group velocity and wavelength: equation (16.5) is in fact the de Broglie relation.

Equation (16.4) is also valid if the potential is piecewise constant. In this case the angular frequency $\omega$ remains constant, otherwise the boundary conditions for $\psi$ could not be satisfied (cf. §16*d*). It follows from equation (16.4) that the quantity $k^2 + \beta\Phi$ remains unchanged when the radiation travels from a region of potential $\Phi_1$ to one of potential $\Phi_2$:

$$k_1^2 + \beta\Phi_1 = k_2^2 + \beta\Phi_2 \quad \text{or} \quad k_1^2 - k_2^2 = \beta(\Phi_2 - \Phi_1) \tag{16.6}$$

The constant $\beta$ therefore connects the change in wavelength of the radiation with the potential difference $\Phi_2 - \Phi_1$. Equation (16.5) represents the theorem of conservation of energy for particles, expressed in terms of wave theory.

The constants $\alpha$ and $\beta$ may therefore be obtained from measurements of wavelengths, group velocities, and potential differences. But $\alpha$ and $\beta$ are in fact the experimental data corresponding to pure wave theory, and in this connection it is somewhat misleading to represent them, as in (16.2), by means of the typical quantities $e$ and $m$, that are associated with particles. However, the charge-mass ratio of the radiation can be determined by deflection experiments and expressed in terms of the constants associated with the wave representation $(e/m = 2\beta/\alpha^2)$.

### (b) Physical quantities and conservation theorems

It is clear that those experiments which are exclusively concerned with the wave-like nature of cathode rays will lead to the formulation of the wave equation (16.3) or (16.1*a*). This equation is not very informative, however, if we do not know the physical meaning of the wave function $\psi$. In order to endow the theoretical pattern with physical meaning, we must understand the relationship between the wave function and the various relevant physical quantities. The typical physical quantities in a theory of the continuum are densities (charge, mass, momentum, and energy densities) and associated current or flow densities. In general, these densities satisfy conservation theorems, one of which we encountered in §14 in the case of the charge density $\rho_e$:

$$\frac{\partial \rho_e}{\partial t} + \text{div} \, \mathbf{j}_e = 0 \qquad (16.7)$$

or in integral form, after integration over a volume $V$ bounded by a surface $S$

$$\frac{d}{dt} \int_V \rho_e \, d\mathbf{r} = - \int_S \mathbf{j}_e . d\mathbf{f} \qquad (16.7a)$$

(Equation (16.7a) signifies that the charge contained in volume $V$ decreases with time because an electric current flows out across the boundary surface $S$.) If the system is self-contained, that is, if the current density $\mathbf{j}_e$ vanishes at the boundary, the total charge remains constant:

$$\frac{d}{dt} \int_V \rho_e \, d\mathbf{r} = 0$$

These conservation theorems must be implicit in the wave equation. We may express this as follows:

"If the wave equation (16.1a) leads to conservation theorems in the form of (16.7), then with certain reservations it is permissible to interpret as physical quantities the densities and current densities occurring in equation (16.7)".

We must now determine whether the conservation theorems are in fact contained in (16.1a), and if so, in what form.

We now state without proof two possible conservation theorems for scalar quantities; it may be shown that these are deducible from equation (16.1a) and its complex conjugate equation.

$$\frac{\partial}{\partial t} \underbrace{\psi^* \psi}_{\rho} + \text{div} \underbrace{\frac{1}{i\alpha} (\psi^* \, \text{grad} \, \psi - \psi \, \text{grad} \, \psi^*)}_{\mathbf{j}} = 0 \qquad (16.8)$$

$$\frac{\partial}{\partial t} \underbrace{(\text{grad} \, \psi^* \, \text{grad} \, \psi + \beta \Phi \psi^* \psi)}_{\tilde{\rho}} +$$
$$+ \text{div} \underbrace{[-(\psi^* \, \text{grad} \, \psi + \psi \, \text{grad} \, \psi^*)]}_{\tilde{\mathbf{j}}} = 0 \qquad (16.9)$$

In view of the above theorems, we interpret $\rho$ as charge or mass density, both of which are proportional to each other, since $e/m = 2\beta/\alpha^2$. We may then look upon (16.8) as the equation of continuity for charge or mass, and (16.9) as the corresponding equation for energy. There is still one free factor available in these equations, so that $\rho$ need only be proportional to the charge density. Since this factor is quite arbitrary we can put $\rho_e = e'\psi^*\psi$ for the electric charge density $\rho_e$, where $e'$ represents an arbitrary unit charge. The only reason for choosing $e'$ to be equal to the electron charge $e$ is to provide a simple corpuscular interpretation of the physical quantities occurring in the wave theory. Further, we shall choose the factor in (16.9) in such a manner that the term $\beta\Phi\psi^*\psi$ may be replaced by $\Phi\rho_e$; this latter quantity is the energy density of a charge distribution $\rho_e$ in an electric potential $\Phi(\mathbf{r})$. Hence we have*

| | | |
|---|---|---|
| Charge density $\rho_e = e\rho$ | Electrical current density | $\mathbf{j}_e = e\mathbf{j}$ |
| Mass density $\rho_m = m\rho$ | Mass flow density | $\mathbf{j}_m = m\mathbf{j}$  (16.10) |
| Energy density $u = \dfrac{e}{\beta}\tilde{\rho}$ | Energy flow density | $\mathbf{f} = \dfrac{e}{\beta}\tilde{\mathbf{j}}$ |

It is of interest to deduce the forms of the above expressions in the simplest three-dimensional case of a plane wave $\psi = a\exp i(\mathbf{k}.\mathbf{r} - \omega t)$. This is a solution of the wave equation for constant $\Phi$, if $k^2 + \beta\Phi = a\omega$. In the three-dimensional case the group velocity is $\mathbf{v}_g = \partial\omega/\partial\mathbf{k} = (2/\alpha)\mathbf{k}$. Then

$$\rho_e = e|a|^2 \qquad\qquad \mathbf{j}_e = \rho_e\mathbf{v}_g$$
$$\rho_m = m|a|^2 \qquad\qquad \mathbf{j}_m = \rho_m\mathbf{v}_g \qquad (16.11)$$
$$u = \tfrac{1}{2}\rho_m\mathbf{v}_g^2 + \rho_e\Phi \qquad \mathbf{f} = u\mathbf{v}_g$$

The densities and currents are constant with respect to time and position, and the fact that the currents simply transport the corresponding densities at group velocity $\mathbf{v}_g$ leads to the following statement:

"The plane wave describes a state of constant density $\rho_e$ or $\rho_m$ which moves at a constant velocity $\mathbf{v}_g$."

Clearly, therefore, the first term in the expression for $u$ corresponds to kinetic energy density, and the second to potential energy density.

---

* Note that $j_e$ vanishes when $\psi$ is real.

Furthermore, $\mathbf{j}_m$ is also the wave momentum density* (cf. Exercise 5, p. 108). The corpuscular representation requires that $\rho = |a|^2$ be interpreted as the particle density; the plane wave is a process in which a statistically constant distribution of particles moves with velocity $\mathbf{v}_g$. The momentum per particle is $m\mathbf{v}_g = \hbar\mathbf{k}$, and the kinetic energy, $\frac{1}{2}mv_g^2 = \hbar^2 k^2/2m$.

It is clear from the above considerations that the wave equation may justifiably be written in its original form (16.1); it is important to realize, however, that the equation contains only two essential constants, the third being given by the arbitrary choice $e' = e$.

### (c) The wave packet

In §14a the wave function $\psi(x,t)$ was derived from the function at time $t = 0$ for linear motion in the absence of external forces, for which $\Phi = 0$. The solution for the case of an initially Gaussian distribution was discussed in §13. If we start with the normalized function

$$\psi(x,0) = \frac{1}{(b\sqrt{\pi})^{1/2}} \exp\left(-\frac{x^2}{2b^2} + ik_0 x\right) \qquad (16.12)$$

the necessary integrations can be evaluated by elementary methods and lead to the result (13.14a) for $\psi(x,t)$. The factor $s$ in this expression is replaced below by $\hbar/b\sqrt{2}$.

The resulting densities are

$$\rho(x,0) = |\psi(x,0)|^2 = \frac{1}{b\sqrt{\pi}} \exp -\frac{x^2}{b^2} \qquad (16.13)$$

$$\rho(x,t) = |\psi(x,t)|^2 = \frac{1}{b(t)\sqrt{\pi}} \exp -\frac{(x-v_0 t)^2}{b^2(t)} \qquad (16.14)$$

where $\omega_0$ is the angular frequency associated with $k_0$ ($\alpha\omega_0 = k_0^2$), and $v_0$ is the corresponding group velocity ($v_0 = 2k_0/\alpha$). The "width" or dispersion, $b(t)$, of the density distribution $\rho$ is a function of the time:

$$b(t) = b\left\{1 + \left(\frac{\hbar t}{mb^2}\right)^2\right\}^{1/2} = b\left\{1 + \left(\frac{2t}{\alpha b^2}\right)^2\right\}^{1/2}$$

The above equations represent a charge distribution $\rho(x,t)$, of total charge $e$ and mass $m$, the motion of which is described by the wave equation (16.3). The total momentum is independent of the

---

* There is also a conservation theorem for the momentum density (for constant potential), which we have not stated above.

time: $\int \mathbf{j}_m \, dx = m v_0 = \hbar k_0$; the total energy is also independent of time and includes kinetic energy only: $\int u \, dx = \frac{1}{2} m v_0^2 (1 + 1/2 b^2 k_0^2)$. The solution can of course be multiplied by a factor $a$, in which case the total charge, the mass, the momentum, and the energy are multiplied by $|a|^2$.

If we can neglect the variation of $b(t)$ with time, we obtain the result expressed by (13.14). This is only permissible for small intervals of time, for which $t \ll \frac{1}{2} \alpha b^2$; for large intervals, the dispersion is proportional to the time: $b(t) = 2t/\alpha b$. The sides of the Gauss curve separate at a constant velocity $2/\alpha b = v_0/b k_0$. If $b$ is large compared to the wavelength $\lambda_0 = 2\pi/k_0$, the rate of separation is small in comparison with the mean velocity $v_0$ of the wave packet; if this condition does not apply, the separation rate is the dominant factor. A measure of the dispersion rate of the wave packet is provided by the time $\tau$ at which $b(\tau) = 2b$, from which $\tau = \frac{1}{2} \alpha b^2 \sqrt{3}$. If $b$ is of atomic dimensions $(10^{-8}\,\mathrm{cm})$, $\tau$ is about $10^{-16}\,\mathrm{s}$. This result makes it impossible to describe an electron as a particle and as a wave packet represented by (16.12); if we wish to take account of the corpuscular properties of the electron, we must interpret the wave function in a statistical sense, because, while a probability distribution can become increasingly dispersed, the particle itself certainly cannot.

The behaviour of the wave packet is illustrated in figure 19. It should perhaps be mentioned that the dispersion of the wave packet is not an irreversible process—it is true that (16.12) leads to a dispersion that increases with time, but this is simply the consequence of our choice of initial function. For instance, by choosing suitable phase relations between the individual waves composing the packet, it is possible to make the dispersion decrease initially with time. As a solution of the wave equation, (13.14a) holds for all values of the time; therefore, if we start, say, at time $t = -\tau$, the density distribution is halved after an interval $\tau$ (see figure 19, broken curve).

For a wave packet of form corresponding to (16.12) in three dimensions, the general solution is simply a product of three solutions in the form of (16.13), in which we are still free to choose the components of the width and velocity along each axis.

### (d) *Reflection and transmission at potential barriers* (*wave theory*)

We shall now investigate the effect of the potential on the solution of the wave equation. We restrict ourselves to one dimension, so that $\psi = \psi(x, t)$, $\Phi = \Phi(x)$, and the wave equation assumes the form (16.3).

We shall consider only stationary states, which are defined as those states in which the densities and currents are independent of the time. Although the wave function $\psi(x,t)$ contains a time factor $e^{-i\omega t}$, it is clear that the densities and currents given by (16.9) and (16.10) are

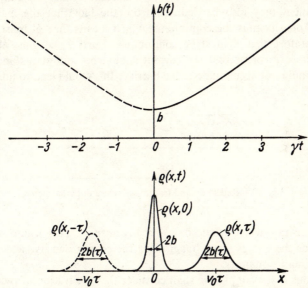

Fig. 19.—Density distribution $\rho(x,t)$ and width $b(t)$ of wave packet, from (16.14)

not functions of the time, since they include only factors of the form $\psi^*\psi$. We have already discussed the special case of the representation of a stationary state by means of a plane wave; the corresponding densities and currents are given by (16.11). If we now put $\psi = \phi(x)e^{-i\omega t}$ in (16.3), we obtain

$$-\phi'' + \beta\Phi\phi = \alpha\omega\phi \qquad (16.15)$$

For constant potential $\Phi$, the general solution* of this equation is

$$\phi(x) = a\,e^{ikx} + b\,e^{-ikx} \qquad (16.16)$$

where $a$ and $b$ are arbitrary complex quantities, and

$$k^2 + \beta\Phi = \alpha\omega \qquad (16.16a)$$

---

* When $\omega$ and $k$ are positive (which we shall always assume to be the case), the first term in (16.16) represents a wave travelling to the right, while the second term represents one travelling to the left.

If the potential $\Phi$ is a continuous function of position, it follows from (16.15) that $\phi''$ must exist; $\phi$, $\phi'$, and $\phi''$ are also continuous functions in this case. This is not so if potential steps are present, so that the potential is only piecewise continuous. At the step, $\phi$ and $\phi'$ are still continuous, while $\phi''$ itself is obviously discontinuous. The continuity requirement may also be deduced from the fact that the mass (or charge) density must be continuous, which means that $\phi'$ must exist, and therefore that $\phi$ must be continuous. Further, $\phi'$ must also be continuous, in order that the current should be continuous and that mass should not accumulate at the barrier; this would lead to infinitely

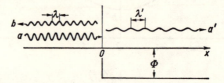

Fig. 20.—Reflection and transmission at a potential step

great densities, which would be physically absurd. This argument also holds for the energy flow, but not for the energy density, which is a discontinuous quantity.

As our first example, we shall consider the penetration of radiation through a potential barrier; the latter may be taken as a potential difference concentrated into a very small interval. We can picture this as follows. Let an initial plane wave $ae^{i(kx-\omega t)}$ be incident on the barrier from the left. At the barrier the wave will be partly reflected, partly transmitted; finally a steady state will be reached. The reason for the present investigation is that any given waveform, such as a wave packet incident from the left, can be created by superposing different stationary solutions with various angular frequencies.

When the system has reached a stationary state, we should expect to find an incident wave $ae^{ikx}$ and a reflected wave $be^{-ikx}$ to the left of the barrier, and a wave $a'e^{ik'x}$ to its right (figure 20). We wish to find the amplitudes $b$ and $a'$ in terms of the amplitude $a$ of the incident wave. A unique solution is obtained if we impose the condition, on physical grounds, that no wave $b'e^{-ik'x}$ reaches the barrier from the right, although a wave travels towards the left from this point. Dropping the time factor $e^{-i\omega t}$, which is common to all the waves, we may now summarize the expressions for the wave functions and the mass flow in the regions to the right and left of the barrier:

Left of the barrier
$x < 0$
$$\begin{cases} \phi = a\,e^{ikx} + b\,e^{-ikx} & k^2 = \alpha\omega \\ j_m = |a|^2\,2k/\alpha - |b|^2\,2k/\alpha \end{cases} \qquad (16.17)$$

Right of the barrier
$x > 0$
$$\begin{cases} \phi = a'\,e^{ik'x} & k'^2 = \alpha\omega - \beta\Phi = k^2 - \beta\Phi \\ j_m = |a'|^2\,2k'/\alpha \end{cases} \quad (16.17a)$$

We see from (16.17) that the mass flow consists of an incident and a reflected component. The reflection coefficient $R$ is defined as the ratio of the reflected to the incident flow density; the transmission coefficient $D$ is the ratio of the transmitted to the incident flow density:*

$$R = \left|\frac{b}{a}\right|^2 \qquad D = \left|\frac{a'}{a}\right|^2 \frac{k'}{k} \qquad (16.18)$$

As a result of the conservation of mass,

$$R + D = 1 \qquad (16.18a)$$

$\phi$ and $\partial\phi/\partial x$ must be continuous at the point $x = 0$, hence

$$a + b = a' \qquad k(a - b) = k'a' \qquad (16.19)$$

and

$$\frac{b}{a} = \frac{1 - k'/k}{1 + k'/k} \qquad R = \left|\frac{1 - k'/k}{1 + k'/k}\right|^2 \qquad (16.20)$$

or

$$R = \left|\frac{1 - n}{1 + n}\right|^2 \qquad (16.20a)$$

where the "refractive index" $n = k'/k$ is introduced, by analogy with optics; $n$ is the ratio of the wavelengths in the two regions separated by the potential barrier.

From (16.17a)

$$n = (1 - \beta\Phi/k^2)^{1/2} = (1 - V/E_{kin})^{1/2} \qquad (16.21)$$

The quantity $\beta\Phi/k^2$ may be replaced by $V/E_{kin}$, where $e\Phi = V$ is the jump in potential energy for an electron and $E_{kin} = \hbar^2 k^2/2m$ is the kinetic energy per electron in the incident wave. Figure 21 illustrates

---

* The quantities $j_m$ and $D$ deduced from (16.17a) and (16.18) are valid only for real values of $k'$. For pure imaginary values of $k'$, $j_m = 0$ ($D = 0$ and $R = 1$).

the variation of the reflection coefficient with the kinetic energy of the electron.  In the region for which the kinetic to barrier potential energy ratio lies between 0 and 1, $n$ and $k'$ are pure imaginaries and total reflection takes place.*  Expressed in terms of the corpuscular interpretation: the kinetic energy is insufficient to enable the electron to surmount the potential barrier $V$.  At greater values of the kinetic energy $R$ decreases gradually to zero.  The full curve in figure 21 shows the reflection coefficient for positive values of $\beta\Phi$ or $V$, the broken

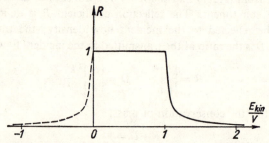

Fig. 21.—Reflection coefficient $R$ at a potential step, shown as a function of $E_{kin}/V$.  The broken line illustrates the variation of $R$ for negative values of $V$.

curve, the coefficient at negative values.  It is evident that, apart from the region of total reflection, we always find values of $R$ and $D$ that are different from zero.

According to classical corpuscular theory there are only two possible alternatives: if $E_{kin} > V$, the electron surmounts the barrier; if $E_{kin} < V$, it recoils.  Classical wave theory, on the other hand, states that a certain fraction of the wave is reflected and that the remainder is transmitted through the barrier†.  In order to describe particles and waves by means of a single model it is necessary once again to have recourse to the statistical interpretation: we may say, for instance, that

---

* When $x > 0$, $j_m \equiv 0$.

† In this connection, a remarkable situation appears to exist when $V$ is negative and $|V| \gg E_{kin}$: this is the case when the radiation is highly accelerated by a large potential difference. According to the corpuscular theory, $R = 0$; according to wave theory, $R \approx 1$ because $n \gg 1$. The almost complete reflection predicted by wave theory appears to be strongly at variance with the classical behaviour of particles, and is not confirmed experimentally when cathode rays are accelerated. The answer is that such a sharp potential step cannot be realized experimentally; the solution under discussion applies only to the case where the wavelength of the radiation is large compared to the dimensions of the potential step. When we reflect that, according to (13.9), the wavelength of this radiation is generally less than $10^{-8}$cm, we can appreciate that such dimensions cannot be realized in practice. For light waves the situation is different; in this case it is possible to arrange for the refractive index to change from one medium to the other within a small fraction of the wavelength.

$R$ is the probability of recoil and $D$ the probability of penetrating the barrier, or that on the average a fraction $R$ of the incident particles is reflected and a fraction $D$ transmitted.

Another important type of potential barrier is illustrated in figure 22. In this case there are four amplitudes, $b$, $a'$, $b'$, $a''$, which must be

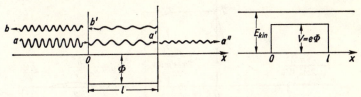

Fig. 22.—Amplitudes at a potential barrier

evaluated in terms of the amplitude $a$ of the incident wave; for this purpose there are two boundary conditions at our disposal at each of the potential steps. The calculations are similar to those which we performed for the case of the potential step, and are omitted. The results are:

$$R = \left| \frac{\frac{1}{2}(n-1/n)\sin(nkl)}{\cos(nkl)-\frac{1}{2}i(n+1/n)\sin(nkl)} \right|^2 \qquad (16.22)$$

where $n$ is again defined by (16.21). If $E_{kin} > V$, $n$ is real, and we obtain

$$R = \frac{\frac{1}{4}(n-1/n)^2 \sin^2(nkl)}{1+\frac{1}{4}(n-1/n)^2 \sin^2(nkl)}$$

$$D = \frac{1}{1+\frac{1}{4}(n-1/n)^2 \sin^2(nkl)} \qquad (16.22a)$$

If $E_{kin} > V$, $n = i\bar{n}$ is a pure imaginary, and

$$R = \frac{\frac{1}{4}(\bar{n}+1/\bar{n})^2 \sinh^2(\bar{n}kl)}{1+\frac{1}{4}(\bar{n}+1/\bar{n})^2 \sinh^2(\bar{n}kl)}$$

$$D = \frac{1}{1+\frac{1}{4}(\bar{n}+1/\bar{n})^2 \sinh^2(\bar{n}kl)} \qquad (16.22b)$$

Figure 23 illustrates the manner in which the reflection coefficient $R$ depends on the ratio $E_{kin}/V$. It is of interest to note that $R$ becomes zero at the points for which $nkl = \pi, 2\pi, 3\pi, \ldots$: this corresponds to the familiar interference phenomena in optics.

In classical physics, equation (16.22$b$) corresponds to the case of total reflection at the barrier; in the present treatment, this equation predicts almost total reflection provided that the width $l$ of the barrier

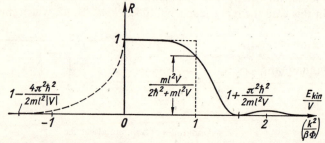

Fig. 23.—Reflection coefficient $R$ at a potential barrier, shown as a function of $E_{kin}/V$. The figure is drawn for the case $\pi^2 h^2/ml^2|V| = 5/4$. The broken line illustrates the variation of $R$ for negative values of $V$.

is sufficiently great (i.e. $\bar{n}kl \gg 1$). For smaller values of $\bar{n}kl$, however, we find that the barrier is appreciably "transparent" to the radiation, although according to classical corpuscular theory it should be insurmountable.

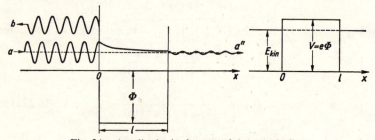

Fig. 24.—Amplitudes in the case of the tunnel effect

This result is easily explained in terms of the wave theory. If we restrict ourselves to the case for which $\bar{n}kl \gg 1$, then

$$D = \left(\frac{4\bar{n}}{\bar{n}^2+1}\right)^2 e^{-2\bar{n}kl} \ll 1$$

This implies that the wave amplitude decreases exponentially within the barrier according to the factor $e^{-\bar{n}kx}$; a small amplitude is still present at the right-hand boundary, however, and this gives rise to an outward-bound wave to the right of the barrier (figure 24).

The quantum theory again requires a statistical interpretation of this result: the transmission coefficient $D$ represents the chance that an incident particle can pass through the barrier. A large number of phenomena are explained by this so-called "tunnel effect", according to which a particle has a chance of penetrating a barrier which, on classical theory, would be insurmountable. If we imagine the particle to be confined within two similar potential barriers then, according to classical mechanics, it cannot leave this region. If we now take into account the tunnel effect, we can estimate the escape probability approximately as follows. The particle strikes the wall of the enclosure at a definite frequency $v$ which depends on its velocity and on the enclosure dimensions. At each impact the chance that the particle will escape is $D$; the probability in a time $dt$ is therefore equal to the number of attempts $v\,dt$ multiplied by $D$. If we now consider a system consisting of $N$ confined particles, the average number escaping during the interval $dt$ is $NvD\,dt$; this is equal to the decrease $dN$ in the number of confined particles, therefore $dN/dt = -NvD$. It follows that the average number of remaining particles at time $t$ is $N(t) = N(0)e^{-vDt}$, where $N(0)$ is the number of particles present initially. This is the law of radioactive disintegration.

More precise quantum-mechanical calculations lead to a similar result, and afford a particularly satisfactory explanation of α-particle emission when the appropriate constants are introduced. For electrons, the tunnel effect is of importance in connection with

1. the ionization probabilities of atoms in an electric field (cf. §39),
2. the emission of electrons from metals in an electric field, in the absence of thermal excitation.

### (e) The "quantum conditions" applicable to the classical wave theory

Bohr's hypothesis (10.1, 10.2), as applied to corpuscular theory, consists essentially of the following two postulates:

1. Stationary orbits exist, possessing energy values $E_n$.
2. The radiated frequencies are $\omega_{ns} = (E_n - E_s)/\hbar$.

It is clear that the condition of stationary non-radiating states is satisfied by the wave theory. As we have indicated in previous sections (in the case of the linear oscillator, for instance), every state $\psi(\mathbf{r}, t)$ may be developed in terms of a set of physically significant stationary solutions:

$$\psi(\mathbf{r}, t) = \sum_n a_n \phi_n(\mathbf{r}) e^{-i\omega_n t} \qquad (16.23)$$

In order to determine the emission of electric waves for a state represented by (16.23) let us investigate the motion of the centre of charge $er_s$, as in the case of radiation from a dipole:

$$ex_s(t) = \int \rho_e(\mathbf{r}) x \, d\mathbf{r} = e \int x \, |\psi(\mathbf{r}, t)|^2 \, d\mathbf{r}$$

or, from (16.23)

$$ex_s(t) = e \sum_{n,m} a_n^* a_m x_{nm} e^{i(\omega_n - \omega_m)t} \qquad (16.24)$$

where the "matrix element" $x_{nm}$ is defined as

$$x_{nm} = \int \phi_n^* x \phi_m \, d\mathbf{r} \qquad (16.24a)$$

The radiation therefore contains the frequencies $\omega_{nm} = \omega_n - \omega_m$, provided $x_{nm} \neq 0$, in accordance with Bohr's theory.

The existence of the elementary charge $e$ must clearly be allowed for in wave theory by the requirement* that $\int \rho \, d\mathbf{r}$ should be an integral number. In the case of the single electron this requirement implies that the wave function $\psi$ be normalized. If the functions $\phi_n$ are normalized, it may be shown that the total energy $E_n = \int u \, d\mathbf{r}$ in the $\phi_n$ state is equal to $\hbar \omega_n$; the second Bohr postulate is therefore satisfied. In the single electron case, therefore, the quantum condition for wave theory is simply that the total charge be normalized to $e$. We then have an exact method for the treatment of single-electron problems.

There is one important matter, however, that we have not so far taken into consideration. The potential $\Phi$ appears in the wave equation; since the electron wave is electrically charged we must allow for the potential of its charge distribution. $\Phi$ then consists of a potential due to the external fields and a potential $\Phi_s$ due to the charge distribution $\rho_e$:

$$\Phi_s(\mathbf{r}) = \int \frac{\rho_e(\mathbf{r}')}{|\mathbf{r} - \mathbf{r}'|} \, d\mathbf{r}' \qquad (16.25)$$

If the wave is very tenuous $\Phi_s$ may be neglected, as we have tacitly assumed up to now. However, if we were to calculate again the energy

---

* If we had allowed the normalization charge $e'$ to remain arbitrary, we should now have to postulate that $\int \rho_e d\mathbf{r} = e' \int |\psi|^2 d\mathbf{r}$ be an integral multiple of the elementary charge $e$.

levels of the hydrogen atom, using $\Phi_s$, we should obtain a completely incorrect result,* since in this case $\Phi_s$ is by no means negligibly small. Clearly, we must postulate that $\Phi_s$ should vanish. This is the second quantum condition for wave theory—there can be no interaction of the electron with itself. Here again the statistical interpretation of $\psi$ comes to the rescue, for the postulate would be quite incomprehensible if $\rho = |\psi|^2$ were a real density. The quantum postulates in the particle and wave theories are complementary, and the task of the quantum theory consists in finding a theoretical pattern that will take into account both the corpuscular and the wave concepts.

## Exercises

### 1. *Planck's radiation formula interpolation*

The spectral distribution of radiation from an enclosure may be expressed in terms of the mean thermal energy $\varepsilon(v,T)$ of an oscillator whose frequency is $v$. Before the discovery of Planck's radiation formula it was necessary to distinguish between two spectral regions, each characterized by a different mathematical relation: the Rayleigh-Jeans region for frequencies well below the maximum of the spectral distribution, and the Wien region lying well above this maximum.

$$\varepsilon(v,T) = \begin{cases} kT, & \text{Rayleigh-Jeans, "small" } v \\ \text{const.} \, v \exp -\dfrac{\alpha v}{T}, & \text{Wien, "large" } v \end{cases}$$

From thermodynamics, the mean square deviation $\Delta$ of the energy $E$ of any system at temperature $T$ is $\Delta = kT^2 \partial \bar{E} / \partial T$.

Express the mean square deviation in the Rayleigh-Jeans region, $\Delta_{RJ}$, and in the Wien region, $\Delta_W$, in terms of $\varepsilon$ and $v$. Verify that the sum $\Delta = \Delta_{RJ} + \Delta_W$ approximately represents the deviation in each region. This expression may be used as an interpolation formula for the mean square deviation throughout the whole spectral region. If this formula is assumed to be strictly correct, and equated to $\Delta = kT^2 \partial \varepsilon / \partial T$, a differential equation is obtained for $\varepsilon$, the solution of which is Planck's formula.

### 2. *Derivation of the hydrogen terms from the correspondence principle*

According to the correspondence principle, $dE/dn = \hbar \omega(E)$ for large quantum numbers $n$; $\omega(E)$ is the frequency of the classical motion. This is a differential equation for the function $E(n)$ in which $n$ is to be taken as a continuously variable quantity. The possible energy values according to quantum theory are those for which $n$ is a whole number. Calculate $\omega(E)$ and hence $E(n)$ for a linear oscillator, and for the hydrogen atom orbits.

### 3. *First approximation to the energy of an anharmonic linear oscillator*

Let the potential energy be $V(x) = \frac{1}{2}m\omega^2 x^2 + \lambda x^4$, where $\lambda$ is "small". Calculate the energy values $E_n$ from (10.4) to the first degree of approximation, by expanding in powers of $\lambda$. Neglect terms in $\lambda$ of the second and higher degree.

---

* In the case of problems involving many electrons it is often possible to replace the strict quantum-mechanical treatment by wave theory, taking into account $\Phi_s$. This corresponds somewhat to Hartree's "self-consistent field" treatment.

### 4. Vibrations of solid bodies at the absolute zero of temperature

When a linear oscillator is in its lowest state, at the absolute zero of temperature, the mean square deviation of its amplitude is $\overline{x^2} = \hbar/2M\omega$, where $M$ is the mass of the oscillator and $\omega$ its angular frequency; this follows because the average values of the potential and kinetic energies are equal. This also applies in the case of atomic vibrations in a solid body. When calculating the vibration of an atom we may assume that the neighbouring atoms remain fixed, as in Einstein's model of a solid body. An atom in a cubical crystal lattice vibrates like a harmonic oscillator with a small amplitude in a definite direction. If the lattice constant (the distance between neighbouring atoms) is $a$, $\omega$ may be expressed approximately in terms of the velocity of sound $c$, thus $\omega \approx c/a$. This follows because the time $a/c$ required for the transfer of an impulse from one atom to the next must be roughly equal to the period of an oscillation $1/\omega$. Evaluate the ratio $\overline{x^2}/a^2$, the root of which is a measure of the linear displacement of the atoms from their equilibrium positions at absolute zero. For aluminium, $a \approx 2 \times 10^{-8}$ cm, $c = 5 \times 10^3$ m/s, $M = 6 \times 10^{-23}$ g.

### 5. Rate of change of total momentum in wave theory

If $\mathbf{j}_m$ is correctly interpreted by (16.10), the rate of change of total momentum must be equal to the total force exerted by the electric field $\mathbf{E} = -\operatorname{grad}\Phi$ on the charge distribution $\rho_e$:

$$\frac{d}{dt}\int \mathbf{j}_m \, d\mathbf{r} = \int \rho_e \mathbf{E} \, d\mathbf{r}$$

Deduce the above equation from the equation of motion (16.1) on the assumption that $\psi(\mathbf{r}, t)$ vanishes at infinity.

### 6. Motion of the centre of gravity of a wave packet

The centre of gravity $x_s$ of a wave packet is defined by

$$x_s = \int x\rho_m \, dx \Big/ \int \rho_m \, dx$$

Prove the following relation for motion in one dimension in the absence of external forces:

$$\dot{x}_s = \int j_m \, dx \Big/ \int \rho_m \, dx = v_0$$

(Total mass × velocity of centre of gravity = total momentum.) Show also that $v_0$ is the mean value of all the group velocities in the wave packet:

$$v_0 = \int \frac{d\omega}{dk} \, |g(k)|^2 \, dk \Big/ \int |g(k)|^2 \, dk$$

*Hint*: $\psi(x, t) = \int g(k, t) e^{ikx} \, dk/(2\pi)^{1/2}$, where $g(k, t) = g(k) e^{-i\omega(k)t}$. For any two functions $\phi_1(x) = \int g_1(k) e^{ikx} \, dk/(2\pi)^{1/2}$ and $\phi_2(x) = \int g_2(k) e^{ikx} \, dk/(2\pi)^{1/2}$:

$$\int \phi_1^*(x)\phi_2(x) \, dx = \int g_1^*(k) g_2(k) \, dk$$

### 7. Energy according to wave theory, and the Hamiltonian operator

Using integration by parts, prove that $\int u \, d\mathbf{r} = \int \psi^* \mathscr{H} \psi \, d\mathbf{r}$. $u$ is the wave-theory energy density, $\mathscr{H}$ is the Hamiltonian operator of particle theory.

# CHAPTER BII

## The general basis of quantum mechanics

### §17. Vectors and operators in Hilbert space

The separate lines of development of the quantum theory which were described in the last chapter lead to a common pattern, which we shall now examine.

To begin with, it is clear that both Heisenberg's and Schrödinger's quantum-mechanical formulations are formally similar to each other. For instance, Heisenberg's postulates (12.13) may also be expressed as follows: matrices must first be found, designated by $p_{\mu\nu}$ and $x_{\mu\nu}$, for which

$$(px - xp)_{\mu\nu} = \frac{\hbar}{i}\delta_{\mu\nu}$$

These matrices are then used to form the matrix

$$\mathscr{H}_{\mu\nu} = \left\{\frac{p^2}{2m} + V(x)\right\}_{\mu\nu}$$

which is then diagonalized by means of a "rotation". It is known from matrix algebra that diagonalization is equivalent to determining the eigenvectors $\phi(\nu)$ of the matrix $\mathscr{H}$:

$$\sum_{\nu} \mathscr{H}_{\mu\nu}\phi_n(\nu) = E_n\phi_n(\mu) \tag{17.1}$$

On the other hand, if we take the classical Hamiltonian function and replace the momentum $p$ by the operator $-i\hbar\,\partial/\partial x$, we obtain the Schrödinger equation (14.12). Then for every differentiable $\phi(x)$:

$$(px - xp)\phi(x) = \frac{\hbar}{i}\phi(x)$$

The eigenvalues $E_n$ are then obtained from the solution of the boundary-value equation

$$\mathscr{H}\phi_n(x) = E_n\phi_n(x) \quad \text{for} \quad \int \phi_n^*(x)\phi_n(x)\,dx \text{ finite} \tag{17.2}$$

109

It will now be shown that equations (17.1) and (17.2) are two equivalent forms of a single eigenvalue equation

$$\mathcal{H}\phi = E\phi$$

This means that the above equation may be represented in two different "coordinate systems" which can be transformed into each other by means of a "rotation" in so-called "Hilbert space".

Before this equivalence can be demonstrated certain preliminary mathematical concepts and propositions must be presented.

Hilbert space consists of a set of "elements" or "vectors", $\phi$, $\chi, \psi, \ldots$, possessing the following properties:

(1). If $a$ and $b$ are any two complex numbers, $a\phi + b\psi$ is also an element in Hilbert space.

(2). Associated with any two elements $\phi$ and $\psi$ there is a (generally complex) number $(\phi, \psi)$, which is termed the scalar or inner product of $\phi$ and $\psi$. The following relation is always true:

$$(\phi, \psi) = (\psi, \phi)^* \tag{17.3}$$

Further, if $a$ is an arbitrary number, then

$$(\phi, a\psi) = a(\phi, \psi) \text{ and hence } (a\phi, \psi) = a^*(\phi, \psi) \tag{17.4}$$

In particular, if $(\phi, \psi) = 0$, $\phi$ and $\psi$ are said to be *orthogonal* to each other.

(3). For any three elements $\phi$, $\chi$, $\psi$,

$$(\phi + \psi, \chi) = (\phi, \chi) + (\psi, \chi)$$

(4). From (17.3), we see that $(\phi, \phi)$ is always a real number. We shall also stipulate that $(\phi, \phi) \geqq 0$. Elements for which $(\phi, \phi)$ would be infinite are not admissible to Hilbert space.

The quantity $(\phi, \phi)$ is called the norm of $\phi$. If $(\phi, \phi) = 1$, the vector $\phi$ is said to be normalized. If $(\phi, \phi) = 0$, $\phi$ is said to be a null element of Hilbert space; $\phi = 0$.

If $\phi$ and $\psi$ are given and are not null elements, then $\phi$ may be uniquely resolved into an element $a\psi$, proportional to $\psi$, and an element $\chi$ orthogonal to $\psi$, by means of the identity

$$\phi = \underbrace{\psi \frac{(\psi, \phi)}{(\psi, \psi)}}_{a} + \underbrace{\left\{ \phi - \psi \frac{(\psi, \phi)}{(\psi, \psi)} \right\}}_{\chi} \tag{17.5}$$

Obviously, $(\psi, \chi) = 0$. If we now form the norm on both sides of (17.5), bearing in mind (17.4) and the fact that $(\psi, \chi) = 0$, we get

$$(\phi, \phi) = \frac{|(\phi, \psi)|^2}{(\psi, \psi)} + (\chi, \chi)$$

This leads to Schwarz's inequality*

$$(\phi, \phi)(\psi, \psi) \geqq |(\phi, \psi)|^2 \tag{17.6}$$

in which the equality sign only applies if $\chi = 0$, when $\phi = a\psi$.

The elements $\phi_1, \ldots, \phi_r$ are said to be linearly dependent if any of the $\phi$ can be expressed as a linear combination of the others, i.e. if coefficients $c_1, \ldots, c_r$ exist that are not all zero, such that

$$c_1 \phi_1 + \ldots + c_r \phi_r = 0 \tag{17.7}$$

If this is not the case the elements are said to be linearly independent.

A system of elements $\alpha_1, \ldots, \alpha_n, \ldots$ in Hilbert space is said to be *normalized* if $(\alpha_n, \alpha_n) = 1$ for each $\alpha_n$. It is *orthogonal*, if $(\alpha_n, \alpha_m) = 0$ for all $n \neq m$. It is *complete* if each Hilbert space element $\phi$ may be expressed as a linear combination of the $\alpha_n$:

$$\phi = \sum_n a_n \alpha_n \tag{17.8}$$

If the $\alpha_n$ fulfil all three conditions, the system is said to be *complete and orthogonal*. In future we shall employ only basic systems of this type. Then

$$(\alpha_n, \alpha_m) = \delta_{nm}$$

and from (17.8)

$$a_n = (\alpha_n, \phi) \equiv \phi(n)$$

We therefore have the following identity for every $\phi$:

$$\phi = \sum_n (\alpha_n, \phi)\alpha_n = \sum_n \phi(n)\alpha_n \tag{17.8a}$$

We may say that the $\alpha_n$ span the whole Hilbert space, or that every element of Hilbert space may be represented by a linear combination of the $\alpha_n$. The elements $\alpha_n$ are therefore said to form a basis $\alpha$ in Hilbert

---

* Equation (17.6) ensures the validity of postulate (I): if the norms of $\phi$ and $\psi$ exist, so does the norm of any linear combination of these two elements.

space, and the components $\phi(n)$ of the vector $\phi$ in this coordinate system constitute the representation of $\phi$ referred to the basis $\alpha$.

If
$$\phi = \sum_n \phi(n)\alpha_n \qquad \psi = \sum_n \psi(n)\alpha_n$$

the inner product is defined as

$$(\phi, \psi) = \sum_{n,m} \phi^*(n)\psi(m)(\alpha_n, \alpha_m) = \sum_n \phi^*(n)\psi(n) \qquad (17.9)$$

where
$$(\alpha_n, \alpha_m) \equiv \alpha_m(n) = \delta_{mn}$$

The numbers $m$ are often continuous variables; in this case the summations in the above formulae are replaced by integrals. If $\alpha_a$ are the orthogonal basis vectors, then

$$\phi = \int (\alpha_a, \phi)\alpha_a \, da = \int \phi(a)\alpha_a \, da$$

If we now form the scalar product with $\alpha_{a'}$, we get

$$\phi(a') = \int \phi(a)(\alpha_{a'}, \alpha_a) \, da \qquad (17.10)$$

The function $\delta_{aa'} = (\alpha_{a'}, \alpha_a)$ is called the Dirac delta function; it corresponds in the continuous case to the Kronecker symbol $\delta_{nm}$ that occurs in the discrete case. This function has the property* that $\int \phi(a)\delta_{aa'} \, da = \phi(a')$ for all $\phi(a)$. It may be pictured as a function vanishing when $a \neq a'$, but tending to infinity in such a manner at $a = a'$ that $\int \delta_{aa'} \, da = 1$.

The Dirac function is obviously not a function in the ordinary sense, just as the vectors $\alpha_a$ do not really belong to Hilbert space because $(\alpha_a, \alpha_a) = \delta_{aa} = \infty$; however, we shall include it in our treatment for the sake of completeness of presentation. We shall encounter the function in (20.11) (cf. also Exercises 1 and 2, p. 137).

The eigenvectors $\phi_n(v)$ are a typical example of the representation of discrete vectors in Hilbert space. Examples of continuous quantities are provided by the wave functions $\phi(x)$ or their Fourier transforms $g(p)$, the scalar products of which are $(\phi, \psi) = \int \phi^*(x)\psi(x) \, dx$ and $(g_1, g_2) = \int g_1^*(p)g_2(p) \, dp$. In these cases Hilbert

---

* This property may be taken as the definition of the $\delta$-function. Then clearly
$$\int \phi(a)\delta_{a+d, a'+d} \, da = \int \phi(a-d)\delta_{a, a'+d} \, da = \phi(a')$$
i.e. $\delta_{a+d, a'+d} = \delta_{a, a'}$. Then $\delta_{a, a'}$ only depends on the difference $a - a'$; for this reason, the function is also written as $\delta(a - a')$.

space consists of all functions $\phi(x)$ the squares of whose absolute magnitudes are integrable.

The beginner should familiarize himself with these concepts in three-dimensional space, bearing in mind that Hilbert space possesses an infinite number of dimensions. In vector calculus it is usual to represent a complete and orthonormal system by means of the unit vectors $\mathbf{i}$, $\mathbf{j}$, $\mathbf{k}$, parallel to the axes $x$, $y$, $z$. Any vector $\mathbf{a}$ can then be represented as

$$\mathbf{a} = a_x\mathbf{i} + a_y\mathbf{j} + a_z\mathbf{k} \tag{17.10a}$$

The norm of the vector $\mathbf{a}$ is equal to the square of its magnitude $a$. The scalar product of two vectors $\mathbf{a}$, $\mathbf{b}$ is

$$(\mathbf{a}, \mathbf{b}) \equiv \mathbf{a}.\mathbf{b} = ab \cos \phi$$

where $a$ and $b$ are the magnitudes of the vectors and $\phi$ is the angle they make with each other. The inequality (17.6) becomes simply

$$a^2b^2 \geqq (ab \cos \phi)^2, \quad \text{i.e. } \cos^2 \phi \leqq 1$$

The analogue of equation (17.9) is

$$(\mathbf{a}, \mathbf{b}) = a_xb_x + a_yb_y + a_zb_z$$

If $\mathbf{e}$ is a normalized vector, then $(\mathbf{a}, \mathbf{e}) = a \cos \phi$ is the projection of $\mathbf{a}$ in the direction of $\mathbf{e}$.

The importance of the Hilbert space vectors lies in the fact that they can be used to represent the physical states of quantum-mechanical systems. This means that, in principle, it is possible to calculate the result of any measurement on a system in a given state if we know the vector $\phi$ associated with that state. It remains to be seen how this calculation may be accomplished.

We saw in the last chapter that operators, such as matrices and differential operators, are associated in quantum theory with physically observable quantities such as position, momentum or energy.

In Hilbert space, an operator $A$ represents a rule for transforming any vector $\phi$ into another vector $\psi$: $A\phi = \psi$. $A$ is said to be linear if

$$A(a\phi + b\psi) = aA\phi + bA\psi \tag{17.11}$$

for any two elements $\phi$, $\psi$, and complex numbers $a$ and $b$. In what follows we shall only be concerned with linear operators.

The sum of two operators is defined as follows:

$$(A + B)\phi = A\phi + B\phi \tag{17.12}$$

and the product as

$$(AB)\phi = A(B\phi) \tag{17.13}$$

in which $B$ is first applied to $\phi$, then $A$ to $B\phi$. The adjoint operator to $A, A^+$, is defined by the relation

$$(\phi, A\psi) = (A^+\phi, \psi) \tag{17.14}$$

for all $\phi$ and $\psi$. It follows immediately that

$$(A^+)^+ = A, \quad (aA)^+ = a^*A^+, \quad (AB)^+ = B^+A^+ \tag{17.15}$$

An operator is said to be Hermitian or self-adjoint if

$$A^+ = A \tag{17.16}$$

The unitary operator is defined by

$$U^+U = UU^+ = 1 \tag{17.17}$$

We shall see later that Hermitian operators always correspond to physical quantities. Unitary operators are important for such purposes as the representation of the variation of a state $\phi$ with respect to time.

We shall now describe the matrix representation of the above operator relations. The representation in the basis $\{\alpha_1, \ldots, \alpha_n, \ldots\}$ of the equation

$$A\phi = \psi$$

is obtained by forming the inner product with $\alpha_m$:

$$(\alpha_m, A\phi) = \sum_n (\alpha_m, A\alpha_n)\phi(n) = \sum_n A_{mn}\phi(n) = \psi(m) \tag{17.18}$$

The operation of $A$ on $\phi$ is represented in matrix form by multiplying $\phi(n)$ by the matrix $A_{mn} = (\alpha_m, A\alpha_n)$. The matrix $A_{mn}$ is said to represent the operator $A$. The matrix representation of $A + B$ is

$$(A+B)_{mn} = A_{mn} + B_{mn} \tag{17.19}$$

The product of two operators is represented by the product of two matrices:

$$(AB)_{mn} = \sum_l (\alpha_m, AB_{ln}\alpha_l) = \sum_l A_{ml} B_{ln} \tag{17.20}$$

Finally, the matrix of the adjoint operator is the conjugate complex of the original transposed matrix:

$$(A^+)_{mn} = (\alpha_m, A^+\alpha_n) = (A\alpha_m, \alpha_n) = A_{nm}^* \tag{17.21}$$

For a Hermitian operator, therefore,

$$A_{mn} = A_{nm}^*$$

The analogue of matrix representation in the continuous case is provided by integral operators: for instance, if $\phi$ is represented by wave functions $\phi(x)$, $A\phi = \psi$ is represented by

$$\int A(x, x')\phi(x')\,dx' = \psi(x)$$

Differential operators are of greater importance, however; among these, we encountered the momentum operator $-i\hbar\,\partial/\partial x$ in §14. Multiplication by $x$ (or by a function* $V(x)$) represents another simple and important operation.

We shall now derive a relation between two different matrix representations of the same Hilbert space vector $\phi$: if these are designated by $\phi(n)$ and $\tilde{\phi}(v)$ referred to bases $\alpha$ and $\beta$ respectively, we have

$$\phi = \sum_n \phi(n)\alpha_n = \sum_v \tilde{\phi}(v)\beta_v \tag{17.22}$$

Forming the inner products with $\alpha_n$ and $\beta_v$:

$$\phi(n) = \sum_v (\alpha_n, \beta_v)\tilde{\phi}(v) = \sum_v U_{nv}\tilde{\phi}(v) \tag{17.23a}$$

$$\tilde{\phi}(v) = \sum_n (\beta_v, \alpha_n)\phi(n) = \sum_n U_{nv}^*\phi(n) \tag{17.23b}$$

The transformation of one matrix representation into another is therefore effected by means of a matrix $U_{nv}$, which is unitary because

$$\sum_v U_{nv} U_{mv}^* = \sum_v (\alpha_n, \beta_v)(\beta_v, \alpha_m)$$

$$= \sum_v \alpha_n^*(v)\alpha_m(v) = (\alpha_n, \alpha_m) = \delta_{nm} \tag{17.24a}$$

and

$$\sum_m U_{m\mu}^* U_{mv} = \delta_{\mu v} \tag{17.24b}$$

The transformation of matrix elements is most simply effected by first expressing the elements $\beta_v$ in terms of the $\alpha_n$:

$$\beta_v = \sum_n (\alpha_n, \beta_v)\alpha_n = \sum_n \beta_v(n)\alpha_n = \sum_n U_{nv}\alpha_n$$

---

* When the Dirac $\delta$-function is introduced, multiplication and differentiation become equivalent to integration:

$$V(x)\phi(x) = \int V(x')\delta_{xx'}\phi(x')\,dx' \quad \text{and} \quad \frac{\partial\phi(x)}{\partial x} = \int \delta_{xx'}\frac{\partial\phi(x')}{\partial x'}\,dx' = -\int \frac{\partial\delta_{xx'}}{\partial x'}\phi(x')\,dx'$$

If this expression is now introduced into $\tilde{A}_{\mu\nu} = (\beta_\mu, A\beta_\nu)$ we obtain:

$$\tilde{A}_{\mu\nu} = \sum_{mn} U^*_{m\mu} A_{mn} U_{n\nu}$$

similarly
$$A_{mn} = \sum_{\mu\nu} U_{m\mu} \tilde{A}_{\mu\nu} U^*_{n\nu} \qquad (17.25)$$

Since the elements $\alpha_n$ and $\beta_\nu$ each constitute an orthonormal basis in Hilbert space we may say that they transform into each other by means of a rotation, since the relations between them are similar to those governing the transformation of coordinate systems in three-dimensional space.

We shall still retain the concept of a rotation even when one of the two orthogonal systems, say $\beta_\nu$, is continuous. It is true that in this case we cannot say strictly that each vector $\alpha_n$ belonging to the discrete basis is transformed by a unitary operation in Hilbert space into a corresponding vector $\beta_\nu$ in the continuous basis. The matrix $U_{n\nu}$ is now replaced by a generalized unitary "matrix" $U_{n\nu}$, possessing one discrete and one continuously variable index. The unitary relations then become

$$\sum_m U^*_{m\nu} U_{m\mu} = \delta(\mu - \nu) \qquad \int U_{n\mu} U^*_{m\mu}\, d\mu = \delta_{mn} \qquad (17.26)$$

The above relations are similar to those expressed by (17.24) and suggest, therefore, that we may also refer to this class of transformation as a "rotation".

This obliteration of the difference between continuous and discrete manifolds is of somewhat doubtful mathematical validity; it is on a par with our introduction of the Dirac $\delta$-function and is justified in so far as it is impossible to discriminate physically between a continuous basis $\beta_\nu$ and a discrete set $\beta_{\nu_n}$ if the $\nu_n$ lie close enough to each other. (A relevant example is provided by the frequent cases in which a Fourier integral is replaced by a summation.)

## §18. Average value and standard deviation

The concepts so far introduced provide an extremely concise means of formulating the quantum theory, and one that is independent of the particular form of representation selected. We shall now describe in some detail the relations between vectors, operators, states, and physical quantities.

Let us measure a quantity $A$ which is an attribute of a large number of identical systems, all of which are in the state $\phi$. If the mean of all the measurements is formed, this is the expectation value of $A$, denoted by $\bar{A}$. This value may be calculated from the associated Hilbert space vector $\phi$ and the operator $A$ according to the formula

$$\bar{A} = \frac{(\phi, A\phi)}{(\phi, \phi)} \tag{18.1}$$

Thus a relation has been established for the first time between the abstract quantities $A, \phi$ and a directly measurable quantity $\bar{A}$. Equation (18.1) obviously remains unchanged if $\phi$ is replaced by $a\phi$, where $a$ is any complex number; $\phi$ and $a\phi$ represent the same state. It is often useful, but not essential, to assume $\phi$ to be normalized. Then the expectation value is simply

$$\bar{A} = (\phi, A\phi)$$

The significance of equation (18.1) may be made clearer if we can show that it provides a statistical interpretation of the normalized wave function $\phi(x)$ and its Fourier transform $g(p)$, which were discussed in §13. If for instance we take $\phi$ as the wave function $\phi(x)$ and $A$ as the operator "multiply by $x$", we obtain

$$\bar{x} = \int x \, | \phi(x) |^2 \, dx \tag{18.2}$$

This is of course just what we should put for the average value of $x$ if we interpreted $| \phi(x) |^2 \, dx$ to be the probability of finding the particle between $x$ and $x + dx$. Similarly for the momentum:

$$\bar{p} = \int p \, | g(p) |^2 \, dp \tag{18.3}$$

If we wish to determine $\bar{p}$, we might proceed by first finding the Fourier transform of $\phi(x)$ and using this to evaluate the integral (18.3). Equation (18.1) relieves us of this necessity, however: we require only the representation of the operator $p$ in the coordinate system in which the state vector is represented as $\phi(x)$, and can then put

$$\bar{p} = \int \phi^*(x) p \phi(x) \, dx \tag{18.4}$$

From §14, we should expect that the momentum $p$ could be replaced in (18.4) by the operator $-i\hbar\partial/\partial x$. We shall leave it to the reader to show that, since $\phi(x) = \int g(p)e^{ipx/\hbar}dp/\sqrt{(2\pi\hbar)}$, (18.3) is identical with (18.4).

Since the operators $A$ correspond to physical quantities, they must obviously be of such a form that $\bar{A}$ is always real; if $(\phi,\phi) = 1$, this means that

$$\bar{A} = \bar{A}^* = (\phi, A\phi)^* = (A\phi, \phi) = (\phi, A^+\phi)$$

Since this must be true for all states $\phi, A$ must be equal to $A^+$; in other words, Hermitian operators must always correspond to the physical quantities.

If $A$ is an observable quantity in the state $\phi$, where $(\phi,\phi) = 1$, the dispersion $\Delta A$ is defined as the mean square deviation

$$(\Delta A)^2 = \overline{(A-\bar{A})^2} = \overline{A^2} - \bar{A}^2 \tag{18.5}$$

where $\overline{A^2} = (\phi, A^2\phi)$. If we now define a vector $\chi$ by

$$A\phi - \bar{A}\phi = \chi \tag{18.6}$$

it follows immediately from the fact that $A$ is Hermitian that

$$\overline{A^2} - \bar{A}^2 = (\chi, \chi) \tag{18.7}$$

Therefore $(\Delta A)^2$ vanishes only when $\chi = 0$, and the dispersion of $A$ in the state $\phi$ is only zero if

$$A\phi = \bar{A}\phi = a\phi$$

where $a$ is real. In this case $\phi$ is said to be an eigenvector of $A$, and the corresponding eigenvalue is $a$. If the system is therefore in an eigenstate of $A$ for which the corresponding eigenvalue is $a$, each measurement of $A$ will always yield the value $a$.

The measured values of other quantities relating to the system in state $\phi$ will in general be dispersed about their mean values.

As we have already seen, it is impossible to find states in which position and momentum each have a definite value at the same time. We shall now prove the generalized uncertainty relation $\Delta A \Delta B \geq \frac{1}{2}\hbar$ for any canonically conjugate operators $A$ and $B$:

Two Hermitian operators $A$ and $B$ are connected by the relation

$$AB - BA = \hbar/i \tag{18.8}$$

with which we are already familiar for the case when $A = p$ and $B = x$. We shall show that

$$(A\phi, A\phi)(B\phi, B\phi) \geqq |(A\phi, B\phi)|^2 \geqq \tfrac{1}{4}\hbar^2 \qquad (18.9)$$

for all $\phi$, where $(\phi, \phi) = 1$. The first inequality follows directly from (17.6) when $\phi$ is replaced by $A\phi$ and $\psi$ by $B\phi$; the equality only holds when

$$A\phi = \lambda B\phi \qquad (18.10)$$

The second inequality in (18.9) follows from (18.8). If we put

$$(A\phi, B\phi) = (\phi, AB\phi) = a + ib$$

where $a$ and $b$ are real, then

$$(\phi, BA\phi) = a - ib$$

Subtracting: $2b = -\hbar$, in view of (18.8), hence $|(\phi, AB\phi)|^2 = a^2 + b^2 = a^2 + \tfrac{1}{4}\hbar^2$. The second equality sign in (18.9) holds only if $a = 0$, when $(A\phi, B\phi)$ is a pure imaginary quantity. In order to satisfy both the possible equalities in (18.9), therefore, $\lambda$ in (18.10) must be a pure imaginary, and

$$A\phi = i\gamma B\phi, \text{ where } \gamma \text{ is real} \qquad (18.11)$$

If $A$ and $B$ are replaced by their deviations from the expectation values, $A - \bar{A}$ and $B - \bar{B}$, the commutation relations (18.8) are preserved, and since $(A\phi, A\phi) = (\phi, A^2\phi)$, etc., it follows from (18.9) that

$$\overline{(A - \bar{A})^2(B - \bar{B})^2} \geqq \tfrac{1}{4}\hbar^2, \text{ i.e. } \Delta A \Delta B \geqq \tfrac{1}{2}\hbar \qquad (18.12)$$

where the equality holds if

$$(A - \bar{A})\phi = i\gamma(B - \bar{B})\phi \qquad (18.13)$$

In particular, if we now put $A = p = -i\hbar \partial/\partial x$ and $B = x$, (18.13) gives the differential equation for the function of $\phi$ with the minimum uncertainty product,

$$\frac{\hbar}{i}\frac{\partial \phi}{\partial x} - \bar{p}\phi = i\gamma(x - \bar{x})\phi$$

the solution of which is

$$\phi(x) = \exp\left(\frac{i}{\hbar}\bar{p}x\right)\exp\left(-\frac{\gamma}{2\hbar}(x - \bar{x})^2\right) \qquad (18.14)$$

This is the equation of the wave packet* that was discussed in §§13 and 16. The constant $\gamma$ shows how the indeterminacy is distributed between $x$ and $p$:

$$\Delta x = \sqrt{\frac{\hbar}{2\gamma}} \qquad \Delta p = \sqrt{\left(\tfrac{1}{2}\gamma\hbar\right)}$$

---

* In the previous treatment, the uncertainty was found to be a minimum at time $t = 0$; this corresponds to the equality condition in (18.12).

## §19. Eigenvalues and eigenvectors

### (a) *Determination of the eigenvalues and eigenvectors*

If a vector $\phi$ remains unchanged apart from a numerical factor $a$ when acted on by an operator $A$, i.e.

$$A\phi = a\phi \tag{19.1}$$

$\phi$ is said to be an eigenvector of $A$, and $a$ is the associated eigenvalue.

An equation for the possible eigenvalues of $A$ is obtained if (19.1) is expressed in any orthogonal system $\alpha_1, \alpha_2, \ldots$:

$$\sum_n A_{mn}\phi(n) = a\phi(m) \quad \text{for} \quad m = 1, 2, 3, \ldots \tag{19.2}$$

If the Hilbert space consists of a finite number of dimensions only (say $N$),* (19.2) contains $N$ linear homogeneous equations for the $N$ coefficients $\phi(1), \ldots, \phi(N)$. A solution of these equations exists if the determinant

$$\begin{vmatrix} A_{11} - a & A_{12} & A_{13} & \cdots \\ A_{21} & A_{22} - a & A_{23} & \cdots \\ \vdots & \vdots & \vdots & \end{vmatrix} \tag{19.3}$$

vanishes. This is an algebraic equation of the $N$th degree in $a$, known as the secular equation. The $N$ roots of this equation are therefore the eigenvalues of $a$; their existence follows from the fundamental theorem of algebra. Hilbert space often has an infinite number of dimensions, however; in this case the investigation of the existence of the eigenvalues is a much more difficult problem, for which we shall not give a general treatment.

In the infinite case, the eigenvalues might perhaps be determined as successive approximations by breaking off the infinite determinant (19.3) at a given finite value of $N$, equating to zero, and solving; $N$ would then be increased and the procedure repeated in the hope that it would lead to convergent values.

Generally, however, more elegant procedures exist for the solution of the eigenvalue problems that are of importance in physics. In many cases the latter may be expressed as boundary conditions associated with differential equations (e.g. Schrödinger's eigenvalue equation in

---

* Examples of such space occur in §26 (perturbation theory), §36 (Zeeman effect), and §39 (Stark effect).

§14); alternatively they may be solved independently of any particular basis as a result of special commutation properties of the operator $A$ (e.g. in the treatment of the linear oscillator given in §15, in which the eigenvalues of $\mathscr{H}$ were determined from the commutation relations of $b^+$ and $b$; further examples are provided in §24 (angular momentum) and §53 (quantization of black-body radiation)).

We shall now consider the two important special cases, in which $A$ is unitary or Hermitian.

*The eigenvalues u of a unitary operator U have absolute value unity.*

The scalar product of the equation $U\psi = u\psi$ with itself gives $(U\psi, U\psi) = u^*u(\psi, \psi)$; hence $(U\psi, U\psi) = (\psi, U^+U\psi) = (\psi, \psi)$. If $(\psi, \psi) \neq 0$, then $u^*u = 1$, as stated.

*The eigenfunctions belonging to different eigenvalues of a unitary operator are orthogonal to each other.*

If the scalar product of the two equations $U\psi_1 = u_1\psi_1$ and $U\psi_2 = u_2\psi_2$ is formed, it follows that

$$(\psi_1, \psi_2) = u_1^* u_2 (\psi_1, \psi_2) \tag{19.4}$$

But $u_1^* = 1/u_1$, therefore either $u_1 = u_2$ or $(\psi_1, \psi_2) = 0$.

*The eigenvalues of a Hermitian operator A are real.*

It follows from $A\psi = a\psi$ that $a = (\psi, A\psi)$. Hence from (17.14) and (17.16) $a^* = (A\psi, \psi) = (\psi, A\psi)$, and therefore $a = a^*$.

*The eigenfunctions belonging to different eigenvalues of a Hermitian operator are orthogonal to each other.*

It follows from $A\phi_1 = a_1\phi_1$ and $A\phi_2 = a_2\phi_2$ that

$$(A\phi_1, \phi_2) - (\phi_1, A\phi_2) = (a_1 - a_2)(\phi_1, \phi_2) \tag{19.5}$$

When $A$ is Hermitian the left-hand side is equal to zero; therefore either $a_1 = a_2$ or $(\phi_1, \phi_2) = 0$.

In the case of a continuous spectrum of eigenvalues

$$A\phi_a = a\phi_a$$

It follows from equation (19.5) that $(a - a')(\phi_a, \phi_{a'}) = 0$, i.e. $(\phi_a, \phi_{a'}) = 0$ for $a \neq a'$. The behaviour at $a = a'$ depends on the normalization, which can be so chosen that

$$(\phi_a, \phi_{a'}) = \delta(a - a')$$

It also occurs very frequently that the eigenvalue spectrum is partly discrete and partly continuous. It is left to the reader to generalize the corresponding formulae.

## (b) Degeneracy

An eigenvalue is said to be "*r*-fold degenerate" if there are *r* linearly independent eigenvectors corresponding to it. In the case of two-fold degeneracy, for instance,

$$A\phi_1 = a\phi_1 \quad \text{and} \quad A\phi_2 = a\phi_2$$

In this case every linear combination $\alpha\phi_1 + \beta\phi_2$ is an eigenfunction corresponding to the same eigenvalue $a$; in other words, all elements of the subspace spanned by $\phi_1$ and $\phi_2$ are eigenfunctions of $A$ corresponding to $a$. This subspace may also be spanned by two orthogonal elements $\chi_1$ and $\chi_2$, say

$$\chi_1 = \phi_1, \quad \chi_2 = \phi_2 - (\phi_1, \phi_2)\phi_1 \tag{19.6}$$

where $\phi_1$ and $\phi_2$ are assumed to be normalized. Then $(\chi_1, \chi_2) = 0$.

The "orthogonalization" process assumed in (19.6) may be performed for any number $r$ of linearly independent elements $\phi_1, \phi_2, \ldots, \phi_r$. We may therefore state the following result:

*In the case of an r-fold degenerate eigenvalue, the r eigenvectors may always be chosen in such a manner that they are mutually orthogonal.*

If $\phi_1, \ldots, \phi_r$ are the original eigenvectors, there always exist linear combinations

$$\chi_k = \sum_{l=1}^{r} c_{kl} \phi_l \qquad k = 1, 2, \ldots, r \tag{19.7}$$

such that the $\chi_k$ are orthogonal and normalized. When the eigenvectors of a Hermitian operator $A$ are chosen in this manner they all form an orthogonal system.

In addition, this orthogonal system is complete—a fact that we state without proof.

## (c) Commuting operators

Let $AB = BA$. Then since $A\psi = a\psi$, it follows that $AB\psi = BA\psi = aB\psi$. We may express this result as follows:

If $B$ and $A$ commute, then both $\psi$ and $B\psi$ are eigenfunctions of $A$ corresponding to the same eigenvalue $a$.

If the eigenvalue $a$ is non-degenerate, $B\psi$ must be identical with $\psi$ apart from a numerical factor; therefore $\psi$ is also an eigenfunction of $B$. On the other hand, if the eigenvalue is $r$-fold degenerate, with orthonormal eigenvectors $\phi_1, \ldots, \phi_r$, we can only infer that all the $B\phi_j$ belong to the subspace spanned by the $\phi_1, \ldots, \phi_r$:

$$B\phi_j = \sum_{k=1}^{r} B_{kj} \phi_k \quad \text{where} \quad B_{kj} = (\phi_k, B\phi_j) \qquad (19.8)$$

In this case, however, we can find a linear combination

$$\chi = \sum_{k=1}^{r} d_k \phi_k \qquad (19.9)$$

such that $\chi$ is also an eigenfunction of $B$, i.e.

$$B\chi = b\chi$$

It follows that

$$\sum_{k=1}^{r} d_k B\phi_k = b \sum_{j=1}^{r} d_j \phi_j$$

Forming the scalar product with $\phi_j$ gives the result

$$\sum_{k=1}^{r} B_{jk} d_k = b \, d_j \qquad (19.10)$$

The above expression represents $r$ linear homogeneous equations for the coefficients $d_k$ introduced in (19.9); a solution only exists if the determinant

$$\begin{vmatrix} B_{11} - b & B_{12} & \ldots \\ B_{21} & B_{22} - b & \ldots \\ \vdots & \vdots & \end{vmatrix} \qquad (19.11)$$

vanishes. To each root $b^{(s)}$ of this equation of the $r$th degree there corresponds a solution $d^{(s)}_1, \ldots, d^{(s)}_r$ of the equations (19.10), and a linear combination

$$\chi_s = \sum_{k=1}^{r} d_k^{(s)} \phi_k \qquad s = 1, 2, \ldots, r \qquad (19.12)$$

which is an eigenfunction both of $A$ and of $B$, for which the eigenvalues are $a$ and $b^{(s)}$ respectively. This result may be expressed as follows:

If $A$ and $B$ commute, there always exists an orthonormal system of eigenfunctions that is common to both operators.

With regard to the roots $b^{(s)}$ of (19.11), there are two cases to be distinguished:

1. The roots are all different. Then, when $a$ and $b$ are given, the eigenfunction $\chi_s$ is uniquely determined. The degeneracy is said to be completely removed.

2. The roots are not all different. Then there is certainly another operator $C$ that commutes with $A$ and $B$, and there is an orthogonal system, the elements of which are eigenfunctions of $A$, $B$, and $C$, with associated eigenvalues $a$, $b^{(s)}$, $c^{(n)}$. The degeneracy is considered to be removed if the eigenfunction is uniquely determined by the three eigenvalues.

This method of establishing a unique orthogonal system, by means of commuting Hermitian or unitary operators, is of great importance in connection with the description of atomic spectra.

### (d) *Functions of Hermitian and unitary operators*

A function $f(A)$ of an operator $A$ is defined as follows: firstly, $A^2, A^3, \ldots$ are interpreted as repeated applications of $A$. Hence $f(A)$ is defined, provided $f(x)$ is developed as a power series $\sum_n c_n x^n$. Then

$$f(A) = \sum_n c_n A^n$$

A general definition proceeds from the orthogonal system associated with $A$,

$$A\phi_n = a_n \phi_n$$

where the $a_n$ are ordinary numbers. The operator $f(A)$ can then be defined by the requirement that

$$f(A)\phi_n = f(a_n)\phi_n$$

This defines $f(A)$ for all elements of the basis of the $\phi_n$, and therefore for any linear combination $\psi = \sum_n c_n \phi_n$.

We should also note that, if $A$ is Hermitian and possesses eigenvalues $a_n$, then $e^{iA} = U$ is unitary with eigenvalues $e^{ia_n}$. We shall require this result in connection with the integration of the Schrödinger equation in §22.

## §20. The correspondence between matrix and wave mechanics

It was shown in the last section that a Hilbert space basis may be formed from the eigenfunctions of a Hermitian operator; we shall now illustrate this procedure in connection with some important physical examples.

In the first instance we shall consider the problem in one dimension: a particle with coordinate $x$. In quantum theory, this coordinate corresponds to an operator which we shall designate by $X$. The eigenstates of $X$, denoted by $\alpha_x$, satisfy the equation

$$X\alpha_x = x\alpha_x \qquad (20.1)$$

We shall now show that, in consequence of the commutation relation

$$PX - XP = \frac{\hbar}{i} \qquad (20.2)$$

the eigenvalues $x$ and $p$ corresponding to the operators $X$ and $P$ extend over all real numbers. If (20.2) is multiplied by $-P/\hbar^2$ from the right and from the left, and the two products are added, the result is

$$\left(\frac{i}{\hbar}P\right)^2 X - X\left(\frac{i}{\hbar}P\right)^2 = 2\frac{i}{\hbar}P = \frac{\hbar}{i}\frac{d}{dP}\left(\frac{i}{\hbar}P\right)^2$$

If this procedure is continued, we obtain

$$\left(\frac{i}{\hbar}P\right)^n X - X\left(\frac{i}{\hbar}P\right)^n = \frac{\hbar}{i}\frac{d}{dP}\left(\frac{i}{\hbar}P\right)^n$$

If we now multiply by $(-x')^n/n!$ and sum over all $n$, the following result is obtained:

$$\exp\left(-\frac{i}{\hbar}Px'\right)X - X\exp-\frac{i}{\hbar}Px' = -x'\exp-\frac{i}{\hbar}Px' \quad (20.3)$$

If this equation is applied to $\alpha_x$, then from (20.1):

$$X\left\{\exp\left(-\frac{i}{\hbar}Px'\right)\alpha_x\right\} = (x+x')\exp\left(-\frac{i}{\hbar}Px'\right)\alpha_x \quad (20.4)$$

$\exp[-(i/\hbar)Px']\alpha_x$ is therefore an eigenfunction of $X$, with the eigenvalue $x+x'$. Since $x'$ can be any real number, our statement is proved for $x$; the proof for $p$ is identical.

An eigenfunction $\alpha_x$ is derived from $\alpha_0$ by putting

$$\alpha_x = f(x) \exp\left(-\frac{i}{\hbar} Px\right)\alpha_0$$

where $f(x)$ is an arbitrary normalization factor for which $f(0) = 1$; this factor may be chosen in such a manner that $(\alpha_x, \alpha_{x'}) = \delta(x - x')$. Since $f(x)$ is a pure number it commutes with $P$; hence $f(x) = \exp i\gamma(x)$, where $\gamma(x)$ is an initially arbitrary real function of $x$ for which $\gamma(0) = 0$.

Any vector $\phi$ may be represented by

$$\phi = \int \phi(x)\alpha_x \, dx \tag{20.5}$$

$\phi(x)$ is termed the representation of $\phi$ in the basis spanned by the eigenfunctions of $X$, or simply, the $x$-representation of $\phi$. The $x$-representation of $X$ is obtained by applying the operator $X$ to (20.5):

$$X\phi = \int \phi(x)X\alpha_x \, dx = \int x\phi(x)\alpha_x \, dx \tag{20.6}$$

In the $x$-representation, therefore, the application of $X$ to $\phi$ simply means the multiplication of $\phi(x)$ by $x$. In order to obtain the simplest possible representation of $P$, we make the stipulation that the arbitrary function $\gamma(x)$ shall vanish identically. Then, since $P$ commutes with the number $x$:

$$\alpha_x = \exp\left(-\frac{i}{\hbar} Px\right)\alpha_0 \quad \text{therefore} \quad \frac{\hbar}{i}\frac{\partial \alpha_x}{\partial x} = -P\alpha_x$$

Hence

$$P\phi = \int \phi(x)P\alpha_x \, dx = -\int \phi(x)\frac{\hbar}{i}\frac{\partial \alpha_x}{\partial x} \, dx = \int \frac{\hbar}{i}\frac{\partial \phi(x)}{\partial x} \alpha_x \, dx$$

when $\phi(x)$ vanishes at infinity. The $x$-representation of $P$ is therefore the same as the Schrödinger form of the momentum operator

$$P\phi(x) = \frac{\hbar}{i}\frac{\partial}{\partial x} \phi(x) \tag{20.7}$$

Thus in principle we know the effect of all operators of the form $f(P, X)$ in their $x$-representation. In particular, the representation of the eigenvalue equation $\mathscr{H}(P, X)\psi_n = E_n\psi_n$ is found to be the Schrödinger equation

$$\mathscr{H}\left(\frac{\hbar}{i}\frac{\partial}{\partial x}, x\right)\psi_n(x) = E_n\psi_n(x) \qquad (20.8)$$

The eigenvectors $\psi_n$ of $\mathscr{H}$ may of course be taken as a basis instead of the $\alpha_x$. The representation of an arbitrary vector $\phi$ then becomes

$$\phi = \sum_n \phi(n)\psi_n$$

(assuming a discrete spectrum for the sake of simplicity). This is called the $E$-representation. Just as the $x$-representation of $X$ implies the multiplication of $\phi(x)$ by $x$ when $X$ is applied to $\phi$, so the $E$-representation of $\mathscr{H}$ means the multiplication of $\phi(n)$ by $E_n$. The representations of the operators $X$ and $P$ now become matrices

$$X_{mn} = (\psi_m, X\psi_n), \qquad P_{mn} = (\psi_m, P\psi_n) \qquad (20.9)$$

which satisfy both (20.2) and

$$\sum_n \mathscr{H}_{mn}(P, X)\phi(n) = E_m\phi(m) \quad \text{for all} \quad \phi(n)$$

from which it follows that

$$\mathscr{H}_{mn} = E_m\delta_{mn} \qquad (20.10)$$

This is precisely Heisenberg's quantum-mechanical formulation (cf. §12).

Since the elements $\phi(x)$ and $\phi(n)$ represent the same vector $\phi$ in two different orthogonal systems $\alpha_x$ and $\psi_n$, each set may be converted into the other by means of the transformation formulae (17.23) or (17.26):

$$\phi(x) = \sum_n U_{nx}^*\phi(n) = \sum_n (\alpha_x, \psi_n)\phi(n) = \sum_n \psi_n(x)\phi(n) \qquad (20.11)$$

$$\phi(n) = \int U_{nx}\phi(x)\,dx = \int (\psi_n, \alpha_x)\phi(x)\,dx = \int \psi_n^*(x)\phi(x)\,dx$$

The correspondence between matrix and wave mechanics is now apparent: the Heisenberg and Schrödinger formulations correspond respectively to the energy and position representations of vectors in Hilbert space. The Schrödinger energy eigenfunctions $\psi_n(x)$ constitute a unitary matrix that affords a means of transforming from one representation to another. In particular, they serve to derive the matrix elements $X_{mn} = x_{mn}$ which are of importance in radiation theory:

$$x_{mn} = (\psi_m, X\psi_n) = \int \psi_m^*(x) x \psi_n(x)\, dx \qquad (20.12)$$

The momentum representation is also important: its relation to the $x$-representation is given in the accompanying table, in which the verification of the results is left to the reader (cf. also Exercise 3, p. 138). In the table are included the $x$- and $p$-representations of the operators $X$ and $P$, an arbitrary vector $\phi$, and the eigenfunctions $\phi_{x'}$ and $\psi_{p'}$ of $X$ and $P$.

|  | $x$-representation | $p$-representation |
|---|---|---|
| $X$ | $x$ | $i\hbar\, \partial/\partial p$ |
| $P$ | $-i\hbar\, \partial/\partial x$ | $p$ |
| $\phi$ | $\phi(x)$ | $g(p)$ |
| $\phi_{x'}$ | $\delta(x-x')$ | $\exp(-ipx'/\hbar)/\sqrt{(2\pi\hbar)}$ |
| $\psi_{p'}$ | $\exp(ip'x/\hbar)/\sqrt{(2\pi\hbar)}$ | $\delta(p-p')$ |

It should be noted that the expressions $\delta(x-x')$ in the $x$-representation and $\exp(-ipx'/\hbar)/\sqrt{(2\pi\hbar)}$ in the $p$-representation, which are analytically quite different, represent the same physical situation, namely, a particle situated at $x'$; on the other hand, the analytically similar expressions $\exp(-ipx'/\hbar)/\sqrt{(2\pi\hbar)}$ and $\exp(ip'x/\hbar)/\sqrt{(2\pi\hbar)}$ denote physically different states. Physical quantities are denoted by abstract elements in Hilbert space, which can be represented by different real or complex numbers according to the choice of coordinate system.

From the above table, we may derive a possible representation of the $\delta$-function:

$$(\phi_x, \phi_{x'}) = \frac{1}{2\pi\hbar} \int_{-\infty}^{\infty} \exp \frac{i}{\hbar} p(x-x')\, dp = \delta(x-x')$$

Let us now integrate, not from $-\infty$ to $\infty$, but from $-A\hbar$ to $A\hbar$, after which $A$ is allowed to increase without limit:

$$\delta(x-x') = \lim_{A \to \infty} \frac{1}{\pi} \frac{\sin A(x-x')}{(x-x')} = \lim_{A \to \infty} s_A(x-x') \qquad (20.13)$$

In order to verify this, let us take an arbitrary function $f(x)$ and form the integral

$$\int_{-\infty}^{+\infty} f(x) s_A(x) dx = \frac{1}{\pi} \int_{-\infty}^{+\infty} f(x) \frac{\sin Ax}{x} dx$$

Substituting $x = z/A$, we obtain

$$\int_{-\infty}^{+\infty} f(x) s_A(x) dx = \frac{1}{\pi} \int_{-\infty}^{+\infty} f\left(\frac{z}{A}\right) \frac{\sin z}{z} dz$$

If $A$ is now allowed to tend to infinity, $z/A$ tends to 0. Then since $\int_{-\infty}^{+\infty} \frac{\sin z}{z} = \pi$

$$\lim_{A \to \infty} \int f(x) s_A(x) dx = f(0) \qquad (20.14)$$

The above result is all that it is intended to express by defining the $\delta$-function as

$$\lim_{A \to \infty} \frac{1}{2\pi} \int_{-A}^{+A} e^{ikx} dk = \delta(x)$$

since the only significant property of this "function" is that

$$\int_{-\infty}^{+\infty} f(x) \delta(x) dx = f(0) \quad \text{for any } f(x)$$

The derivative of the $\delta$-function can be derived in a similar manner as

$$\delta'(x) = \lim_{A \to \infty} s'_A(x)$$

hence

$$\int f(x) \delta'(x) dx = -f'(0)$$

Other representations and properties of the $\delta$-function will be found in Exercises 1 and 2, p. 137.

*Note*: For the sake of completeness, we must mention a notation due to P. A. M. Dirac, which is described in his book *The Principles of Quantum Mechanics*, and which is frequently employed in the literature on the subject. In this notation vectors are denoted by the brackets $|\rangle$ or $\langle|$, depending upon whether they occur on the right or the left side of the scalar product. The symbol $\langle|$ is called a "bra" vector, and $|\rangle$, a "ket" vector. The suffixes required to designate the vectors $\phi$ are written inside the brackets. The scalar product $(\phi_m, \phi_n)$ is written $\langle m|n\rangle$, and $A\phi_m$ is represented as $A|m\rangle$.

Dirac's notation possesses no advantages over that which has been used so far in this book. Its usefulness first appears in connection with the completeness relation for an orthonormal system $|m\rangle$, which it expresses very concisely. The projection operator $|m\rangle\langle m|$ is introduced for this purpose: the condition for completeness is then denoted by $\sum_m |m\rangle\langle m| = 1$.

If this equation is multiplied by $|\rangle$, we obtain $|\rangle = \sum_m |m\rangle\langle m|\rangle$, which is the representation of $|\rangle$ in terms of the $\langle m|\rangle$. Similarly, an operator $A$ may be represented in terms of its eigenvalues $a_m$ and eigenvectors $|m\rangle$: $A = \sum_m a_m |m\rangle\langle m|$.

## §21. Probability and the quantum theory

We saw in §18 that if the state $\phi$ of a system is known, the average value and dispersion of any set of observations can be calculated. We now wish to ascertain the probability of any given measurement. When an observable $A$ is measured, the system passes from its original state to an eigenstate of $A$; therefore the only possible results of such measurements are the eigenvalues of $A$. We may therefore pose the following question: given a normalized state $\phi$, what is the probability $w(a_n)$ of finding the eigenvalue $a_n$ of $A$ when the observable $A$ is measured, or in the case of a continuous spectrum, what is the probability $w(a)\,da$ that the measured value will lie between $a$ and $a + da$?

To find the answer, we express the equation $\bar{A} = (\phi, A\phi)$ in the $A$-representation:

$$\bar{A} = (\phi, A\phi) = \sum_n a_n \,|\phi(n)|^2 \tag{21.1}$$

where

$$A\psi_n = a_n \psi_n \qquad \phi(n) = (\psi_n, \phi)$$

Similarly, for a continuous spectrum:

$$\bar{A} = \int a\,|\phi(a)|^2\,da \tag{21.1a}$$

where

$$A\psi_a = a\psi_a \qquad \phi(a) = (\psi_a, \phi)$$

We obtain in the same way

$$\overline{A^2} = \sum_n a_n^2\,|\phi(n)|^2 \quad \text{and} \quad \overline{A^2} = \int a^2\,|\phi(a)|^2\,da \tag{21.2}$$

and for an arbitrary function of $A$

$$\overline{f(A)} = \sum_n f(a_n)\,|\phi(n)|^2 \quad \text{and} \quad \overline{f(A)} = \int f(a)\,|\phi(a)|^2\,da$$

Therefore, if $\phi$ is resolved into components $\phi(n)$ or $\phi(a)$ along the eigenvectors $\psi_n$ or $\psi_a$ of $A$, the required probabilities can be calculated from the formula

$$w(a_n) = |\phi(n)|^2 \quad \text{or} \quad w(a)\,da = |\phi(a)|^2\,da \tag{21.3}$$

If a measurement of $A$ yields the value $a_n$ or $a$, then after the measurement the system is in an eigenstate $\phi$; this means that a further measurement of $A$ would yield the same result. If $B$ is an operator that commutes with $A$, and $a_n$ or $a$ are not degenerate, then $\phi$ is also an eigenvector of $B$, and the result of a subsequent measurement of the observable $B$ is also uniquely determined. When degeneracy is present, it is only possible to say that $\phi$ lies somewhere in the Hilbert subspace spanned by the eigenvectors of $a_n$ or $a$. After the measurement of $B$, $\phi$ is also an eigenfunction of the operator $B$, and if it is no longer degenerate with regard to this operator, then it is uniquely determined. If this is not the case, it is necessary to seek another operator that commutes with $A$ and $B$. Thus in simple cases it is possible in principle to determine a state uniquely by means of measurements on a complete set of operators.

A much more difficult problem is inherent in the determination of the original state of a system. It is obviously insufficient to determine by many measurements the probability distribution $w(x_1, \ldots)dx_1 \ldots dx_f$ of the eigenvalues of a complete system of operators $X_1, \ldots, X_f$, because this does not yield the phase $S(x_1, \ldots)$ of the wave function $\phi = \sqrt{(w)}e^{iS}$, which is necessary to determine the mean values $\bar{p}$, $\overline{p^2}$, etc. It is not even enough to know the momentum distribution $\tilde{w}(p_1, \ldots)$ in addition to the distribution $w(x_1, \ldots)$. This is apparent in the example of the wave packet discussed in §16: in this case $\phi(x, t)$ and $\phi(x, -t)$ have the same distributions $w(x)$ and $\tilde{w}(p)$, but correspond to different states, as is clear from their representation in terms of time ($\phi(x, -t)$ converges, $\phi(x, t)$ diverges). In the case of a particle, $\phi(x, t)$ can be determined if $w(x, t)$ and $w(x, t+dt)$ are measured, since, from the continuity equation

$$\dot{w} = -\frac{\partial}{\partial x}\frac{\hbar}{2mi}\left(\phi^*\frac{\partial\phi}{\partial x} - \phi\frac{\partial\phi^*}{\partial x}\right) = -\frac{\hbar}{m}\frac{\partial}{\partial x}\left(w\frac{\partial S}{\partial x}\right)$$

from which the phase $S(x)$ is obtained by integrating twice.

If $B$ and $A$ do not commute, then in general an indeterminate result is obtained when $B$ is measured after $A$. An eigenstate of $A$ has several non-zero components in the basis spanned by the eigenvectors of $B$; the squares of their magnitudes give the probability of finding the corresponding eigenvalue of $B$, by (21.1 or 21.1$a$). In the case of canonically conjugate operators, an accurate measurement of one quantity necessarily involves a corresponding indeterminacy in the other, in accordance with the uncertainty relation.

In most cases, of course, it is impossible in practice to determine a state by a series of measurements, particularly for systems with many degrees of freedom, such as macroscopic bodies. For instance, if we assume that we have only measured the energy of a macroscopic system, and that we knew that it lay between $E$ and $E+\Delta E$, then we only know in the first instance that the state $\phi$ of the system lies in the Hilbert subspace spanned by the eigenstates $\phi_\nu$ of the energy shell $\Delta E$. From statistical mechanics we then know that there is an equal probability $w_\nu$ of finding any given state $\phi_\nu$ out of all the states in the energy shell.

A system is called an *ensemble* if it consists of a set of states $\phi_\nu$, each of which occurs with a corresponding probability $w_\nu$; this is in contrast to the so-called pure case, in which all the $w_\nu$ vanish except for one. If we assume that the $\phi_\nu$ constitute a complete system, then when $A$ is measured, the probability of finding the eigenvalue $a_n$ is the product of the two probabilities, first, of finding $\phi_\nu$, and secondly, given $\phi_\nu$, of finding $a_n$, summed over all $\nu$:

$$w(a_n) = \sum_\nu w_\nu \,|\,\phi_\nu(n)\,|^2 \tag{21.4}$$

i.e.
$$\bar{A} = \sum_\nu w_\nu (\phi_\nu, A\phi_\nu) \tag{21.5}$$

There is an essential difference between the various states $\phi_\nu$ of an ensemble and the same states in a superposition $\phi = \sum_\nu c_\nu \phi_\nu$; in the latter case, the probability $w_\nu$ of finding the state $\phi_\nu$ in $\phi$ is $|\,c_\nu\,|^2$. But from (21.3) the probability of finding the measured value $a_n$ when the system is in the state $\phi$ is

$$w(a_n) = |\,\phi(n)\,|^2 = \sum_{\nu\mu} c_\nu c_\mu^* \,\phi_\nu(n)\phi_\mu^*(n)$$

$$= \sum_\nu w_\nu \,|\,\phi_\nu(n)\,|^2 + \sum_{\nu \neq \mu} c_\nu c_\mu^* \,\phi_\nu(n)\phi_\mu^*(n)$$

Comparison with (21.4) indicates that in the pure case the $\phi_\nu$ can interfere with each other, whereas the interference terms are absent in the case of the ensemble.

## §22. The equation of motion in quantum mechanics

When we come to consider the problem of the time variation of a system, a new essential physical element appears in the concept of Hilbert space that we have used hitherto. If we refer back to Chapter BI, we see two solutions given which appear to be completely different, even though the Hamiltonian operator $\mathscr{H}$ appears in each of them.

In §12, the principles of Heisenberg's matrix mechanics were developed from the correspondence principle, and the following relation was obtained for the time variation of every quantum mechanical quantity that is represented by a matrix:

$$\left(\frac{dA}{dt}\right)_{nm} = \frac{i}{\hbar}(\mathcal{H}A - A\mathcal{H})_{nm} \tag{22.1}$$

This may be expressed as an operator equation:

$$\dot{A} = \frac{i}{\hbar}(\mathcal{H}A - A\mathcal{H}) \tag{22.2}$$

In §14, on the other hand, the Schrödinger wave equation for a state function $\psi$ was established; in virtue of (14.10) and (14.11), it may be expressed in the form

$$-\frac{\hbar}{i}\dot{\psi} = \mathcal{H}\psi \tag{22.3}$$

where the operators corresponding to physical quantities such as $x$, $p_x$, etc., are introduced as time-independent quantities such as "multiplication by $x$", $-i\hbar\,\partial/\partial x$, etc.

It is important to realize that these two representations, known respectively as the Heisenberg and the Schrödinger picture (not to be confused with the $E$- and $x$-representations), lead to the same physical results in spite of their different viewpoints and their different initial equations.

The reason for this is that neither the vectors nor the operators $A$ have any direct physical significance; this is a property of the expectation values $(\psi, A\psi)$ alone. The latter remain unchanged, however, if an arbitrary (and possibly time-dependent) unitary transformation $\psi \to \psi' = U\psi$, $A \to A' = UAU^+$ is applied to $\psi$ and $A$:

$$(\psi', A'\psi') = (U\psi, UAU^+U\psi) = (\psi, A\psi) \tag{22.4}$$

For this reason, only the time dependence of the expectation values is physically defined; that of the vectors and operators themselves is not.*

In the Schrödinger picture, the operators $A$ are assumed to be independent of time, and the vectors are expressed in terms of time by

---

* For the sake of brevity we shall assume here that the Hamiltonian operator $\mathcal{H}$ is independent of time. In the general case, when $\mathcal{H}$ depends explicitly on time, we have the situation described on p. 156.

integrating (22.3), in which $\mathscr{H}$ is taken to be constant with respect to time:

$$\psi(t) = \exp\left(-\frac{i}{\hbar}\mathscr{H}t\right)\psi(0) \tag{22.5}$$

For time-independent expectation values, therefore,

$$\{\psi(t), A\psi(t)\} = \left\{\psi(0), \exp\left(\frac{i}{\hbar}\mathscr{H}t\right)A\exp\left(-\frac{i}{\hbar}\mathscr{H}t\right)\psi(0)\right\} \tag{22.6}$$

It is easy to see that the operators

$$A(t) = \exp\left(\frac{i}{\hbar}\mathscr{H}t\right)A\exp\left(-\frac{i}{\hbar}\mathscr{H}t\right) \tag{22.7}$$

satisfy the "equations of motion" (22.2). Hence we have the following relationships between the Heisenberg and the Schrödinger pictures; in the former we are dealing with time-independent vectors $\psi = \psi(0)$, and the dependence upon time of the operators $A(t)$ is given by (22.7). The transition from the Schrödinger to the Heisenberg picture is effected by the unitary transformation

$$U = \exp\frac{i}{\hbar}\mathscr{H}t$$

The difference between the two pictures disappears in the case of vectors which may be represented by time-dependent bases. For instance, if we consider the identity

$$\psi(x, t) = \{\alpha_x, \psi(t)\} = \left\{\alpha_x, \exp\left(-\frac{i}{\hbar}\mathscr{H}t\right)\psi(0)\right\}$$

$$= \left\{\exp\left(\frac{i}{\hbar}\mathscr{H}t\right)\alpha_x, \psi(0)\right\}$$

$$= \{U\alpha_x, \psi(0)\} = \{\alpha_x(t), \psi(0)\} \tag{22.8}$$

we may say that

(1) $\psi(x,t)$ is the representation of the Schrödinger vector $\psi(t)$ in the system constituted by the eigenfunctions $\alpha_x$ of $X$ (for which $X\alpha_x = x\alpha_x$), or

(2) $\psi(x,t)$ is the representation of the Heisenberg vector $\psi(0)$ in the system of eigenfunctions $\alpha_x(t)$ of $X(t)$ (for which $X(t)\alpha_x(t) = x\alpha_x(t)X(t) = UXU^+$).

We shall now investigate the variation with respect to time of the momentum and position expectation values. In the Heisenberg picture, from (22.2):

$$\dot{x} = \frac{i}{\hbar}(\mathscr{H}x - x\mathscr{H}) \qquad \dot{p} = \frac{i}{\hbar}(\mathscr{H}p - p\mathscr{H}) \qquad (22.9)$$

Since $px - xp = \hbar/i$, it follows that for every function $F(x,p)$ of the operators $x$ and $p$

$$Fx - xF = \frac{\hbar}{i}\frac{\partial F}{\partial p} \qquad Fp - pF = -\frac{\hbar}{i}\frac{\partial F}{\partial x} \qquad (22.10)$$

To prove the above relations, let us expand $F(x,p)$ in the first equation (22.10) in powers of $p$, noting that

$$p^n x - x p^n = \frac{\hbar}{i} n p^{n-1} = \frac{\hbar}{i}\frac{\partial p^n}{\partial p} \qquad (22.11)$$

for all $n$, as may be confirmed by proceeding step by step from the familiar commutation relation (for which $n = 1$). The second equation of (22.10) is proved in a similar manner from the commutation of $x^n$ with $p$. Hence when $F \equiv \mathscr{H}$, (22.9) becomes the quantum-mechanical analogue of the Hamiltonian equations of motion

$$\dot{x} = \frac{\partial \mathscr{H}}{\partial p} \qquad \dot{p} = -\frac{\partial \mathscr{H}}{\partial x} \qquad (22.12)$$

The equations (22.9) only assume physical significance in connection with the determination of the expectation values. In particular, if we put $p^2/2m + V(x)$ for $\mathscr{H}$, then for any state $\psi$ in the Heisenberg picture

$$\frac{d}{dt}(\psi, x\psi) = \frac{1}{m}(\psi, p\psi) \qquad \frac{d}{dt}(\psi, p\psi) = -\left(\psi, \frac{dV}{dx}\psi\right) \qquad (22.13)$$

from which we obtain the so-called Ehrenfest theorem

$$m\frac{d^2}{dt^2}(\psi, x\psi) = -\left(\psi, \frac{dV}{dx}\psi\right), \quad \text{i.e.} \quad m\ddot{\bar{x}} = -\overline{\left(\frac{\partial V}{\partial x}\right)} \qquad (22.14)$$

This simple expression may be immediately extended to the case of motion in three dimensions.

The centre of gravity of a wave packet therefore moves in accordance with Newton's equation, if the acting force $-\operatorname{grad} V$ is averaged over the wave packet. In particular, if the field of force is uniform (as in the deflection experiments for the measurement of $e/m$, for instance), the centre of gravity of the charge distribution moves exactly in accordance with the laws of classical physics.

The above statements only have obvious meaning if the wave packet remains fairly concentrated. For instance, if we consider a wave packet at a potential step, part of it is reflected and part transmitted. The packet divides into two separate portions travelling in different directions. The centre of gravity still moves according to (22.14), but it lies somewhere between the two parts and can no longer suggest that the charge is concentrated in its neighbourhood.

When magnetic fields are present the Hamiltonian operator no longer assumes the simple form $p^2/2m + V(x)$; it also contains product terms of the type $\mathbf{pA}(\mathbf{r}, t)$, where $\mathbf{A}$ is the vector potential. Although calculation is not quite as simple as in (22.13) it can still be performed, and gives the corresponding result to (22.14) (cf. Exercise 6, p. 138)

$$m\ddot{\mathbf{r}} = \bar{\mathbf{F}}$$

where $\mathbf{F}$ is the operator of the Lorentz force,

$$\mathbf{F} = e\mathbf{E} + \frac{e}{2c}(\mathbf{v} \times \mathbf{H} - \mathbf{H} \times \mathbf{v})$$

$\mathcal{H}$ and $\mathbf{F}$ must be Hermitian operators; it is therefore necessary to pay attention to the included non-commuting operators, such as $\mathbf{p}$ and $\mathbf{A}$ in $\mathcal{H}$, $\mathbf{v}$ in (22.9), and $\mathbf{H}$ in $\mathbf{F}$.

In connection with Ehrenfest's principle, it is useful to give the quantum-mechanical form of the virial theorem. In classical mechanics this is obtained as follows: the equation of motion is first multiplied by $\mathbf{r}$: $m\ddot{\mathbf{r}}\mathbf{r} = -\mathbf{r}\,\mathrm{grad}\,V$.

The identity

$$\ddot{\mathbf{r}}\mathbf{r} = \frac{d}{dt}\dot{\mathbf{r}}\mathbf{r} - \dot{\mathbf{r}}^2$$

is now used to substitute for $\ddot{\mathbf{r}}\mathbf{r}$. If $\dot{\mathbf{r}}\mathbf{r}$ is finite, as is always the case for stationary atomic orbits, then on taking the time average:

$$\overline{2E_{kin}} = m\overline{\dot{\mathbf{r}}^2} = \overline{\mathbf{r}\,\mathrm{grad}\,V} \qquad (22.15)$$

A similar result is obtained in quantum mechanics if the Schrödinger equation for stationary states, $\mathcal{H}\phi = E\phi$, is differentiated with respect to $\mathbf{r}$:

$$\mathcal{H}\,\mathrm{grad}\,\phi + \phi\,\mathrm{grad}\,V = E\,\mathrm{grad}\,\phi$$

Since

$$\mathcal{H}(\mathbf{r}\phi^*) = \mathbf{r}\mathcal{H}\phi^* - \frac{\hbar^2}{m}\text{grad }\phi^*$$

if we multiply by $\phi^*\mathbf{r}$ and integrate with respect to $\mathbf{r}$, we obtain

$$-\frac{\hbar^2}{m}\int \text{grad }\phi \text{ grad }\phi^* \, d\mathbf{r} + \int |\phi|^2 \mathbf{r} \text{ grad } V \, d\mathbf{r} = 0$$

after integration by parts. Now, since

$$\frac{\hbar^2}{2m}\int \text{grad }\phi \text{ grad }\phi^* \, d\mathbf{r} = -\int \phi^* \frac{\hbar^2}{2m}\nabla^2\phi \, d\mathbf{r} = \overline{E_{kin}}$$

we obtain the same result as in (22.15):

$$2\overline{E_{kin}} = \overline{\mathbf{r} \text{ grad } V} \qquad (22.16)$$

although the average values in both equations are formed according to different rules (cf. Exercise 8, p. 138).

## Exercises

1. *Various representations of the δ-function*

Represent the function $\delta(x)$ by means of the expression $\lim_{b \to 0} C_b S(x/b)$, in which the following functions $S$ are to be used:

(1) $S = \exp -\dfrac{x^2}{b^2}$  (2) $S = \dfrac{1}{1+x^2/b^2}$  (3) $S = \begin{cases} 0 & \text{for } |x| > b \\ 1 & \text{for } |x| < b \end{cases}$  (4) $S = \dfrac{\sin^2 x/b}{x^2/b^2}$

The quantity $b$ is taken to be the width of the distribution ($b > 0$). Determine the normalization constant $C_b$ in each case so that the integral over the range $-\infty \leqq x \leqq +\infty$ gives the value unity. Illustrate the functions graphically. Show also that

$$\lim_{b \to 0} \int_{-a}^{a'} f(x) C_b S(x/b) \, dx = f(0)$$

provided that the integration takes place about the point $x = 0$, i.e. that $a', a > 0$.

2. *δ-function relations*

Prove the following relations:

(1) $\delta(Cx) = \dfrac{1}{C}\delta(x) \qquad C = \text{const}$

(2) $\delta\{(x-a)(x-b)\} = \dfrac{1}{|a-b|}\{\delta(x-a) - \delta(x-b)\} \qquad a \neq b \text{ const}$

(3) $\displaystyle\prod_{i=1}^{N}\delta\left(\sum_{k=1}^{N}\alpha_{ik}x_k\right) = \dfrac{1}{|\det \alpha_{ik}|}\prod_{k=1}^{N}\delta(x_k)$ where $\det \alpha_{ik} \neq 0$

Verify for the case of two variables $x$ and $y$:

$$\delta(x+y)\delta(x-y) = \tfrac{1}{2}\delta(x)\delta(y)$$

### 3. *The Schrödinger equation in momentum space*

Express the Schrödinger equation $\mathscr{H}\phi = i\hbar\dot{\phi}$ in momentum space (a) as an integral equation, (b) as a differential equation where $x = -i\hbar\partial/\partial p$ and $V(x)$ is assumed to be given as a power series. (c) What is the form taken by the equation for the linear oscillator?

### 4. *One-dimensional potential well*

Let a particle move parallel to the $x$-axis in a potential field

$$V(x) = \begin{cases} 0 & \text{for } |x| > l \\ -V_0 & \text{for } |x| < l \end{cases}$$

Since $\mathscr{H} = p^2/2m + V(x)$ commutes with the reflection operator $P$ (defined by $P\phi(x) = \phi(-x)$), the eigenfunctions of $\mathscr{H}$ can also be chosen as eigenfunctions of $P$, i.e. as odd or even functions. It will therefore be sufficient to consider the functions in the range $x \geqq 0$. For the even functions we shall put

$$\phi_g = \begin{cases} a\cos qx & \text{for } 0 < x < l \\ S_g(k)\,e^{-ikx} - e^{ikx} & \text{for } x > l \end{cases}$$

and for the odd functions,

$$\phi_u = \begin{cases} a\sin qx & \text{for } 0 < x < l \\ S_u(k)\,e^{-ikx} - e^{ikx} & \text{for } x > l \end{cases}$$

We also have

$$\frac{\hbar^2 k^2}{2m} + V_0 = \frac{\hbar^2 q^2}{2m}$$

Determine the functions $S_g(k)$ and $S_u(k)$ from the boundary conditions that $\phi$ and $\phi'$ are continuous at $x = l$. Where do the zero points of $S$ lie along the positive imaginary axis of $k$? What is their significance?

### 5. *Potential in δ-function form*

Determine the eigenfunctions and eigenvalues for the potential $-\dfrac{\hbar^2}{2m\lambda}\delta(x)$ (a) from the previous exercise by passing to the limits

$$V_0 \to \infty, \quad l \to 0, \quad \text{but with } 2V_0 l = \frac{\hbar^2}{2m\lambda}$$

(b) directly from the boundary condition for $x = 0$.
Are there any stationary states?

### 6. *Ehrenfest's theorem and the magnetic field*

Determine the equation for the time rate of change of the operators **r** and **p** in the presence of a magnetic field.

### 7. *Current and magnetic field*

Extend the continuity equation (14.15) to the case in which a magnetic field is present.
What is the current density **j** in this case?

### 8. *The virial theorem*

Discuss the virial equation (22.16) when the potential is (a) of the Coulomb type, (b) that of an oscillator.

# CHAPTER BIII

## Linear and angular momentum

### §23. Translation and rotation as unitary operators

*(a) Definitions*

We saw in §20 that the operator $T_a = \exp(-iP_x a/\hbar)$ converts a Hilbert space basis vector $\alpha_x$ to $\alpha_{x+a}$, and therefore displaces a state designated by $x$, by an amount equal to $a$. In the three-dimensional case, the unitary operator $T_{\mathbf{a}} = \exp(-i\mathbf{Pa}/\hbar)$ effects a similar translation of the states $\alpha_{\mathbf{r}}$ by an amount equal to the vector $\mathbf{a}$.

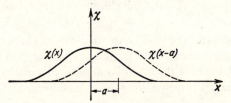

Fig. 25.—$\chi(x-a)$ represents a state $\chi(x)$ displaced by an amount $a$

In the $\mathbf{r}$-representation $\chi(\mathbf{r})$, $T_{\mathbf{a}}$ takes the form

$$T_{\mathbf{a}} = \exp\left(-\frac{i}{\hbar}\mathbf{pa}\right), \quad \text{where } \mathbf{p} = \frac{\hbar}{i}\frac{\partial}{\partial\mathbf{r}} = \frac{\hbar}{i}\,\text{grad} \qquad (23.1)$$

Then

$$T_{\mathbf{a}}\chi(\mathbf{r}) = \chi(\mathbf{r}-\mathbf{a}) \qquad (23.1a)$$

Equation (23.1*a*) can be immediately verified: if $T_{\mathbf{a}}$ is expanded in powers of $\mathbf{a}$, (23.1*a*) is clearly identical with the Taylor expansion of $\chi(\mathbf{r}-\mathbf{a})$. In particular, for the displacement in the $x$-direction (fig. 25):

$$T_a = \exp\left(-\frac{i}{\hbar}p_x a\right) = \exp\left(-a\frac{\partial}{\partial x}\right)$$

$$T_a\chi(x, y, z) = \chi(x-a, y, z) \qquad (23.1b)$$

Similarly, the operator $\mathbf{M}$ of the angular momentum may be associated with a rotation $D_t(\alpha)$ through an angle $\alpha$ about an axis $\mathbf{t}$ (where $\mathbf{t}^2 = 1$) through the origin of coordinates. In the $\mathbf{r}$-representation, the operator that rotates a state in this manner is

$$D_t(\alpha) = \exp\left(-\frac{i}{\hbar}\mathbf{M}\mathbf{t}\alpha\right), \quad \text{where } \mathbf{M} = \mathbf{r} \times \mathbf{p} = \frac{\hbar}{i}\mathbf{r} \times \text{grad} \qquad (23.2)$$

Then

$$D_t(\alpha)\chi(\mathbf{r}) = \chi(\bar{\mathbf{r}}), \quad \text{where } \bar{\mathbf{r}} = D_t(-\alpha)\mathbf{r} \qquad (23.2a)$$

The vector $\bar{\mathbf{r}}$ is produced from $\mathbf{r}$ by a rotation through the angle $-\alpha$, that is, by the reciprocal operation to $D_t(\alpha)$. If we take the particular case of a rotation about the $z$-axis, and express this in polar coordinates:

$$\frac{i}{\hbar}M_z = (\mathbf{r} \times \text{grad})_z = x\frac{\partial}{\partial y} - y\frac{\partial}{\partial x} = \frac{\partial}{\partial \phi}$$

Therefore

$$D_z(\alpha) = \exp\left(-\alpha\frac{\partial}{\partial \phi}\right), \quad D_z(\alpha)\psi(r, \theta, \phi) = \psi(r, \theta, \phi - \alpha) \qquad (23.2b)$$

Thus (23.2) is shown to be valid, since the $z$-axis can have any direction.

   The above expressions for the translation and rotation operators are generally valid, even for the case of a number of particles, when $\mathbf{p}$ and $\mathbf{M}$ are the operators corresponding to the total linear and total angular momentum.

## (b) Eigenvalues and eigenfunctions

   Since $\mathbf{p}$ and $\mathbf{M}$ are Hermitian operators, $T_a$ and $D_t(\alpha)$ are unitary, and the eigenvalues of $T_a$ and $D_z(\alpha)$ therefore have absolute value unity. In addition, $(T_a)^2 = T_{2a}$ and $[D_z(\alpha)]^2 = D_z(2\alpha)$. Further, if $k$ and $m$ are real numbers, and $\chi_k$ and $\psi_m$ are eigenfunctions of $T_a$ and $D_z(\alpha)$ respectively, then

$$T_a\chi_k = e^{-ika}\chi_k \quad \text{and} \quad D_z(\alpha)\psi_m = e^{-im\alpha}\psi_m \qquad (23.3)$$

Comparison with (23.1b) shows that the function $\chi_k(x)$ satisfies the relation

$$\chi_k(x-a) = e^{-ika}\chi_k(x)$$

The solution of this functional equation is

$$\chi_k(x) = e^{ikx}u_k(x), \quad \text{where } u_k(x) = u_k(x-a) \qquad (23.4)$$

$u_k(x)$ is therefore periodic, with period $a$. Similarly, $\psi_m$ may be written

$$\psi_m(r, \theta, \phi) = e^{im\phi} w_m(r, \theta, \phi)$$

where
$$w_m(r, \theta, \phi) = w_m(r, \theta, \phi - \alpha) \qquad (23.5)$$

In order that $\psi_m$ should be unique, $m$ must be a whole number and $\alpha = 2\pi/n$ where $n$ is an integer.

An infinitesimal translation or rotation is produced if $a$ or $\alpha$ is made infinitesimally small. Then in virtue of (23.1$b$) and (23.2$b$) we may define

$$T_{inf} = \left(\frac{T_a - 1}{a}\right)_{a \to 0} = -\frac{i}{\hbar} p_x, \quad (D_z)_{inf} = \left\{\frac{D_z(\alpha) - 1}{\alpha}\right\}_{\alpha \to 0} = -\frac{i}{\hbar} M_z$$

The relation (23.4) implies that when $a$ is infinitesimal, $u_k$ is quite independent of $x$. Similarly, $w_m$ is independent of $\phi$ when $\alpha$ is infinitesimal. Then from (23.4) and (23.5) the eigenfunctions of $T_{inf}$ and $(D_z)_{inf}$ are respectively

$$e^{ikx} u_k(y, z) \qquad e^{im\phi} w_m(r, \theta) \qquad (23.6)$$

The functions (23.6) are of course also eigenfunctions of $p_x$ and $M_z$, corresponding to eigenvalues $\hbar k$ and $\hbar m$ respectively.

## (c) Commutability with $\mathcal{H}$

If the Hamiltonian operator $\mathcal{H}$ commutes with a unitary operator $U$, then $\phi$ and $U\phi$ are both eigenfunctions of $\mathcal{H}$ corresponding to the same eigenvalue $E$. If $E$ is non-degenerate, then $\phi$ is also an eigenfunction of $U$. On the other hand, if $E$ is, say, $s$-fold degenerate with eigenfunctions $\phi_1, \ldots, \phi_s$, then $U\phi_j$ ($j = 1, 2, 3, \ldots, s$) must be a linear combination of these eigenfunctions. In this case, however, new functions $\chi_j = \sum_{r=1}^{s} a_{jr} \phi_r$, $j = 1, \ldots, s$ may be formed, which are eigenfunctions of $U$; hence $U\chi_j = u_j \chi_j$. Each $\chi_j$ then belongs to the eigenvalue $E$ of $\mathcal{H}$ and to the eigenvalue $u_j$ of $U$. If a number of operators $U', U'', \ldots$ commute with $\mathcal{H}$ and with each other, then without restriction of generality the eigenfunctions of $\mathcal{H}$ can always be chosen so that they are also eigenfunctions of $U', U'', \ldots$ We shall now consider two applications of this fundamental theorem, which we have already discussed in §19$c$.

## 1.  *Single electron functions in a metallic crystal lattice*

An electron moving through a metal is acted on by a spatially periodic potential due to the metal ions and the other conduction electrons in the metal lattice. If the sides of the elementary cell in the lattice are represented by the vectors $\mathbf{a}_1, \mathbf{a}_2, \mathbf{a}_3$, then the potential energy $V(\mathbf{r})$ of the electron takes the form

$$V(\mathbf{r}) = V(\mathbf{r} + n_1\,\mathbf{a}_1 + n_2\,\mathbf{a}_2 + n_3\,\mathbf{a}_3) \qquad (23.7)$$

where $n_1$, $n_2$, $n_3$ are whole numbers. The Hamiltonian operator

$$\mathcal{H} = -\frac{\hbar^2}{2m}\nabla^2 + V(\mathbf{r}) \qquad (23.8)$$

then commutes with the translation operators $T_{\mathbf{a}_1}$, $T_{\mathbf{a}_2}$, $T_{\mathbf{a}_3}$. Indeed for any function $f(\mathbf{r})$

$$T_{\mathbf{a}_1}V(\mathbf{r})f(\mathbf{r}) = V(\mathbf{r}-\mathbf{a}_1)f(\mathbf{r}-\mathbf{a}_1) = V(\mathbf{r})f(\mathbf{r}-\mathbf{a}_1)$$
$$= V(\mathbf{r})T_{\mathbf{a}_1}f(\mathbf{r})$$

the commutability of $T$ and $\nabla^2$ is trivial.

Since the translation operators $T_{\mathbf{a}_1}$, $T_{\mathbf{a}_2}$, $T_{\mathbf{a}_3}$ also commute with each other, we may assume that the eigenfunctions $\phi$ of (23.8) are also eigenfunctions of the $T_{\mathbf{a}_j}$, and therefore, that (corresponding to (23.3))

$$T_{\mathbf{a}_j}\phi_{\mathbf{K}} = \exp(-i\mathbf{K}.\mathbf{a}_j)\,\phi_{\mathbf{K}} \qquad (j = 1, 2, 3)$$

As in (23.4), the $\phi_{\mathbf{K}}$ may be written in the form

$$\phi_{\mathbf{K}} = \exp(i\mathbf{K}.\mathbf{r})\,u_{\mathbf{K}}(\mathbf{r})$$

where the $u_{\mathbf{K}}$ and $V(\mathbf{r})$ exhibit the period of the lattice.

## 2.  *Electron in a central field of force*

In the case under consideration, the potential energy depends only on the distance $r$ from a fixed centre; therefore

$$\mathcal{H} = -\frac{\hbar^2}{2m}\nabla^2 + V(r) \qquad (23.9)$$

$\mathcal{H}$ commutes with any rotation about this fixed centre: for instance, for $D_z(\alpha)$

$$D_z(\alpha)V(r)f(r,\theta,\phi) = V(r)f(r,\theta,\phi-\alpha) = V(r)D_z(\alpha)f(r,\theta,\phi)$$

Since this property of commutability also holds when $\alpha$ is infinitesimal, $M_x$, $M_y$, and $M_z$ also commute with $\mathscr{H}$, and so does the sum of the squares

$$\mathbf{M}^2 = M_x^2 + M_y^2 + M_z^2$$

However, the individual components of $\mathbf{M}$ do not commute with each other; if we make use of the relations $p_x x - x p_x = \hbar/i$, etc., the following commutation relations may easily be verified:

$$M_x M_y - M_y M_x = i\hbar M_z$$
$$M_y M_z - M_z M_y = i\hbar M_x \qquad (23.10)$$
$$M_z M_x - M_x M_z = i\hbar M_y$$

On the other hand, each component of $\mathbf{M}$ commutes with $\mathbf{M}^2$, e.g.

$$M_z \mathbf{M}^2 - \mathbf{M}^2 M_z = 0 \qquad (23.11)$$

In the case of a central field, therefore, there are two operators, say $\mathbf{M}^2$ and $M_z$, that commute with $\mathscr{H}$ and with each other. It is therefore possible to choose the eigenfunctions of $\mathscr{H}$ in such a manner that they are also eigenfunctions of $\mathbf{M}^2$ and $M_z$.

We shall determine the eigenvalues and eigenfunctions of $\mathbf{M}^2$ and $M_z$ in the next section, for which we shall require the following operators:

$$\vec{\Lambda} = \frac{1}{\hbar}\mathbf{M}, \quad \text{i.e.} \quad \begin{cases} \Lambda_x = \dfrac{1}{i}\left(y\dfrac{\partial}{\partial z} - z\dfrac{\partial}{\partial y}\right) \\[2mm] \Lambda_y = \dfrac{1}{i}\left(z\dfrac{\partial}{\partial x} - x\dfrac{\partial}{\partial z}\right) \\[2mm] \Lambda_z = \dfrac{1}{i}\left(x\dfrac{\partial}{\partial y} - y\dfrac{\partial}{\partial x}\right) \end{cases} \qquad (23.12)$$

and

$$\Lambda_+ = \Lambda_x + i\Lambda_y \qquad \Lambda_- = \Lambda_x - i\Lambda_y \qquad (23.13)$$

Since the $M_i$ and $\Lambda_i$ are Hermitian operators, $\Lambda_+$ and $\Lambda_-$ are adjoint to each other:

$$\Lambda_+ = \Lambda_-^+$$

Equations (23.10) and (23.11) then become

$$\Lambda_x \Lambda_y - \Lambda_y \Lambda_x = i\Lambda_z, \ldots, \ldots \qquad \vec{\Lambda}^2 \Lambda_z - \Lambda_z \vec{\Lambda}^2 = 0 \quad (23.14)$$

In addition, we have the following commutation rules for $\Lambda_+$ and $\Lambda_-$:

$$\Lambda_z\Lambda_+ - \Lambda_+\Lambda_z = \Lambda_+ \qquad \Lambda_z\Lambda_- - \Lambda_-\Lambda_z = -\Lambda_- \quad (23.15)$$

$$\Lambda_+\Lambda_- = \overset{\rightarrow 2}{\Lambda} + \Lambda_z - \Lambda_z^2 \qquad \Lambda_-\Lambda_+ = \overset{\rightarrow 2}{\Lambda} - \Lambda_z - \Lambda_z^2 \quad (23.16)$$

Equipped with these relations, we shall now determine the eigenvalues of $\Lambda^2$ and $\Lambda_z$ from the commutation rules alone, without making use of (23.12).

## §24. Eigenvalues of the angular momentum

We shall now consider an eigenfunction $Y_\lambda$ corresponding to a fixed eigenvalue of $\overset{\rightarrow 2}{\Lambda}$. Since $(Y_\lambda, \overset{\rightarrow 2}{\Lambda} Y_\lambda) = \sum_k (\Lambda_k Y_\lambda, \Lambda_k Y_\lambda) \geqq 0$, this eigenvalue cannot be negative. It may be written without restriction of generality in terms of a real number $\lambda \geqq 0$, in the form $\lambda(\lambda+1)$:

$$\overset{\rightarrow 2}{\Lambda} Y_\lambda = \lambda(\lambda+1)Y_\lambda$$

In general, the eigenvalue $\lambda(\lambda+1)$ will be degenerate, since an electron with a given total angular momentum can have very different values for the components. We shall take the $z$-component as an example: since $\overset{\rightarrow 2}{\Lambda}$ and $\Lambda_z$ commute, we may assume that $Y_\lambda$ is also an eigenfunction of $\Lambda_z$, say with eigenvalue $\mu$. Then we may write

$$\Lambda_z Y_{\lambda\mu} = \mu Y_{\lambda\mu} \qquad (24.1)$$

$$\overset{\rightarrow 2}{\Lambda} Y_{\lambda\mu} = \lambda(\lambda+1)Y_{\lambda\mu} \qquad (24.2)$$

Since

$$(Y_{\lambda\mu}, \Lambda_z^2 Y_{\lambda\mu}) \leqq (Y_{\lambda\mu}, \overset{\rightarrow 2}{\Lambda} Y_{\lambda\mu})$$

$$\mu^2 \leqq \lambda(\lambda+1)$$

Hence for any given $\lambda$ there is always a maximum $z$-component $\bar{\mu}$ and a minimum $\underline{\mu}$.

We now multiply (24.1) by $\Lambda_+$ and make use of (23.15); then

$$\Lambda_z(\Lambda_+ Y_{\lambda\mu}) = (\mu+1)(\Lambda_+ Y_{\lambda\mu})$$

in other words $\Lambda_+ Y_{\lambda\mu}$ either belongs to the eigenvalue $\mu+1$ or it vanishes.

In particular, we must have

$$\Lambda_+ Y_{\lambda\bar{\mu}} = 0$$

otherwise $\Lambda_+ Y_{\lambda\bar{\mu}}$ would belong to the eigenvalue $\bar{\mu}+1$, and $\bar{\mu}$ would not be the greatest value. It then follows from (23.16) that

$$\overset{\rightarrow 2}{\Lambda} Y_{\lambda\bar{\mu}} = \lambda(\lambda+1)Y_{\lambda\bar{\mu}} = \{\Lambda_-\Lambda_+ + \Lambda_z(\Lambda_z+1)\} Y_{\lambda\bar{\mu}} = \bar{\mu}(\bar{\mu}+1)Y_{\lambda\bar{\mu}}$$

therefore, since $\lambda \geqq 0$,

$$\bar{\mu} = \lambda \qquad (24.3)$$

In classical mechanics, the $z$-component of the electron's angular momentum would be a maximum if the latter vector were directed along the positive $z$-axis: we should then expect $M_x^2$ and $M_y^2$ to be zero, and $\mathbf{M}^2 = M_z^2 = \hbar^2\bar{\mu}^2$. The quantum-theory result, on the other hand, is $\hbar^2\bar{\mu}(\bar{\mu}+1) = \hbar^2\bar{\mu}^2 + \hbar^2\bar{\mu}$. The reason is that, since $M_x$, $M_y$, and $M_z$ do not commute, an eigenfunction of $M_z$ cannot also be an eigenfunction of $M_x$ and $M_y$. It is true that when $M_z = \hbar\bar{\mu}$ the expectation values of $M_x$ and $M_y$ are zero, but deviations occur about these mean values, the order of magnitude of which is $\overline{M_x^2} \approx \overline{M_y^2} \approx \hbar^2\bar{\mu}$. (Cf. Exercise 2, p. 152.) In the case of large quantum numbers, the relative deviation $\overline{M_x^2}/\mathbf{M}^2 \approx \dfrac{1}{\bar{\mu}}$ becomes vanishingly small and the classical formula is then valid.

It follows by a similar argument that $\Lambda_- Y_{\lambda\mu}$ either belongs to the eigenvalue $\mu-1$ of $\Lambda_z$ or vanishes. If the operator $\Lambda_-$ is applied to $Y_{\lambda\bar{\mu}}$ a number of times in succession, each successive eigenvalue of $\Lambda_z$ is one less than the previous value. After a finite number of steps the eigenvalue $\mu$ is reached, for which

$$\Lambda_- Y_{\lambda\mu} = 0$$

Hence, in view of (23.16)

$$\overset{\rightarrow 2}{\Lambda} Y_{\lambda\mu} = \lambda(\lambda+1)Y_{\lambda\mu} = \{\Lambda_+\Lambda_- + \Lambda_z(\Lambda_z-1)\} Y_{\lambda\mu} = \mu(\mu-1)Y_{\lambda\mu}$$

and therefore

$$\mu = -\lambda \qquad (24.4)$$

Now, since $Y_{\lambda\mu}$ has been formed from $Y_{\lambda\bar{\mu}}$ by a finite number of operations in which successive eigenvalues are reduced by unity, $\bar{\mu}-\mu = 2\lambda$ must be an integer:

$$\bar{\mu}-\mu = 2\lambda = 0, 1, 2, 3, \ldots$$

In the case of the special angular-momentum operators (23.12), $\lambda$ and $\mu$ must be whole numbers (cf. §25). However, by proceeding from the general commutation rules (23.14), half-integral values of $\lambda$ and $\mu$ are also obtained; this case is important in connection with electron spin, although $\Lambda_x$, $\Lambda_y$, $\Lambda_z$ can no longer be taken as differential operators or $Y_{\lambda\mu}$ as angle functions.

In describing the effect of the operators $\Lambda_+$ and $\Lambda_-$ on the eigenfunctions we shall assume the $Y_{\lambda\mu}$ to be normalized. Then

$$\Lambda_+ Y_{\lambda\mu} = \alpha_{\lambda\mu} Y_{\lambda,\mu+1} \tag{24.5}$$

and by forming the inner product, remembering that $\Lambda_+ = \Lambda_-^+$, we obtain

$$|\alpha_{\lambda\mu}|^2 = (\Lambda_+ Y_{\lambda\mu}, \Lambda_+ Y_{\lambda\mu}) = (Y_{\lambda\mu}, \Lambda_- \Lambda_+ Y_{\lambda\mu})$$

$$= (Y_{\lambda\mu}, [\overset{\to}{\Lambda}{}^2 - \Lambda_z - \Lambda_z^2] Y_{\lambda\mu}) = \lambda(\lambda+1) - \mu(\mu+1)$$

We may therefore write

$$\Lambda_+ Y_{\lambda\mu} = \sqrt{[\lambda(\lambda+1) - \mu(\mu+1)]}\, Y_{\lambda,\mu+1}$$

$$= \sqrt{[(\lambda+\mu+1)(\lambda-\mu)]}\, Y_{\lambda,\mu+1} \tag{24.6}$$

and similarly

$$\Lambda_- Y_{\lambda\mu} = \sqrt{[\lambda(\lambda+1) - \mu(\mu-1)]}\, Y_{\lambda,\mu-1}$$

$$= \sqrt{[(\lambda-\mu+1)(\lambda+\mu)]}\, Y_{\lambda,\mu-1} \tag{24.7}$$

We next consider the case when $\lambda = \frac{1}{2}$: the only eigenfunctions that occur are

$$Y_{\frac{1}{2},\frac{1}{2}} = \alpha \quad \text{and} \quad Y_{\frac{1}{2},-\frac{1}{2}} = \beta$$

These two functions constitute a basis which spans the whole Hilbert space of the angular-momentum functions. Then from (24.6), (24.7), and (24.1)

$$\Lambda_+ \alpha = 0 \qquad \Lambda_+ \beta = \alpha$$

$$\Lambda_- \alpha = \beta \qquad \Lambda_- \beta = 0$$

$$\Lambda_z \alpha = \tfrac{1}{2}\alpha \qquad \Lambda_z \beta = -\tfrac{1}{2}\beta$$

A vector $u$ in this Hilbert space is determined by two numbers $a$ and $b$ as follows:

$$u = \begin{pmatrix} a \\ b \end{pmatrix} = a\alpha + b\beta \quad \text{where} \quad \alpha = \begin{pmatrix} 1 \\ 0 \end{pmatrix}, \quad \beta = \begin{pmatrix} 0 \\ 1 \end{pmatrix}$$

The effect of the operators $\Lambda_+$, $\Lambda_-$, $\Lambda_z$ on the vector $u$ is represented by the following matrices:

$$\Lambda_+ = \begin{pmatrix} 0 & 1 \\ 0 & 0 \end{pmatrix} \qquad \Lambda_- = \begin{pmatrix} 0 & 0 \\ 1 & 0 \end{pmatrix} \qquad \Lambda_z = \frac{1}{2}\begin{pmatrix} 1 & 0 \\ 0 & -1 \end{pmatrix}$$

In the theory of electron spin, for which the above formalism will be required, the angular-momentum operator is written

$$\mathbf{M} = \hbar\vec{\Lambda} \equiv \mathbf{s} = \tfrac{1}{2}\hbar\vec{\sigma} \tag{24.8}$$

It follows from (23.13) that the components of the vector operator $\vec{\sigma}$ are

$$\sigma_x = \begin{pmatrix} 0 & 1 \\ 1 & 0 \end{pmatrix} \qquad \sigma_y = \begin{pmatrix} 0 & -i \\ i & 0 \end{pmatrix} \qquad \sigma_z = \begin{pmatrix} 1 & 0 \\ 0 & -1 \end{pmatrix} \tag{24.9}$$

These are the Pauli spin matrices, which we shall meet later. It can be verified by multiplying out that $\mathbf{s}$ is proportional to the unit matrix:

$$\mathbf{s}^2 = \tfrac{1}{4}\hbar^2(\sigma_x^2 + \sigma_y^2 + \sigma_z^2) = \tfrac{3}{4}\hbar^2\begin{pmatrix} 1 & 0 \\ 0 & 1 \end{pmatrix}$$

The constant of proportionality has the value $\hbar^2\lambda(\lambda+1)$ where $\lambda = \tfrac{1}{2}$.

We next consider the case in which the $\Lambda$-operators are given by the differential expressions (23.12). As we shall see in §25, $\lambda$ and $\mu$ are whole numbers, usually designated by $l$ and $m$; the eigenfunctions $Y_{lm} = Y_{lm}(\theta, \phi)$ are functions of the polar angles $\theta$ and $\phi$. They are termed *spherical harmonics*.

## §25. Spherical harmonics

Expressed in polar coordinates, (23.12) and (23.13) lead to the following results:

$$\Lambda_z = \frac{1}{i}\frac{\partial}{\partial\phi} \qquad \Lambda_+ = e^{i\phi}\left(\frac{\partial}{\partial\theta} + i\cot\theta\,\frac{\partial}{\partial\phi}\right)$$

$$\Lambda_- = e^{-i\phi}\left(-\frac{\partial}{\partial\theta} + i\cot\theta\,\frac{\partial}{\partial\phi}\right) \tag{25.1}$$

$$\vec{\Lambda}^2 = -\frac{1}{\sin^2\theta}\left(\sin\theta\,\frac{\partial}{\partial\theta}\sin\theta\,\frac{\partial}{\partial\theta} + \frac{\partial^2}{\partial\phi^2}\right) \tag{25.2}$$

We first determine $Y_{ll}$ from the equations

$$\Lambda_+ Y_{ll} = 0 \qquad \Lambda_z Y_{ll} = lY_{ll}$$

From (25.1), the solution is

$$Y_{ll} = \alpha_l \, e^{il\phi} \sin^l \theta \tag{25.3}$$

as $\phi$ increases by $2\pi$ the solution is single-valued only for integral values of $l$.

The constants $\alpha_l$ are determined by the condition for normalization:

$$(Y_{ll}, Y_{ll}) = \int Y_{ll}^* \, Y_{ll} \, d\Omega$$

$$= \alpha_l^2 \int_0^\pi \int_0^{2\pi} \sin^{2l} \theta \sin \theta \, d\theta \, d\phi = 1 \tag{25.3a}$$

(It is assumed that $\alpha_l$ is a positive real number.) Evaluation of the integral by elementary methods yields the following result:

$$\alpha_l^{\ 2} = \frac{(2l+1)!}{2^{2l+1}(l!)^2 2\pi} \quad \text{or} \quad \alpha_l = \frac{1}{2^l l!} \sqrt{\frac{(2l+1)!}{4\pi}} \tag{25.3b}$$

The remaining spherical harmonics $Y_{lm}$ are obtained from $Y_{ll}$ by the repeated application of $\Lambda_-$, using (24.7). As a result of (24.1), $Y_{lm}$ and hence $\Lambda_- Y_{lm}$ depend on $\phi$ through the factors $e^{im\phi}$ and $e^{i(m-1)\phi}$ respectively. Therefore

$$\sqrt{[(l-m+1)(l+m)]} \, Y_{l,m-1} = \Lambda_- Y_{lm} = e^{-i\phi}\left(-\frac{\partial}{\partial\theta} - m \cot \theta\right) Y_{lm}$$

$$= \frac{e^{-i\phi}}{\sin^{m-1}\theta} \frac{\partial}{\partial \cos\theta}(\sin^m \theta \, Y_{lm})$$

If $Y_{l,l-1}, Y_{l,l-2}, \ldots,$ are calculated successively from $Y_{ll}$, we obtain the expression

$$Y_{lm} = \frac{1}{2^l l!} \left(\frac{2l+1}{4\pi}\right)^{1/2} \left(\frac{(l+m)!}{(l-m)!}\right)^{1/2} \frac{1}{\sin^m \theta} \frac{d^{l-m}(\sin^{2l}\theta)}{d\cos\theta^{l-m}} e^{im\phi} \tag{25.4}$$

In particular:

$$Y_{lo} = \frac{1}{2^l l!} \left(\frac{2l+1}{4\pi}\right)^{1/2} \frac{d^l \sin^{2l}\theta}{d\cos\theta^l} \tag{25.4a}$$

and

$$Y_{l,-l} = \frac{(-1)^l}{2^l l!} \left(\frac{(2l+1)!}{4\pi}\right)^{1/2} \sin^l \theta \, e^{-il\phi} \tag{25.4b}$$

Apart from the factor $(-1)^l$, it would have been possible to obtain this last formula in the same way as (25.3), directly from the equation $\Lambda_- Y_{l,-l} = 0$, subject to the corresponding normalization condition. Conversely, we could have found the $Y_{lm}$ from $Y_{l,-l}$ or from $Y_{lo}$ by (24.6), by repeated application of $\Lambda_+$. In this case

$$\sqrt{[(l+m+1)(l-m)]}\, Y_{l,m+1} = \Lambda_+ Y_{lm} = -e^{i\phi}\sin^{m+1}\theta \frac{d}{d\cos\theta}\frac{Y_{lm}}{\sin^m\theta}$$

and from the above expression together with (25.4a) or (25.4b) we obtain

$$Y_{lm} = \frac{(-1)^m}{2^l l!}\left(\frac{2l+1}{4\pi}\right)^{1/2}\left(\frac{(l-m)!}{(l+m)!}\right)^{1/2}\sin^m\theta \frac{d^{l+m}\sin^{2l}\theta}{d\cos\theta^{l+m}}e^{im\phi} \quad (25.5)$$

It can be verified by differentiation that the above formula agrees with (25.4).

The above relations constitute the usual definition of the spherical harmonics as given in the literature. Putting $\cos\theta = \eta$, the ordinary Legendre polynomials are defined as

$$P_l(\eta) = \frac{1}{2^l l!}\frac{d^l(\eta^2-1)^l}{d\eta^l}$$

and the associated Legendre functions as

$$P_l^m(\eta) = (-1)^m\sin^m\theta \frac{d^m}{d\eta^m}P_l(\eta) = \frac{(-\sin\theta)^m}{2^l l!}\frac{d^{l+m}(\eta^2-1)^l}{d\eta^{l+m}}$$

the latter expression is also valid for negative values of $m$. Formulae (25.4) and (25.5) may now be expressed in the following simplified forms:

$$Y_{lm} = (-1)^{l-m}\left(\frac{2l+1}{4\pi}\right)^{1/2}\left(\frac{(l+m)!}{(l-m)!}\right)^{1/2}P_l^{-m}e^{im\phi}$$

or

$$= (-1)^l\left(\frac{2l+1}{4\pi}\right)^{1/2}\left(\frac{(l-m)!}{(l+m)!}\right)^{1/2}P_l^m e^{im\phi}$$

As a result of the equivalence of these two expressions we have the familiar relation for associated Legendre functions:

$$P_l^{-m} = (-1)^m\frac{(l-m)!}{(l+m)!}P_l^m$$

The spherical harmonics corresponding to $l = 0, 1, 2, 3, 4, \ldots$ are designated successively as the $s-, p-, d-, f-, g-, \ldots$ functions. Since $m$ can assume a total of $2l+1$ different values between $-l$ and $+l$, there are one $s$-function, three $p$-functions, five $d$-functions, etc. The $s$- and $p$-functions are given below:

$$Y_{00} = \frac{1}{(4\pi)^{1/2}}$$

$$Y_{11} = \left(\frac{3}{8\pi}\right)^{1/2} \sin\theta\, e^{i\phi}; \quad Y_{10} = -\left(\frac{3}{4\pi}\right)^{1/2} \cos\theta; \quad Y_{1,-1} = -\left(\frac{3}{8\pi}\right)^{1/2} \sin\theta\, e^{-i\phi}$$

The probability distributions $w(\theta,\phi) = |Y_{l,m}(\theta,\phi)|^2$ are of some interest, and are shown as polar diagrams in figure 26, apart from their normalization factors.

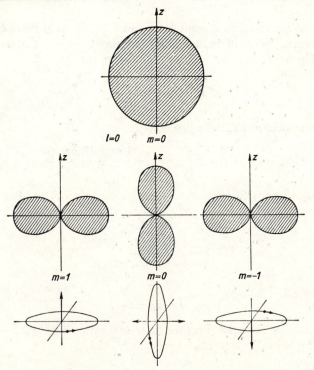

Fig. 26.—Probability distributions of the $s$-function and the three $p$-functions, and the three corresponding classical orbits

Since the effect of the operators $\Lambda_+$, $\Lambda_-$ on the functions $Y_{lm}$ is now known, the matrix elements $(Y_{l'm'}, \mathbf{M}Y_{lm})$ can be easily deduced, using (23.12). We now give another method of calculating the important matrix elements $(Y_{l'm'}, \mathbf{r}Y_{lm})$. For this purpose, we put

$$\mathbf{r}/r = \mathbf{a} = (\sin\theta\cos\phi, \sin\theta\sin\phi, \cos\theta)$$

and

$$a_\pm = a_x \pm ia_y = \sin\theta\, e^{\pm i\phi} \tag{25.6}$$

We can then easily verify the following commutation relations:

$$\Lambda_+ a_+ - a_+ \Lambda_+ = 0 \tag{25.7a}$$

$$\Lambda_- a_- - a_- \Lambda_- = 0 \tag{25.7b}$$

$$\Lambda_+ a_- - a_- \Lambda_+ = -(\Lambda_- a_+ - a_+ \Lambda_-) = 2a_z \tag{25.7c}$$

Also, from (25.4)

$$a_- Y_{ll} = -\left(\frac{2}{(2l+1)(2l+3)}\right)^{1/2} Y_{l+1,l-1} + \left(\frac{2l}{2l+1}\right)^{1/2} Y_{l-1,l-1} \tag{25.8}$$

Since $a_-$ and $\Lambda_-$ commute, if the latter operator is applied $(l-m)$ times to (25.8) we obtain

$$a_- Y_{lm} = -\left(\frac{(l-m+1)(l-m+2)}{(2l+3)(2l+1)}\right)^{1/2} Y_{l+1,m-1}$$
$$+ \left(\frac{(l+m)(l+m-1)}{(2l+1)(2l-1)}\right)^{1/2} Y_{l-1,m-1} \tag{25.9}$$

Similarly, when $\Lambda_+$ is applied to $a_+ Y_{l,-l}$,

$$a_+ Y_{lm} = \left(\frac{(l+m+1)(l+m+2)}{(2l+3)(2l+1)}\right)^{1/2} Y_{l+1,m+1}$$
$$- \left(\frac{(l-m)(l-m-1)}{(2l+1)(2l-1)}\right)^{1/2} Y_{l-1,m+1} \tag{25.10}$$

Finally, from (25.9) and (24.6) and the commutation relation (25.7c):

$$a_z Y_{lm} = -\left(\frac{(l+m+1)(l-m+1)}{(2l+3)(2l+1)}\right)^{1/2} Y_{l+1,m}$$
$$- \left(\frac{(l+m)(l-m)}{(2l+1)(2l-1)}\right)^{1/2} Y_{l-1,m} \tag{25.11}$$

We now have all the quantities required for the calculation of the angle-dependent portions of the matrix elements $\mathbf{r}_{l'm', lm}$. We should particularly note that these elements only differ from zero if $\Delta l = l' - l = \pm 1$ and $\Delta m = m' - m = 0, \pm 1$. These "selection rules" are extremely important in connection with the optical spectra of atoms. We might have guessed the condition for $m$ immediately because the quantities $a_\pm$ include a factor $e^{\pm i\phi}$ and $a_z = \cos\theta$ is independent of $\phi$. Hence in the calculation of the matrix elements integration over the range of $\phi$ always gives the value zero, provided that $\Delta m$ is not equal to 0 or $\pm 1$.

## Exercises

1. *Parity of the spherical harmonics*

Let $P\phi(\mathbf{r}) = \phi(-\mathbf{r})$ represent the operation of reflection in the origin. Show that     (1) The parity operator $P$ has the eigenvalues $+1$ and $-1$;
    (2) $P Y_{lm}(\theta, \phi) = (-1)^l Y_{lm}(\theta, \phi)$.

2. *Angular momentum matrices*

Using (24.6) and (24.7), express the matrices $M_x$, $M_y$, $M_z$, and $\mathbf{M}^2$ in the $l$-, $m$-representation, for the subspace $l = 1$ ($p$-functions). Use the results to calculate the root mean square deviations $\Delta M_x$ and $\Delta M_y$, and explain these deviations by means of the classical model of a precessing top.

3. *Eigenfunctions of the electron spin*

Determine the eigenvectors of the component of s in the direction $\theta$, $\phi$,
    (a) when the eigenvectors of

$$s_{\theta,\phi} = s_x \sin\theta \cos\phi + s_y \sin\theta \sin\phi + s_z \cos\theta$$

are directly determined by the relation

$$s_{\theta,\phi} \begin{pmatrix} a \\ b \end{pmatrix} = \hbar\lambda \begin{pmatrix} a \\ b \end{pmatrix}$$

in which the eigenvalues $\lambda$ are naturally independent of the direction and equal to $\pm\frac{1}{2}$;

    (b) when the two eigenstates $\alpha = \begin{pmatrix} 1 \\ 0 \end{pmatrix}$ and $\beta = \begin{pmatrix} 0 \\ 1 \end{pmatrix}$ are rotated into the direction $\theta$, $\phi$ in accordance with (23.2), in which $\mathbf{M} = \mathbf{s}$. (The first rotation is about the $y$-axis through an angle $\theta$, the second about the $z$-axis through an angle $\phi$.)
    (c) Compare the two results.

4. *Relation between $\Lambda_+$ and $\Lambda_-$*

Prove the relation $\Lambda_+ = \Lambda_-^+$, using (25.1) and the definition (25.3a) for the scalar product.

5. *A spherical harmonic relation*

Prove that
$$\sum_{m=-l}^{l} |Y_{lm}|^2 = \frac{2l+1}{4\pi}$$

# CHAPTER B IV

## Approximation methods

### §26. Schrödinger's perturbation theory

There are not many cases in quantum theory for which exact solutions can be obtained, and it is therefore important to be able to have recourse to approximation procedures. The first to be discussed will be Schrödinger's method for the approximative determination of eigenfunctions and eigenvalues. For this purpose we assume that the Hamiltonian operator may be written in the form $\mathcal{H}_0 + W$, in which the solution of the eigenvalue problem

$$\mathcal{H}_0 \phi_m = \varepsilon_m \phi_m \qquad (26.1)$$

is known, and $W$ is a small "perturbation" of $\mathcal{H}_0$; $\mathcal{H}_0$ and $W$ are Hermitian. The permissible magnitude of $W$ will be more precisely specified later in this discussion. It is further assumed that the $\phi_m$ constitute a complete orthonormal system in which the required eigenfunctions $\psi_\mu$ of

$$(\mathcal{H}_0 + W)\psi_\mu = E_\mu \psi_\mu \qquad (26.2)$$

can be expressed as follows:

$$\psi_\mu = \sum_m c_{m\mu} \phi_m \qquad (26.3)$$

Provided that the term $\varepsilon_m$ is non-degenerate and that $W$ is not too great, there will be just one term in the "disturbed" (or exact) problem that can be said to have been generated from $\varepsilon_m$ by the perturbation $W$. More precisely, the solution $\phi_m(\lambda)$ of

$$(\mathcal{H}_0 + \lambda W)\phi_m(\lambda) = \varepsilon_m(\lambda)\phi_m(\lambda)$$

is a continuous function of $\lambda$ for which

$$\phi_m(0) = \phi_m \qquad \varepsilon_m(0) = \varepsilon_m$$
$$\phi_m(1) = \psi_\mu \qquad \varepsilon_m(1) = E_\mu$$

$\psi_\mu$ may then be denoted by the old quantum number $m$. In the degenerate case, however, when two or more terms $\varepsilon_m$ almost or completely coincide,* it is no longer possible to find a relation of this sort between individual unperturbed and perturbed states; it is solely the group of coincident $\varepsilon_m$ (denoted below by the index $m'$) that is transformed into the corresponding group of perturbed terms $E_\mu$. This is formally represented as follows:

$$\psi_\mu = \sum_{m'} c_{m'\mu} \phi_{m'} + P\psi_\mu \qquad (26.4)$$

The operator $P$ eliminates the terms in $m'$ and is defined as follows:

$$P\psi = \sum_{m(\neq m')} \phi_m(\phi_m, \psi) \qquad (26.4a)$$

The operator $P$ is a projector, since it projects each state in the subspace of the $\phi_m$ for which $m \neq m'$: $P\phi_m = \phi_m$, $P\phi_{m'} = 0$. This operator commutes with $\mathscr{H}_0$. We now introduce the Schrödinger equation (26.2) in the form $\qquad \psi_\mu = \dfrac{1}{E_\mu - \mathscr{H}_0} W\psi_\mu$

into the second term of (26.4), and obtain the equation

$$\psi_\mu = \sum_{m'} c_{m'\mu} \phi_{m'} + \frac{P}{E_\mu - \mathscr{H}_0} W\psi_\mu \qquad (26.5)$$

the formal solution of which is

$$\psi_\mu = \left(1 - \frac{P}{E_\mu - \mathscr{H}_0} W\right)^{-1} \sum_{m'} c_{m'\mu} \phi_{m'} = \sum_{m'} c_{m'\mu} \Omega\phi_{m'} \qquad (26.6)$$

The operator $\Omega$ may be expressed as a power series:

$$\Omega = \left(1 - \frac{P}{E_\mu - \mathscr{H}_0} W\right)^{-1}$$

$$= 1 + \frac{P}{E_\mu - \mathscr{H}_0} W + \frac{P}{E_\mu - \mathscr{H}_0} W \frac{P}{E_\mu - \mathscr{H}_0} W + \dots \qquad (26.7)$$

The necessary condition for the convergence of this series is that

$$\left| \left( \phi_m, \frac{P}{E_\mu - \mathscr{H}_0} W\phi_{m'} \right) \right| = \left| \frac{W_{mm'}}{E_\mu - \varepsilon_m} \right| < 1$$

---

* A criterion will be given later for the case in which two terms are to be considered as "almost" coincident.

for all $m \neq m'$. It is clear that the expansion (26.7) is only of value if $|W_{mm'}| \ll |E_\mu - \varepsilon_m|$, since only in this case can we restrict ourselves to the first few terms of the series.

From (26.7) it follows that

$$(\phi_{m'}, \Omega\phi_{n'}) = \delta_{m'n'}$$

In order to determine the coefficients $c_{m'\mu}$ we introduce (26.6) in (26.2) and form the scalar product with $\phi_{m'}$. We then obtain the equation

$$\sum_{n'} \{\varepsilon_{m'} \delta_{m'n'} + (W\Omega)_{m'n'}\} c_{n'\mu} = E_\mu c_{m'\mu} \qquad (26.8)$$

This is a finite secular equation, the degree of which is determined by the number of coincident terms $\varepsilon_{m'}$.

It is disturbing at first sight to find that $\Omega$ depends on $E_\mu$ and that the solution of the secular equation does not therefore give $E_\mu$ explicitly. However, if the perturbation $W$ is sufficiently small, $\Omega$ depends only slightly upon $E_\mu$, and the latter quantity can be determined from (26.8) by iteration; the zero-order approximation $\varepsilon_{m'}$ is introduced in the left-hand side of (26.8), $E_\mu$ is determined as a first approximation and introduced in turn in the left-hand side, and so on.

If we introduce the zero-order approximation of (26.7), $\Omega \approx 1$, into (26.8), this equation simply becomes

$$\sum_{n'} (\varepsilon_{m'} \delta_{m'n'} + W_{m'n'}) c_{n'\mu} = E_\mu c_{m'\mu} \qquad (26.8a)$$

The $E_\mu$ are therefore the solutions of the algebraic equation

$$\det \{(\varepsilon_{m'} - E_\mu) \delta_{m'n'} + W_{m'n'}\} = 0 \qquad (26.9)$$

In the non-degenerate case we obtain

$$E_\mu = \varepsilon_{m'} + W_{m'm'} = \varepsilon_{m'} + (\phi_{m'}, W\phi_{m'}) \qquad (26.10)$$

in other words, the energy perturbation of first order in $W$ is equal to the mean value of the perturbation operator $W$ in the undisturbed state.

If we go one step further and introduce the zero-order approximation $\varepsilon_{m'}$ in (26.7), we obtain the following expressions for $\psi_\mu$ and $E_\mu$:

$$\psi_\mu = \phi_{m'} + \frac{P}{\varepsilon_{m'} - \mathcal{H}_0} W\phi_{m'} = \phi_{m'} + \sum_{m(\neq m')} \frac{W_{mm'}}{\varepsilon_{m'} - \varepsilon_m} \phi_m \qquad (26.11)$$

$$E_\mu = \varepsilon_{m'} + W_{m'm'} + \sum_{m(\neq m')} \frac{|W_{mm'}|^2}{\varepsilon_{m'} - \varepsilon_m} \qquad (26.12)$$

It is evident that in the lowest state the second approximation to the energy perturbation is always negative, since $\varepsilon_m > \varepsilon_{m'}$.

The eigenfunctions of zero-order approximation belonging to each eigenvalue are obtained by determining the coefficients $c_{m'\mu}$ from (26.8a). Their evaluation is often facilitated if a Hermitian or unitary operator $A$ exists that commutes with $W$ and $\mathcal{H}_0$, and the eigenfunctions and eigenvalues $a_l$ of which are known. In this case we choose the eigenfunctions of zero-order approximation in such a manner that they are also eigenfunctions of $\mathcal{H}_0$ and $A$. It can then be shown that $W_{m'n'} = 0$ when $\varepsilon_{m'} = \varepsilon_{n'}$, provided that $a_{m'} \neq a_{n'}$. The proof depends on the postulate that $AW - WA = 0$, from which it follows that

$$(\phi_{m'}, (AW - WA)\phi_{n'}) = (a_{m'} - a_{n'})W_{m'n'} = 0$$

The results of this section may be summarized as follows. If no degeneracy is present, to the zero-order approximation, a perturbation leads to a displacement of the original energy levels in accordance with (26.6). In the degenerate case, on the other hand, there are energy values that are originally equal, but which correspond to different "correct" eigenfunctions of zero-order approximation. These are in general displaced by different amounts; in other words, the perturbation causes the originally degenerate term to split. We shall encounter examples of such splitting in later chapters.

## §27. Dirac's perturbation calculation

Dirac's procedure deals with time-dependent processes. The starting point is the time-dependent Schrödinger equation

$$\mathcal{H}\psi = i\hbar\dot{\psi} \tag{27.1}$$

in which the Hamiltonian operator $\mathcal{H}$ is resolved into an unperturbed operator $\mathcal{H}_0$ that is independent of time, and a small perturbation operator $W$ that may be time-dependent:

$$\mathcal{H} = \mathcal{H}_0 + W \tag{27.1a}$$

The object is to determine the state $\psi(t)$ at time $t$ from a given initial state $\psi(0)$ at time $t = 0$, assuming that the change with respect to time produced by $W$ can be treated as a small perturbation.

Since $\mathcal{H}$ is a Hermitian operator, from (27.1) all scalar products remain constant with respect to time:

$$\frac{d}{dt}(\phi, \psi) = i\hbar\{(\phi, \mathcal{H}\psi) - (\mathcal{H}\phi, \psi)\} = 0$$

If $\psi(t)$ is related to $\psi(0)$ by means of an operator $U$:

$$\psi(t) = U(t)\psi(0),$$

hence
$$(\phi(t), \psi(t)) = (\phi(0), U^*U\psi(0)) \qquad (27.2)$$

and $U$ must consequently be unitary.

The operator $U(t)$ was introduced in §22 for the case in which $\mathscr{H}$ is independent of time. When $\mathscr{H}(t)$ depends on time we obtain an expression which is equally simple, at any rate in form:

$$U(t) = \qquad\qquad\qquad\qquad\qquad\qquad\qquad\qquad\qquad (27.3)$$

$$\sum_{\nu=0}^{\infty} \left(\frac{1}{i\hbar}\right)^\nu \int_{t \geqq \theta_1 \geqq \theta_2 \geqq \ldots \geqq \theta_\nu \geqq 0} d\theta_1 \, d\theta_2 \ldots d\theta_\nu \, \mathscr{H}(\theta_1)\,\mathscr{H}(\theta_2)\ldots\mathscr{H}(\theta_\nu)$$

or written out in full,

$$U(t) = 1 + \frac{1}{i\hbar}\int_0^t d\theta_1 \, \mathscr{H}(\theta_1) + \frac{1}{(i\hbar)^2}\int_0^t d\theta_1 \int_0^{\theta_1} d\theta_2 \, \mathscr{H}(\theta_1)\,\mathscr{H}(\theta_2) + \ldots$$
$$(27.3a)$$

Differentiation of (27.3) with respect to time gives $i\hbar\dot{U} = \mathscr{H}U$, and $U(0) = 1$; hence (27.2) satisfies the Schrödinger equation and the initial condition.

The transformation may be expressed symbolically in a somewhat simpler form:

$$U(t) = T\exp\left(\frac{1}{i\hbar}\int_0^t \mathscr{H}(\theta)\,d\theta\right) \qquad (27.4)$$

In the above expression the exponential operator is defined as a power series expansion of the exponential function:

$$\exp\left(\frac{1}{i\hbar}\int_0^t \mathscr{H}(\theta)\,d\theta\right) = \sum_{\nu=0}^{\infty} \frac{1}{\nu!}\left\{\frac{\int_0^t \mathscr{H}(\theta)\,d\theta}{i\hbar}\right\}^\nu =$$

$$= \sum_{\nu=0}^{\infty} \left(\frac{1}{i\hbar}\right)^\nu \frac{1}{\nu!}\int_0^t d\theta_1 \int_0^t d\theta_2 \ldots \int_0^t d\theta_\nu \, \mathscr{H}(\theta_1)\,\mathscr{H}(\theta_2)\ldots\mathscr{H}(\theta_\nu) \quad (27.4a)$$

The symbol $T$ in (27.4) means that the operators $\mathscr{H}(\theta_1),\ldots,\mathscr{H}(\theta_\nu)$ are so arranged in terms of time that $\theta_1 \geqq \theta_2 \geqq \ldots \geqq \theta_\nu$. The complete range of integration $t \geqq \theta_1, \theta_2,\ldots,\theta_\nu \geqq 0$ can then be divided into ranges of the form $t \geqq \theta_1 \geqq \theta_2 \geqq \ldots \geqq \theta_\nu \geqq 0$. There are $\nu!$ such ranges, corresponding to the $v!$ permutations of the $\theta_1,\ldots,\theta_v$ among themselves. When the $\mathscr{H}(\theta)$ are arranged according to the conditions imposed by $T$ each partial range of integration provides the same contribution, so that (27.4) is identical with the expression (27.3).

The symbol $T$ is of no importance if the operators $\mathscr{H}(\theta)$ and $\mathscr{H}(\theta')$ commute:

$$U(t) = \exp\left(\frac{1}{i\hbar}\int_0^t \mathscr{H}(\theta)\,d\theta\right), \text{ if } [\mathscr{H}(\theta), \mathscr{H}(\theta')] = 0 \quad (27.5)$$

The simplest case of this sort occurs when $\mathscr{H}$ is independent of the time:

$$U(t) = \exp\frac{1}{i\hbar}\mathscr{H}t \text{ for } \dot{\mathscr{H}} = 0 \qquad (27.5a)$$

If the perturbation $W = A f(t)$ consists of a time-dependent numerical factor and a time-independent operator $A$ that commutes with $\mathscr{H}_0$:

$$U(t) = \exp\left[\frac{1}{i\hbar}\left\{\mathscr{H}_0 t + A\int_0^t f(\theta)\,d\theta\right\}\right]$$

$$= \exp\left(\frac{1}{i\hbar}\mathscr{H}_0 t\right)\exp\left(\frac{1}{i\hbar}A\int_0^t f(\theta)\,d\theta\right) \qquad (27.5b)$$

In many cases it is expedient to separate the effect of the unperturbed operator $\mathscr{H}_0$ in order to show the variation due to the perturbation $W$ alone. For this purpose we put

$$\psi = U_0(t)\tilde{\psi}, \text{ where } U_0(t) = \exp\frac{\mathscr{H}_0 t}{i\hbar} \qquad (27.6)$$

Then

$$\dot{\psi} = \dot{U}_0\tilde{\psi} + U_0\dot{\tilde{\psi}} = \frac{1}{i\hbar}\mathscr{H}_0 U_0\tilde{\psi} + U_0\dot{\tilde{\psi}}$$

If no perturbation were present, $\tilde{\psi}$ would be constant with respect to time and equal to $\psi(0)$; the variation of $\tilde{\psi}$ with time is only due to the perturbation. If we introduce (27.6) into the Schrödinger equation we obtain

$$WU_0(t)\tilde{\psi} = i\hbar U_0(t)\dot{\tilde{\psi}} \qquad (27.7)$$

or

$$\tilde{W}\tilde{\psi} = i\hbar\dot{\tilde{\psi}} \qquad (27.7a)$$

where $\tilde{W} = U_0^+(t) W U_0(t) = \exp\left(-\frac{\mathscr{H}_0 t}{i\hbar}\right)W\exp\left(\frac{\mathscr{H}_0 t}{i\hbar}\right) = \tilde{W}^+$

(Note that if $W$ and $\mathcal{H}_0$ do not commute, $\tilde{W}$ is time-dependent even though $W$ is not a function of time.) This equation must be solved for the initial condition $\tilde{\psi}(0) = \psi(0)$; the result is

$$\tilde{\psi}(t) = T \exp\left(\frac{1}{i\hbar} \int_0^t \tilde{W}(\theta)\, d\theta\right) \psi(0) \qquad (27.7b)$$

$U$ has thus been split into two factors, only one of which contains the perturbation:

$$U(t) = \exp\left(\frac{1}{i\hbar} \mathcal{H}_0 t\right) T \exp\left(\frac{1}{i\hbar} \int_0^t \tilde{W}(\theta)\, d\theta\right) \qquad (27.8)$$

Dirac's approximation consists in expanding the perturbation factor and retaining only the lowest powers. If we include only the linear terms in $W$ or $\tilde{W}$, we have

$$U = U_0 \left\{ 1 + \frac{1}{i\hbar} \int_0^t \tilde{W}(\theta)\, d\theta \right\} \qquad (27.8a)$$

If the terms of the second degree are included,

$$U = U_0 \left\{ 1 + \frac{1}{i\hbar} \int_0^t \tilde{W}(\theta)\, d\theta + \frac{1}{(i\hbar)^2} \int_0^t d\theta_1 \int_0^{\theta_1} d\theta_2 \, \tilde{W}(\theta_1)\, \tilde{W}(\theta_2) \right\}$$
$$(27.8b)$$

Accordingly, when $\psi(t)$ is obtained from (27.2) by the use of (27.8a), it is represented by an expansion only to terms of the first order of the perturbation potential; the approximation (27.8b), on the other hand, provides an expansion to terms of the second order. If the function of first approximation $\psi^{(1)}(t)$ is used to obtain expectation values, it is of course permissible to retain terms only up to the first order of the perturbation potential; if the expectation values are required to the second degree of approximation they must be calculated from the second-order wave functions $\psi^{(2)}(t)$.

The above approximations may only be used if

$$\left( \psi(0), \left\{ \frac{1}{\hbar} \int_0^t \tilde{W}(\theta)\, d\theta \right\}^2 \psi(0) \right) \ll 1 \qquad (27.9)$$

This means that they are in general only valid for comparatively small time intervals. (Cf. however p. 162 and §41.)

In general equation (27.8) is expressed in the representation in which $\mathscr{H}_0$ is diagonal. $\psi(t)$ is therefore resolved into eigenfunctions $\phi_m$ of $\mathscr{H}_0$, the amplitudes of which are time-dependent:

$$\psi(t) = \sum_m a_m(t)\phi_m \qquad (27.10)$$

In the above expression

$$a_m(t) = \sum_n U_{mn}(t)a_n(0)$$

where

$$U_{mn}(t) = \{\phi_m, U(t)\phi_n\} \qquad (27.11)$$

Since $U_0(t)$ is also diagonal along with $\mathscr{H}_0$, i.e.

$$\mathscr{H}_{0mn} = E_m\delta_{mn} \quad \text{and} \quad U_{0mn}(t) = \exp\left(\frac{1}{i\hbar}E_m t\right)\delta_{mn}$$

it follows that

$$\widetilde{W}_{mn}(\theta) = \exp\left(-\frac{1}{i\hbar}E_m\theta\right)W_{mn}\exp\left(\frac{1}{i\hbar}E_n\theta\right)$$

Hence $U_{mn}(t)$ is given by the expression

$$U_{mn}(t) = \exp\left(\frac{1}{i\hbar}E_m t\right)\left\{\delta_{mn} + \frac{1}{i\hbar}\int_0^t d\theta_1\, W_{mn}(\theta_1)\times\right.$$

$$\times \exp\left(-\frac{1}{i\hbar}(E_m - E_n)\theta_1\right) +$$

$$+\frac{1}{(i\hbar)^2}\int_0^t d\theta_1\int_0^{\theta_1} d\theta_2 \sum_{m'} W_{mm'}(\theta_1)\times$$

$$\times \exp\left(-\frac{1}{i\hbar}(E_m - E_{m'})\theta_1\right)W_{m'n}(\theta_2)\times$$

$$\left. \times \exp\left(-\frac{1}{i\hbar}(E_{m'} - E_n)\theta_2\right) + \ldots\right\} \qquad (27.12)$$

The integration is easily performed in the particular case where the perturbation $W$ is independent of time. The result is expressed in a form which enables the extension to higher orders of approximation to be easily deduced:*

$$U_{mn}(t) = \exp\left(\frac{1}{i\hbar}E_m t\right)\delta_{mn} + W_{mn}\left\{\frac{\exp\frac{1}{i\hbar}E_m t}{E_m - E_n} + \frac{\exp\frac{1}{i\hbar}E_n t}{E_n - E_m}\right\} +$$

$$+ \sum_{m'} W_{mm'} W_{m'n}\left\{\frac{\exp\frac{1}{i\hbar}E_m t}{(E_m - E_{m'})(E_m - E_n)} + \frac{\exp\frac{1}{i\hbar}E_{m'} t}{(E_{m'} - E_m)(E_{m'} - E_n)} + \right.$$

$$\left. + \frac{\exp\frac{1}{i\hbar}E_n t}{(E_n - E_m)(E_n - E_{m'})}\right\} + \ldots \qquad (27.12a)$$

If we proceed from the state "0" of the unperturbed operator $\mathcal{H}_0$, for which $a_m(0) = \delta_{m0}$, then the amplitude $a_m(t)$ is equal to $U_{m0}(t)$, and $|U_{m0}(t)|^2$ is the probability of encountering the state $\phi_m$ at time $t$. If we include terms up to the second degree in $W$, then

$$|U_{m0}(t)|^2 = |W_{m0}|^2 \frac{\sin^2\{\frac{1}{2}(E_m - E_0)t/\hbar\}}{\frac{1}{4}(E_m - E_0)^2} \quad \text{for } m \neq 0 \qquad (27.13a)$$

$$|U_{00}(t)|^2 = 1 - \sum_{m \neq (0)} |W_{m0}|^2 \frac{\sin^2\{\frac{1}{2}(E_m - E_0)t/\hbar\}}{\frac{1}{4}(E_m - E_0)^2} \qquad (27.13b)$$

In equation (27.13a) the only contribution is from the linear term in $W$, shown in equation (27.12a), since the first term of this latter equation vanishes; in (27.13b), on the other hand, the quadratic term of the expansion must be taken into account. Equations (27.13) simply express the fact that $U$ is unitary to the second-order terms in $W$, provided that the quadratic terms in $U$ itself are taken into account.† $(\sum_m |U_{m0}(t)|^2 = 1$ or $\sum_m |a_m(t)|^2 = 1$.) This approximation is obviously only permissible provided that $\sum_{m(\neq 0)} |U_{m0}(t)|^2$ in (27.13a) is small

---

* The individual terms of the expansion contain no singularities, although it appears at first sight that infinite terms would occur on the vanishing of differences such as $E_m - E_{m'}$. It may easily be verified that in all these cases the transition to the limit, $E_m \to E_{m'}$, produces a finite and correct result according to (27.12). Indeed, we can see immediately from this expression that any equal values of $E_m$ that may occur simply give rise to a factor $t$.

† The relation (27.13b) can therefore be obtained from the unitariness property, without using the second approximation in (27.12a).

compared to 1 (see (27.9)), and that the initial state $\phi_0$ is therefore still most likely to obtain.

As an example of Dirac's approximation method, we shall treat the problem of the scattering of a particle by a potential $W(\mathbf{r})$. The Hamiltonian operator of the free particle is $\mathscr{H}_0 = -(\hbar^2/2m)\nabla^2$. The treatment will be facilitated if we imagine the particle to be contained in a cubical enclosure of side $L$ and volume $V = L^3$; this is done simply to obtain discrete eigenfunctions of $\mathscr{H}_0$. This restriction may be eliminated in the final result by allowing $V$ to tend to infinity. The normalized eigenfunctions of $\mathscr{H}_0$ are

$$\phi_{\mathbf{k}} = \frac{1}{\sqrt{V}}\exp i\mathbf{k}\mathbf{r}$$

where $\mathbf{k} = 2\pi\mathbf{n}/L$ and $\mathbf{n}$ is any vector the components of which have integral values. The energy values are $E_{\mathbf{k}} = \hbar^2\mathbf{k}^2/2m$. If the volume is large, the permissible values of $\mathbf{k}$ lie very close together; the number of possible values in an element $d\mathbf{k} = dk_x dk_y dk_z$ is $V d\mathbf{k}/(2\pi)^3$.

If we proceed from an initial state $\phi_{\mathbf{k}_0}$, in which a particle of momentum $\hbar\mathbf{k}_0$ is present in the enclosure, then from (27.13) the probability $w_{\mathbf{k}}(t)$ of finding a particle of momentum $\hbar\mathbf{k}$ after an interval $t$ is

$$w_{\mathbf{k}}(t) = |W_{\mathbf{k}\mathbf{k}_0}|^2 \frac{\sin^2\left[\frac{1}{2}(E_{\mathbf{k}} - E_{\mathbf{k}_0})t/\hbar\right]}{\frac{1}{4}(E_{\mathbf{k}} - E_{\mathbf{k}_0})^2} \quad \text{for } \mathbf{k} \neq \mathbf{k}_0 \quad (27.14)$$

in which the matrix element $W_{\mathbf{k}\mathbf{k}_0}$ is given by the expression

$$W_{\mathbf{k}\mathbf{k}_0} = \frac{1}{V}\int_V \exp\left[i(\mathbf{k}_0 - \mathbf{k})\mathbf{r}\right] W(\mathbf{r})\, d\mathbf{r} = \frac{1}{V}\overline{W}_{\mathbf{k}\mathbf{k}_0} \quad (27.14a)$$

It is always possible to ensure that $\sum_{\mathbf{k}(\neq \mathbf{k}_0)} w_{\mathbf{k}}(t)$ is small compared to 1 by taking the volume $V$ to be sufficiently large. In this case, therefore, the Dirac approximation is always applicable, however long the time interval; its value depends, however, on the extent to which it is possible to neglect the higher terms of the expansion.

The probability of finding the particle with values corresponding to $\mathbf{k}$ at time $t$ in the element $d\mathbf{k}$ is clearly $w(\mathbf{k},t)\,d\mathbf{k} = \left[V/(2\pi)^3\right] w_{\mathbf{k}}(t)\,d\mathbf{k}$. We can now define a differential effective cross-section $dq$ for scattering in the element $d\mathbf{k}$, as follows. Let the current of particles associated with the primary wave, $\mathbf{j}_0 = \hbar\mathbf{k}_0/mV$, pass through an element of area

$dq$, normal to $\mathbf{j}_0$. The size of this element is such that all particles passing through it in the time $t$ are scattered into the element $d\mathbf{k}$ of k-space:*

$$\frac{\hbar k_0}{mV} t \, dq = w(\mathbf{k}, t) \, d\mathbf{k} \qquad (27.15)$$

It is not immediately evident that $dq$ is independent of time, since at first sight $w(\mathbf{k}, t)$ is not proportional to the time; this difficulty is only apparent, however. If the time-dependent component of $w(\mathbf{k}, t)$ is plotted as a function of $E_k$ (figure 27), we see that for large values of the time there is a maximum of magnitude $(t/\hbar)^2$ at the point

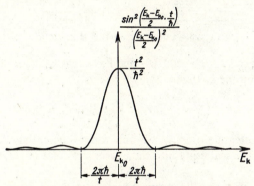

Fig. 27.—Relation between the probability $w_k(t)$ and the energy, according to (27.14)

$E_k = E_{k_0}$, and that the interval in which this component is appreciable is of order of magnitude $|E_k - E_{k_0}| \approx \hbar/t$. Therefore over a sufficiently long interval the only scattered particles to appear are those for which $E_k \approx E_{k_0}$; hence $k \approx k_0$. The particles are scattered elastically, without change of energy. If the width $\hbar/t$ is small enough for $W_{kk_0}$ to remain virtually unchanged within this region, the time factor in (27.14) may be replaced by a $\delta$-function of argument $E_k - E_{k_0}$ (cf. Exercise 1, p. 137), multiplied by the value of the integral

$$\int_{-\infty}^{+\infty} \frac{\sin^2 \left[ \frac{1}{2}(E_k - E_{k_0})t/\hbar \right]}{\frac{1}{4}(E_k - E_{k_0})^2} \, dE_k = \frac{2\pi}{\hbar} t$$

---

* Since the wave function in the volume $V$ is normalized to unity $\mathbf{j}_0$ is a probability current. The argument remains unchanged if we proceed from the initial function $(N/V)^{1/2} \exp i k_0 r$ representing $N$ particles of momentum $\hbar k_0$. $N\mathbf{j}_0$ is the particle flux, and $Nw(\mathbf{k}, t)d\mathbf{k}$ the number of scattered particles of class $(\mathbf{k}, d\mathbf{k})$.

Hence

$$dq = \frac{mV}{\hbar k_0 t} w(\mathbf{k}, t)\, d\mathbf{k} = \frac{m}{(2\pi\hbar)^2} \cdot \frac{1}{k_0} |\overline{W}_{\mathbf{k}\mathbf{k}_0}|^2\, \delta(E_\mathbf{k} - E_{\mathbf{k}_0})\, d\mathbf{k}$$

The differential cross-section will now be expressed in terms of polar coordinates in k-space. If we take $k^2 = 2mE_k/\hbar^2$, then $d\mathbf{k} = k^2\, dk\, d\Omega = (mk/\hbar^2)\, dE_k\, d\Omega$, where $d\Omega$ is the element of solid angle.

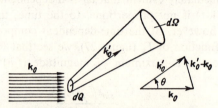

Fig. 28.—Effective scattering cross-section

Introducing a further vector $\mathbf{k}_0'$ parallel to the element of solid angle and of magnitude $k_0$ (figure 28), we obtain

$$dq = \left(\frac{m}{2\pi\hbar^2}\right)^2 |\overline{W}_{\mathbf{k}_0'\mathbf{k}_0}|^2 \cdot \delta(E_k - E_{k_0})\, dE_k\, d\Omega$$

If we now integrate over $E_k$, we obtain the differential effective cross-section $dQ$ for scattering into the element of solid angle $d\Omega$:

$$dQ = \left(\frac{m}{2\pi\hbar^2}\right)^2 |\overline{W}_{\mathbf{k}_0'\mathbf{k}_0}|^2\, d\Omega \qquad (27.16)$$

The particles of the primary wave striking $dQ$ are scattered into the element of solid angle $d\Omega$; $\mathbf{k}_0'$ is determined by the direction from which the particles are observed. For a spherically symmetrical potential $W(r)$,

$$\overline{W}_{\mathbf{k}_0'\mathbf{k}_0} = \int \exp\left[i(\mathbf{k}_0' - \mathbf{k}_0)\mathbf{r}\right] W(r)\, d\mathbf{r}$$

$$= \frac{4\pi}{|\mathbf{k}_0' - \mathbf{k}_0|} \int_0^\infty dr \cdot r W(r) \sin(|\mathbf{k}_0' - \mathbf{k}_0| r)$$

depends only on $|\mathbf{k}_0' - \mathbf{k}_0|$; since $|\mathbf{k}_0' - \mathbf{k}_0| = 2k_0 \sin\frac{1}{2}\theta$ according to figure 28, $W_{\mathbf{k}_0'\mathbf{k}_0}$ depends only on the angle between the direction of observation and the direction of incidence.

The total effective cross-section is obtained by integrating throughout the complete range of solid angles. An example of the calculation is given in Exercise 4, p. 176.

For the Dirac approximation to be applicable, it must be possible to treat $W$ as a small perturbation in comparison with the energy of the primary particles. The approximation procedure employed in the case of scattering is therefore typical of the treatment applicable to high energies. What is meant by "high energies" or "small $W$" can really be decided only from a knowledge of the higher approximations, whose further contribution must be small compared with the first approximations derived above.

In the Dirac approximation, the occurrence of the factor $\delta(E_m - E_0)$ for large time values is often taken to be an instance of the conservation of energy. This is very misleading, particularly as the impression is created that the conservation theorem holds only as an approximation for large values of the time, whereas in fact when the Hamiltonian operator is independent of time this theorem is strictly valid ($\dot{\mathscr{H}} = 0$). In fact, the $\delta$-function expresses much more; it implies that the energy associated with the unperturbed operator is conserved over long time intervals. This may be expressed in another way: when the perturbation is small it is permissible to resolve the state $\psi(t)$ into the eigenfunctions of $\mathscr{H}_0$; it then contains functions practically all of which have the same unperturbed energy. If we now consider the wave aspect of the scattering process previously discussed, its significance is easily perceived. The scattering is detectable because the plane wave associated with the incident beam produces a spherical wave propagated from the scattering centre. While this spherical wave is within the effective range of the scattering potential it contains values of $\mathbf{k}$ that differ markedly from the value $\mathbf{k}_0$ for the plane wave. The situation is different when the spherical wave has travelled outwards a distance of many wavelengths, and only a small fraction lies within the effective range of the potential. In this case nearly all the k-values occurring in the outer regions have the magnitude $k_0$, and the small portion within effective range of the potential may be neglected.*

*Appendix:* **The Born approximation for the time-independent treatment of scattering.**

In the scattering case which we discussed above, a stationary state will set in if we wait long enough. This problem may also be treated by means of Born's "time-free" approximation procedure. Putting $E = \hbar^2 k_0^2/2m$ for the energy, the Schrödinger equation becomes

$$\nabla^2 \psi + k_0^2 \psi = \frac{2m}{\hbar^2} W \psi \qquad (27.17)$$

We shall now deduce the corresponding integral equation, which is easier to deal with. Since

$$(\nabla^2 + k_0^2)\frac{e^{ikr}}{r} = -4\pi\,\delta(\mathbf{r})$$

---

* In classical mechanics the scattered particles also regain their initial energy after passing through the scattering potential.

equation (27.17) is satisfied by

$$\psi(\mathbf{r}) = \psi_0(\mathbf{r}) - \frac{m}{2\pi\hbar^2} \int W(\mathbf{r}')\psi(\mathbf{r}') \frac{\exp(ik_0|\mathbf{r}-\mathbf{r}'|)}{|\mathbf{r}-\mathbf{r}'|} \, d\mathbf{r}' \quad (27.18)$$

if $\psi_0(\mathbf{r})$ represents a solution of the equation in the absence of the perturbation potential,

$$\nabla^2\psi_0 + k_0^2\psi_0 = 0 \tag{27.19}$$

It is true that the transition from (27.17) to (27.18) is not unique, since we could have put $\exp(-ik_0|\mathbf{r}-\mathbf{r}'|)$ instead of $\exp(+ik_0|\mathbf{r}-\mathbf{r}'|)$. The choice of the $+$ sign resulted from the boundary condition that, at a great distance from the scattering centre, $\psi(\mathbf{r})$ must consist of a plane wave and a spherical wave diverging from the centre. The latter is provided for by the sign of the exponential function in (27.18), since the following approximation is valid when the distance $r$ is large:

$$\frac{\exp(ik_0|\mathbf{r}-\mathbf{r}'|)}{|\mathbf{r}-\mathbf{r}'|} \rightarrow \frac{\exp\left[ik_0 r - ik_0\left(\mathbf{r}'\frac{\mathbf{r}}{r}\right)\right]}{r}$$

$$= \frac{\exp(ik_0 r)}{r}\exp(-i\mathbf{k}_0'\mathbf{r}') \quad \text{for } |\mathbf{r}| \gg |\mathbf{r}'|$$

$$(27.18a)$$

In the above expression $\mathbf{k}_0' = k_0\mathbf{r}/r$. The first condition is satisfied by putting

$$\psi_0 = a \exp ik_0\mathbf{r} \tag{27.19a}$$

Born's approximation consists in the iterative solution of the integral equation (27.18); the $\psi$-function under the integral sign is replaced by the whole right-hand side of (27.18), the $\psi$-function occurring again is replaced in turn, the procedure being continued as far as is necessary. If terms up to the second approximation are included we obtain the following result:

$$\psi(\mathbf{r}) = \psi_0(\mathbf{r}) - \frac{m}{2\pi\hbar^2} \int W(\mathbf{r}')\psi_0(\mathbf{r}') \frac{\exp(ik_0|\mathbf{r}-\mathbf{r}'|)}{|\mathbf{r}-\mathbf{r}'|} \, d\mathbf{r}' +$$

$$+ \left(\frac{m}{2\pi\hbar^2}\right)^2 \iint W(\mathbf{r}')W(\mathbf{r}'')\psi_0(\mathbf{r}'') \times$$

$$\times \frac{\exp(ik_0|\mathbf{r}-\mathbf{r}'|)}{|\mathbf{r}-\mathbf{r}'|} \cdot \frac{\exp(ik_0|\mathbf{r}'-\mathbf{r}''|)}{|\mathbf{r}'-\mathbf{r}''|} \cdot d\mathbf{r}' \, d\mathbf{r}'' + \dots \quad (27.20)$$

We shall now write down the first approximation, taking due account of (27.18*a*) and (27.19*a*) at great distances from the scattering centre. Then, making use of the abbreviation (27.14*a*), we obtain

$$\psi(\mathbf{r}) \to a \left\{ \exp i\mathbf{k}_0\mathbf{r} - \frac{m}{2\pi\hbar^2} \frac{\exp ik_0 r}{r} \overline{W}_{\mathbf{k}_0'\mathbf{k}_0} \right\} \qquad (27.21)$$

The first term of the above expression obviously represents the primary wave, and the second, the spherical wave diverging from the scattering centre.

The differential cross-section $dQ$ is the ratio of the particle current through $d\Omega$ and the primary current across unit area:

$$dQ = \left( \frac{m}{2\pi\hbar^2} \right)^2 |\overline{W}_{\mathbf{k}_0'\mathbf{k}_0}|^2 d\Omega$$

This agrees exactly with the formula (27.16) which was obtained by means of Dirac's "time-dependent" approximation.

## §28. The variational method for the determination of eigenvalues

The eigenvalues and eigenfunctions of a Hamiltonian operator $\mathscr{H}$ are determined by the equation

$$\mathscr{H}\phi = E\phi \qquad (28.1)$$

Equation (28.1) possesses solutions, the norm of which $(\phi, \phi)$ exists only for certain definite values of $E$. If $\phi$ is normalized,

$$(\phi, \phi) = 1 \qquad (28.2)$$

then the energy

$$E = \overline{\mathscr{H}} = (\phi, \mathscr{H}\phi) \qquad (28.2a)$$

is equal to the expectation value of $\mathscr{H}$ in the state $\phi$. If $\phi$ is not normalized, then

$$E = \overline{\mathscr{H}} = \frac{(\phi, \mathscr{H}\phi)}{(\phi, \phi)} \qquad (28.3)$$

Equation (28.1) may be formulated as the solution of a variational problem. The latter is defined as follows: $\phi$ is determined in such a manner that, when $\mathscr{H}$ is given by (28.3), or by (28.2*a*) and the auxiliary condition (28.2), $E$ possesses an extreme value. Let $\phi$ be a solution, and $\phi + \alpha\eta$ a small variation of the "correct" function $\phi$, where $\alpha$ is a small complex number and $\eta$ an arbitrary function, the norm of which

$(\eta, \eta)$ exists. Then if $\mathscr{H}(\alpha)$ is formed from $\phi + \alpha\eta$ according to (28.3), it must possess an extreme value when $\alpha = 0$. This is expressed as follows:

$$\frac{\partial \mathscr{H}(\alpha)}{\partial \alpha} = \frac{\partial}{\partial \alpha} \frac{[(\phi + \alpha\eta), \mathscr{H}(\phi + \alpha\eta)]}{[(\phi + \alpha\eta), (\phi + \alpha\eta)]} \to 0 \quad \text{as} \quad \alpha \to 0 \quad (28.4)$$

If $\mathscr{H}(\alpha)$ is expanded in powers of $\alpha$ for small $\alpha$, then

$$\mathscr{H}(\alpha) = E + \frac{1}{(\phi, \phi)} \{ \alpha^* [\eta, (\mathscr{H} - E)\phi] + \alpha [\phi, (\mathscr{H} - E)\eta] \} \quad (28.5)$$
$$+ \text{ higher terms in } \alpha \text{ and } \alpha^*.$$

In the above expression $E$ is introduced as an abbreviation for $(\phi, \mathscr{H}\phi)/(\phi, \phi)$. The condition (28.4) implies that the terms of the first degree in $\alpha$ and $\alpha^*$ must vanish in (28.5). When $\alpha$ is real, we have the result

$$[\eta, (\mathscr{H} - E)\phi] + [\phi, (\mathscr{H} - E)\eta]$$
$$= [\eta, (\mathscr{H} - E)\phi] + [\eta, (\mathscr{H} - E)\phi]^* = 0 \quad (28.6a)$$

and when $\alpha$ is a pure imaginary quantity,

$$[\eta, (\mathscr{H} - E)\phi] - [\eta, (\mathscr{H} - E)\phi]^* = 0 \quad (28.6b)$$

It follows that for all $\eta$

$$[\eta, (\mathscr{H} - E)\phi] = 0 \quad (28.6c)$$

Since, however, the functions $\eta$ are entirely arbitrary, and can for instance cover all functions of a complete system, it follows that $(\mathscr{H} - E)\phi$ must vanish, and therefore that $\phi$ is a solution of equation (28.1).

If we use the definitions (28.2, 28.2a), the auxiliary condition (28.2) implies that $\eta$ must be orthogonal to $\phi$: $(\eta, \phi) = 0$. The extremum condition for (28.2a) leads to the result $(\eta, \mathscr{H}\phi) = 0$ for all $\eta$, with $(\eta, \phi) = 0$. Then $\mathscr{H}\phi$ must be proportional to $\phi$, and we again obtain (28.1).

We may express the above facts as follows. If we proceed from the exact eigenfunctions, then in the first approximation the energy value $E = \mathscr{H}$ remains unchanged if $\mathscr{H}$ is formed from a slightly modified function $\phi + \alpha\eta$. The energy value itself is relatively unaffected by variations of $\phi$; it is therefore possible to obtain good eigenvalues with comparatively poor eigenfunctions. Actually, this is just the way that the energy values behave in an ordinary perturbation calculation (§26).

The energy values are always better than the eigenfunctions by one degree of approximation; the energy in first approximation is $\mathscr{H}$, formed from the unperturbed eigenfunctions, while the energy in second approximation is determined from the eigenfunctions of the first approximation, and so on.

The extreme values of $\mathscr{H}$ do not necessarily have to be either minima, maxima, or saddle-points (cf. Exercise 3, p. 47). However, the lowest energy value, which is the energy $E_0$ of the ground state, is always the absolute minimum of $\mathscr{H}$, since the contribution of eigenfunctions with higher energy values can only result in $\mathscr{H}$ being greater than $E_0$.

The extremal property of $\mathscr{H}$ can thus be used to determine the energy of the ground state, as follows. In cases where an exact solution is impossible or difficult, we choose an initial function $\phi$ which can be presumed to represent the ground state approximately. Let this function contain one or more parameters $\beta_i$. Then $\mathscr{H}(\beta_1, \beta_2, \ldots)$ is determined as a function of the parameters, which are chosen in such a manner that $\mathscr{H}$ is a minimum. This minimum value approximately represents the energy of the ground state, and it is the optimum approximation over the function range employed. The calculated energy is always greater than the exact value, since the latter is in fact the absolute minimum. The energies calculated in this manner are generally very good approximations, although the function used to determine $\mathscr{H}$ may still be a relatively poor approximation to the eigenfunction of the ground state. There is of course an art in choosing physically suitable initial functions. The value of different types of approximation can only be judged by the results. The best approximations are obtained for the lowest energy values. Examples of the applications of approximation procedures are given in §§43 and 44.

For excited states, the application of the above method is not so simple. If the energy of the first excited state is to be determined in this manner, for instance, it is necessary to ensure that the chosen functions are orthogonal to the exact ground-state function, since any contribution from the ground state lowers the minimum energy, and it is then by no means clear whether the calculated energy is greater or less than the exact value. In many cases, however, the possibility of any contribution from the lower states can be excluded for reasons of symmetry.*

---

* The case of the atom in an electric field, described overleaf, provides another example in which the calculated state does not give the lowest value of the energy: when the distance $x$ is very large, $W$ is also very large, and negative. When the electron is removed to a great distance from the nucleus and $W$ is negative, the corresponding state certainly has a lower energy. Actually, there are no stationary bound states when a homogeneous electric field is present; the calculation given overleaf refers to a metastable state corresponding to a relative minimum (cf. remarks in §39).

A simple example may serve to illustrate the applicability of the method of variations: we shall determine the ground-state energy of a hydrogen atom in an electric field $F$ directed along the $x$-axis. The Hamiltonian operator in this case is

$$\mathscr{H} = \underbrace{\frac{\mathbf{p}^2}{2m} - \frac{e^2}{r}}_{\mathscr{H}_0} - \underbrace{eFx}_{W} \qquad (28.7)$$

The electric field is assumed to be small enough for it to be permissible to neglect terms in $F$ of higher degree than the second, in the expression for the energy. The eigenfunction $\phi_0$ of the ground state in the absence of the field is given by

$$\mathscr{H}_0 \phi_0 = E_0 \phi_0, \quad \phi_0 = \frac{e^{-r/a}}{\sqrt{(\pi a^3)}} \quad \text{where} \quad a = \frac{\hbar^2}{me^2}, \quad E_0 = -\frac{e^2}{2a} \quad (28.8)$$

The calculation of $\phi_0$ and the proof that it is the eigenfunction of the ground state are given in the next section. It may easily be verified by differentiation that $\phi_0$ is in fact an eigenfunction of $\mathscr{H}_0$ corresponding to the eigenvalue* $E_0$.

The electron charge density $\rho = e\phi_0^2$ is distributed about the nucleus with spherical symmetry and decreases exponentially with distance. The situation is similar to Thomson's model, though in this case the approximately homogeneous electron charge is distributed throughout a sphere of radius $a$ while the point-like nucleus lies at the centre of this distribution. We can determine the polarizability and the energy change in the electric field if we can assume that the charge distribution remains unaltered when the field is applied (as in Thomson's model), and that it merely moves as a whole through a fixed distance $\tilde{x}$ relative to the nucleus.

For the Thomson model (§4) in a state of equilibrium we have the following relations:

$$\frac{e^2}{a^3} \tilde{x} = eF, \quad e\tilde{x} = a^3 F; \quad \text{therefore} \quad \alpha = a^3, \quad E = E_0 - \tfrac{1}{2}\alpha F^2 \qquad (28.9)$$

In the above expressions, $a$ is the radius of the model sphere, $e\tilde{x}$ is the dipole moment, $\alpha$ is the polarizability, and $E - E_0$ is the energy change

---

* $\phi_0$ is normalized. If we take $\phi_0$ as an approximate solution and $a$ as parameter, then since $(\phi_0, \mathscr{H}\phi_0) = \hbar^2/2ma^2 - e^2/a$ must be a minimum for the best estimate of $a$, it follows that $a$ has the value $\hbar^2/me^2$.

caused by the field (i.e. the change in the electron's electrostatic energy as a result of the displacement $\bar{x}$).

We shall now calculate the polarizability from the displacement of the rigid charge distribution relative to the nucleus, just as we did for the Thomson model; in this case, however, the charge is no longer homogeneously distributed. Nevertheless, for small displacements the result depends on the field strength and the charge density at short distances from the centre of the distribution; the latter is approximately constant in this region, and equal to $e/\pi a^3$. In the Thomson model the corresponding charge density is $\frac{3}{4}e/\pi a^3$. If we therefore replace $a^3$ in (28.9) by $\frac{3}{4}a^3$, we obtain the following values for the polarizability and the energy of the present model:

$$\alpha = \tfrac{3}{4}a^3 \qquad E = E_0 - \tfrac{1}{2}\alpha F^2 \qquad (28.10)$$

where $a$ is now the Bohr radius given by (28.8). Taking $a = 0.53 \times 10^{-8}$ cm, we obtain the value $\frac{3}{4} \times 0.15 \times 10^{-24} \text{cm}^3$ for the polarizability of the hydrogen atom; this result does not agree very well with the value $0.66 \times 10^{-24} \text{cm}^3$ found by Wentzel and Waller[*] from a second-order perturbation calculation. The discrepancy is not surprising, however, in view of the crude approximations on which the present calculation is based.[†]

At this point the variational method will be found to be of assistance. So far, we have assumed that the charge distribution is rigid; this implies that the wave function $\phi_0(x,y,z)$ is replaced by $\phi_0(x - \beta_1, y, z)$, where $\beta_1$ is the displacement of the distribution. For small displacements this is equivalent to putting

$$\phi = \left(1 - \frac{i}{\hbar}\beta_1 p_x\right)\phi_0 \qquad (28.11)$$

If the above expression is used to calculate $\mathscr{H}(\beta_1)$, and $\beta_1$ is determined from the minimum of $\mathscr{H}$, the result, as might be expected, is identical with (28.10).

As has been shown by P. Gombàs,[‡] the trial solution

$$\phi = (1 + \beta_2 W)\phi_0 \qquad (28.12)$$

[*] G. Wentzel, *Z. f. Phys.* **38** (1927), 518; J. Waller, *Z. f. Phys.* **38** (1927), 635.

[†] The somewhat uncertain experimental value is given as $0.66 \times 10^{-24} \text{cm}^3$ (Landolt-Börnstein I/1, Berlin 1955).

[‡] Gombàs, P. *Theorie und Lösungsmethoden des Mehrteilchenproblems der Wellenmechanik* (Basel, 1950).

frequently leads to very good results in perturbation problems. In this case it leads to the better approximation $4a^3$ for $\alpha$. Finally, if we combine both solutions in the form

$$\phi = \left(1 - \frac{i}{\hbar}\beta_1 p_x + \beta_2 W\right)\phi_0 \qquad (28.13)$$

we obtain

$$\alpha = \tfrac{3}{4}a^3 = 0.11 \times 10^{-24}\,\mathrm{cm}^3 \text{ with } \beta_1 \text{ alone} \qquad (28.11a)$$

$$\alpha = 4a^3 = 0.60 \times 10^{-24}\,\mathrm{cm}^3 \text{ with } \beta_2 \text{ alone} \qquad (28.12a)$$

$$\alpha = (4+\tfrac{1}{8})a^3 = 0.62 \times 10^{-24}\,\mathrm{cm}^3 \text{ with } \beta_1 \text{ and } \beta_2 \quad (28.13a)$$

It is certain that the true polarizability is greater and the true energy value lower than the calculated value. The results show that the variational method is almost as good as the second-order perturbation calculation of Waller and Wentzel. The calculations using (28.13) are very simple, containing elementary integrals and merely requiring the solution of a linear system of equations for $\beta_1$ and $\beta_2$. When choosing trial functions it is of course necessary to ensure that they will lead to simple and completely integrable expressions.

The method of calculation will now be briefly described. The expectation value of the energy is formed from (28.3), using (28.13):

$$E = \frac{((1+O)\phi_0, \mathscr{H}(1+O)\phi_0)}{((1+O)\phi_0, (1+O)\phi_0)} = \frac{(\phi_0, (1+O^+)\mathscr{H}(1+O)\phi_0)}{(\phi_0, (1+O^+)(1+O)\phi_0)} \qquad (28.14)$$

in which the abbreviations

$$O = \beta_2 W - \frac{i}{\hbar}\beta_1 p_x \qquad O^+ = \beta_2 W + \frac{i}{\hbar}\beta_1 p_x \qquad (28.15)$$

have been introduced. ($\beta_1$ and $\beta_2$ are taken to be real numbers, while $W$ and $p_x$ are Hermitian operators.) Mean values over $\phi_0$ are denoted by bars. Many of the terms in (28.14) vanish because of the radial symmetry of $\phi_0$, e.g. $\overline{O}$, $\overline{O^+}$, $\overline{W}$, $\overline{O^+\mathscr{H}_0}$, $\overline{\mathscr{H}_0 O}$, $\overline{O^+WO}$. Hence from (28.14)

$$E = \frac{(\phi_0, (\mathscr{H}_0 + O^+W + WO + O^+\mathscr{H}_0 O)\phi_0)}{(\phi_0, (1+O^+O)\phi_0)} = \frac{E_0 + \overline{O^+W} + \overline{WO} + \overline{O^+\mathscr{H}_0 O}}{1+\overline{O^+O}} \qquad (28.14a)$$

The terms containing $O$ or $W$ are linear in terms of the perturbation, and therefore proportional to the field strength $F$. It will appear later that $\beta_1$ is proportional to the field strength, while $\beta_2$ is independent of it. The expression (28.14a) contains quadratic terms only. If we require the expansion only as far as terms in $F^2$, the denominator can be expanded, and we obtain

$$E = E_0 + \overline{O^+W} + \overline{WO} + \overline{O^+\mathscr{H}_0 O} - E_0\overline{O^+O} \qquad (28.14b)$$

The last two terms can be combined in the form $\overline{(O^+\mathcal{H}_0 - \mathcal{H}_0 O^+)O}$. If we now insert the expressions in (28.15) corresponding to $O$ and $O^+$, we obtain

$$E = E_0 - \beta_1 eF + \beta_2 . 2e^2F^2\overline{x^2}$$

$$- \beta_1{}^2 \frac{ie^2}{\hbar}\overline{\frac{x}{r^3}p_x} + \beta_2{}^2 . \frac{e^2F^2i\hbar}{m}\overline{p_x x} - \beta_1\beta_2 \left\{ \overline{e^3F\frac{x^2}{r^3}} + \overline{\frac{eF}{m}p_x{}^2} \right\} \qquad (28.14c)$$

The required mean values are as follows, in view of the spherical symmetry of $\phi_0$:

$$\overline{r^2} = 3a^2 \qquad \overline{r} = \frac{3}{2}a \qquad \overline{\frac{1}{r}} = \frac{1}{a} \qquad \overline{\frac{1}{r^2}} = \frac{2}{a^2}$$

Then

$$E = E_0 - \beta_1 eF + \beta_2 . 2e^2F^2a^2 + \beta_1{}^2\frac{2e^2}{3a^3} + \beta_2{}^2\frac{e^2F^2\hbar^2}{2m} - \beta_1\beta_2 \left\{\frac{e^3F}{3a} + \frac{eF\hbar^2}{3ma^2}\right\}$$
$$(28.14d)$$

If we deal with each parameter separately, then at the minimum

$$\beta_1 = \frac{3a^3F}{4e} \quad \text{when } \beta_2 = 0, \text{ therefore } E = E_0 - \tfrac{1}{2}\times\tfrac{3}{4}a^3F^2, \quad \alpha = \tfrac{3}{4}a^3 \qquad (28.15a)$$

$$\beta_2 = -\frac{2ma^2}{\hbar^2} \quad \text{when } \beta_1 = 0, \text{ therefore } E = E_0 - \tfrac{1}{2}\times 4a^3F^2, \quad \alpha = 4a^3 \qquad (28.15b)$$

These are the results given in (28.11a) and (28.12a). If we now take $\beta_1$ and $\beta_2$ together and determine the lowest value, we obtain a linear system of equations which serves to determine the optimum values:

$$\beta_1\frac{4e^2}{3a^3} - \beta_2\frac{2e^3F}{3a} = eF \qquad \beta_1\frac{2e^3F}{3a} - \beta_2\frac{e^2F^2\hbar^2}{m} = 2e^2F^2a^2 \quad (28.16)$$

The solution gives the lowest energy:

$$\beta_1 = -\frac{3Fa^3}{8e} \qquad \beta_2 = -\frac{9ma^2}{4\hbar^2}$$

therefore $\qquad E = E_0 - \frac{1}{2}\left(4 + \frac{1}{8}\right)a^3F^2 \qquad \alpha = \left(4 + \frac{1}{8}\right)a^3 \qquad (28.15c)$

From the above example it is clear that the choice of initial function is of decisive importance in the final result. If we refer back to the beginning of the previous analysis we can see that the assumption of a rigid unchanged electron structure was an unfortunate one. Judging from the satisfactory result provided by the trial function (28.12) it is evident that the outer part of the electron distribution is much more strongly influenced than the inner portion. (28.12) corresponds in effect to an additional term $x\phi_0$, whereas the corresponding quantity for (28.11) is $(x/r)\phi_0$. The two expressions therefore differ by the factor $r$ which ensures that due weight is given to the perturbation in the outer regions where it is strongest.

When choosing trial functions it is also important to make sure that no un-necessary parameters are introduced, since they would have no effect on the result. In the above example, for instance, we might have introduced functions that were proportional to $y\phi_0$ or $\partial\phi_0/\partial y$. As we can see from the fact that the symmetry is cylindrical about the $x$-axis after the field is applied, the corresponding parameters would have vanished when the minimum value was derived, and we should have expended time and trouble for nothing. It is also evident that a variation of $a$, though possible in principle, would only provide contributions proportional to $F^4$, because $E_0(a)$ is already a minimum.

In many cases a superposition of independent linear functions $\phi_i (i = 1, 2, \ldots, n)$ can be used as a trial function:

$$\phi = \sum_{i=1}^{n} c_i \phi_i$$

The initial functions do not need to be either normalized or orthogonal. The optimum values of the coefficients $c_i$ can then be determined by the method of variations.

Putting $\mathscr{H}_{ik} = (\phi_i, \mathscr{H}\phi_k)$ and $S_{ik} = (\phi_i, \phi_k)$, we obtain

$$\overline{\mathscr{H}}(c_1, \ldots, c_n) = \frac{\sum\limits_{i,k} c_i^* c_k \mathscr{H}_{ik}}{\sum\limits_{i,k} c_i^* c_k S_{ik}}$$

The derivatives with respect to the $c_i$ and $c_i^*$ must vanish. This gives the result

$$\sum_k \mathscr{H}_{ik} c_k = E \sum_k S_{ik} c_k$$

where the value of $\overline{\mathscr{H}}$ at the extremum is denoted by $E$. There is only one solution to this linear homogeneous equation if the determinant vanishes:

$$\det\left[\mathscr{H}_{ik} - ES_{ik}\right] = 0$$

The solution of this equation of the $n$th degree yields $n$ quantities $E^{(s)}$ ($s = 1, \ldots, n$) which are the stationary energy values in the available function range. Associated with each value $E^{(s)}$ are coefficients $c_i^{(s)}$ and a function $\phi^{(s)}$, the expectation value of which is simply $E^{(s)}$:

$$(\phi^{(s')}, \mathscr{H}\phi^{(s)}) = E^{(s)} \delta_{ss'}$$

The functions $\phi^{(s)}$ can be chosen in such a manner that they are normal-ized and mutually orthogonal.

The above procedure is called Ritz's method after its originator. If the system of functions is complete, it yields all the exact eigenvalues. If this is not the case approximations are obtained, the degree of accuracy of which depends once more on a skilful choice of the functions $\phi_i$.

A trial function of this type is often suitable as a starting-point for a perturbation calculation when degenerate terms are present. It is evident that this method gives energy values of first approximation and eigenfunctions of zero-order approximation, provided that we admit only orthonormal functions of the unperturbed Hamiltonian operator (for which $S_{ik} = \delta_{ik}$) corresponding to one definite energy value. These functions are in fact defined by the condition that $(\phi^{(s')}, \mathcal{H}\phi^{(s)})$ should have only diagonal elements, which are identical with the energy values of first approximation (cf. §26). However, the method of variations gives the optimum energy values even if the functions are not orthonormal, or if they are not eigenfunctions of a common unperturbed Hamiltonian operator. A relevant example is provided by the treatment of the hydrogen molecule in §44, in which the two functions chosen for the calculation of the ground state have approximately the same energy, but do not possess a common unperturbed operator.

If the functions $\phi_i$ contain additional parameters which are also varied, the energy values $E^{(s)}$ depend on these parameters as well. Since the extremum conditions are given by

$$\frac{\partial \overline{\mathcal{H}}(\ldots c_i \ldots, \ldots \beta_\lambda \ldots)}{\partial c_i} = 0 \quad \text{and} \quad \frac{\partial \mathcal{H}(\ldots c_i \ldots, \ldots \beta_\lambda \ldots)}{\partial \beta_\lambda} = 0$$

we obtain the further equations $\partial E^{(s)}/\partial \beta_\lambda = 0$

## Exercises

1. *First-order perturbation calculation for an anharmonic oscillator*

Calculate the energy to the first degree of approximation in the case of an anharmonic oscillator for which $V(x) = \frac{1}{2}m\omega^2 x^2 + \lambda x^4$. Evaluate the matrix elements $(x^4)_{nn}$ according to the method given in §15. Compare the result with the "classical" calculation in Exercise 3, p. 107; the results should agree for high quantum numbers.

2. *Quadratic secular equation*

Let the Hamiltonian operator consist of the following matrix:

$$\mathcal{H} = \begin{pmatrix} \varepsilon_1 + \lambda W_{11} & \lambda W_{12} \\ \lambda W_{21} & \varepsilon_2 + \lambda W_{22} \end{pmatrix}$$

Determine the eigenvalues (*a*) exactly, by solving the quadratic equation $\text{Det}(\mathcal{H} - \varepsilon.1) = 0$; (*b*) by perturbation theory, assuming that $|\lambda W_{12}/(\varepsilon_2 - \varepsilon_1)| \ll 1$.

(*c*) Expand the exact solution in powers of $\lambda$ and compare it with (*b*). (*d*) Discuss the degenerate case when $\varepsilon_1 = \varepsilon_2$.

### 3. *A primitive model of the electron in a metal*

Assume that an electron moves freely along the *x*-axis, and that its wave function is

$$\phi(x) = e^{ikx}/\sqrt{l} \qquad k = 2\pi n/l \qquad (n \text{ integer})$$

Investigate the effect of a small perturbation in the form of a periodic potential

$$W(x) = W(x+a) = \sum_n W_{\frac{2\pi n}{a}} \exp\left(\frac{2\pi i n}{a} x\right)$$

using second-order perturbation theory. For which neighbourhoods does the perturbation theory fail? Find the correct functions of zero-order approximation at these points. What is the form of the energy spectrum in this case? (Answer: It is split into bands.) The calculation is facilitated by noting that

$$\mathscr{H} = -(\hbar^2/2m)\,\partial^2/\partial x^2 + V(x)$$

commutes with the displacement operator $T_a$ of §23.

### 4. *Effective scattering cross-section for a screened Coulomb potential*

Calculate the differential and total cross-sections for a scattering potential $-(e^2/r)\exp(-r/a)$, using Dirac's approximation. This so-called "screened Coulomb potential" represents the interaction between an electron and a neutral hydrogen atom. The potential of the nucleus, $-e^2/r$, is screened by the electrons of the atom; *a* is of the order of magnitude of the Bohr radius. If *a* is allowed to tend to infinity a pure Coulomb potential is obtained. Compare the differential cross-section calculated in this manner with the classical scattering formula of Rutherford.

### 5. *Polarizability of a spherically symmetrical system*

Calculate the polarizability of a spherically symmetrical atom with $N$ electrons (representing an inert gas), using the trial function $\phi = (1+\beta W)\phi_o$, where $W = -eF \sum_{i=1}^{N} x_i$. $\overline{W}$ vanishes because of the spherical symmetry, since $\Sigma \overline{x_i} = 0$. $\mathscr{H}_0$ contains kinetic and Coulomb energies only. Let $\phi_0$ be the exact eigenfunction corresponding to $H_0$ and the eigenvalue $E_0$. Express the polarizability in terms of mean values corresponding to the state $\phi_0$.

# C

---

Problems
involving
one
electron

## Stationary states

### §29. The electron in a central field of force

We shall now describe the motion of an electron in the central field of force of an atom. Strictly speaking, such a field exists only in the cases of the hydrogen atom and of hydrogen-like ions, since all other atoms contain more than one electron. However, the atoms of many monovalent elements (in particular, the alkali and precious metals) may be represented as though they consisted of a rigid "hull" formed from the nucleus and the firmly bound inner electrons, around which a single valence electron revolves. This outer electron therefore moves in the Coulomb potential of the nucleus, although this potential is screened by the inner electrons. It is spherically symmetrical, acting at a great distance from the nucleus like the potential due to a nucleus with a single charge. At shorter distances the potential approaches that due to a nucleus of charge $Z$, where $Z$ is the atomic number of the element in question. The energy levels of the valence electron are therefore determined from a Schrödinger equation of the form

$$\left\{ -\frac{\hbar^2}{2m}\nabla^2 + V(r) \right\} \phi(\mathbf{r}) = E\phi(\mathbf{r}) \tag{29.1}$$

Since the potential depends only on the magnitude of $\mathbf{r}$, the eigenfunctions may be chosen in accordance with §§24 and 25 in such a manner that they are eigenfunctions of the square of the angular momentum $\mathbf{M}$ and also of its $z$-component.

Equation (29.1) can be expressed in the following form, as may easily be verified:*

$$-\frac{\hbar^2}{2m}\frac{1}{r}\frac{\partial^2(r\phi)}{\partial r^2} + \left(\frac{\mathbf{M}^2}{2mr^2} + V(r)\right)\phi = E\phi \tag{29.2}$$

* In polar coordinates $r$, $\vartheta$, $\phi$,
$$\nabla^2 = \frac{1}{r}\frac{\partial^2}{\partial r^2}r + \frac{1}{r^2\sin^2\theta}\left\{ \sin\theta\frac{\partial}{\partial\theta}\sin\theta\frac{\partial}{\partial\theta} + \frac{\partial^2}{\partial\phi^2} \right\}$$

Cf. expression (25.2) for $\mathbf{M}^2$, and Vol. I, equation (13.4$b$).

Since $\phi$ must be an eigenfunction of $\mathbf{M}^2$, it follows, if we put $r\phi = \chi(r)Y_{lm}(\theta, \phi)$, that

$$-\frac{\hbar^2}{2m}\chi'' + \left(\frac{l(l+1)\hbar^2}{2mr^2} + V(r)\right)\chi = E\chi \qquad (29.3)$$

In this equation the term added to $V(r)$ comes from the centrifugal force $F_r$:

$$F_r = -\frac{\partial}{\partial r}\frac{l(l+1)\hbar^2}{2mr^2} = \frac{l(l+1)\hbar^2}{mr^3} = m\omega^2 r$$

if the associated classical angular velocity $\omega = \hbar\sqrt{[l(l+1)]}/mr^2$ is substituted for the angular momentum $\hbar\sqrt{[l(l+1)]}$.

We shall first consider equation (29.3) for the situations in which $r$ is very small, and very large, distinguishing between the cases $E < 0$ and $E > 0$. When $r$ is small, $V$ and $E$ are small compared with the potential of the centrifugal force $1/r^2$. Hence, approximately,

$$-\chi'' + \frac{l(l+1)}{r^2}\chi = 0$$

There are two independent linear solutions of the above equation:

$$\chi = ar^{l+1} \quad \text{and} \quad \chi = br^{-l}$$

The second solution must be excluded, since it leads to a singularity of the radial part of the wave function, $R = \chi/r$, as $r$ tends to 0. Hence for small values of $r$ we must have*

$$\chi = ar^{l+1} \qquad (29.4)$$

For sufficiently large values of $r$, the terms in brackets in equation (29.3) become negligible compared with $E$, and we obtain the solution†

$$\chi = \begin{cases} A\sin kr + B\cos kr & \text{for } \dfrac{\hbar^2 k^2}{2m} = E > 0 \\[2ex] A e^{\beta r} + B e^{-\beta r} & \text{for } -\dfrac{\hbar^2\beta^2}{2m} = E < 0 \end{cases} \qquad (29.5)$$

---

* Equation (29.4) is also valid for $l = 0$; in this case the singularity of the excluded solution is weak enough for the wave function to remain normalizable. The solution must still be excluded, however, since in this case $\nabla^2\phi$ in (29.1) would exhibit a singularity similar to that of the $\delta$-function whereas $V\phi$ has a singularity of order $1/r^2$ only.

† When the Coulomb potential is expanded asymptotically the factor $1/r$ must still be retained: the relation (29.5) remains correct, in so far as $A$ and $B$ are now proportional to powers of $r$.

In the first case, $E > 0$, and the particle possesses sufficient energy to escape to infinity; this corresponds to the classical case in which a particle approaches from and flies away to infinity. By a suitable choice of the parameters $A$ and $B$ for each value of $k$ we should expect to obtain a continuation of equation (29.4). The energy is therefore not quantized, the spectrum is continuous, and the eigenfunctions cannot be normalized.

In the second case, $E < 0$, and we obtain the closed orbits of classical theory, to which correspond the spatially limited bound states of quantum theory. It is necessary that $A = 0$ in order that the eigenfunctions should remain finite when $r$ tends to infinity. A connection with equation (29.4) is established by means of the parameter $B$ alone, but only for certain definite values of $E$; the energy is quantized, and the spectrum is discrete.

In the first instance we shall consider the case of the hydrogen atom, which provides a prototype of such a spectrum; the potential $V(r) = -e^2/r$. It is convenient to replace $r$ and $E$ by the dimensionless quantities

$$\rho = \frac{r}{\hbar^2/m\,e^2} = \frac{r}{a_0} \quad \text{and} \quad \varepsilon = \frac{E}{e^2/2a_0} \tag{29.6}$$

in other words, lengths are measured in terms of the Bohr radius

$$a_0 = \frac{\hbar^2}{m\,e^2} = 0\cdot53\,\text{A.U.}$$

and energies are expressed in terms of the Rydberg unit

$$\frac{e^2}{2a_0} = 13\cdot55\,\text{eV}$$

Then, from (29.3),

$$\frac{d^2\chi}{d\rho^2} + \left\{ \varepsilon + \frac{2}{\rho} - \frac{l(l+1)}{\rho^2} \right\} \chi = 0 \tag{29.7}$$

Using (29.4) and (29.5), we can put $\chi$ in the form

$$\chi = \rho^{l+1} \left( \sum_{\nu=0}^{\infty} c_\nu \rho^\nu \right) e^{-\lambda\rho}, \quad \text{where } \lambda = \sqrt{(-\varepsilon)}$$

We then introduce the above expression into (29.7) and determine the coefficients of $\rho^{\nu+l}$:

$$c_{\nu+1}\left[(\nu+l+2)(\nu+l+1) - l(l+1)\right] - 2c_\nu\left[\lambda(\nu+l+1) - 1\right]$$

This expression must vanish for all values of $v$ in order to satisfy (29.7):

$$c_{v+1} = 2\frac{\lambda(v+l+1)-1}{(v+l+2)(v+l+1)-l(l+1)}c_v \qquad (29.8)$$

The above relation serves to determine the coefficients $c_v$ in terms of $c_0$. The value of the latter coefficient is arbitrary, and is determined by the normalization condition. We may now draw the following alternative conclusions:

1. $\lambda = 1/(n_r+l+1) = 1/n$, where $n_r = 0,1,2,\ldots$ Then $c_v = 0$ when $v \geq n_r+1$, i.e. the series terminates at $v = n_r$.

2. $n_r$ is not an integer: in this case the series does not terminate. For large values of $v$, (29.8) gives

$$c_{v+1} \approx \frac{2\lambda}{v+1}c_v$$

therefore asymptotically,

$$c_v \approx \frac{(2\lambda)^v}{v!}c_0$$

hence

$$\sum_v c_v \rho^v \approx c_0 e^{2\lambda\rho}$$

As $\rho \to \infty$, $\chi$ would increase like $e^{\lambda\rho}$ and would not be normalizable; this case must therefore be excluded.

The result is, therefore, that the energy levels are given by the following relation:

$$E_n = -\frac{e^2}{2a_0}\frac{1}{n^2}, \text{ where } n \geq l+1 \text{ is an integer} \qquad (29.9)$$

We must now investigate the degree of degeneracy of $E_n$. Firstly, if we take the relation $n = n_r+l+1$, we always obtain the same value if we put in succession $l = 0,1,\ldots,n-1$ and $n_r = n-1,n-2,\ldots,1,0$.

From §25, we know that each value of $l$ is associated with $2l+1$ different eigenfunctions $Y_{l,m}$, denoted by the quantum numbers $l$ and $m = -l,\ldots,0,\ldots,l$. Thus the total degree of degeneracy is

$$\sum_{l=0}^{n-1}(2l+1) = n^2$$

If we denote the radial component by $R_{nl}(r) = \chi_{nl}/r$, corresponding to energy $E_n$ and angular-momentum quantum number $l$, then

$$\phi_{nlm}(\mathbf{r}) = R_{nl}(r)Y_{lm}(\theta,\phi) \qquad (29.10)$$

is the complete eigenfunction of $\mathcal{H}$, $\mathbf{M}^2$, and $M_z$. The first three radial components, with their appropriate normalization factors, are as follows:

$$R_{10} = \frac{2}{a_0^{3/2}} \exp -\frac{r}{a_0}$$

$$R_{20} = \frac{1}{(2a_0)^{3/2}} \left(2 - \frac{r}{a_0}\right) \exp -\frac{r}{2a_0} \tag{29.11}$$

$$R_{21} = \frac{1}{(24a_0^3)^{1/2}} \frac{r}{a_0} \exp -\frac{r}{2a_0}$$

In order to illustrate the above components we shall consider the functions

$$w_{nl}(r)\,dr = r^2 R_{nl}^2(r)\,dr$$

which represent the probability of finding the electron in a spherical shell of thickness $dr$ at a distance $r$ from the nucleus, when the atom is in the state $R_{nl}$ (figure 29.) The maxima of these functions move outwards with each increase in the principal quantum number, and occur

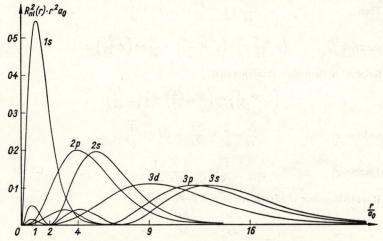

Fig. 29.—Radial probability distributions for the hydrogen atom

| $w_{10}$: 1s-state | $w_{21}$: 2p-state | $w_{31}$: 3p-state |
| $w_{20}$: 2s-state | $w_{30}$: 3s-state | $w_{32}$: 3d-state |

near the Bohr orbits of the old quantum theory.  The maximum of the function

$$w_{10} = Cr^2 \exp -\frac{2r}{a_0}$$

occurs at $r = a_0$, the radius of the smallest Bohr orbit.

The mean values $\overline{1/r}$, $\overline{1/r^2}$, $\overline{1/r^3}$ are determined by means of the functions $w_{nl}$:

$$\overline{\frac{1}{r^\nu}} = \int_0^\infty R^2_{nl} \frac{1}{r^\nu} \, r^2 dr = \left( \chi_{nl}, \frac{1}{r^\nu} \chi_{nl} \right)$$

The mean value $\overline{1/r}$ follows from the virial theorem (22.16) for Coulomb forces: $\overline{V} = 2E = -e^2/n^2a = -e^2/r$, from which

$$\overline{\frac{1}{r}} = \frac{1}{n^2} \frac{1}{a_0} \tag{29.12}$$

The mean value $\overline{1/r^2}$ may be determined by making use of the fact that the definition of the eigenvalues of (29.7) is applicable to integral and non-integral values of $l$. This equation may therefore be differentiated with respect to $l$. For this purpose it may be written as

$$\mathscr{H} \chi_{nl} = \varepsilon \chi_{nl}$$

where $\quad \mathscr{H} = \left\{ -\frac{d^2}{d\rho^2} + \frac{l(l+1)}{\rho^2} - \frac{2}{\rho} \right\} \quad$ and $\quad \varepsilon = -\frac{1}{(n_r + l + 1)^2} = -\frac{1}{n^2}$

Then

$$\frac{\partial \mathscr{H}}{\partial l} \chi + \mathscr{H} \frac{\partial \chi}{\partial l} = \frac{\partial \varepsilon}{\partial l} \chi + \varepsilon \frac{\partial \chi}{\partial l}$$

therefore

$$\left( \chi, \frac{\partial \mathscr{H}}{\partial l} \chi \right) + \left( \chi, \mathscr{H} \frac{\partial \chi}{\partial l} \right) = \frac{\partial \varepsilon}{\partial l} + \varepsilon \left( \chi, \frac{\partial \chi}{\partial l} \right)$$

Since $\mathscr{H}$ is Hermitian, it follows that

$$\left( \chi, \mathscr{H} \frac{\partial \chi}{\partial l} \right) = \left( \mathscr{H} \chi, \frac{\partial \chi}{\partial l} \right) = \varepsilon \left( \chi, \frac{\partial \chi}{\partial l} \right)$$

and hence

$$\frac{\partial \varepsilon}{\partial l} = \frac{2}{n^3} = \overline{\frac{\partial \mathscr{H}}{\partial l}} = (2l+1) \overline{\frac{1}{\rho^2}}$$

Therefore

$$\overline{\frac{1}{r^2}} = \frac{2}{(2l+1)n^3} \frac{1}{a_0^2} \tag{29.13}$$

If we differentiate (29.7) with respect to $\rho$, we obtain

$$\frac{\partial \mathscr{H}}{\partial \rho} \chi + \mathscr{H} \frac{\partial \chi}{\partial \rho} = \varepsilon \frac{\partial \chi}{\partial \rho}$$

Forming the scalar product with $\chi$:

$$\left( \chi, \frac{\partial \mathscr{H}}{\partial \rho} \chi \right) = 0$$

Since
$$\frac{\partial \mathcal{H}}{\partial \rho} = -\frac{2l(l+1)}{\rho^3} + \frac{2}{\rho^2}$$

it follows that
$$\overline{\frac{1}{\rho^2}} = \frac{\overline{l(l+1)}}{\rho^3}$$

hence, when $l \neq 0$*

$$\overline{\frac{1}{r^3}} = \frac{2}{l(l+1)(2l+1)n^3} \frac{1}{a_0^3} \tag{29.14}$$

We have now determined all the mean values of interest to us in connection with later calculations.

In the case of other monovalent atoms, equation (29.3) cannot be exactly integrated. However, by means of Schrödinger's perturbation theory it is possible to deduce the extent of the displacement of the energy levels with respect to the hydrogen terms. For this purpose we shall put

$$V(r) = -\frac{e^2}{r} + S(r)$$

treating the deviation $S(r)$ of the potential $V(r)$ from the hydrogen atom potential as a perturbation. We then obtain the hydrogen terms as the zero-order approximation and, using the methods of §26, the following perturbed terms as a first-order approximation:†

$$E_{nl} = E_n + \int_0^\infty R_{nl}^2(r) r^2 S(r) \, dr \tag{29.15}$$

$S(r)$ is only appreciably different from zero in the neighbourhood of $r = 0$, and its sign is always negative. The perturbed energy terms are therefore lower than the corresponding values of $E_n$; the deviation increases with increased concentration of the wave function $R_{ln}$ at small values of $r$. This energy diminution decreases as $l$ increases, $n$ being fixed, and as $n$ increases at a fixed value of $l$.

In figure 30 the spectra of lithium and sodium are illustrated; these provide the simplest examples of spectra of monovalent atoms other than hydrogen. The term $n = 1$ does not appear in the case of lithium, because it is occupied by the electrons of the inner shell, and cannot

---

* For the $s$-term (for which $l = 0$), $\overline{1/r^3}$ is divergent.

†This is a degenerate case; however, $R_{nl}$ and $Y_{lm}$ are the correct initial functions of zero-order approximation, since they are eigenfunctions corresponding to different eigenvalues of the operators $M^2$ and $M_z$, which commute with $\mathcal{H}$ and with $S$.

therefore be occupied by the outer electron in virtue of the Pauli exclusion principle. The same situation applies in the case of sodium for the terms $n = 1$ and $n = 2$.

The characteristic difference between the hydrogen spectrum and the spectra of the other monovalent elements consists in the removal of

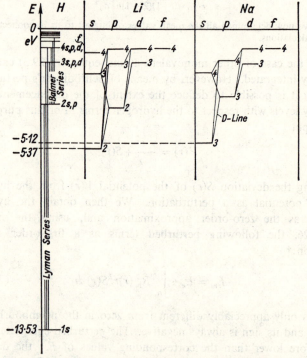

Fig. 30.—Term systems of hydrogen, lithium, and sodium

the $l$-degeneracy; for hydrogen, the terms can be designated by a "principal quantum number" $n$, whereas for the other elements the terms corresponding to the same value of $n$ but different values of $l$ are separated, and both quantum numbers are required for their designation. There are, for instance, $2s$-terms corresponding to $n = 2$, $l = 0$, $2p$-terms corresponding to $n = 2$, $l = 1$, and similarly $3s$-, $3p$-, and $3d$-terms with different energy values.

The intensity of the emitted radiation is given by the matrix elements of **r**. The results obtained in §25 imply that the only spectral lines that occur are those corresponding to the transitions $\Delta l = \pm 1$ and $\Delta m = 0$, $\pm 1$. In the case of the hydrogen atom, transitions are possible between all terms, since for any two terms there are always two corresponding functions that do not violate the above selection rules. In the case of lithium and sodium the $l$-degeneracy is removed, and it is necessary to take account of the selection rule $\Delta l = \pm 1$. The only transitions occurring are those between $p$- and $s$-states, for instance, or between $p$- and $d$-states, as shown in figure 30. The selection rule $\Delta m = 0, \pm 1$ comes into effect only when the degeneracy is completely removed.

## §30. The normal Zeeman effect

When a magnetic field is switched on, the $m$-degeneracy vanishes. In order to see this, let us express the Hamiltonian operator of the atom in a magnetic field as follows:*

$$\mathcal{H} = \frac{\left(\mathbf{p} - \frac{e}{c}\mathbf{A}\right)^2}{2m_0} + V(r) \approx \frac{\mathbf{p}^2}{2m_0} - \frac{e}{m_0 c}(\mathbf{A}, \mathbf{p}) + V(r) \quad (30.1)$$

The magnetic field is assumed to be homogeneous, and to be represented by the vector potential

$$\mathbf{A} = \tfrac{1}{2}\mathbf{H} \times \mathbf{r} \quad (30.2)$$

then, from (30.1)

$$\mathcal{H} = \frac{\mathbf{p}^2}{2m_0} - \frac{e}{2m_0 c}(\mathbf{M}, \mathbf{H}) + V(r)$$

The middle term represents the energy, in the field $\mathbf{H}$, of a magnetic moment

$$\vec{\mu} = \frac{e}{2m_0 c}\mathbf{M}$$

associated with an angular momentum† $\mathbf{M}$. If we take the $z$-axis of

---

* We omit the quadratic term in A; this is permissible for sufficiently small magnetic fields (i.e. for those fields that can be realized in practice), provided that the contribution of the first-degree term in A is different from zero. The electron mass will be denoted here by $m_0$ in order to avoid confusion with the quantum number $m$. The charge of the electron, $e$, is equal to the negative elementary charge $- e_0$, and is therefore a minus quantity. Since div $\mathbf{A} = 0$, $(\mathbf{p}, \mathbf{A}) = (\mathbf{A}, \mathbf{p})$, from which it follows that $\mathcal{H}$ is Hermitian.

† It is necessary to distinguish between the "canonical angular momentum" $\mathbf{M} = \mathbf{r} \times \mathbf{p}$ and the "mechanical angular momentum"

$$m\mathbf{r} \times \dot{\mathbf{r}} = \mathbf{M} - \mathbf{r} \times \frac{e}{c}\mathbf{A}$$

Strictly speaking, the magnetic moment is proportional to the mechanical angular momentum, but this can be neglected in the present approximation.

our coordinate system to be parallel to **H** we obtain the Schrödinger equation

$$\left\{-\frac{\hbar^2}{2m_0}\nabla^2 + V(r) - \frac{e}{2m_0 c}M_z H\right\}\psi = E\psi \qquad (30.3)$$

The solutions are the same as those in the absence of a magnetic field; the only difference is that the $m$-degeneracy is removed. We obtain

$$E_{nlm} = E_{nl} - \frac{e\hbar}{2m_0 c}mH \qquad (30.4)$$

where $E_{nl}$ is the energy of the atom in the absence of the magnetic field. The energy perturbation due to the magnetic field can be expressed as

$$\Delta E = \mu_B mH = -\mu_z H$$

where $\mu_B = -e\hbar/2m_0 c$ appears as the smallest unit of the magnetic moment, termed the *Bohr magneton*.

According to the Bohr frequency condition, the possible emission and absorption frequencies of the spectrum (30.4) are

$$\omega_{nlm,n'l'm'} = \frac{E_{nl} - E_{n'l'}}{\hbar} - \frac{eH}{2m_0 c}(m - m')$$

$$= \omega_0 + \omega_L(m - m') \qquad (30.5)$$

where

$$\omega_L = -\frac{e}{2m_0 c}H$$

is the "Larmor frequency" (cf. §35).

The intensity and polarization of the spectral lines are obtained from the "correspondence" postulates of §11. These lead to the result that the radiation corresponding to the transition $nlm \rightarrow n'l'm'$ can be considered to be emitted by a Hertzian dipole of moment

$$e\mathbf{r}(t) = e(\tilde{\mathbf{r}} e^{i\omega t} + \tilde{\mathbf{r}}^* e^{-i\omega t})$$

where

$$\omega = \omega_0 + \omega_L(m - m')$$

and

$$\tilde{\mathbf{r}} = (\phi_{nlm}, \mathbf{r}\phi_{n'l'm'})$$

is the transition element, with components $(\tilde{x}, \tilde{y}, \tilde{z})$. If $(x, y, z)$ are replaced by the quantities

$$x \pm iy = r\sin\theta\, e^{\pm i\phi} \quad \text{and} \quad z = r\cos\theta$$

then, putting $\phi_{nlm} = R_{nl} Y_{lm}$,

$$\tilde{x} \pm i\tilde{y} = (R_{nl}, rR_{n'l'})(Y_{lm}, \sin\theta \, e^{\pm i\phi} Y_{l'm'}) \qquad (30.6a)$$

$$\tilde{z} = (R_{nl}, rR_{n'l'})(Y_{lm}, \cos\theta \, Y_{l'm'}) \qquad (30.6b)$$

We now consider the angular components of the wave function. According to the recurrence formulae (25.9) to (25.11) for the spherical harmonics, the latter vanish except when $l - l' = \pm 1$ and $m - m' = 0, \pm 1$. We then have the following selection rule: for the electric dipole radiation of an electron in a central field of force the following conditions always apply,

$$\Delta l = \pm 1 \qquad \Delta m = 0, \pm 1$$

When $m = m'$, it follows from (25.9) and (25.10) that $\tilde{x} = \tilde{y} = 0$, $\tilde{z} \neq 0$, and hence that

$$\bar{\mathbf{r}}(t) = A(0, 0, \cos\omega_0 t)$$

where $A$ is a constant.

In this case the dipole moment oscillates parallel to the $z$-axis with unchanged angular frequency $\omega_0$; it therefore produces no radiation in the direction of the $z$-axis, and linearly polarized light in the plane normal thereto.

When $m = m' + 1$, $\tilde{x} + i\tilde{y} \neq 0$, $\tilde{x} - i\tilde{y} = \tilde{z} = 0$, therefore

$$\mathbf{r}(t) = B\left[\cos(\omega_0 + \omega_L)t, \, \sin(\omega_0 + \omega_L)t, 0\right]$$

where $B$ is a constant. In this case the dipole moment rotates with increased frequency in a right-handed sense about the $z$-axis. The light emitted in the direction of the $z$-axis is circularly polarized in the right-handed sense; at right angles, in the $xy$-plane, the light is linearly polarized.

Similarly, when $m = m' - 1$,

$$\mathbf{r}(t) = C\left[\cos(\omega_0 - \omega_L)t, \, -\sin(\omega_0 - \omega_L)t, 0\right]$$

where $C$ is a constant. Therefore, if we observe the system along a direction parallel to the $z$-axis and the magnetic field, we find two circularly polarized components of angular frequencies $\omega_0 \pm \omega_L$; this is the longitudinal Zeeman effect. If the system is observed at right angles to the field we obtain the transverse Zeeman effect, which comprises one component (called the $\pi$-component) of frequency $\omega_0$, linearly polarized parallel to the field, and two components (termed the $\sigma$-components) of frequencies $\omega_0 \pm \omega_L$, linearly polarized at right angles to the field (figure 31).

This phenomenon is the familiar Lorentz triplet of classical physics. In practice, the splitting of the spectral lines in the magnetic field is found to be much more complicated. Equation (30.3) cannot therefore be regarded as the correct representation of the electron motion in a

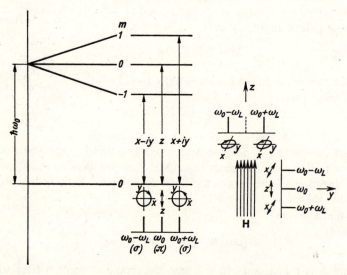

Fig. 31.—Normal Zeeman effect: term scheme and line pattern for transitions between a *p*-term and an *s*-term

magnetic field; the reason for its failure is that the electron, in addition to its electric charge, also possesses a magnetic moment that has not yet been taken into account (see chapter C II).

Finally, it should be mentioned that there are no selection rules for the radial component of (30.6), and hence for $n$. The evaluation of the integrals for these components is extremely laborious; the reader is referred to the literature.*

---

* W. Gordon, *Ann. Phys.* 2 (1929), 1031.

## Exercises

1. *Energy terms and degree of degeneracy for a three-dimensional oscillator* *

Calculate the energy terms of a three-dimensional oscillator of potential $V(r) = \frac{1}{2}m\omega^2 r^2$, in the same manner as was done for the case of the hydrogen atom. Determine the behaviour at short and at long distances from the centre, for which $\chi = r^{l+1}$ and $\exp -\frac{1}{2}\alpha r^2$ respectively; $\alpha = m\omega/\hbar$. For any distance, represent $\chi$ by the following general expression:

$$\chi = r^{l+1}\left[\exp -\tfrac{1}{2}\alpha r^2\right]\sum_{v=0}^{\infty} c_v r^v$$

The Schrödinger equation leads to a recurrence formula for the coefficients $c_v$, from which it appears that the power series must terminate. Give the formula for the energy terms and represent each term by its appropriate spectroscopic symbol ($s, p, d, f, g, h$). Deduce a general formula for the degrees of degeneracy.

2. *Hermitian nature of the Hamiltonian operator expressed in polar coordinates*

Determine the adjoint operator $p_r^+$ of $p_r = -i\hbar\partial/\partial r$, and express the radial part of $(-\hbar^2/2m)\nabla^2$,

$$-\frac{\hbar^2}{2m}\cdot\frac{1}{r^2}\frac{\partial}{\partial r}r^2\frac{\partial}{\partial r} = -\frac{\hbar^2}{2m}\frac{1}{r}\frac{\partial^2}{\partial r^2}r$$

in terms of $p_r$ and $p^+$, so that its Hermitian nature is evident.

3. *Probability function and Bohr orbits of the hydrogen atom*

Show that, in the case of a large principal quantum number $n$, and maximum angular-momentum quantum number $l = n-1$, an electron is to be found at a mean distance $\bar{r}$ from the nucleus with a corresponding small root mean square deviation $\Delta r$ given by

$$(\Delta r)^2 = \overline{(r-\bar{r})^2} = \overline{r^2} - \bar{r}^2$$

$\bar{r}$ being given by the radius of the corresponding Bohr orbit. Compare the result with (29.12) and (29.13).

---

* In nuclear theory the term scheme of the three-dimensional oscillator is often employed for the representation of the nucleon terms. The degree of degeneracy is closely connected with the so-called "magic numbers" of the nuclear shell model. (cf. Mayer-Jensen, *Elementary Theory of Nuclear Shell Structure* (New York, 1955).

# CHAPTER CII

## Electron spin

### §31. The Stern-Gerlach experiment

It was stated at the end of §30 that equation (30.3) fails to explain the experimental results of the Zeeman effect. The inadequacy of this equation becomes even more apparent when attempting to interpret the Stern-Gerlach experiment for the determination of atomic magnetic moments. An outline of this experiment is given below.

If a beam of neutral atoms of moment $\mu$ is passed through a homogeneous magnetic field $\mathbf{H_0}$, each atom is subjected to a couple $\bar{\mu} \times \mathbf{H_0}$, but not to any deflecting force. The magnetic moments therefore

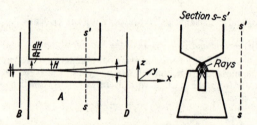

Fig. 32.—The Stern-Gerlach experiment

precess about the direction of the field in such a manner that their components in this direction remain constant; the motion of the atoms through the field, however, remains rectilinear.

The situation is different when the atomic beam is passed through an inhomogeneous magnetic field, as in the arrangement shown in figure 32. In the classical sense the magnetic moments still precess about the field direction; in addition, however, a force

$$\mathbf{F} = (\bar{\mu} . \nabla)\mathbf{H} = \left( \mu_x \frac{\partial}{\partial x} + \mu_y \frac{\partial}{\partial y} + \mu_z \frac{\partial}{\partial z} \right)\mathbf{H} \qquad (31.1)$$

arises, which produces a deviation of the beam (cf. Vol. I, §31).

In order to deduce the magnetic moment of the individual atoms from the observed deflection, we must pass the beam through the magnetic field using the simplest possible geometrical arrangement. We take the $x$-axis parallel to the initial beam direction, the $z$-axis in the plane of symmetry of the magnet, and the $y$-axis perpendicular to it (figure 32). Then $H_x = 0$, and all derivatives with respect to $x$ vanish. Hence from (31.1)

$$F_x = 0 \qquad F_y = \mu_y \frac{\partial H_y}{\partial y} + \mu_z \frac{\partial H_y}{\partial z} \qquad F_z = \mu_y \frac{\partial H_z}{\partial y} + \mu_z \frac{\partial H_z}{\partial z}$$

In particular, if the beam lies in the symmetry plane $y = 0$, $H_y = 0$ and $\partial H_y / \partial z = \partial H_z / \partial y = 0$; in the latter case because the magnetostatic field is irrotational. Further, since the field is solenoidal in the region of the beam, $\partial H_y / \partial y + \partial H_z / \partial z = 0$, and the components of the force are

$$F_x = 0 \qquad F_y = \mu_y \frac{\partial H_y}{\partial y} = -\mu_y \frac{\partial H_z}{\partial z} \qquad F_z = \mu_z \frac{\partial H_z}{\partial z}$$

We see, therefore, that deflecting components of the force exist in both the $z$- and $y$-directions.

Now the magnetic moment precesses about the direction of the field at any point, which is virtually parallel to the axis of $z$. The component $\mu_z$ is therefore practically constant, whereas $\mu_y$ oscillates about zero. If the magnetic field is so strong that the atomic magnet executes many precessional cycles during the flight of the particle, the mean value of $F_y$ vanishes, and the only non-zero component of the force is

$$F_z = \mu_z \frac{\partial H_z}{\partial z} \qquad (31.2)$$

If the atoms in question are in a state corresponding to an azimuthal quantum number $l$, this force should cause the original beam to split into $2l+1$ constituent beams, corresponding to the $2l+1$ possible values of $\mu_z$; each constituent beam would be associated with a definite value of the magnetic-moment component

$$\mu_z = -\mu_B m = \frac{e\hbar}{2m_0 c} m \qquad (31.3)$$

In particular, no splitting should occur in the case of $s$-electrons. In fact, however, when beams of alkali and noble-metal atoms are observed it is found that they split into two component beams.

Since there is certainly no orbital angular momentum associated with the electrons of these atoms in the ground state, the observed moment can only be ascribed to the electron itself, possibly because it rotates about its own axis. The value of the magnetic moment is found from measurements to be one Bohr magneton.

The experimental fact of this splitting into two component beams, together with our analysis of the angular momentum in §24 (for the case $\lambda = \frac{1}{2}$), suggests the hypothesis that the magnetic moment is connected with a spin angular momentum, the $z$-component of which can assume only the values $\pm\hbar/2$. This hypothesis was first put forward by S. A. Goudsmit and G. E. Uhlenbeck; W. Pauli was responsible for the mathematical analysis of this concept, and in particular for the representation of the angular momentum operators by means of the matrices $(\sigma_x, \sigma_y, \sigma_z)$ (§24).

The ratio of the magnetic moment to the spin angular momentum is

$$\frac{e\hbar}{2m_0 c}\bigg/\frac{\hbar}{2} = \frac{e}{m_0 c} \tag{31.4}$$

whereas the ratio of this moment to the orbital angular momentum is $e/2m_0 c$. This result makes it difficult to put forward a simple explanation of the magnetic moment in terms of electron spin; such an assumption would lead to the ratio $e/2m_0 c$ (cf. §35).

This discrepancy affords an interpretation of the magneto-mechanical anomaly in ferromagnetic materials, in the case of the Einstein-de Haas effect (Vol. III). In the study of this effect the ratio (31.4) was found experimentally; this leads necessarily to the assumption that ferromagnetism originates from electron spin, and not from the orbital angular momentum.

It will appear later that the assumption of electron spin provides an explanation of the multiplet structure (§34) and of the anomalous Zeeman effect (§36). It is found, for instance, that the energy levels of the alkali atoms, apart from the $s$-states, are split into two levels, called *doublets*, lying close together. A well-known example is the splitting of the yellow sodium D line into the two lines $D_1 = 5890$ A.U. and $D_2 = 5896$ A.U., which correspond to the transition $3p \rightarrow 3s$: this arises as a result of the separation of the $3p$-level into two neighbouring terms. The phenomenon may be explained as follows: the electrons orbiting the nucleus create a magnetic field with respect to which the spin takes up a parallel or an anti-parallel position, thus producing two different energy levels.

We shall go into the subject more thoroughly in §34, and will deal next with the motion of the spin in external fields.

### §32. Spin in a steady magnetic field

According to the previous section the operators $(s_x, s_y, s_z) = \frac{1}{2}\hbar(\sigma_x, \sigma_y, \sigma_z)$ are associated with the spin angular momentum of the electron. The spin state of the electron is represented by a function $\chi$, lying in a two-dimensional Hilbert space in which the operators $(\sigma_x, \sigma_y, \sigma_z)$ act, and which according to §24 may be considered to be spanned by the two eigenfunctions of $\sigma_z$, denoted by $\alpha, \beta$. We may therefore write

$$\chi = a\alpha + b\beta \qquad (32.1)$$

where $a$ and $b$ are complex numbers. $\chi$ is normalized if $(\chi, \chi) = |a|^2 + |b|^2 = 1$; $|a|^2$ and $|b|^2$ then represent the respective probabilities that the electron spin is parallel to the positive $z$-axis or in the opposite direction.

If we put $\alpha = \kappa_{+1}$, $\beta = \kappa_{-1}$, where $\sigma_z \kappa_s = s\kappa_s$, and $s = \pm 1$, then (32.1) may be written

$$\chi = \sum_{s=\pm 1} \kappa_s \chi(s), \text{ where } \chi(s) = (\kappa_s, \chi).$$

The spin function (or spin coordinate) $\chi(s)$ represents a combination of the two numbers $(a, b)$, and can assume only the two values $\pm 1$. An operator $A$ may then be constructed from the Pauli matrices and represented by a two-row matrix $A_{ss'}$. Such an operator acts as follows:

$$A\chi(s) = \sum_{s'=\pm 1} A_{ss'}\chi(s')$$

The normalization condition is

$$(\chi, \chi) = \sum_s \chi^*(s)\chi(s) = 1$$

The state of an electron is completely represented by a function of the position and spin coordinates, $\chi(\mathbf{r}, s, t)$, or alternatively by the pair of functions $\chi(\mathbf{r}, 1, t) = a$ and $\chi(\mathbf{r}, -1, t) = b$. The position and spin probabilities are then determined by the product $\chi^*(\mathbf{r}, s, t)\chi(\mathbf{r}, s, t)d\mathbf{r}$. To begin with we shall neglect the dependence on position and concern ourselves only with the time relationship. For this purpose we shall assume that the spatial relationship for both spin settings is represented at any moment by a comparatively broad wave packet, the dispersion of which may be neglected, and which in other respects obeys the laws of motion of classical mechanics (cf. Exercise 1, p. 206, and 2, p. 207).

The magnetic moment associated with the spin is

$$\vec{\mu} = \frac{e\hbar}{2m_0 c}\vec{\sigma} = -\mu_B \vec{\sigma}$$

and its energy in a magnetic field **H** is $-\vec{\mu}\mathbf{H}$. Hence

$$-(\vec{\mu}\mathbf{H})\chi = -\frac{\hbar}{i}\dot{\chi} \tag{32.2}$$

Taking the field to be parallel to the z-axis, and using (24.9) to represent the Pauli matrix $\sigma_z$, we may write (32.2) in terms of its components:

$$\dot{a} = -i\omega_L a \qquad \dot{b} = i\omega_L b \qquad \text{where} \qquad \omega_L = -\frac{eH}{2m_0 c} \tag{32.3}$$

The solution of these equations is

$$a(t) = a_0 e^{-i\omega_L t} \qquad b(t) = b_0 e^{i\omega_L t} \tag{32.4}$$

The normalization condition $|a_0|^2 + |b_0|^2 = 1$ is always satisfied if we put

$$a_0 = e^{i\gamma}\cos\tfrac{1}{2}\theta \qquad b_0 = e^{i\delta}\sin\tfrac{1}{2}\theta$$

It may easily be verified that
$(\chi, \vec{\sigma}\chi) = (\bar{\sigma}_x, \bar{\sigma}_y, \bar{\sigma}_z)$

$$= [\sin\theta\cos(2\omega_L t - \gamma + \delta), \sin\theta\sin(2\omega_L t - \gamma + \delta), \cos\theta] \tag{32.5}$$

The mean value of the angular momentum therefore precesses about the field direction, with twice the Larmor frequency, on account of the magneto-mechanical spin anomaly. The states $a_0 = 1$, $b_0 = 0$ and $a_0 = 0$, $b_0 = 1$ are eigenfunctions with energies

$$E = \mu_B H = \hbar\omega_L \quad \text{and} \quad E = -\mu_B H = -\hbar\omega_L \text{ respectively} \tag{32.6}$$

The Larmor precession (32.5) may perhaps be better understood if we consider the time-dependent operators

$$\vec{\sigma}(t) = \exp\left(\frac{i}{\hbar}\mathscr{H}t\right)\vec{\sigma}\exp\left(-\frac{i}{\hbar}\mathscr{H}t\right)$$

and their time derivatives

$$\frac{d\vec{\sigma}}{dt} = \frac{i}{\hbar}(\mathscr{H}\vec{\sigma} - \vec{\sigma}\mathscr{H}) \tag{32.7}$$

For instance, the $x$-component of (32.7) is

$$\frac{d\sigma_x}{dt} = -\frac{ie}{2m_0 c}[(\mathbf{H}\vec{\sigma})\sigma_x - \sigma_x(\mathbf{H}\vec{\sigma})]$$

If we write out the scalar product in terms of the components and make use of the commutation rules

$$\sigma_x\sigma_y - \sigma_y\sigma_x = 2i\sigma_z, \ldots, \ldots,$$

we obtain

$$\frac{d\vec{\sigma}}{dt} = 2\vec{\omega}_L \times \vec{\sigma} \quad \text{where} \quad \vec{\omega}_L = -\frac{e}{2m_0 c}\mathbf{H} \tag{32.8a}$$

or

$$\frac{d\mathbf{s}}{dt} = \vec{\mu} \times \mathbf{H} \quad \text{since} \quad \mathbf{s} = \tfrac{1}{2}\hbar\vec{\sigma} \tag{32.8b}$$

The latter equation may be expressed as in classical mechanics: the rate of change of angular momentum is equal to the mechanical torque. Since the equations (32.8) are linear as regards the operators $\sigma_i$, they may also serve to express the expectation values $(\chi, \sigma_i\chi)$ in the same form.

## §33. Spin resonance in an oscillating magnetic field

The existence of the Larmor precession affords a means of determination of the magnetic moments of atoms and of atomic nuclei that is more accurate than that provided by the original Stern-Gerlach method. According to equation (32.6), a particle with spin $\tfrac{1}{2}\hbar$ and magnetic moment $\mu$ in a magnetic field of magnitude $H$ is capable of assuming two different states, of energies

$$E_0 = \mu H = -\hbar\omega_L \qquad E_1 = -\mu H = \hbar\omega_L$$

If the spin and the magnetic moment are anti-parallel as in the case of the electron, then $\mu$ is negative and $E_0$ is the energy of the lower state.

If the particle is brought into a weak oscillating magnetic field normal to the constant field, transitions between the ground and the excited states can be produced, similar to those that occur in the case of an atom in the radiation field. These transitions are most easily brought about when the angular frequency $\omega_0$ of the oscillating field satisfies the Bohr frequency condition $\hbar\omega_0 = E_1 - E_0 = 2\hbar\omega_L$.

This condition obviously implies that the oscillating field resonates with the spin precessional frequency in the constant magnetic field, since, according to (32.5), the latter is equal to twice the Larmor frequency. At resonance the vertex angle of the cone of precession is continuously increased by the oscillating field, and the spin is finally "switched round" from its original direction to the opposite one, provided that it is not impeded by any damping effects.

In atomic beams, in which the spin precesses practically undamped, the onset of resonance can be directly determined by measuring the

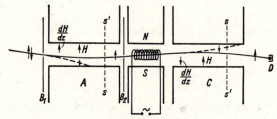

Fig. 33a.—Rabi's experiment

number of "switched spins" (I. I. Rabi, 1939). In fluids and solids strong damping forces exist as a result of interaction between spins and vibrations of the crystal lattice. For these materials, the oscillating field merely causes a precession of the spin about the constant field at the frequency of oscillation, and produces the usual resonance phenomena of a damped oscillator, such as a large increase in the amplitude of precession and a rise in absorption in the neighbourhood of the resonance frequency. Both these phenomena have been observed (F. Bloch, 1946; E. Purcell, 1946). Figure 33a shows the experimental arrangement of Rabi's method in diagrammatic form. A beam of atoms or molecules passes through the slit $B_1$ and enters the region of the inhomogeneous field due to the magnet A. A single direction of spin is selected by the screen $B_2$. In the succeeding homogeneous field NS, the field due to the coil excites additional Larmor precessions and "switches" spins. After a further splitting of the beam in the inhomogeneous magnetic field C, the unswitched spins are recorded by the detector D.

We shall now make a closer study of the Larmor precession and the deflection of the spin axis. For this purpose we shall again make use of equation (32.2), but with a magnetic field given by

$$H_x = H' \cos \omega_0 t \qquad H_y = H' \sin \omega_0 t \qquad H_z = H$$

We have assumed a rotating field instead of an oscillating one since this facilitates calculation and has little effect on the result; any oscillating field can be taken as a superposition of two rotating fields, the directions of rotation being clockwise and anti-clockwise respectively. If one of the rotating fields is in resonance with the Larmor precession, then as we shall see, the other is always far from resonance, and therefore hardly affects the precession at all. We shall take the solution of (32.2) to be

$$\chi = \begin{cases} a(t) \exp - i\omega_L t \\ b(t) \exp i\omega_L t \end{cases}$$

then, since $\hbar\omega_L = -\mu H$, $\hbar\omega'_L = -\mu H'$,

$$\dot{a} = -i\omega'_L \exp i(2\omega_L - \omega_0)t \cdot b$$

$$\dot{b} = -i\omega'_L \exp -i(2\omega_L - \omega_0)t \cdot a \qquad (33.1)$$

If we differentiate the first equation and insert the value of $b$ from the second, we obtain

$$\ddot{a} - i(2\omega_L - \omega_0)\dot{a} + \omega'^2_L a = 0$$

If we put $a \sim e^{i\omega t}$, we obtain a quadratic equation in $\omega$, the two solutions of which are

$$\omega_{1,2} = \omega_L - \tfrac{1}{2}\omega_0 \pm \sqrt{[(\omega_L - \tfrac{1}{2}\omega_0)^2 + \omega'^2_L]} \equiv \omega_r \pm \omega_K \quad (33.2)$$

We are interested in the case in which the spins are parallel to the z-axis on leaving the slit $B_2$ at the instant $t = 0$; accordingly, we put $a(0) = 1$, $b(0) = 0$. Then at later instants of time

$$a = \left( \cos \omega_K t - i \frac{\omega_r}{\omega_K} \sin \omega_K t \right) \exp i\omega_r t$$

$$b = -i \frac{\omega'_L}{\omega_K} \sin \omega_K t \exp - i\omega_r t \qquad (33.3)$$

Let $\tau$ be the time of transit through the oscillating field $H'$: then at the end of the field region we find a fraction

$$|b|^2 = \frac{\omega'^2_L}{(\omega_L - \tfrac{1}{2}\omega_0)^2 + \omega'^2_L} \sin^2 \left[ \sqrt{\{(\omega_L - \tfrac{1}{2}\omega_0)^2 + \omega'^2_L\}} \, \tau \right]$$

of the atoms of the original beam with altered spins. The relation of $|b|^2$ to $\omega_0$ exhibits the character of a resonance curve.

In order to change as many spins as possible at the resonance point $\omega_0 = 2\omega_L$, the parameters of the oscillating field are chosen to make $\tau\omega_L' = \frac{1}{2}\pi$. Rabi's method is chiefly used for the determination of nuclear magnetic moments $\mu \approx e\hbar/M_{\text{prot}}c$, when the moment of the electron shell vanishes. For $H' \approx 1\,\text{Oe}$ and a beam of atoms with

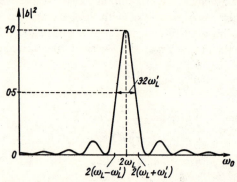

Fig. 33*b*.—Resonance curve obtained in Rabi's experiment

thermal velocities $v \approx 10^5\,\text{cm/s}$, this corresponds to a field dimension of a few centimetres. Figure 33*b* shows the form of the resonance curve. Its width is of order of magnitude $\omega_L'$; since $\omega_L'$ is chosen to be much smaller than $\omega_L$, the resonance point is very sharply defined.

In general, however, the lines are broadened by other effects, chief among which may be mentioned:

1. Dispersion of $\omega_L$ as a result of inhomogeneities in $H$, partly because of the dispersion of the external magnetic field, and also (in the case of solid and liquid materials) because $H$ does not represent the external field alone, but also includes the field of the neighbouring atoms, which naturally varies from point to point.

2. In the Rabi method, the dispersion of $\tau$ as a result of the distribution of velocities in the atomic beam.

3. For solids and liquids, there is a further damping effect due to the transfer of spin precessional energy to neighbouring atoms through spin-spin or spin-lattice interaction. (In the frequency region of interest, radiation damping may be completely neglected.) This broadening is analogous to the normal broadening of the resonance curve of a forced oscillator as a result of the damping terms.

## §34. Spin-orbit interaction and doublet splitting

Spin-orbit interaction is in itself a relativistic effect, the full treatment of which will be deferred until Section F. We shall content ourselves here with a semi-classical derivation, which is correct except for terms of order $v^2/c^2$. In the first place, we know from Vol. I, §87, that a magnetic dipole of moment $\vec{\mu}$, moving with velocity $\mathbf{v}$, possesses an electric dipole moment

$$\vec{\pi} = \frac{\mathbf{v}}{c} \times \vec{\mu}$$

For the ordinary permanent electrostatic dipole moment the potential energy in a field $\mathbf{E}$ would be equal to $-\vec{\pi}.\mathbf{E}$, that is, equal to the work required to turn the dipole away from the direction of the field. In the present case, however, the dipole is associated with an angular momentum which produces a precessional motion that we must consider somewhat more closely. The equation for this motion is

$$\frac{d\mathbf{s}}{dt} = \vec{\mu} \times \mathbf{H} + \left(\frac{\mathbf{v}}{c} \times \vec{\mu}\right) \times \mathbf{E}$$

This equation is not obtained by adding the terms $-\vec{\mu}\mathbf{H} - \vec{\pi}\mathbf{E} = -\vec{\mu}(\mathbf{H} + \mathbf{E} \times \mathbf{v}/c)$ to the Hamiltonian function; according to §32, this would lead to the equation

$$\frac{d\mathbf{s}}{dt} = \vec{\mu} \times \left(\mathbf{H} + \mathbf{E} \times \frac{\mathbf{v}}{c}\right) = \vec{\mu} \times \mathbf{H} + \left(\frac{\mathbf{v}}{c} \times \vec{\mu}\right) \times \mathbf{E} + (\vec{\mu} \times \mathbf{E}) \times \frac{\mathbf{v}}{c}$$

which differs from the previous equation by the presence of a third term that is certainly not negligibly small. The first equation is obtained in an approximate manner, however, by adding the terms $-\vec{\mu}\mathbf{H} - \frac{1}{2}\vec{\pi}\mathbf{E}$ to the Hamiltonian function. These additional terms lead in the first place to the equation for spin

$$\frac{d\mathbf{s}}{dt} = \vec{\mu} \times \left(\mathbf{H} + \frac{1}{2}\mathbf{E} \times \frac{\mathbf{v}}{c}\right)$$

which differs from the initial equation by the expression

$$\left(\frac{\mathbf{v}}{c} \times \vec{\mu}\right) \times \mathbf{E} - \frac{1}{2}\vec{\mu} \times \left(\mathbf{E} \times \frac{\mathbf{v}}{c}\right) = \frac{1}{2}\left\{\frac{\mathbf{v}}{c} \times (\vec{\mu} \times \mathbf{E}) + \mathbf{E} \times \left(\vec{\mu} \times \frac{\mathbf{v}}{c}\right)\right\}$$

In our present approximation, we may take the equation of motion of the electron to be $m\dot{\mathbf{v}} = e\mathbf{E}$; the above expression may then be written in the form

$$\frac{m}{2ec}\left\{\mathbf{v} \times (\vec{\mu} \times \dot{\mathbf{v}}) + \dot{\mathbf{v}} \times (\vec{\mu} \times \mathbf{v})\right\} = \frac{m}{2ec}\frac{d}{dt}\{\mathbf{v} \times (\vec{\mu} \times \mathbf{v})\} - \frac{1}{2}\frac{\mathbf{v}}{c} \times \left(\dot{\mathbf{s}} \times \frac{\mathbf{v}}{c}\right)$$

In the present approximation this expression may be neglected, because the mean value of the time derivative over one cycle vanishes, and because the frequency of the spin precession is small compared with the orbital frequency of the electron.

The Hamiltonian function associated with the spin equation is therefore approximately

$$\mathcal{H} = \frac{\mathbf{p}^2}{2m} + V(\mathbf{r}) - \vec{\mu}\,\mathbf{H} - \tfrac{1}{2}\vec{\mu}\left(\mathbf{E} \times \frac{\mathbf{v}}{c}\right)$$

In particular, when $\mathbf{H} = 0$ and the electric field is centrally symmetrical

$$\mathbf{E} \times \frac{\mathbf{v}}{c} = -\frac{1}{e}\frac{dV(r)}{dr}\frac{\mathbf{r}}{r} \times \frac{\mathbf{v}}{c} = -\frac{1}{emc}\frac{dV(r)}{dr} \cdot \frac{\mathbf{L}}{r}$$

and

$$\mathcal{H} = \frac{\mathbf{p}^2}{2m} + V(\mathbf{r}) + \frac{1}{2m^2c^2}\frac{1}{r}\frac{dV}{dr}(\mathbf{Ls}) \tag{34.1}$$

In order to facilitate the integration of the Schrödinger equation we first look for all operators that commute with $\mathcal{H}$. As we may see immediately, these are $\mathbf{Ls}, \mathbf{L}^2, \mathbf{s}^2$, and hence $\mathbf{J}^2 = (\mathbf{L}+\mathbf{s})^2 = \mathbf{L}^2 + \mathbf{s}^2 + 2\mathbf{Ls}$.

We thus have an operator $\mathbf{J}$, the total angular momentum, the magnitude of which is constant with respect to time (i.e. commutable with $\mathcal{H}$). Furthermore, it may be verified that the individual components of $\mathbf{J}$ also commute with $\mathcal{H}$.

Using a similar procedure to that employed in the case of the central field alone, the spin and angle components can be separated; the radial part, for which $\mathbf{Ls}$ is a pure number, may then be treated by itself. We shall therefore look for functions that are eigenfunctions of $\mathbf{J}^2, J_z, \mathbf{L}^2$ or $\mathbf{s}^2$.

To illustrate the procedure, we shall consider two operators $\mathbf{L}_1$ and $\mathbf{L}_2$, together with the total angular momentum $\mathbf{J} = \mathbf{L}_1 + \mathbf{L}_2$. We shall investigate the eigenfunctions of $\mathbf{J}^2$ and $J_z$, knowing the common eigenfunctions of $\mathbf{L}_1^2$ and $L_{1z}$, and of $\mathbf{L}_2^2$ and $L_{2z}$ respectively:

$$\mathbf{L}_1^2 Y_{l_1 m_1} = \hbar^2 l_1(l_1+1)Y_{l_1 m_1} \qquad L_{1z}Y_{l_1 m_1} = \hbar m_1 Y_{l_1 m_1}, \text{ where } |m_1| \le l_1$$

$$\mathbf{L}_2^2 Y_{l_2 m_2} = \hbar^2 l_2(l_2+1)Y_{l_2 m_2} \qquad L_{2z}Y_{l_2 m_2} = \hbar m_2 Y_{l_2 m_2}, \text{ where } |m_2| \le l_2$$

We now form the $(2l_1+1)(2l_2+1)$ products $Y_{l_1 m_1} Y_{l_2 m_2}$, corresponding to given $l_1$ and $l_2$; these are obviously also eigenfunctions of the operator $J_z$:

$$J_z Y_{l_1 m_1} Y_{l_2 m_2} = \hbar m_j Y_{l_1 m_1} Y_{l_2 m_2}, \text{ where } m_j = m_1 + m_2$$

In general, these products are not eigenfunctions of $\mathbf{J}^2$. However, since the application of a component of $\mathbf{L}_1$ or $\mathbf{L}_2$ to one of the products yields a function that still lies within the $(2l_1+1)(2l_2+1)$-dimensional Hilbert space that they span, the application of $\mathbf{J}^2$ must also lead to a function in this space, and since $\mathbf{J}^2$ and $J_z$ commute, this function will be associated with the same value of $m_j$ as that corresponding to the original product function.

Since there is only one function $Y_{l_1 l_1} Y_{l_2 l_2}$ corresponding to the maximum value of $m_j$ ($m_j = l_1 + l_2$), it must be an eigenfunction of the operator $\mathbf{J}^2$ corresponding to the eigenvalue $\hbar^2 j(j+1)$, where $j = l_1 + l_2$. Conversely, according to the general properties of angular-momentum operators there must be a total of $2j+1$ eigenfunctions corresponding to the values $j, j-1, \ldots, -j$ of $m_j$; these are derived by repeated application of the operator $J_- = L_{1-} + L_{2-}$ to the function $Y_{l_1 l_1} Y_{l_2 l_2}$. We shall now consider the two eigenfunctions of the operator $J_z$ belonging to the eigenvalue $m_j = j-1$: these are $Y_{l_1 l_1 - 1} Y_{l_2 l_2}$ and $Y_{l_1 l_1} Y_{l_2 l_2 - 1}$. A linear combination of these two functions, $J_- Y_{l_1 l_1} Y_{l_2 l_2}$, belongs to the value $j = j$; a second linear combination, orthogonal to the first one, must then belong to $j = j-1$, $m_j = j = j-1$. Similarly, linear combinations corresponding to $j = j$ and $j = j-1$ may be formed from the three functions for which $m_j = j-2$; the third combination, orthogonal to the others, must then correspond to $j = j-2$. In this manner it is possible to separate the required functions.

We thus obtain the values $l_1 + l_2$, $l_1 + l_2 - 1, \ldots$, $l_1 - l_2$, if $l_1 \geqq l_2$. The number of these functions,

$$\sum_{j=l_1-l_2}^{l_1+l_2} (2j+1) = (l_1 + l_2 + 1)^2 - (l_1 - l_2)^2$$

is equal to the number of the initial functions $(2l_1+1)(2l_2+1)$. The determination of the eigenvalues of $\mathbf{J}^2$ and their degree of degeneracy can be accomplished without any knowledge of the corresponding eigenfunctions; the explicit formulation of the latter is only necessary for the calculation of intensities (cf. Exercise 2, p. 238).

In the present case we are concerned with the quantum numbers $l_1 = l$ and $l_2 = s = \frac{1}{2}$, associated respectively with the orbital and spin angular momenta. For the total angular momentum we therefore obtain the two quantum numbers $j = l + \frac{1}{2}$ and $j = l - \frac{1}{2}$ when $l \neq 0$, and a single quantum number $j = s = \frac{1}{2}$ for $l = 0$.

The wave function $\psi$ can therefore be defined by four quantum numbers $n$, $l$, $j$, $m_j$, designated as follows:

$n$: principal quantum number,

$l$: orbital angular momentum,

$j$: total angular momentum,

$m_j$: z-component of the total angular momentum.

Then

$$(\mathbf{Ls})\psi = \tfrac{1}{2}(\mathbf{J}^2 - \mathbf{L}^2 - \mathbf{s}^2)\psi = \tfrac{1}{2}\hbar^2 \left[ j(j+1) - l(l+1) - s(s+1) \right] \psi$$

The equation for $\chi(r) = rR(r)$ is therefore

$$\left( -\frac{\hbar^2}{2m}\frac{d^2}{dr^2} + \frac{\hbar^2 l(l+1)}{2mr^2} + V + \right.$$
$$\left. + \frac{\hbar^2}{(2mc)^2}\frac{1}{r}\frac{dV}{dr}\left\{ j(j+1) - l(l+1) - s(s+1) \right\} \right)\chi = E\chi \quad (34.2)$$

Spin-orbit interaction therefore vanishes in the case of s-terms, for which $l = 0$,* and when $l \neq 0$ it produces a small perturbation of $V(r)$. In the latter case the doublet energy levels can be represented as the sum of the unperturbed energy and the expectation value of the perturbation energy:†

$$E = E_{nl} + \frac{\hbar^2}{(2mc)^2}\frac{1}{r}\overline{\frac{dV}{dr}} \cdot \left\{ \begin{array}{l} l \text{ for } j = l+\tfrac{1}{2} \\ -(l+1) \text{ for } j = l-\tfrac{1}{2} \end{array} \right. \quad (34.3)$$

If the difference between $V(r)$ and the Coulomb potential $-e^2/r$ is neglected in the spin-orbit correction term, then, using (29.14), we obtain

$$E = E_{nl} + \frac{e^2}{2a_0 n^2}\frac{\alpha^2}{(2l+1)n} \cdot \left\{ \begin{array}{l} \dfrac{1}{l+1} \text{ for } j = l+\tfrac{1}{2} \\ -\dfrac{1}{l} \text{ for } j = l-\tfrac{1}{2} \end{array} \right. \quad (34.3a)$$

In the above expression $\alpha = e^2/\hbar c \approx 1/137$ is the dimensionless Sommerfeld fine-structure constant. The energy difference due to spin-orbit

---

* Dirac's relativistic theory of the electron provides a correction term (see §60, the term with div E) that ensures that, in the case of the hydrogen atom, the formula (34.3a) is also valid for $l = 0$ when $j = l+\tfrac{1}{2}$.

† The functions designated by $n$, $l$, $j$, $m_j$ are the correct initial functions for the perturbation calculation.

interaction is therefore of the order of $\alpha^2 \approx 10^{-4}$ times smaller than $E_{nl}$. Theory predicts that the doublet energy separation should be proportional to $1/n^3$, and this is in good agreement with experiment for alkali atoms; the dependence on $l$ is not very well confirmed, because the theoretical value is much affected by the deviation of $V(r)$ from the Coulomb potential $-e^2/r$.

In addition to the splitting of the energy levels as a result of spin-orbit interaction there is a further correction term of order of magnitude $\alpha^2$, the relativistic variation of mass. If the relativistic formula for the energy

$$E_{kin} = mc^2 \sqrt{\left(1 + \frac{\mathbf{p}^2}{m^2 c^2}\right)} - mc^2$$

is expanded as far as the fourth power of $p$, we obtain

$$E_{kin} = \frac{\mathbf{p}^2}{2m} - \frac{1}{2mc^2}\left(\frac{\mathbf{p}^2}{2m}\right)^2$$

The second term again represents a small perturbation. The energy perturbation is obtained by forming the expectation value from the unperturbed eigenfunctions:

$$-\frac{1}{2mc^2}\overline{\left(\frac{\mathbf{p}^2}{2m}\right)^2} = -\frac{1}{2mc^2}\overline{[E_{nl} - V(r)]^2}$$

In the case of the alkali atoms the term separations due to $l$ are already large, and their values are imperceptibly altered by this effect. In the case of the hydrogen atom, on the other hand, the perturbation produces a separation of the originally degenerate terms corresponding to the same value of $n$. In order to determine the magnitude of this separation we introduce the appropriate values for hydrogen for $E_{nl}$ and $V$ in the above equation:

$$\Delta E_n = -\frac{1}{2mc^2}\left\{E_n^2 + 2E_n e^2 \overline{\frac{1}{r}} + e^4 \overline{\frac{1}{r^2}}\right\}$$

Then, using (29.12, 29.13) and the relation $e^2/a_0 = mc^2\alpha^2$:

$$\Delta E_n = E_n\left(\frac{2}{2l+1} - \frac{3}{4n}\right)\frac{\alpha^2}{n} \qquad (34.4)$$

We now combine (34.3) and (34.4) and replace $l$ by $j$. Then strangely enough the terms with $j = l + \frac{1}{2}$ and $j = l - \frac{1}{2}$ coincide; i.e. in the case of hydrogen, terms corresponding to the same values of $n$ and $j$ are degenerate. The total energy, including the rest mass energy, is

$$E_{nj} = mc^2 \left\{ 1 - \frac{\alpha^2}{2n^2} - \frac{\alpha^4}{2n^4} \left( \frac{n}{j+\frac{1}{2}} - \frac{3}{4} \right) + \ldots \right\} \qquad (34.5)$$

According to our method of derivation, the above approximation formula should only be valid for the case $l \neq 0$. However, if we compare it with the strictly correct Sommerfeld fine-structure formula,* which can be derived from Dirac's relativistic theory of the electron (Chapter F II), and which has been experimentally confirmed as far as terms of order $mc^2\alpha^5$, we find that (34.5) can also be employed in the case of $s$-terms. The Sommerfeld formula is as follows:

$$E_{nj} = mc^2 \left\{ 1 + \frac{\alpha^2}{[n - (j+\frac{1}{2}) + \sqrt{\{(j+\frac{1}{2})^2 - \alpha^2\}}]^2} \right\}^{-\frac{1}{2}} \qquad (34.6)$$

If this formula is expanded in terms of $\alpha^2$ and the series is terminated after the term in $\alpha^4$, we obtain precisely the formula (34.5). The first term is obviously the rest-mass energy of the electron, the second term gives the Rydberg formula, and the third is derived from spin-orbit interaction and the relativistic variability of mass.

The third term in the approximation formula only amounts to a correction, in the case of the hydrogen terms. It is significantly greater, however, for the X-ray terms of the heavy elements. These result from the displacement of a valence electron that jumps into a previously produced gap in the lowest electron shell. The electrons in the inner shells move approximately in the Coulomb potential due to the nucleus of charge $Z$. The energy levels are then given by (34.5), in which $e^2$ is replaced by $Ze^2$, or $\alpha^2$ by $Z^2\alpha^2$. The spin-orbit perturbation is therefore magnified relative to the Rydberg terms by a factor $Z^2$; for $Z \approx 30$ the factor is therefore as much as $10^3$.

## Exercises

1. *Detailed description of the Stern-Gerlach experiment for a neutral particle of spin $\frac{1}{2}$*

For the description of the experiment we assume that the wave function at time $t = 0$ can be separated into position and spin components: $\Psi_0 = \phi_0(\mathbf{r})\chi_0$. Let the function of the coordinates $\phi_0(\mathbf{r})$ be a broad wave packet, the motion of which can be calculated by classical mechanics. Let $\chi_0$ be an arbitrary spin function, e.g. an eigenfunction corresponding to $\sigma_z$ or $\sigma_y$. Both functions are assumed to be normalized.

---

* Deviations from the Sommerfeld formula (e.g. the Lamb shift) occur as a result of interaction of the electron with the zero-point oscillations of the electromagnetic radiation field; this interaction is not taken into account in the treatment of the hydrogen atom according to Dirac's theory. See Lamb, W. E., and Retherford, R. C., *Phys. Rev.* 72 (1947), 241.

(a) Let the magnetic field be parallel to the z-axis, and let its inhomogeneity be represented by $dH/dz$, as in figure 32. The velocity of the wave packet is assumed to be parallel to the x-axis; the packet therefore travels along the ray paths shown in section in figure 32. How does the separation into two beams arise as a result of the equation of motion $\mathscr{H}\Psi = i\hbar\dot{\Psi}$? What are the probabilities of finding the particle in each beam?

(b) How are conditions altered if the apparatus is rotated through an angle $\theta$ about the beam direction? Consider the special cases $\theta = 0$ (case (a)) and $\theta = \frac{1}{2}\pi$, corresponding to a field direction parallel to the y-axis.

*Hint*: If H lies in the direction $\theta$, then $\mathbf{H}\vec{\sigma} = H\sigma_\theta$, where $\sigma_\theta$ is the component of $\vec{\sigma}$ in the direction $\theta$. $\Psi$ may therefore be resolved in terms of the eigenfunctions $\kappa_{s'}{}^\theta$ of $\sigma_\theta$:

$$\Psi = \Sigma\phi_{s'}(\mathbf{r}, t)\kappa_{s'}{}^\theta \text{ with } \sigma_\theta\kappa_{s'}{}^\theta = s' \kappa_{s'}{}^\theta, s' = \pm 1$$

(cf. Exercise 3, p. 152). The equation of motion $\mathscr{H}\Psi = i\hbar\dot{\Psi}$ is thus separated into two equations corresponding to $s' = \pm 1$. We finally obtain two separate wave packets, each associated with a given spin setting.

## 2. *Multiple Stern-Gerlach experiment*

If a beam is first passed through a magnetic field parallel to z, it is split into two component beams. If these are now analysed by means of a further Stern-Gerlach arrangement in which the field is inclined at an angle $\theta$, each component beam is again split into two. Calculate the relative beam intensities, using the results of the previous exercise. Discuss the conditions in the case of a triple arrangement, in which the field is first parallel to z, then to x, and finally to z again. How does this affect the measurement of two non-commuting operators $\sigma_x$ and $\sigma_z$?

## 3. *Spin precession in a magnetic field*

Give the general solution of (32.2) in operator form for a constant magnetic field, and deduce the rate of change of the state function by comparison with (23.2).

## 4. *Fine structure of the hydrogen spectrum*

Taking account of the relativistic corrections, give a qualitative diagram of the hydrogen spectrum as far as $n = 3$, and an accurate diagram of the fine structure of the $H_\alpha$ line ($n = 3 \rightarrow n = 2$).

# CHAPTER CIII

## The atom in the electromagnetic field

### §35. The classical treatment of the motion of atomic electrons in a constant magnetic field

The effect of a superimposed magnetic field on the motion of the electrons in an atom constitutes the basis of an understanding of the Zeeman effect and diamagnetism. Both these phenomena are governed by a theorem due to Larmor.

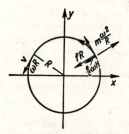

Fig. 34.—Forces acting on an elastically bound electron moving in an orbit normal to the magnetic field

We shall first consider the special case of an electron describing an orbit, which is elastically bound to an equilibrium position (figure 34). Let $-f\mathbf{r}$ be the elastic force of attraction directed towards the centre of the orbit. Then the angular frequency $\omega_0$ follows from the condition that the attractive and centrifugal forces should be equal:

$$-f\mathbf{r} + m\omega_0^2 \mathbf{r} = 0, \text{ therefore } \omega_0 = \sqrt{\frac{f}{m}} \qquad (35.1)$$

If there is a magnetic field of strength $H$ normal to the orbital plane, there arises an additional force $(e/c)\mathbf{v} \times \mathbf{H}$. In the present case the vector $\mathbf{v} \times \mathbf{H}$ also lies in the direction of $\mathbf{r}$, and is equal to $\omega H \mathbf{r}$ if $\omega$ is taken to be positive for an electron motion about the field direction in the sense of a right-handed screw. When the magnetic field is present we therefore obtain the following equation for the angular frequency $\omega$:

$$-f+\frac{e}{c}\omega H+m\omega^2 = 0 \qquad (35.2)$$

or in terms of the Larmor frequency $\omega_L = -eH/2mc$,

$$-\omega_0^2-2\omega_L\omega+\omega^2 = 0$$

Now let the magnetic field be so weak that $\omega_L$ is vanishingly small compared to $\omega_0$. (In atomic problems this condition is satisfied except for the strongest fields that can be achieved in practice.) Then the two solutions of the quadratic equation in $\omega$ are

$$\omega_1 = -\omega_0+\omega_L \qquad \omega_2 = \omega_0+\omega_L \qquad (35.3)$$

The electron rotating in a clockwise direction therefore experiences an increase $\omega_L$ in its orbital frequency, while the electron rotating in the opposite direction is retarded by the same amount. Therefore, if (35.1) represents a possible motion in the absence of the field, then (35.3) represents a possible motion when the field is present. We have not yet stated, however, whether in fact equation (35.1) is transformed into (35.3) when the field $H$ is applied.

In order to answer this question we must take account of the following facts. We know that the force exerted by a magnetic field on an electron is always normal to the orbit. The field **H** can perform no work on the electron, and the kinetic energy of the latter therefore cannot change. In fact, however, the kinetic energy of the motion corresponding to $\omega_2$, for instance, is greater than that of the initial motion, by an amount

$$\tfrac{1}{2}mr^2\{(\omega_0+\omega_L)^2-\omega_0^2\} \approx mr^2\omega_0\omega_L$$

How does this increase in kinetic energy arise? In answer, we must take into consideration the fact that when the magnetic field is applied an electric field must be created according to the equation of the law of induction

$$\operatorname{curl}\mathbf{E} = -\frac{1}{c}\dot{\mathbf{H}}$$

This electric field is capable of performing work. If the work performed by this field during one cycle is obtained by integrating over the period $\tau \approx 2\pi/\omega_0$, then, from Stokes's theorem,

$$\oint \mathbf{F} \cdot d\mathbf{s} = e \oint \mathbf{E} \cdot d\mathbf{s} = e \int (\operatorname{curl} \mathbf{E})_n \, dS = -\frac{e}{c} \dot{H} \pi r^2$$

The work done per second is therefore

$$-\frac{e\dot{H}}{2c} r^2 \omega_0$$

and the total work performed as the field increases from 0 to $H$ is

$$-\frac{eH}{2mc} mr^2 \omega_0 = mr^2 \omega_0 \, \omega_L$$

This is just the energy increase that we found above.

In the above derivation it is assumed that the orbital radius is almost constant over the period of one cycle. This condition is satisfied only if the rate of growth of the magnetic field is sufficiently slow to enable many cycles to occur during the interval; in this case the motion previously expressed by (35.1) is now represented by (35.3). Conversely, if the growth interval were short compared to the orbital period, completely different results would be obtained.

The two particular cases of a clockwise and an anticlockwise orbit are represented by (35.3), and may be summarized as follows. The influence of a slowly applied magnetic field on the motion of an electron is such that the latter possesses the same motion with respect to a co-ordinate system rotating with angular velocity $\omega_L$ as it had with respect to a system at rest before the field was applied. This is Larmor's theorem.

We must now verify that we have dealt with the general case of the elastically bound electron as a result of our previous consideration of the two orbits. For this purpose we shall consider the equation of motion of an elastically bound electron that is also subjected to a magnetic field of magnitude $H$, directed along the $z$-axis. Then the equation of motion contains not only the elastic force, but the Lorentz force as well:

$$m(\ddot{\mathbf{r}} + \omega_0^2 \mathbf{r}) = \frac{e}{c} \mathbf{v} \times \mathbf{H} \tag{35.4}$$

or in terms of the coordinates,

$$\ddot{x}+\omega_0^2 x = \frac{eH}{mc}\dot{y} \qquad \ddot{y}+\omega_0^2 y = -\frac{eH}{mc}\dot{x} \qquad \ddot{z}+\omega_0^2 z = 0$$

If the second equation is multiplied by $i$ and added to the first, and the complex number $\zeta = x+iy$ is introduced,

$$\ddot{\zeta}+\omega_0^2\zeta = -i\frac{eH}{mc}\dot{\zeta} = 2i\omega_L\dot{\zeta} \qquad (35.4a)$$

When $|\omega_L| \ll \omega_0$, the general solution of this equation is

$$\zeta = e^{i\omega_L t}(A e^{i\omega_0 t}+B e^{-i\omega_0 t}) \qquad (35.5)$$

where $A$ and $B$ are two arbitrary complex numbers.

We may immediately perceive the accuracy of the previous theorem: an oscillation in the direction of the magnetic field (the $z$-axis) is completely unaffected, whereas the motion in the plane normal to the field differs by a factor $e^{i\omega_L t}$ from the motion that takes place when $H=0$ (and therefore $\omega_L = 0$). This factor represents a rotation of the coordinate system through an angle $\omega_L t$. *Therefore, when the field is applied, a constant rotation of angular frequency $\omega_L$ about the field direction is superimposed upon the original motion.*

From the above considerations, we should expect that, if the light emitted by an electron were spectrally analysed, it would be found that the magnetic field caused an originally single line of angular frequency $\omega_0$ to be split into three lines of frequencies

$$\omega_0-\omega_L \qquad \omega_0 \qquad \omega_0+\omega_L \qquad (35.6)$$

This separation,
$$\Delta\omega = \pm\omega_L = \mp\frac{eH}{2mc} \qquad (35.7)$$

was first observed by Zeeman (somewhat qualitatively in the first place, and without a complete separation of the individual components); H. A. Lorentz immediately interpreted the phenomenon in the manner described above. The magnitude of the separation agrees with the result of the analysis given in §30.

Our analysis of the motion in the magnetic field enables us to predict the polarization of the three Zeeman components. The undisplaced vibration corresponds to a polarization of the electric vector parallel to the direction of the field, while each displaced component corresponds to a circularly polarized wave rotating respectively clockwise and anti-clockwise in a plane normal to the magnetic field. Therefore, if the light emitted by a radiating atom is observed along the direction of the lines of force, two lines are found, separated by an interval $|eH/mc|$, which are circularly polarized in a right- and left-hand sense: this is the longitudinal Zeeman effect. The undisplaced component is absent, because an oscillating dipole emits no light along the direction of its vibrations. In the transverse Zeeman effect, the light is observed from a direction at right angles to the field, and there is found to be an undisplaced line polarized parallel to the field, on either side of which at an interval $eH/2mc$ is a line polarized at right angles to the field (cf. §30). It should be mentioned that this line-splitting, known as the normal Zeeman effect, occurs only exceptionally in the form described. Most lines exhibit a more complicated separation pattern (the anomalous Zeeman effect), the interpretation of which is only possible with the help of quantum theory, including electron spin.

The Larmor theorem may also be demonstrated for the case of several electrons. If the forces on the electrons are due to a potential,* and if the latter is cylindrically symmetrical about the field direction, the possible motions in the magnetic field are obtained by superimposing a rotation upon the motion in the absence of the field, involving a rotation about **H** at an angular velocity $\omega_L$. The forces due to the potential are not affected by the additional motion of rotation, while the additional Coriolis forces $2m\vec{\omega} \times \mathbf{v}$ are compensated by the forces due to the magnetic field,† $(e/c)\mathbf{v} \times \mathbf{H}$, when $\vec{\omega} = \vec{\omega}_L = -(e/2mc)\mathbf{H}$.

In general, it is not possible to determine whether any given orbit is converted on application of the field into the corresponding orbit possessing the additional rotation; this only occurs with certainty when the field is applied slowly.

Since all electrons possess charges of the same sign, the rotational motion that sets in when the field is applied slowly must have an effect on the surroundings that is comparable to that of a current circulating about the atomic centre. Such a current produces a magnetic field that may be represented as due to a magnetic dipole situated in the atom. Therefore, if the atom possesses no magnetic moment in the absence of the field, it will exhibit such a moment when the latter is applied, as a result of the Larmor precession; the direction of this moment will be opposed to that of the field. The property of diamagnetism, which we shall now consider in further detail, is due to this moment which is induced by the magnetic field.

We start from the formula derived in Vol. I, §47 for the magnetic moment of an atom in the interior of which the current density **j** is continuously distributed:

$$\mathbf{p}^{mag} = \frac{1}{2c} \int \mathbf{r}' \times \mathbf{j}(\mathbf{r}') \, d\mathbf{r}' \tag{35.8}$$

---

* The potential includes that due to the nucleus and the electrostatic interaction between the electrons.

† It is again assumed that $\omega_L$ is small enough to enable the centrifugal force of the Larmor precession to be neglected in comparison with the Lorentz force.

If instead of the continuous current density $\mathbf{j}$ we introduce quasi point charges $e_i$ situated at the points $\mathbf{r}_i$ and possessing velocities $\mathbf{v}_i$, the integral (35.8) is reduced to a sum of integrals taken over the individual point charges. The first electron makes a contribution $e_1 \mathbf{r}_1 \times \mathbf{v}_1/2c$; the total result is therefore

$$\mathbf{p}^{mag} = \frac{1}{2c} \sum_i e_i \overline{\mathbf{r}_i \times \mathbf{v}_i} \tag{35.9}$$

where the bar denotes a time average over the electron cycles.

We shall now investigate the manner in which this moment varies when a magnetic field is applied, as a result of the Larmor rotation. The vector $(\vec{\omega}_L \times \mathbf{r}_i)$ due to this rotation must be added to the velocities $\mathbf{u}_i$ of the electrons in the rotating coordinate system; the velocities of the electrons in the stationary system are therefore

$$\mathbf{v}_i = \mathbf{u}_i + \vec{\omega}_L \times \mathbf{r}_i \tag{35.9a}$$

If we insert these values in equation (35.9) we obtain

$$\mathbf{p}^{mag} = \frac{1}{2c} \sum_i e_i \overline{\mathbf{r}_i \times (\mathbf{u}_i + \vec{\omega}_L \times \mathbf{r}_i)} = \mathbf{p}^0 + \frac{1}{2c} \sum_i e_i \overline{\mathbf{r}_i \times (\vec{\omega}_L \times \mathbf{r}_i)}$$

where $\mathbf{p}^0$ denotes the magnetic moment of the atom in the absence of the field. If we take the direction of the field to be parallel to the $z$-axis, the component of $\mathbf{p}^{mag}$ in this direction is

$$p_z^{mag} = p_z^0 + \frac{1}{2c} \sum_i e_i \overline{(x_i^2 + y_i^2)} \omega_L$$

If we insert the value of the Larmor precession $\omega_L = -eH/2mc$, then (if all $e_i = e$)

$$p_z^{mag} = p_z^0 - \frac{e^2 H}{4mc^2} \sum_i \overline{(x_i^2 + y_i^2)}$$

For a spherically symmetrical electron distribution the time averages are

$$\overline{x_i^2} = \overline{y_i^2} = \overline{z_i^2} = \overline{r_i^2}/3, \qquad \overline{x_i y_i} = \overline{y_i z_i} = \overline{z_i x_i} = 0$$

hence in this case

$$p_z^{mag} = p_z^0 - \frac{e^2 H}{6mc^2} \sum_i \overline{r_i^2} \tag{35.10}$$

In contrast, the components at right angles to the field are not altered: the $x$-component, for instance, is

$$p_x^{mag} = p_x^0 + \frac{e\omega_L}{2c} \sum_i \overline{x_i z_i} = p_x^0$$

The above formulae for the dipole moment in the presence of a magnetic field are also correct on the average for atoms that are not spherically symmetrical, provided that the individual atoms are randomly orientated.

If the atoms do not possess a permanent magnetic moment $\mathbf{p}^0$, a material containing $N$ atoms per cubic centimetre will acquire a moment

$$\left( -\frac{Ne^2}{6mc^2} \sum_i \overline{r_i^2} \right) H$$

when the field is applied. The factor occurring with $H$ is termed the diamagnetic susceptibility per unit volume. As the formula indicates, its value is directly proportional to the mean square distance of the electrons from the centre of the atom. If we denote the number of electrons in the atom by $Z$ and the mean square of their distance from the centre by $a^2$, then

$$\sum \overline{r_i^2} = Za^2$$

and the diamagnetic susceptibility is

$$\chi = -Ne^2 Za^2/6mc^2$$

See §38 for experimental values of $\chi$.

We must refer once more to our remarks in §8, when we pointed out that, strictly speaking, formula (35.10) is completely valueless in classical theory, since the thermal average of $\mathbf{p}^{mag}$ vanishes. Obviously, therefore, the term $\mathbf{p}^0$ must compensate the diamagnetism exactly; this may be verified by calculating the thermal average

$$\overline{\mathbf{p}^0} = \frac{e}{2c} \int \mathbf{r} \times \mathbf{u} \exp(-\mathcal{H}/kT) \, d\mathbf{p} \, d\mathbf{r} \Big/ \int \exp(-\mathcal{H}/kT) \, d\mathbf{p} \, d\mathbf{r}$$

where, according to (7.5) and (35.9a),

$$\mathcal{H} = \left( \mathbf{p} - \frac{e}{c} \mathbf{A} \right)^2 \Big/ 2m + V(r) \qquad \mathbf{p} = m\mathbf{v} + \frac{e}{c}\mathbf{A} = m\mathbf{u} \qquad \mathbf{A} = \mathbf{H} \times \mathbf{r}/2$$

In order that $\overline{\mathbf{p}^0}$ should contain terms of the first degree in $\mathbf{H}$, the exponential function is expanded in terms of $\mathbf{H}$ and the canonical angular momentum $\mathbf{M} = \mathbf{r} \times \mathbf{p}$ is introduced (cf. the second footnote on p. 187). Then

$$\overline{\mathbf{p}^0} = \frac{e^2}{2m^2c^2kT} \overline{(\mathbf{r} \times \mathbf{p})(\mathbf{p}\mathbf{A})} = \frac{e^2}{4m^2c^2} \frac{\overline{\mathbf{M}(\mathbf{HM})}}{kT}$$

The mean values are formed by means of the weighting function $\exp(\mathscr{H}_0/kT)$, where $\mathscr{H}_0$ is the Hamiltonian function in the absence of the magnetic field.

Putting $\overline{p_x^2} = \overline{p_y^2} = \overline{p_z^2} = mkT$ (from the law of equipartition) and $\overline{p_x p_y} = \overline{p_y p_z} = \overline{p_z p_x} = 0$, we obtain exactly the diamagnetic term, but with changed sign. In classical theory, therefore, paramagnetism is necessarily associated with diamagnetism, and exactly compensates the latter at the point of thermal equilibrium.

The situation is quite different in quantum theory. In this case the value of $\overline{\mathbf{M}(\mathbf{HM})}$ is not determined by the law of equipartition; on the contrary, the $z$-component and the square of $\mathbf{M}$ are quantized. At sufficiently low temperatures $\overline{\mathbf{M}(\mathbf{HM})}$ is independent of temperature, and $\overline{\mathbf{p}^0} = CH/T$. This is Curie's law, which we shall consider in greater detail in §37.

In conclusion, we should mention another important phenomenon that follows from (35.8) and that is capable of being verified experimentally. The electric current $\mathbf{j}$ must clearly be associated with a particle current of magnitude $(m/e)\mathbf{j}$; the current system therefore possesses an angular momentum

$$\mathbf{L} = \frac{m}{e} \int \mathbf{r} \times \mathbf{j}(\mathbf{r}) \, d\mathbf{r} = \frac{2mc}{e} \mathbf{p}^{mag} \qquad (35.11)$$

This mechanical effect may be detected in the course of the demagnetization of macroscopic specimens, i.e. when $\mathbf{p}^{mag}$ varies with respect to time; it was first observed by A. Einstein and J. de Haas. The first measurements resulted in a ratio of the angular momentum to the magnetic moment of the same order of magnitude as that predicted by (35.11). More precise measurements, in particular those due to S. J. Barnett, yielded the value $mc/e$ for this ratio; as we have already indicated in §31, this points to the fact that ferromagnetism is due to electron spin.

## §36. The anomalous Zeeman effect

In an external magnetic field both the orbital and the spin magnetic moments possess potential energy. Hence from (30.2), (32.2), and (34.1) we obtain the following expression for the Hamiltonian operator $\mathscr{H}_1$:

$$\mathscr{H}_1 = \mathscr{H}_0 + \frac{1}{2m^2c^2} \frac{1}{r} \frac{dV}{dr} (\mathbf{L}.\mathbf{s}) - \frac{e}{2mc} (\mathbf{H}.\mathbf{L}) - \frac{e}{mc} (\mathbf{H}.\mathbf{s}) \qquad (36.1)$$

If the field $\mathbf{H}$ is taken parallel to the $z$-axis,

$$\mathscr{H}_1 = \mathscr{H}_0 + \frac{1}{2m^2c^2} \frac{1}{r} \frac{dV}{dr} (\mathbf{L}.\mathbf{s}) - \frac{eH}{2mc} (J_z + s_z) \qquad (36.2)$$

The operator $\mathcal{H}_1$ no longer commutes with $\mathbf{J}^2$, because of the term in $s_z$; however, it still commutes with $\mathbf{L}^2$, $\mathbf{s}^2$, and $J_z$. Strictly speaking, therefore, we can use the quantum numbers $n$, $l$, $s$, and $m_j$ only for the characterization of the states. However, the eigenfunctions of $\mathcal{H}_1$ are represented by a superposition of the solutions associated with the same value of $m_j$ in the absence of the magnetic field. When $s = 1/2$ they may be written in the form*

$$\phi_{nlm_j} = R_{nl} \{a Y_{l,m_j-1/2} \cdot \alpha + b Y_{l,m_j+1/2} \cdot \beta\} = R_{nl} \tilde{Y}_{lm_j} \quad (36.3)$$

The coefficients $a$ and $b$ depend on $H$, and are determined from the eigenvalue equation

$$\left\{ E_{nl} + \left( R_{nl}, \frac{1}{2m^2c^2} \frac{1}{r} \frac{dV}{dr} R_{nl} \right) (\mathbf{L} \cdot \mathbf{s}) - \frac{eH}{2mc} (J_z + s_z) \right\} \tilde{Y}_{lm_j} = E \tilde{Y}_{lm_j} \quad (36.4)$$

Two limiting cases are of importance, and are easy to treat in the general case when $s$ is free to assume any value.

(a) *The potential energy in the external field is small compared to the spin-orbit interaction*

In this case we may put $H = 0$ in (36.2) as an approximation of zero order; the energy perturbation is the expectation value of $(-eH/2mc)(J_z + s_z)$, formed from the unperturbed eigenfunctions $\tilde{Y}_{ljm_j}$:

$$E_{nljm_j} = E_{nlj} + \mu_B m_j H(1 + \bar{s}_z/\hbar m_j) = E_{nlj} + \mu_B m_j H g \quad (36.5)$$

The factor in brackets is called the Landé splitting factor, $g$, or the Landé $g$-factor. In order to determine it, we require the expectation value of $s_z$, which is obtained as follows. Firstly, we require the following commutation relations, which may easily be verified.

$$\mathbf{L}(\mathbf{L} \cdot \mathbf{s}) - (\mathbf{L} \cdot \mathbf{s})\mathbf{L} = i\hbar \mathbf{s} \times \mathbf{L}, \qquad \mathbf{s}(\mathbf{L} \cdot \mathbf{s}) - (\mathbf{L} \cdot \mathbf{s})\mathbf{s} = -i\hbar \mathbf{s} \times \mathbf{L}$$

Since $\mathbf{J} = \mathbf{L} + \mathbf{s}$, it follows by addition that $\mathbf{J}$ commutes with $(\mathbf{L} \cdot \mathbf{s})$. Hence, if we multiply the second of the above relations vectorially from the right by $\mathbf{J}$, the result may be written

$$(\mathbf{s} \times \mathbf{J})(\mathbf{L} \cdot \mathbf{s}) - (\mathbf{L} \cdot \mathbf{s})(\mathbf{s} \times \mathbf{J}) = -i\hbar(\mathbf{s} \times \mathbf{L}) \times \mathbf{J}$$

$$= -i\hbar \{\mathbf{L}(\mathbf{s} \cdot \mathbf{J}) - \mathbf{s}(\mathbf{L} \cdot \mathbf{J})\} \quad (36.6)$$

$$= i\hbar \{\mathbf{s} \cdot \mathbf{J}^2 - \mathbf{J}(\mathbf{s} \cdot \mathbf{J})\}$$

---

* The functions $Y_{lm}$ are functions of angle alone; the functions $\tilde{Y}_{lm}$ also contain the spin.

The last relation is due to the fact that **L** and **s** commute and that **L** = **J** − **s**.

If we now form the expectation value of (36.6) in the state $\tilde{Y}_{ljm_j}$ the left side vanishes, because (**L**, **s**) is Hermitian and the $\tilde{Y}_{ljm_j}$ are the associated eigenfunctions. Hence

$$\overline{\mathbf{s}\mathbf{J}^2} = \overline{\mathbf{J}(\mathbf{s},\mathbf{J})}, \text{ and in particular, } \overline{s_z\mathbf{J}^2} = \overline{J_z(\mathbf{s},\mathbf{J})} \quad (36.7)$$

Since $\tilde{Y}_{ljm_j}$ is also an eigenfunction of the operators $\mathbf{J}^2$, $J_z$, and $(\mathbf{s},\mathbf{J}) = \tfrac{1}{2}(\mathbf{J}^2 + \mathbf{s}^2 - \mathbf{L}^2)$ with corresponding eigenvalues $\hbar^2 j(j+1)$, $\hbar m_j$, and $\tfrac{1}{2}\hbar^2[j(j+1)+s(s+1)-l(l+1)]$, it follows from (36.5) that the Landé factor is

$$g = 1 + \frac{j(j+1)+s(s+1)-l(l+1)}{2j(j+1)} \quad (36.8)$$

In particular, when $s = \tfrac{1}{2}$,

$$g = \begin{cases} 1 + \dfrac{1}{2l+1} = \dfrac{2l+2}{2l+1} & \text{for } j = l+\tfrac{1}{2} \\[2ex] 1 - \dfrac{1}{2l+1} = \dfrac{2l}{2l+1} & \text{for } j = l-\tfrac{1}{2} \end{cases} \quad (36.9)$$

When these values of $g$ are introduced into the expression (36.5) for the energy, we obtain the term separation for single electron systems applicable to the anomalous Zeeman effect in weak magnetic fields. Figures 35 to 37 show the splitting of the sodium D lines both in the term scheme and on the frequency diagram of the possible transitions, including the polarizations observed transversally.

Equation (36.7) may be interpreted in terms of classical mechanics in a very simple manner. The spin-orbit interaction causes the vectors **L** and **s** to precess about the vector **J** = **L** + **s**, the time rate of change of which is very small when the field is weak (figure 38). Therefore, if the vector **s** is resolved into components parallel and perpendicular to **J**,

$$\mathbf{s} = \frac{(\mathbf{s},\mathbf{J})}{\mathbf{J}^2}\mathbf{J} + \left(\mathbf{s} - \frac{(\mathbf{s},\mathbf{J})}{\mathbf{J}^2}\mathbf{J}\right) = \mathbf{s}_{||} + \mathbf{s}_{\perp}$$

the mean value of $\overline{\mathbf{s}_{\perp}}$ over one precessional period is zero. Hence $\overline{s_z} = (\mathbf{s},\mathbf{J})J_z/\mathbf{J}^2$, which corresponds exactly to (36.7).

(*b*) *The potential energy in the external field is large compared to the spin-orbit interaction*

In this case we obtain the approximation of zero order by eliminating the term in $(\mathbf{L}, \mathbf{s})$ in (36.4); $\mathbf{L}^2$, $L_z$, $\mathbf{s}^2$, and $s_z$ then commute with $\mathscr{H}_1$, and $l$, $m_l$, $s$, $m_s$ may therefore be chosen as the quantum numbers. The energy is

$$E_{nlm_lm_s} = E_{nl} + \mu_B H(m_l + 2m_s) \tag{36.10}$$

As in the case of equations (30.6*a*, 30.6*b*), it may be shown that the selection rules $\Delta l = \pm 1$ and $\Delta m_l = 0, \pm 1$ apply, together with the

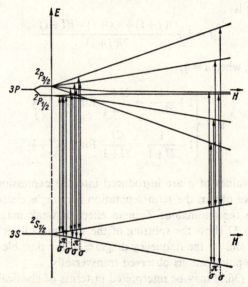

Fig. 35.—Displacement and splitting of the energy terms of the sodium D lines in the magnetic field. Each group of transitions refers to a single field strength, but for the sake of clarity the individual transitions are shown side by side.

additional rule $\Delta m_s = 0$. As a result of this last rule, only the normal Lorentz triplet (30.5) is observed.

This transition from the complicated pattern of the anomalous Zeeman effect in weak fields to the simple Lorentz triplet in strong fields has been verified experimentally and is termed the *Paschen-Back effect*.

In order to analyse the transition quantitatively it is necessary to solve the eigenvalue equation (36.4). When $m_j = \pm(l+\frac{1}{2})$, and hence $m_s = \pm\frac{1}{2}$, only one of the functions $Y_{lm}$ in (36.3) is different from zero.

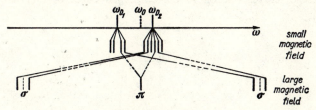

Fig. 36.—Zeeman effect: structure of the sodium D-lines in weak and strong magnetic fields

The expression for the energy is obtained directly from (36.4) by forming the scalar product with this function:

$$E = E_{nl} + (\Delta E_{nl})_{j=l+\frac{1}{2}} \pm \mu_B H(l+1) \quad \text{for} \quad m_j = \pm(l+\frac{1}{2}) \qquad (36.11)$$

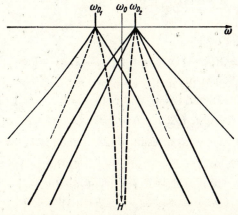

Fig. 37.—Zeeman effect: line shift and intensity as a function of field strength. The broken lines refer to the $\pi$-components. The thickness of the lines indicates the intensity

In the above expression $(\Delta E_{nl})_{j=l+\frac{1}{2}}$ denotes the energy separation when $H = 0$, due to spin-orbit interaction, as calculated in §34, and including the additional effects mentioned there. When $m_j = \pm(l+\frac{1}{2})$, therefore, the displacement of the energy terms is proportional to $H$;

this result agrees with (36.5) and (36.10), and is valid in particular for all $s$-terms, for which $l = 0$, $m_j = m_s = \pm\frac{1}{2}$ (cf. fig. 35).

When $|m_j| < l+\frac{1}{2}$, both the functions $Y_{lm}$ in (36.3) are different from zero, corresponding to the combined effect of the two states

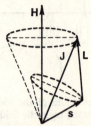

Fig. 38.—Vector model of the angular momenta

$j = l+\frac{1}{2}$ and $j = l-\frac{1}{2}$ possessing the same value of $m_j$. In this case forming the scalar product with $\phi_1 = Y_{l,m_j-\frac{1}{2}}\cdot\alpha$ and $\phi_2 = Y_{l,m_j+\frac{1}{2}}\cdot\beta$ yields a secular equation of the form

$$\begin{vmatrix} (\mathscr{H}_{11}-E) & \mathscr{H}_{12} \\ \mathscr{H}_{21} & (\mathscr{H}_{22}-E) \end{vmatrix} = 0$$

the solutions of which are

$$E_{1,2} = \tfrac{1}{2}(\mathscr{H}_{11}+\mathscr{H}_{22}) \pm [\tfrac{1}{4}(\mathscr{H}_{11}-\mathscr{H}_{22})^2 + |\mathscr{H}_{12}|^2]^{1/2}$$

In order to calculate the matrix elements $\mathscr{H}_{ik} = (\phi_i, \mathscr{H}\phi_k)$ we make use of (23.12), (23.13), and (24.8), and put $\sigma_\pm = \frac{1}{2}(\sigma_x \pm i\sigma_y)$. Then

$$2(\mathbf{L},\mathbf{s}) = \hbar^2(\Lambda_x\sigma_x+\Lambda_y\sigma_y+\Lambda_z\sigma_z) = \hbar^2(\Lambda_+\sigma_-+\Lambda_-\sigma_++\Lambda_z\sigma_z)$$

from which it follows as a result of (24.6) and (24.7) that

$$\mathscr{H}_{11} = E_{nl}+K_{nl}(m_j-\tfrac{1}{2})+\mu_B H(m_j+\tfrac{1}{2})$$

$$\mathscr{H}_{22} = E_{nl}-K_n(m_j+\tfrac{1}{2})+\mu_B H(m_j-\tfrac{1}{2})$$

$$\mathscr{H}_{12} = \mathscr{H}_{21} = +K_{nl}\sqrt{[(l+\tfrac{1}{2})^2-m_j^2]}$$

where $\qquad K_{nl} = \dfrac{\hbar^2}{(2mc)^2}\displaystyle\int_0^\infty R_{nl}\frac{1}{r}\frac{dV}{dr}R_{nl}r^2\,dr > 0$

We finally obtain the following expression for the energy for the case in which $|m_j| \neq l+\frac{1}{2}$:

$$E_{1,2} = E_{nl}-\tfrac{1}{2}K_{nl} \pm \sqrt{[K_{nl}^2(l+\tfrac{1}{2})^2 + K_{nl}\mu_B H m_j + (\tfrac{1}{2}\mu_B H)^2]} + \\ + \mu_B H m_j \quad (36.12)$$

It may be verified that the energy expression (36.11) tends to (34.3) as $H$ becomes vanishingly small; when $H$ is small, it is converted into (36.5), where $g$ is represented by (36.9); when $H$ is very large the expression tends to (36.10) (cf. figs. 35, 36, 37). For strong fields, the doublet structure is preserved in the case of the $\sigma$-components, but vanishes in the case of the $\pi$-component (figure 36 and 37).

## §37. Quantum theory of paramagnetism and diamagnetism

When a material is magnetizable, a magnetic moment is produced in it under the influence of a magnetic field. In the case of diamagnetism this moment is opposed to the field, while for paramagnetic materials it lies parallel to the field direction.

When calculating the magnetizability it is necessary to distinguish whether or not the individual molecules or atoms already possess a magnetic moment. In the latter case the magnetization arises as a result of the moment induced by the field in the atoms, as we have already seen in the classical treatment of §35. This induced moment always produces a diamagnetic effect; the latter is unaffected by the existence of any permanent moment, and is therefore a basic phenomenon occurring in all materials. However, when the individual molecules possess a permanent magnetic moment the resultant paramagnetism is generally much stronger, and the coexisting diamagnetism may therefore be neglected.

We give below a treatment of paramagnetism and diamagnetism of single-electron systems, or more precisely, of atoms with one valence electron which is almost entirely responsible for the magnetism. Such atoms possess a permanent magnetic moment as a result of the electron spin. There are also elements, such as boron in the ground state $^2P_{1/2}$, that possess a non-vanishing orbital moment. We shall, however, perform the analysis in a sufficiently general manner to enable the results to be easily extended to atoms with more than one electron, in which the permanent moment may vanish.*

The Hamiltonian operator for the atom in a constant external field is

$$\mathscr{H} = \mathscr{H}_0 - \frac{e}{mc}(\mathbf{A}, \mathbf{p}) + \frac{e^2}{2mc^2}\mathbf{A}^2 - \frac{e}{mc}(\mathbf{s}, \mathbf{H}) \quad \text{where} \quad \mathbf{A} = \tfrac{1}{2}\mathbf{H} \times \mathbf{r} \quad (37.1)$$

---

* This may occur in the case of the so-called Russell-Saunders or L-S coupling, in which the spin-orbit interaction is so small that the orbital and spin angular momenta are each more strongly coupled among themselves than is the total spin with the total orbital moment. The resultant orbital, spin, and total angular momenta are represented by $L$, $S$, $J$ in place of $l$, $s$, $j$.

In the above expression, $\mathscr{H}_0$ includes the kinetic, potential and spin-orbit energy terms that are independent of **H**. The magnetic moment consists of

the spin component $\quad \bar{\mu}_s = -\mu_0 \bar{\vec{\sigma}}$ and

the orbital component $\quad \bar{\mu}_l = (e/2c)\overline{(\mathbf{r} \times \dot{\mathbf{r}})}$ (cf. 35.9).

The averages represent the quantum-mechanical expectation values, and where applicable, mean values taken over the temperature distribution.

It may easily be verified that, since

$$\dot{\mathbf{r}} = \frac{i}{\hbar}(\mathscr{H}\mathbf{r} - \mathbf{r}\mathscr{H}) = \left(\mathbf{p} - \frac{e}{c}\mathbf{A}\right)\bigg/ m$$

the expectation value of the moment for the state $\phi_i$ and energy $E_i$ is given by

$$\bar{\mu}_z^i = \bar{\mu}_{lz}^i + \bar{\mu}_{sz}^i = -\overline{\frac{\partial \mathscr{H}}{\partial H_z}}^i = -\frac{\partial E_i}{\partial H_z} \qquad (37.2)$$

with corresponding expressions for $\bar{\mu}_x$ and $\bar{\mu}_y$; in the following analysis, however, we shall put $H_x = H_y = 0$. Then by averaging over the temperature distribution we obtain

$$\bar{\mu}_z = \frac{\sum_i \dfrac{\partial E_i}{\partial H} \exp - \beta E_i}{\sum_i \exp - \beta E_i} \quad \text{where } \beta = \frac{1}{kT} \qquad (37.3)$$

We shall restrict ourselves in the first instance to those cases in which the average thermal energy $kT$ is large compared with the Zeeman energy separation. However, the temperature is assumed to be sufficiently low to enable practically all but the lowest multiplet state and its Zeeman separation to be neglected in cases where the ground state of the atom possesses an orbital moment and a consequent term separation as a result of spin-orbit interaction (figure 39).

The summation in (37.3) then merely extends over the $2j+1$ states of this separation. In addition, since we have assumed weak fields and comparatively high temperatures, we may restrict ourselves to terms of the first degree in $\beta$:

$$\bar{\mu}_z = - \sum_{m_j = -j}^{j} \frac{\partial E_{m_j}}{\partial H}(1 - \beta E_{m_j}) \bigg/ \sum_{m_j = -j}^{j} (1 - \beta E_{m_j}) \qquad (37.4)$$

In the above expression, the quantities $E_{m_j}$ represent the portions of the multiplet energies $E_{lsjm_j}$ that are dependent on $H$.

We are interested only in terms of the first degree in $H$; in the expression for $E_{m_j}$, therefore, we merely require terms as far as the

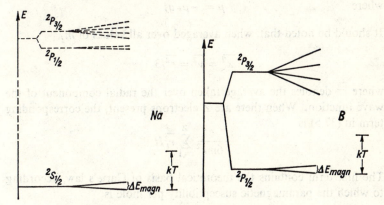

Fig. 39.—Ground-state thermal energy and Zeeman splitting, for sodium and boron

second degree in this quantity. These result firstly from the term separation calculated in §36: for the two cases $j = l + \frac{1}{2}$ and $j = l - \frac{1}{2}$ we have the following result, which follows from (36.11) and from (36.12) after expansion as far as terms of the second degree in $H$,

$$E_{m_j}^{(1)} = \mu_B H g m_j \pm \frac{\mu_B^2 H^2}{4 K_{nl}(2l+1)} \left(1 - \frac{4m_j^2}{(2l+1)^2}\right)$$

where the Landé factor $g$ is given by (36.9).* Secondly, from the term in $\mathbf{A}^2$ in $\mathscr{H}$, which we have hitherto neglected,

$$E_{m_j}^{(2)} = \frac{e^2}{2mc^2}\overline{\mathbf{A}^2}^{m_j} = \frac{e^2}{8mc^2}\overline{(x^2+y^2)}^{m_j} H^2$$

Since $\sum m_j = 0$, $\sum m_j^2 = \frac{1}{3}j(j+1)(2j+1)$, after performing the summation in (37.4) we obtain the following expression for the mean

---

* When $j = l + \frac{1}{2}$, the term in $H^2$ obviously vanishes for $|m_j| = l + \frac{1}{2}$, in agreement with (36.11); the above formula is therefore also valid for s-terms. For $l > 0$, the doublet term with $j = l - \frac{1}{2}$ is the lowest, since $K_{nl} > 0$, and is the only one that requires to be considered.

magnetic moment in the direction of the field, for the cases $j = l \pm \frac{1}{2}$:

$$\bar{\mu}_z = \frac{\mu^2}{kT} \frac{j+1}{3j} H \mp \frac{\mu_B^2(2-g)}{3K_{nl}(2l+1)} H - \frac{e^2}{6mc^2} \overline{r^2} H \qquad (37.5)$$

where
$$\mu = -\mu_B g j$$

It should be noted that, when averaged over all values of $m_j$,

$$\overline{x^2} = \overline{y^2} = \overline{r^2}/3$$

where $\overline{r^2}$ denotes the average taken over the radial component of the wave function. When there are $Z$ electrons present, the corresponding term in (37.5) is

$$-\frac{e^2}{6mc^2} \sum_{i=1}^{z} \overline{r_i^2} H$$

The first term contains the theoretical basis of Curie's law, according to which the paramagnetic susceptibility per mole is

$$\chi_M = \frac{N\bar{\mu}_z}{H} = \frac{C}{T}$$

where $N$ is Avogadro's number, and

$$C = \frac{\mu^2}{k} \frac{(j+1)}{3j} N$$

is the Curie constant.

The second term is strictly absent in the case of the $s$-terms (for which $g = 2$). In systems comprising a single electron $j$ is never zero: in these cases the last two terms may be neglected in comparison with the first (for instance, the ratio of the first term to the second is of the same order of magnitude as that of the interaction term $K_{nl}$ to $kT$). The first term may vanish in the case of systems with more than one electron, and the first and second terms vanish in particular for closed shells—a fact that we state without proof. In this last case the third term represents the normal diamagnetism (cf. §35). The reader is referred to §38 for experimental values of diamagnetic susceptibilities.

We shall now briefly investigate the situation in which $\mu H$ and $kT$ are of comparable magnitude. There are two limiting cases which we shall consider; in doing so we shall neglect the diamagnetic effect.

In the limiting case of classical mechanics, in which **J** may assume any orientation, the energy of a dipole of moment $\mu$, making an angle

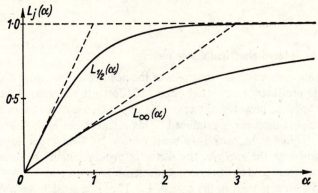

Fig. 40.—The Langevin functions $L_1(\alpha)$ and $L_\infty(\alpha)$

$\theta$ with the field direction, is $E = -\mu H \cos\theta$. Hence from Boltzmann's distribution law the mean moment is

$$\bar{\mu}_z = \frac{\int_0^\pi \mu \cos\theta \, e^{\beta\mu H \cos\theta} \sin\theta \, d\theta}{\int_0^\pi e^{\beta\mu H \cos\theta} \sin\theta \, d\theta} = \mu\left(\coth\alpha - \frac{1}{\alpha}\right) = \mu L_\infty(\alpha)$$

where
$$\alpha = \frac{\mu H}{kT}$$

Secondly, we consider the case $j = \frac{1}{2}$, for which

$$\bar{\mu}_z = \frac{\mu e^{\beta\mu H} - \mu e^{-\beta\mu H}}{e^{\beta\mu H} + e^{-\beta\mu H}} = \mu \tanh\alpha = \mu L_{1/2}(\alpha)$$

All the curves $L_j(\alpha) = \overline{\mu_z(j)}/\mu(j)$, where $j$ is arbitrary and $\mu(j) = -\mu_B g j$, lie between the two limiting curves $L_\infty(\alpha)$ and $L_{1/2}(\alpha)$ (figure 40). The curves $L_j(\alpha)$ are called the Langevin functions.

In experimental work on paramagnetic materials it is usual to operate in the initial linear region of the Langevin curves (figure 40). In the measurements of susceptibility undertaken by Kammerlingh

Onnes and his collaborators on gadolinium sulphate, it was possible to cover a large part of the Langevin curve only by reducing the temperature to that of liquid helium and increasing the field strength to 22,000 gauss.

## §38. The induced electrical dipole moment

An atom in which the centroids of the positive and negative charges coincide produces no external electrical effect. If an external electric field is applied, however, the centroids of the charges are drawn apart and a dipole moment is produced. This moment lies in the direction of the applied field, which we shall denote by $\mathbf{F}$ in order to avoid confusion with the energy. The factor of proportionality between the dipole moment and the field is called the polarizability $\alpha$.

In §4 we saw that, in the case of the Thomson model, the polarizability is proportional to the atomic volume; the only effect of quantum mechanics is to alter the numerical factor (§28). This result is also approximately valid in the case of more complicated atoms and molecules. In the present context we shall confine ourselves to mentioning that the atomic volume enters into several very different effects, such as the volume correction term in the van der Waals equation, and the free path (and therefore the viscosity) in gases. A satisfactory agreement is found to exist between the values of the atomic volumes as determined from polarizability, the gas equation, and viscosity; this is particularly true in all cases in which a dipole moment exists only in the presence of a field.

In the special classical case of the Thomson model of §4, in which the electron is elastically bound and possesses an angular frequency $\omega_0$, the dipole moment $\mathbf{p}^{el}$ induced in an atom by a field $\mathbf{F} = (F,0,0)$ is

$$p_x^{el} = ex = \frac{e^2}{m}\frac{1}{\omega_0^2}F = '\alpha F, \qquad \alpha = a_0^3 \tag{38.1}$$

It will be shown in the next paragraph that, for the general case of an atom posessing one valence electron, quantum mechanics leads to the very similar formula

$$e\bar{x} = \frac{e^2}{m}\sum_{j\neq 0}\frac{f_{j0}}{\omega_{j0}^2}F \quad \text{where} \quad \hbar\omega_{j0} = E_j - E_0 \tag{38.2}$$

In the above expression the angular frequency $\omega_0$ of the classical oscillator is replaced by a summation over all the Bohr transition frequencies from the ground state to the highest excited states. The so-called "oscillator strengths" $f_{j0}$ indicate how much each frequency $\omega_{j0}$ contributes to the total dipole moment; it will be shown in the next paragraph that these strengths have the values

$$f_{j0} = \frac{2m\omega_{j0}}{\hbar} |x_{j0}|^2, \text{ and in particular, } f_{00} = 0 \quad (38.3)$$

The oscillator strengths also occur in the theory of dispersion, absorption, and emission, and are accordingly considered in detail in §§41 and 50. In §41 we give the proof of the summation theorem:

$$\sum_j f_{j0} = 1 \quad (38.4)$$

The polarizability may easily be evaluated with the help of this theorem. In (38.2) we put

$$\sum_{j \neq 0} \frac{f_{j0}}{\omega_{j0}^2} = \frac{1}{\overline{\omega^2}} \sum_j f_{j0} = \frac{1}{\overline{\omega^2}}$$

where $\overline{\omega^2}$ is an appropriate mean value of the $\omega_{j0}^2$. In the case of the hydrogen atom the $f$-values decrease comparatively slowly as $j$ increases, so that an appreciable fraction (about 40 per cent) of the sum in (38.4) lies in the continuous part of the spectrum.

We therefore tentatively replace $\sqrt{(\overline{\omega^2})}$ by the Rydberg frequency $\omega_R = e^2/2\hbar a_0$, and obtain

$$e\bar{x} = 4a_0^3 F, \text{ i.e. } \alpha = 4a_0^3 \quad (38.5)$$

A somewhat better estimate was given in §28 with the help of the method of variations. This method also provided comparatively simple expressions for systems containing more than one electron (cf. Exercise 5, p. 176 and Exercise 6, p. 238). In order to compare the theory with experimental results we shall neglect the correlation terms, i.e. the mean values $\overline{x_i x_k}$ for $i \neq k$, that occur in the above-mentioned formulae, since they may be expected to be small compared to $\overline{x_i^2}$. (If the wave function were separable into a product of functions of a single particle they would even vanish.) Hence for spherically symmetrical atoms

$$\alpha = \frac{4}{9} \frac{1}{N a_0} \left( \sum_i \overline{r_i^2} \right)^2 \quad (38.6)$$

Since the wave functions of complex atoms are generally unknown, the $r_i^2$ cannot be calculated. However, we recall that the same quantity occurs in the expression for the diamagnetic susceptibility (§37):

$$\chi = -e^2 \sum_i \overline{r_i^2}/6mc^2$$

We thus have a relationship between $\chi$ and $\alpha$ in which only experimentally verifiable quantities occur:

$$\chi = -\frac{e^2}{4mc^2}\sqrt{(N\alpha a_0)} \qquad (38.7)$$

### ELECTRIC AND MAGNETIC POLARIZABILITY OF THE INERT GASES

| $N$ | $\alpha(10^{-24}\text{cm}^3)$ observed* | $-\chi(10^{-29}\text{cm}^3)$ observed** | $-\chi$ calculated |
|---|---|---|---|
| He   2 | 0·216 | 0·316 | 0·33 |
| Ne  10 | 0·398 | 1·2 | 1·0 |
| Ar  18 | 1·63 | 3·22 | 2·8 |
| Kr  36 | 2·48 | 4·65 | 4·8 |
| Xe  54 | 4·01 | 7·15 | 7·5 |

\* Landolt-Börnstein I. 1 401 (1950).
\*\* Landolt-Börnstein I. 1 394 (1950).

In the above table numerical values are given for the inert gases which show that (38.6) is a very good approximation.

### §39. The Stark effect

Using quantum-mechanical methods, we shall calculate the effect of an external electric field on the energy levels of an atom (the Stark effect). If we take the $x$-axis of our coordinate system to be parallel to the field, the Hamiltonian operator is

$$\mathscr{H} = \mathscr{H}_0 - eFx \qquad (39.1)$$

In the above expression $\mathscr{H}_0$ is the Hamiltonian operator of the unperturbed atom, with eigenfunctions $\phi_j$ and eigenvalues $E_j$. The expression for the dipole moment in the direction of $x$ is similar to the formula (37.2) for the magnetic moment:

$$e\bar{x} = -\frac{\overline{\partial \mathscr{H}}}{\partial F} = -\frac{\partial \varepsilon}{\partial F} \qquad (39.2)$$

We shall determine the energy $\varepsilon$ by means of Schrödinger's perturbation method. We are interested only in the terms of $\bar{x}$ that are of the first degree in $F$, and consequently need to carry through the perturbation calculation only to the second-order approximation. Using (26.12), and noting that $W_{00} = -ex_{00}F = 0$, we obtain the following expression for the energy of the ground state, in the case of a spherically symmetrical charge distribution:

$$\varepsilon = E_0 + \sum_{j \neq 0} \frac{e^2 |x_{j0}|^2}{E_0 - E_j} F^2 \tag{39.3}$$

In view of the relation (39.2), differentiation of the above expression with respect to $F$ gives equation (38.2), if $f_{j0}$ is substituted from (38.3).

In the derivation of (39.3) it is important that the unperturbed energy value $E_j$ should not be degenerate. This is generally the case for the ground state, but not for the excited states.

Consider, for instance, the first excited level of the hydrogen atom, for which $n = 2$, and which is fourfold degenerate. The quantum numbers $(l, m)$ have the values $(0,0)$, $(1,0)$, $(1,1)$, $(1,-1)$.* For these states, all elements $z_{jk}$ except $z_{(0,0),(1,0)} = z_{(1,0),(0,0)}$ vanish (cf. (25.11) regarding the selection rules $\Delta l = \pm 1$, $\Delta m = 0$ for the $z$-component of the matrix element). This gives

$$-eF z_{(0,0),(1,0)} = -eF \int \phi_{210}(\mathbf{r}) z \phi_{200}(\mathbf{r}) \, d\mathbf{r}$$

$$= -\frac{eF}{16a_0^4} \int_0^\infty \int_{-1}^1 r^4 \left(2 - \frac{r}{a_0}\right) \exp\left(-\frac{r}{a_0}\right) \cos^2 \theta \, d(\cos \theta) \, dr$$

$$= 3ea_0 F$$

In the subspace of the functions $(0,0)$ and $(1,0)$ we therefore have to solve another secular equation of the form of (26.9), in order to remove the degeneracy. In the present case we obtain

$$\begin{vmatrix} H_{11} - \varepsilon & H_{12} \\ H_{21} & H_{22} - \varepsilon \end{vmatrix} = \begin{vmatrix} E_1 - \varepsilon & 3ea_0 F \\ 3ea_0 F & E_1 - \varepsilon \end{vmatrix} = 0$$

the two solutions of which are

$$\varepsilon_1 = E_1 - 3ea_0 F \qquad \varepsilon_2 = E_1 + 3ea_0 F \tag{39.4}$$

* We temporarily take $F$ to be parallel to the $z$-axis (although the latter is more usually employed as the polar axis); the field $F$ is assumed to be strong enough to enable the term separation due to spin-orbit interaction to be neglected.

When no degeneracy is present, therefore, the energy merely decreases proportionately to the square of $F$ (the quadratic Stark effect); in the degenerate case, however, the energy separation is directly proportional* to $F$, as though the atom possessed a permanent dipole moment (linear Stark effect) (see figure 41).

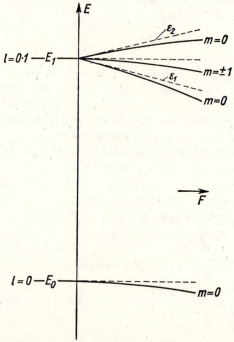

Fig. 41.—The Stark effect for the hydrogen atom. The broken lines illustrate the linear Stark effect; the full lines show the quadratic corrections. (These corrections are magnified by a factor of 10 relative to the linear terms, and are therefore really much smaller than shown.) The ground state and the state $l=1$, $m=\pm1$ exhibit only the quadratic Stark effect

The correct eigenfunctions of zero-order approximation corresponding to the energy values (39.4) are $\phi_{200}\pm\phi_{210}$. These two functions give a charge density with a dipole moment parallel to the positive or the negative $z$-axis. The electrostatic interaction of this dipole moment with the external field gives the additional terms in (39.4). If the $l$-degeneracy is removed when the field is absent (e.g. in the case of sodium), only the quadratic Stark effect is obtained.

* In strong fields the quadratic Stark effect is naturally superimposed on the linear effect.

The validity of the previous perturbation calculation might be open to question, for the following reason. The potential energy, plotted as a function of $x$, exhibits a form as shown in figure 67, p. 384. Now the potential barrier can be penetrated by the electron in virtue of the tunnel effect (§16); the atom cannot therefore possess any stationary bound states, although these are clearly predicted by equation (39.3).

The use of (39.3) is justifiable, however, in spite of the above considerations, provided that the field is not too strong and that the atom is not too highly excited. The Hamiltonian operator (39.1) refers to a literally permanent field, whereas in fact in any experiment the field is switched on and off. If the atom is in the ground state and the field is not too strong, the probability of the electron tunnelling through the potential hill is so small that it remains in the bound state until the field is switched off. The situation is different for strong fields and excited states; in such cases, as we might expect on classical grounds (cf. Exercise 3, p. 238), the atom can be ionized much sooner as a result of the tunnel effect.

## §40. The classical theory of dispersion

We refer again to the concept of the elastically bound atomic electron, which we have already used on a number of occasions. The equation of motion of such an electron, under the influence of an electric field $F_x = F \cos \omega t$, is

$$m(\ddot{x} + \omega_0 x) = eF \cos \omega t \qquad (40.1)$$

if we neglect the effect of damping for the time being. The general solution of this equation is

$$x(t) = A \cos \omega_0 t + B \sin \omega_0 t + \frac{e}{m} F \frac{\cos \omega t}{\omega_0^2 - \omega^2}$$

If we take the initial conditions $x = 0$, $\dot{x} = 0$ at $t = 0$, the dipole moment is

$$p_x^{el}(t) = ex = \frac{e^2 F}{m} \frac{\cos \omega t - \cos \omega_0 t}{\omega_0^2 - \omega^2} \qquad (40.1a)$$

Owing to the damping which is always present (but which has been neglected here for the sake of simplicity), the term in $\cos \omega_0 t$ resulting from the initial conditions soon decays; in the condition of "forced oscillation" the dipole moment is therefore

$$p_x^{el}(t) = \frac{e^2 F}{m(\omega_0^2 - \omega^2)} \cos \omega t \qquad (40.2)$$

In the case of a not too highly compressed gas possessing $N$ atoms per cubic centimetre, the electric displacement in an alternating field of angular frequency $\omega$ is

$$D_x = F_x + 4\pi N p_x^{el} = \left(1 + \frac{4\pi e^2 N}{m(\omega_0^2 - \omega^2)}\right) F_x$$

The refractive index $n = \sqrt{\varepsilon}$ is therefore given by

$$n^2 = 1 + \frac{4\pi N e^2}{m(\omega_0^2 - \omega^2)}$$

If $N$ is so small that $|n^2 - 1| \ll 1$, the dispersion formula becomes

$$n - 1 = 2\pi \frac{N e^2}{m(\omega_0^2 - \omega^2)} \qquad (40.3)$$

Apart from a numerical factor and an additional term, the above formula represents the function $n(\omega)$ in the resonance region $\omega \sim \omega_0$, in a manner that is very satisfactorily confirmed by experiment. When equation (40.1) is supplemented by a frictional term, it leads to the relation for $n(\omega)$ within the absorption line that is termed "anomalous dispersion".

The dispersion formula (40.3) soon proved to be inadequate in one respect: even in the simplest substances (such as a gas consisting of hydrogen atoms or sodium vapour, in which only one electron per atom can be concerned with dispersion), the absorption spectrum does not show a single line $\omega_0$, but a very large number. If the atom is in the quantum state $s$, corresponding to an energy $E_s$, then on Bohr's theory all possible frequencies

$$\omega_{vs} = \frac{E_v - E_s}{\hbar} \qquad v = 0, 1, 2, 3, \ldots$$

can occur as absorption frequencies. Each of these frequencies contributes to the dispersion in the manner prescribed by the expression

(40.3). On the other hand, before the advent of quantum mechanics the oscillator model provided the only means of giving a fairly correct representation of dispersion phenomena. The idea thus came into being that the atom did not merely contain one single oscillator $\omega_0$, but a whole series with characteristic frequencies $\omega_{vs}$, and that it was necessary to add the contributions of all these oscillators when calculating the quantities $p_x^{el}$ and $n-1$. A simple summation of expressions of the form of (40.2) is out of the question, however, just because this would lead to abnormally large values of $p_x^{el}(t)$. To each *ad hoc* oscillator we must be prepared to assign a temporarily unknown "oscillator strength" $f_{vs}$ which indicates the extent to which it contributes to the dispersion.

The dispersion formula then becomes

$$n^2 - 1 = \frac{4\pi N e^2}{m} \sum_v \frac{f_{vs}}{\omega_{vs}^2 - \omega^2} \tag{40.4}$$

and for the case in which $\left| n^2 - 1 \right| \ll 1$

$$n - 1 = \frac{2\pi N e^2}{m} \sum_v \frac{f_{vs}}{\omega_{vs}^2 - \omega^2} \tag{40.5}$$

Using this model, we obtain the following expression for the dipole moment $p_x^{el}$ of an atom in the state $s$:

$$p_x^{el}(t) = \frac{e^2 F}{m} \sum_v \frac{f_{vs}}{\omega_{vs}^2 - \omega^2} \cos \omega t \tag{40.6}$$

Soon after this formula was derived, Thomas and Kuhn put forward a hypothesis about the quantities $f_{vs}$. If the frequency $\omega$ is taken to be so high that $\omega^2$ is large compared with all frequencies $\omega_{vs}^2$ (as in the case of X-rays, for instance), then (40.6) tends to the expression

$$p_x^{el}(t) = -\frac{e^2 F}{m\omega^2} \left( \sum_v f_{vs} \right) \cos \omega t \tag{40.7}$$

The characteristic frequencies $\omega_{vs}$ do not appear in the above expression. We may assume that the old classical formula (40.2), in which

$$p_x^{el} = -\frac{e^2 F}{m\omega^2} \cos \omega t$$

is valid in this limiting case; this formula simply describes the behaviour of a free electron. This can only be true if the relation

$$\sum_{\nu} f_{\nu s} = 1 \quad \text{for all} \quad s \tag{40.8}$$

is satisfied. Equation (40.8) is termed the "summation theorem". The number 1 occurring in this theorem expresses the fact that we are dealing with only one electron in total; the individual quantities $f_{\nu s}$ indicate how the unit oscillator strength of the classical bound electron is to be distributed among the different "equivalent oscillators".

In the next paragraph we shall prove formula (40.6) using the methods of the quantum theory, and shall also investigate the actual numerical values of the quantities $f_{\nu s}$ in simple examples.

## §41. The quantum-mechanical theory of dispersion

Let the unperturbed atom be described by the Hamiltonian operator $\mathcal{H}_0$, with associated eigenvalues $E_\nu$ and eigenfunctions $\phi_\nu$:

$$\mathcal{H}_0 \phi_\nu = E_\nu \phi_\nu \tag{41.1}$$

Let the effect of the light wave be represented by a time-dependent operator $W$. In the simplest particular case of an alternating electric field $F_x = F\cos\omega t$ parallel to the $x$-axis we may put*

$$W = -\tfrac{1}{2}xeF\,(\exp i\omega t + \exp -i\omega t) \tag{41.2}$$

We now take the Schrödinger equation

$$(\mathcal{H}_0 + W)\psi = -\frac{\hbar}{i}\dot{\psi} \tag{41.3}$$

and expand $\psi$ in terms of the eigenfunctions $\phi_\lambda$ of the unperturbed equation:

$$\psi = \sum_{\lambda} a_\lambda(t)\exp\left(-\frac{i}{\hbar}E_\lambda t\right)\phi_\lambda \tag{41.3a}$$

---

* Strictly speaking, in place of $-exF$ we should use the expression

$$-\frac{e}{mc}p_x A_x \quad \text{where} \quad A_x = -\frac{c}{\omega}F\sin\omega t \tag{41.2a}$$

However, the difference between these two quantities is unimportant, provided that the vector potential $A_x$ is nearly constant throughout the volume of the atom, so that a dipole approximation may be used. This may be directly verified from the expression (41.6), if we take into consideration the relation $p_{\nu s} = im\omega_{\nu s}x_{\nu s}$.

If we then form the scalar product of (41.3) and $\phi_v$ we obtain

$$\dot{a}_v(t) = -\frac{i}{\hbar}\sum_\lambda W_{v\lambda}\exp\left(\frac{i}{\hbar}(E_v-E_\lambda)t\right)a_\lambda(t) \qquad (41.4)$$

where
$$W_{v\lambda} = (\phi_v, W\phi_\lambda) \qquad (41.5)$$

As in the case of the Dirac approximation described in §27, the integration of (41.4) is subject to the initial condition that the perturbation $W$ is applied at time $t = 0$, and that until that moment the atom is in the state $s$, i.e. $a_v(0) = \delta_{sv}$. Then when $v \neq s$ we obtain

$$a_v(t) = \frac{eFx_{vs}}{2\hbar}\left\{\frac{\exp i(\omega_{vs}+\omega)t-1}{\omega_{vs}+\omega} + \frac{\exp i(\omega_{vs}-\omega)t-1}{\omega_{vs}-\omega}\right\} \qquad (41.6)$$

It is clear from the above expression that, whenever $\omega$ is in the neighbourhood of one of the resonance frequencies $\omega_{vs}$, the approximation assumed in the Dirac method ($|a_v| \ll 1$) is no longer satisfied even when $F$ is small. We must therefore restrict ourselves in the present instance to frequencies for which the approximation is valid; the reader is referred to §50 for the discussion of the case for which $\omega \approx \omega_{vs}$. Using (41.3a) and (41.6) we obtain the following expression for $\psi$:

$$\psi = \exp\left(-\frac{i}{\hbar}E_s t\right)\phi_s + \sum_{v \neq s} a_v \exp\left(-\frac{i}{\hbar}E_v t\right)\phi_v \qquad (41.7)$$

We are concerned with the expectation value of the dipole moment, which is calculated by means of (41.7):

$$\overline{p_x^{el}(t)} = e(\psi, x\psi) \qquad (41.8)$$

(The mean-value symbol over $p_x^{el}$ denotes a spatial, not a time average.) When calculating (41.8) we shall restrict ourselves in the approximation to terms of the first degree in $a_v$. Then

$$(\psi, x\psi) = x_{ss} + \sum_v (a_v \exp(i\omega_{sv}t)x_{sv} + a_v^* \exp(-i\omega_{sv}t)x_{sv}^*)$$

If we now introduce the quantities $a_v$ as given by the expression (41.6), we obtain the following expression for the dipole moment in place of formula (40.1a), which was calculated from classical theory:

$$\overline{p_x^{el}(t)} = ex_{ss} + \frac{e^2F}{\hbar}\sum_v \frac{2\omega_{vs}|x_{vs}|^2}{\omega_{vs}^2-\omega^2}(\cos\omega t - \cos\omega_{vs}t) \qquad (41.9)$$

We shall omit the contribution of $\cos \omega_{sv} t$ in the above expression, as we did in the case of the corresponding term in the classical treatment of §40. The first term in (41.9), $ex_{ss}$, denotes a constant dipole moment corresponding to an atom in the state $s$, and contributing nothing to the dispersion. The remainder of (41.9) corresponds exactly to formula (40.4), which was derived on semi-classical grounds; it possesses the great advantage, however, that the quantities $f_{vs}$ can now be specified by a fundamental relation. Comparison of (40.6) and (41.9) gives (cf. (38.3))

$$f_{vs} = \frac{m}{\hbar} 2\omega_{vs} |x_{vs}|^2 \qquad (41.10)$$

This result enables us to prove the summation theorem $\sum_v f_{vs} = 1$. We first write (41.10) in the form

$$f_{vs} = \frac{m}{\hbar} (\omega_{vs} x_{vs} x_{sv} - \omega_{sv} x_{sv} x_{vs})$$

The matrix elements of the momentum are

$$p_{vs} = im\omega_{vs} x_{vs}$$

hence $\qquad\qquad f_{vs} = \frac{1}{i\hbar} (x_{sv} p_{vs} - p_{sv} x_{vs})$

Now the summation theorem requires that the following relation should hold for all $s$:

$$\sum_v (p_{sv} x_{vs} - x_{sv} p_{vs}) = \frac{\hbar}{i}$$

This is just the matrix form of the term $r = s$ in the fundamental commutation relation

$$(px - xp)_{rs} = \frac{\hbar}{i} \delta_{rs}$$

the latter may therefore be considered as in a sense a consequence of the summation theorem for $f$.

According to (41.10), the sign of $f_{vs}$ is the same as that of $\omega_{vs} = (E_v - E_s)/\hbar$. If $s$ represents the ground state of the atom, then $E_v$ is always greater than $E_s$, and $f_{vs}$ is positive for all values of $v$; this is the usual situation in the case of dispersion measurements. If $s$ represents an excited state, however, negative values of $f_{vs}$ occur when

$E_v$ is less than $E_s$. A "negative dispersion" of this type may be observed, for instance, in the case of excited (metastable) neon atoms in a neon arc.[*]

The behaviour of a linear harmonic oscillator is of interest. In this case, $E_s = \hbar\omega_0(s+\frac{1}{2})$. The only non-zero matrix elements $x_{vs}$ occur when $v = s+1$ and $v = s-1$:

$$x_{s+1,s} = (s+1)^{1/2}\left(\frac{\hbar}{2m\omega_0}\right)^{1/2} \quad \text{and} \quad x_{s-1,s} = s^{1/2}\left(\frac{\hbar}{2m\omega_0}\right)^{1/2}$$

Hence from (41.10), putting

$$\omega_{s+1,s} = \omega_0 \quad \text{and} \quad \omega_{s-1,s} = -\omega_0:$$
$$f_{s+1,s} = s+1 \quad \text{and} \quad f_{s-1,s} = -s$$

The sum of the contributions of the positive and negative dispersion is unity; hence the value of the dipole moment due to forced oscillations is the same as that given by the classical formula (40.2), for all $s$ including $s = 0$.

As an example, the values of $f$ for the principal series of lithium and sodium are calculated from (41.10), and compared with the observed values; the results are extracted from the excellent compilation contained in A. Unsöld, *Physik der Sternatmosphären*, Berlin (1938), pp. 191 ff., 205 ff.

| Lithium principal series | $\lambda(\text{Å})$ | $f$(calc.) | $f$(obs.) | |
|---|---|---|---|---|
| $n = 2$ | 6708 | 0·750 | 0·750 | Absolute measurement |
| $n = 3$ | 3233 | 0·0055 | 0·0055 ⎫ | |
| $n = 4$ | 2741 | 0·0052 | 0·0048 ⎬ | Relative measurements referred to $n = 2$ |
| $n = 5$ | 2563 | 0·0025 | 0·0032 ⎭ | |

| Sodium principal series | $\lambda(\text{Å})$ | $f$(calc.) | $f$(obs.) | |
|---|---|---|---|---|
| $n = 3$ | 5893 | 0·975 | 1·00 | Absolute measurement |
| $n = 4$ | 3303 | 0·0144 | 0·0144 ⎫ | |
| $n = 5$ | 2853 | 0·00241 | 0·00211 ⎬ | Relative measurements |
| $n = 6$ | 2680 | 0·00098 | 0·00065 ⎭ | |

[*] Cf. for instance H. Kopfermann and R. Ladenburg, *Z. Phys.* **65**, (1930), 167.

It should be noted that only the product $N f_{vs}$ can be obtained from dispersion measurements in the neighbourhood of an absorption line, using the relation

$$n-1 = 2\pi N \frac{e^2}{m} \sum_v f_{vs}/(\omega_{vs}^2 - \omega^2)$$

Considerable uncertainty is involved in the determination of $N$ (the number of atoms per cubic centimetre), and relative measurements are therefore often preferred. Further, the theoretical determination of the quantities $f_{vs}$ is strictly possible in the case of the hydrogen atom alone; all other atoms can only be treated by somewhat laborious approximation procedures, such as Hartree's method. The errors in the calculated transition probabilities are estimated to be 20 per cent. See Exercise 4 below for the calculation of the oscillator strengths in the case of the hydrogen atom.

## Exercises

1. *Zeeman effect for the sodium D-line*

Is it possible to produce the Paschen-Back effect in the case of the sodium D-line, having regard to the fields that can be achieved in practice?

2. *Doublet eigenfunctions*
   Determine the eigenfunctions

$$Y_{l j m_j} = a Y_{l, m_j - 1/2} \alpha + b Y_{l, m_j + 1/2} \beta$$

and use them to calculate the relative intensities of the Zeeman components of the sodium D-lines in a weak magnetic field.

3. *Ionization due to the tunnel effect*

Make a rough estimate of the lifetime of a hydrogen atom in the ground state, acted on by an electric field of order of magnitude $10^6$ V/cm.

4. *Spectral line intensities*

Calculate the oscillator strengths for the transition $2p \to 1s$ in the hydrogen atom. What is the lifetime of the excited $2p$ states?

5. *Oscillator strengths for the principal series of sodium*

In the table on p. 237 the sum of the experimental values of the oscillator strengths for sodium is obviously greater than 1. Does this conflict with the summation theorem for $f$? What is the sum in the case of lithium? See figure 30, p. 186 for the lithium and sodium terms entering into the sums of the $f$-values.

6. *Polarizability of the alkali atoms*

In the alkali atoms the oscillator strengths for the resonance lines corresponding to the transition from the ground state to the first excited state are nearly equal to 1. The polarizability is practically entirely due to the valence electron. Since the ground-state configurations are spherically symmetrical, the energy in an electric field $F$ parallel to $x$ is

$$E = E_0 + \sum_{i \neq 0} \frac{|W_{i0}|^2}{E_0 - E_i}$$

as in the case of the hydrogen atom ($W = -eFx$). The eigenfunctions of the first-order approximation are

$$\phi = \phi_0 + \sum_{i \neq 0} \frac{W_{i0}}{E_0 - E_i} \phi_i$$

Making the assumption that $W_{i0}$ differs from zero only for transitions to the first excited state, show that

(1) the perturbed function $\phi$ is produced by a displacement of $\phi_0$ by an amount $\beta$:

$$\phi = \phi_0 - \beta \frac{\partial}{\partial x} \phi_0$$

(2) the polarizability $\alpha$ is determined from $\beta$; it is expressed in terms of universal constants and the excitation energy $E_a$.

(3) Compare the calculated values of $\alpha$ with the values determined experimentally from the Stark effect.*

|  | Li | Na | K | Rb | Cs |
|---|---|---|---|---|---|
| $E_a$ (eV) | 1·84 | 2·10 | 1·61 | 1·56 | 1·39 |
| $\alpha_{exp}$ ($10^{-24}$ cm$^3$) | 27 | 27 | 46 | 50 | 61 |

## 7. *Refractive index of free electrons*

Determine the refractive index $n(\omega)$ of a rarefied "gas" consisting of free electrons, neglecting radiation damping and ohmic resistance. If $n$ is imaginary, total reflection occurs. For what wavelengths are alkali-metal foils transparent, assuming one free electron per atom? (The specific volumes of the body-centred alkali metals are given by $1/N = \frac{1}{2}a^3$, where $a = 3\cdot5$, $4\cdot3$, $5\cdot3$, $5\cdot6$, $6\cdot1$ respectively for Li, Na, K, Rb, Cs.) What is the density of the free electrons in the ionosphere, if total reflection occurs when $\nu \approx 10 \,\mathrm{Mc/s} = 10^7 \mathrm{s}^{-1}$?

---

* Landolt-Börnstein, I/1, 6th ed., Berlin, 1950.

# D

---

**Problems
involving
several
electrons**

## CHAPTER DI

### Problems involving several electrons

### §42. Pauli's exclusion principle

So far we have considered systems possessing only a single electron. The treatment of systems containing more than one electron requires new concepts in order to make the properties of such systems comprehensible.

If we omit the effect of the electron spin and the electrostatic interaction of the electrons, the Hamiltonian operator for $n$ electrons,

$$\mathcal{H} = \sum_{i=1}^{n} \left( \frac{1}{2m} \mathbf{p}_i^2 + V(\mathbf{r}_i) \right) = \sum_{i=1}^{n} \mathcal{H}_i \qquad (42.1)$$

consists of the sum of the operators $\mathcal{H}_i$, each of which acts on a single set of electron coordinates $\mathbf{r}_i$. If the eigenfunctions $\phi_q$ of $\mathcal{H}_i$ are known,*

$$\mathcal{H}_i \phi_\mathbf{q} = \varepsilon_\mathbf{q} \phi_\mathbf{q} \qquad (42.2)$$

then the complete set of eigenfunctions of $\mathcal{H}$ can be given:

$$\mathcal{H} \Phi_Q(\mathbf{r}_1, \mathbf{r}_2, \ldots, \mathbf{r}_n) = E_Q \Phi_Q \qquad (42.3)$$

where $\quad \Phi_Q = \phi_{\mathbf{q}_1}(\mathbf{r}_1) \phi_{\mathbf{q}_2}(\mathbf{r}_2) \ldots \phi_{\mathbf{q}_n}(\mathbf{r}_n)$ and $E_Q = \sum_{i=1}^{n} \varepsilon_{\mathbf{q}_i} \quad (42.3a)$

The validity of equation (42.3) may easily be verified. Since $\mathcal{H}_i$ acts only on $\phi_{\mathbf{q}_i}(\mathbf{r}_i)$ and leaves the remaining functions unchanged, $\mathcal{H}_i \Phi_Q = \varepsilon_{\mathbf{q}_i} \Phi_Q$, from which (42.3) follows directly. Thus all solutions may be obtained, since the functions $\Phi_Q$ constitute a complete orthogonal system in the coordinate space $\mathbf{r}_1, \mathbf{r}_2, \ldots, \mathbf{r}_n$, if this is true for $\phi_\mathbf{q}(\mathbf{r})$ in the coordinate space of $\mathbf{r}$.

When dealing with atoms possessing more than one electron, it is generally possible as a first approximation to neglect the spin energy,

---

* The symbol q represents an abbreviation for a set of three quantum numbers, such as $n_r$, $l$, and $m$ in the case of a central potential.

which is small. The electrostatic interaction of the electrons can be represented by taking $V(\mathbf{r}_i)$ in the $i$th equation to be the sum of the potential due to the nucleus and that due to the effect of the other electrons in the atom. The Hamiltonian operator then takes the form (42.1). If the system is further specified by the spin coordinates and spin quantum numbers, and if $x$ is used to denote the space and spin coordinates $\mathbf{r}$ and $s$, then the eigenfunctions corresponding to the energy $E_\Lambda = \sum_i \varepsilon_{\lambda_i}$

$$\Psi_\Lambda(x_1, \ldots, x_n) = \Psi_{\lambda_1}(x_1)\Psi_{\lambda_2}(x_2)\ldots\Psi_{\lambda_n}(x_n) \qquad (42.4)$$

where $\lambda$ represents the space quantum numbers $\mathbf{q}$ and the spin quantum numbers $m_s$. Each energy value $E_\Lambda$ therefore exhibits spin degeneracy in addition to spatial degeneracy.

Investigation of the properties of atoms with more than one electron shows that the ground states are quite different from what we would expect in consequence of the above description. In the lowest energy state, all electrons should clearly be in the state $1s$. The corresponding spatial quantum numbers of all electrons would be equal.* In fact, even the simplest atoms behave quite differently. Spectroscopic observation shows that, whereas in helium the electrons occupy two $1s$ states, in lithium two $1s$ states and one $2s$ state are filled. Spectroscopic examination of the elements, and the shell-like electron structure of atoms, expressed by the periodic system, both lead inevitably to the hypothesis that each state can be occupied by only a single electron. This is Pauli's principle.

In consequence of this principle, equal sets of quantum numbers $\lambda_i$ should not occur in the wave function $\Psi_\Lambda$. Using this assumption, it is possible to explain the construction of the periodic system of the elements. The ground state of an atom is obtained by filling the lowest states successively in the equivalent potential $V$, taking due account of the spin degeneracy.

For many reasons, however, the above formulation of Pauli's principle is untenable. For one thing, it is based on the assumption that the interaction of the electrons can be represented by an average potential. In general, however, individual sets of quantum numbers cannot be specified for each electron. Further, it may easily be shown that a perturbation gives rise to forbidden states produced from the

---

* Spin-spin and spin-orbit coupling would be absent. The energy would then be independent of the spin quantum numbers and the ground state would be $2^n$-fold degenerate.

permitted ones. Finally, the above form of the Pauli exclusion principle can be only provisional, since it is based on the special properties of the Hamiltonian operator.

Heisenberg has shown how the Pauli principle may be expressed in general form. In order to understand the principle as expressed in Heisenberg's form we must first deal with the general properties of the Hamiltonian operator for many particles.

The Hamiltonian operator of a system of $n$ electrons may depend in a very complicated manner on the operators $\mathbf{p}_i$, $\mathbf{r}_i$, $\vec{\sigma}_i$, but it possesses one simple property of decisive importance: it is symmetrical with respect to interchanges of the indices $i$, and is invariant for any such interchange. This is because the electrons are physically equivalent particles which are indistinguishable in all respects. All potentials, including the electron interaction, appear in the Hamiltonian operator in symmetrical form. (This may be seen in the case of the Hamiltonian operator in the absence of spin, which consists of the expression (42.1) together with the interaction terms

$$\frac{1}{2}\sum_{i\neq i'}^{n}\frac{1}{|\mathbf{r}_i-\mathbf{r}_{i'}|})$$

If $P$ represents a permutation of the numbers $1,\ldots,n$ and $P_i$, the number replacing $i$ after the permutation has been effected, this property of $\mathscr{H}$ may be represented as follows:

$$\mathscr{H}(\ldots,\mathbf{p}_i,\mathbf{r}_i,\vec{\sigma}_i,\ldots)=\mathscr{H}(\ldots,\mathbf{p}_{P_i},\mathbf{r}_{P_i},\vec{\sigma}_{P_i},\ldots) \qquad (42.5)$$

This may also be described by means of a permutation operator $P$:

$$P\Psi(x_1,\ldots,x_i,\ldots,x_n)=\Psi(x_{P_1},\ldots,x_{P_i},\ldots,x_{P_n}) \qquad (42.6)$$

$\mathscr{H}$ and $P$ then commute:

$$P\mathscr{H}\Psi=\mathscr{H}P\Psi \quad \text{or} \quad P\mathscr{H}-\mathscr{H}P=0 \qquad (42.7)$$

Therefore if $\Psi$ is an eigenfunction of $\mathscr{H}$, $P\Psi$ is also an eigenfunction, belonging to the same eigenvalue.

The simplest relations are those for the case of two particles, where the only permutation is the interchange of the numbers 1 and 2, denoted by $P_{12}$. If $\Psi(x_1,x_2)$ is an eigenfunction of $\mathscr{H}$, then so is $P_{12}\Psi(x_1,x_2)=\Psi(x_2,x_1)$. Since the double application of $P_{12}$ gives the identity ($P_{12}.P_{12}=1$), the only possible eigenvalues of $P_{12}$ are $\pm 1$. Then the eigenfunctions of $\mathscr{H}$, that are also eigenfunctions of $P_{12}$, are evidently symmetric and antisymmetric functions $\Psi^S$ and $\Psi^A$; when acted upon by $P_{12}$, these functions either remain unaltered or suffer a change of sign:

$$\mathscr{H}\Psi = E\Psi, \quad \mathscr{H}\Psi^A = E\Psi^A, \quad \mathscr{H}\Psi^S = E\Psi^S$$

$$P_{12}\Psi^A = -\Psi^A, \quad P_{12}\Psi^S = \Psi^S \tag{42.8}$$

where $\qquad \Psi^A = \Psi - P_{12}\Psi \quad$ and $\quad \Psi^S = \Psi + P_{12}\Psi$

The symmetric or antisymmetric nature of a state does not change in the course of time; if a state is antisymmetric at time $t = 0$, it remains so for all time. This is because the unitary transformation that converts $\Psi(0)$ to $\Psi(t)$ (cf. §27) only contains the operator $\mathscr{H}$; the transformation operator therefore commutes with $P$ even when $\mathscr{H}$ contains arbitrary (but necessarily symmetric) perturbations that are functions of time.

In the present problem, involving two particles, the following conditions apply. A symmetric and an antisymmetric eigenfunction are associated with each energy value $E$ denoted by $\lambda_1$ and $\lambda_2$, where $\lambda_1 \neq \lambda_2$; when $\lambda_1 = \lambda_2$, however, there is only one symmetric function. We may therefore forbid the occurrence of a doubly occupied state by admitting antisymmetric functions only. This leads to the postulate that the wave function must be antisymmetric in the coordinates of both electrons. This form of Pauli's principle is free from objection; it reduces to the original form for the case in which no interaction is present. It does not depend on any special form of the Hamiltonian operator, but imposes a condition on the wave function. Since by § 19 any arbitrary quantum numbers may be introduced by resolving in terms of the corresponding eigenfunction system of commuting operators, the prohibition of identical sets of quantum numbers is quite general. In the Fourier expansion, for instance, no two wave-number vectors can be equal when the spin is the same in each case; the amplitude vanishes when $x_1 = x_2$, as does the antisymmetric wave function $\Psi(x_1, x_2)$.

For the case of several electrons, it is clearly necessary only to postulate that the wave function for each pair of electron coordinates is antisymmetric in order to express Pauli's principle in its general form. If $P_{ik}$ represents the interchange of two electron coordinates, the following relation must hold:

$$P_{ik}\Psi(x_1, \ldots, x_n) = -\Psi(x_1, \ldots, x_n) \text{ for any } i \neq k \tag{42.9}$$

Now if we have an eigenfunction corresponding to a given eigenvalue,

$$\mathscr{H}\Psi = E\Psi \qquad (42.10)$$

then in general, $n!$ functions $P\Psi$ belong to the same eigenvalue, since there are $n!$ different permutations. Since every permutation can be expressed as a product of interchanges of two quantities, the anti-symmetric function $\Psi^A$ associated with $E$ can be represented as follows:

$$\Psi^A = \sum_P (-1)^P P\Psi, \quad \text{where} \quad \mathscr{H}\Psi^A = E\Psi^A \qquad (42.11)$$

In the above expression the summation is to be taken over all permutations; the value of $(-1)^P$ is $-1$ when $P$ contains an odd number of interchanges, and $+1$ when this number is even. Then $\Psi^A$ is an anti-symmetric eigenfunction, though it vanishes whenever the procedure for generating an antisymmetric function yields the value 0, as in the case, for instance, when $\Psi$ is completely symmetrical. The energy values remain unchanged by this procedure; however, some states are excluded in virtue of Pauli's principle, and the degree of degeneracy of others is appreciably reduced.

According to Slater, the procedure for making the wave function (42.4) antisymmetric may be expressed in a simple manner as a determinant, known as the Slater determinant:

$$\Psi_\Lambda^A = \begin{vmatrix} \Psi_{\lambda_1}(x_1) & \Psi_{\lambda_1}(x_2)...\Psi_{\lambda_1}(x_n) \\ \Psi_{\lambda_2}(x_1) & \vdots \\ \vdots & \\ \Psi_{\lambda_n}(x_1)..............\Psi_{\lambda_n}(x_n) \end{vmatrix} \qquad (42.4a)$$

It is clear that the interchange of two coordinates (or two quantum numbers) changes the sign of the above expression, since this corresponds to the interchange of two columns (or two rows) in the determinant. A single eigenfunction is associated with each set of quantum numbers $\lambda_1 \neq \lambda_2 \neq \lambda_3 ... \neq \lambda_n$.

As in the case of two particles, Pauli's principle is completely valid; the antisymmetry property expressed by (42.9) does not change in the course of time, and naturally still holds good when expressed in terms of other coordinates.

It is important to realize that Pauli's principle does not complicate the situation, but that on the contrary it represents a very great simplification. If the principle were not valid, it would for instance be possible to have completely symmetric functions,* $P_{ik}\Psi^s = \Psi^s$. However, the symmetric nature of the Hamiltonian operator implies that transitions between symmetric and antisymmetric states are strictly forbidden and can never take place. In the case of two particles, for instance, the Hilbert space is resolved into two completely independent components, one of which is symmetric, and the other antisymmetric. As a result, we should expect to find a symmetric and an antisymmetric "world", between which there would be no connection. It follows that any theory of thermodynamic statistics would be impossible, since it is assumed in all such statistics that every state is possible, starting from the ground state. We should therefore be compelled to treat the symmetric and antisymmetric components separately in the statistical theory. Fortunately, only one simple function group exists in nature, that of the antisymmetric states.

In the next sections we shall show by means of simple examples how fundamental the Pauli principle is to the physics of atoms and molecules.

The electron coordinates always appear in symmetrical form in physical quantities. For instance

$$|\Psi^A(\mathbf{r}_1, s_1, \mathbf{r}_2, s_2, \ldots)|^2 \, d\mathbf{r}_1 \ldots d\mathbf{r}_n$$

is the probability of finding electron 1 in the element of space $d\mathbf{r}_1$ and with spin coordinate $s_1$, electron 2 in $d\mathbf{r}_2$ with spin $s_2$, etc.; this quantity is symmetric in the electron coordinates $x_1, \ldots, x_n$ and independent of the numbering of the electrons. In this case there is no possibility of distinguishing between the individual electrons; it would therefore be more correct to say that $|\Psi^A|^2 \, d\mathbf{r}_1 \ldots d\mathbf{r}_n$ represents the probability of finding an electron in $d\mathbf{r}_1$ with spin $s_1$, another in $d\mathbf{r}_2$ with spin $s_2$, etc. In equation (42.4) the state $\Psi_\Lambda$ represents electron 1 in state $\lambda_1$, electron 2 in state $\lambda_2$, etc.; the physical interpretation of $\Psi_\Lambda{}^A$, on the other hand, is that there is an electron in state $\lambda_1$, another in state $\lambda_2$, etc. Pauli's principle therefore implies that the physical equivalence of the electrons is expressed by the equivalence of the different states when the electrons are permuted.

In principle, all the electrons of the universe should be handled together in a single antisymmetric wave function; this would of course be an impossibly complicated matter. In the case of spatially separate

---

* When there are more than two electrons, other groups of functions exist in addition to antisymmetric and symmetric functions; they are associated with other forms of representation of the permutation group.

physical systems, however, it may easily be shown that the process of obtaining the antisymmetric wave function has no outside effect; when discussing the hydrogen atom, for instance, it is permissible to consider a single electron. The situation is perhaps best illustrated by taking the case of two electrons bound to two widely separated protons $a$ and $b$; the states of the electrons are denoted by $\lambda_a$ and $\lambda_b$. Then $\Psi = \Psi_{\lambda_a}(x_1)\Psi_{\lambda_b}(x_2)$ and $\Psi^A = (1/\sqrt{2})[\Psi_{\lambda_a}(x_1)\Psi_{\lambda_b}(x_2) - \Psi_{\lambda_a}(x_2)\Psi_{\lambda_b}(x_1)]$ are both normalized functions, if $(\Psi_{\lambda_a}, \Psi_{\lambda_a}) = (\Psi_{\lambda_b}, \Psi_{\lambda_b}) = 1$, and if $(\Psi_{\lambda_a}, \Psi_{\lambda_b})$ vanishes as it certainly does when the separation is great. Physically, $\Psi$ and $\Psi^A$ are quite indistinguishable, and all physical quantities, which must necessarily be symmetric in $x_1$, $x_2$, are the same whether calculated from $\Psi$ or $\Psi^A$. The expectation values of all operators $A$ are the same, whether formed from $\Psi$ or $\Psi^A$, but the degree of degeneracy may well be reduced; two different $\Psi$ functions are associated with $\lambda_a$ and $\lambda_b$, but only one function $\Psi^A$. It is therefore immaterial whether $\Psi$ or $\Psi^A$ is employed, provided that $\Psi$ is initially chosen so that, when $x_1 = x_2$ and the Pauli principle would take effect, this function vanishes simply on account of the large interval between the two systems.

Pauli's principle applies not only to electrons, but to all elementary particles that possess half-integral spin, such as protons and neutrons whose spin is $\frac{1}{2}$; for all these particles, the principle requires that the corresponding wave functions should be antisymmetric. Such particles are called Fermi particles or fermions, and are subject to Fermi-Dirac statistics, so-called because E. Fermi was the first to develop a relation between the term selection given by Pauli's principle and statistical thermodynamics.

Finally, we must refer to the behaviour of composite particles. We may consider these as single particles of appropriate spin, provided that the binding energy is high, that the interaction is not too great, and that the inner state of the compound structure remains unchanged. The Hamiltonian operator corresponding to such particles is symmetric, and the particles themselves are indistinguishable. With regard to the behaviour of the wave function, it is necessary to distinguish between particles composed of an even number of fermions, e.g.

| | |
|---|---|
| Deuteron: | 1 neutron, 1 proton, spin 1 |
| $\alpha$-particle: | 2 neutrons, 2 protons, spin 0 |
| $He^4$ atom: | 2 neutrons, 2 protons, 2 electrons, spin 0 |

and those composed of an odd number, such as

> Triton:          2 neutrons, 1 proton, spin $\frac{1}{2}$
>
> He$^3$ nucleus:   1 neutron, 2 protons, spin $\frac{1}{2}$
>
> He$^3$ atom:      1 neutron, 2 protons, 2 electrons, spin $\frac{1}{2}$

In the first of these groups, the sign of the wave function remains the same when two identical particles are interchanged; in the second group, however, the sign changes when this occurs. If $X_i$ are the co-ordinates of the centre of mass and the spin, then

$$P_{ik}\Psi(\ldots X_i \ldots) = \Psi \quad \text{for an even number of fermions} \quad (42.12a)$$

$$P_{ik}\Psi(\ldots X_i \ldots) = -\Psi \text{ for an odd number of fermions} \quad (42.12b)$$

Particles whose wave functions behave according to (42.12b) obey the Fermi-Dirac statistics, while those whose wave functions are given by (42.12a) are said to be governed by the Bose-Einstein statistics. Since all particles behaving according to (42.12b) possess half-integral values of the spin, while the spin of those governed by (42.12a) is an integral number or 0, the behaviour of the corresponding wave functions can be characterized by the spin; Fermi-Dirac and Bose-Einstein statistics apply respectively in the cases of half-integral and integral values of the spin. The wave functions obeying Bose-Einstein statistics are completely symmetric with respect to permutations of the co-ordinates. (Light quanta also belong to the family of Bose particles, or bosons.) It is important to note, however, that the above description is only valid when all the composite particles are in the ground state and excited states are excluded; an excited helium atom, for instance, would be physically distinguishable from such an atom in the ground state.

## §43. The helium atom

### (a) General properties of the eigenfunctions

In order to provide a simple illustration of the application of the Pauli principle, we shall discuss the properties of a system of two electrons in the Coulomb field of a nucleus of charge $Ze_0 = -Ze$. For the helium atom $Z = 2$; we shall use the general value of $Z$ at first, however, so as to be able to treat simultaneously the cases of the negative hydrogen ion ($Z = 1$) and the positive lithium ion ($Z = 3$).

If we disregard the small contribution of the spin in the Hamiltonian operator, then

$$\mathscr{H} = \frac{1}{2m}(\mathbf{p}_1^2 + \mathbf{p}_2^2) - Ze^2\left(\frac{1}{r_1} + \frac{1}{r_2}\right) + \frac{e^2}{r_{12}} \qquad (43.1)$$

$\mathscr{H}$ is symmetric in $\mathbf{r}_1$ and $\mathbf{r}_2$; the spatial components of the eigenfunctions of $\mathscr{H}$ can therefore be chosen to be symmetric and antisymmetric:

$$P_{12}\Phi^S(\mathbf{r}_1, \mathbf{r}_2) = \Phi^S(\mathbf{r}_2, \mathbf{r}_1) = \Phi^S(\mathbf{r}_1, \mathbf{r}_2) \qquad (43.2a)$$

$$P_{12}\Phi^A(\mathbf{r}_1, \mathbf{r}_2) = \Phi^A(\mathbf{r}_2, \mathbf{r}_1) = -\Phi^A(\mathbf{r}_1, \mathbf{r}_2) \qquad (43.2b)$$

Since the entire wave function must be antisymmetric in all the electron coordinates, the spin component must be chosen to be symmetric for $\Phi^A$ [$\chi^S(s_1, s_2)$, where $P_{12}\chi^S = \chi^S$], and antisymmetric for $\Phi^S$ [$\chi^A(s_1, s_2)$, where $P_{12}\chi^A = -\chi^A$].* Thus the only total wave functions to occur are the two combinations

$$\Psi(x_1, x_2) = \Phi^S\chi^A \qquad (43.3a)$$

$$\Psi(x_1, x_2) = \Phi^A\chi^S \qquad (43.3b)$$

(b) *Properties of the spin functions $\chi^S$ and $\chi^A$*

One antisymmetric and three symmetric functions can be formed from the two normalized and mutually orthogonal spin functions $\alpha$ and $\beta$ (cf. §§24, 32) that are available for an electron:†

| | $s_z$ | $s^2$ | |
|---|---|---|---|
| $\chi^A = \dfrac{1}{\sqrt{2}}\{\alpha(s_1)\beta(s_2) - \alpha(s_2)\beta(s_1)\}$ | 0 | 0 | (43.4a) |
| $\chi_1^S = \alpha(s_1)\alpha(s_2)$ | $\hbar$ | $2\hbar^2$ | |
| $\chi_0^S = \dfrac{1}{\sqrt{2}}\{\alpha(s_1)\beta(s_2) + \alpha(s_2)\beta(s_1)\}$ | 0 | $2\hbar^2$ | (43.4b) |
| $\chi_{-1}^S = \beta(s_1)\beta(s_2)$ | $-\hbar$ | $2\hbar^2$ | |

* The spin and spatial components can be separated only in the case of systems comprising two electrons, since there is then only one permutation operator $P_{12}$. A further condition is that it should be possible to neglect the spin-orbit interaction.
† The factors in (43.4) are so chosen that the spin functions are normalized to unity. In addition they are mutually orthogonal since they belong to different eigenvalues of $s^2$ and $s_z$, as can be immediately verified.

It may readily be verified that the above functions are both eigen-functions of the square of the angular-momentum operator for a total spin $\mathbf{s}^2 = (\mathbf{s}_1 + \mathbf{s}_2)^2$ as well as eigenfunctions of the latter's $z$-component $s_z = s_{1z} + s_{2z}$: for this purpose we make use of the representation of the operator $\mathbf{s}$ given in §24, and note that $\mathbf{s}_1$ acts only on functions of $s_1$.

It follows that $\chi^A$ corresponds to a total spin of value 0, whereas in $\chi^S$ the spins of the two electrons, each of value $\frac{1}{2}$, combine to give a total spin of value 1. In the states represented by (43.4a), therefore, the angular momentum and the associated magnetic moment are due solely to the orbits of the two electrons.

Transitions between the states represented by (43.3a) and (43.3b) are very rare. Since $(\Phi^S, \Phi^A) = (\chi^S, \chi^A) = 0$, any perturbation $W$ that could produce such transitions must contain spatial and spin com-ponents (e.g. spin-orbit interaction), in order that the matrix element $(\Phi^S \chi^A, W \Phi^A \chi^S)$ should be different from zero. Transitions in the optical region (for which $W$ would be approximately proportional to $\mathbf{r}_1 + \mathbf{r}_2$) do not occur. From the optical point of view the term systems corresponding to (43.3a) and (43.3b) behave as though they were almost completely separate. The prohibition of transitions in the optical region is admittedly not a strict one, since as a result of the spin-orbit interaction the exact eigenfunctions do contain small com-ponents of both groups of functions.* No prohibition exists in electron collision processes, since the addition of a further electron to the two atomic electrons completely alters the symmetry conditions.

The helium atom states (43.3a), in which the spins are "antiparallel" $(\mathbf{s}^2 = 0)$ are called *parhelium*, and the states (43.3b) in which the spins are "parallel" $(\mathbf{s}^2 = 2\hbar^2)$ are termed *orthohelium*; the reason for this nomenclature is that it was originally sought to explain the different optical behaviour of the two types of helium by means of two different modifications of the helium atom. The terms belonging to the parhelium system are also called *singlet* terms, those belonging to orthohelium, *triplet* terms, since each ortho term can generally split into three terms as a result of the spin-orbit interaction.

---

* Thus there is for instance a weak line due to intercombination between the $2p$ ortho- and the $1s$ para-state (cf. figure 42).

*(c) The spatial components of the eigenfunctions in the presence of a perturbation $e^2/r_{12}$*

If we treat the electrostatic interaction of the electrons $e^2/r_{12}$ as a perturbation, then the eigenfunctions of zero-order approximation are the products of eigenfunctions in the Coulomb field, which were treated in detail in §29 for the case of the hydrogen atom, and which may be taken to be real functions:

$$\Phi_Q = \phi_{q_1}(\mathbf{r}_1)\,\phi_{q_2}(\mathbf{r}_2) \tag{43.5}$$

In the above expression, $\mathbf{q}$ represents the three quantum numbers $n$, $l$, and $m$ occurring in the hydrogen problem. Neglecting the perturbation, the energy then consists of the sum of the two components:

$$E_Q = \varepsilon_{q_1} + \varepsilon_{q_2}, \quad \text{where} \quad \varepsilon_q = -\frac{m(Ze^2)^2}{2\hbar^2 n^2} = -\frac{Z^2 I_H}{n^2} \tag{43.6}$$

($I_H$ is the energy of ionization of the hydrogen atom.)

In the above approximation, the ground state of helium would possess the energy $E_0 = -8I_H$ ($n_1 = n_2 = 1$); the energy of ionization would be $4I_H$, which is the energy difference between the helium ground state and the energy ($-4I_H$) of the singly ionized helium atom in the ground state, characterized by $n_1 = 1$, $n_2 = \infty$ (or $n_1 = \infty$, $n_2 = 1$). States in which both electrons are excited (such as $n_1 = n_2 = 2$, for which $E_Q = -2I_H$) are situated above the helium ionization limit and may be disregarded in any discussion of the discrete spectrum of eigenvalues. In what follows, therefore, we shall consider only those states for which one of the $n$-values is unity; the corresponding wave function $\phi_{100}$ will be denoted by $\phi_0$.

$$\Phi_{nlm} = \phi_0(\mathbf{r}_1)\phi_{nlm}(\mathbf{r}_2) \tag{43.5a}$$

The symmetric and antisymmetric functions formed from (43.5a)

$$\Phi_{nlm}^S = \frac{1}{\sqrt{2}}\{\phi_0(\mathbf{r}_1)\phi_{nlm}(\mathbf{r}_2) + \phi_0(\mathbf{r}_2)\phi_{nlm}(\mathbf{r}_1)\} \tag{43.6a}$$

$$\left.\right\} \quad n = 2, 3, 4, \ldots$$

$$\Phi_{nlm}^A = \frac{1}{\sqrt{2}}\{\phi_0(\mathbf{r}_1)\phi_{nlm}(\mathbf{r}_2) - \phi_0(\mathbf{r}_2)\phi_{nlm}(\mathbf{r}_1)\} \tag{43.6b}$$

$$\Phi_{100}^S = \phi_0(\mathbf{r}_1)\phi_0(\mathbf{r}_2) \quad \text{for} \quad n = 1 \tag{43.7}$$

are the correct normalized initial eigenfunctions for the perturbation; they yield the following first-order approximation for the total energy:

$$E_{nl}^{S} = (\Phi_{nlm}^{S}, \mathscr{H}\Phi_{nlm}^{S}) \tag{43.8a}$$

$$E_{nl}^{A} = (\Phi_{nlm}^{A}, \mathscr{H}\Phi_{nlm}^{A}) \tag{43.8b}$$

This follows because $\mathscr{H}$, as given by (43.1), commutes with $P_{12}$, and with the components of the total angular momentum $\mathbf{L} = \mathbf{L}_1 + \mathbf{L}_2$, and therefore also with $\mathbf{L}^2$. The same applies to the perturbation $e^2/r_{12} = \mathscr{H} - \mathscr{H}_0$, since the unperturbed operator commutes in a trivial manner with these three operators. The functions (43.6) are chosen in such a manner that they are eigenfunctions of $L_z$, $\mathbf{L}^2$, and $P_{12}$:

|  | $L_z$ | $\mathbf{L}^2$ | $P_{12}$ |
|---|---|---|---|
| $\Phi_{nlm}^{S}$ | $\hbar m$ | $\hbar^2 l(l+1)$ | $+1$ |
| $\Phi_{nlm}^{A}$ | $\hbar m$ | $\hbar^2 l(l+1)$ | $-1$ |

$$\tag{43.9}$$

(Note that each component of $\mathbf{L}_1$ when applied to $\phi_0(\mathbf{r}_1)$ gives the result 0.) The different functions corresponding to the same initial energy (i.e. a fixed value of $n$) belong to different eigenvalues of operators that commute with $\mathscr{H}$ and $e^2/r_{12}$ (cf. §26). This implies the vanishing of the perturbation matrix elements composed of the different functions; the latter are therefore the correct initial functions for the perturbation calculation. The factors are so chosen that the functions are normalized, using the same criteria as for the hydrogen functions.

It can be shown that the energies given by (43.8) do not depend on $m$. The proof is most simply effected by introducing operators applicable to two particles that are the counterparts of (24.6) or (24.7); such operators change $m$ by one unit and commute with $\mathscr{H}$.

The ground state is the parhelium state (43.7a) with zero spin, since there is no antisymmetric spatial function corresponding to it. This state is spherically symmetrical and possesses neither spin nor orbital angular momentum. Apart from this, the unperturbed energies are the same for the parhelium and orthohelium states; however, the ortho-helium states possess the triple degeneracy of the corresponding par-helium states since three independent spin functions are associated with each orthohelium term.

The effect of the perturbation can easily be seen qualitatively. Since the perturbation $e^2/r_{12}$ is positive, the first-order perturbation energy is also positive, and all the unperturbed energy values are displaced upwards. The perturbation effect is stronger for $\Phi^S$ than for $\Phi^A$, because $\Phi^A$ vanishes in the region for which $r_{12} \approx 0$, where the perturbation is strongest. The parhelium energy terms are generally

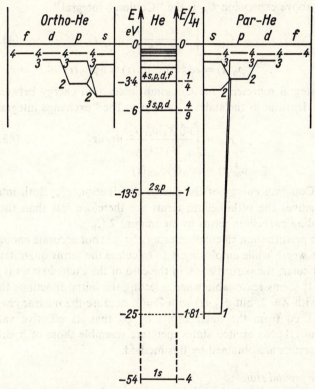

Fig. 42.—Term scheme of the helium atom. The centre column shows the unperturbed energies (43.6) up to the ionization level. On either side are shown the experimental term schemes of ortho- and parhelium

higher than the corresponding energy terms for orthohelium. Figure 42 shows the unperturbed energy values in comparison with the experimental term values. The deviation from the unperturbed energies is quite considerable, which suggests that the method of treating the electron interaction as a perturbation is not a particularly good approximation.

According to (43.6, 43.8) the perturbation energy $(\Phi, (e^2/r_{12})\Phi) = \delta E$ consists of two parts:

$$\delta E_{nl}^S = C_{nl} + A_{nl} \qquad (43.10a)$$

$$\delta E_{nl}^A = C_{nl} - A_{nl} \qquad (43.10b)$$

In the above expression $C_{nl}$ is the "Coulomb integral"

$$C_{nl} = \int \frac{\rho_0(\mathbf{r}_1)\rho_{nl}(\mathbf{r}_2)}{|\mathbf{r}_1 - \mathbf{r}_2|} d\mathbf{r}_1 \, d\mathbf{r}_2 \qquad (43.11)$$

where $\qquad \rho_0(\mathbf{r}) = e\phi_0^2(\mathbf{r}), \quad \rho_{nl}(\mathbf{r}) = e\phi_{nl0}^2(\mathbf{r})$

This integral represents the Coulomb interaction energy between the charge densities in the states $\rho_0$ and $\rho_{nl}$. The "exchange integral"*

$$A_{nl} = \int \frac{\rho_{nl}^A(\mathbf{r}_1)\rho_{nl}^A(\mathbf{r}_2)}{|\mathbf{r}_1 - \mathbf{r}_2|} d\mathbf{r}_1 \, d\mathbf{r}_2 \qquad (43.12)$$

where $\qquad \rho_{nl}^A(\mathbf{r}) = e\phi_0(\mathbf{r})\phi_{nl0}(\mathbf{r})$

is the Coulomb energy of the charge distribution $\rho_{nl}^A$. Both integrals are positive: the orthohelium terms are therefore less than the corresponding parhelium terms by an amount $2A_{nl}$.

The perturbation calculation using $e^2/r_{12}$ is not accurate enough for it to be worth while employing it to calculate the terms quantitatively. In particular, the perturbation in the case of the excited states is fairly large. It seems reasonable when selecting the initial functions to take $\phi_0(\mathbf{r})$ with $Z = 2$, but $\phi_{nlm}(\mathbf{r})$ with $Z = 1$, because the nuclear charge is so screened from the outer orbits by $\phi_0$ that its effective value is practically 1; the excited states therefore resemble those of hydrogen. Better results are obtained by this method.

### (d) The ground state

The calculation of the ground-state energy needs to be discussed in some detail. The first-order approximation for the energy calculated by the perturbation method is

$$E_0^{(1)} = E_0 + C_0 \qquad (43.13)$$

where the unperturbed energy $E_0 = -2Z^2 I_H$, and the Coulomb integral $C_0$ is given by

$$C_0 = \int \frac{\rho_0(\mathbf{r}_1)\rho_0(\mathbf{r}_2)}{|\mathbf{r}_1 - \mathbf{r}_2|} d\mathbf{r}_1 \, d\mathbf{r}_2 \qquad (43.14)$$

---

* The physical meaning and nomenclature of this integral are discussed in the following paragraphs.

From (29.11),

$$\phi_0 = \frac{1}{\sqrt{(\pi a^3)}} \exp -\frac{r}{a} \quad \text{where} \quad a = \frac{1}{Z} \frac{\hbar^2}{me^2} = \frac{a_0}{Z}$$

hence
$$\rho_0(\mathbf{r}) = \frac{e}{\pi a^3} \exp -\frac{2r}{a} \tag{43.15}$$

depends only on the magnitude of $r$, as does $\Phi(\mathbf{r})$, the potential of the charge distribution $\rho_0(\mathbf{r})$. Instead of calculating $\Phi$ from the usual integral

$$\Phi(\mathbf{r}) = \int \frac{\rho_0(\mathbf{r}')}{|\mathbf{r}-\mathbf{r}'|} d\mathbf{r}' \tag{43.16}$$

let us determine this quantity from the equation for the potential

$$\nabla^2\Phi = -4\pi\rho_0, \quad \text{hence} \quad \frac{1}{r}\frac{d^2}{dr^2} r\Phi(r) = -4\pi\rho_0(r) \tag{43.17}$$

$\Phi$ must vanish like $e/r$ at infinity. Integration of (43.17) gives

$$\Phi = \frac{e}{r}\left\{1 - \left(1 + \frac{r}{a}\right)\exp -\frac{2r}{a}\right\}$$

then
$$C_0 = \int_0^\infty \rho_0(r)\Phi(r)4\pi r^2\, dr = \frac{5}{8}\frac{e^2}{a}$$

hence
$$C_0 = \frac{5}{8} \times 4I_H \quad \text{for} \quad Z = 2 \tag{43.14a}$$

The ionization energy would therefore be $1\cdot5I_H$, instead of $4I_H$ as in the case of the initial approximation; the experimental value is $1\cdot82I_H$. The value calculated by the first approximation is therefore still about $0\cdot3I_H$ or 4 eV too high; however, in view of rough approximation used, the agreement is surprisingly good.

A better approximation is obtained if the perturbation $W$ is kept as small as possible. If it can be arranged that the first-order perturbation energy actually disappears, the effect of the perturbation is particularly small and the approximation procedure is optimal. For this purpose the Hamiltonian operator (43.1) is artificially split into two parts,

$$\mathscr{H} = \underbrace{\frac{1}{2m}(\mathbf{p}_1^2 + \mathbf{p}_2^2) - Z'e^2\left(\frac{1}{r_1} + \frac{1}{r_2}\right)}_{\mathscr{H}_0} + \underbrace{e^2\left\{\frac{1}{r_{12}} - (Z-Z')\left(\frac{1}{r_1} + \frac{1}{r_2}\right)\right\}}_{W} \tag{43.18}$$

by introducing an effective nuclear charge $Z'$ of such magnitude that, when the average value of $W$ is formed from the eigenfunctions of $\mathcal{H}_0$, it vanishes. The eigenfunctions and eigenvalues of $\mathcal{H}_0$ are those

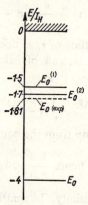

Fig. 43.—Comparison of the various energy values found for the ground state of the helium atom.

$E_0$(exp): experimental value
$E_0$: unperturbed value
$E_0^{(1)}$: value with perturbation $e^2/r_{12}$
$E_0^{(2)}$: value obtained from method of variations with optimum $Z'$

occurring in the analysis of the hydrogen atom, but with nuclear charge $Z'$ and with $a' = a_0/Z'$. The unperturbed energy is

$$E_0 = -2Z'^2 I_H \qquad (43.19a)$$

and the mean value of the perturbation energy is

$$\overline{W} = \frac{5}{8}\frac{e^2}{a'} - (Z-Z')\frac{2e^2}{a'} \qquad (43.19b)$$

It follows from the requirement that $\overline{W} = 0$ that the best choice of $Z'$ is

$$Z' = Z - \tfrac{5}{16} \qquad (43.20)$$

The above approximation leads to a value of about $1\cdot7 I_H$ for the ionization energy of helium; this is appreciably nearer the experimental value than the results of the previous approximation (figure 43).

The same choice for $Z'$ can also be substantiated on other grounds. From (43.19), it is clear that $E_0 + \overline{W}$ is simply the expectation value of $\mathscr{H}$, formed from the hydrogen functions with a nuclear charge $Z$:

$$\mathscr{H}(Z') = -\frac{e^2}{a_0}\{Z'^2 - \tfrac{5}{8}Z' + 2Z'(Z-Z')\} \qquad (43.21)$$

$Z'$ is a parameter entering into the wave functions. In accordance with the procedure of §28, the best approximation for the ground state is obtained by choosing $Z'$ so that $\overline{\mathscr{H}}(Z')$ is a minimum: this leads once more to the value for $Z'$ given by (43.20). The energy found by a variational method of this sort must be higher than the exact value. We can also see that the perturbation calculation using $e^2/r_{12}$ must yield an even higher value, since the total energy found by this procedure is $\overline{\mathscr{H}}(Z)$, which is the expectation value of $\mathscr{H}$ when the wave function is not an optimum. If we include still more parameters (say 9 in total) in the trial solution used in the variational method, and in particular, terms containing both $r_1$ and $r_2$, it is possible to calculate the position of the ground state with almost spectroscopic accuracy.[*] The result given by (43.20) can also be expressed as follows: the distributed charge of an electron screens the field of the nucleus to such an extent that the effective nuclear charge for the other electron is just $Z'e$.

From (43.19a) and (43.20), the ionization energy $I_Z$ of a nucleus of charge $Z$ with two electrons (i.e. the work of separation of an electron from an ion with positive charge $Z-2$) is

$$I_Z = I_H(2Z'^2 - Z^2) = I_H\{Z^2 - \tfrac{5}{4}Z + 2(\tfrac{5}{16})^2\} \qquad (43.22)$$

A comparison between theoretical and experimental values is given in the following table.

Ionization energy of helium-like ions, from equation (43.22)

| Ion | $Z$ | $I/I_H$ theor. | $I/I_H$ exp.[†] |
|-----|-----|----------------|-----------------|
| H⁻ | 1 | −0·055 | +0·053 |
| He | 2 | +1·70 | 1·81 |
| Li⁺ | 3 | 5·45 | 5·57 |
| Be⁺⁺ | 4 | 11·20 | 11·32 |
| B³⁺ | 5 | 18·95 | 19·08 |
| C⁴⁺ | 6 | 28·70 | 28·84 |
| N⁵⁺ | 7 | 40·45 | 40·61 |
| O⁶⁺ | 8 | 54·20 | 54·39 |
| F⁷⁺ | 9 | 69·95 | 70·12 |

* F. A. Hylleraas, *Z. f. Phys.* 54 (1929), 347.    † Landolt-Börnstein, 6th ed. (1950), Vol. I, part 1, p. 211.

Apart from the case of the negative hydrogen ion the agreement is good. The negative sign of $I_1$ would mean that the negative hydrogen ion could not exist, since the calculated energy of its ground state is higher than the energy of the hydrogen atom in the ground state. In the present approximation, the electrostatic repulsion exceeds the Coulomb attraction of the single nuclear charge. However, this approximation is not sufficiently accurate to enable the existence of the negative hydrogen ion to be excluded. This ion does in fact exist, and possesses a comparatively small ionization energy* of $0 \cdot 05 I_H$.

## §44. The hydrogen molecule

The Hamiltonian operator for the hydrogen molecule may be expressed as follows:

$$\mathscr{H} = \frac{\mathbf{P}_a^2 + \mathbf{P}_b^2}{2M} + \frac{\mathbf{p}_1^2 + \mathbf{p}_2^2}{2m} + e^2 \left\{ \frac{1}{R} + \frac{1}{r_{12}} - \frac{1}{r_{a_1}} - \frac{1}{r_{a_2}} - \frac{1}{r_{b_1}} - \frac{1}{r_{b_2}} \right\} \quad (44.1)$$

In the above expression, $a$ and $b$ refer to the nuclei, 1 and 2 to the electrons, and $R = r_{ab}$ is the distance between the two nuclei (figure 44). Since we have neglected the spin-orbit interaction, the wave function may be expressed as the product of the spin and orbital functions:†

$$\Psi = \psi(\mathbf{R}_a, \mathbf{R}_b, \mathbf{r}_1, \mathbf{r}_2) \chi(s_a, s_b) \eta(s_1, s_2) \quad (44.2)$$

Since $\mathscr{H}$ commutes with the permutation operators $P_{ab}$ and $P_{12}$, the eigenfunctions of $\mathscr{H}$ can be simultaneously chosen to be eigenfunctions of $P_{ab}$ and $P_{12}$. Therefore $\psi(\mathbf{R}_a, \ldots)$ is either symmetric or antisym-

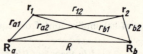

Fig. 44.—Distances between electrons and nuclei in the hydrogen molecule

metric in $a, b$ and $1, 2$; in virtue of Pauli's principle, the spin functions $\chi$ and $\eta$ must always have the opposite symmetry to $\psi$. As in the case of the helium atom, therefore, singlet and triplet states exist due to both nuclear and electron spin.

---

* The quantity $I_1$ relating to the negative hydrogen ion is called the *electronic affinity*, by analogy with other atoms such as the halogens, which form negative ions. This affinity is the energy obtained when an electron is picked up by the neutral atom.

† This is possible because there are only two similar particles. When there are several particles the functions (44.2) must be made antisymmetric in their spin and orbital components together.

It is not possible to solve the Schrödinger equation $\mathscr{H}\psi = E\psi$ exactly, but good approximate solutions exist.

We shall now consider the adiabatic approximation due to M. Born and R. Oppenheimer (1927). It is based on the great difference in mass between electrons and protons, as a result of which the latter may be considered to move slowly relative to the electrons in stationary bound states, according to the classical concept of the atom. We shall therefore assume that at each instant the electrons are disposed according to the instantaneous positions of the nuclei, and accordingly express the wave function as follows:

$$\psi(\mathbf{R}_a, \mathbf{R}_b, \mathbf{r}_1, \mathbf{r}_2) = \kappa(\mathbf{R}_a, \mathbf{R}_b)\phi(\mathbf{r}_1, \mathbf{r}_2; \mathbf{R}_a, \mathbf{R}_b) \qquad (44.3)$$

In the above expression $\phi$ is the wave function for fixed values of $\mathbf{R}_a$ and $\mathbf{R}_b$, and $|\kappa(\mathbf{R}_a, \mathbf{R}_b)|^2 d\mathbf{R}_a d\mathbf{R}_b$ is the probability of finding $\mathbf{R}_a$ and $\mathbf{R}_b$ in $d\mathbf{R}_a$ and $d\mathbf{R}_b$ respectively. $\phi$ must therefore satisfy the equation

$$\mathscr{H}_{el}\phi = \left\{ \frac{\mathbf{p}_1^2 + \mathbf{p}_2^2}{2m} + e^2 \left( \frac{1}{r_{12}} - \frac{1}{r_{a_1}} - \frac{1}{r_{a_2}} - \frac{1}{r_{b_1}} - \frac{1}{r_{b_2}} \right) \right\}\phi \qquad (44.4)$$

$$= E_{el}(R)\phi$$

In this equation $E_{el}(R)$ is the energy of the electrons expressed as a function of the positions of the nuclei: for reasons of symmetry this quantity must be a function of the interval $|\mathbf{R}_a - \mathbf{R}_b| = R$ alone.

If we insert (44.3) in the Schrödinger equation $\mathscr{H}\psi = E\psi$, and then multiply by $\phi^*$ and integrate over $\mathbf{r}_1$ and $\mathbf{r}_2$, we obtain

$$\int \phi^* \left\{ \frac{\mathbf{P}_a^2 + \mathbf{P}_b^2}{2M} + \frac{e^2}{R} + \mathscr{H}_{el} \right\} \phi\kappa \, d\mathbf{r}_1 \, d\mathbf{r}_2 = E\kappa \int |\phi|^2 \, d\mathbf{r}_1 \, d\mathbf{r}_2$$

$\phi$ may be assumed to be real and normalized without loss of generality. Then since

$$\mathbf{P}_a^2 \kappa\phi = \phi\mathbf{P}_a^2\kappa + \kappa\mathbf{P}_a^2\phi + 2(\mathbf{P}_a\kappa)(\mathbf{P}_a\phi)$$

and

$$\int \phi^* \mathbf{P}_a \phi \, d\mathbf{r}_1 \, d\mathbf{r}_2 = \mathbf{P}_a \tfrac{1}{2} \int \phi^2 \, d\mathbf{r}_1 \, d\mathbf{r}_2 = \frac{\hbar}{2i}\frac{\partial}{\partial \mathbf{R}_a}1 = 0$$

we obtain from (44.4) the following result:

$$\left\{ \frac{\mathbf{P}_a^2 + \mathbf{P}_b^2}{2M} + \int \phi^* \left( \frac{\mathbf{P}_a^2 + \mathbf{P}_b^2}{2M} \right) \phi \, d\mathbf{r}_1 \, d\mathbf{r}_2 + E_{el}(R) + \frac{e^2}{R} \right\}\kappa = E\kappa \qquad (44.5)$$

We shall see later that $\phi$ depends essentially only on the differences $R_a - r_1$, etc. The order of magnitude of the second term in (44.5) is therefore $(\phi, [(\mathbf{p}_1^2 + \mathbf{p}_2^2)/2M] \phi)$, i.e. smaller than the kinetic energy $(\phi, [(\mathbf{p}_1^2 + \mathbf{p}_2^2)/2m] \phi)$ of the electrons by a factor $m/M$; this term can therefore be neglected in comparison with $E(R)$. Hence from (44.5)

$$\left\{ \frac{\mathbf{P}_a^2 + \mathbf{P}_b^2}{2M} + E_{el}(R) + \frac{e^2}{R} \right\} \kappa = E\kappa \qquad (44.6)$$

It is evident that the total energy of the electrons appears as the potential energy of the nuclei. The solution of the problem reduces approximately to the solution of equations (44.4) and (44.6).

The translational and rotational motions can be separated in the usual way in equation (44.6). We introduce the coordinates $\mathbf{R}_s = \frac{1}{2}(\mathbf{R}_a + \mathbf{R}_b)$ and $\mathbf{R} = \mathbf{R}_a - \mathbf{R}_b$, and put

$$\kappa = \exp(i\mathbf{K}\mathbf{R}_s) Y_{lm}(\Theta, \Phi) \frac{u(R)}{R}$$

Substituting for $\kappa$ in (44.6) yields the following equation for $u$:*

$$-\frac{\hbar^2}{M} u'' + \left\{ \frac{\hbar^2 l(l+1)}{MR^2} + E_{el}(R) + \frac{e^2}{R} \right\} u = \left( E - \frac{\hbar^2 \mathbf{K}^2}{4M} \right) u \quad (44.7)$$

In order that the molecule should possess a bound state, the function $V(R) = E_{el}(R) + e^2/R$ must possess a minimum corresponding to a definite equilibrium interval $a$. In the lowest energy states of the molecule only small vibrations can occur about the equilibrium position. We may therefore expand the expression in curly brackets in (44.7) about $a$:

$$\frac{\hbar^2 l(l+1)}{MR^2} + V(R) = \frac{\hbar^2 l(l+1)}{Ma^2} + V(a) + \tfrac{1}{4}M\omega^2(a - R)^2$$

If we neglect the displacement of the minimum of $V(R)$ due to the centrifugal force, we have the following relations:

$$\frac{dV(R)}{dR}\bigg|_{R=a} = 0; \quad \tfrac{1}{4}M\omega^2 = \frac{1}{2}\frac{d^2}{dR^2} V(R)\bigg|_{R=a}$$

Then from (44.7) we obtain the equation of a harmonic oscillator with energy levels $E_{osc} = \hbar\omega(n + \frac{1}{2})$.

---

* Note that the reduced mass $\frac{1}{2}M$ occurs in place of $M$ in the equation of relative motion.

Thus we arrive at the result that the total energy of the molecule may be divided as follows:

$$E = E_{trans} + E_{rot} + E_{osc} + V(a) \qquad (44.8)$$

We wish to compare the order of magnitude of each of the above terms:

$E_{trans} = \dfrac{\hbar^2}{4M}K^2 \approx \dfrac{\hbar^2}{M\lambda^2}$,    where $\lambda$ is the de Broglie wavelength corresponding to the translational momentum.

$E_{rot} = \dfrac{\hbar^2 l(l+1)}{Ma^2} \approx \dfrac{\hbar^2}{Ma^2}$,    where $a$ is the molecular equilibrium distance.

$E_{osc} = \hbar\omega(n+\tfrac{1}{2}) \approx \hbar\sqrt{\dfrac{E''_{el}(a)}{M}} \approx \left(\dfrac{M}{m}\right)^{1/2}\dfrac{\hbar^2}{Ma^2}$

$V(a) \approx \dfrac{\hbar^2}{ma^2} = \dfrac{M}{m}\dfrac{\hbar^2}{Ma^2}$

(We have assumed that $V(a)$ is of the same order of magnitude as the kinetic energy, since the potential energy is approximately equal to this quantity. $E''_{el}(a)$ is of the same order of magnitude as $\hbar^2/ma^4$.) In a gas consisting of diatomic molecules, translational motion is first produced, followed by rotational motion, and then by molecular vibrations, as the temperature is increased. With a further increase in temperature electron transitions may be excited; however, dissociation often occurs first.

The dissociation energy is given by $D = 2\varepsilon_0 - E_0$, where $\varepsilon_0$ is the energy of a free atom and the energy of the ground state of the molecule is $E_0 = V(a) + \tfrac{1}{2}\hbar\omega$.

There is a further point with regard to Pauli's principle, in connection with the rotational spectrum. The wave function $Y_{lm}(\Theta,\Phi)$ is symmetric in $\mathbf{R}_a$ and $\mathbf{R}_b$ if $l$ is even, and antisymmetric if $l$ is odd (cf. Exercise 1, p. 152). Since $\mathbf{R}_s$ is symmetric in $\mathbf{R}_a$ and $\mathbf{R}_b$ and the same is true of the wave function $\phi(\mathbf{r}_1,\mathbf{r}_2; \mathbf{R}_a,\mathbf{R}_b)$ of the bound states (as we shall see in §45), nuclear-spin function triplets correspond to odd values of $l$, and singlets to the even values; this is a consequence of Pauli's principle. The nuclear triplet states are called *orthohydrogen*, and the singlets *parahydrogen*, by analogy with the terminology used for helium. At high temperatures molecular hydrogen is a mixture of ortho- and parahydrogen in the ratio of 3 to 1; in this case all rotation states are excited with equal probability and the ratio depends only on the relative weights of the spin functions. At low temperatures no rotation terms are excited, and the hydrogen molecule should therefore occur in the para form. However, as a result of the almost total prohibition of singlet to triplet transitions, it takes many days before the pure gas reaches the new state of equilibrium after its temperature is lowered. Once this equilibrium has been achieved and the temperature is raised once more, it takes equally long for orthohydrogen to appear, which it does as a result of the weak nuclear spin interactions during molecular collisions.

The difference between the rotational terms of ortho- and parahydrogen is particularly noticeable experimentally, both in the rotation spectrum and in the rotational contribution to the specific heat. The specific heat of a 3:1 mixture of ortho- and parahydrogen is obtained by cooling below room temperature; heating pure parahydrogen from very low temperatures gives the specific heat of this form alone. It is clear from the above account that Pauli's principle can entail quite surprising and far-reaching experimental results.

## §45. Chemical bonds

In §44 we assumed that the total potential energy $V(R)$ of the nucleus possessed a minimum; we now wish to know how this comes about.

While many questions concerning the heteropolar or ionic bond had been successfully dealt with before the advent of the quantum theory, the homopolar or covalent bond had obstinately resisted every attempt at explanation in terms of classical physics. We shall see that the quantum theory, without requiring any additional assumptions, provides both qualitative and quantitative answers to the problem of the nature of the homopolar bond.

Firstly, we must consider equation (44.4): an exact solution is not available, but approximations to the lowest energy states may be formed from the functions of the hydrogen atom, which have the following form:

$$\phi_0(\mathbf{r} - \mathbf{R}_a), \quad \phi_0(\mathbf{r} - \mathbf{R}_b), \quad \text{where} \quad \phi_0(\mathbf{r}) = \frac{1}{\sqrt{(\pi a_0^3)}} \exp -\frac{r}{a_0}$$

We shall make use of the abbreviations

$$a(1) = \phi_0(\mathbf{r}_1 - \mathbf{R}_a), \quad b(1) = \phi_0(\mathbf{r}_1 - \mathbf{R}_b)$$

together with similar abbreviations $a(2)$ and $b(2)$.

Four functions of two particles can be constructed from these single-particle functions:

$$a(1)a(2), \quad a(1)b(2), \quad b(1)a(2), \quad b(1)b(2)$$

The complete wave function $\phi(\mathbf{r}_1, \mathbf{r}_2; \mathbf{R}_a, \mathbf{R}_b)$ is expressed as a linear combination of these functions.

The form of these linear combinations is largely determined by the condition that they must be eigenfunctions of $P_{ab}$ and $P_{12}$. It may easily be verified that this condition is satisfied only by the following combinations:*

$$\phi_A = c_1 \{a(1)b(2) - a(2)b(1)\} \quad P_{ab}\phi_A = P_{12}\phi_A = -\phi_A \quad (45.1)$$

$$\phi_S = c_2 \{a(1)b(2) + a(2)b(1)\} + c_3 \{a(1)a(2) + b(1)b(2)\}$$
$$P_{ab}\phi_S = P_{12}\phi_S = \phi_S \quad (45.2)$$

$$\phi_{AS} = c_4 \{a(1)a(2) - b(1)b(2)\}$$
$$P_{ab}\phi_{AS} = -\phi_{AS} \quad P_{12}\phi_{AS} = \phi_{AS} \quad (45.3)$$

The subscripts $S$ and $A$ mean symmetric and antisymmetric respectively.

The coefficients $c_3$ and $c_4$ multiply functions representing two electrons associated with each nucleus: the corresponding states are termed "polar", although in contrast to the heteropolar or ionic bond the charge of the two electrons is always symmetrically distributed about both nuclei.

We shall omit any consideration of these states in the first instance, and shall investigate the lowest energy state of the two remaining combinations. For this purpose we shall require the mean particle density

$$\rho(\mathbf{r}) = \int |\phi(\mathbf{r}, \mathbf{r}')|^2 \, d\mathbf{r}'$$

for the two states

$$\phi_A = c_1 \{a(1)b(2) - a(2)b(1)\}, \quad \phi_S = c_2 \{a(1)b(2) + a(2)b(1)\} \quad (45.4)$$

The result is

$$\rho_A(1) = \frac{\rho_a(1) + \rho_b(1) - 2S\rho_{ab}(1)}{2(1 - S^2)}, \quad \rho_S(1) = \frac{\rho_a(1) + \rho_b(1) + 2S\rho_{ab}(1)}{2(1 + S^2)} \quad (45.5)$$

---

* It is not possible to construct a function $\phi_{SA}$ from the above two-particle functions.

The coefficients $c_1$ and $c_2$ are chosen so as to normalize the functions $\phi_A$ and $\phi_S$. Further,

$$\rho_a(1) = a^2(1) \qquad \rho_b(1) = b^2(1) \qquad \rho_{ab}(1) = a(1)b(1)$$

$$S = \int \rho_{ab}(1)\, d\mathbf{r}_1$$

Figure 45 illustrates the variation of the "overlap" integral $S$ with the separation between the nuclei.

In classical physics the expected result would be simply $\rho = \tfrac{1}{2}(\rho_a + \rho_b)$; the expressions (45.5), on the other hand, exhibit additional terms containing the "exchange density" $\rho_{ab}$. These are due to the requirement

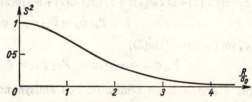

Fig. 45.—The square of the "overlap integral" $S$ as a function of distance from the nucleus

that the stationary states should be eigenfunctions of the symmetry operators $P_{ab}$ and $P_{12}$; they have no analogue in classical physics. Since the function $\rho_{ab}(\mathbf{r})$ is greatest along the line joining the nuclei $a$, $b$, the function $\rho_A$ represents a reduction of the charge density between the nuclei in comparison with the classical value, while $\rho_S$ represents an increase (cf. figure 46). The effect on the nuclei is that $\rho_A$ involves an additional force of repulsion and $\rho_S$ an additional force of attraction. We should therefore expect to find that $\phi_S$ relates to the bound state of the molecule, and this is in fact the case.

We are now provided with an interpretation of the chemical valency symbols in the case of covalent bonds. Owing to a symmetry effect of quantum mechanics* the electron charge accumulates between the nuclei in the case of those states possessing antisymmetric electron spin; it then acts electrostatically as a "cement" holding the nuclei together.

We now return to the "polar" states. These involve a reduction of the kinetic energy, since they enable each electron to spread over the

---

* Not to be confused with Pauli's principle.

whole molecule independently of the other. This can perhaps be seen most clearly if we put $c_3 = c_2$ in (45.2); we can then express this equation as follows:

$$\bar{\phi}_S = c_2 [a(1) + b(1)] [a(2) + b(2)] \qquad (45.6)$$

This means that both electrons move around the two nuclei completely independently of each other; this involves an increase in the de Broglie wavelengths of the individual electrons, corresponding to a reduction

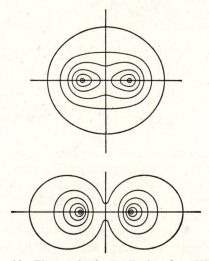

Fig. 46.—Electron density distribution, from (45.5)

of the kinetic energy. On the other hand the potential energy is increased as a result of the Coulomb repulsion between the electrons, since in comparison with $\phi_S$ as given by (45.4), there is a higher probability of finding two electrons associated with the same nucleus.

If the constants $c_2$ and $c_3$ are determined so as to make the energy a minimum, a comparatively small admixture of polar states to $\phi_S$ is obtained ($c_3 = 0{\cdot}26c_2$). For heavier molecules the proportion of the admixture is higher; in particular, the polar states provide appreciable contributions in the case of metallic bonds.

We shall neglect the polar states in our further discussion of the hydrogen molecule, in which we shall now consider the approximate treatment of W. Heitler and F. London (1927). The energy $E_{el}(R)$ is equal to the expectation value of $\mathscr{H}_{el}$ in the states $\phi_A$ and $\phi_S$. Using (44.4) we obtain

$$E_S(R) = 2\varepsilon_0 + \frac{C+A}{1+S^2} \qquad (45.7)$$

$$E_A(R) = 2\varepsilon_0 + \frac{C-A}{1-S^2} \qquad (45.8)$$

in which

$$C(R) = e^2 \int \left( \frac{1}{r_{12}} - \frac{1}{r_{a2}} - \frac{1}{r_{b1}} \right) \rho_a(1)\,\rho_b(2)\,d\mathbf{r}_1\,d\mathbf{r}_2$$

is the Coulomb integral,

$$A(R) = e^2 \int \left( \frac{1}{r_{12}} - \frac{1}{r_{a2}} - \frac{1}{r_{b1}} \right) \rho_{ab}(1)\,\rho_{ab}(2)\,d\mathbf{r}_1\,d\mathbf{r}_2$$

the "exchange integral", and

$$\varepsilon_0 = \left( a(1), \left( \frac{\mathbf{p}_1^2}{2m} - \frac{e^2}{r_{a1}} \right) a(1) \right)$$

the energy of the ground state of the hydrogen atom.

As regards the term "exchange integral", it should be realized that it is meaningless to speak of an exchange of electrons, because such an interchange cannot be observed. On the other hand it is quite permissible to say that the two atoms exchange their spins; the frequency of this exchange is actually given by $A/\hbar$ (cf. Exercise 5, p. 272).

For the simple but rather long calculation of the integrals $C$ and $A$ the reader is referred to the original literature;[*] the results are given in figure 47. As we should expect, the symmetric state provides a bond as a result of the negative sign of $A(R)$.

Agreement with experiment is quite good, but can be appreciably improved by refinements in the expression for the wave function. Firstly, we can treat the nuclear charge number as a parameter in the method of variations (45.4), as we did in the case of helium (§43). This gives an energy level of about 17 per cent below the simple Heitler-London function, when the effective atomic number $Z'$ is 1·166.[†] In addition, the constant $c_3$ occurring in (45.2) can also be included as a parameter; this implies that allowance is made for the polar states. We then obtain a further reduction in the energy of 10 per cent, when

---

[*] Y. Sugiura, *Z. Phys.* **45** (1927), 484.

[†] Note that in the present case $Z' > Z = 1$, in contrast to the situation for the helium atom.

$Z' = 1\cdot19$ and $c_3 = 0\cdot256c_2$. The remaining 25 per cent difference was accounted for by H. M. James and A. S. Coolidge: by making a more accurate allowance for the correlation between the electrons and the

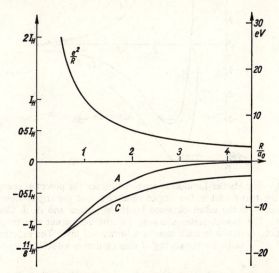

Fig. 47.—Coulomb integral $C$, exchange integral $A$, and electrostatic energy of the nuclei $e^2/R$. When $R \gtrsim 1\cdot5a_0$, $-C \approx e^2/R$

factors in the wave function depending explicitly on $r_{12}$, they obtained values for the energy of dissociation and the inter-nuclear equilibrium distance that agreed with experimental data to within 1 per cent (cf. figure 48).

We now refer to another way of expressing equations (45.7) and (45.8), of importance in connection with ferromagnetism. Owing to the Pauli principle, the symmetric spin function of the electron is associated with $\phi_A$ and the antisymmetric function with $\phi_S$. We may therefore say that the potential energy of the nuclei depends on the setting of the electron spins. Equations (45.7) and (45.8) are put into a new form to make this more apparent:

$$P^s_{12}\eta_S(s_1,s_2) = \eta_S, \qquad P^s_{12}\eta_A(s_1,s_2) = -\eta_A \qquad (45.9)$$

In the above equations $P^s_{12}$ represents the permutation operator of the spin variables, and $\eta_S$ and $\eta_A$ are the symmetric and antisymmetric spin functions. Equations (45.7) and (45.8) can therefore be combined as follows:

$$\mathscr{H}^s(R)\eta = \tfrac{1}{2}\{(1+P^s_{12})E_A+(1-P^s_{12})E_S\}\eta$$

$$= 2\varepsilon_0\eta+\left\{\frac{C-AS^2}{1-S^4}-P^s_{12}\frac{A-CS^2}{1-S^4}\right\}\eta = E_{el}(R)\eta \qquad (45.10)$$

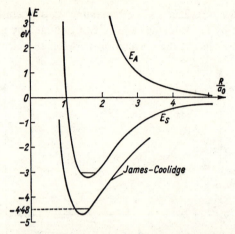

Fig. 48.—The Heitler-London approximation for the potential energy between the two nuclei in the singlet state $(E_S)$ and the triplet state $(E_A)$, compared with the result obtained by H. M. James and A. S. Coolidge for $E_S$. The lowest state, allowing for the zero-point energy of the oscillation, is shown in each case by a horizontal line. The experimental value for the energy of dissociation is $4·48\,\text{eV}$

$\eta_A$ and $\eta_S$ are eigenfunctions of $\mathcal{H}^s$, with respective eigenvalues $E_S$ and $E_A$. In (45.9) we can replace $P^s_{12}$ by $\frac{1}{2}\{1+(\vec{\sigma}_1, \vec{\sigma}_2)\}$, where $\vec{\sigma}_1$ and $\vec{\sigma}_2$ are the Pauli spin operators acting on $s_1$ and $s_2$ (cf. Exercise 6, p. 272). Then

$$\mathcal{H}^s(R) = K(R) - \tfrac{1}{2}J(R)(\vec{\sigma}_1, \vec{\sigma}_2) \qquad (45.11)$$

where

$$K(R) = 2\varepsilon_0 + \frac{C-AS^2}{1-S^4} - \frac{1}{2}\frac{A-CS^2}{1-S^4}, \qquad J(R) = \frac{A-CS^2}{1-S^4} < 0 \qquad (45.12)$$

Since $\eta_A$ and $\eta_S$ are eigenfunctions of $\mathcal{H}^s(R)$, the correct spin functions and energies are obtained by solving the eigenvalue equation (45.10) for the spin function $\eta$.

Equation (45.10) implies that, for a given internuclear distance $R$, there is an interaction between the spins that tends to set them antiparallel, since $\eta_A$ is the ground state. This is clearly expressed by the term $-\tfrac{1}{2}J(\vec{\sigma}_1, \vec{\sigma}_2)$ in (45.11).

This interaction is purely electrostatic in character and occurs indirectly as a result of Pauli's principle. It has no connection with the magnetic spin-spin interaction (which is also present). Heisenberg advanced the hypothesis (in 1928) that ferromagnetism was due to an exchange interaction of this sort between the spins. According to this theory, the difference between the "ferromagnetic electrons" of iron, cobalt, nickel, etc. (i.e. the electrons of the incomplete inner shells) and the electrons of the hydrogen molecule lies in the fact that $J$ is positive for the former at the equilibrium distance, which is largely determined by the valence electrons (i.e. the electrons of the outer shell). It is not yet possible to decide conclusively whether the situation is really as described, but it is assumed that Heisenberg's theory is correct in its essentials.

## Exercises

### 1. *Energy of an ideal Fermi "gas"*

The spatial components of the eigenfunctions corresponding to the particles of an ideal gas enclosed in a periodicity volume $V = L^3$ are $e^{i\mathbf{k}\mathbf{r}}/\sqrt{V}$, where $\mathbf{k} = 2\pi\mathbf{n}/L$ and $n_x$, $n_y$, $n_z$ are integers. These functions may also contain a spin function $\alpha$ or $\beta$ if the particles possess a spin of value $\frac{1}{2}$. The energy is $E = \sum\limits_{j=1}^{n} \hbar^2 \mathbf{k}_j{}^2/2m$, where $n$ is the total number of particles and $\mathbf{k}_1, \ldots, \mathbf{k}_n$ are the propagation vectors represented in the eigenfunctions. In consequence of Pauli's exclusion principle, the $\frac{1}{2}n$ values of $\mathbf{k}$ with the lowest energies contain both $\alpha$ and $\beta$ spins. What is the maximum value of the magnitude of $\mathbf{k}$, denoted by $k_0$? What is the energy of the ground state? (Since the $\mathbf{k}$-values lie very close together when the volume is large, the summation should be replaced by integrals.)

### 2. *Correlations for the ideal Fermi gas*

The single particle functions for the ideal Fermi gas are $\Psi_\lambda(x) = e^{i\mathbf{k}\mathbf{r}}\kappa_s/\sqrt{V}$, where $x$ represents the spatial and spin coordinates and $\lambda$ the quantum numbers $\mathbf{k}$ and $s$, and

$$\kappa_s = \begin{cases} \alpha & \text{when } s = 1 \\ \beta & \text{when } s = -1 \end{cases}$$

The functions $\Psi_\lambda(x)$ are normalized and orthogonal. The ground state of the gas is

$$\Psi_\Lambda{}^A = \frac{1}{\sqrt{(n!)}} \Sigma(-1)^P \Psi_{\lambda_{P_1}}(x_1) \ldots \Psi_{\lambda_{P_n}}(x_n)$$

The indices $\lambda_i$ extend over all values of $\mathbf{k}$ corresponding to those of Exercise 1 and the values $\pm 1$ of $s$. $\Psi_\Lambda{}^A$ is normalized (see below). By integrating over all $x_i$ except $x_1$ and $x_2$, we obtain the probability $w(\mathbf{r}_1, s_1, \mathbf{r}_2, s_2) d\mathbf{r}_1 d\mathbf{r}_2$ of finding an electron with spin $s_1$ in $d\mathbf{r}_1$ and an electron with spin $s_2$ in $d\mathbf{r}_2$:

$$w = \frac{1}{n!} \int dx_3 \ldots dx_n \Sigma(-1)^P(-1)^{P'} \Psi^*_{\lambda_{P'_1}}(x_1) \ldots \Psi^*_{\lambda_{P'_n}}(x_n) \Psi_{\lambda_{P_1}}(x_1) \ldots \Psi_{\lambda_{P_n}}(x_n)$$

Since the functions $\Psi_\lambda$ are orthogonal, the only contributing permutations are those for which $P_i = P_i'$ where $i \geq 3$. The permutations $P$ and $P'$ can therefore differ only in the case of the first two indices: this means that $P$ and $P'$ are either identical or that they simply differ through an interchange $P_1' = P_2$, $P_2' = P_1$. In the first case $(-1)^P(-1)^{P'} = 1$, in the second, $-1$. (It follows that $\Psi_\Lambda$ is normalized. When integrating over all $x_i$ we are only concerned with the case $P = P'$; this gives $n!$ times 1.) Hence

$$w = \frac{1}{n!} \Sigma_P \left( \Psi^*_{\lambda_{P_1}}(x_1) \Psi^*_{\lambda_{P_2}}(x_2) - \Psi^*_{\lambda_{P_2}}(x_1) \Psi^*_{\lambda_{P_1}}(x_2) \right) \Psi_{\lambda_{P_1}}(x_1) \Psi_{\lambda_{P_2}}(x_2)$$

The summation $\Sigma\limits_{P}$ may be replaced by a factor $(n-2)!$ corresponding to the number of permutations that leave $P_1$ and $P_2$ unchanged, and by a summation over the range of quantum numbers:

$$w = \frac{1}{n(n-1)} \Sigma_{\lambda,\lambda'} \left( \Psi^*_\lambda(x_1) \Psi^*_{\lambda'}(x_2) - \Psi^*_{\lambda'}(x_1) \Psi^*_\lambda(x_2) \right) \Psi_\lambda(x_1) \Psi_{\lambda'}(x_2)$$

(*a*) Calculate $w$, replacing the summations over **k** by integrals.

(*b*) $w$ is a function of the interval $|\mathbf{r}_1-\mathbf{r}_2| = r$. If $\mathbf{r}_1$, $s_1$ are fixed, then apart from a normalization factor, $wd\mathbf{r}$ is the probability of finding an electron with spin $s_2$ at a distance **r** from a given electron with spin $s_1$. Multiplication by the number $\frac{1}{2}n$ of electrons with spin $s_2$ gives the density $\rho_{s_1 s_2}(\mathbf{r})$ at the distance **r** from the electron with spin $s_1$; in this derivation the small difference between $n$ and $n-1$ is neglected.

$$\rho_{s_1 s_2}(\mathbf{r}) = \tfrac{1}{2}nw(\mathbf{r}_1, s_1; \mathbf{r}_1+\mathbf{r}, s_2)/\int wd\mathbf{r} \approx 2Vnw$$

Discuss the densities $\rho_{1,1}$ and $\rho_{1,-1}$, and the total density $\rho_{1,1}+\rho_{1,-1}$ as observed from an electron. Illustrate them diagrammatically.

### 3. *Polarizability of helium*

Using the formula given in Exercise 5, p. 176, evaluate the polarizability of a helium atom by deriving an approximation to the exact eigenfunction of the ground state in the form of a product of hydrogen functions with an optimum value of $Z$.

### 4. *Rotation terms of the deuterium molecule*

The heavy hydrogen isotope, deuterium, has a spin of value 1, and therefore obeys Bose-Einstein statistics. Give the relative weights of the rotation states of $D_2$ for the different values of $l$, using a similar argument to that employed in the case of the hydrogen molecule.

### 5. *The exchange integral and spin interchange in the hydrogen molecule*

A state consisting of the combination

$$\phi = c\{[a(1)b(2)-a(2)b(1)][\alpha(1)\beta(2)+\alpha(2)\beta(1)]+$$
$$+[a(1)b(2)+a(2)b(1)][\alpha(1)\beta(2)-\alpha(2)\beta(1)]\}$$
$$= f_A+f_S$$

at time $t = 0$ depends on time as follows (cf. §45):

$$\phi(t) = f_A \exp(-iE_A t/\hbar)+f_S \exp(-iE_S t/\hbar)$$

Show that this wave function represents the oscillation of an $\alpha$ and a $\beta$ spin to and fro between the atoms $a$ and $b$, and that the frequency of this exchange of spin is approximately given by $A/\hbar$, where $A$ is the exchange integral.

### 6. *Dirac's identity for the spin vector*

Let $P^s\chi(s_1,s_2) = \chi(s_2,s_1)$. Show that

$$P^s \equiv \tfrac{1}{2}\{1+(\vec{\sigma}_1 \vec{\sigma}_2)\}$$

### 7. *Eigenfunctions of* $S^2$

Using the identity proved in the previous exercise, determine the eigenfunctions and eigenvalues of

$$S^2 = \tfrac{1}{4}\hbar^2(\vec{\sigma}_1+\vec{\sigma}_2)^2$$

### 8. *Van der Waals potential for the Thomson atomic model*

According to §44, the potential energy between the two atoms of the hydrogen molecule decreases exponentially and is negligibly small at distances of a few Bohr radii. The energy calculated in that paragraph is essentially a first-order approximation, in which allowance is made only for the ground states of the hydrogen atoms. At great distances the main contribution is due to the second-order approximation, in which excited states are taken into account in the trial solution for the wave function; this contribution takes the form of a decreasing potential proportional to $1/R^6$ (the van der Waals potential). This potential is to be evaluated for a simple molecular model. Each atom is represented by a Thomson model of radius $R_A$ (figure 49). Electron 1 ($r_1$) is in nucleus $a$, electron 2 ($r_2$) in nucleus $b$. Symmetrizing effects are unimportant when the inter-nuclear interval $R$ is large.

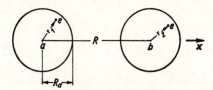

Fig. 49.—Two Thomson atoms

(*a*) Set up the classical equations of motion for each electron. (Since the displacements $r_1$ and $r_2$ are assumed to be small, it does not matter whether the field of dipole $b$ is calculated at $a$ or at $r_1$.) The equation of motion for $r_1$ contains a component proportional to $r_1$ and $r_2$, and similarly for $r_2$. We thus obtain a system of differential equations for coupled oscillators.

(*b*) Calculate the eigenfrequencies $\omega_\nu$.

(*c*) Find the energy of the ground state, $V(R)$. It is the sum of the zero-point energies $\sum_{\nu=1}^{6} \tfrac{1}{2}\hbar\omega_\nu$. Derive an approximate expression for $V(R)$ for great distances.

(*d*) Obtain the eigenfunction corresponding to the ground state, and calculate the mean values of the individual coordinates and their products. Discuss the significance of the sign of the correlation product $\overline{x_1 x_2}$.

### 9. *Forces between two helium atoms*

Calculate the potential energy arising between two helium atoms, using the same method as was employed in the case of the hydrogen molecule. The ground state of the atom is assumed to be represented by products of the hydrogen functions $a$, $b$ with the optimum value of $Z$. The present calculation is simpler because the ground state is not degenerate. The initial function $\Psi^A$ is the antisymmetric function

$$\Psi^A = \sum_P (-1)^P P\Psi$$

where $\quad \Psi = a(1)\alpha(1)a(2)\beta(2)b(3)\alpha(3)b(4)\beta(4)$
Hence
$$E_{el}(R) = (\Psi^A, \mathscr{H}_{el}\Psi^A)/(\Psi^A, \Psi^A) = (\Psi^A, \mathscr{H}_{el}\Psi)/(\Psi^A, \Psi)$$

In addition, there is the electrostatic interaction of the two nuclei. Express $E_{el}$ in terms of the Coulomb and exchange integrals that were used in the case of the hydrogen molecule; in order to simplify the calculation, assume that the overlap integral $S$ may be neglected except in the exchange integral itself. As a result of this assumption $(\Psi^A, \Psi) \approx 1$; it should also be noted that, in addition to the identical permutation, only $P_{13}$, $P_{24}$, and $P_{13}P_{24}$ (with appropriate sign) occur in $(\Psi^A, \mathscr{H}_{el}\Psi)$ because the spin functions are orthogonal.

Fig. 45—Two Thomson atoms

# E

---

The
theory
of
radiation

# CHAPTER EI

## Black-body radiation

### §46. Thermodynamics of black-body radiation

#### (a) Kirchhoff's proof of the existence of a universal function $u_\omega(\omega,T)$

If the walls of a completely evacuated enclosure are brought to a
definite temperature $T$, electromagnetic radiation is produced in its
interior. A state of equilibrium is attained when the material of the
walls absorbs as much radiation energy per unit time as it emits. This
radiation is represented by its energy density $u$; since the radiation is
electromagnetic this density has the value

$$u = \frac{1}{8\pi}(\mathbf{E}^2 + \mathbf{H}^2) \tag{46.1}$$

The spectral distribution of the radiation energy is denoted by a
function $u_\omega$ of $\omega$; $u_\omega d\omega$ represents the portion of the energy density in
the elementary angular frequency range $\omega, \omega + d\omega$. Naturally, the
following relation is always valid:

$$u = \int_0^\infty u_\omega \, d\omega \tag{46.2}$$

It was found by G. R. Kirchhoff that, at a given temperature, the
function $u_\omega$ does not depend at all on the nature of the material forming
the enclosure; $u_\omega$ is determined solely by the temperature, and is
completely independent of the material constants of the walls.

Kirchhoff's proof of this law is based on the following argument.
Let us assume that there were two enclosures A and B formed within
different materials, in which different values of $u_\omega$ arose at some point
of the spectrum when they were each brought into contact with a heat
source of temperature $T$. Then we could make use of this situation to
produce a finite temperature difference between the two heat sources

that were originally at the same temperature $T$, without performing any work.  To do this, we should have to place enclosure A in contact with one source and enclosure B in contact with the other; we should then have to reflect (by means of a suitable optical arrangement) a small aperture constructed in A into a similar small aperture in B.  The apertures would be covered by filters passing only the "colour" $\omega$ for which there is a difference between the two values of $u_\omega$.  If for instance $u_\omega$ were greater in A than in B, these measures would lead to more energy being radiated into B from A than would be radiated back.  In consequence, energy would be abstracted from enclosure A; the temperature of the heat source surrounding A would therefore drop, and that of the source in contact with B would rise.  These changes would continue until the value of $u_\omega$ was the same for each enclosure.  The temperature difference so created could be used to perform mechanical work by means of a heat engine.  A difference between the values of $u_\omega$ in A and B would therefore permit the construction of a *perpetuum mobile* of the second kind.  Therefore by the second law of thermodynamics the value of $u_\omega$ in each enclosure must be the same at all frequencies.  Since $u_\omega$ depends only on $\omega$ and on the temperature $T$, there must be a universal function $u_\omega(\omega, T)$ which gives the spectral distribution of the energy of the black-body radiation at temperature $T$.

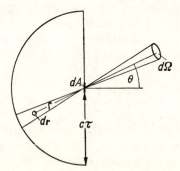

Fig. 50.—Calculation of the radiation intensity

Kirchhoff assigned to physical research the task of discovering this universal function; the stage-by-stage solution of the problem is associated with the names of L. Boltzmann, W. Wien, and M. Planck. When putting forward his law, Kirchhoff did not need to make any assumption regarding the physical nature of the black-body radiation.

Boltzmann and Wien merely used the fact that, for electromagnetic radiation, energy and momentum differ by the factor $c$. The complete solution was first produced by Planck with the help of his quantum hypothesis.

For further thermodynamic analysis, we need to know the relation between the radiation pressure $P$ and intensity $K$, and the radiation energy density. Let us consider the radiation energy incident during a small time interval $\tau$ on a small element of surface $dA$ of the enclosure boundary (figure 50). Since the energy is propagated with the velocity of light, all points for which $r \leq c\tau$ contribute to the radiation through $dA$. A certain fraction $u\,d\mathbf{r}$ of the radiation energy present in the small element of volume $d\mathbf{r}$ will strike $dA$. For isotropic radiation, which we shall assume, this fraction is given by the ratio of the solid angle $dA\cos\theta/r^2$ subtended by the element of surface at $d\mathbf{r}$, divided by the total solid angle $4\pi$:

$$u\,d\mathbf{r}\frac{dA\cos\theta}{4\pi r^2}$$

The radiation incident on $dA$ from a given solid angle $d\Omega$ is obtained by integrating over the volume of the cone $d\Omega$ contained in the hemisphere $r < c\tau$:

$$u\frac{c}{4\pi}dA\cos\theta\,d\Omega\tau = K\,dA\cos\theta\,d\Omega\tau, \quad \text{i.e.} \quad K = u\frac{c}{4\pi} \quad (46.3)$$

The total radiation incident on $dA$ is obtained by integrating over the complete hemisphere:

$$u\frac{c}{4}dA\tau = S\,dA\tau, \quad \text{i.e.} \quad S = u\frac{c}{4} \quad (46.4)$$

$K\cos\theta$ is the energy radiated from the solid angle $d\Omega$, per unit time and per unit area of the boundary wall; $S$ is the corresponding total energy. If $dA$ is a small aperture in the enclosure $K$ and $S$ denote the energy emitted into $d\Omega$ and the total energy, respectively. Equations (46.3) and (46.4) are naturally also valid for each spectral component:

$$K_\omega = u_\omega\frac{c}{4\pi}, \qquad S_\omega = u_\omega\frac{c}{4} \quad (46.5)$$

Since the radiation energy and momentum differ by a factor $1/c$, we can now derive the momentum transferred to the wall, and hence the radiation pressure $P$. If we assume that the boundary wall is an ideal reflector, the radiation incident at an angle $\theta$ will be reflected at the same angle $\theta$ to the normal, in accordance with the law of reflection

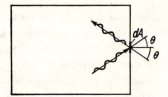

Fig. 51.—Ideal reflection at cavity wall

(figure 51). The momentum in the direction of the normal that is transferred to $dA$ during the interval $\tau$ is clearly

$$dA\,\tau \int \frac{K}{c}\cos\theta\,.\,2\cos\theta\,d\Omega = dA\,\tau\frac{4\pi K}{3c} = dA\,\tau\frac{u}{3}$$

The radiation pressure is therefore

$$P = \frac{u}{3} \tag{46.6}$$

This relation can also be derived from the hypothesis of light quanta (cf. Exercise 1, p. 296).

### (b) The Stefan-Boltzmann law and Wien's law

The laws of Stefan-Boltzmann and of Wien are necessary consequences of the second law of thermodynamics.

In order to derive the Stefan-Boltzmann law we make use of the fact that a reversible heat engine working between two heat sources at temperature $T$ and $T-\delta T$ possesses an efficiency $\delta T/T$. We wish to perform work by making use of the radiation pressure $P$; we know from (46.6) and Kirchhoff's law that this quantity is a function of temperature alone. The enclosure is provided with a frictionless moving piston and brought into contact with the heat source $T$. If the volume $V$ is now increased by an amount $v$ as a result of the slow motion outwards of the piston, the radiation pressure performs an amount of work $Pv$. This results in the extraction of a quantity of heat $Q$ from the heat source, consisting of the energy equivalent of the work performed, $Pv$, and the increase in the energy content of the enclosure, $uv$. In total, therefore, $Q = (P+u)V$. We now remove the enclosure from the heat source and lower its temperature by an amount $\delta T$ by means of an adiabatic expansion of amount $\delta V$. We now compress the system isothermally at temperature $T - \delta T$ to the point from which a succeeding adiabatic compression brings it back to the initial state. The net external work performed in this cycle is found from the $P - V$ diagram* (figure 52) to be $\delta P . V$, where $\delta P$ is the difference in pressure at temperatures $T$ and $T - \delta T$ $[\delta P = (dP/dT) \delta T]$. From the second law of thermodynamics, this net work must be equal to $Q \, \delta T/T$; hence

$$\frac{dP}{dT} = \frac{P+u}{T}$$

If we put $P = \frac{1}{3}u$ (from (46.6)) in the above equation, integration gives

$$u = \text{constant} \times T^4 \qquad (46.7)$$

which expresses the Stefan-Boltzmann law.

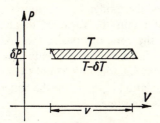

Fig. 52.—*P–V* diagram for the Stefan-Boltzmann law. The work performed is equal to the shaded area

The proof of Wien's law is based on the following argument. We imagine a cavity with a movable piston, and with walls composed of perfectly reflecting material. Only reflections can take place in such a cavity; there can be no absorption or emission processes. When any radiation has once been introduced into the cavity, therefore, the corresponding spectral distribution of energy would exist indefinitely. However, as soon as we introduce a "black" particle capable of absorbing and emitting energy, it will exchange energy with the radiation until the Kirchhoff distribution $u_\omega(\omega, T)$ is established, $T$ being determined by the original total energy of the radiation and the particle.

---

* The difference between the area shown in figure 52 and the expression $\delta P.v$ is a small quantity of higher order.

The heat capacity of the latter is assumed to be vanishingly small compared with that of the enclosure, so that the total radiation energy $uV$ remains unchanged by this process of producing a "black" distribution. After the Kirchhoff distribution $u_\omega$ has thus been established in the enclosure, the particle is removed and the radiation in the cavity is compressed by pressing in the piston with a constant and very small velocity $v$. This naturally alters the spectral distribution of the energy, since by the Doppler principle a light wave of frequency $\omega$ incident on the piston at an angle $\theta$ is reflected with an increased frequency

$$\tilde{\omega} = \omega\alpha(\theta) = \omega\left(1 + \frac{2v}{c}\cos\theta\right) \tag{46.8}$$

The energy is also increased as a result of reflection at the moving piston, since work is performed by the piston on the radiation to the extent of the factor $\alpha(\theta)$, as we can see from the argument leading to equation (46.6).

The second law of thermodynamics requires that the radiation in the enclosure should retain the form of the Kirchhoff distribution during the course of the compression, as the following argument will show. Let us assume that a radiation density $u'(\omega)$ were to exist in the enclosure after the compression. We could then compare this density with the Kirchhoff radiation $u_\omega(\omega, T')$ at the temperature for which the total energy density is the same, i.e.

$$\int_0^\infty u'(\omega)\,d\omega = \int_0^\infty u_\omega(\omega, T')\,d\omega$$

If $u'(\omega)$ is not equal to $u_\omega(\omega, T')$ at all frequencies, there must be at least one frequency $\omega_1$ for which $u'(\omega_1) > u_\omega(\omega_1, T')$, and another frequency $\omega_2$ for which $u'(\omega_2) < u_\omega(\omega_2, T')$. Now let A and B once more represent two enclosures filled with "black" radiation at temperature $T$ and placed in a heat source at temperature $T'$. We could then bring A and B into contact with our working enclosure as far as the respective "colours" $\omega_1$ and $\omega_2$ are concerned, using our previous optical arrangement together with coloured filters; by this means, energy would be introduced into enclosure A and extracted from enclosure B. By a suitable choice of exposure times we could always ensure that the total energy in the working enclosure remained unaltered, merely suffering a change in the spectral distribution. Then the pressure inside the working enclosure would also remain unchanged, so that after removal of the optical contact we should obtain exactly the same amount of work from an adiabatic expansion that we previously expended on the compression. The final result would simply be that we would have produced a temperature difference between the two heat sources in contact with A and B, and this is impossible to accomplish without performing work.

We shall now calculate the change in the spectral distribution of the energy $U_\omega d\omega = V u_\omega d\omega$ in the enclosure in the interval $\tau$, when a piston of area $A$ is inserted with velocity $v$. If we consider a fixed frequency interval $d\omega = \Delta$, the change in $U_\omega \Delta$ is

$$dU_\omega \Delta = -A\tau\pi K_\omega \Delta + \int_{\omega \le \tilde{\omega}' \le \omega+\Delta} A\tau K_{\omega'}\, d\omega'\cos\theta\, d\Omega\, \alpha(\theta) \qquad (46.9)$$

The first term is the total energy in the interval that is incident on the surface $A$, in accordance with (46.4); as a result of the Doppler effect it will be converted into a different frequency and $U_\omega$ will be reduced. The integrand of the second term contains the energy contributions from the various intervals $d\omega'$ that are converted into radiation of frequency $\tilde{\omega}'$, their frequency and energy being altered by a factor $\alpha(\theta)$ on reflection. The fraction of this radiation for which $\omega \le \tilde{\omega}' \le \omega+\Delta$ produces an increase in $U_\omega$. When $\Delta$ is small we have, from (46.9),

$$dU_\omega = -\pi A\tau K_\omega + \int A\tau K\left(\frac{\omega}{\alpha}\right)\cos\theta\, d\Omega \qquad (46.9a)$$

If we expand

$$K\left(\frac{\omega}{\alpha}\right) = K\left(\omega \middle/ \left(1+\frac{2v}{c}\cos\theta\right)\right) \approx K(\omega) - \frac{\partial K_\omega}{\partial\omega}\omega\frac{2v}{c}\cos\theta$$

and neglect the terms of the second degree in $v$, the first term of (46.9a) cancels, and after integrating over the solid angle $2\pi$ there remains

$$dU_\omega = -\frac{4\pi}{3}\frac{\omega v A\tau}{c}\frac{\partial K_\omega}{\partial\omega} \qquad (46.10)$$

Since $vA\tau$ represents the decrease in volume $-dV$ as the piston is moved inwards during the interval $\tau$, (46.10) may also be expressed in the form

$$dU_\omega = \frac{\omega}{3}\frac{\partial U_\omega}{\partial\omega}\frac{dV}{V} = \frac{\omega}{3}\frac{\partial u_\omega}{\partial\omega}dV \qquad (46.11)$$

using the relation (46.5). Equation (46.11) can be looked upon as a partial differential equation of the first order that must be satisfied by the function $U_\omega(V)$:

$$\frac{\omega}{3V}\left(\frac{\partial U_\omega}{\partial \omega}\right)_V = \left(\frac{\partial U_\omega(V)}{\partial V}\right)_\omega \tag{46.12}$$

This implies that $U_\omega(V)$ can only be a function $\psi(\omega^3 V)$, arbitrary in the first instance, of the argument $\omega^3 V$:

$$U_\omega(V) = V u_\omega = \psi(\omega^3 V) \tag{46.13}$$

or putting $\qquad \psi(\xi) = \xi \phi(\xi)$

$$u_\omega = \omega^3 \phi(\omega^3 V) \tag{46.13a}$$

From this equation we can deduce the form of the function of interest to us, $u_\omega(T)$. From the first law of thermodynamics, $dU = -P\,dV$ for the adiabatic compression; this equation also follows immediately from (46.11) if we integrate over all $\omega$ and put $P = \frac{1}{3}u$. Then since $u = \text{const.}\,T^4$ it follows that

$$d(VT^4) + \tfrac{1}{3}T^4\,dV = 0 \quad \text{or} \quad VT^3 = \text{const.}$$

Therefore, if we replace $V$ in (46.13a) by $\text{const.}/T^3$, we obtain

$$u_\omega(T) = \omega^3 f(\omega/T) \tag{46.14}$$

This is Wien's equation, by means of which Kirchhoff's problem is reduced to the determination of the function $f(\omega/T)$ of the single variable $\omega/T$. Integration of (46.14) over the range of $\omega$ yields the Stefan-Boltzmann law (46.7); the equation also leads to Wien's displacement law, which states that $\omega_{max}/T$ or $\lambda_{max}T$ must possess a fixed universal numerical value at the maxima of the spectral distributions.

The Stefan-Boltzmann law and Wien's equation follow purely as a result of the general laws of thermodynamics and the electromagnetic nature of the radiation. In the next section we shall be concerned with the additional concepts required for the complete evaluation of the function $u_\omega(T)$.

## §47. Mathematical model for black-body radiation

According to Kirchhoff's law, the function $u_\omega(T)$ representing the spectral distribution of the radiation energy density in the enclosure is independent of the nature of the enclosure walls and of the manner in which contact is established with the temperature bath $T$. We are therefore free to choose both in such a way as to make the theoretical treatment as clear as possible. We shall adopt the following model.

The bounding surface of the cavity is assumed to consist of perfectly reflecting walls. Inside the cavity there is assumed to be a linear oscillator in the form of a particle of charge $e$ and mass $m$, capable of performing elastic oscillations parallel to the $x$-axis, about an equilibrium position. If $E_x$ is the $x$-component of the electrical field strength at the particle, and

$$\gamma = \frac{2e^2\omega^2}{3mc^3} \tag{47.1}$$

is the radiation damping (cf. §5), then the equation of motion of the particle is

$$\ddot{x} + \gamma\dot{x} + \omega_0^2 x = \frac{e}{m}E_x \tag{47.2}$$

In order to define the temperature of the system we assume that the cavity contains an electrically neutral gas; this gas interacts with the boundary surface which is heated to the temperature $T$ and thus "imposes the temperature $T$" on the oscillator by means of molecular collisions with it. On its side, the oscillator maintains contact with the cavity radiation as a result of the properties described by (47.1) and (47.2).

When integrating (47.2) we must take account of the fact that both $x(t)$ and $E_x(t)$ vary greatly and irregularly with respect to time. We are not interested in the values of these functions at any given moment, but in their statistical representation. The most appropriate way of treating such functions is to represent them by means of a Fourier integral:

$$E_x(t) = \frac{1}{\sqrt{(2\pi)}} \int_{-\infty}^{\infty} C(\omega)\, e^{i\omega t}\, d\omega, \quad \text{where} \quad C(-\omega) = C^*(\omega) \tag{47.3}$$

Substituting this expression for $E_x(t)$ in (47.2), the solution of that equation may be expressed in the form

$$x(t) = \frac{1}{\sqrt{(2\pi)}} \frac{e}{m} \int_{-\infty}^{\infty} \frac{C(\omega)}{-\omega^2 + \omega_0^2 + i\gamma\omega}\, e^{i\omega t}\, d\omega \tag{47.4}$$

In order to evaluate the above relations we have to make an assumption about the behaviour of the function $E_x(t)$ for large intervals of time. We stipulate that $E_x(t)$ shall only differ from 0 in a finite interval extending from $t = -\frac{1}{2}t_0$ to $t = +\frac{1}{2}t_0$; $t_0$ can then be arbitrarily large, though finite. Using this assumption, we obtain the following expression for the time average of $E_x^2$, from (47.3):

$$\overline{E_x^2} = \frac{1}{t_0}\frac{1}{2\pi}\iiint_{-\infty}^{\infty} C(\omega)\,e^{i\omega t}\,C^*(\omega')\,e^{-i\omega' t}\,d\omega\,d\omega'\,dt$$

Since

$$\frac{1}{2\pi}\int_{-\infty}^{\infty} e^{i(\omega-\omega')t}\,dt = \delta(\omega-\omega')$$

it follows that

$$\overline{E_x^2} = \frac{2}{t_0}\int_0^{\infty} |C(\omega)|^2\,d\omega$$

Using the abbreviation

$$E_\omega^2 = \frac{2}{t_0}|C(\omega)|^2 \qquad (47.5)$$

we obtain the spectral distribution of the field strength

$$\overline{E_x^2(t)} = \int_0^{\infty} E_\omega^2\,d\omega \qquad (47.6)$$

Similarly, from (47.4) and (47.5) we have

$$\overline{x^2} = \frac{e^2}{m^2}\int_0^{\infty} \frac{E_\omega^2\,d\omega}{(\omega^2-\omega_0^2)^2+\gamma^2\omega^2} \qquad (47.7)$$

Now in general $\gamma \ll \omega_0$; hence the integral in (47.7) has such a steep maximum at $\omega = \omega_0$ that we may replace $E_\omega^2$ by $E_{\omega_0}^2$. If we take $\mu = \omega - \omega_0$ as the variable of integration, the denominator becomes

$$(\omega-\omega_0)^2(\omega+\omega_0)^2+\gamma^2\omega^2 \approx 4\omega_0^2\mu^2+\gamma^2\omega_0^2$$

The approximate integration can then be performed between the limits $-\infty$ and $+\infty$; the result is

$$\overline{x^2} = E_{\omega_0}^2\frac{e^2\pi}{m^2\omega_0^2\,2\gamma} \qquad (47.8)$$

To obtain the relation between the mean oscillator energy $\overline{E_{osc}}$ and the spectral energy density $u_\omega\,d\omega$ in the cavity we must introduce into

(47.8) the value of $\gamma$ given by (47.1), and the value $m\omega_0^2 x^2$ for $\overline{E_{osc}}$. For isotropic black-body radiation

$$u_\omega \, d\omega = \frac{1}{8\pi} E_\omega^2 \cdot 6 \, d\omega$$

where the factor 6 is due to the fact that on average the squares of the three components of **E** and the three components of **H** are all equal. Since the relation (47.8) must hold for every oscillator frequency $\omega$, it follows that for isotropic black-body radiation

$$\overline{E_{osc}} = u_\omega \frac{\pi^2 c^3}{\omega^2}$$

or
$$u_\omega \, d\omega = \overline{E_{osc}} \frac{\omega^2}{\pi^2 c^3} \, d\omega \qquad (47.9)$$

We may observe the following relation, which we shall require later. From §5, the energy radiated per second by a linear oscillator is

$$S = \frac{2e^2\omega^2}{3mc^3} \overline{E_{osc}} \qquad (47.9a)$$

Since the oscillator is in equilibrium with the black-body radiation, it must absorb the same amount of energy per second from the latter, at its characteristic frequency. From (47.9), therefore, the energy absorbed per second is

$$A = \frac{2\pi^2 e^2}{3m} u_\omega \qquad (47.9b)$$

If we put $\overline{E_{osc}} = kT$ in (47.9), as required by classical statistical mechanics, we obtain the Rayleigh-Jeans formula for black-body radiation:

$$u_\omega(T) \, d\omega = kT \frac{\omega^2}{\pi^2 c^3} \, d\omega \qquad (47.10)$$

This result is obtained from classical physics alone. It leads to the absurd statement that the spectral distribution of the energy density $u_\omega$ increases without limit as the frequency increases, and hence that the energy density

$$u(T) = \int_0^\infty u_\omega(T) \, d\omega$$

is infinitely great for all finite values of $T$. This is the celebrated "ultra-violet catastrophe".

The difficulty was overcome by the Quantum Hypothesis, put forward by Max Planck in the year 1900. If $\overline{E_{osc}}$ is replaced in (47.9) by the value required by the quantum theory,*

$$\overline{E_{osc}} = \frac{\hbar\omega}{\exp\dfrac{\hbar\omega}{kT} - 1} \qquad (47.11)$$

(cf. §9), the result is Planck's formula, which remains finite even when $\omega$ tends to infinity:

$$u_\omega(T)\,d\omega = \frac{\hbar\omega}{\exp\dfrac{\hbar\omega}{kT} - 1}\frac{\omega^2}{\pi^2 c^3}\,d\omega \qquad (47.12)$$

The above derivation is admittedly open to the objection that we have essentially retained formula (47.9), which was deduced from the arguments of classical physics, in spite of the fact that the inadequacy of these arguments is demonstrated by the mere necessity for the postulate represented by (47.11). The rigour of this objection may be somewhat modified by considering a totally different form of (47.9). This is obtained from a consideration of the characteristic electromagnetic oscillations in a cavity (such as a cube of side $L$), and in particular, the number $z(\omega)\,d\omega$ of the oscillations contained in the frequency interval $\omega, \omega + d\omega$. A simple calculation shows that

$$z(\omega)\,d\omega = L^3 \frac{\omega^2}{\pi^2 c^3}\,d\omega \qquad (47.13)$$

Equation (47.9) may therefore be written as

$$L^3 u_\omega\,d\omega = \overline{E_{osc}}\,z(\omega)\,d\omega \qquad (47.14)$$

In the above expression, the left-hand side denotes the mean energy of all oscillations taking place inside the volume $L^3$ and lying within the

---

\* Quantum theory also yields the zero-point energy, in addition to (47.11). Strictly speaking, making allowance for this energy also leads to an "ultra-violet catastrophe". However, this is not so important as the difficulty inherent in the classical theory; we must remember that any experimental statement regarding emission and radiation pressure refers only to differences between the radiation cavity and its external surroundings. In the "balance-sheet", the zero-point energy drops out in each frequency interval, and therefore in total. $u_w(T)$ as given by (47.12) is the energy density which may be used in conjunction with (46.2) and (46.4) to give the radiation and the pressure in a cavity at temperature $T$, in comparison with a vacuum at the temperature of absolute zero.

interval $d\omega$. Equation (47.14) therefore simply states that the mean energy of any characteristic oscillation in the cavity is equal to the mean energy $\overline{E_{osc}}$ of a material oscillator with the same frequency.

To provide a strict proof of the above statement, we must first put Maxwell's equations for the cavity into their Hamiltonian form and apply to them the established propositions of statistical mechanics and the quantum theory. We then find that each characteristic oscillation behaves just as if it were a linear oscillator of the corresponding frequency; this result will be proved in detail in §51.

We shall now discuss Planck's formula (47.12) in greater detail, and establish a connection between it and the two familiar laws of black-body radiation, Wien's displacement law and the Stefan-Boltzmann law of total radiation.

### Wien's displacement law

If the spectral distribution of the energy is expressed in terms of the wavelength $\lambda$ instead of the frequency $\omega$, (47.12) gives the following result:

$$u_\lambda \, d\lambda = \frac{16\pi^2 \hbar c}{\lambda^5} \frac{1}{\exp \dfrac{2\pi \hbar c}{\lambda kT} - 1} \, d\lambda \qquad (47.15)$$

We wish to determine the wavelength $\lambda_m$ for which $u_\lambda$ is a maximum for any given value of $T$. The condition for this is $\partial u_\lambda / \partial \lambda = 0$; taking this into account, and using the abbreviation $y = 2\pi \hbar c / \lambda_m kT$, we obtain the equation

$$5(1 - e^{-y}) = y$$

The root of this transcendental equation may be seen to lie in the neighbourhood of $y = 5$. If we put $y = 5 - \eta$ we then obtain the relation

$$5e^{-5}e^\eta = \eta$$

Since $\eta$ is a small number we can replace $e^\eta$ by $1 + \eta$; we then obtain the approximation

$$\eta = \frac{1}{\frac{1}{5}e^5 - 1} = 0 \cdot 035$$

The constant of Wien's displacement law is therefore given by

$$\lambda_m T = \frac{2\pi \hbar c}{ky} = \frac{2\pi \hbar}{k} \frac{c}{4 \cdot 965} \qquad (47.16)$$

Simultaneous measurements of $\lambda_m$ and $T$ on a black body will therefore give the ratio $\hbar/k$ of the two universal constants, named after Planck and Boltzmann. The fact that the maximum for solar radiation lies in the green region, for which $\lambda \approx 0.5 \times 10^{-4}$ cm, and that the corresponding value of $T$ is about $6000°$ K, enable us to determine an approximate figure for the ratio $\hbar/k$.

### The Stefan-Boltzmann law of total radiation

The total energy density of black-body radiation is obtained by integrating (47.12):

$$u = \frac{\hbar}{\pi^2 c^3} \int_0^\infty \frac{\omega^3 \, d\omega}{\exp\dfrac{\hbar\omega}{kT} - 1}$$

Introducing $x = \hbar\omega/kT$ as a new variable,

$$u = \frac{\hbar}{\pi^2 c^3} \left(\frac{kT}{\hbar}\right)^4 \int_0^\infty \frac{x^3 \, dx}{e^x - 1}$$

The numerical value of this integral is $\pi^4/15$; this gives

$$u = \frac{\hbar}{\pi^2 c^3} \frac{\pi^4}{15} \left(\frac{kT}{\hbar}\right)^4 \tag{47.17}$$

We have thus determined the constant appearing in the Stefan-Boltzmann law as given by equation (46.7). As a rule direct measurements are not made on the energy density, but on the energy $S$ emitted per second per unit area of the surface of a black body on one side (i.e. into a solid angle $2\pi$). This quantity is related to $u$ by equation (46.4):

$$S = \frac{c}{4} u$$

The energy emitted per second per square centimetre of a black body is therefore

$$S = \sigma T^4, \quad \text{where} \quad \sigma = \frac{\pi^2 k^4}{60 c^2 \hbar^3} \tag{47.18}$$

The measurement of $S$ at a known temperature gives us the ratio $k^4/\hbar^3$; thus, merely by measuring the maximum and the total radiation we can determine the two fundamental constants $\hbar$ and $k$.

The experimental results are as follows:

$$\lambda_m T = 0 \cdot 290 \, \text{cm deg},$$

$$\sigma = 5 \cdot 68 \times 10^{-5} \, \text{erg s}^{-1} \, \text{cm}^{-2} \, \text{deg}^{-4}$$

$$\hbar = 1 \cdot 05 \times 10^{-27} \, \text{erg s},$$

$$k = 1 \cdot 38 \times 10^{-16} \, \text{erg deg}^{-1}$$

A knowledge of Boltzmann's constant enables us to determine Avogadro's number $N$ and the elementary electric charge $e$:

$$N = \frac{R}{k} = \frac{8 \cdot 31 \times 10^7}{1 \cdot 38 \times 10^{-16}} = 6 \cdot 02 \times 10^{23}$$

$$e = \frac{F}{N} = \frac{96,500}{6 \cdot 02 \times 10^{23}} \text{C} = 1 \cdot 60 \times 10^{-19} \text{C} = 4 \cdot 80 \times 10^{-10} \text{e.s.u.}$$

This method of Planck's for determining the elementary electric charge from radiation measurements alone deserves particular consideration, being by far the most accurate at the time (1900). It was only much later that it was surpassed in accuracy by the measurements of Millikan, which were described in Vol. I.

Planck's formula (47.12) may be expressed in a somewhat simpler form when the ratio $\hbar\omega/kT$ is either very large or very small compared with 1. The results for these two limiting cases are as follows:

$$\hbar\omega \gg kT: \quad u_\omega = \frac{\hbar\omega^3}{\pi^2 c^3} \exp - \frac{\hbar\omega}{kT} \qquad (47.19)$$

$$\hbar\omega \ll kT: \quad u_\omega = \frac{\omega^2}{\pi^2 c^3} kT \qquad (47.20)$$

Formula (47.19), known as Wien's radiation formula, had been established by W. Wien before Planck had put forward his hypothesis; however, it contained an undetermined numerical factor. The formula represents quite accurately the ultra-violet part of the energy distribution including the maximum. The other limiting case covered by Planck's formula, (47.20), is identical with the Rayleigh-Jeans radiation law (47.10); this is only to be expected in view of the method of derivation. When Planck's formula was put forward, therefore, it provided a connection between the experimentally verified equation (47.19) and the equation (47.20), which had been deduced from classical theory (cf. figure 53).

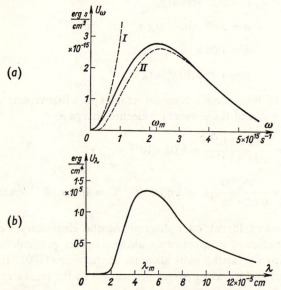

Fig. 53.—Spectral distribution of black-body radiation for $T = 6000°$K
(a) Planck's distribution, Rayleigh-Jeans law (I), and Wien radiation formula (II), as functions of the angular frequency $\omega$.
(b) Planck's distribution as a function of the wavelength $\lambda$.

## §48. Einstein's derivation of Planck's formula

Einstein, whose process of reasoning was of decisive importance in connection with the concepts of the quantum theory, successfully advanced a self-consistent proof of Planck's formula.* It is characteristic of the derivation that it demands a minimum number of basic assumptions. These are set out below.

1. If $E_s$ and $E_n$ are two non-degenerate energy levels of an atom ($E_n > E_s$), the latter reacts with the black-body radiation by emitting or absorbing light quanta of magnitude

$$\hbar\omega_{ns} = E_n - E_s \qquad (48.1)$$

depending on whether it is in the state $n$ or the state $s$.

* A. Einstein, *Phys. Z.* 18 (1917), 121.

2. Let there be a number of such atoms (e.g. free gas atoms) in a cavity at temperature $T$. In particular,

$N_s$ = number of atoms in the state $s$,

$N_n$ = number of atoms in the state $n$.

Then when the system is in thermal equilibrium it follows from a basic formula of statistical mechanics that

$$\frac{N_n}{N_s} = \frac{\exp - E_n/kT}{\exp - E_s/kT} \qquad (48.2a)$$

and therefore

$$N_n = N_s \exp - \frac{\hbar\omega_{ns}}{kT} \qquad (48.2b)$$

3. In the limiting case for which $\hbar\omega_{ns} \ll kT$ the laws of classical physics and hence the Rayleigh-Jeans formula

$$u(\omega_{ns}) \rightarrow \frac{\omega_{ns}^2}{\pi^2 c^3} kT \qquad (48.3)$$

are valid.

We now consider in turn the number $Z_{sn}$ of transitions from $s$ to $n$ taking place in the time $dt$, and the number $Z_{ns}$ of opposite processes. At equilibrium, both numbers must be the same. We take the number of absorption processes $Z_{sn}$ to be proportional to $N_s$ and to the energy density $u(\omega_{ns})$ of the radiation at the corresponding frequency, and include a constant $B_{sn}$ that is characteristic of the absorption:

$$Z_{sn} = N_s B_{sn} u(\omega_{ns}) \, dt \qquad (48.4)$$

The number of emission processes is proportional to $N_n$. We now have to distinguish between two types of emission. The first is spontaneous emission, which is independent of $u$, and which is represented by the probability $A_{ns} \, dt$ of a spontaneous transition from $n$ to $s$ in the interval $dt$. The second type is "forced" emission, proportional to $u(\omega_{ns})$, and expressed as $B_{ns} u(\omega_{ns})$. In total, therefore,

$$Z_{ns} = N_n \left[ A_{ns} + B_{ns} u(\omega_{ns}) \right] dt \qquad (48.5)$$

The forced emission introduced here corresponds in classical physics to a vibrating oscillator which can extract energy from an incident electric wave, or convey energy to it, depending on the phase angle between the vibration and the electric field of the wave. Now it follows from the condition $Z_{ns} = Z_{sn}$, bearing in mind (48.2), that

$$\exp\frac{\hbar\omega_{ns}}{kT}B_{sn}u = A_{ns} + B_{ns}u$$

from which

$$u(\omega_{ns}) = \frac{A_{ns}}{B_{sn}\exp\dfrac{\hbar\omega}{kT} - B_{ns}} \qquad (48.6)$$

where $\omega_{ns} = \omega$.

In order to apply our third postulate we take $\hbar\omega/kT \ll 1$. Then

$$u(\omega_{ns}) \rightarrow \frac{A_{ns}}{B_{sn} - B_{ns} + B_{sn}\dfrac{\hbar\omega}{kT}}$$

In order that this formula should be identical with (48.3) we must have

$$B_{sn} = B_{ns} \qquad (48.7)$$

that is, the forced emission and the absorption constants must be equal. In addition we must have

$$\frac{A_{ns}}{\hbar\omega B_{sn}}kT = \frac{\omega^2}{\pi^2 c^3}kT$$

i.e. the following relationship must exist between $A_{ns}$ and $B_{ns}$:

$$\frac{A_{ns}}{B_{ns}} = \frac{\hbar\omega^3}{\pi^2 c^3} \qquad (48.8)$$

When the relations (48.7) and (48.8) between the initially arbitrary coefficients $A$ and $B$ are introduced into (48.6), Planck's formula

$$u(\omega) = \frac{\omega^2}{\pi^2 c^3}\frac{\hbar\omega}{\exp\hbar\omega/kT - 1}$$

is obtained.

The characteristic feature of the description of absorption and emission given by (48.4) and (48.5) is that it assumes that the occurrence of each elementary process is due to chance. Thus $A_{ns}dt$ is simply the probability that an atom in state $n$ changes to the state $s$ as a result of the spontaneous emission of a quantum $\hbar\omega_{ns}$, during the interval $dt$.

Most physicists were at first inclined to look upon this representation as a provisional expedient. It was only the further development of the quantum theory that showed that the physical situation was in a sense completely described by the statistical postulates (48.4) and (48.5). We shall see in §53 how the coefficients $A_{ns}$ and $B_{ns}$ can actually be calculated by means of the quantum theory.

Even without such a calculation, however, Einstein's relations (48.7) and (48.8) enable an important connection to be established between the coefficients $A_{ns}$ and the oscillator strengths $f_{ns}$ introduced in the theory of dispersion; this was first demonstrated by R. Ladenburg. We shall calculate the energy absorbed by an atom obeying the dispersion formula

$$n - 1 = N \sum_s \frac{2\pi e^2}{m} \frac{f_{ns}}{\omega_{ns}^2 - \omega^2}$$

in the neighbourhood of a given line $\omega_{ns}$ and over an interval $dt$, when it is in an isotropic radiation field of spectral energy density $u_\omega$. From (47.9b) this energy is

$$f_{ns} \frac{2\pi^2 e^2}{3m} u(\omega_{ns}) \, dt$$

Using Einstein's representation of the absorption process, the same energy is given by

$$\hbar \omega_{ns} B_{ns} u(\omega_{ns}) \, dt$$

If the above expressions are equated to each other and the relation (48.8) is introduced, we obtain

$$A_{ns} = \frac{2}{3} \frac{e^2 \omega_{ns}^2}{mc^3} f_{ns} = \gamma f_{ns} \qquad (48.9)$$

This equation shows that it is possible to obtain quantitative estimates of spontaneous transition probabilities from measurements of dispersion. This method is much more exact and reliable than the one based on the duration of the luminosity of canal rays in Wien's well-known experiment.

If we introduce into (48.9) the value of $f_{ns}$ given by (41.10),

$$f_{ns} = \frac{2m |x_{ns}|^2 \omega_{ns}}{\hbar} \qquad (48.10)$$

we obtain equation (11.9) for the probability of spontaneous emission, which we had previously inferred from the correspondence principle.

For the three-dimensional case we must add the contributions $|y_{ns}|^2$ and $|z_{ns}|^2$ due to the other two vibrational directions; the formula then becomes identical with (11.9a).

This agreement provides strong support for the validity of the argument of §11, based on the correspondence principle, and leading to equation (48.9). However, a self-consistent proof of this equation can be given only in terms of the quantum theory of the electromagnetic field (§53).

### Exercises

1. *Pressure and energy density in an ideal gas*

A gas consisting of atoms of rest mass $m$ is present in a container. The density of the atoms is $n$, and the distribution of the momentum is given by $f(\mathbf{p})d\mathbf{p}$, which is the number of atoms per cubic centimetre possessing a momentum in the range $(\mathbf{p}, d\mathbf{p})$ ($\int f d\mathbf{p} = n$). The pressure $P$ is defined as the momentum transferred to the container walls by collisions with the atoms, per unit time and per unit area. The collisions are assumed to be elastic, and the distribution $f(\mathbf{p})$ is isotropic. Show that

$$P = \frac{2}{3}n\overline{E_{kin}} \text{ in the non-relativistic limit,}$$

$$P = \frac{1}{3}n\overline{E} \text{ in the limit } \overline{E} \gg mc^2.$$

$E$ is the total energy, $E_{kin}$ the kinetic energy. The mean values are defined by

$$\overline{F} = \frac{\int F(\mathbf{p})f(\mathbf{p})d\mathbf{p}}{\int f(\mathbf{p})d\mathbf{p}} = \frac{1}{n}\int F(\mathbf{p})f(\mathbf{p})d\mathbf{p}$$

The first relation gives the ideal gas equation if we put $\overline{E_{kin}} = \frac{3}{2}kT$; the second relation gives the connection between pressure and energy density (equation (46.6)) for a gas consisting of light quanta, for which $m = 0$.

2. *The solar constant*

The temperature of the sun's surface is about 6000°K. Using (47.17), calculate the solar constant $S$, i.e. the solar energy radiated per minute on to $1\,\text{cm}^2$ of the surface on the earth normal to the incident radiation. (The angle $\delta$ subtended by the sun at the earth is about $\frac{1}{2}°$.)

# CHAPTER EII

## Absorption and emission

### §49. The classical treatment of absorption and anomalous dispersion

The theory of dispersion that was given in §40 applies only to the transparent region outside the spectral lines (i.e. $\omega \neq \omega_0$). In order to describe the behaviour within the lines, we must include the damping in the equation of motion for the electron. This equation then becomes

$$\ddot{\mathbf{r}} + \gamma \dot{\mathbf{r}} + \omega_0^2 \mathbf{r} = \frac{e}{m} \mathbf{E} \tag{49.1}$$

If $\mathbf{E}$ depends on time through the factor $e^{i\omega t}$, the solution of (49.1) is given by

$$\mathbf{r}(t) = \frac{e}{m} \frac{1}{\omega_0^2 - \omega^2 + i\gamma\omega} \mathbf{E} \tag{49.2}$$

We shall first consider the balance of energy in the system. The work done by the field on the electron in the time interval $dt$ is $e\mathbf{E}\dot{\mathbf{r}}\,dt$. From (49.1) the mean power due to $\mathbf{E}$ is therefore

$$e\overline{\mathbf{E}\dot{\mathbf{r}}} = m\gamma\overline{\dot{\mathbf{r}}^2} \tag{49.3}$$

since the mean values of

$$\ddot{\mathbf{r}}\dot{\mathbf{r}} = \frac{1}{2}\frac{d}{dt}\dot{\mathbf{r}}^2 \quad \text{and} \quad \mathbf{r}\dot{\mathbf{r}} = \frac{1}{2}\frac{d}{dt}(\mathbf{r}^2)$$

are equal to zero for periodic motion. This power is to be compared with the mean radiated power of the electron, which is

$$S = \frac{2}{3}\frac{e^2}{c^3}\overline{\ddot{\mathbf{r}}^2} = \frac{2}{3}\frac{e^2}{c^3}\omega^2\overline{\dot{\mathbf{r}}^2} \tag{49.4}$$

Now (49.3) represents the energy withdrawn from the field per second, while equation (49.4) gives the energy which is returned to the field in the form of radiation. Both these quantities are equal when $\gamma$ has the usual value for the radiation damping

$$\gamma_{rad} = \frac{2}{3}\frac{e^2\omega^2}{mc^3}$$

If $\gamma$ is greater than $\gamma_{rad}$, part of the power $eE\dot{r}$ is converted into heat, e.g. into kinetic energy of the atoms.

The order of magnitude of the ratio $\gamma_{rad}/\omega$ is of some interest: it is

$$\gamma_{rad}/\omega = \frac{2}{3}\frac{e^2/mc^2}{c/\omega} \approx \frac{R_{el}}{\lambda}$$

In the above expression, $R_{el}$ is the classical electron radius ($\approx 10^{-13}$cm) and $\lambda$ the wavelength employed ($\approx 10^{-5}$cm).

If we make use of (49.2) to express the current density $\mathbf{j} = Ne\dot{r}$, and substitute in the first of Maxwell's equations,

$$\operatorname{curl}\mathbf{H} = \frac{4\pi}{c}\mathbf{j} + \frac{1}{c}\dot{\mathbf{E}}$$

we obtain

$$\operatorname{curl}\mathbf{H} = \frac{\varepsilon}{c}\dot{\mathbf{E}}, \text{ where } \varepsilon = 1 + 4\pi N\frac{e^2}{m}\frac{1}{\omega_0^2 - \omega^2 + i\gamma\omega} \quad (49.5)$$

We then obtain the following equations for a wave travelling parallel to the $z$-axis, and polarized parallel to the $x$-axis:

$$E_x = a\exp i\omega\left(t - \frac{\sqrt{\varepsilon}}{c}z\right), \qquad H_y = a\sqrt{\varepsilon}\exp i\omega\left(t - \frac{\sqrt{\varepsilon}}{c}z\right) \quad (49.6)$$

If we put

$$\sqrt{\varepsilon} = n - i\kappa \quad (49.7)$$

where $n$ and $\kappa$ are real, then the real part of the solution is

$$E_x = a\exp\left(-\frac{\omega\kappa}{c}z\right)\cos\omega\left(t - \frac{n}{c}z\right)$$

$$H_y = a(n^2 + \kappa^2)^{1/2}\exp\left(-\frac{\omega\kappa}{c}z\right)\cos\left[\omega\left(t - \frac{n}{c}z\right) - \psi\right] \quad (49.8)$$

where $\tan\psi = \kappa/n$, $\sqrt{\varepsilon} = (n^2 + \kappa^2)^{1/2}e^{-i\psi}$.

The time average of the Poynting vector $\mathbf{S} = (c/4\pi)\mathbf{E} \times \mathbf{H}$ is

$$S_z = S = \frac{c}{8\pi} n a^2 \exp -\frac{2\omega\kappa}{c} z$$

The reduction in intensity over the distance $dz$ is thus

$$dS = -\frac{2\omega\kappa}{c} S\, dz \qquad (49.9)$$

$n$ is termed the (real) refractive index, and $\kappa$ the extinction coefficient. If the gas is sufficiently rarefied it can be assumed that $|\varepsilon - 1| \ll 1$; then $\sqrt{\varepsilon} = 1 + \frac{1}{2}(\varepsilon - 1)$ approximately, whence

$$n = 1 + 2\pi N \frac{e^2}{m} \frac{\omega_0^2 - \omega^2}{(\omega_0^2 - \omega^2)^2 + \gamma^2\omega^2}$$

$$\kappa = 2\pi N \frac{e^2}{m} \frac{\gamma\omega}{(\omega_0^2 - \omega^2)^2 + \gamma^2\omega^2} \qquad (49.10)$$

In order to see how the above quantities vary in the neighbourhood of a spectral line, we shall put

$$\omega = \omega_0 + \mu$$

When $|\mu| \ll \omega_0$, $\omega_0^2 - \omega^2 = -2\mu\omega_0$. Elsewhere we may replace $\omega$ by $\omega_0$. Then

$$n = 1 + 2\pi N \frac{e^2}{m\omega_0} \frac{-2\mu}{4\mu^2 + \gamma^2}$$

$$\kappa = 2\pi N \frac{e^2}{m\omega_0} \frac{\gamma}{4\mu^2 + \gamma^2} \qquad (49.11)$$

The manner in which $\kappa$ and $n$ vary with $\omega = \mu + \omega_0$ is shown in figure 54. Omitting the factor $2\pi N e^2/m\omega_0$ which is common to $n-1$ and $\kappa$, the latter quantity has a maximum of height $1/\gamma$ at $\mu = 0$, and half this value at $\mu = \pm\frac{1}{2}\gamma$. Now $n-1$ has a maximum at $\mu = -\frac{1}{2}\gamma$ and a minimum at $\mu = \frac{1}{2}\gamma$, the magnitudes of which are $1/2\gamma$. Between these two values, that is, within the spectral line, the refractive index decreases with increasing frequency, in contrast to its behaviour in the transparent region outside the spectral line. This property is termed *anomalous dispersion*.

The curve of $\kappa$ becomes higher and narrower as $\gamma$ decreases, but the area included under it is independent of the damping, and is equal to

$$\int_{-\infty}^{\infty} \frac{\gamma \, d\mu}{4\mu^2 + \gamma^2} = \tfrac{1}{2}\pi$$

Therefore

$$\int_{-\infty}^{\infty} \kappa \, d\omega = 2\pi N \frac{e^2}{m\omega_0} \frac{\pi}{2} \tag{49.12}$$

Using this result, the absorption in the case of a continuous spectrum may easily be calculated. Provided that the intensity (i.e. the energy

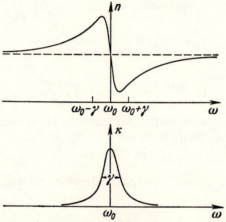

Fig. 54.—Refractive index and extinction coefficient in the neighbourhood of a spectral line

per second per square centimetre) of the incident radiation is not in the form of a sharp line, but extends over a continuous spectrum,

$$S = \int_0^{\infty} I(\omega) \, d\omega \tag{49.13}$$

we should expect from (49.9) that the absorption over an interval $dz$ would be

$$dS = -dz \int \frac{2\omega\kappa}{c} I(\omega) \, d\omega$$

Since $\kappa$ has a sharp maximum at $\omega = \omega_0$, $\omega I(\omega)$ can be replaced by

$\omega_0 I(\omega_0)$ in the integrand provided that $I(\omega)$ remains nearly constant within the line width. Then from the last equation and (49.12)

$$dS = -N \, dz \frac{2\pi^2}{c} \frac{e^2}{m} I(\omega_0) \qquad (49.14)$$

Equation (49.14) shows that the total energy absorbed by a line from the continuous spectrum is independent of the line width.

This equation may also be interpreted as follows. $N \, dz$ is the number of atoms contained in a layer of thickness $dz$ and $1 \, cm^2$ cross-section. Equation (49.14) therefore states that the energy absorbed per second by each oscillator from the spectrum represented by (49.13) is

$$\frac{2\pi^2 e^2}{cm} I(\omega_0) \qquad (49.15)$$

If we represent the incident radiation by its energy density $u$(ergs per $cm^3$) instead of by its intensity $S$(ergs per second per $cm^2$), we have the following relation:

$$u(\omega) = \frac{1}{c} I(\omega)$$

The spectral energy density of the radiation field being $u(\omega)$, the energy absorbed from the field by a three-dimensional oscillator is therefore

$$\frac{2\pi^2 e^2}{m} u(\omega) \qquad (49.15a)$$

A linear oscillator which is capable only of motion parallel to the $x$-axis can only react with the $x$-component of the field. For such an oscillator only one-third of the radiant energy of an isotropic field is effective, hence:

Energy absorbed by a linear oscillator is

$$\frac{2\pi^2 e^2}{3m} u(\omega) \qquad (49.15b)$$

The damping constant does not appear in the above formulae for absorption in a continuous spectrum.

In view of the later treatment of the subject by means of the quantum theory, it is of interest to note that the same formulae for the absorption of energy by an oscillator can be obtained without introducing the damping. In §40, we gave the solution of the equation

$$\ddot{\mathbf{r}} + \omega_0^2 \mathbf{r} = \frac{e}{m} \mathbf{F} \cos \omega t$$

(where $\mathbf{F}$ is the amplitude of the electric field), for initial conditions $\mathbf{r} = \dot{\mathbf{r}} = 0$ at $t = 0$:

$$\mathbf{r}(t) = \frac{e}{m} \mathbf{F} \frac{\cos \omega t - \cos \omega_0 t}{\omega_0^2 - \omega^2} \qquad (49.16)$$

(We may no longer omit the term $\cos \omega_0 t$, since we wish to perform the calculation when the damping is zero.) Using the identity $\cos \alpha - \cos \beta = 2 \sin \frac{1}{2}(\beta + \alpha) \sin \frac{1}{2}(\beta - \alpha)$ we may express (49.16) as follows:

$$\mathbf{r}(t) = -\frac{e}{m} \mathbf{F} \frac{2 \sin \frac{1}{2}(\omega + \omega_0)t \sin \frac{1}{2}(\omega - \omega_0)t}{(\omega_0 + \omega)(\omega_0 - \omega)} \qquad (49.17)$$

The potential energy of the oscillator is $\frac{1}{2} m \omega_0^2 \mathbf{r}^2$, which is equal to its mean kinetic energy; the total energy is therefore $m \omega_0^2 \overline{\mathbf{r}^2}$. For the magnitude of this energy to be appreciable, $\omega$ must be nearly equal to $\omega_0$. We shall therefore put $\omega_0 + \omega \approx 2\omega_0$ in (49.17), and since the term $\sin^2 \omega_0 t$ oscillates rapidly we shall replace it in the following expression by its mean value $\overline{\sin^2 \omega_0 t} = \frac{1}{2}$. Then the energy absorbed by the oscillator in time $t$ is

$$m \omega_0^2 \mathbf{r}^2 = \frac{e^2}{m} \mathbf{F}^2 \frac{1}{2} \frac{\sin^2 \frac{1}{2}(\omega - \omega_0)t}{(\omega_0 - \omega)^2}$$

The radiation intensity corresponding to a field amplitude $\mathbf{F}$ is

$$S = \frac{c}{8\pi} \mathbf{F}^2$$

If this intensity is distributed according to (49.13),

$$m \omega_0^2 \overline{\mathbf{r}^2} = \frac{4\pi e^2}{mc} \int_0^\infty I(\omega) \frac{\sin^2 \frac{1}{2}(\omega - \omega_0)t}{(\omega_0 - \omega)^2} \, d\omega$$

Putting $\omega - \omega_0 = \mu$ once more, we have for the function next to $I(\omega)$ in the integral

$$\phi(\mu) = \frac{\sin^2 \frac{1}{2}\mu t}{\mu^2}$$

which we have already encountered in §27. When $\mu = 0$ it has the value $\frac{1}{4}t^2$, and its first zero is at $\mu_0 = 2\pi/t$. The area under the curve is approximately a triangle of base $2\mu_0$ and height $\frac{1}{4}t^2$, and is therefore equal to $\frac{1}{2}\pi t$. It may in fact easily be verified that

$$\int_{-\infty}^{\infty} \phi(\mu)\,d\mu = \frac{1}{2}\pi t$$

exactly. Now if $t$ is so large that $I(\omega)$ does not change appreciably when $\omega$ changes by an amount $\pi/t$, we may take $I(\omega_0)$ out of the integral. We then have

$$m\omega_0^2\,\overline{\mathbf{r}^2} = \frac{2\pi^2 e^2}{mc} I(\omega_0)t$$

in agreement with (49.15).

## §50. The probability of an induced transition

In §41, an atom subject from time $t = 0$ to the effect of a perturbation was represented by the equation

$$\psi = \exp\left(-\frac{i}{\hbar}E_s t\right)\phi_s + \sum_{v \neq s} a_v \exp\left(-\frac{i}{\hbar}E_v t\right)\phi_v \qquad (50.1)$$

$|a_v(t)|^2$ is the probability of finding the atom in the state $v$ at time $t$, given the fact that it was in the state $s$ at time $t = 0$. Provided that $|a_v| \ll 1$, it is permissible to say that $|a_v|^2$ represents the probability that a transition from $s$ to $v$ took place during the interval $t$. In (41.6) we gave the following expression for $a_v(t)$ in the case of an alternating field $E_x = F\cos\omega t$:

$$a_v(t) = \frac{eFx_{vs}}{2\hbar}\left\{\frac{\exp i(\omega + \omega_{vs})t - 1}{\omega + \omega_{vs}} + \frac{\exp i(\omega - \omega_{vs})t - 1}{\omega - \omega_{vs}}\right\} \qquad (50.2)$$

The condition $|a_v| \ll 1$ is no longer satisfied when $\omega$ is nearly equal to $\omega_{vs}$; to deal with this case it is really necessary to take account of the damping. However, in the classical treatment of absorption it was shown that the total energy absorbed per second could be correctly deduced without taking the damping into account. In the present discussion we shall therefore limit ourselves to the calculation of the total absorption, making no allowance for damping.

We shall again consider the case of a continuous spectrum of incident radiation. The only frequencies contributing appreciably to

$|a_v(t)|^2$ are those for which one of the two denominators, $\omega_{vs}+\omega$ or $\omega_{vs}-\omega$, is very small. In particular, if $\phi_s$ denotes the ground state of the atom, $\omega_{vs} > 0$, and only the term with $\omega_{vs}-\omega$ as denominator contributes significantly. On the other hand, if $\phi_s$ represents an excited state, $\omega_{vs}$ can be less than 0, and $|a_v(t)|^2$ then represents the probability that the radiation field induces a transition from $\phi_s$ to a state of lower energy $\phi_v$; this is termed *forced*, or *stimulated, emission*.

We now consider the case when $\omega_{vs} > 0$. When monochromatic radiation is incident,

$$|a_v|^2 = \frac{e^2 |x_{vs}|^2}{4\hbar^2} F^2 \frac{\sin^2\left[\frac{1}{2}(\omega_{vs}-\omega)t\right]}{\frac{1}{4}(\omega_{vs}-\omega)^2} \tag{50.3}$$

(In the above equation we have used the relation $|e^{i\beta}-1|^2 = 4\sin^2\frac{1}{2}\beta$.) When the incident radiation is in the form of a continuous spectrum $F^2$ is replaced by $(8\pi/c)I_\omega d\omega$, and

$$|a_v|^2 = \frac{e^2 |x_{vs}|^2}{4\hbar^2} \frac{8\pi}{c} \int I(\omega) \frac{\sin^2\left[\frac{1}{2}(\omega_{vs}-\omega)t\right]}{\left[\frac{1}{2}(\omega_{vs}-\omega)\right]^2} d\omega \tag{50.4}$$

When $t$ is sufficiently great we can remove $I(\omega_{vs})$ from the integral, as we did in the previous paragraph; we then have

$$|a_v(t)|^2 = \frac{e^2 |x_{vs}|^2}{\hbar^2} \frac{4\pi^2}{c} I(\omega_{vs})t \tag{50.5}$$

The energy of the atom increases by an amount $\hbar\omega_{vs}$ in the case of a transition from $\phi_s$ to $\phi_v$. The increase in energy per second is therefore

$$\frac{1}{t}\hbar\omega_{vs}|a_v(t)|^2 = \omega_{vs}\frac{e^2 |x_{vs}|^2}{\hbar} \frac{4\pi^2}{c} I(\omega_{vs}) \tag{50.6}$$

This increase in energy may be compared with that found for the classical oscillator in (49.15), when the latter expression is multiplied by its appropriate $f$-value, as given by (48.10). In this case the increase in energy per second is

$$f_{vs}\frac{2\pi^2}{c}\frac{e^2}{m}I(\omega_{vs}) = \omega_{vs}\frac{e^2 |x_{vs}|^2}{\hbar}\frac{4\pi^2}{c}I(\omega_{vs}) \tag{50.7}$$

The above expression is in complete agreement with the result given by equation (50.6).

## §51. The Hamiltonian form of the Maxwell equations for a vacuum

Before we can make use of the quantum theory in the treatment of the radiation field, we must first put Maxwell's equations in their Hamiltonian form. The methods so far established cannot be directly employed because, unlike the position and momentum of a particle, the electromagnetic field is not represented by discrete numbers but by continuous functions of position. These methods become applicable, however, if we consider the fields inside a cavity such as an enclosure with reflecting walls.

As a result of the boundary conditions, which require the vanishing of the appropriate tangential and normal components of the fields, only a discrete number of plane waves can exist inside the enclosure, in contrast to the situation in infinite space. The fields can therefore be represented by means of a discrete set of wave amplitudes, dependent on time. Lagrangian or Hamiltonian differential equations may be derived for the latter, which can easily be made compatible with the postulates of the quantum theory.

We may obtain the results for continuous fields by proceeding to the limiting case of an infinitely large enclosure. In doing so, we may reasonably expect that the boundary conditions will not affect the final result; we are therefore free to choose them in the most appropriate manner for calculation. Instead of postulating ideally reflecting walls we shall therefore assume that all field quantities are periodic in the enclosure; for simplicity, we shall take the enclosure to be a cube, the sides of which are of length $L$ and are parallel to the coordinate axes. More precisely, if $f(x,y,z,t)$ is any field quantity, then

$$f(x, y, z, t) - \\ = f(x+L, y, z, t) = f(x, y+L, z, t) = f(x, y, z+L, t) \quad (51.1)$$

The advantage of this procedure is that it permits the use of the more convenient exponential functions in the Fourier expansions occurring in the calculation, instead of sine and cosine waves.

In the first instance we shall consider fields in empty enclosures. The Maxwell equations

$$\operatorname{curl} \mathbf{E} = -\frac{1}{c}\dot{\mathbf{H}}, \quad \operatorname{div} \mathbf{E} = 0$$

$$\operatorname{curl} \mathbf{H} = \frac{1}{c}\dot{\mathbf{E}}, \quad \operatorname{div} \mathbf{H} = 0$$

are solved by putting

$$H = \operatorname{curl} A, \quad E = -\frac{1}{c}\dot{A}, \quad \operatorname{div} A = 0 \qquad (51.2)$$

giving the following equation for the vector potential $A$:

$$\nabla^2 A - \frac{1}{c^2}\ddot{A} = 0 \qquad (51.3)$$

$A$ is expanded in a Fourier series:

$$A(r, t) = \sum_{k,\lambda} A_{k\lambda}(t)\, s_{k\lambda}\, \frac{e^{ikr}}{L^{3/2}} \qquad (51.4)$$

The quantities

$$s_{k\lambda} = s_{-k\lambda} \qquad (51.4a)$$

are linearly independent unit vectors defining the polarization of the corresponding wave. Since $\operatorname{div} A = 0$, $(s_{k\lambda} k) = 0$, which implies that the waves are transverse. The summation over $\lambda$ therefore extends only over the values 1, 2. The two vectors $s_{k1}$ and $s_{k2}$, which are normal to $k$, will also be chosen to be orthogonal to each other.

As a result of the postulate (51.1) $k$ assumes only the discrete values

$$k = \frac{2\pi}{L}(n_x, n_y, n_z), \quad \text{where} \quad n_x, n_y, n_z = 0, \pm 1, \pm 2, \ldots \quad (51.5)$$

The summation over $k$ in (51.4) therefore represents a triple summation over all integers $n_x, n_y, n_z$.

The proper variables to employ for the representation of the field are the amplitudes $A_{k\lambda}(t)$. Since $A$ is a real quantity,

$$A_{k\lambda} = A_{-\lambda k}^{*} \qquad (51.4b)$$

in virtue of (51.4a).* From (51.3) we obtain the "equation of motion"

$$\ddot{A}_{k\lambda} + \omega_k^2 A_{k\lambda} = 0, \quad \text{where} \quad \omega_k = c\,|\,k\,| \qquad (51.6)$$

This is the equation of a linear oscillator, for which the Hamiltonian formalism and quantization characteristics are already known. The equations (51.6) are not mutually independent, however, as a result of the accessory condition (51.4b), and the amplitudes are complex; we

---

* The relation of $s_{k\lambda}$ to $s_{-k\lambda}$ given by (51.4a) is arbitrary. Once it has been assumed, however, (51.4b) follows necessarily.

must therefore give further consideration to the problem of the Hamiltonian representation of the radiation field.

We shall assume that, as in the mechanics of particles, the total energy $E$ of the field assumes the role of the Hamiltonian function. Then, taking account of the relation

$$\frac{1}{L^3} \int_0^L \int_0^L \int_0^L \exp i(\mathbf{k}-\mathbf{k}')\mathbf{r} \, dx \, dy \, dz = \begin{cases} 1 & \text{when } \mathbf{k}=\mathbf{k}' \\ 0 & \text{when } \mathbf{k} \neq \mathbf{k}' \end{cases} = \delta_{\mathbf{k}\mathbf{k}'}$$

and the fact that $(\mathbf{k}\mathbf{s}_{\mathbf{k}\lambda})=0$, we obtain from (51.2), (51.4), and (51.6):

$$E = \frac{1}{8\pi} \int (\mathbf{E}^2+\mathbf{H}^2) \, dV = \frac{1}{8\pi c^2} \sum_{\mathbf{k}\lambda} (\dot{A}_{\mathbf{k}\lambda}^* \dot{A}_{\mathbf{k}\lambda} + \omega_{\mathbf{k}}^2 A_{\mathbf{k}\lambda}^* A_{\mathbf{k}\lambda}) \qquad (51.7)$$

We must now form canonically conjugate variables $q_j$ and $p_j$ from the $A_{\mathbf{k}\lambda}$ and $\dot{A}_{\mathbf{k}\lambda}$ such that, when the field energy $E$ is expressed in them, it may be employed as the Hamiltonian function $\mathscr{H}(p_j, q_j)$ in a canonical representation. This implies that the equations of motion (51.6) follow from the equations

$$\dot{p}_j = -\frac{\partial \mathscr{H}}{\partial q_j} \qquad \dot{q}_j = \frac{\partial \mathscr{H}}{\partial p_j} \qquad (51.8)$$

Now the amplitudes $A_{\mathbf{k}\lambda}$ themselves cannot be employed as coordinates in this canonical representation, because of the restriction imposed by (51.4b). However, if new amplitudes are introduced, defined by

$$a_{\mathbf{k}\lambda} = \frac{1}{2}\left(A_{\mathbf{k}\lambda} + \frac{i}{\omega_{\mathbf{k}}}\dot{A}_{\mathbf{k}\lambda}\right) \qquad a_{-\mathbf{k}\lambda}^* = \frac{1}{2}\left(A_{\mathbf{k}\lambda} - \frac{i}{\omega_{\mathbf{k}}}\dot{A}_{\mathbf{k}\lambda}\right) \quad (51.9)$$

then

$$A_{\mathbf{k}\lambda} = a_{\mathbf{k}\lambda} + a_{-\mathbf{k}\lambda}^* \qquad (51.10)$$

Condition (51.4b) is thus automatically satisfied, while the amplitudes $a_{\mathbf{k}\lambda}$ and $a_{\mathbf{k}\lambda}^*$ are free from any further restriction. It would thus appear that the $a_{\mathbf{k}\lambda}$ could be employed as coordinates in the required representation; it can be shown immediately that this is in fact possible.

From (51.9) and (51.6) we first derive the new equations of motion:

$$\dot{a}_{\mathbf{k}\lambda} + i\omega_{\mathbf{k}} a_{\mathbf{k}\lambda} = 0 \qquad \dot{a}_{\mathbf{k}\lambda}^* - i\omega_{\mathbf{k}} a_{\mathbf{k}\lambda}^* = 0$$

Thus two differential equations of the first order have replaced the single equation of the second order (51.6). In addition,

$$\dot{A}_{\mathbf{k}\lambda} = -i\omega_{\mathbf{k}}(a_{\mathbf{k}\lambda} - a_{-\mathbf{k}\lambda}^*) \qquad (51.11)$$

The field energy $E$ may now be expressed in terms of the $a_{k\lambda}$ and $a_{k\lambda}^*$ alone, in view of the definition (51.9):

$$E = \frac{1}{4\pi c^2} \sum_{k\lambda} \omega_k^2 (a_{k\lambda}^* a_{k\lambda} + a_{-k\lambda}^* a_{-k\lambda})$$

$$= \frac{1}{2\pi c^2} \sum_{k\lambda} \omega_k^2 a_{k\lambda}^* a_{k\lambda} \qquad (51.12)$$

If the $a_{k\lambda}$ are now identified with the canonical coordinates $q_j$, and the quantities $i\omega_k a_{k\lambda}^*/2\pi c^2$ with the conjugate momenta $p_j$, the energy may be expressed as the Hamiltonian function in the form

$$\mathcal{H} = -i \sum_{k\lambda} \omega_k q_j p_j \qquad (51.12a)$$

The equations (51.11) then follow directly from the canonical equations of motion (51.8), thereby establishing the required connection with the canonical form of representation.*

In order that we may later form the necessary commutation relations as simply as possible, we shall replace the $a_{k\lambda}$ by the quantities

$$b_{k\lambda} = \frac{1}{c} \left( \frac{\omega_k}{2\pi\hbar} \right)^{1/2} a_{k\lambda} \qquad (51.13)$$

following a similar procedure to that which we adopted in the case of the individual linear oscillator in §15. Then (51.12) becomes

$$E = \sum_{k\lambda} \hbar\omega_k b_{k\lambda}^* b_{k\lambda} \qquad (51.14)$$

and the corresponding canonically conjugate coordinates and momenta are

$$b_{k\lambda} \quad \text{and} \quad i\hbar b_{k\lambda}^*$$

As before, it may easily be verified that equations (51.11) follow from (51.9).

We now express the vector potential (51.4) in terms of the $b_{k\lambda}$. Using (51.10) and (51.13), we first obtain

$$\mathbf{A}(\mathbf{r}, t) = \frac{c}{L^{3/2}} \sum_{k\lambda} \left( \frac{2\pi h}{\omega_k} \right)^{1/2} \mathbf{s}_{k\lambda} (b_{k\lambda} + b_{-k\lambda}^*) \exp i\mathbf{k}\mathbf{r}$$

---

* In the canonical equations of motion (51.8) the coordinates and momenta are complex conjugates, apart from a factor; however, they may be regarded as independent of each other in partial differentiation. This may be proved by resolving them into real and imaginary parts, and treating the latter as independent quantities.

If we replace $-\mathbf{k}$ by $\mathbf{k}$ in the second summation, and put

$$b_{\mathbf{k}\lambda} = b_{\mathbf{k}\lambda}^0 \exp - i\omega_{\mathbf{k}}t \qquad b_{\mathbf{k}\lambda}^* = b_{\mathbf{k}\lambda}^{0*} \exp i\omega_{\mathbf{k}}t$$

in virtue of the relation (51.11), the expression becomes

$$\mathbf{A}(\mathbf{r}, t) = \frac{c}{L^{3/2}} \sum_{\mathbf{k},\lambda} \left( \frac{2\pi h}{\omega_{\mathbf{k}}} \right)^{1/2} \mathbf{s}_{\mathbf{k}\lambda} \times$$

$$\times \{ b_{\mathbf{k}\lambda}^0 \exp i(\mathbf{kr} - \omega_{\mathbf{k}}t) + b_{\mathbf{k}\lambda}^{0*} \exp - i(\mathbf{kr} - \omega_{\mathbf{k}}t) \} \quad (51.15)$$

This is the sum of real waves of amplitude $|b_{\mathbf{k}\lambda}^0|$, travelling in the direction of $\mathbf{k}$. The result may be used to derive the number of waves occurring in the interval $d\omega$, which has already been given in (47.13). From (51.5) and the relation $\omega^2 = c^2 k^2$

$$\left( \frac{L}{2\pi} \right)^2 \frac{\omega^2}{c^2} = n_x^2 + n_y^2 + n_z^2$$

In the space of the $n_x$, $n_y$, $n_z$, the number of "lattice points" corresponding to integral values for which $n_x^2 + n_y^2 + n_z^2 \leqq R^2$ is equal to the volume of the sphere of radius $R$, namely $(4\pi/3)R^3$, when $R$ is large. Since two directions of polarization ($\lambda = 1, 2$) are associated with each value of $\mathbf{k}$, the total number of waves of angular frequency less than $\omega$ is

$$s(\omega) = 2 \frac{4\pi}{3} \left( \frac{L\omega}{2\pi c} \right)^3$$

The number of waves occurring in the interval $d\omega$ is therefore

$$z(\omega)\, d\omega = \frac{ds}{d\omega} d\omega = L^3 \frac{\omega^2}{\pi^2 c^3} d\omega$$

as stated in (47.13).

## §52. The quantum theory of the radiation field

Now that Maxwell's equations have been expressed in canonical form with coordinates $b$ and momenta $i\hbar b^*$, we may proceed to the quantum theory, replacing the numbers $b^*$ and $b$ by operators $b^+$ and $b$ subject to the commutation relations†

$$i\hbar b_{\mathbf{k}\lambda}^+ b_{\mathbf{k}'\lambda'} - b_{\mathbf{k}'\lambda'}\, i\hbar b_{\mathbf{k}\lambda}^+ = \frac{\hbar}{i} \delta_{\mathbf{k}\mathbf{k}'} \delta_{\lambda\lambda'} \qquad b_{\mathbf{k}\lambda} b_{\mathbf{k}'\lambda'} - b_{\mathbf{k}'\lambda'} b_{\mathbf{k}\lambda} = 0$$

---

† The fact that $b^*$ must be replaced in quantum theory by $b^+$ is best seen by resolving $b$ into real and imaginary parts; these parts must become real numbers in Hermitian operators.

In other words, all operators $b$, $b^+$ commute except those with the same values of $\mathbf{k}$ and $\lambda$, for which

$$b_{\mathbf{k}\lambda} b_{\mathbf{k}\lambda}^+ - b_{\mathbf{k}\lambda}^+ b_{\mathbf{k}\lambda} = 1 \tag{52.1}$$

In terms of these operators, the Hamiltonian operator is

$$\mathscr{H} = \sum_{\mathbf{k}\lambda} \hbar\omega_{\mathbf{k}} b_{\mathbf{k}\lambda}^+ b_{\mathbf{k}\lambda} \tag{52.2}$$

The transition to quantum theory is not quite free from ambiguity, since the Hamiltonian could equally well have been expressed as $\sum \hbar\omega_{\mathbf{k}} b_{\mathbf{k}\lambda} b_{\mathbf{k}\lambda}^+$. The expression (52.2) implies a particular choice of energy scale, corresponding in fact to a ground state of zero energy.

The vector potential now becomes the operator

$$\mathbf{A} = \sum_{\mathbf{k}\lambda} c \left(\frac{2\pi\hbar}{L^3}\right)^{1/2} \frac{\mathbf{s}_{\mathbf{k}\lambda}}{\sqrt{\omega_{\mathbf{k}}}} \{b_{\mathbf{k}\lambda} e^{i\mathbf{k}\mathbf{r}} + b_{\mathbf{k}\lambda}^+ e^{-i\mathbf{k}\mathbf{r}}\} \tag{52.3}$$

We can now construct a Hilbert space from the eigenfunctions of $\mathscr{H}$, just as we did in the case of the linear oscillator in §15. If $\Phi$ is an eigenvector of $\mathscr{H}$ (cf. (15.11)) corresponding to the eigenvalue $E$,

$$\sum_{\mathbf{k}} \hbar\omega_{\mathbf{k}} b_{\mathbf{k}}^+ b_{\mathbf{k}} \Phi = E\Phi \tag{52.4}$$

(The index $\lambda$ has been temporarily omitted.) If we form the scalar product with $\Phi$, taking account of the fact that $b^+$ is adjoint to $b$, we obtain

$$\sum_{\mathbf{k}} \hbar\omega_{\mathbf{k}} (b_{\mathbf{k}}\Phi, b_{\mathbf{k}}\Phi) = E \tag{52.4a}$$

All the terms on the left are positive or zero; therefore $E$ is positive, and only equal to 0 if $b_{\mathbf{k}}\Phi$ is 0 for all $\mathbf{k}$. We now apply a particular operator $b_{\mathbf{k}'}$ to (52.4); it then follows from (52.1) that

$$\sum_{\mathbf{k}} \hbar\omega_{\mathbf{k}} b_{\mathbf{k}}^+ b_{\mathbf{k}} (b_{\mathbf{k}'}\Phi) = (E - \hbar\omega_{\mathbf{k}'}) b_{\mathbf{k}'}\Phi \tag{52.4b}$$

If the operator $b_{\mathbf{k}'}$ is applied $n$ times, the following result is obtained: either $b_{\mathbf{k}'}^n \Phi$ belongs to the eigenvalue $E - n\hbar\omega_{\mathbf{k}'}$, or it is equal to 0. The latter case must occur once for a finite value of $n$, since the eigenvalue certainly cannot be negative. Let the number $n_{\mathbf{k}'}$ be defined such that $b_{\mathbf{k}'}^{n_{\mathbf{k}'}} \Phi$ corresponds to the eigenvalue $E - n_{\mathbf{k}'} \hbar\omega_{\mathbf{k}'}$ but that $b_{\mathbf{k}'}^{n_{\mathbf{k}'}+1} \Phi = 0$. Now let us apply the same procedure to the equation

$$\sum_{\mathbf{k}} \hbar\omega_{\mathbf{k}} b_{\mathbf{k}} b_{\mathbf{k}}^+ (b_{\mathbf{k}'}^{n_{\mathbf{k}'}} \Phi) = (E - n_{\mathbf{k}'} \hbar\omega_{\mathbf{k}'}) b_{\mathbf{k}'}^{n_{\mathbf{k}'}} \Phi$$

forming the scalar product $n_{\mathbf{k}''}$ times with another operator $b_{\mathbf{k}''}$. Then the result,

$$\Psi = b_{\mathbf{k}''}^{n_{\mathbf{k}''}} b_{\mathbf{k}'}^{n_{\mathbf{k}'}} \Phi$$

corresponds to the eigenvalue

$$E - n_{\mathbf{k}'} \hbar\omega_{\mathbf{k}'} - n_{\mathbf{k}''} \hbar\omega_{\mathbf{k}''}$$

and

$$b_{\mathbf{k}''} \Psi = 0$$

We continue in this manner, taking each $b_{\mathbf{k}}$ in succession. We finally obtain an element

$$\Phi_0 = \left(\prod_{\mathbf{k}} b_{\mathbf{k}}^{n_{\mathbf{k}}}\right) \Phi$$

corresponding to the eigenvalue

$$E - \sum_{\mathbf{k}} \hbar\omega_{\mathbf{k}} n_{\mathbf{k}}$$

for which $b_{\mathbf{k}} \Phi_0 = 0$ for all $\mathbf{k}$. Then from (52.4a) the eigenvalue is equal to 0; the eigenvalue $E$ in (52.4) must therefore have the form

$$E = \sum_{\mathbf{k}} \hbar\omega_{\mathbf{k}} n_{\mathbf{k}} \tag{52.5}$$

If the discrete sequence of $\mathbf{k}$ values is numbered in some way, say $\mathbf{k}_1, \mathbf{k}_2, \mathbf{k}_3, \ldots$, we may denote the eigenvector corresponding to the eigenvalue (52.5) as follows:

$$\Phi n_{\mathbf{k}_1}, n_{\mathbf{k}_2}, n_{\mathbf{k}_3}, \ldots \tag{52.6}$$

We can now take the rest of the argument directly from our earlier treatment of the linear oscillator. As in (15.12), if $\Phi$ is an eigenfunction corresponding to $E$, then $b_{\mathbf{k}}^+ \Phi$ is an eigenfunction corresponding to $E + \hbar\omega_{\mathbf{k}}$. We can therefore construct the whole Hilbert space by means of the repeated application of the $b_{\mathbf{k}}^+$ to $\Phi_{0,0,0,\ldots} = \Phi_0$.* After once more introducing $\lambda$ and normalizing, we obtain

$$\Phi \ldots, n_{\mathbf{k}\lambda}, \ldots = \prod_{\mathbf{k}', \lambda'} \frac{b_{\mathbf{k}'\lambda'}^{+ n_{\mathbf{k}'\lambda'}}}{\sqrt{(n_{\mathbf{k}'\lambda'}!)}} \Phi_{0,0,0,\ldots}$$

---

* The ground state represented by $\Phi_0$ corresponds to a vacuum; it still contains zero-point vibrations, however, as in the case of the linear oscillator (example 4, p. 108). Since we are dealing with an infinite number of oscillators the mean square values of the fields, $\overline{E^2}$, $\overline{H^2}$, must also be infinitely great. A completely satisfactory treatment of this anomaly does not yet exist (see p. 319).

The $n_{k\lambda}$ form an infinite set of positive integers, and define the number of light quanta with wave number $k$ and polarization $\lambda$ that are present in the state $\Phi \ldots, n_{k\lambda}, \ldots$ The energy corresponding to $\Phi$ is

$$E = \sum_{k\lambda} \hbar\omega_k n_{k\lambda} \tag{52.7}$$

The operators $b$ and $b^+$ have the same effect on the eigenfunctions as in the case treated in §15:

$$
\begin{aligned}
b_{k\lambda}^+ \Phi \ldots, n_{k\lambda}, \ldots &= \sqrt{(n_{k\lambda}+1)}\, \Phi \ldots, n_{k\lambda}+1, \ldots \\
b_{k\lambda} \Phi \ldots, n_{k\lambda}, \ldots &= \sqrt{(n_{k\lambda})}\, \Phi \ldots, n_{k\lambda}-1, \ldots
\end{aligned}
\tag{52.8}
$$

They can therefore be designated as the photon generating and annihilating operators.

This completes our description of the vectors forming the Hilbert space basis, and of the effect of the application of the operators $b_{k\lambda}^+$ and $b_{k\lambda}$ to them.

### §53.  Quantum mechanics of the atom in the radiation field

We shall now consider the case in which radiation and an atom with one valence electron are both present within the container. The stationary states $\phi_i$ of the electron satisfy a Schrödinger equation $\mathscr{H}_{el}\phi_i = \varepsilon_i\phi_i$; for the sake of simplicity they will be assumed to be non-degenerate. The index $i$ represents the complete set of electron quantum numbers. The stationary states of the complete system, consisting of the electron and the radiation field, and neglecting their interaction, are given by

$$\Phi_\mu = \phi_i \Phi \ldots, n_{k\lambda}, \ldots \text{ with energy } E_\mu = \varepsilon_i + \sum \hbar\omega_k n_{k\lambda} \tag{53.1}$$

(Greek subscripts represent the quantum numbers of the complete system, e.g. $\mu$ stands for the set of numbers $i, \ldots, n_{k\lambda}, \ldots$)

The interaction* between the field and the electron is

$$W = -\frac{e}{mc}\left(\mathbf{A}, \frac{\hbar}{i}\,\mathrm{grad}\right) = -\frac{e}{mc}(\mathbf{A}, \mathbf{p})$$

The states $\Phi_\mu$ are no longer stationary, but functions of the time. Their dependence on time is determined by the Hamiltonian operator

$$\mathscr{H} = \mathscr{H}_0 + W$$

---

* The term in $A^2$ is omitted, since it makes no contribution to the processes discussed below.

where $\mathcal{H}_0 = \mathcal{H}_{el} + \sum \hbar\omega_k b_{k\lambda}^+ b_{k\lambda}$

and $\quad W = -\frac{e}{m}\left(\frac{2\pi\hbar}{L^3}\right)^{1/2} \sum_{k\lambda} \frac{(s_{k\lambda}, p)}{\sqrt{\omega_k}} \{b_{k\lambda} \exp ikr + b_{k\lambda}^+ \exp -ikr\}$

If we take the general expression for a state to be

$$\Phi = \sum_\nu c_\nu(t) \exp\left(-\frac{i}{\hbar}E_\nu t\right)\Phi_\nu$$

then, since $\mathcal{H}\Phi = i\hbar\dot{\Phi}$, it follows that

$$\dot{c}_\mu(t) = -\frac{i}{\hbar}\sum_\nu (\Phi_\mu W\Phi_\nu) \exp(i\omega_{\mu\nu}t) c_\nu(t), \text{ where } \hbar\omega_{\mu\nu} = E_\mu - E_\nu$$

Therefore if the system is in the state $\Phi = \Phi_\nu$ at time $t = 0$, there is a probability $|c_\mu(t)|^2$ of finding other states $\Phi_\mu$ at time $t$. This probability is found from Dirac's perturbation theory (§27) to be

$$|c_\mu|^2 = w_{\mu\nu} t = 2\pi \frac{|(\Phi_\mu W\Phi_\nu)|^2}{\hbar^2} \frac{\sin^2 \frac{1}{2}\omega_{\mu\nu}t}{\frac{1}{2}\pi\omega_{\mu\nu}^2 t} \; t$$

where $w_{\mu\nu}$ is the "transition probability".

Since the vector potential and the operators $b^+$ and $b$ only occur in linear form in the interaction $W$, the matrix element $W_{\mu\nu} = (\Phi_\mu, W\Phi_\nu)$ only differs from 0 if a single one of the $n_{k\lambda}$ of the final state $\Phi_\mu$ differs by unity from its corresponding value in the initial state $\Phi_\nu$. For instance: from (52.8),

$$(\Phi_\mu b_{k\lambda}^+ \Phi_\nu) = (\Phi \ldots, n_{k\lambda}+1, \ldots, b_{k\lambda}^+\Phi \ldots, n_{k\lambda}, \ldots) = \sqrt{(n_{k\lambda}+1)}$$

when $\Phi_\mu$ and $\Phi_\nu$ agree in all quantum numbers except $n_{k\lambda}$. This means that, in the first-order approximation of Dirac's perturbation theory, the only transitions to occur are those in which a single light quantum is produced (i.e. emitted) or annihilated (absorbed).

When $t$ is large enough, the function $\sin^2 \frac{1}{2}\omega t / \frac{1}{2}\pi\omega^2 t$ differs from 0 only in the immediate neighbourhood of $\omega = 0$. Since

$$\int_{-\infty}^\infty \frac{\sin^2 \frac{1}{2}\omega t}{\frac{1}{2}\pi\omega^2 t} d\omega = 1$$

the function may be replaced by $\delta(\omega)$ in all further integrations over $\omega$. This implies that the only transitions to occur are those for which

$\omega_\mu \approx \omega_\nu$. Therefore, if $\omega_k$ denotes the frequency of the emitted or absorbed quantum:

$$\underbrace{\varepsilon_i + \hbar\omega_k = \varepsilon_j}_{E_{initial}} \qquad \text{Absorption} \quad (\varepsilon_j > \varepsilon_i)$$
$$\varepsilon_i = \underbrace{\varepsilon_j + \hbar\omega_k}_{E_{final}} \qquad \text{Emission} \quad (\varepsilon_j < \varepsilon_i) \qquad (53.2)$$

We shall now consider the case of emission. Using the relations (52.8), the transition probability is

$$w_{k\lambda}^{i\to j} = \frac{(2\pi)^2}{L^3} \frac{e^2}{\hbar m^2} \frac{n_{k\lambda}+1}{\omega_k} |(\phi_j, \mathbf{s}_{k\lambda}\, \mathbf{p} \exp(i\mathbf{k}\mathbf{r})\, \phi_i)|^2\, \delta(\omega_{ij} - \omega_k)$$

$$(53.3)$$

This may be expressed in words as follows: if an atom in the state $\phi_i$ is present at time $t = 0$ in the radiation field $\Phi \ldots, n_{k\lambda}, \ldots$, then $w_{k\lambda}^{i\to j}\, dt$ is the probability that after a time $dt$ the atom should be in the state $\phi_j$, having emitted a light quantum of wave number $\mathbf{k}$ and polarization $\lambda$. This probability consists of two components. One component contains the factor $n_{k\lambda}$, and is therefore proportional to the intensity of the radiation field. It proves to be identical with the probability of the induced transition, which was introduced in §§48 and 50. The other term contains the factor 1, and is therefore quite independent of the radiation field, existing even when the field vanishes; this component gives the probability of the spontaneous emission. The formula that was derived in §48 on the basis of the correspondence principle is deducible in the present treatment as a natural consequence of the quantization of the radiation field. It is also apparent from the relations (52.8) and from the argument on which the construction of the Hilbert space was based that the figure 1 occurring next to $n_{k\lambda}$ in (53.3) results directly from the commutation relations (52.1).

In order to deduce the familiar formulae (48.9) or (11.9) for the probability of spontaneous emission, we shall investigate the probability of a transition from $i$ to $j$ accompanied by the simultaneous emission of radiation into an element of solid angle $d\Omega$. In order to do this we must sum (53.3) over both polarizations and over all the $\mathbf{k}$ whose directions are contained in $d\Omega$. At the same time we proceed to the limit $L \to \infty$; as a result, the permitted values of $\mathbf{k}$ are packed closer and closer together, and the summation becomes an integral:

$$\sum_k \Rightarrow \left(\frac{L}{2\pi}\right)^3 d\Omega \int k^2\, dk, \text{ where } k = |\mathbf{k}| = \frac{\omega}{c}$$

Owing to the $\delta$-function in (53.3), the only value of $\mathbf{k}$ remaining after integration has the magnitude $|\mathbf{k}| = \omega_{ij}/c$.

In order to make the comparison with formula (11.9) complete we shall restrict ourselves to the case in which the wavelength of the emitted light is large compared with the diameter of the atom; we may then use the dipole approximation $e^{i\mathbf{k}\mathbf{r}} = 1$. In addition,

$$(\phi_i, \mathbf{p}\phi_j) = im\omega_{ij}(\phi_i, \mathbf{r}\phi_j) = im\omega_{ij}\mathbf{r}_{ij}$$

We now take $\mathbf{s}_{k1}$ to lie in the plane defined by the real or imaginary part of $\mathbf{r}_{ij}$ and by $\mathbf{k}$, and $\mathbf{s}_{k2}$ normal to it. Then $(\mathbf{s}_{k2}, \mathbf{r}_{ij}) = 0$, and there remains only

$$\frac{1}{2\pi} \frac{e^2 \omega_{ij}^3}{\hbar c^3} |(\mathbf{s}_{k1}, \mathbf{r}_{ij})|^2 \, d\Omega$$

From figure 55 we see that the scalar product $(\mathbf{s}_{k1}, \mathbf{r}_{ij})$ contains the factor $\sin\theta$; hence after integrating over $d\Omega$ we obtain

$$w_{spont}^{i \to j} = \frac{4}{3} \frac{e^2 \omega_{ij}^3}{\hbar c^3} |\mathbf{r}_{ij}|^2 \tag{53.4}$$

The energy emitted in connection with the transition $i \to j$ is obtained by multiplying (53.4) by $\hbar\omega_{ij}$. The results agree with the formulae based on the correspondence principle, (11.8, 11.9).

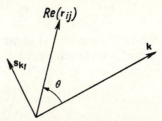

Fig. 55.—Polarization vector $\mathbf{s}_{k_1}$, dipole moment Re $(\mathbf{r}_{ij})$, and propagation vector $\mathbf{k}$. The polarization vector $\mathbf{s}_{k_2}$ is perpendicular to the plane of the diagram. To integrate over $[\mathbf{s}_{k\lambda}, \text{Im } (\mathbf{r}_{ij})]^2$, $\mathbf{s}_{k_1}$ is chosen to lie in the plane defined by Im $(\mathbf{r}_{ij})$ and $\mathbf{k}$

We shall now deal with the term containing the factor $n_{k\lambda}$ in equation (53.3), in order to obtain the corresponding formula for the stimulated emission. In view of its application to the treatment of temperature radiation in a cavity we shall restrict ourselves to the case in which the radiation distribution is isotropic; in other words, we assume that the $n_{k\lambda}$ depend only on the magnitude of **k**, but not on its direction or on the polarization $\lambda$.

We introduce once more a spectral energy density $u(\omega)$ such that $u(\omega)\,d\omega$ is the energy density of the radiation field in the frequency interval $d\omega$. Then in this interval there are $u(\omega)\,d\omega/\hbar\omega$ light quanta per unit volume of the cavity, each half of which is associated with one of the two independent polarization directions. We thus obtain the following expression for the probability of an induced dipole transition from $i$ to $j$:

$$w_{induced}^{i\to j} = \sum_\lambda \int \frac{(2\pi)^2}{L^3} \frac{e^2}{m^2\hbar\omega} |(\phi_i, \mathbf{s}_{k\lambda}\mathbf{p}\phi_j)|^2 \delta(\omega_{ij}-\omega) \frac{L^3 u(\omega)}{2\hbar\omega} d\omega \frac{d\Omega}{4\pi}$$

This integral is evaluated as above, and gives

$$w_{induced}^{i\to j} = \frac{4\pi^2 e^2}{3} \frac{1}{\hbar^2} u(\omega_{ij}) |\mathbf{r}_{ij}|^2$$

which agrees with the result (50.6) when we put $I(\omega) = cu(\omega)$ in that equation.

The above analysis also covers the case of absorption, since the transition probability has the same form as (53.3), with the factor $(n_{k\lambda}+1)$ replaced by $n_{k\lambda}$. We thus have the familiar relation

$$w_{induced}^{i\to j} = w_{absorp}^{j\to i}$$

A comparison with the coefficients $A$ and $B$ introduced in Einstein's derivation of the Planck radiation formula yields the following relations:

$$A_{i\to j} = w_{spont}^{i\to j} = \frac{4}{3}\frac{e^2\omega_{ij}^3}{\hbar c^3} |\mathbf{r}_{ij}|^2$$

$$B_{i\to j} = \frac{4\pi^2 e^2}{3}\frac{1}{\hbar^2} |\mathbf{r}_{ij}|^2 = B_{j\to i}$$

$$B_{i\to j} u(\omega_{ij}) = w_{induced}^{i\to j}$$

As we can see, the relation (48.8)

$$A_{i\to j} = B_{j\to i}\frac{\hbar\omega_{ij}^3}{\pi^2 c^3}$$

is in fact satisfied.

*The natural width of spectral lines*

In §5 we saw that the spectral line emitted by a classical oscillator possesses a certain width due to the radiation damping of the electron. The line intensity distribution is given by

$$I_\omega \, d\omega = \frac{\gamma_{cl}}{2\pi} \frac{I_0 \, d\omega}{(\omega - \omega_0)^2 + \frac{1}{4}\gamma_{cl}^2} \tag{53.5}$$

for which the half-value width is

$$\gamma_{cl} = \frac{2e^2\omega_0^2}{3mc^3}$$

On the other hand, the immediate result given by Dirac's perturbation calculation for the line width $\Delta\omega$ is the relation $\Delta\omega \cdot t \approx 1$, from which it would appear that the quantum theory predicted infinitely sharp spectral lines in the limiting case of very large time intervals. This conclusion is false, however, because the Dirac approximation is not valid for arbitrary values of the time; on the contrary, it assumes that in the expansion

$$\Phi(t) = c_0(t)\Phi_0 + \sum_{\nu \neq 0} c_\nu(t)\Phi_\nu$$

the probability $|c_0|^2$ for the occurrence of the initial state is nearly 1. In view of the normalization condition $\sum_{\nu \neq 0} |c_\nu|^2 + |c_0|^2 = 1$, this implies that for a sufficiently weak radiation field $\sum_{\nu \neq 0} |c_\nu(t)|^2 \approx w_{\text{spont}} \cdot t \ll 1$, where $w_{\text{spont}}$ is the probability of a spontaneous transition from the initial state to some lower excited state of the atom. For greater time intervals, however, $c_0(t)$ will decrease as $t$ increases; we shall attempt to allow for this in the perturbation calculation by making the initial assumption

$$c_0(t) = \exp -\tfrac{1}{2}\gamma t \tag{53.6}$$

instead of $c_0(t) \approx 1$.

For simplicity we shall restrict ourselves to the case of an atom in the first excited state $\phi_a$ in the absence of the radiation field; we shall neglect processes in which more than one light quantum is emitted, and shall use the dipole approximation. Then the light quantum wave number **k** is sufficient to denote uniquely the final states represented by the atom in the ground state $\phi_g$ and by one light quantum of frequency $\omega_k$ in the radiation field. We thus have the following equation for the amplitudes $c_k(t)$ of these states:

$$-\frac{\hbar}{i}\dot{c}_k(t) = W_{k0}\exp\left[i(\omega_k-\omega_0)t\right]c_0(t) \qquad (53.7)$$

where

$$\hbar\omega_0 = E_a - E_g$$

and

$$W_{k0} = -\frac{e}{m}\left(\frac{2\pi h}{L^3}\right)^{1/2}\frac{im\omega_0}{\sqrt{\omega_k}}(s_k, r_{ag})$$

The vector $s_k$ is normal to $k$ in the plane defined by $k$ and the matrix element $r_{ag}$ (see figure 55).

Substituting (53.6) and (53.7) and integrating,

$$c_k(t) = W_{k0}\frac{\exp\left[i(\omega_k-\omega_0)t-\frac{1}{2}\gamma t\right]-1}{\hbar(\omega_0-\omega_k)-\frac{1}{2}i\hbar\gamma}$$

When $\gamma t \gg 1$, the probability distribution of the emitted quanta is

$$|c_k(\infty)|^2 = \frac{|W_{k0}|^2}{\hbar^2}\frac{1}{(\omega_0-\omega_k)^2+\frac{1}{4}\gamma^2}$$

The number of quanta present in the frequency interval $d\omega = c\,dk$ is

$$n(\omega)\,d\omega = \left(\frac{L}{2\pi}\right)^3 k^2\,dk\int|c_k(\infty)|^2\,d\Omega$$

The above expression is integrated in the same manner as that leading to (53.4); the result is

$$n(\omega)\,d\omega = w_{spont}^{a\to g}\frac{\omega}{\omega_0}\frac{1}{2\pi}\frac{d\omega}{(\omega_0-\omega)^2+\frac{1}{4}\gamma^2} \qquad (53.8)$$

The normalization condition is now

$$e^{-\gamma t}+\sum_k|c_k(t)|^2 = 1$$

Since this condition must hold good for all values of the time it follows that

$$\sum_k|c_k(\infty)|^2 = \int_0^\infty n(\omega)\,d\omega = 1$$

Since $\gamma \ll \omega_0$, the only appreciable contributions to this integral come from the range in the neighbourhood of $\omega = \omega_0$; its value is therefore

$$\int_0^\infty\frac{\omega}{\omega_0}\frac{d\omega}{(\omega_0-\omega)^2+\frac{1}{4}\gamma^2} \approx \int_{-\infty}^\infty\frac{d\omega}{(\omega_0-\omega)^2+\frac{1}{4}\gamma^2} = \frac{2\pi}{\gamma}$$

Then from (53.8) and the normalization condition we obtain the following relation for the lifetime of the excited state $\phi_a$:

$$\gamma = w_{spont}^{a \to g} \tag{53.9}$$

Since the initial assumption $c_0 = \exp -\frac{1}{2}\gamma t$ is not a strict solution of the Schrödinger equation, the normalization condition is not strictly fulfilled at all times. Further, $\gamma$ is determined by the normalization condition only if it does not contain an imaginary part. It is therefore somewhat more satisfactory to determine $\gamma$ from the equation of motion for $c_0(t)$,

$$-\frac{\hbar}{i} \dot{c}_0 = \sum_k W_{0k} \exp\left[i(\omega_0 - \omega_k)t\right] c_k(t)$$

Using this procedure, we obtain the previous result (53.9) for the real part of $\gamma$, and an imaginary part which represents a displacement of the spectral line $\omega_0$. Unfortunately, this imaginary part is a divergent integral.*

This divergence has for long been an insuperable difficulty of the quantum theory; it has not yet been completely overcome, but has been ingeniously circumvented through the concept of the mass renormalization of the electron (Kramers, 1945). The discussion of these developments lies outside the scope of this book. The reader is referred to such a book as Jauch and Rohrlich, *The Theory of Photons and Electrons*, Cambridge, Mass., 1955.

We give without proof the result for the case in which the electron is in some higher state $\phi_i$; $w^{i \to j}$ denotes the probability of spontaneous transition to some lower state $\phi_j$. From the correspondence principle we would expect the width $\gamma_{ij}$ of the line $\omega_{ij}$ to be equal to $w^{i \to j}$. This is not the case, however; the result is

$$\gamma_{ij} = \gamma_i + \gamma_j = \sum_{k<i} w^{i \to k} + \sum_{k<j} w^{j \to k}$$

Each term $\omega_i$ is associated with a given width $\gamma_i$ dependent on the probabilities of the transitions to lower terms; in the case of the transition $\omega_{ij}$ the widths of the terms $\gamma_i$ and $\gamma_j$ are to be added to give the width of the line $\omega_{ij}$.

## Exercises

1. *Fourier representation of Hertz's solution*

Integrate the equations

$$\nabla^2 A - \ddot{A}/c^2 = -4\pi j/c \qquad \nabla^2 \phi - \ddot{\phi}/c^2 = -4\pi\rho$$

using the relations

$$A(r,t) = \int A_{k,\omega} \exp i(kr - \omega t)\, dk\, d\omega \qquad \phi(r,t) = \int \phi_{k,\omega} \exp i(kr - \omega t)\, dk\, d\omega$$

---

* This divergence is directly associated with the infinite field oscillations discussed on p. 311. See Weisskopf and Wigner, E., *Z. Phys.* 63 (1930); Heitler, *Quantum Theory of Radiation*, third ed., p. 181 (Oxford, 1954).

Perform the integration with respect to $\omega$ in the complex plane. How must the path of integration be taken with respect to the poles of $A_{k,\omega}$ and $\phi_{k,\omega}$, in order to obtain the solutions of the retarded potentials?

## 2. *Total radiation momentum*

The momentum density of electromagnetic radiation is

$$\mathbf{g} = \frac{1}{4\pi c} \mathbf{E} \times \mathbf{H}, \quad \text{or} \quad \mathbf{g} = -\frac{1}{4\pi c^2} \dot{\mathbf{A}} \times \text{curl } \mathbf{A}$$

Expressed in component form,

$$g_i = -\frac{1}{4\pi c^2} \sum_k \dot{A}_k \frac{\partial A_k}{\partial x_i} + \frac{1}{4\pi c^2} \sum_k \dot{A}_k \frac{\partial A_i}{\partial x_k} = g_i^{(1)} + g_i^{(2)}$$

Only the first term contributes to the total momentum $\mathbf{G} = \int \mathbf{g} d\mathbf{r}$. The contribution of the second term vanishes—this may be seen by integrating by parts, where the term outside the integral vanishes because of the periodicity assumed in (51.1), and the remaining integrand vanishes because $\text{div } \mathbf{A} = 0$. Represent $\mathbf{G}$ in terms of $b_{k\lambda}$ and $b_{k\lambda}^*$, then proceed to the quantum theory and determine the eigenvalues of the operator $\mathbf{G}$.

Note the relations (51.4a) and (51.12), and express the result with $b^+$ to the left of $b$, as shown for the Hamiltonian operator (52.2).

## 3. *Angular momentum of the radiation*

The angular momentum $\mathbf{M}$ of electromagnetic radiation is $\mathbf{M} = \int \mathbf{r} \times \mathbf{g} d\mathbf{r}$, where $\mathbf{g}$ is the momentum density. If $\mathbf{g}$ is resolved into the two components $\mathbf{g}^{(1)} + \mathbf{g}^{(2)}$ of the previous exercise, the component $\mathbf{g}^{(2)}$ may be transformed by integration by parts, and we obtain

$$\mathbf{M} = \int \mathbf{r} \times \mathbf{g}^{(1)} d\mathbf{r} + \frac{1}{4\pi c^2} \int \mathbf{A} \times \dot{\mathbf{A}} d\mathbf{r}$$

The first term is of no interest; it clearly depends upon the choice of the origin of coordinates, and corresponds as it were to the "orbital angular momentum" of the radiation. It can always be made to vanish by a suitable choice of origin. In contrast, the second term is independent of the coordinate system, and corresponds to an intrinsic angular momentum, or "spin", of the radiation.

Represent 
$$\mathbf{M} = \frac{1}{4\pi c^2} \int \mathbf{A} \times \dot{\mathbf{A}} d\mathbf{r}$$

in terms of the operators $b_{k\lambda}^+$ and $b_{k\lambda}$, taking due account of (51.9), and the remarks at the end of the previous exercise. As in the case of $\mathscr{H}$ and $\mathbf{G}$, $\mathbf{M}$ may be represented as a sum of terms, each of which is associated with a given value of $\mathbf{k}$: $\mathbf{M} = \sum_\mathbf{k} \mathbf{M}^{(\mathbf{k})}$. In contrast to $\mathscr{H}^{(\mathbf{k})}$ and $\mathbf{G}^{(\mathbf{k})}$, however, the two polarizations $\lambda = 1, 2$ are still combined.

Introduce new operators in order to separate the polarizations in $\mathbf{M}^{(\mathbf{k})}$. What are the eigenvalues of $\mathbf{M}$? What classical waves correspond to the eigenstates? Express $\mathscr{H}$ and $\mathbf{G}$ in terms of the new operators. Describe the Hilbert space constructed from the eigenstates of $\mathscr{H}$, $\mathbf{G}$, and $\mathbf{M}$.

**F**

---

The
relativistic
theory
of
the
electron

# CHAPTER FI

## Relativistic classical mechanics

### §54. The equations of motion of the electron

We shall assume a knowledge of the principles of the theory of relativity that were derived in the first volume; the results required for our present purpose are summarized below.

The non-relativistic equation of motion of an electron in an electromagnetic field $\mathbf{E}$, $\mathbf{H}$ is

$$\frac{d}{dt} m\mathbf{v} = e \left( \mathbf{E} + \frac{\mathbf{v}}{c} \times \mathbf{H} \right) \tag{54.1}$$

The energy equation is obtained by forming the scalar product with $\mathbf{v}$:

$$\frac{d}{dt} \tfrac{1}{2} m\mathbf{v}^2 = e(\mathbf{E}\mathbf{v}) \tag{54.1a}$$

The relativistic analogue of the velocity $v_i = dx_i/dt$ is the four-velocity $u_\nu = dx_\nu/d\tau$, where $x_\nu$ stands for $x, y, z, ict$, and $d\tau = \sqrt{(1 - v^2/c^2)}\, dt$ is the differential of the proper time. Since the $x_\nu$ constitute a four-vector and the proper time $d\tau$ is a scalar quantity, the $u_\nu$ also form a four-vector:

$$(u_1, u_2, u_3) = \frac{\mathbf{v}}{\sqrt{(1 - v^2/c^2)}} \qquad u_4 = \frac{ic}{\sqrt{(1 - v^2/c^2)}} \tag{54.2}$$

The length of this vector is

$$\sum_{\nu=1}^{4} u_\nu^2 = -c^2 \tag{54.2a}$$

In the relativistic case the field strengths* $\mathbf{E}$, $\mathbf{H}$, can be combined in an antisymmetric tensor of the second order:

$$F_{\mu\nu} = \left\{ \begin{array}{c|cccc} {}_\mu\!\diagdown^\nu & 1 & 2 & 3 & 4 \\ \hline 1 & 0 & H_z & -H_y & -iE_x \\ 2 & -H_z & 0 & H_x & -iE_y \\ 3 & H_y & -H_x & 0 & -iE_z \\ 4 & iE_x & iE_y & iE_z & 0 \end{array} \right\} \tag{54.3}$$

* Strictly, the magnetic induction $\mathbf{B}$ should be used instead of $\mathbf{H}$, since $\mathbf{E}$ and this quantity together form a tensor. In what follows, however, we are only concerned with a vacuum, where $\mathbf{H}$ and $\mathbf{B}$ are identical.

If we form the vector of the four-force $K_\mu$ from $F_{\mu\nu}$ and $u_\nu$,

$$K_\mu = \frac{e}{c} \sum_\nu F_{\mu\nu} u_\nu \qquad (54.4)$$

then, substituting for $u_\nu$ according to (54.2), we obtain:

$$(K_1, K_2, K_3) = e\left\{\mathbf{E} + \frac{\mathbf{v}}{c} \times \mathbf{H}\right\}\Big/ \sqrt{(1 - v^2/c^2)} \qquad (54.5)$$

$$K_4 = \frac{ie}{c} \mathbf{Ev}\Big/ \sqrt{(1 - v^2/c^2)}$$

The above equations represent the relativistic generalization of the right-hand side of (54.1). The left-hand side becomes $dmu_\nu/d\tau$; the relativistic equation of motion therefore assumes the form

$$\frac{d}{d\tau} mu_\nu = \frac{e}{c} \sum_\nu F_{\mu\nu} u_\nu \qquad (54.6)$$

Expressed in components:

$$\frac{d}{dt} \frac{m\mathbf{v}}{\sqrt{(1 - v^2/c^2)}} = e\left(\mathbf{E} + \frac{\mathbf{v}}{c} \times \mathbf{H}\right) \qquad (54.7)$$

$$\frac{d}{dt} \frac{mc^2}{\sqrt{(1 - v^2/c^2)}} = e(\mathbf{Ev}) \qquad (54.7a)$$

The spatial components (54.7) are the experimentally verified equations of motion that have already been discussed in §3. The fourth component (54.7a) provides the relativistic generalization of the theorem of the conservation of energy. Equations (54.6) therefore summarize and generalize equations (54.1) and (54.1a).

The energy of the electron is

$$E = mc^2 / \sqrt{(1 - v^2/c^2)} \qquad (54.8)$$

For small velocities

$$E = mc^2 + \tfrac{1}{2}mv^2 + \dots$$

We thus obtain for $E$ the kinetic energy of non-relativistic mechanics, apart from an additive constant which is the rest energy $mc^2$. It is apparent that in the non-relativistic limiting case, for which $v/c \ll 1$, the four equations (54.6) tend to the four equations (54.1, 54.1a).

If $u_v$ is multiplied by the electron mass $m$, a new four-vector is obtained,

$$p_v = m u_v, \text{ where } (p_1, p_2, p_3) = \frac{m\mathbf{v}}{\sqrt{(1-v^2/c^2)}} = \mathbf{p}, \quad p_4 = \frac{i}{c}E \quad (54.9)$$

the spatial parts of which are the components of the momentum. Then from (54.2a) we obtain the following relation between energy and momentum:

$$\mathbf{p}^2 - \frac{E^2}{c^2} = -m^2c^2 \qquad E = \sqrt{(p^2c^2 + m^2c^4)} \qquad (54.10)$$

## §55. The Lagrangian and Hamiltonian functions

The relativistic Hamiltonian function can be obtained in its most concise form from the quantity $G = \int L\,dt$ which was introduced in (6.6). In non-relativistic mechanics the vanishing of the variation of $G$ leads to a formulation of the equations of motion that is independent of the coordinates. $G$ is a scalar, and if the relativistic equations of motion are based on a similar principle of the extremum, then $G$ must be a scalar relativistic invariant.

Recollecting that the differential of the proper time $d\tau = dt\sqrt{(1-v^2/c^2)}$ is a scalar invariant, although $dt$ is not, we may put

$$L\,dt = L'\,d\tau \qquad (55.1)$$

where $L'$ is also to be invariant. Comparison with (7.3) suggests a form for $L'$ which is deduced as follows. We consider the two four-vectors

$$\text{four-potential } (\Phi_1, \Phi_2, \Phi_3, \Phi_4) = (\mathbf{A}, i\phi) \qquad (55.2)$$

and $\qquad$ four-velocity $(u_1, u_2, u_3, u_4)$.

We assume that $L'$ contains the scalar product of these two four-vectors, plus a scalar invariant. The latter must be equal to $-mc^2$, since in the non-relativistic case and in the absence of the field

$$L = L'\sqrt{(1-v^2/c^2)} \approx L'(1-v^2/2c^2)$$

This must reduce to the non-relativistic Lagrange function $L = \text{const} + \frac{1}{2}mv^2$. Hence for $\Phi_v \equiv 0$, $L'$ must be equal to $-mc^2$. We therefore try the following solution:

$$L' = -mc^2 + \frac{e}{c}\sum_v u_v \Phi_v \qquad (55.3)$$

Our new Lagrange function $L$ follows from this trial solution together with (55.1) and (55.2):

$$L = -mc^2 \sqrt{(1-v^2/c^2)} + \frac{e}{c}(\mathbf{v}\mathbf{A}) - e\phi \qquad (55.4)$$

Hence the relativistic canonical momentum is

$$p_x = \frac{\partial L}{\partial \dot{x}} = \frac{m\dot{x}}{\sqrt{(1-v^2/c^2)}} + \frac{e}{c}A_x, \text{ i.e. } \mathbf{p} = \frac{m\mathbf{v}}{\sqrt{(1-v^2/c^2)}} + \frac{e}{c}\mathbf{A} \qquad (55.5)$$

It may easily be verified that the Lagrangian form of the equations of motion

$$\frac{dp_x}{dt} = \frac{\partial L}{\partial x} \qquad \frac{dp_y}{dt} = \frac{\partial L}{\partial y} \qquad \frac{dp_z}{dt} = \frac{\partial L}{\partial z} \qquad (55.6)$$

agrees with (54.7).

For the Hamiltonian function $\mathbf{p}\mathbf{v} - L$ it follows from (55.4) and (55.5) that

$$\frac{mv^2}{\sqrt{(1-v^2/c^2)}} + mc^2 \sqrt{(1-v^2/c^2)} + e\phi = \frac{mc^2}{\sqrt{(1-v^2/c^2)}} + e\phi$$

We must now express the velocities in terms of the momenta, according to (7.4). Then from

$$\left(\mathbf{p} - \frac{e}{c}\mathbf{A}\right)^2 = \frac{m^2 v^2}{1-v^2/c^2} = m^2 c^2 \left(\frac{1}{1-v^2/c^2} - 1\right)$$

it follows that

$$\mathcal{H}(\mathbf{r}, \mathbf{p}, t) = c \sqrt{\left[\left(\mathbf{p} - \frac{e}{c}\mathbf{A}\right)^2 + m^2 c^2\right]} + e\phi \qquad (55.7)$$

This is the relativistic Hamiltonian function for the motion of a particle of charge $e$ and mass $m$ in an electromagnetic field represented by the potentials $\mathbf{A}$ and $\phi$.

This result may be expressed in the following form,

$$\left(\mathbf{p} - \frac{e}{c}\mathbf{A}\right)^2 - \left(\frac{\mathcal{H}}{c} - \frac{e}{c}\phi\right)^2 = -m^2 c^2$$

which brings the relativistic invariance into prominence when the four-vectors

$$(p_1, p_2, p_3, p_4) = \left(\mathbf{p}, \frac{i}{c}\mathcal{H}\right) \qquad (\Phi_1, \Phi_2, \Phi_3, \Phi_4) = (\mathbf{A}, i\phi)$$

are introduced. This gives the following result:

$$\sum_\nu \left( p_\nu - \frac{e}{c}\Phi_\nu \right)^2 = \sum_\nu (mu_\nu)^2 = -m^2 c^2 \qquad (55.8)$$

When the external fields vanish (55.8) is identical with (54.10) when the energy $E$ is replaced by the Hamiltonian function.

## Exercises

1. *The energy spectrum of electrons in β-disintegration, according to Fermi*

In the course of the β-disintegration of a nucleus, the latter emits an electron $e$ and a neutrino $\nu$. The energy of disintegration $E_m$ is distributed between the two emitted particles. The only observable particles, the electrons, therefore possess different energies, depending on the share of the energy associated with each neutrino. In order to calculate the energy distribution of the electrons it is necessary to make an assumption concerning the distribution of the disintegration energy between the two particles. One obvious assumption drawn from statistical mechanics is that the number of disintegrations $n(\mathbf{p}_e, \mathbf{p}_\nu) d\mathbf{p}_e d\mathbf{p}_\nu$, for which the momenta of the two particles lie in the intervals $(\mathbf{p}_e, d\mathbf{p}_e)$ and $(\mathbf{p}_\nu, d\mathbf{p}_\nu)$, is proportional to $d\mathbf{p}_e d\mathbf{p}_\nu$. The nucleus itself may be assumed to be practically at rest in view of its large mass, and it is therefore unnecessary to take account of the conservation of momentum. The momenta are subject to the further condition that the energy of the two particles $\mathscr{H}_e(\mathbf{p}_e) + \mathscr{H}_\nu(\mathbf{p}_\nu)$ is equal to the energy of disintegration $E_m$, with an associated infinitesimal uncertainty $\Delta E_m$ which is introduced purely for convenience of calculation. Hence

$$n(\mathbf{p}_e, \mathbf{p}_\nu) d\mathbf{p}_e d\mathbf{p}_\nu = C d\mathbf{p}_e d\mathbf{p}_\nu, \quad \text{where} \quad E_m \leqq \mathscr{H}_e + \mathscr{H}_\nu \leqq E_m + \Delta E_m$$

The constant of normalization is determined from the total number $N$ of observed disintegrations:

$$\int_{E_m \leqq \mathscr{H}_e + \mathscr{H}_\nu \leqq E_m + \Delta E_m} n \, d\mathbf{p}_e d\mathbf{p}_\nu = N$$

The number of electrons in the interval $(\mathbf{p}_e, d\mathbf{p}_e)$ is obtained by integrating over $\mathbf{p}_\nu$, subject to the energy condition; the number of electrons $n(E_e) dE_e$ with energies in the interval $(E_e, dE_e)$ is found by a further integration over $\mathbf{p}_e$ with $E_e \leqq \mathscr{H}_e \leqq E_e + dE_e$.

Evaluate $n(E_e)$: the calculation should be performed using the relativistic Hamiltonian function, first for a finite neutrino rest mass $m_\nu$, then for the case $m_\nu = 0$.

2. *The relativistic Hamiltonian equations*

Derive the equations of motion (54.7) from the Hamiltonian function (55.7).

# CHAPTER FII

## Relativistic quantum mechanics

### §56. Relativistic wave equations

In non-relativistic quantum theory momentum and energy are represented by operators as follows:

$$\mathbf{p} = \frac{\hbar}{i}\frac{\partial}{\partial \mathbf{r}} \qquad E = -\frac{\hbar}{i}\frac{\partial}{\partial t} \tag{56.1}$$

These operators act on a wave function $\psi$. Then, retaining the classical relation between $\mathscr{H}$ and $\mathbf{p}$, and applying the energy theorem $\mathscr{H} = E$, we obtain the Schrödinger equation

$$\mathscr{H}(\mathbf{p},\mathbf{r})\psi = i\hbar\dot\psi \tag{56.2}$$

The relations (56.1) are also retained in the relativistic theory; they may be summarized by the four-vector $p_\mu$ in the form

$$p_\mu = \frac{\hbar}{i}\frac{\partial}{\partial x_\mu} \tag{56.1a}$$

which exhibits their invariant character.

We shall now attempt to establish a relativistic Schrödinger equation for the electron. The equation employed so far is not relativistically invariant. It contains derivatives of the second order with respect to the space coordinates, but a derivative of the first order with respect to time; the relativistically equivalent derivatives $\partial/\partial x_\mu$ are not treated in a symmetrical manner. If we wish to obtain a relativistic generalization of the original equation, there are two possibilities: the equation must either be of the second order in $\partial/\partial \mathbf{r}$ and so in $\partial/\partial t$ as well, or it remains linear in $\partial/\partial t$ and we must arrange to make it linear in $\partial/\partial \mathbf{r}$ also.

The first alternative together with (55.8) leads to the Klein-Gordon equation*

$$\sum_\mu \left(\frac{\hbar}{i}\frac{\partial}{\partial x_\mu} - \frac{e}{c}\Phi_\mu\right)^2 \psi = -m^2c^2\psi \tag{56.3}$$

or

$$\left\{\left(\mathbf{p} - \frac{e}{c}\mathbf{A}\right)^2 - \frac{1}{c^2}\left(\frac{\hbar}{i}\frac{\partial}{\partial t} + e\phi\right)^2\right\}\psi = -m^2c^2\psi \tag{56.3a}$$

* Klein, O., *Z. Phys.* **37** (1926), 895; Gordon, W., *Z. Phys.* **40** (1926), 117.

When the external forces vanish (i.e. $\Phi_\mu = 0$) the equation becomes

$$-\hbar^2 \nabla^2 \psi + \frac{\hbar^2}{c^2} \ddot{\psi} = -m^2 c^2 \psi \tag{56.3b}$$

If we now consider the second alternative, then following Dirac,* we can attempt to linearize the force-free form of equation (56.3) by assuming the following solution, which is known as Dirac's equation:

$$\left\{ \sum_\mu \gamma_\mu p_\mu \right\} \psi \equiv \left\{ \sum_\mu \gamma_\mu \frac{\hbar}{i} \frac{\partial}{\partial x_\mu} \right\} \psi = imc\psi \tag{56.4}$$

In the above expression the quantities $\gamma_\mu$ are constant operators that have yet to be determined. The manner of linearization must be such that the solutions of (56.4) also satisfy the force-free Klein-Gordon equation (56.3b), thus preserving the previous relation between energy and momentum (i.e. between the propagation vector and the frequency in the case of plane waves). If the operation $\sum_\mu \gamma_\mu p_\mu$ is again performed on (56.4) the result is

$$\sum_{\mu, \nu} \gamma_\mu \gamma_\nu p_\mu p_\nu \psi = -m^2 c^2 \psi \tag{56.5}$$

If this equation is to agree with (56.3b), this clearly requires that†

$$\tfrac{1}{2}(\gamma_\mu \gamma_\nu + \gamma_\nu \gamma_\mu) = \delta_{\mu\nu} \tag{56.6}$$

This equation cannot be satisfied by ordinary numbers $\gamma_\mu$, but as we shall see, it can be by square matrices $\gamma_{\mu, mn}$ of at least four rows. This means that in (56.4) $\psi$ must be a wave function consisting of a number of components: $\psi = (\psi_1, \psi_2, \ldots, \psi_n, \ldots)$. This is very satisfactory since the electron must be associated with additional degrees of freedom to take account of the spin. The $\gamma_\mu$ act on the components $\psi_n$ in the manner with which we are already familiar from the discussion of spin:

$$(\gamma_\mu \psi)_m = \sum_n \gamma_{\mu, mn} \psi_n$$

Later on we shall encounter special representations of the $\gamma_\mu$.

In the presence of external fields (56.4) is generalized to

$$\sum_\mu \gamma_\mu \left( p_\mu - \frac{e}{c} \Phi_\mu \right) \psi = imc\psi \tag{56.7}$$

---

* Dirac, P. A. M., *Proc. Roy. Soc.* **117** (1928), 610; **118**, 351.

† Only the symmetric part (56.6) occurs in (56.5), since $p_\mu p_\nu$ is symmetric in $\mu$ and $\nu$.

While (56.4) together with the relation (56.6) leads to the Klein-Gordon equation, in the case of (56.7) the same procedure yields additional terms:

$$\sum_{\mu,\nu} \gamma_\mu \gamma_\nu \left(p_\mu - \frac{e}{c}\Phi_\mu\right)\left(p_\nu - \frac{e}{c}\Phi_\nu\right)\psi = -m^2c^2\psi \qquad (56.8)$$

or after a simple transformation, taking account of (56.6),

$$\left\{\sum_\mu \left(p_\mu - \frac{e}{c}\Phi_\mu\right)^2 - \sum_{\mu,\nu} \frac{\hbar e}{4ic}(\gamma_\mu\gamma_\nu - \gamma_\nu\gamma_\mu)F_{\mu\nu}\right\}\psi = -m^2c^2\psi \qquad (56.8a)$$

where

$$F_{\mu\nu} = \frac{\partial \Phi_\nu}{\partial x_\mu} - \frac{\partial \Phi_\mu}{\partial x_\nu}$$

is the field strength tensor (54.2). Equation (56.8$a$) differs from (56.3) through an additional term that may be interpreted as the interaction of the spin with the external field.

In order to illustrate this more clearly, we introduce the vectors

$$\vec{\sigma}' = (\sigma'_1, \sigma'_2, \sigma'_3) \quad \text{and} \quad \vec{\alpha} = (\alpha_1, \alpha_2, \alpha_3)$$

the components of which are

$$\begin{array}{llll}
& \sigma'_1 = i\gamma_3\gamma_2, & \sigma'_2 = i\gamma_1\gamma_3, & \sigma'_3 = i\gamma_2\gamma_1 \\
\text{and} & \alpha_1 = i\gamma_4\gamma_1, & \alpha_2 = i\gamma_4\gamma_2, & \alpha_3 = i\gamma_4\gamma_3
\end{array} \qquad (56.9)$$

and confine ourselves to the non-relativistic case. We now assume a solution $\psi = \tilde{\psi}\exp - imc^2t/\hbar$ together with the approximation $|\dot{\tilde{\psi}}| \ll |mc^2\tilde{\psi}|$. We can then neglect $\dot{\tilde{\psi}}$ and consequently obtain

$$\left\{\frac{1}{2m}\left(\mathbf{p} - \frac{e}{c}\mathbf{A}\right)^2 + e\phi - \frac{e\hbar}{2mc}\vec{\sigma}'\mathbf{H} + i\frac{e\hbar}{2mc}\vec{\alpha}\mathbf{E}\right\}\psi = i\hbar\dot{\psi} \qquad (56.10)$$

This is the correct representation of the interaction between spin and field postulated in Pauli's theory of spin. The operators $\sigma'_i$ are very similar to the Pauli spin matrices $\sigma_i$; this is shown by the fact that they satisfy the same commutation relations

$$\sigma'_1\sigma'_2 - \sigma'_2\sigma'_1 = 2i\sigma'_3, \quad \text{etc., and} \quad \sigma'^2_k = 1 \qquad (56.11)$$

(as may be verified from (56.6)), and that the three components of $\sigma'$ can therefore have only the eigenvalues $+1$ and $-1$. In §60 we shall show that, apart from additional terms, (56.10) is actually identical with Pauli's spin theory.

It can be regarded as a great advance over the Schrödinger equation that the electron spin, including the commutation relations of the Pauli matrices, and the correct value of the magnetic moment follow automatically as it were from the relations (56.6) (cf. magneto-mechanical anomaly, p.194).

Nevertheless, serious difficulties are encountered in connection with the physical interpretation of the Klein-Gordon and Dirac equations as quantum-mechanical equations for the electron; that is, in taking $\psi$ to represent a state vector, as in the non-relativistic theory. We shall give a qualitative discussion of these difficulties in the following pages.

The Klein-Gordon equation does not possess the established form $\mathcal{H}\psi = i\hbar\dot{\psi}$; on the contrary, it is of the second order in $\partial/\partial t$. A knowledge of the state at a given instant is therefore insufficient to determine the situation at a later time; for this purpose we require the initial values of $\psi$ and $\dot{\psi}$. Again, it is not possible to define a probability density $\rho$ (represented by $\psi^*\psi$ in the non-relativistic case): to see this, we shall attempt to determine $\rho$ by establishing the relativistic generalization of the continuity equation $\dot{\rho} + \text{div } \mathbf{j} = 0$ (which implies the maintenance of the normalization $\int \rho \, d\tau$). Expressed in terms of the four-current* $j_\mu$, the continuity equation is

$$\sum_\mu \partial j_\mu / \partial x_\mu = 0$$

The only possible equation of this type is obtained by a similar procedure to that used in the case of the non-relativistic Schrödinger equation in the absence of external forces:

$$j_\mu = \frac{\hbar}{2mi}\left\{\psi^*\frac{\partial\psi}{\partial x_\mu} - \psi\frac{\partial\psi^*}{\partial x_\mu}\right\}$$

It may easily be verified that the continuity equation is valid in virtue of (56.3b). The resultant density

$$\rho = j_4/ic = \frac{i\hbar}{2mc^2}\{\psi^*\dot{\psi} - \dot{\psi}^*\psi\}$$

cannot be interpreted as a probability density because it can assume positive and negative values, and negative probability densities are obviously absurd.

This difficulty cannot be removed by restricting $\psi$ and $\dot{\psi}$ to those values for which $\rho$ is positive, because this positive value of $\rho$ generally

* Cf. Vol. I, §81.

gives rise in the course of time to a density which is still negative in places. We cannot therefore make use of the Klein-Gordon equation to describe a particle.

However, this equation possesses meaning when considered as a classical wave equation in which $e\rho$ appears as a charge density (where the charges may be positive or negative). The normal quantization methods may then be applied to it, similar to the procedure employed in the case of Maxwell's equations (§§51, 52). Particles associated in quantum mechanics with the "Klein-Gordon field" possess a mass $m$, charge $\pm e$, and no spin; $\pi$-mesons are an example.

The above difficulties do not appear in the Dirac equation. This equation is linear in $\partial/\partial t$; further, the corresponding equation of continuity enables a value of $\rho$ to be defined that is always positive (cf. following sections). On the other hand there is another difficulty, as we shall see later. It is found that the energy levels of the stationary states vary between $-\infty$ and $+\infty$; there is no state of lowest energy.* This means that it is impossible to find a completely positive expression for the energy density when the Dirac equation is interpreted as a classical wave equation; nor can the difficulty be eliminated by excluding the states of negative energy from consideration on the grounds that they are physically useless, because positive energy states can become states of negative energy in the presence of external fields.

Nevertheless, the Dirac equation may be "quantized" to provide a many-particle theory which at first sight also includes particles with negative energy possessing physically absurd properties (see §58). However, since the electrons satisfy Pauli's exclusion principle, Dirac was able to eliminate this difficulty by assuming that all the negative energy states were already filled and that therefore no transitions could take place between positive and negative states. This situation represents the vacuum state, that is, the state of lowest energy with zero charge and mass. An additional electron can assume only the unoccupied energy values, and a lowest state exists.†

The vacuum state is not empty; it possesses a lower stratum of occupied negative-energy levels which entail a number of physical

---

* The situation is different in the case of the Klein-Gordon equation. The energy density of the field theory is positive definite; in the quantized theory, therefore, the energy spectrum contains positive values only (in contrast to the frequency spectrum, which extends from $-\infty$ to $+\infty$).

† It may be shown directly that the occurrence of negative energies in the case of Dirac particles necessarily entails quantization in accordance with Pauli's principle. To some extent this proves the relation between spin and the statistics that was described in pp. 249-50.

properties. This can be most clearly seen from the fact that it is possible to produce electron and positron pairs. A light quantum of sufficient energy can raise an electron with negative energy into a positive energy state; the ground state is then short of a particle with negative energy and charge. What we should therefore be able to observe besides the excited electron is a new physical entity which, according to Dirac, behaves like a particle with positive energy and charge! It was in 1930 that Dirac advanced the hypothesis of the existence of "positrons"* on the strength of his bold and at first sight extremely surprising assumption with regard to the vacuum state. Such particles were in fact first observed two years later by Anderson in cosmic radiation.

Fundamentally, the Dirac and Klein-Gordon equations are similar in that each necessarily leads to the consideration of many particles. This is a general characteristic of all relativistic theories, which are represented in physical terms by quantized fields with many particles.

We shall, however, look upon the Dirac equation in an approximate manner as the Schrödinger equation for an electron, provided that we note that the states of negative energy are filled, and provided that we may neglect transitions from positive to negative energy states.

## §57. The physical interpretation of Dirac's equation

We shall first put the Dirac equation

$$\left\{\sum_\mu \gamma_\mu \left(p_\mu - \frac{e}{c}\Phi_\mu\right) - imc\right\}\psi = 0 \tag{57.1}$$

into the usual form $\mathscr{H}\psi = i\hbar\dot\psi$. This is achieved by applying $ic\gamma_4$ to (57.1), since $\gamma_4\gamma_4 = 1$. Using the definitions (56.9) together with

$$\beta = \gamma_4 \qquad \alpha_l = i\beta\gamma_l \qquad (l = 1, 2, 3) \tag{57.2}$$

or

$$\vec{\alpha} = i\beta\vec{\gamma} \tag{57.2a}$$

we obtain the following equation:

$$\left\{\beta mc^2 + c\vec{\alpha}\left(\mathbf{p} - \frac{e}{c}\mathbf{A}\right) + e\phi\right\}\psi = i\hbar\dot\psi \tag{57.3}$$

---

* Dirac first thought that there was a connection between the "holes" in the ground state and the proton, which was then the only other particle known besides the electron. However, it was soon shown by R. Oppenheimer (*Phys. Rev.* 35 (1930), 461) and H. Weyl (1931) that the mass of a "hole" must be equal to the mass of the electron.

Thus in this expression the Hamiltonian operator is given by

$$\mathscr{H} = \beta mc^2 + c\vec{\alpha}\left(\mathbf{p} - \frac{e}{c}\mathbf{A}\right) + e\phi \qquad (57.3a)$$

$\mathscr{H}$ must be Hermitian in order to possess real eigenvalues. We therefore choose the matrices $\beta$ and $\alpha_l$ to be Hermitian; the $\gamma_\mu$ are also Hermitian in consequence of the definitions (57.2):

$$\beta^+ = \beta \qquad \alpha_l^+ = \alpha_l \qquad \gamma_\mu^+ = \gamma_\mu \qquad (57.4)$$

or expressed in terms of the matrix elements,

$$\beta_{mn}^* = \beta_{nm} \qquad \alpha_{l,mn}^* = \alpha_{l,nm} \qquad \gamma_{\mu,mn}^* = \gamma_{\mu,nm} \qquad (57.4a)$$

In virtue of (57.2) and (56.6), the $\alpha_l$ and $\beta$ satisfy the relations

$$\beta^2 = 1 \qquad \alpha_l\beta + \beta\alpha_l = 0 \qquad \alpha_l\alpha_k + \alpha_k\alpha_l = 2\delta_{kl} \qquad (57.5)$$

Hence in view of (57.4) and (57.5) the $\alpha_l, \beta, \gamma_\mu$ are unitary as well as Hermitian matrices.

We can now establish an equation of continuity, using a similar method to that employed in connection with the non-relativistic Schrödinger equation. Equation (57.3) is first expressed in terms of the components:

$$\sum_n \left\{ \beta_{mn} mc^2 + c\vec{\alpha}_{mn}\left(p - \frac{e}{c}A\right) + e\phi\,\delta_{mn} \right\} \psi_n = i\hbar\dot{\psi}_m$$

This is multiplied by $\psi_m^*$, summed over $m$, and the conjugate complex removed. Then in view of the relations (57.4a) the result is

$$\frac{\partial\rho}{\partial t} + \text{div}\,\mathbf{j} = 0$$

where
$$\rho = \sum_m \psi_m^* \psi_m \qquad \mathbf{j} = \sum_{m,n} c\vec{\alpha}_{mn} \psi_m^* \psi_n \qquad (57.6)$$

This continuity equation enables $\rho(\mathbf{r},t)\,d\mathbf{r}$ to be interpreted as the probability of the occurrence of the electron in the volume element $d\mathbf{r}$.

We can now calculate the centroid $\bar{\mathbf{r}} = \int \mathbf{r}\rho\,d\mathbf{r}$ of a wave packet together with its time rate of change $\dot{\mathbf{r}}$. It is apparent from §22 that we merely need to form the operator $\dot{\mathbf{r}} = (i/\hbar)[\mathscr{H}, \mathbf{r}]$, when $\dot{\bar{\mathbf{r}}} = \dot{\mathbf{r}}$. Then using (57.3a) and the usual commutation relations for $\mathbf{p}$ and $\mathbf{r}$ we obtain

$$\dot{\mathbf{r}} = \frac{i}{\hbar}[\mathscr{H}, \mathbf{r}] = c\vec{\alpha} \qquad (57.7)$$

Although this operator has exactly the form that we would expect from the continuity equation ($\bar{\mathbf{r}} = \int \mathbf{j}(\mathbf{r}, t)\, d\mathbf{r}$), it has some unusual properties which we shall now describe.

Since the $\alpha_l$ are Hermitian and unitary, they can have only the eigenvalues $\pm 1$; an exact measurement of a velocity component would therefore always yield the value $+c$ or $-c$. Further, the components $\dot{x}$, $\dot{y}$, $\dot{z}$ do not commute with each other or with $\mathscr{H}$ in the absence of the external forces $\Phi_\mu$. Thus there are no free particles with sharply defined velocity.

These properties of the operator $\dot{\mathbf{r}}$ indicate that it has very little in common with the corresponding operator $\{\mathbf{p} - (e/c)\mathbf{A}\}/m$ occurring in the non-relativistic theory, particularly since in experimental determinations of the velocity components any values may be found between $-c$ and $+c$.

We shall see in §60, however, that there is an operator $\mathbf{v}$ that is essentially different from $c\bar{\alpha}$, and which becomes the ordinary velocity operator in the non-relativistic case; this operator is directly connected with the experimentally determined velocity.

We shall now consider the time rate of change of the operator $\{\mathbf{p} - (e/c)\mathbf{A}\}$; since $c\bar{\alpha} = \dot{\mathbf{r}}$, this is

$$\frac{d}{dt}\left(\mathbf{p} - \frac{e}{c}\mathbf{A}\right) = \frac{i}{\hbar}\left[\mathscr{H}, \mathbf{p} - \frac{e}{c}\mathbf{A}\right] - \frac{e}{c}\frac{\partial \mathbf{A}}{\partial t} = e\mathbf{E} + \frac{e}{c}\dot{\mathbf{r}} \times \mathbf{H} \quad (57.8)$$

Although the form of this equation is similar to that of the classical equation of motion (54.7) its significance is not the same, because of the difference mentioned above between the mean momentum and the velocity of the centroid multiplied by $m$ (cf. §60).

For a free particle (i.e. one for which $\mathbf{A} = \phi = 0$):

$$\frac{d\mathbf{p}}{dt} = \frac{i}{\hbar}[\mathscr{H}, \mathbf{p}] = 0$$

Therefore although there are no free particles with a sharp value of $\dot{\mathbf{r}}$, such particles do exist with sharply defined values of $\mathbf{p}$; we shall consider these in more detail in the next paragraph.

We shall now investigate the conservation law for the orbital angular momentum of a free particle, $\mathbf{L} = \mathbf{r} \times \mathbf{p}$. Since $\mathscr{H}$ and $\mathbf{p}$ commute we have

$$\dot{\mathbf{L}} = \frac{i}{\hbar}[\mathscr{H}, \mathbf{L}] = \frac{i}{\hbar}[\mathscr{H}, \mathbf{r}] \times \mathbf{p} = \dot{\mathbf{r}} \times \mathbf{p} \quad (57.9)$$

The above equation shows that the orbital angular momentum of a free particle is not constant in time. This is also true of the spin, for which we may tentatively assign the value $s = \frac{1}{2}\hbar\vec{\sigma}'$, where $\vec{\sigma}'$ is defined by (56.9). Then from (57.5) and the relations $\sigma'_1 = i\alpha_3\alpha_2$, etc., we obtain

$$\dot{s} = \frac{i}{\hbar}[\mathcal{H}, s] = -\dot{r} \times p \qquad (57.10)$$

If we compare this result with (57.9) we see that the total angular momentum $J = L + s$ is constant in time (cf. Exercise 1, p. 359).

We shall now derive the form of the matrices $\gamma_\mu$, $\alpha_l$, and $\beta$. These are merely required to satisfy the relations (56.6) and (57.5) and the Hermitian condition (57.4). It follows in the first instance that the matrices are determined except for unitary transformations $U\gamma_\mu U^+$ which leave (56.6) and (57.5) unchanged.

We can first choose $U$ so that $\beta$ is diagonal. Since $\beta$ is unitary and Hermitian it can have only the eigenvalues $\pm 1$, which can be so arranged that

$$\beta = \begin{pmatrix} 1 & 0 \\ 0 & -1 \end{pmatrix} = \begin{pmatrix} 1 & & & & \vdots & 0......0 \\ & 1 & & & \vdots & \vdots \\ & & \ddots & & \vdots & \vdots \\ & & & 1 & \vdots & 0 \quad\quad 0 \\ \cdots & \cdots & \cdots & \cdots & & \cdots\cdots \\ 0......0 & & & & \vdots & -1 \\ \vdots & & & & \vdots & \ddots \\ 0......0 & & & & \vdots & & -1 \end{pmatrix} \qquad (57.11)$$

In the first matrix the figure 1 represents unit matrices, and the figure 0 stands for null matrices possessing an initially unknown number of rows and columns. Now it follows from the relations $\alpha_k\beta + \beta\alpha_k = 0$ and $\alpha_k\alpha_k^+ = 1$ that

$$\alpha_k \beta \alpha_k^+ = -\beta$$

Then, forming the matrix traces, we have

$$\mathrm{Tr}(-\beta) = \mathrm{Tr}\,\alpha_k\beta\alpha_k^+ = \mathrm{Tr}\,\alpha_k^+\alpha_k\beta = \mathrm{Tr}\,\beta, \text{ hence } \mathrm{Tr}\,\beta = 0$$

This means that, in (57.11), 1 and $-1$ have the same number of columns, and 0 is a square matrix. It then follows from the relations $\alpha_k\beta + \beta\alpha_k = 0$ and $\alpha_k = \alpha_k^+$ that $\alpha_k$ must have the form

$$\alpha_k = \begin{pmatrix} 0 & \alpha^{(k)} \\ \alpha^{(k)+} & 0 \end{pmatrix} \qquad (57.12)$$

with corresponding square matrices $\alpha^{(k)}$ and 0. From (56.9) the $\sigma'_k$ assume the form

$$\sigma'_k = \begin{pmatrix} \sigma_k & 0 \\ 0 & \tilde{\sigma}_k \end{pmatrix} \tag{57.13}$$

From (56.11) we know that the quantities $\frac{1}{2}\hbar\sigma'_k$ satisfy the familiar commutation relations for the angular momentum; so do $\frac{1}{2}\hbar\sigma_k$ and $\frac{1}{2}\hbar\tilde{\sigma}_k$ in view of (57.13). They must therefore be the usual angular-momentum matrices, in accordance with §24. Further, we know by definition that $\sigma'^2_k = 1$, so that $\sigma^2_k = \tilde{\sigma}^2_k = 1$ also; therefore the only quantum number for the angular momentum arising is $l = \frac{1}{2}$. $\sigma_k$ and $\tilde{\sigma}_k$ may then be put into the form of the Pauli spin matrices by means of a unitary transformation of $\sigma'_k$

$$U = \begin{pmatrix} U_1 & 0 \\ 0 & U_2 \end{pmatrix}$$

which leaves $\beta$ unchanged: then*

$$\vec{\sigma}' = \begin{pmatrix} \vec{\sigma} & 0 \\ 0 & \vec{\sigma} \end{pmatrix} \tag{57.14}$$

The $\alpha_l$ may now be determined by introducing the matrix

$$\alpha_0 = -i\alpha_1 \alpha_2 \alpha_3 = \begin{pmatrix} 0 & \alpha^{(0)} \\ \alpha^{(0)+} & 0 \end{pmatrix} \tag{57.15}$$

This matrix is also unitary and Hermitian, and commutes with all $\alpha_l$ and $\sigma_l$, as may be shown from (56.9) and (57.5). Hence $\alpha^{(0)}$ can only have the form $e^{i\phi}\begin{pmatrix} 1 & 0 \\ 0 & 1 \end{pmatrix}$ (cf. Exercise 3, p. 360). Then by means of a unitary transformation

$$U = \begin{pmatrix} e^{i\phi/2} \cdot 1 & 0 \\ 0 & e^{-i\phi/2} \cdot 1 \end{pmatrix}$$

which leaves $\beta$ and $\tilde{\alpha}$ unchanged, $\alpha_0$ may be put into the form

$$\alpha_0 = \begin{pmatrix} 0 & 1 \\ 1 & 0 \end{pmatrix} \tag{57.15a}$$

---

* Trivial generalizations of the form

$$\vec{\sigma}' = \begin{bmatrix} \vec{\sigma} & 0 & & \\ 0 & \vec{\sigma} & & \\ & & \vec{\sigma} & 0 \\ & & 0 & \vec{\sigma} \end{bmatrix}$$

would be conceivable; these would simply lead to two completely similar unconnected Dirac equations.

From (56.9) and (57.15) we finally obtain

$$\alpha_l = \alpha_0 \, \sigma_l' = \begin{pmatrix} 0 & \sigma_l \\ \sigma_l & 0 \end{pmatrix}$$

which completes our analysis.

We thus have the following result. The $\alpha_l$, $\beta$, and $\gamma_\mu$ are uniquely determined by means of the relations (56.6) and (57.5), apart from unitary transformations (see footnote, p. 337). They are $4 \times 4$ matrices, and may be put into the form

$$\beta = \begin{pmatrix} 1 & 0 \\ 0 & -1 \end{pmatrix} \qquad \vec{\alpha} = \begin{pmatrix} 0 & \vec{\sigma} \\ \vec{\sigma} & 0 \end{pmatrix} \qquad (57.16)$$

where $\vec{\sigma}$ represents the usual Pauli matrices (24.9).

In connection with later applications, it is useful to introduce the following designations. Matrices in the form of $\beta$ and $\sigma_l'$, which contain null matrices in their secondary diagonal, are called $D$-matrices, while those having the form of $\alpha_l$ with null matrices in the principal diagonal are termed $A$-matrices.* The following rules apply to the products of such matrices:

$$\beta A + A\beta = 0 \qquad \beta D - D\beta = 0 \qquad (57.17)$$

$D \cdot D'$ gives a $D$ matrix, $A \cdot A'$ gives a $D$ matrix, $A \cdot D$ gives an $A$ matrix.

## §58. The Dirac equation in the absence of an external field

We shall now look for solutions of the equation

$$(c\vec{\alpha}\mathbf{p} + \beta mc^2)\psi = i\hbar\dot{\psi} \qquad (58.1)$$

We first investigate the stationary states $\psi = \phi e^{-i\omega t}$:

$$(c\vec{\alpha}\mathbf{p} + \beta mc^2)\phi = \hbar\omega\phi \qquad (58.2)$$

Since $\mathbf{p}$ and $\mathcal{H}$ commute, as we have seen, $\phi$ may also be assumed to be an eigenfunction of $\mathbf{p}$, or expressed in terms of components:

$$\phi_n(\mathbf{k}, \mathbf{r}) = a_n(\mathbf{k}) \, e^{i\mathbf{k}\mathbf{r}} \qquad (58.3)$$

In the above expression $\hbar\mathbf{k}$ is the eigenvalue of $\mathbf{p}$ and $\hbar\omega = E$ is the eigenvalue of $\mathcal{H}$.

When (58.3) is inserted in (58.2) we obtain the eigenvalue equation

$$\sum_n \{\vec{\alpha}c\hbar\mathbf{k} + \beta mc^2\}_{mn} a_n(\mathbf{k}) = Ea_m(\mathbf{k}) \qquad (58.4)$$

* Translator's note: The German for "secondary diagonal" is "Ausserdiagonal".

Using the relations (57.16), this equation may be expressed in terms of components as follows:

$$
\begin{aligned}
(E-mc^2)a_1 - c\hbar k_3\, a_3 - c\hbar(k_1-ik_2)a_4 &= 0 \\
(E-mc^2)a_2 - c\hbar(k_1+ik_2)a_3 + c\hbar k_3\, a_4 &= 0 \\
(E+mc^2)a_3 - c\hbar k_3\, a_1 - c\hbar(k_1-ik_2)a_2 &= 0 \\
(E+mc^2)a_4 - c\hbar(k_1+ik_2)a_1 + c\hbar k_3\, a_2 &= 0
\end{aligned}
\tag{58.5}
$$

The relativistic energy formula

$$
E^2 = (c\hbar k)^2 + m^2c^4 \tag{58.6}
$$

results for the eigenvalues $E$, if the determinant of the above system of equations is set equal to 0, or if the operator $\{c\tilde{\alpha}\hbar\mathbf{k} + \beta mc^2\}$ is simply applied again to (58.4), and the relations (57.5) are employed. Hence

$$
E = \pm E(\hbar\mathbf{k}) = \pm\sqrt{\{(c\hbar k)^2 + m^2c^4\}} \tag{58.7}
$$

Equating the determinant to zero yields an equation of the fourth degree in $E$; there are therefore four solutions, of which two pairs coincide.[*] These are designated (figure 56)

$$
E_1 = E_2 = E(\hbar\mathbf{k}) \qquad E_3 = E_4 = -E(\hbar\mathbf{k})
$$

Since the determinantal equation possesses double roots we are free to choose any two of the components $a_n$ as we wish, the other two then being determined by (58.5). One possible choice is

$$
\begin{pmatrix} a_1^1 \ldots\ldots a_1^4 \\ \vdots \qquad \vdots \\ a_4^1 \ldots\ldots a_4^4 \end{pmatrix} = a
\begin{pmatrix} 1 & 0 & & ? \\ 0 & 1 & & \\ & & 1 & 0 \\ ? & & 0 & 1 \end{pmatrix}
$$

where the $a_n^s$ are the components associated with $E_s$. The as yet undetermined factor $a$ is chosen so as to normalize $a_n^s$ (i.e. $\sum\limits_n |a_n^s|^2 = 1$). Then the solution of (58.5) gives

$$
a_n^s = \frac{1}{\sqrt{\{2E(\hbar\mathbf{k})[E(\hbar\mathbf{k})+mc^2]\}}} b_n^s
$$

---

[*] From (58.7) there are only two different eigenvalues $E_i$. In addition, $\sum\limits_{i=1}^{4} E_i = 0$, because the trace of the secular matrix vanishes (Tr $\beta$ = Tr $\alpha$ = 0).

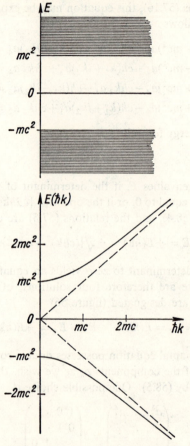

Fig. 56.—The energy spectrum of the Dirac equation for a free particle

where the values of $b_n^s$ are given in the following table:

| $n$ \ $s$ | 1 | 2 | 3 | 4 |
|---|---|---|---|---|
| 1 | $E(\hbar k) + mc^2$ | 0 | $-c\hbar k_3$ | $-c\hbar(k_1 - ik_2)$ |
| 2 | 0 | $E(\hbar k) + mc^2$ | $-c\hbar(k_1 + ik_2)$ | $c\hbar k_3$ |
| 3 | $c\hbar k_3$ | $c\hbar(k_1 - ik_2)$ | $E(\hbar k) + mc^2$ | 0 |
| 4 | $c\hbar(k_1 + ik_2)$ | $-c\hbar k_3$ | 0 | $E(\hbar k) + mc^2$ |

The components $a_n^s$ may be expressed more concisely, using the relations (57.16):

$$a_n^s(\hbar\mathbf{k}) = \left\{ \frac{c\tilde{\alpha}\hbar\mathbf{k} + \beta\left[E(\hbar\mathbf{k}) + mc^2\right]}{\sqrt{\{2E(\hbar\mathbf{k})\left[E(\hbar\mathbf{k}) + mc^2\right]\}}} \beta \right\}_{ns}$$

$$= \frac{\{\mathscr{H}\beta + E(\hbar\mathbf{k})\}_{ns}}{\sqrt{\{2E(\hbar\mathbf{k})\left[E(\hbar\mathbf{k}) + mc^2\right]\}}} \qquad (58.8)$$

The $a_n^s$ clearly form an orthogonal system, since the matrix (58.8) is unitary because

$$[\mathscr{H}\beta + E(\hbar\mathbf{k})][\mathscr{H}\beta + E(\hbar\mathbf{k})]^+ = \mathscr{H}\beta^2\mathscr{H} + (\mathscr{H}\beta + \beta\mathscr{H})E(\hbar\mathbf{k}) + E^2(\hbar\mathbf{k})$$

$$= 2E(\hbar\mathbf{k})[E(\hbar\mathbf{k}) + mc^2]$$

Further, we can see from (58.8) that the $a_n^s$ satisfy equation (58.4), since

$$\mathscr{H}[(\mathscr{H}\beta + E(\hbar\mathbf{k})] = E^2(\hbar\mathbf{k})\beta + \mathscr{H}E(\hbar\mathbf{k}) = [\mathscr{H}\beta + E(\hbar\mathbf{k})]\beta E(\hbar\mathbf{k})$$

which is the same as (58.4) except for the factor

$$\sqrt{\{2E(\hbar\mathbf{k})[E(\hbar\mathbf{k}) + mc^2]\}}$$

For our further calculations we shall again assume a periodicity box* of volume $V = L^3$. We can then represent all functions $\psi(\mathbf{r},t)$ in this box in terms of the complete system

$$\phi_n^s(\mathbf{k},\mathbf{r}) = a_n^s(\mathbf{k})\frac{e^{i\mathbf{k}\mathbf{r}}}{\sqrt{V}}, \quad \text{where} \quad \mathbf{k} = \frac{2\pi}{L}\mathbf{n} \qquad (58.9)$$

$$\psi_n(\mathbf{r},t) = \sum_{\mathbf{k},s} b_s(\mathbf{k},t)\,\phi_n^s(\mathbf{k},\mathbf{r}) \qquad (58.10)$$

The dependence on time of the coefficients $b_s(\mathbf{k},t)$ is found by introducing (58.10) into the Dirac equation. In the absence of external fields we obtain the following result:

$$E_s(\hbar\mathbf{k})b_s(\mathbf{k},t) = i\hbar\dot{b}_s(\mathbf{k},t)$$

or $\qquad\qquad b_s(\mathbf{k},t) = b_s(\mathbf{k},0)\,e^{-i\omega_s(\mathbf{k})t} \qquad (58.11)$

---

* This box distinguishes the system of reference in which it is at rest from all other systems. However, since we are not interested in the properties of the quantities subject to Lorentz transformations, this need not concern us any further.

We give below the expressions for the expectation values of the energy, momentum, charge density, and total charge, corresponding to (58.10):

$$\tilde{\mathscr{H}} = \int \sum_{m,n} \psi_m^* \{c\tilde{\alpha}\mathbf{p} + \beta mc^2\}_{mn} \psi_n \, d\mathbf{r} = \sum_{\mathbf{k},s} \hbar\omega_s(\mathbf{k}) b_s^*(\mathbf{k}) b_s(\mathbf{k}) \quad (58.12a)$$

$$\bar{\mathbf{G}} = \int \sum_m \psi_m^* \mathbf{p}\psi_m \, d\mathbf{r} = \sum_{\mathbf{k},s} \hbar\mathbf{k} b_s^*(\mathbf{k}) b_s(\mathbf{k}) \quad (58.12b)$$

$$\tilde{\rho} = e\sum_m \psi_m^* \psi_m = e \sum_{\substack{m,\mathbf{k},\mathbf{k}' \\ s,s'}} b_s^*(\mathbf{k}) b_{s'}(\mathbf{k}') \phi_m^{*s}(\mathbf{k},\mathbf{r}) \phi_m^{s'}(\mathbf{k}',\mathbf{r}) \quad (58.12c)$$

$$\tilde{Q} = e\int \sum_m \psi_m^* \psi_m \, d\mathbf{r} = e\sum_{\mathbf{k},s} b_s^*(\mathbf{k}) b_s(\mathbf{k}) \quad (58.12d)$$

We may summarize the foregoing analysis as follows. In the absence of external forces the solutions of the Dirac equation are superimposed plane waves. Four linear independent solutions are associated with a given value of $\mathbf{k}$, of which two correspond to positive energy and two to negative energy: the spectrum is illustrated in figure 56. The negative energy states cannot be ignored, because they are definitely part of the complete system represented by (58.9), and when external fields are present transitions can take place between states of positive and negative energy.

In spite of the above remarks, the negative energy states are physically impossible systems, as will be seen from the following example. If we consider a wave packet consisting of such states, with wave numbers in the neighbourhood of $\mathbf{k}_0$, it moves with a group velocity $\mathbf{v}_0 = [\partial\omega_s(\mathbf{k})/\partial\mathbf{k}]_0$; this means that when $\mathbf{k}_0$ is small, $\mathbf{v}_0 = -\hbar\mathbf{k}_0/m$, and the momentum of the wave packet, $\hbar\mathbf{k}_0$, has the opposite direction to $\mathbf{v}$. If we consider a collision between the wave train and ordinary matter of mass $M$, at rest, the energy and momentum conservation laws yield the following equations for the velocities $\mathbf{v}$ and $\mathbf{V}$ of the wave packet and the mass (considering motion in one dimension only, and using the non-relativistic approximation):

$$-mv_0^2 = MV^2 - mv^2 \qquad -mv_0 = MV - mv$$

Hence

$$V = -\frac{2m}{M-m} v_0 \qquad v = -\frac{M+m}{M-m} v_0$$

This means that after the collision $v$ is always greater than $v_0$ while $M$ is now in motion. The wave packet is therefore accelerated by each collision with matter at rest. We indicated in §56 how Dirac was able to circumvent this difficulty, and shall now give a more precise account of his ideas on the subject of the quantization of the Dirac equation.

## §59. The quantum theory of Dirac's field equations

If we interpret the Dirac equation as the classical field equation of a wave function $\psi$ rather than as a quantum-mechanical equation representing a particle (as we did in the case of the wave equation in §16), it can then be quantized in the same way as Maxwell's equations, and the properties of the particles associated with the field may be studied. We are indeed forced to resort to this procedure, since only by so doing can we obtain a consistent representation of many particles required to eliminate the difficulty of the negative energy states. According to this interpretation, the expectation values given by (58.12) are the total energy, momentum, charge density, and charge of the $\psi$ field, and the normalization condition $\int \rho \, d\mathbf{r} = e$ does not apply. Since the procedure was fully discussed in §52 for the Maxwell equations we can deal with it more briefly here.

The quantization is accomplished for the Dirac field alone, without taking any account of electromagnetic fields. For this purpose we make use of the Dirac field expansion given by (58.10). The equations of motion (58.11) for the coefficients of the expansion indicate that $\mathscr{H}$ in (58.12a) may be interpreted as a Hamiltonian function if $i\hbar b_s^*(\mathbf{k})$ is the momentum canonically conjugate to $b_s(\mathbf{k})$.

In proceeding to the quantum theory the time-dependent variables $b_s^*(\mathbf{k})$ and $b_s(\mathbf{k})$ of the field theory become the Hermitian conjugate operators $b_s^+(\mathbf{k})$ and $b_s(\mathbf{k})$ in the quantum field theory. Similarly, the physical quantities specified in (58.12) acquire the character of operators:

$$\tilde{\mathscr{H}} = \sum_{\mathbf{k},s} \hbar\omega_s(\mathbf{k}) \, b_s^+(\mathbf{k}) \, b_s(\mathbf{k}) \tag{59.1a}$$

$$\tilde{\mathbf{G}} = \sum_{\mathbf{k},s} \hbar\mathbf{k} b_s^+(\mathbf{k})_s \, b(\mathbf{k}) \tag{59.1b}$$

$$\tilde{\rho} = e \sum_{\mathbf{k},\mathbf{k}';\, s,s':n} b_s^+(\mathbf{k}) b_{s'}(\mathbf{k}) \, \phi_n^{*s}(\mathbf{k},\mathbf{r}) \phi_n^{s'}(\mathbf{k}',\mathbf{r}) \tag{59.1c}$$

$$\tilde{Q} = e \sum_{\mathbf{k},s} b_s^+(\mathbf{k}) \, b_s(\mathbf{k}) \tag{59.1d}$$

If we were to employ the usual commutation relations between canonically conjugate quantities,

$$b_{s'}(\mathbf{k}') \, b_s^+(\mathbf{k}) - b_s^+(\mathbf{k}) \, b_{s'}(\mathbf{k}') = \delta_{ss'} \delta_{\mathbf{k}\mathbf{k}'} \tag{59.2}$$

$$b_{s'}(\mathbf{k}') \, b_s(\mathbf{k}) - b_s(\mathbf{k}) \, b_{s'}(\mathbf{k}') = b_{s'}^+(\mathbf{k}') \, b_s^+(\mathbf{k}) - b_s^+(\mathbf{k}) \, b_{s'}^+(\mathbf{k}')$$
$$= 0$$

we could construct the Hilbert space in which the operators act in the same way as we did in the case of light. As a result of the commutation relations the quantities $b_s^+(\mathbf{k}) b_s(\mathbf{k})$ would have the integral eigenvalues $n_s(\mathbf{k}) = 0, 1, 2, 3, \ldots$ These would represent the occupation numbers of the states designated by $\mathbf{k}$ and $s$: at any given time there would be $n_s(\mathbf{k})$ particles of type $s$, $\mathbf{k}$ in a state $\psi_{\ldots n_s(\mathbf{k}) \ldots}$

We can see at once, however, that this quantization procedure is useless for two reasons. For one thing the particles arising from it do not satisfy the Fermi statistics that must be postulated for electrons, and in which the only possible occupation numbers are $n_s(\mathbf{k}) = 0$ and 1. Further, as in the case of the solutions of the Dirac equation there is no state of lowest energy

$$E = \sum_{\mathbf{k}, s = 1 \ldots 4} \hbar \omega_s(\mathbf{k}) n_s(\mathbf{k})$$

since the $\omega_s(\mathbf{k})$ can assume positive and negative values and the $n_s(\mathbf{k})$ can become arbitrarily great. In fact, the quantization procedure that we have described leads to the Bose-Einstein statistics. This can most easily be perceived by constructing the elements of Hilbert space by means of generating operators. The "vacuum state" $\tilde{\Psi}_0$, in which no particles are present, is defined by

$$b_s(\mathbf{k}) \tilde{\Psi}_0 = 0 \text{ for all } s, \mathbf{k} \tag{59.3}$$

The general element

$$b_{s_1}^+(\mathbf{k}_1) \ldots b_{s_n}^+(\mathbf{k}_n) \tilde{\Psi}_0 \tag{59.4}$$

arises from the application of the generation operators to $\tilde{\Psi}_0$ and is symmetric in the quantum numbers $s_i$, $\mathbf{k}_i$ in virtue of the commutation relations (59.2).

Assuming the validity of the Fermi statistics, the general element (59.4) should be antisymmetric. Following P. Jordan and E. Wigner (1928), this can be achieved by substituting a plus sign for the minus sign everywhere in the commutation relations (59.2), which are thus radically altered:*

$$b_{s'}(\mathbf{k}') b_s^+(\mathbf{k}) + b_s^+(\mathbf{k}) b_{s'}(\mathbf{k}') = \delta_{ss'} \delta_{\mathbf{k}\mathbf{k}'} \tag{59.5}$$

$$b_{s'}(\mathbf{k}') b_s(\mathbf{k}) + b_s(\mathbf{k}) b_{s'}(\mathbf{k}') = b_{s'}^+(\mathbf{k}') b_s^+(\mathbf{k}) + b_s^+(\mathbf{k}) b_{s'}^+(\mathbf{k}')$$
$$= 0$$

---

* Note that in this case $b_s(\mathbf{k}) b_s(\mathbf{k}) = 0$ and $b_s^+(\mathbf{k}) b_s^+(\mathbf{k}) = 0$.

It can be seen that this arrangement is self-consistent from the fact that a Hilbert space can again be constructed in a similar manner to before. Equations (59.3) and (59.4) remain unchanged (see Exercise 4, p. 360). The element (59.4) is clearly antisymmetric; it differs from zero only for $s_1, \mathbf{k}_1 \neq s_2, \mathbf{k}_2 \neq \ldots \neq s_n, \mathbf{k}_n$, and the non-zero elements are normalized with $\tilde{\Psi}_0$. As before, $b_s^+(\mathbf{k})$ and $b_s(\mathbf{k})$ act as particle generating and annihilating operators, and the eigenvalues of the "particle number operator" $b_s^+(\mathbf{k}) b_s(\mathbf{k})$ are $n_s(\mathbf{k}) = 0, 1$. (Cf. Exercise 4, p. 360.)

The difficulties associated with the states of negative energy may now be avoided by means of Dirac's postulate that these states are filled when the system is in the ground state. The ground or vacuum state $\Psi_0$ is then no longer defined by (59.3), but as follows:*

$$b_s(\mathbf{k})\Psi_0 = 0 \quad \text{for} \quad s = 1, 2 \text{ and all } \mathbf{k} \tag{59.6a}$$

$$b_s^+(\mathbf{k})\Psi_0 = 0 \quad \text{for} \quad s = 3, 4 \text{ and all } \mathbf{k} \tag{59.6b}$$

In the state $\Psi_0$, the expectation or eigenvalues of the operators (59.1) are undefined: for instance, we should obtain a "vacuum value" $-\infty$ for $\tilde{\mathscr{H}}$. On the other hand, if we wish to interpret $\Psi_0$ as the vacuum state, we must postulate that in this state the energy, momentum, and charge all vanish. In order to express this mathematically we subtract the appropriate vacuum expectation values from the operators (59.1), thus obtaining new operators, the values of which vanish in the state† $\Psi_0$. We shall illustrate the procedure by forming the new Hamiltonian operator $\mathscr{H}$, which we obtain from $\tilde{\mathscr{H}}$ by subtracting the vacuum expectation value $\sum_{\mathbf{k}, s=3,4} \hbar\omega_s(\mathbf{k})$:

$$\mathscr{H} = \tilde{\mathscr{H}} - \sum_{\mathbf{k}, s=3,4} \hbar\omega_s(\mathbf{k}) = \sum_{\mathbf{k}, s=1,2} \hbar\omega_s(\mathbf{k}) b_s^+(\mathbf{k}) b_s(\mathbf{k}) +$$
$$+ \sum_{\mathbf{k}, s=3,4} \hbar\omega_s(\mathbf{k}) \{b_s^+(\mathbf{k}) b_s(\mathbf{k}) - 1\}$$

---

* The relation between $\Psi_0$ and $\tilde{\Psi}_0$ is given symbolically by

$$\Psi_0 = \prod_{\substack{\mathbf{k} \\ s=3,4}} b_s^+(\mathbf{k}) \tilde{\Psi}_0$$

in which the product extends over all values of $\mathbf{k}$. If $\Psi_0$ is acted on by the operator $b_{s''}^+(\mathbf{k}'')$, where $s'' = 3, 4$, the result must be zero in view of the commutation relations. (Note that $b_s^+(\mathbf{k}) b_s^+(\mathbf{k}) = 0$.)

† Since the vacuum values are mathematically undefined divergent expressions, this subtraction procedure is open to objection. It should be considered merely as a heuristic method for establishing a physically-rational relativistically-invariant system of equations (59.9) for many particles (cf. renormalization, p. 319).

Applying the commutation relation (59.5), we obtain

$$\mathcal{H} = \sum_{k,s=1,2} \hbar\omega(\mathbf{k})\, b_s^+(\mathbf{k})\, b_s(\mathbf{k}) + \sum_{k,s=3,4} \hbar\omega(\mathbf{k})\, b_s(\mathbf{k})\, b_s^+(\mathbf{k}) \qquad (59.7a)$$

where

$$\omega(\mathbf{k}) = \omega_{1,2}(\mathbf{k}) = -\omega_{3,4}(\mathbf{k})$$

We similarly obtain

$$\mathbf{G} = \sum_{k,s=1,2} \hbar\mathbf{k}\, b_s^+(\mathbf{k})\, b_s(\mathbf{k}) - \sum_{k,s=3,4} \hbar\mathbf{k}\, b_s(\mathbf{k})\, b_s^+(\mathbf{k}) \qquad (59.7b)$$

$$\rho = e \sum_{\substack{kk' \\ n}} \left\{ \sum_{\substack{s=1,2 \\ s'=1,2}} + \sum_{\substack{s=1,2 \\ s'=3,4}} + \sum_{\substack{s=3,4 \\ s'=1,2}} \right\} b_s^+(\mathbf{k})\, b_{s'}(\mathbf{k}')\, \phi_n^{*s}(\mathbf{k})\, \phi_n^{s'}(\mathbf{k}') -$$

$$- e \sum_{\substack{k,k' \\ s,s'=3,4 \\ n}} b_{s'}(\mathbf{k}')\, b_s^+(\mathbf{k})\, \phi_n^{*s}(\mathbf{k})\, \phi_n^{s'}(\mathbf{k}') \qquad (59.7c)$$

$$Q = e \sum_{k,s=1,2} b_s^+(\mathbf{k})\, b_s(\mathbf{k}) - e \sum_{k,s=3,4} b_s(\mathbf{k})\, b_s^+(\mathbf{k}) \qquad (59.7d)$$

All the operators (59.7) now have zero expectation or eigenvalues in the state $\Psi_0$; in addition, the expectation value of $\mathcal{H}$ is always positive, and unlike $\tilde{Q}$, the new total charge operator $Q$ can assume positive and negative values.*

We may simplify the above expressions by introducing new operators $c_{1,2}(\mathbf{k})$ in place of $b_{3,4}(\mathbf{k})$. These operators are defined as follows:

$$b_{3,4}^+(-\mathbf{k}) = c_{1,2}(\mathbf{k}) \qquad (59.8)$$

Then

$$\mathcal{H} = \sum_{k,s=1,2} \hbar\omega(\mathbf{k})\{b_s^+(\mathbf{k})\, b_s(\mathbf{k}) + c_s^+(\mathbf{k})\, c_s(\mathbf{k})\}$$

$$\mathbf{G} = \sum_{k,s=1,2} \hbar\mathbf{k}\,\{b_s^+(\mathbf{k})\, b_s(\mathbf{k}) + c_s^+(\mathbf{k})\, c_s(\mathbf{k})\} \qquad (59.9)$$

$$Q = e \sum_{k,s=1,2} \{b_s^+(\mathbf{k})\, b_s(\mathbf{k}) - c_s^+(\mathbf{k})\, c_s(\mathbf{k})\}$$

The definition of $\Psi_0$ now becomes

$$b_s(\mathbf{k})\Psi_0 = 0 \qquad c_s(\mathbf{k})\Psi_0 = 0 \text{ for } s = 1,2 \text{ and all } \mathbf{k} \qquad (59.10)$$

The states $b_s^+(\mathbf{k})\Psi_0$ and $c_s^+(\mathbf{k})\Psi_0$ (for $s=1,2$) are eigenstates of the energy and momentum operators $\mathcal{H}$ and $\mathbf{G}$ with respective eigenvalues $\hbar\omega(\mathbf{k})$ and $\hbar\mathbf{k}$, and belong respectively to the eigenvalues $e$ and $-e$ of the charge operator $Q$. The operators $b_s^+(\mathbf{k})$ and $c_s^+(\mathbf{k})$ therefore create electrons and positrons respectively. Similarly, $b_s^+(\mathbf{k})b_s(\mathbf{k}) = n_s^e(\mathbf{k})$ and $c_s^+(\mathbf{k})c_s(\mathbf{k}) = n_s^{\bar{e}}(\mathbf{k})$ are the particle number operators of the

---

* This is not surprising; when an electron of charge $e$ is removed from a negative energy state the resultant charge is $-e$.

electrons $(e)$ and positrons $(\bar{e} = -e)$. The total number of particles is associated with the operator

$$N = \sum_{\mathbf{k}, s = 1, 2} \{ n_s^e(\mathbf{k}) + n_s^{\bar{e}}(\mathbf{k}) \} \tag{59.11}$$

$N\Psi_0 = 0$ then defines the vacuum state.

The previous difficulties have now disappeared; the energy is always positive, but on the other hand we now have two sorts of particles with charges of different sign. The subtraction procedure employed is only possible for systems obeying Fermi statistics; it also satisfies the relativity requirements, but we shall not prove this.

The most general state is given by an expansion of the form

$$\Psi = F_0 \Psi_0 + \sum_{\mathbf{k}_1, s_1 = 1, 2} F_{1,0}(\mathbf{k}_1, s_1)\, b_{s_1}^+(\mathbf{k}_1)\, \Psi_0 +$$
$$+ \sum_{\mathbf{k}_1', s_1' = 1, 2} F_{0,1}(\mathbf{k}_1', s_1')\, c_{s_1'}^+(\mathbf{k}_1')\, \Psi_0 + \tag{59.12}$$
$$+ \sum_{\substack{\mathbf{k}_1, \mathbf{k}_1' \\ s_1, s_1' = 1, 2}} F_{1,1}(\mathbf{k}_1, s_1; \mathbf{k}_1', s_1')\, b_{s_1}^+(\mathbf{k}_1')\, c_{s_1'}^+(\mathbf{k}_1')\, \Psi_0 + \ldots$$

The Hilbert space elements in terms of which $\Psi$ is expanded in (59.12) are normalized if $\Psi_0$ is assumed to be normalized (cf. Exercise 4, p. 360). Then $\left| F_{n,n'}(\mathbf{k}_1, s_1 \ldots; \ldots, \mathbf{k}_{n'}', s_{n'}') \right|^2$ is the probability of finding $n$ electrons in the state $\Psi$ with quantum numbers $k_1, s_1, \ldots$ and $n'$ positrons with numbers $k_1', s_1', \ldots$ For free particles* $N$ commutes exactly with $\mathcal{H}$, while for slowly varying external fields it does so approximately; the number of particles is therefore constant. We can then limit ourselves to the consideration of states for which (e.g.) only $F_{1,0}$ (one electron) or only $F_{0,1}$ (one positron) differs from zero.

To proceed from the representation in momentum space to coordinate space representation, we introduce the Fourier transforms of the $F(\mathbf{k}, s)$, which we denote by $G(\mathbf{r}, s)$:

$$G_0 = F_0$$
$$G_{1,0}(\mathbf{r}, s) = \sum_{\mathbf{k}} F_{1,0}(\mathbf{k}, s) \frac{e^{i\mathbf{k}\mathbf{r}}}{\sqrt{V}}$$
$$F_{1,0}(\mathbf{k}, s) = \int_V G_{1,0}(\mathbf{r}, s) \frac{e^{-i\mathbf{k}\mathbf{r}}}{\sqrt{V}}\, d\mathbf{r}$$

---

* In the quantization of the Dirac field we have not only neglected the external electromagnetic fields but the electromagnetic interactions of the field charges as well. This latter simplification is justifiable only if the particle densities are sufficiently low.

The general state then becomes

$$\Psi = G_0 \Psi_0 + \sum_{s=1,2} \int_V d\mathbf{r}\, G_{1,0}(\mathbf{r}, s) \sum_{\mathbf{k}} b_s^+(\mathbf{k}) \frac{e^{-i\mathbf{k}\mathbf{r}}}{\sqrt{V}} \Psi_0 + \ldots \qquad (59.13)$$

The norm of $\Psi$ is

$$(\Psi, \Psi) = |F_0|^2 + \sum_{\mathbf{k}, s=1,2} |F_{1,0}(\mathbf{k}, s)|^2 + \ldots$$

$$= |G_0|^2 + \sum_{s=1,2} \int_V d\mathbf{r}\, |G_{1,0}(\mathbf{r}, s)|^2 + \ldots \qquad (59.14)$$

The expansion functions $F$ and $G$ are functions of time as well as of the variables shown above. Their equation of motion is obtained from the general equation $\mathscr{H}\Psi = i\hbar\dot{\Psi}$, which is valid for free particles because of the commutability of the particle number operators for each term of (59.12) and (59.13). A simple calculation gives the following results:

$$\dot{F}_0 = 0, \qquad \hbar\omega(\mathbf{k}) F_{1,0}(\mathbf{k}, s) = i\hbar\dot{F}_{1,0}, \ldots$$

$$\dot{G}_0 = 0, \qquad E(\mathbf{p}) G_{1,0}(\mathbf{r}, s) = i\hbar\dot{G}_{1,0}, \ldots, \qquad (59.15)$$

where $E(\mathbf{p})$ represents the operator

$$E(\mathbf{p}) = \{\mathbf{p}^2 c^2 + m^2 c^4\}^{1/2} = mc^2 + \frac{\mathbf{p}^2}{2m} - \frac{\mathbf{p}^4}{8m^3 c^2} + \ldots \qquad (59.16)$$

and

$$\mathbf{p} = \frac{\hbar}{i} \frac{\partial}{\partial \mathbf{r}}$$

These are relativistic Schrödinger equations for two-component "Schrödinger functions" $G_{1,0}(\mathbf{r}, s)$, where $s = 1, 2$.

What is the physical significance of these functions? At first sight we are inclined to assume that $|G_{1,0}|^2$ is the probability density for an electron; this is not the case, however, as we shall see immediately.

Let us examine the expectation value $\bar{\rho}(\mathbf{R})$ of the density $\rho(\mathbf{R})$ in a state in which only $G_{1,0}$ or $F_{1,0}$ is non-zero, and which therefore contains just one electron. The only terms in (59.7c) contributing to $\bar{\rho}$ are those for which $s, s' = 1, 2$; it may easily be verified that the expectation values of the remaining terms vanish. Then $\bar{\rho}$ is the expectation value of

$$e \sum_{\substack{\mathbf{k}, \mathbf{k}'n \\ s,s'=1,2}} b_s^+(\mathbf{k}) b_{s'}(\mathbf{k}') \phi_n^{*s}(\mathbf{k}, \mathbf{R}) \phi_n^{s'}(\mathbf{k}', \mathbf{R})$$

in the state

$$\Psi = \sum_{k,s=1,2} F(k,s) b_s^+ (k) \Psi_0 = \sum_{s=1,2} \int_V dr\, G(r,s) \sum_k b_s^+ (k) \frac{e^{-ikr}}{\sqrt{V}} \Psi_0$$

We obtain

$$\bar{\rho}(R) = \sum_{\substack{k k' k'' k''' n \\ s s' s'' s''' = 1,2}} F^*(k'',s'') F(k''',s''') e \phi_n^{*s}(k,R) \phi_n^{s'}(k',R) \times$$

$$\times \left( b_{s''}^+ (k'') \Psi_0, b_s^+ (k) b_{s'}(k') b_{s'''}^+ (k''') \Psi_0 \right) \qquad (59.17)$$

It follows from the commutation relations (59.5), the definition (59.6a) and the fact that $(\Psi_0, \Psi_0) = 1$, that the second factor is $\delta_{s's'''} \delta_{k'k'''} \delta_{ss''} \delta_{kk''}$. Hence

$$\bar{\rho}(R) = e \sum_{\substack{k,k'n \\ s,s'=1,2}} \phi_n^{*s}(k,R) \phi_n^{s'}(k',R) F^*(k,s) F(k',s') \qquad (59.17a)$$

or inserting the values for $\phi_n^s$ given in (58.9),

$$\bar{\rho}(R) = \frac{e}{V^2} \sum_{\substack{k,k'n \\ s,s'=1,2}} \int dr\, dr'\, a_n^{*s}(\hbar k) a_n^{s'}(\hbar k') \times$$

$$\times e^{ik'(R-r') - ik(R-r)} G^*(r,s) G(r',s') \qquad (59.17b)$$

This result may be put into the form

$$\bar{\rho}(R) = e \sum_{\substack{s,s'=1,2 \\ n=1...4}} \int dr\, dr'\, G^*(r,s) G(r',s') K_n^{*s}(R-r)\, K_n^{s'}(R-r') \qquad (59.17c)$$

The functions $K_n^s$ in the above expression are defined by

$$K_n^s(r) = \sum_k a_n^s(\hbar k) \frac{e^{ikr}}{V} \qquad (59.18)$$

or when $V$ becomes infinite,

$$K_n^s(r) = \frac{1}{(2\pi)^3} \int dk\, a_n^s(\hbar k) e^{ikr} \qquad (59.18a)$$

When (59.18a) applies, the integrations in (59.17c) extend over all space, thus eliminating the arbitrarily introduced periodicity volume. The density is therefore not given by $|G|^2$, but is dispersed by a superposition of integral transforms in terms of the functions $K_n^s$. When the total charge is calculated by integrating over the range of $\mathbf{R}$ the $K_n^s$ drop out, because the $\phi_n^s$ form an orthonormal system of functions.

In connection with the calculation of the charge distribution from $G(\mathbf{r}, s)$ a number of relativistic modifications of the previous rules are encountered.

(*a*) It is no longer possible to associate an operator $\mathbf{R}$ with the position coordinate $\underline{\mathbf{R}}$ such that

$$\int \mathbf{R}\bar{\rho}(\mathbf{R})\, d\mathbf{R} = e(G, \underline{\mathbf{R}}G)$$

and

$$\int f(\mathbf{R})\, \bar{\rho}(\mathbf{R})\, d\mathbf{R} = e(G, f(\underline{\mathbf{R}})\, G)$$

This is because the functions $\phi_n^s$ do not constitute a complete system when $s$ ranges over the values $1, 2$ only.

In field theory, "position" does not correspond to any observable in the usual sense, but occurs as a parameter, as does the time. The observable quantities are all densities, such as energy, momentum, and charge density.

(*b*) The centroid of charge is displaced in a characteristic manner relative to the centroid of $|G|^2$. If we calculate the mean value

$$e\bar{\mathbf{R}}^\rho = \int \mathbf{R}\bar{\rho}(\mathbf{R})\, d\mathbf{R}$$

then after some intermediate calculations (see Appendix on p. 352) we obtain the non-relativistic approximation

$$\begin{aligned}
\bar{\mathbf{R}} &= \int \sum_{s=1}^{2} G^*(\mathbf{r}, s)\, \mathbf{r}\, G(\mathbf{r}, s)\, d\mathbf{r} - \\
&\quad - \int \sum_s G^*(\mathbf{r}, s)\frac{\hbar}{4m^2c^2}\left(\vec{\sigma} \times \frac{\hbar}{i}\operatorname{grad}\right)G(\mathbf{r}, s)\, d\mathbf{r} \\
&= \bar{\mathbf{R}}^G - \frac{\hbar}{4m^2c^2}\overline{(\vec{\sigma} \times \mathbf{p})}^G
\end{aligned} \tag{59.19}$$

This result is very closely connected with the magnetic moment of the electron. We indicated in §34 that a moving magnetic moment behaves like an electric dipole. Now if we separate the Dirac current $\mathbf{j}_v = iec\psi^*\beta\gamma_v\psi$ into a "spin component" and an "orbital component" (see Exercise 5, p. 360), the spin portion $\bar{\rho}_s$ of the density $\bar{\rho}$ may be written in the form $\bar{\rho}_s = -\operatorname{div} \mathbf{\bar{P}}$. This may be regarded as the density of the "dielectric polarization" $\mathbf{\bar{P}}$ of the electron. The mean value

$$e\overline{\mathbf{R}^{\rho_s}} = -\int \mathbf{R} \operatorname{div} \mathbf{\bar{P}} \, d\mathbf{R} = \int \mathbf{\bar{P}} \, d\mathbf{R}$$

agrees with the above displacement, while the orbital portion $\bar{\rho}_b$ coincides with $\sum\limits_{s=1,2} |G|^2$ in the non-relativistic limiting case. This displacement appears experimentally in the form of the spin-orbit interaction (§§34 and 60).

(c) The Dirac density is not only displaced relative to $\sum\limits_{s=1,2} |\bar{G}|^2$, but is also "blurred". We can best see what is meant by this by comparing the quantities $\overline{\mathbf{R}^2}^{\rho}$ and $\overline{\mathbf{R}^2}^{G}$. The result is (see Appendix)

$$\overline{\mathbf{R}^2}^{\rho} = \overline{\mathbf{R}^2}^{G} + \frac{3}{4}\frac{\hbar^2}{m^2 c^2} \tag{59.20}$$

The mean square deviation of the Dirac density is therefore greater than the Schrödinger-Pauli density by approximately the square of the Compton wavelength of the electron.

This result is associated with the fact that we have restricted ourselves to a pure electron state. It can readily be seen that, as in classical field theory, a sharply defined charge density must be represented by a superposition of solutions containing both positive and negative frequencies.

The effect on measurements would be as follows. If we attempt to determine the "position" of an electron, say with a microscope, to within less than the Compton wavelength, we must use $\gamma$-rays of wavelength $\lambda \ll \hbar/mc$, i.e. of energy $\varepsilon \gg mc^2$. Since this energy is sufficient to create electron-positron pairs, positrons will thus appear in any such accurate measurement of position.

The greater dispersion of the actual charge density compared with the Schrödinger density $|G|^2$ gives rise to an additional term in Pauli's equation that is experimentally observable (see following section).

(d) So far we have only considered states containing one particle. In the case where a number of particles are present, the densities of the individual particles are not simply additive; additional terms arise as a result of interference effects between electron and positron states (cf. the sums $\sum\limits_{\substack{s'=1,2 \\ s=3,4}}$ in (59.7d)). It would therefore be a mistake to think that, in the absence of external fields, quantization of the wave equation would lead to a system that was separated with regard to electrons and positrons. It is true that when the electromagnetic fields can be neglected the equations of motion are separated, but physical quantities such as the charge density are not.

## Appendix

Calculation of the mean values of functions of position in the single electron state of the Dirac theory.

The most convenient representation of the expectation value of the charge density $\bar{\rho}(\mathbf{R})$ in the state described by the Schrödinger function $G(\mathbf{R},s)$ is

$$\bar{\rho}(\mathbf{R}) = e \sum_{\substack{s,s'=1,2 \\ n}} \{a_n{}^s(\mathbf{p})\,G(\mathbf{R},s)\}^* \{a_n^{s'}(\mathbf{p})\,G(\mathbf{R},s')\} \tag{A.1}$$

where $\mathbf{p} = \dfrac{\hbar}{i}\dfrac{\partial}{\partial \mathbf{R}}$

This expression is obtained by passing to the limiting case of an infinite volume in (59.17b), expressing the arguments $\hbar\mathbf{k}$ of the components $a_n{}^s$ in terms of derivatives of the exponential functions, and transferring these derivatives to the functions $G$. The remaining integrals over $\mathbf{k}$ and $\mathbf{k}'$ give $\delta(\mathbf{R}-\mathbf{r})\,\delta(\mathbf{R}-\mathbf{r}')$, so that the integrations can be performed over $\mathbf{r}$ and $\mathbf{r}'$.

From (A.1), the mean value of any function $f(\mathbf{R})$ of weight $\bar{\rho}$ is

$$\bar{f^\rho} = \sum_{\substack{s,s'=1,2 \\ n=1\ldots 4}} \int d\mathbf{R}\,G^*(\mathbf{R},s)\,a_n^{+s}(\mathbf{p})\,f(\mathbf{R})\,a_n^{s'}(\mathbf{p})\,G(\mathbf{R},s') \tag{A.2}$$

The representation of the operators dependent on $\mathbf{p}$ is obtained from (58.8):

$$a_n{}^s(\mathbf{p}) = \frac{c\{\vec{a}\beta\}_{ns}\,\mathbf{p}+\delta_{ns}\,[E(\mathbf{p})+mc^2]}{\sqrt{\{2E(\mathbf{p})\,[E(\mathbf{p})+mc^2]\}}}$$

$$a_n^{+s}(\mathbf{p}) = \frac{c\{\beta\vec{a}\}_{sn}\,\mathbf{p}+\delta_{sn}\,[E(\mathbf{p})+mc^2]}{\sqrt{\{2E(\mathbf{p})\,[E(\mathbf{p})+mc^2]\}}} \tag{A.3}$$

In addition

$$\sum_n a_n^{+s'}(\mathbf{p})\,a_n{}^s(\mathbf{p}) = \delta_{s's}$$

In order to deal with the non-relativistic limiting case we first expand the quantities in (A.3) in terms of powers of $1/mc$:

$$E(\mathbf{p}) = mc^2\left(1+\frac{\mathbf{p}^2}{2m^2c^2}+\ldots\right)$$

$$a_n{}^s(\mathbf{p}) = \{\vec{a}\beta\}_{ns}\,\frac{\mathbf{p}}{2mc}\left(1-\frac{3\mathbf{p}^2}{8m^2c^2}\ldots\right)+\delta_{ns}\left(1-\frac{\mathbf{p}^2}{8m^2c^2}\ldots\right) \tag{A.4}$$

with a corresponding expression for $a_n^{+s}(\mathbf{p})$. If we include terms to the second power of $1/mc$, we then have from (A.2):

$$\bar{f^\rho} = \sum_{\substack{s,s'=1,2 \\ n}} \int d\mathbf{R}\,G^*(\mathbf{R},s) \times \tag{A.2a}$$

$$\times\,[\delta_{ns}\delta_{ns'}f-\delta_{ns}\delta_{ns'}(\mathbf{p}^2f+f\mathbf{p}^2)/8m^2c^2+\{\beta\vec{a}\}_{s'n}\,\mathbf{p}f\{\vec{a}\beta\}_{ns}\,\mathbf{p}/4m^2c^2]\,G(\mathbf{R},s')$$

No term of the first degree in $1/mc$ appears, because $\tilde{a}\beta$ is a secondary diagonal matrix whose components vanish for the range $s, s' = 1, 2$. The first term in the square brackets gives $f^G$; the other two terms give the deviations from the mean value to the required degree of approximation. The summation over $n$ can then be performed; $\delta_{ss'}$ occurs in the first two terms, while in the third term it is necessary to take account of the relations (57.5). We then obtain

$$f^p = f^G + \sum_{s, s' = 1, 2} \int d\mathbf{R}\, G^*(\mathbf{R}, s) \times$$
$$\times \left[ \sum_{i, l = 1, 2, 3} \{\alpha_l \alpha_i\}_{ss'}\, p_l\, fp_i / 4m^2 c^2 - \delta_{ss'}(\mathbf{p}^2 f + f\mathbf{p}^2)/8m^2 c^2 \right] G(\mathbf{R}, s')$$
(A.2b)

The second term in square brackets in (A.2b) may be conveniently transformed as follows:

$$\mathbf{p}^2 f + f\mathbf{p}^2 \to 2\mathbf{p}f\mathbf{p} + [\mathbf{p}, [\mathbf{p}, f]] = \sum_{i = 1 \ldots 3} \{2p_i\, fp_i + [p_i, [p_i, f]]\}$$

We now combine the first term in the above expression with the term in $\alpha_l \alpha_i$ in

$$\sum_{i, l = 1 \ldots 3} (\{\alpha_l \alpha_i\}_{ss'} - \delta_{il}\delta_{ss'}) p_l\, fp_i / 4m^2 c^2$$

From (57.5) we can express this as

$$\sum_{i, l} \tfrac{1}{2} \{\alpha_l \alpha_i - \alpha_i \alpha_l\}_{ss'}\, p_l\, fp_i / 4m^2 c^2$$

Since $fp_i = p_i f + i\hbar\, \partial f / \partial X_i$, this expression becomes

$$\sum_{i, l = 1 \ldots 3} (\alpha_l \alpha_i - \alpha_i \alpha_l) p_l\, \frac{\partial f}{\partial X_i}\, i\hbar / 8m^2 c^2$$

because the contribution of $p_l p_i f$ vanishes owing to the symmetry in $i, l$. We then finally obtain

$$f^p = f^G + \overline{\sum_{i, l = 1 \ldots 3} (\alpha_l \alpha_i - \alpha_i \alpha_l) p_l\, \frac{\partial f}{\partial X_i}\, i\hbar / 8m^2 c^2}^G$$
$$- \overline{\sum_{i = 1 \ldots 3} [p_i, [p_i, f]]/8m^2 c^2}^G$$
(A.2c)

In the calculation of the mean value of the position, $f = X_k$, the double commutation drops out, and since $\partial X_k / \partial X_i = \delta_{ik}$ we have

$$X_k^p = X_k^G + \overline{\sum_{l = 1 \ldots 3} (\alpha_l \alpha_k - \alpha_k \alpha_l) p_l\, i\hbar / 8m^2 c^2}^G$$
(A.5)

In view of (57.16), this is identical with equation (59.19),

$$X_k^p = X_k^G - \overline{(\vec{\sigma} \times \mathbf{p})_k \cdot \hbar / 4m^2 c^2}^G$$
(A.5a)

If the complete calculation is strictly performed, we obtain

$$\bar{\mathbf{R}}^p = \bar{\mathbf{R}}^G - \overline{\frac{(\vec{\sigma} \times \mathbf{p})\hbar c^2}{2E(\mathbf{p})[E(\mathbf{p}) + mc^2]}}^G$$
(A.6)

which tends to (A.5a) in the non-relativistic approximation.

If we calculate the mean square deviation $(f = \mathbf{R}^2)$ for a $G$-function that is spherically symmetrical about $\mathbf{R}=0$ ($\overline{\mathbf{R}^\rho}=\overline{\mathbf{R}^G}=0$, $\overline{\mathbf{p}}=0$, $\overline{\mathbf{p}\times\mathbf{R}^G}=0$), the second term in (A.2c) contributes nothing. This follows either from the symmetry in $l, i$ of the averaged expression $p_l \mathbf{R}^2 p_i$, or from the fact that $\overline{\mathbf{p}\times\mathbf{R}}=0$ for the spherically symmetrical state, because only the components of the "orbital angular momentum" $\mathbf{p}\times\mathbf{R}$ contribute to the second term. The value of the double commutation term is a constant, $6\hbar^2/m^2c^2$. In total we therefore obtain the result that was given in §59,

$$\overline{\mathbf{R}^2}^\rho = \overline{\mathbf{R}^2}^G + \frac{3}{4}\frac{\hbar^2}{m^2c^2} \tag{A.7}$$

This result does not apply to arbitrarily small values of $\overline{\mathbf{R}^2}^G$; on the contrary, a necessary condition is that $\overline{\mathbf{R}^2}^G \gg \hbar^2/m^2c^2$, since if it were not, the expansion (A.4) would be meaningless.

## §60. The Pauli spin theory as an approximation

In the last paragraph we saw how the equations (59.15) for two-component functions can be set up, proceeding from the quantization of the Dirac field in the absence of external forces. In the non-relativistic case, for which $v/c \ll 1$, these equations become the ordinary Schrödinger equations for a free particle. When external fields are present and the interaction between particles must be taken into account, it is no longer possible to establish two-component equations of this sort, because electron-positron pairs can be produced by the external field. The operator $N$ for the total number of particles no longer commutes with $\mathscr{H}$, and we cannot confine ourselves to a state in which only a single function $G_{nn'}(\mathbf{r}_1 \ldots; \mathbf{r}_1' \ldots)$ differs from zero.

However, if the probability of pair production is small, we can attempt to treat the transitions between states with different numbers of particles as a perturbation of the independent two-component equations. The prerequisites for this procedure exist when the external field is almost constant* over a distance of the order of magnitude of the Compton wavelength $\hbar/mc$:

$$\frac{\hbar}{mc}\left|\frac{\partial \mathbf{E}}{\partial x}\right| \ll |\mathbf{E}| \qquad \frac{\hbar}{mc}\left|\frac{\partial \mathbf{H}}{\partial x}\right| \ll |\mathbf{H}| \tag{60.1}$$

In deriving the two-component equation it proves to be unnecessary to adopt the roundabout procedure involved in the quantization of the

---

*This results from a more exact discussion of the Dirac equation in the presence of external fields. The condition is plausible, since it states that the field contains only Fourier components of wavelength $\lambda \gg \hbar/mc$, and hence that the energies of the corresponding quanta $\hbar\omega \approx hc/\lambda$ are small compared with $2mc^2$, the minimum energy necessary for the production of electron-positron pairs.

Dirac equation, provided that we restrict ourselves to correction terms in the Schrödinger equation that are of the second degree in $\hbar/mc$ and $v/c$. It is then sufficient to treat the original Dirac equation itself. As an illustration we shall again consider the case of a free particle.

We seek an equation that is equivalent to the Dirac equation

$$\mathscr{H}\psi = \{c\tilde{\alpha}\mathbf{p} + \beta mc^2\}\psi = i\hbar\dot{\psi}$$

in which the posititive energy states are represented by functions $G(\mathbf{r},s)$ of two components ($s = 1, 2$). This is obviously equivalent to finding a unitary transformation $U\psi = G$ in which the transformed Hamiltonian operator $\mathscr{H}' = U\mathscr{H}U^+$ contains only $D$-type matrices (where $D$ is defined by (57.17)), because $A$-type matrices mix the components $(1,2)$ of the functions $\psi_n(\mathbf{r},t)$ with the components $(3,4)$.

For a free particle, the transformation $U$ can be derived from (58.8):

$$U = \frac{c\beta\tilde{\alpha}\mathbf{p} + E(\mathbf{p}) + mc^2}{\sqrt{\{2E(\mathbf{p})[E(\mathbf{p}) + mc^2]\}}} \qquad (60.2)$$

Hence

$$\mathscr{H}' = \beta E(\mathbf{p})$$

and the eigensolutions for positive values of the energy are

$$G(\mathbf{r},s) = \delta_{ss_z}\frac{e^{i\mathbf{k}\mathbf{r}}}{\sqrt{V}} \qquad s_z = 1, 2 \qquad (60.3)$$

For $U\mathbf{r}U^+$ etc., we obtain comparatively involved expressions (see table on p. 356). The transformation of $\mathbf{r}$ was effectively performed in the previous section.* Those operators that transform to $\mathbf{r}$, $\dot{\mathbf{r}}$, etc., do not have the same physical significance as these quantities in their previous sense; the only exception is $\mathbf{p}$, because $U\mathbf{p}U^+ = \mathbf{p}$. In order to express this difference we shall designate the previous Dirac operators by the terms electron position $\mathbf{R}_{el}$, electron velocity $\dot{\mathbf{R}}_{el}$, etc., which indicates that they are associated with the electric charge ($e\overline{\mathbf{R}}_{el} = \int \mathbf{r}\bar{\rho}_{el}d\mathbf{r}$). In the case of the new operators, which possess a simple representation in the Schrödinger-Pauli theory, we shall retain the designations of position, velocity, etc. (see table on p. 356). We saw in the last paragraph that the "electron position" is both displaced and dispersed with respect to "position" as defined above. External fields do not "act"

---

* It is $(U\mathbf{r}U^+)_{ss'} = \Sigma_n a_n^{+s}(\mathbf{p})\mathbf{r}a_n^{s'}(\mathbf{p})$ for $s,s' = 1 \ldots 4$; see equation (A.2) in §59.

at the latter, but at the electron position. The centre of charge is given by $\overline{\mathbf{R}}_{el}$. On the other hand it is the operators $\mathbf{R}$, etc. which in the non-relativistic case become the corresponding operators $x$, $y$, $z$, etc., of the Schrödinger-Pauli theory. For instance, in the new representation

$$\mathscr{H}' = \beta E(\mathbf{p}) = \beta\left(mc^2 + \frac{\mathbf{p}^2}{2m} + \ldots\right)$$

$$\dot{\mathbf{r}} = \frac{i}{\hbar}[\mathscr{H}', \mathbf{r}] = \beta\frac{\partial E(\mathbf{p})}{\partial \mathbf{p}} = \beta\left(\frac{\mathbf{p}}{m} + \ldots\right)$$

$$\dot{\vec{\sigma}}' = \frac{i}{\hbar}[\mathscr{H}', \vec{\sigma}'] = 0$$

These are just the properties of the operators in the non-relativistic theory. In particular, $\dot{\mathbf{R}} = \mathbf{v}$ for the positive energy states,* and the state of a free particle can be chosen as an eigenfunction of $\mathbf{p}$ and $\vec{\sigma}$.

| Quantity | Symbol | Representation according to | |
|---|---|---|---|
| | | Dirac | Schrödinger-Pauli |
| Electron position | $\mathbf{R}_{el}$ | $\mathbf{r}$ | $\mathbf{r} - \dfrac{\hbar c^2\vec{\sigma}' \times \mathbf{p} - i\hbar c^3\beta(\vec{\alpha}\mathbf{p})\mathbf{p}/E(\mathbf{p})}{2E(\mathbf{p})[E(\mathbf{p})+mc^2]} - \dfrac{i\hbar c}{2E(\mathbf{p})}\beta\vec{\alpha}$ |
| Position | $\mathbf{R}$ | $U^+\mathbf{r}U$ | $\mathbf{r}$ |
| Electron spin | $\vec{\Sigma}_{el}$ | $\vec{\sigma}'$ | $\vec{\sigma}' + \dfrac{i\beta\vec{\alpha}\times\mathbf{p}c}{E(\mathbf{p})} - \dfrac{\mathbf{p}\times(\vec{\sigma}'\times\mathbf{p})c^2}{E(\mathbf{p})[E(\mathbf{p})+mc^2]}$ |
| Spin | $\vec{\Sigma}$ | $U^+\vec{\sigma}'U$ | $\vec{\sigma}'$ |
| State | $\psi$ | $\psi_n(\mathbf{r})$ | $\sum_n U_n{}^s\psi_n(\mathbf{r}) = G(\mathbf{r},s)$ |
| Mean values | $\begin{cases}\overline{f(\mathbf{R}_{el})}=\overline{f}{}^\rho \\ \overline{f(\mathbf{R})}=\overline{f}{}^G\end{cases}$ | $\int f(\mathbf{r})\sum_n\lvert\psi_n(\mathbf{r})\rvert^2 d\mathbf{r} = \int\sum_{ss'}f_{ss'}(\mathbf{r})G^*(\mathbf{r},s)G(\mathbf{r},s')d\mathbf{r}$ $\sum_{nn's}\int(U^+fU)_{nn'}^{ss}\psi_n{}^*\psi_{n'}d\mathbf{r} = \int f(\mathbf{r})\sum_s\lvert G(\mathbf{r},s)\rvert^2 d\mathbf{r}$ | |

We shall now consider the case in which external fields are present, and attempt to bring the appropriate Hamiltonian operator into a form in which it contains only $D$-type matrices. Since the transformation that effects this contains the fields and possibly the time explicitly, we

---

* In §57 we found that $\dot{\mathbf{r}} = c\vec{\alpha}$, $\dot{\vec{\sigma}}' = -2c(\vec{\alpha}\times\mathbf{p})/\hbar$.

must assume the following form for the transformed Hamiltonian operator:

$$\mathscr{H}' = U\mathscr{H}U^+ - i\hbar U\frac{\partial U^+}{\partial t} \tag{60.4}$$

This is necessary to ensure that the equation

$$\mathscr{H}U^+G = i\hbar\frac{\partial}{\partial t}(U^+G)$$

takes the form

$$\mathscr{H}'G = i\hbar\dot{G}$$

in the new representation.

We now write

$$\mathscr{H} = \beta mc^2 + e\phi + c\tilde{\alpha}\left(\mathbf{p} - \frac{e}{c}\mathbf{A}\right)$$

in the form

$$\mathscr{H} = \beta mc^2 + D + A$$

where $A$ and $D$ are defined by (57.17), and try the solution*

$$U_1 = \exp\frac{\beta A}{2mc^2} \tag{60.6}$$

We shall use this expression to determine the Hamiltonian operator as given by (60.4). In the final result we shall include only terms as far as $(v/c)^4$, since we are interested only in non-relativistic velocities and in slowly varying fields as defined by (60.1). When (60.6) is expanded and introduced into (60.4), we obtain the following result for $\mathscr{H}'$ after some intermediate calculations:

$$\mathscr{H}' = \beta mc^2 + D + \frac{\beta A^2}{2mc^2} - \frac{1}{8(mc^2)^2}\left[A,[A,D] + i\hbar\frac{\partial A}{\partial t}\right] - \frac{\beta A^4}{8(mc^2)^3} +$$

$$+ \frac{\beta}{2mc^2}\left\{[A,D] + i\hbar\frac{\partial A}{\partial t}\right\} - \frac{A^3}{3(mc^2)^2} + \ldots \tag{60.7}$$

In view of the rules governing $A$- and $D$-matrices, the first line contains $D$-matrices only and the second line $A$-matrices only. Thus $\mathscr{H}'$ now has the form $\mathscr{H}' = \beta mc^2 + D' + A'$, where $A'$ is only of order of magni-

---

* This solution is suggested by the non-relativistic expansion of the corresponding operator (60.2) in the case of the free particle: $U = 1 + \beta\tilde{\alpha}\mathbf{p}/2mc + \frac{1}{2}(\beta\tilde{\alpha}\mathbf{p}/2mc)^2 + \ldots$ This agrees with the expansion of (60.6) apart from terms of the order of magnitude of $(v/c)^2$. Cf. Foldy, L. L., and Wouthuysen, S. A., *Phys. Rev.* **78** (1950), 29.

tude $(v/c)^3$. Using exactly the same procedure, these non-diagonal terms can be reduced by a further order of magnitude by means of another transformation $U_2 = \exp \dfrac{\beta A'}{2mc^2}$, etc. It may be verified that the use of $U_2$ supplies only correction terms to $D'$ of higher order than the fourth in $v/c$; the first line of (60.7) therefore correctly represents the Hamiltonian operator to within terms in $(v/c)^4$.

In order to gain a more precise understanding of the significance of this approximation, we now introduce the quantities $D = e\phi$ and $A = c\tilde{\alpha}\left(\mathbf{p} - \dfrac{e}{c}\mathbf{A}\right)$ into (60.7), when we obtain

$$\mathscr{H}' = \beta mc^2 + e\phi + \frac{\beta}{2m}\left(\mathbf{p} - \frac{e}{c}\mathbf{A}\right)^2 - \frac{e\hbar}{2m c}\beta\vec{\sigma}'\,\mathbf{H} - \frac{\beta}{8m^3c^2}\left(\mathbf{p} - \frac{e}{c}\mathbf{A}\right)^4 -$$

$$- \frac{e\hbar}{8m^2c^2}\vec{\sigma}'\left\{\mathbf{E}\times\left(\mathbf{p} - \frac{e}{c}\mathbf{A}\right) - \left(\mathbf{p} - \frac{e}{c}\mathbf{A}\right)\times\mathbf{E}\right\} -$$

$$- \frac{e\hbar^2}{8m^2c^2}\operatorname{div}\mathbf{E} + \ldots \tag{60.8}$$

The quantities $\mathbf{v}/c$ and $\dfrac{\hbar}{mc}\left|\dfrac{\partial\mathbf{E}}{\partial x}\right|\Big/|\mathbf{E}|$ are independent of each other, but have the same order of magnitude for the lowest states of the hydrogen atom:

$$\overline{|\mathbf{v}|}/c \approx \frac{\hbar}{mc}\overline{\left|\frac{\partial\phi}{\partial x}\right|}\Big/|\phi| \approx \frac{e^2}{\hbar c} \approx 1/137$$

We can see that the two terms in the second line of (60.8) are of the same order of magnitude, since the first is approximately $\dfrac{\hbar}{mc}\dfrac{\partial e\phi}{\partial x}\dfrac{v}{c}$ and the second is $\left(\dfrac{\hbar}{mc}\dfrac{\partial}{\partial x}\right)^2 e\phi$.

If we restrict ourselves to positive energy states and irrotational fields (when curl $\mathbf{E} = 0$, i.e. $-\mathbf{p}\times\mathbf{E} = \mathbf{E}\times\mathbf{p}$), (60.8) finally becomes

$$\mathscr{H}' = mc^2 + e\phi + \frac{1}{2m}\left(\mathbf{p} - \frac{e}{c}\mathbf{A}\right)^2 - \frac{e\hbar}{2mc}\vec{\sigma}\mathbf{H} - \frac{1}{8m^3c^2}\left(\mathbf{p} - \frac{e}{c}\mathbf{A}\right)^4 -$$

$$- \frac{e\hbar}{4m^2c^2}\vec{\sigma}\,\mathbf{E}\times\left(\mathbf{p} - \frac{e}{c}\mathbf{A}\right) - \frac{e\hbar^2}{8m^2c^2}\operatorname{div}\mathbf{E} \tag{60.9}$$

The physical significance of each term occurring in the above expression has been explained in §34, with the exception of the term $-\dfrac{e\hbar^2}{8m^2c^2}\operatorname{div}\mathbf{E}$, the origin of which, however, is quite clear as a result of our previous discussion. We have already seen that the quantity $\mathbf{r}$ occurring in the Schrödinger function $G(\mathbf{r},s)$ is not the point at which the fields act; the latter is displaced and dispersed relative to the Schrödinger "position" as previously defined. Hence if $e\phi(\mathbf{R}_{el})$ is expanded in terms of powers of $\mathbf{R}_{el}-\mathbf{R}$, the spin-field interaction* is given by the dipole term implied by the displacement (59.19). (This interaction is represented by the sixth term in (60.9).) For spherically symmetrical dispersion the quadrupole term is

$$\frac{1}{2}\sum_{ik}\frac{\partial^2 e\phi}{\partial x_i\,\partial x_k}\overline{x_i x_k}=\frac{\overline{\mathbf{r}^2}}{6}\nabla^2 e\phi=-\frac{e}{6}\overline{\mathbf{r}^2}\operatorname{div}\mathbf{E}$$

where $\overline{\mathbf{r}^2}$ is a measure of the dispersion of the electrical charge relative to the electron position. Comparison with (60.9) shows that

$$\overline{\mathbf{r}^2}=\frac{3}{4}\left(\frac{\hbar}{mc}\right)^2$$

in agreement with (59.20); the electron charge is therefore dispersed over a region of the order of magnitude of the Compton wavelength.

For the Coulomb field of the hydrogen nucleus $\operatorname{div}\mathbf{E}=-\nabla^2\phi=4\pi e\,\delta(\mathbf{r})$. The additional term therefore produces an energy perturbation (actually an increase) in the case of those states for which $\left|G(0,s)\right|^2\neq 0$, that is, for all $s$-states (cf. §34). However, when calculating the energy terms for an electron in a pure Coulomb field it is unnecessary to adopt the preceding approximation procedure, since in this case the Dirac equation can be integrated exactly. The result (which we shall not prove) is Sommerfeld's formula (34.6).

## Exercises

### 1. *The spin operator of the Dirac equation*

Equation (57.10) indicated that the spin of a free particle is not constant with respect to time. Show that the component of spin parallel to the momentum is constant, however, and that the spin and electron spin components (using the terminology of §60) in this direction are the same.

---

* Cf. Becker, R.: *Akad. d. Wiss., Math. Phys.*, Göttingen 1945.

## 2. The matrix trace as a unitary invariant

Show that the trace $\Sigma A_{nn}$ of a matrix $A$ remains unchanged in value when subjected to an arbitrary unitary transformation.

Derive an expression for the trace of a Hermitian matrix in terms of the matrix eigenvalues $a_i$.

## 3. Commutation with the Pauli matrices

Show that if a matrix $\begin{pmatrix} a & b \\ c & d \end{pmatrix}$ commutes with all three Pauli matrices $\sigma_x$, $\sigma_y$, $\sigma_z$, it must be a multiple of the unit matrix.

## 4. Hilbert space for Fermi-Dirac statistics

Describe the formation of Hilbert space for the operators $b_k^+$ and $b_k$ which have the following commutation relations:

$$b_k^+ b_{k'} + b_{k'} b_k^+ = \delta_{kk'}, \qquad b_k b_{k'} + b_{k'} b_k = b_k^+ b_{k'}^+ + b_{k'}^+ b_k^+ = 0$$

## 5. The spin and orbital current of the Dirac electron

Equation (57.6) may be written in abbreviated form as follows:

$$\rho = \psi^* \psi \qquad \mathbf{j} = c\psi^* \vec{a}\psi$$

Putting $\rho = \tfrac{1}{2}(\psi^*\psi + \psi^*\psi)$, with a corresponding expression for $\mathbf{j}$, replace $\psi$ and $\psi^*$ once each by their value as given by the expression resulting from the Dirac equation,

$$c\vec{a}\mathbf{p}\psi + \beta mc^2\psi = i\hbar\dot{\psi}$$

i.e.

$$\psi = \frac{\beta}{mc^2}(i\hbar\dot{\psi} - c\vec{a}\mathbf{p}\psi)$$

Compare the two components of $\rho$ and $\mathbf{j}$ with the Schrödinger expressions and with (56.10).

# G

Solutions

# Solutions

## Chapter AI (p. 15)

### 1. *Rutherford's scattering formula*

(a) The three conservation laws are proved by differentiating out and taking account of the equation of motion. The laws of conservation of energy and angular momentum are valid for central forces in general; the third conservation theorem holds good only for a potential of the form $1/r$.

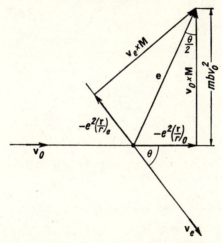

Fig. 57.—Graphical construction for the vector **e**

(b) The graphical construction for **e** is illustrated in figure 57. The relation $\tan \frac{1}{2}\theta = e^2/mv_0^2 b$ is directly deduced from it. Note that the angular momentum **M** is normal to the plane of the diagram, that **M** has the magnitude $mv_0 b$, and hence that $|\mathbf{v}_0 \times \mathbf{M}| = mv_0^2 b$.

(c)
$$dQ = \pi db^2 = \pi \left(\frac{e^2}{mv_0^2}\right)^2 d \cot^2 \tfrac{1}{2}\theta = \left(\frac{e^2}{2mv_0^2}\right)^2 \frac{d\Omega}{\sin^4 \tfrac{1}{2}\theta}$$

### 2. *Aston's mass spectrograph*

The deflection in the electric field is

$$\delta v_z = \frac{e}{m} E \delta t = \frac{eEa}{mv_x} \rightarrow \theta \approx \frac{eEa}{mv^2}$$

The deflection in the magnetic field is

$$\delta v_z = \frac{ev_x}{mc} H \delta t = \frac{eHb}{mc} \rightarrow \phi \approx \frac{eHb}{mvc}$$

The equation for the path when $x > l$ is $z = -\theta x + \phi(x-l)$. Focusing takes place when

$$\left(\frac{\partial z}{\partial v}\right)_{z=z_f} = 0, \text{ i.e. when } z_f = \theta x_f = \frac{e}{mv^2} Eax_f = \frac{Ea}{2V} x_f$$

The photographic plate must therefore be inclined to the $x$-axis at an angle $\theta = Ea/2V$.

# Chapter A II (p. 34)

1. *Broadening of spectral lines*

(a) $\left(\dfrac{\Delta\omega}{\omega}\right)_{\text{radiation}} = \dfrac{2}{3}\dfrac{e^2\omega}{mc^3} \approx \dfrac{e^2/mc^2}{\lambda} \approx 10^{-8}$ for $\lambda \approx 10^{-5}$ cm (visible light)

(b) $\left(\dfrac{\Delta\omega}{\omega}\right)_{\text{collision}} = \dfrac{2}{\tau\omega} \approx \dfrac{\bar{v}}{l}\dfrac{\lambda}{c} \approx \dfrac{\sigma^2\lambda}{V/N}\dfrac{\bar{v}}{c}$

$$\approx \begin{cases} 10^{-7} \text{ for normal atmospheric pressure,} \\ 10^{-10} \text{ for } 10^{-2}\text{mm Hg} \end{cases}$$

$\sigma$ is the atomic radius, $l$ the mean free path, $\bar{v} = \sqrt{(kT/M)}$ the mean velocity of the atoms, $V/N$ the specific volume.

(c) $\omega' = \omega_0\left(1 + \dfrac{v_x}{c}\right)$ is the frequency observed along the $x$-direction. $v_x$ is seldom greater than $\pm\bar{v}$; hence

$$\left(\frac{\Delta\omega}{\omega}\right)_{\text{Doppler}} \approx \frac{\bar{v}}{c} \approx 10^{-5}$$

(d) The broadening due to collisions can be made small compared with the width due to radiation damping by reducing the pressure; this latter width, however, is generally small compared with the Doppler width. Nevertheless, the radiation damping can be measured separately as a result of the different line form produced by the two effects. Taking account of the Maxwell distribution, we obtain the following expression for the spectral energy density in the emitted light:

$$S(\omega) = \frac{\gamma W_0}{2\pi} \int_{-\infty}^{\infty} \frac{\exp -\dfrac{Mv_x^2}{2kT} \, dv_x}{\left(\omega - \omega_0 - \omega_0 \dfrac{v_x}{c}\right)^2 + \tfrac{1}{4}\gamma^2} \Bigg/ \int_{-\infty}^{\infty} \exp -\frac{Mv_x^2}{2kT} \, dv_x$$

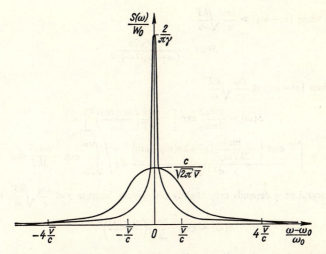

Fig. 58.—$S(\omega)/W_0$ for pure radiation damping and for pure Doppler broadening

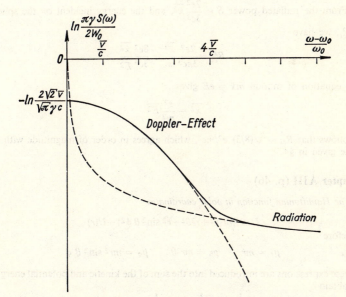

Fig. 59.—$S(\omega)/W_0$ for radiation and Doppler broadening on a logarithmic scale

Hence when $|\omega-\omega_0| \gg \dfrac{\omega_0}{c}\sqrt{\dfrac{kT}{m}}$

$$S(\omega) = \frac{\gamma W_0/2\pi}{(\omega-\omega_0)^2+\frac{1}{4}\gamma^2}$$

and when $|\omega-\omega_0| \ll \dfrac{\omega_0}{c}\sqrt{\dfrac{kT}{m}}$

$$S(\omega) = \frac{\gamma W_0 c}{2\pi\omega_0}\exp\left[-\frac{mc^2}{2kT}\left(\frac{\omega-\omega_0}{\omega_0}\right)^2\right] \times$$

$$\times \int_{-\infty}^{+\infty} \frac{\exp\left\{-\dfrac{mc^2}{2kT\omega_0{}^2}[\eta^2+2\eta(\omega_0-\omega)]\right\}}{\eta^2+\frac{1}{4}\gamma^2}\, d\eta \left/ \int_{-\infty}^{+\infty} \exp-\frac{mv_x{}^2}{2kT}\, dv_x \right.$$

The integral in $\eta$ depends only slightly on $\omega-\omega_0$, because $\gamma \ll \dfrac{\omega_0}{c}\sqrt{\dfrac{kT}{m}}$, hence

$$S(\omega) \sim \exp-\frac{mc^2}{2kT}\left(\frac{\omega-\omega_0}{\omega_0}\right)^2$$

We therefore have practically pure Doppler line form within the Doppler width, while outside this width the line form is almost entirely due to the radiation damping.

2. *Radiation damping and the electron radius*

From the radiated power $S = \dfrac{2e^2}{3c^3}\overline{\ddot{x}^2}$, and the energy incident on the sphere, $s\pi R_{el}^2$, we have

$$R_{el}^2 = \frac{2e^2}{3\pi c^3}\frac{\overline{\ddot{x}^2}}{s} = \frac{8e^2}{3c^4}\frac{\overline{\ddot{x}^2}}{\overline{E^2}}$$

The equation of motion $m\ddot{x} = eE$ gives

$$\overline{\ddot{x}^2} = \frac{e^2}{m^2}\overline{E^2}$$

It follows that $R_{el} = \sqrt{(8/3)}.e^2/mc^2$, which agrees in order of magnitude with the value given in §4.

## Chapter A III (p. 46)

1. *The Hamiltonian function in polar coordinates*

$$L = \tfrac{1}{2}m\{\dot{r}^2+r^2\dot{\theta}^2+r^2\sin^2\theta\,\dot{\phi}^2\}-U(r)$$

therefore

$$p_r = m\dot{r} \qquad p_\theta = mr^2\dot{\theta} \qquad p_\varphi = mr^2\sin^2\theta\,\dot{\phi}$$

If these expressions are introduced into the sum of the kinetic and potential energies, we obtain

$$\mathscr{H} = \frac{1}{2m}\left\{p_r^2+\frac{1}{r^2}p_\theta^2+\frac{1}{r^2\sin^2\theta}p_\varphi^2\right\}+U(r)$$

## 2. *The Lagrangian function and the conservation of momentum*

$$\int_{t_0}^{t_1} L(\mathbf{r}_1 \ldots) dt = \int_{t_0}^{t_1} L(\mathbf{r}_1 + \mathbf{R}_1, \ldots) dt$$

The second integral is evaluated in the same manner as on p. 37:

$$\int_{t_0}^{t_1} L(\mathbf{r}_1 + \mathbf{R}_1, \ldots) dt$$
$$= \int_{t_0}^{t_1} L(\mathbf{r}_1, \ldots) dt + \int_{t_0}^{t_1} dt \sum_i \mathbf{R}_i \left\{ -\frac{d}{dt}\frac{\partial L}{\partial \dot{\mathbf{r}}_i} + \frac{\partial L}{\partial \mathbf{r}_i} \right\} + \left[ \sum_i \mathbf{R}_i \frac{\partial L}{\partial \dot{\mathbf{r}}_i} \right]_{t_1} - \left[ \sum_i \mathbf{R}_i \frac{\partial L}{\partial \dot{\mathbf{r}}_i} \right]_{t_0}$$

Since the initial path $\mathbf{r}_i$ was "correct", the second integral on the right-hand side vanishes, and we obtain the following condition for invariance:

$$\left[ \sum_i \mathbf{R}_i \frac{\partial L}{\partial \dot{\mathbf{r}}_i} \right]_{t_1} = \left[ \sum_i \mathbf{R}_i \frac{\partial L}{\partial \dot{\mathbf{r}}_i} \right]_{t_0} \quad \text{for all } t_1, t_0,$$

or
$$\frac{d}{dt} \sum_i \mathbf{R}_i \frac{\partial L}{\partial \dot{\mathbf{r}}_i} = 0 \qquad \frac{d}{dt} \sum_i \mathbf{R}_i m_i \mathbf{v}_i = 0$$

When $\mathbf{R}_i = \mathbf{R}_0$ we obtain the law of conservation of linear momentum:

$$\frac{d}{dt} \sum_i m_i \mathbf{v}_i = 0$$

$\mathbf{R}_i \rightarrow \mathbf{n} \times \mathbf{r}_i$ yields the law of conservation of angular momentum:

$$\frac{d}{dt} \sum_i \mathbf{r}_i \times m_i \mathbf{v}_i = 0$$

## 3. *Hamilton's principle of the extremum*

(a) In this case we have

$$G\{x+\gamma\} - G\{x\} = \int_{t_0}^{t_1} \tfrac{1}{2} m \dot{\gamma}^2 dt$$

The term of the first degree in $\gamma$ vanishes if $x(t)$ represents the correct path. $G$ is therefore a minimum for this path, since

$$\int_{t_0}^{t_1} \tfrac{1}{2} m \dot{\gamma}^2 dt > 0$$

(b) In this case

$$G\{x+\gamma\} - G\{x\} = \int_{t_0}^{t_1} \tfrac{1}{2} m \{ \dot{\gamma}^2 - \omega^2 \gamma^2 \} dt$$

since the term of the first degree in $\gamma$ vanishes as in case (a). Since $\gamma(t)$ must vanish at $t_0$ and $t_1$ it can be represented by a sine series:

$$\gamma(t) = \sum_{n=1}^{\infty} a_n \sin n\pi \frac{t-t_0}{t_1-t_0} = \sum_{n=1}^{\infty} a_n \sin \omega_n(t-t_0)$$

The integral may then be performed, to give the result

$$G\{x+\gamma\} - G\{x\} = \tfrac{1}{4} m(t_1-t_0) \sum_{n=1}^{\infty} a_n^2 (\omega_n^2 - \omega^2)$$

If all $\omega_n > \omega$, i.e. $t_1 - t_0$ is smaller than the half-period of the oscillation, $\frac{1}{2}T$, $G$ is a minimum. If $t_1 - t_0$ is greater than $\frac{1}{2}T$, angular frequencies $\omega_n$ with small values of $n$ are smaller than $\omega$, while the other $\omega_n$ are greater; $G$ is then indeterminate in its behaviour. If we choose a variation $\gamma$ that contains only small values of $n$, $G$ is a maximum; otherwise it is a minimum. However, the coefficients $a_n$ may be so chosen that $G$ does not change at all.

## Chapter BI (p. 107)

### 1. *Planck's radiation formula interpolation*

$$\Delta = \begin{cases} \varepsilon^2 & \text{Rayleigh-Jeans} \\ \alpha k v \varepsilon & \text{Wien} \end{cases}$$

The representation $\Delta = \varepsilon^2 + \alpha k v \varepsilon$ is a reasonable interpolation for the mean square deviation $\Delta$, since the second term is small compared with the first in the Rayleigh-Jeans region, and vice versa in the Wien region. The differential equation for $\varepsilon$ as a function of $T$ possesses the solution

$$\varepsilon(v, T) = \frac{\alpha v k}{C \exp(\alpha v/T) - 1}$$

In the above expression the constant of integration $C$ must be assumed to be 1, since for small values of $v$ in the Rayleigh-Jeans region we should have $\varepsilon = kT$. When $\alpha = h/k$ we obtain Planck's formula. The mean-square-deviation formula applicable to a small frequency range of the black-body radiation is obtained by multiplying the mean square deviation of the oscillator by the number of oscillations in the frequency interval. In the Wien region these deviations are identical with those of an ideal gas composed of particles of energy $hv$, while in the Rayleigh-Jeans region the deviations can clearly be interpreted in terms of the interference effects of the light waves. Planck's formula thus demonstrates most clearly both the wave and particle nature of light.

### 2. *Derivation of the hydrogen terms from the correspondence principle*

For a linear oscillator $\omega$ is independent of $E$, hence $E(n) = \text{const} + nh\omega$. In a Coulomb field, and for an orbit of radius $R$,

$$E = -\tfrac{1}{2}mv^2 = -e^2/2R = -\tfrac{1}{2}e^2(m\omega^2/e^2)^{1/3}, \text{ hence } \omega = \{-8E^3/me^4\}^{1/2}$$

Therefore $\qquad dE/\omega(E) = -d\{-2E/me^4\}^{-1/2} = \hbar\,dn$

from which $\qquad -me^4/2E = (\text{const} + \hbar n)^2$

and $E = -me^4/2(\hbar n + \text{const})^2$, which gives the hydrogen terms when the constant of integration vanishes.

### 3. *First approximation to the energy of an anharmonic linear oscillator*

$$E_n = nh\omega + \tfrac{3}{2}\lambda \left(\frac{nh}{m\omega}\right)^2$$

**4.** *Vibrations of solid bodies at the absolute zero of temperature*

$$\overline{x^2}/a^2 = \hbar/2mca, \text{ hence } \sqrt{(\overline{x^2}/a^2)} \approx 1/30 \approx 3 \text{ per cent}$$

It should be realized that the disturbance is much greater than the displacements due to normal elastic stresses.

**5.** *Rate of change of total momentum in wave theory*

From (16.8):

$$\frac{d}{dt} \int j_m \, d\mathbf{r} = \frac{m}{i\alpha} \int d\mathbf{r} \{\psi^* \text{ grad } \psi + \psi^* \text{ grad } \dot{\psi} - \text{conjugate complex quantity}\}$$

$$= \frac{m}{i\alpha} \int d\mathbf{r} \left\{ \frac{\beta}{i\alpha} \psi^* \psi \text{ grad } \Phi + \frac{1}{i\alpha} (\nabla^2 \psi^* \text{ grad } \psi - \psi^* \text{ grad} \nabla^2 \psi) - \text{conj. compl.} \right\}$$

using (16.1a). The second term of the integrand vanishes on integrating by parts. Then since $2m\beta \to e\alpha^2$,

$$\frac{d}{dt} \int j_m \, d\mathbf{r} = -\frac{2m\beta}{\alpha^2} \int \psi^* \psi \text{ grad } \Phi \, d\mathbf{r} = \int \rho_e \mathbf{E} \, d\mathbf{r}$$

**6.** *Motion of the centroid of a wave packet*

The first relation is obtained from the equation of motion by integrating by parts. We can deduce from the previous exercise that $v_0$ is constant (since $\mathbf{E} = 0$), and hence that $\int \rho_m \, dx$ does not depend on the time. In addition

$$\psi(x,t) = \frac{1}{\sqrt{(2\pi)}} \int g(k,t) e^{ikx} \, dk,$$

$$x\psi(x,t) = \frac{1}{\sqrt{(2\pi)}} \int g(k,t) \left( \frac{\partial}{\partial k} e^{ikx} \right) dk = \frac{i}{\sqrt{(2\pi)}} \int \left( \frac{\partial}{\partial k} g(k,t) \right) e^{ikx} \, dk$$

integrating by parts once more. Accordingly,

$$x_s = i \int g^*(k,t) \frac{\partial}{\partial k} g(k,t) \, dk \Big/ \int |g(k,t)|^2 \, dk$$

and

$$\dot{x}_s = \int |g(k,t)|^2 \frac{d\omega}{dk} \, dk \Big/ \int |g(k,t)|^2 \, dk$$

using the relation $g(k,t) = g(k,0) e^{-i\omega(k)t}$

**7.** *Energy according to wave theory, and the Hamiltonian operator*

$$\int u \, d\mathbf{r} = \int \left( \frac{e}{\beta} \text{ grad } \psi^* \text{ grad } \psi + e\Phi \psi^* \psi \right) d\mathbf{r}$$

$$= \int \psi^* \left\{ -\frac{e}{\beta} \nabla^2 \psi + e\Phi \psi \right\} d\mathbf{r} = \int \psi^* \mathscr{H} \psi \, d\mathbf{r}$$

## Chapter B II (p. 137)

### 1. *Various representations of the δ-function*

$$\delta(x) = \lim_{b \to 0} \frac{1}{b} S(x/b), \text{ where}$$

(1)  $S(\eta) = \dfrac{\exp -\eta^2}{\sqrt{\pi}}$      (2)  $S(\eta) = \dfrac{1}{\pi(1+\eta^2)}$

(3)  $S(\eta) = \begin{cases} 0 & \text{for } |\eta| > 1 \\ \frac{1}{2} & \text{for } |\eta| \leq 1 \end{cases}$      (4)  $S(\eta) = \dfrac{\sin^2 \eta}{\pi \eta^2}$

$$\int_{-\infty}^{+\infty} \frac{1}{b} S(x/b)\, dx = \int_{-\infty}^{+\infty} S(\eta)\, d\eta = 1$$

in all four representations. In addition:

$$\int_{-a}^{a'} f(x) \frac{1}{b} S(x/b)\, dx = \int_{-a/b}^{a'/b} f(b\eta) S(\eta)\, d\eta \xrightarrow[b \to 0]{} f(0) \int_{-\infty}^{+\infty} S(\eta)\, d\eta = f(0)$$

It can be seen from the graphical representation (figure 60) that the maximum heights are proportional to $1/b$ and that the widths are proportional to $b$; the total area under the curves is therefore independent of $b$.

### 2. *δ-function relations*

(1) This relation is most simply proved if the $\delta$-function is represented by a Gaussian function as in the previous exercise. If $x$ is replaced by $Cx$, the normalization must be adjusted by the factor $C$ in order to yield $\delta(x)$ again.

(2) $\delta[(x-a)(x-b)]$ behaves like a $\delta$-function at $x = a$ and $x = b$. When $x \approx a$, $x-b$ may be replaced by $a-b$ and removed, and similarly when $x \approx b$.

(3) Since $|\det \alpha_{ik}| \neq 0$, the left-hand side of the relation is only zero for $x_k = 0$, and may therefore be represented as the product of simple $\delta$-functions. The factor is the functional determinant of the transformation of $\sum_k \alpha_{ik} x_k$ to the $x_k$; it is obtained by effecting the transformation, noting that

$$\int dx_1 \ldots dx_N \prod_k \delta(x_k) = 1$$

For orthogonal transformations $|\det \alpha_{ik}| = 1$, hence

$$\delta\left[\frac{1}{\sqrt{2}}(x+y)\right] \delta\left[\frac{1}{\sqrt{2}}(x-y)\right] = \delta(x)\,\delta(y)$$

### 3. *The Schrödinger equation in momentum space*

The operator effecting the transformation between position and momentum space is

$$U(p,x) = \frac{1}{\sqrt{(2\pi\hbar)}} \exp(ipx/\hbar)$$

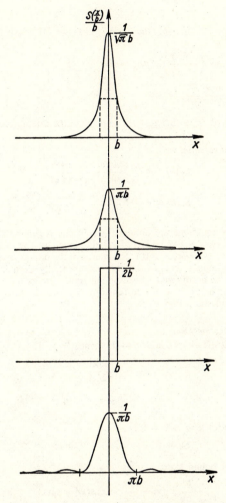

Fig. 60.—Various representations of the $\delta$-function

The representation of $\phi(x,t)$ in momentum space is therefore

$$\psi(p,t) = \int U(p,x)\,\phi(x,t)\,dx$$

Hence $\qquad \mathscr{H}\phi = \int \mathscr{H}\exp(-ipx/\hbar)\frac{\psi(p,t)}{\sqrt{(2\pi\hbar)}}\,dp$

Therefore

(a)   $\dfrac{p^2}{2m}\,\psi(p,t)+\displaystyle\int V_{p-p'}\,\psi(p',t)\,dp' = i\hbar\dot\psi; \qquad V(x) = \displaystyle\int V_p\exp(-ipx/\hbar)\,dp$

(b)   $\dfrac{p^2}{2m}\,\psi(p,t)+V\!\left(\dfrac{-\hbar}{i}\dfrac{\partial}{\partial p}\right)\psi(p,t) = i\hbar\dot\psi$

(c)   $\dfrac{p^2}{2m}\,\psi-\dfrac{\hbar^2 m\omega^2}{2}\dfrac{\partial^2}{\partial p^2}\,\psi = i\hbar\dot\psi$

The form of equation (c) is therefore the same as in the $x$-representation.

### 4. One-dimensional potential well

$$S_g(k) = e^{2ikl}\,\frac{q\sin ql+ik\cos ql}{q\sin ql-ik\cos ql}$$

$$S_u(k) = e^{2ikl}\,\frac{q\cos ql-ik\sin ql}{q\cos ql+ik\sin ql}$$

The roots of these equations are obtained graphically (figure 61) from the intersections of the curves $q^2+\kappa^2 = 2mV_0 l^2/\hbar^2$ and $q\tan ql = \kappa$, $-q\cot ql = \kappa$ in the plane of $q$ and $\kappa$. $(k = i\kappa; \kappa > 0.)$ The energy of the bound states is $-\hbar^2\kappa^2/2m$, since $S(-i\kappa) = 0$ implies that $\phi$ has only exponentially decaying components, and may therefore be normalized. Even bound states also exist for any arbitrarily small potential $V_0$; in the case of odd functions $V_0$ must be greater than $\hbar^2/2ml^2$ in order that bound states may exist.

### 5. Potential in δ-function form

(a) In the limit $q^2l \to 1/2\lambda$ and $ql \to 0$; hence

$$S_g \to \frac{1+2i\lambda k}{1-2i\lambda k}$$

$$\phi_g = \frac{2i}{1-2i\lambda k}\,(2\lambda k\cos kx - \sin k\,|\,x\,|\,)$$

with a bound state for $\kappa = -ik = 1/2\lambda$;

$$S_u \to 1 \qquad \phi_u = -2i\sin kx$$

with no associated bound state.

(b) If

$$-\frac{\hbar^2}{2m}\,\phi''(x) - \frac{\hbar^2}{2m\lambda}\,\delta(x)\,\phi(x) = \varepsilon\phi(x)$$

is integrated over a very small interval containing $x = 0$, it follows that

$$\phi'(+0) - \phi'(-0) + \phi(0)/\lambda = 0$$

The first derivative of $\phi$ therefore has a discontinuity $-\phi(0)/\lambda$ at $x = 0$, while $\phi(x)$ itself is continuous. If we seek once more to find the odd and even eigenfunctions we obtain the same solutions as in (a). $\phi(0) = 0$ for the odd functions; since there is a vanishing probability of finding the particle at the point at which $V(x) \neq 0$, the odd functions are those of a free particle.

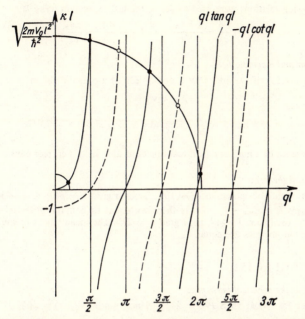

Fig. 61.—Determination of the bound states of the potential well. The intersections of the circular quadrant of radius $\{2mV_0l^2/\hbar\}^{\frac{1}{2}}$ with the curves $ql\tan ql$ and $-ql\cot ql$ give the values of $\kappa$ for the bound states. (Dots correspond to even functions, circles to odd functions.) If the radius is less than $\pi/2$ there is only one even bound state

## 6. *Ehrenfest's theorem and the magnetic field*

In the presence of a magnetic field

$$\mathscr{H} = \frac{1}{2m}\left(\mathbf{p} - \frac{e}{c}\,\mathbf{A}(\mathbf{r},t)\right)^2 + e\phi(\mathbf{r},t);$$

hence the velocity operator is

$$\dot{\mathbf{r}} = \frac{i}{\hbar}(\mathscr{H}\mathbf{r} - \mathbf{r}\mathscr{H}) = \frac{1}{m}\left(\mathbf{p} - \frac{e}{c}\,\mathbf{A}\right)$$

and the *x*-component of $m\ddot{\mathbf{r}}$ is

$$m\ddot{x} = \frac{i}{\hbar}\left[\mathscr{H}\left(p_x - \frac{e}{c}A_x\right) - \left(p_x - \frac{e}{c}A_x\right)\mathscr{H}\right] - \frac{e}{c}\frac{\partial A_x}{\partial t}$$

$$= eE_x + \frac{e}{2c}\left[(\dot{y}H_z - \dot{z}H_y) + (H_z\dot{y} - H_y\dot{z})\right]$$

The following relations were used to derive the last expression (cf. § 7):

$$E_x = -\frac{\partial \phi}{\partial x} - \frac{1}{c}\frac{\partial A_x}{\partial t} \qquad H_x = \frac{\partial A_z}{\partial y} - \frac{\partial A_y}{\partial z}, \ldots$$

### 7. *Current and magnetic field*

As in (14.15), if we add $\phi^* \mathscr{H} \phi$ and $-\phi \mathscr{H} \phi^*$ we again obtain

$$\dot{\rho} + \text{div } \mathbf{j} = 0$$

where $\qquad \rho = \phi^* \phi$ and $\mathbf{j} = \dfrac{\hbar}{2mi}(\phi^* \text{ grad } \phi - \phi \text{ grad } \phi^*) - \dfrac{e}{mc} A \phi^* \phi$

In the presence of a magnetic field real functions also yield a current density $\mathbf{j} \neq 0$.

### 8. *The virial theorem*

For a Coulomb potential $V \sim 1/r$; hence $\mathbf{r}$ grad $V = r\, \partial V/\partial r = -V$. The total energy $E = \overline{E_{kin}} + \overline{V} = \frac{1}{2}\overline{V}$ is therefore equal to half the potential energy.

For the oscillator, $V \sim r^2$, i.e. $\mathbf{r}$ grad $V = 2V$. The mean values of the kinetic and potential energies are equal.

## Chapter BIII (p. 152)

### 1. *Parity of the spherical harmonics*

(1) Let $P\phi = \lambda\phi$: then $P^2\phi = \lambda^2\phi$. Hence, since $P\phi(-\mathbf{r}) = \phi(\mathbf{r})$, $\lambda^2 = 1$.

(2) The relation $PY_{lm}(\theta, \phi) = Y_{lm}(\pi - \theta, \phi - \pi)$ is correct when $m = l$, since $Y_{ll} = \text{const } (\sin \theta)^l e^{il\varphi}$. Since $Y_{lm} = \text{const } \Lambda_-^{l-m} Y_{ll}$ and $P\Lambda_- - \Lambda_- P = 0$, it follows that the relation is valid for all $m$.

### 2. *Angular-momentum matrices*

$$M_x = \frac{\hbar}{\sqrt{2}}\begin{pmatrix} 0 & 1 & 0 \\ 1 & 0 & 1 \\ 0 & 1 & 0 \end{pmatrix} \qquad M_y = \frac{\hbar}{\sqrt{2}}\begin{pmatrix} 0 & -i & 0 \\ i & 0 & -i \\ 0 & i & 0 \end{pmatrix} \qquad M_z = \hbar\begin{pmatrix} 1 & 0 & 0 \\ 0 & 0 & 0 \\ 0 & 0 & -1 \end{pmatrix}$$

$$M^2 = M_x{}^2 + M_y{}^2 + M_z{}^2 = 2\hbar^2\begin{pmatrix} 1 & 0 & 0 \\ 0 & 1 & 0 \\ 0 & 0 & 1 \end{pmatrix}$$

Since $\overline{M_x} = (M_x)_{lm,\,lm} = 0$ for all $m$,

$$(\Delta M_x)^2 = \overline{M_x{}^2} = \sum_{l'm'} (M_x)_{lm,\,l'm'}(M_x)_{l'm',\,lm}$$

Since $M_x$ possesses no matrix elements for which $l' \neq l$ the $\overline{M_x{}^2}$ are directly obtained as the diagonal elements in the squares of the above matrices. Hence

$$\Delta M_x = \Delta M_y = \begin{cases} \hbar/\sqrt{2} & \text{for } m = \pm 1 \\ \hbar & \text{for } m = 0 \end{cases}$$

for a total angular momentum $\hbar\sqrt{2}$.

In the case of a top with total angular momentum $M = \hbar\sqrt{2}$ precessing about the cones shown in figure 62, the time averages of $\overline{M_x{}^2}$ and $\overline{M_y{}^2}$ are equal to the above values.

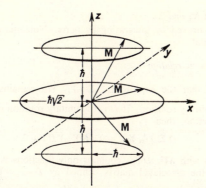

Fig. 62.—Representation of an angular-momentum vector and its dispersion due to a precessing moment about the *z*-axis

### 3. *Eigenfunctions of the electron spin*

(a) The matrix of $s_{\theta, \varphi}$ is

$$s_{\theta, \varphi} = \tfrac{1}{2}\hbar \begin{pmatrix} \cos \theta & \sin \theta \, e^{-i\varphi} \\ \sin \theta \, e^{i\varphi} & -\cos \theta \end{pmatrix}$$

The eigenvalues of this matrix are $\pm \tfrac{1}{2}\hbar$; this may be seen at once because the sum of the eigenvalues (the sum of the diagonal elements) is zero and the product (which is the determinant) is $-\tfrac{1}{4}\hbar^2$. Then the relations

$$\cos \theta . a + \sin \theta \, e^{-i\varphi} . b = \pm a, \qquad \sin \theta \, e^{i\varphi} . a - \cos \theta . b = \pm b$$

give the normalized eigenstates corresponding to the eigenvalues $\pm \tfrac{1}{2}\hbar$:

$$\begin{pmatrix} a_+ \\ b_+ \end{pmatrix} = \begin{pmatrix} \cos \tfrac{1}{2}\theta \, e^{-i\varphi} \\ \sin \tfrac{1}{2}\theta \end{pmatrix} \text{ and } \begin{pmatrix} a_- \\ b_- \end{pmatrix} = \begin{pmatrix} -\sin \tfrac{1}{2}\theta \, e^{-i\varphi} \\ \cos \tfrac{1}{2}\theta \end{pmatrix}$$

(b) The required rotation operator is

$$D = \exp - \frac{i}{\hbar} \, \phi s_z . \exp - \frac{i}{\hbar} \, \theta s_y = \exp -\tfrac{1}{2}i\phi\sigma_z . \exp -\tfrac{1}{2}i\theta\sigma_y$$

Now since $\sigma_y{}^2 = \sigma_z{}^2 = 1$,

$$\exp -\tfrac{1}{2}i\theta\sigma_y = \sum_\nu (-\tfrac{1}{2}i\theta)^\nu \frac{\sigma_y{}^\nu}{\nu!} = \sum_{\nu \, even} (-\tfrac{1}{2}i\theta)^\nu \frac{1}{\nu!} + \sum_{\nu \, odd} (-\tfrac{1}{2}i\theta)^\nu \frac{1}{\nu!} \, \sigma_y$$

$$= \cos \tfrac{1}{2}\theta - i \sin \tfrac{1}{2}\theta . \sigma_y$$

Hence $\qquad D = (\cos \tfrac{1}{2}\phi - i \sin \tfrac{1}{2}\phi\sigma_z)(\cos \tfrac{1}{2}\theta - i \sin \tfrac{1}{2}\theta\sigma_y)$

The application of this operator to $\alpha = \begin{pmatrix} 1 \\ 0 \end{pmatrix}$ and $\beta = \begin{pmatrix} 0 \\ 1 \end{pmatrix}$ gives

$$D\alpha = \begin{pmatrix} a_+ \\ b_+ \end{pmatrix} = \begin{pmatrix} \cos \tfrac{1}{2}\theta \exp -\tfrac{1}{2}i\phi \\ \sin \tfrac{1}{2}\theta \exp \tfrac{1}{2}i\phi \end{pmatrix} \qquad D\beta = \begin{pmatrix} a_- \\ b_- \end{pmatrix} = \begin{pmatrix} -\sin \tfrac{1}{2}\theta \exp -\tfrac{1}{2}i\phi \\ \cos \tfrac{1}{2}\theta \exp \tfrac{1}{2}i\phi \end{pmatrix}$$

(c) The two states thus determined are identical, differing only by a factor $\exp \tfrac{1}{2}i\phi$ of magnitude 1.

#### 4. *Relation between* $\Lambda_+$ *and* $\Lambda_-$

The relation is proved by integrating by parts. Note the factor $\sin \theta$ in the normalization integral.

#### 5. *A spherical harmonic relation*

We first show that the expression $\sum\limits_{m=-l}^{l} Y_{lm}^* Y_{lm}$ remains unchanged when subjected to an arbitrary rotation. If $D$ is the unitary rotation operator and $D Y_{lm}$ is the rotated function, then

$$D \sum Y_{lm}^* Y_{lm} = \sum (D Y_{lm})^* D Y_{lm}$$

Since $D$ commutes with $\mathbf{M}^2$, $D Y_{lm}$ contains only functions with the same value of $l$. If $D_{mm'}$ is the associated matrix defined by $D_{mm'} = (Y_{lm'}, D Y_{lm})$, then

$$D Y_{lm} = \sum_{m'} D_{mm'} Y_{lm'}$$

Making use of the unitariness property of the rotation matrices,

$$\sum D_{mm'}^* D_{mm''} = \delta_{m' m''}$$

we then obtain

$$D \sum_m Y_{lm}^* Y_{lm} = \sum_{mm'm''} D_{mm'}^* Y_{lm'}^* D_{mm''} Y_{lm''} = \sum_{m'} Y_{lm'}^* Y_{lm'}$$

We have thus proved that $\sum Y_{lm}^* Y_{lm}$ is invariant to rotation, and therefore independent of $\theta$ and $\phi$:

$$\sum_m Y_{lm}^* Y_{lm} = C$$

The numerical value of the constant $C$ is obtained by integrating the above relation over a unit sphere, taking account of the normalization condition $\int Y_{lm}^* Y_{lm} d\Omega = 1$:

$$\int \sum_m Y_{lm}^* Y_{lm} d\Omega = \sum_{m=-l}^{l} 1 = 2l+1, \qquad \int C d\Omega = 4\pi C$$

$C$ therefore has the value $(2l+1)/4\pi$.

### Chapter BIV (p. 175)

#### 1. *First-order perturbation calculation for an anharmonic oscillator*

$$E_n = (n + \tfrac{1}{2})\hbar\omega + \frac{3\lambda}{2}\left(\frac{\hbar}{m\omega}\right)^2 \{(n+\tfrac{1}{2})^2 + \tfrac{1}{4}\}$$

Compare the solution of Exercise 3, p. 368.

#### 2. *Quadratic secular equation*

$$E_{1,2} = \tfrac{1}{2}(\varepsilon_1 + \varepsilon_2) + \tfrac{1}{2}\lambda(W_{11} + W_{22}) \pm \{[\tfrac{1}{2}(\varepsilon_1 - \varepsilon_2) + \tfrac{1}{2}\lambda(W_{11} - W_{22})]^2 + \lambda^2 |W_{12}|^2\}^{1/2}$$

$$= \tfrac{1}{2}(\varepsilon_1 + \varepsilon_2) + \tfrac{1}{2}\lambda(W_{11} + W_{22}) \pm \left\{\tfrac{1}{2}(\varepsilon_1 - \varepsilon_2) + \tfrac{1}{2}\lambda(W_{11} - W_{22}) + \lambda^2 \frac{|W_{12}|^2}{\varepsilon_1 - \varepsilon_2}\right\} + \dots$$

Perturbation theory gives the same result.

### 3. *A primitive model of the electron in a metal*

Put

$$\frac{1}{\sqrt{l}} e^{ikx} = \frac{1}{\sqrt{l}} e^{iKx} e^{i\kappa x} = \phi_{K\kappa}, \qquad K = \frac{2\pi}{a} n \qquad -\frac{\pi}{a} \leqq \kappa < \frac{\pi}{a}$$

Then

$$T_a \phi_{K,\kappa} = e^{i\kappa a} \phi_{K,\kappa}$$

Therefore all matrix elements $\mathscr{H}_{K\kappa\,K'\kappa'}$ of $\mathscr{H}$ vanish for $\kappa \neq \kappa'$, and

$$E_{K,\kappa} = \frac{\hbar^2}{2m} (K+\kappa)^2 + V_{00} + \sum_{K' \neq K} \frac{2m|V_{(K-K')}|^2}{\hbar^2\{(K+\kappa)^2 - (K'+\kappa)^2\}}$$

$$V_{K\kappa, K'\kappa} = \frac{1}{l} \int V e^{i(K-K')x} \, dx = V_{(K-K')}$$

This expansion fails when $(K+\kappa)^2 \approx (K'+\kappa)^2$, i.e. when $K+K' \approx -2\kappa$, which only occurs when $\kappa \approx \pm \pi/a, 0$. The degeneracy is removed by setting up the following determinant and solving for $\varepsilon$:

$$\begin{vmatrix} \dfrac{\hbar^2}{2m}(K+\kappa)^2 - \varepsilon & V_{(K'-K)} \\[2mm] V_{(K-K')} & \dfrac{\hbar^2}{2m}(K'+\kappa)^2 - \varepsilon \end{vmatrix} = 0$$

whence

$$\varepsilon_{1,2} = \frac{\hbar^2}{2m} \frac{(K+\kappa)^2 + (K'+\kappa)^2}{2} \pm \sqrt{\left[\hbar^4\left(\frac{(K+\kappa)^2 - (K'+\kappa)^2}{4m}\right)^2 + |V_{(K-K')}|^2\right]}$$

The form of $\varepsilon(\kappa)$ is shown graphically in figure 63.

### 4. *Effective scattering cross-section for a screened Coulomb potential*

$$dQ = \left(\frac{2me^2}{\hbar^2}\right)^2 \frac{d\Omega}{\left(\dfrac{1}{a^2} + |\mathbf{k_0}' - \mathbf{k_0}|^2\right)^2} = \left(\frac{2me^2}{\hbar^2}\right)^2 \frac{d\Omega}{\left(\dfrac{1}{a^2} + 4k_0^2 \sin^2 \tfrac{1}{2}\theta\right)^2}$$

If $a$ is allowed to tend to infinity and the kinetic energy $\hbar^2 k^2/2m$ of the incident particle is represented by $E$ as in the classical formula for scattering, we surprisingly obtain the same formula $dQ = \left(\dfrac{e^2}{4E}\right)^2 \dfrac{d\Omega}{\sin^4 \tfrac{1}{2}\theta}$ as in the classical case. Further, the exact quantum-mechanical calculation always gives the classical result in the case of the Coulomb potential for any value of the energy. The total effective cross-section is

$$Q = \left(\frac{4me^2a^2}{\hbar^2}\right)^2 \frac{\pi}{1 + 4a^2k_0^2}$$

If $ak_0 \gg 1$, $Q$ is proportional to the square of the wavelength. For a pure Coulomb potential the effective cross-section is divergent.

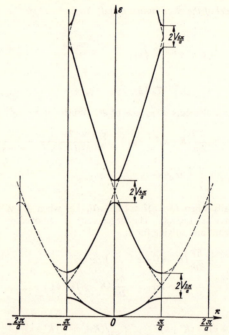

Fig. 63.—Energy perturbation of free electrons in a periodic potential field

## 5. *Polarizability of a spherically symmetrical atom*

The energy expressed as a function of $\beta$ is derived as in §28, and is

$$E(\beta) = E_0 + 2\beta\overline{W^2} + \beta^2\{\overline{W\mathscr{H}_0 W} - E_0 \overline{W^2}\}$$

Since the state $\phi_0$ over which the average is taken satisfies the equation $\mathscr{H}_0 \phi_0 = E_0 \phi_0$, we may replace $E_0 \overline{W^2}$ by $\overline{\mathscr{H}_0 W^2}$, $\overline{W^2 \mathscr{H}_0}$, or $\frac{1}{2}(\overline{\mathscr{H}_0 W^2 + W^2 \mathscr{H}_0})$. The factor of $\beta^2$ then only contains the commutations:

$$\overline{W\mathscr{H}_0 W} - \frac{1}{2}(\overline{\mathscr{H}_0 W^2 + W^2 \mathscr{H}_0}) = \frac{1}{2}\overline{[[W, \mathscr{H}_0], W]} = \frac{e^2 F^2 \hbar^2 N}{2m}$$

Only the kinetic part of $\mathscr{H}_0$ is of importance in $[W, \mathscr{H}_0]$ (i.e. $[W, \mathscr{H}_0] \sim \sum_i p_{x_i}$),

and the remaining commutation then merely contains the simple commutation relations between $p_{x_i}$ and $x_i$. The energy at the minimum is

$$E = E_0 - \frac{\{\overline{W^2}\}^2}{\overline{W\mathscr{H}_0 W} - E_0 \overline{W^2}} = E_0 - \frac{2e^2 m}{\hbar^2 N}\{\overline{(\sum x_i)^2}\}^2 F^2$$

$$= E_0 - \frac{2}{Na}\{\overline{(\sum x_i)^2}\}^2 F^2$$

and the polarizability is thus $\alpha = 4/Na\{\overline{(\Sigma x_i)^2}\}^2$, where $a$ is the Bohr radius. The mixed terms may not be omitted when forming the mean value of $(\Sigma x_i)^2$:

$$\overline{(\Sigma x_i)^2} = \sum_i \overline{x_i^2} + \sum_{i \neq k} \overline{x_i x_k}$$

It is true that $\Sigma \overline{x_i}$ vanishes for spherically symmetrical states, but $\sum_{i \neq k} \overline{x_i x_k}$ does not do so, because the electrons tend to avoid each other as a result of the mutual Coulomb repulsion. This effect is termed correlation, and is referred to in §38.

## Chapter CI (p. 191)

1. *Energy terms and degree of degeneracy for a three-dimensional oscillator*

The Schrödinger equation gives

$$\sum_{v=0}^{\infty} \{r^{v-2}c_v[v(v-1) + 2v(l+1)] + r^v c_v[\beta - 2\alpha(v+l+3/2)]\} = 0$$

where $\alpha = m\omega/\hbar$ and $\beta = 2mE/\hbar^2$. Comparison of the coefficients associated with the same powers of $r$ shows that the odd coefficients $c_1, c_3, c_5, \dots$ must vanish. When $v$ is even and equal to $2n_r$, the power series terminates only when $\beta = 2\alpha(2n_r + l + 3/2)$ or $E = \hbar\omega(2n_r + l + 3/2)$, where $n_r$ and $l$ can extend over all positive integers and zero. The number of terms corresponding to a given $n = 2n_r + l$ is $(n+1)(n+2)/2$. Even values of $l$ are associated with even values of $n$, and similarly for odd values. The results for the first few terms are summarized in the following table:

| Quantum number $n$: | 0 | 1 | 2 | 3 | 4 | 5 |
|---|---|---|---|---|---|---|
| Energy $E$, excluding zero point energy | 0 | $\hbar\omega$ | $2\hbar\omega$ | $3\hbar\omega$ | $4\hbar\omega$ | $5\hbar\omega$ |
| Degeneracy | 1 | 3 | 6 | 10 | 15 | 21 |
| Term symbols | $s$ | $p$ | $s,d$ | $p,f$ | $s,d,g,$ | $p,f,h,$ |

2. *Hermitian nature of the Hamiltonian operator expressed in polar coordinates*

$$\int \psi_1^* (p_r \psi_2) r^2 \sin\theta \, dr \, d\theta \, d\phi = \int \left( \frac{1}{r^2} p_r r^2 \psi_1 \right)^* \psi_2 r^2 \sin\theta \, dr \, d\theta \, d\phi$$

as may be seen by integrating by parts. In order that the integrated part should vanish, the wave functions must vanish sufficiently strongly at infinity and be regular at $r = 0$. Then $p_r^+ = (1/r^2).p_r r^2$, and hence the radial component of the kinetic energy

$$-\frac{\hbar^2}{2m} \frac{1}{r^2} \frac{\partial}{\partial r} r^2 \frac{\partial}{\partial r} = \frac{1}{2m} p_r^+ p_r$$

is a Hermitian operator.

3. *Probability function and Bohr orbits of the hydrogen atom*

When $l = n - 1$ the radial component of the wave function takes the form

$$R = \text{const. } r^{n-1} \exp -\frac{r}{na_0}$$

Therefore   $w(r) = (Rr)^2 = $ const. exp $2n(\log r - r/n^2 a_0)$,   with   a   maximum   at $r_n = n^2 a_0$, the radius of the $n$th Bohr orbit. The deviation is obtained by expanding the exponent about the maximum value:

$$w(r) \approx \text{const. exp} \left[ -(r - r_n)^2/n^3 a_0^2 \right]$$

Hence   $$\Delta r/\bar{r} = 1/\sqrt{(2n)}$$

The relative deviations therefore decrease proportionately to $1/\sqrt{n}$.

Comparison with (29.12) and (29.13): since the deviations are small, we must have

$$\Delta \frac{1}{r} \bigg/ \frac{\bar{1}}{r} \approx \Delta r/\bar{r}$$

this result may also be proved from the fact that

$$\Delta \frac{1}{r} = \sqrt{\left[ \left( \frac{\overline{1}}{r} \right)^2 - \frac{\overline{1}^2}{r} \right]}$$

and   $$\frac{\overline{1}}{r^2} = \frac{2}{(2l + 1)n^3 a_0^2} \approx \frac{1}{n^4 a_0^2} + \frac{1}{2n^5 a_0^2} \text{ for } n \gg 1$$

## Chapter C II (p. 206)

1. *Detailed description of the Stern-Gerlach experiment for a neutral particle of spin* $1/2$

(a) If the trial solution

$$\Psi(\mathbf{r}, t) = \sum_s \phi_s(\mathbf{r}, t) \, \kappa_s^z \text{ whereby } \sigma_z \kappa_s^z = s\kappa_s^z \text{ with } s = \pm 1$$

is introduced into the equation of motion, we obtain

$$\left( \frac{\mathbf{p}^2}{2m} + \mu_B H(z)s \right) \phi_s(\mathbf{r}, t) = i\hbar \dot{\phi}_s(\mathbf{r}, t) \text{ for } s = \pm 1$$

This result may be verified by inserting $\Psi$ into the equation of motion and forming the scalar product with $\kappa_1$ or $\kappa_{-1}$; it follows from the orthogonal property of the two spin functions. In the two equations $\pm \mu_B H(z)$ occurs as the potential energy and $\pm \mu_B \partial H/\partial z$ as the force. The initial conditions are clearly

$$\phi_s(\mathbf{r}, 0) = \phi_0(\mathbf{r})(\kappa_s^z, \chi_0) = \phi_0(\mathbf{r})\chi_0(s)$$

In view of the breadth of the wave packet the resultant motion approximately follows the laws of classical mechanics; therefore, when $\partial H/\partial z$ is negative (as in figure 32), the wave packet $\phi_{+1}$ corresponding to $s = +1$ moves parallel to the positive $x$-axis like a particle of moment $\mu_B$ in this direction, and similarly in the case of $\phi_{-1}$ in the opposite direction. We finally obtain two spatially separated wave packets, each of which has an eigenfunction $\kappa_s^z$ as factor. In view of the equation of motion, $\int |\phi_s|^2 d\mathbf{r}$ is independent of the time and equal to $|(\kappa_s^z, \chi_0)|^2$, since $\phi_0$ is normalized. Therefore $|(\kappa_s^z, \chi_0)|^2 = |\chi_0(s)|^2$ is the probability of finding the particle in the component beam denoted by $s$; in experiments with many particles $|\chi_0(s)|^2$ is proportional to the intensity of the beam.

(b) In this case the two probabilities correspond to $|(\kappa_{s'}^\theta, \chi_0)|^2$. As in Exercise 3, p. 375,   $\kappa_{s'}^{\theta,\phi} = \kappa_{s'}^z \cos \frac{1}{2}\theta - s' \kappa_{-s'}^z \sin \frac{1}{2}\theta \exp is'\phi$ with $\sigma_z \kappa_s^z = s\kappa_s^z$.

In case (*a*), for $\theta = 0$, $\kappa_s^{\theta, \phi}$ becomes $\kappa_s^z$.

In case (*b*), when $\theta = \pi/2$, $\phi = \pi/2$, therefore we obtain the following values for the intensities:

$$\kappa_{s'}^{\pi/2} = (1/\sqrt{2})(\kappa_{s'}^z + i\kappa_{-s'}^z) \text{ and } \tfrac{1}{2}|\chi_0(+1) \pm i\chi_0(-1)|^2 \text{ for } s' = \pm 1$$

The above analysis applies to a hydrogen or alkali atom in the ground state (when the appropriate mass is introduced), since in these cases there is merely a spin moment, and no orbital moment is present. For charged particles such as electrons the Stern-Gerlach splitting is completely masked by the effect of the Lorentz force.

## 2. *Multiple Stern-Gerlach experiment*

From the previous exercise we see that, when the component beam $s_z$ is again split by a magnetic field parallel to $\theta$, the resultant relative intensities are

$$|(\kappa_s^z, \kappa_{s'}^\theta)|^2 = \cos^2 \tfrac{1}{2}\theta \text{ or } \sin^2 \tfrac{1}{2}\theta$$

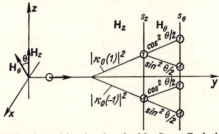

Fig. 64.—Relative intensities in the double Stern-Gerlach experiment

The value of these quantities is derived from the eigenfunctions $\kappa_{s'}^\theta$ of the previous solution (figure 64). In the case of the triple experiment we first have $|\kappa_s^z, \kappa_{s'}^x|^2 = \tfrac{1}{2}$, then $|\kappa_s^x, \kappa_{s'}^z|^2 = \tfrac{1}{2}$ (figure 65). We can see from this experiment that it is clearly impossible to make a simultaneous and precise determination of the values of two non-commuting quantities $\sigma_z$ and $\sigma_x$. In the first experiment we

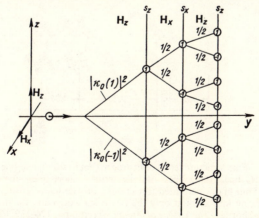

Fig. 65.—Relative intensities in the triple Stern-Gerlach experiment

measure $\sigma_z$, selecting the upper beam, for which $\sigma_z$ is exactly equal to 1; in the second experiment we measure $\sigma_x$, again selecting the upper beam, for which $\sigma_x = 1$; if we now attempt to measure $\sigma_z$ again in a third experiment we do not obtain a sharp result.

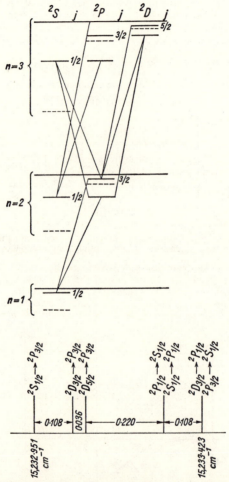

Fig. 66.—Fine-structure terms of hydrogen (not to scale), and fine structure of the $H_\alpha$ line. Long lines—Balmer terms. Short lines—fine-structure terms. Thin lines—permitted transitions. Broken lines—positions of the terms taking into account only the relativistic correction for mass (from Condon-Shortley, *The Theory of Atomic Spectra*, pp. 124–139, Cambridge 1953)

### 3. *Spin precession in a magnetic field*

The solution corresponding to the initial state $\chi(0)$ is

$$\chi(t) = \exp\left\{\frac{i}{\hbar}(\vec{\mu} H)t\right\} \chi(0) = \exp\left\{-\frac{2i}{\hbar}(s\vec{\omega}_L)t\right\} \chi(0)$$

Comparison with (23.2) shows that $\chi(t)$ is produced from $\chi(0)$ by a rotation about H through an angle $2\omega_L t$.

### 4. *Fine structure of the hydrogen spectrum*

Note that $\Delta l = \pm 1$, $\Delta j = 0, \pm 1$. A study of equations (34.3) and (34.5) leads to the term scheme shown in figure 66.

## Chapter C III (p. 238)

### 1. *Zeeman effect for the sodium D-line*

The Zeeman splitting and spin-orbit splitting are of approximately equal magnitude when $\Delta E_{\text{spin-orbit}} \approx \mu_B H$; $\Delta E$ is determined from the separation of the D-lines $\Delta\lambda = 6$ A.U.   $\Delta E = hc\Delta\lambda/\lambda^2$, where $\lambda = 5893$ A.U. This leads to the result $H \approx 10^5$ oersteds.

### 2. *Doublet eigenfunctions*

The eigenfunctions are obtained by applying $(j - m_j)$ times the operator $J_- = L_- + \sigma_-$ to $Y_{ll}\alpha$ (for $j = l + \frac{1}{2}$), and by finding the orthogonal combinations for $j = l - \frac{1}{2}$. The result is:

$$Y_{ljm_j} = \begin{cases} Y_{00}\alpha \text{ and } Y_{00}\beta & \text{for } l=0 \\[2mm] \sqrt{\left(\dfrac{l+m_j+\frac{1}{2}}{2l+1}\right)} Y_{l,\,m_j-\frac{1}{2}}\alpha - \sqrt{\left(\dfrac{l-m_j+\frac{1}{2}}{2l+1}\right)} Y_{l,\,m_j+\frac{1}{2}}\beta & \text{for } l=j-\frac{1}{2} \\[2mm] \sqrt{\left(\dfrac{l-m_j+\frac{1}{2}}{2l+1}\right)} Y_{l,\,m_j-\frac{1}{2}}\alpha + \sqrt{\left(\dfrac{l+m_j+\frac{1}{2}}{2l+1}\right)} Y_{l,\,m_j+\frac{1}{2}}\beta & \text{for } l=j+\frac{1}{2} \end{cases}$$

The angular components of the matrix elements
$(Y_{l'j'm'_j}|\mathbf{r}|Y_{ljm_j}) = r_{al'j'm'_j;\,ljm_j}$ are given for the sodium D-lines in the following table:

| Transition | $m_j'$ | $m_j$ | $18a_x^2$ | $18a_y^2$ | $18a_z^2$ |
|---|---|---|---|---|---|
| $2P_{3/2}$ | 3/2 | 1/2 | 3 | 3 | 0 |
|  | 3/2 | −1/2 | 0 | 0 | 0 |
| ↓ | 1/2 | 1/2 | 0 | 0 | 4 |
| $2S_{1/2}$ | 1/2 | −1/2 | 1 | 1 | 0 |
| $2P_{1/2}$ | 1/2 | 1/2 | 0 | 0 | 2 |
| ↓ $2S_{1/2}$ | 1/2 | −1/2 | 2 | 2 | 0 |

The intensities of the Zeeman components for a weak field are calculated in accordance with §§30 and 36, using the above table. Note that $\Delta m_j = 0, \pm 1$, $\Delta l = \pm 1$, $\Delta j = 0, \pm 1$.

### 3. *Ionization due to the tunnel effect*

From p. 105, the probability of ionization of a hydrogen atom owing to the tunnel effect is

$$w\,dt \approx \omega D\,dt \qquad \hbar\omega = \frac{e^2}{2a_0} \approx \hbar \times 10^{16}\mathrm{s}^{-1}$$

$$D \approx e^{-\kappa l} \qquad \kappa \approx \sqrt{\left(\frac{2m}{\hbar^2}(V_0 - E_0)\right)}$$

From figure 67:

$$l \approx \frac{-e}{2a_0 F} \qquad \kappa \approx \frac{1}{a_0} \to D \approx \exp - \frac{e}{2a_0{}^2 F} \approx 10^{-10^3/F}$$

When $F = 10^6\,\mathrm{V/cm}$ the lifetime is therefore of the order of $10^{1000}\mathrm{s}$ (one year $\approx$ $10^7\mathrm{s}$, age of the earth $\approx 10^9$ years). The state is therefore practically stationary.

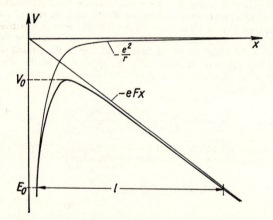

Fig. 67.—Potential of an electron in the electric field

### 4. *Spectral line intensities*

The $x$- and $y$-components of the transition elements, for which $m = \pm 1$, $m' = 0$, are $a_x{}^2 = a_y{}^2 = 1/6$, calculated from formulae (25.9/25.11); the $z$-component (for which $m = 0$, $m' = 0$) is $a_z{}^2 = 1/3$. From (29.11), the radial portion of the eigenfunction is

$$\int R_{10} r R_{21} r^2 dr = \frac{24}{\sqrt{6}}\left(\frac{2}{3}\right)^5 a_0$$

the $x$-, $y$-, and $z$-components of $f$ therefore have the values $f_{m0}$ shown in the table below:

| $m$ | $x$ | $y$ | $z$ |
|-----|------|------|------|
| 1 | 0·21 | 0·21 | 0 |
| 0 | 0 | 0 | 0·42 |
| −1 | 0·21 | 0·21 | 0 |

Hence
$$f(x) = f(y) = f(z) = \Sigma f_{m0} = 0 \cdot 42$$
$$2p,1s \quad 2p,1s \quad 2p,1s$$

The lifetime $\tau$ is given by

$$\frac{1}{\tau} = \frac{2}{3} \frac{e^2\omega^2}{mc^3} \frac{2m}{\hbar} \omega_{01} |\mathbf{r}_{01}|^2$$

and is about $10^{-9}$s for all $p$-states (for which $m = \pm 1, 0$).

### 5. Oscillator strengths for the principal series of sodium

No.

The summation theorem $\Sigma_\nu f_{\nu\mu} = 1$ is valid only when $\nu$ extends over all quantum numbers of a complete system. For sodium, however, the initial state $\mu$ is $3s$. The transition from this state to the $2p$-state is forbidden by Pauli's principle, because the $2p$-level is occupied; this term is therefore absent from the sum of the $f$-values. Since the $2p$-energy level is lower than the $3s$-level this term is negative, and the experimental value of the $f$-sum is therefore greater than unity. For lithium, the $f$-value is zero for the transition of the valence electron to the ground state which is occupied by the electrons of the inner shell, as a result of the selection rule $\Delta l \neq 0$; in this case, therefore, the sum of the $f$-values must be unity.

### 6. Polarizability of the alkali atoms

$$\phi = \phi_0 + \underset{i \neq 0}{\Sigma} \frac{W_{i0}(E_i - E_0)}{(E_0 - E_i)(E_i - E_0)} \phi_i$$

From the above sum the square of the excitation energy can be removed from the denominator, and the summation can then be effected over all $i$:

$$\phi = \phi_0 + \frac{eF}{E_a^2} \Sigma_i x_{i0}(E_i - E_0)\phi_i$$

Since
$$x_{i0}(E_i - E_0) = (\mathcal{H}x - x\mathcal{H})_{i0} = -\frac{i\hbar}{m} p_{xi0}$$

$$\phi = \phi_0 - \frac{eFi\hbar}{E_a^2 m} \Sigma_i p_{xi0}\phi_i = \phi_0 - \frac{eFi\hbar}{E_a^2 m} p_x \phi_0$$
$$= \phi_0 - \beta \frac{\partial}{\partial x} \phi_0$$

where the displacement $\beta$ is given by $\beta = eF\hbar^2/E_a^2 m$.

The polarizability $\alpha$ is the factor of $F$ in the dipole moment $e\beta$:

$$\alpha = \frac{e^2\hbar^2}{mE_a^2} = 4a^3 \left(\frac{I_H}{E_a}\right)^2$$

where $a$ is the Bohr radius and $I_H$ the ionization energy of the hydrogen atom. The experimental and the calculated values agree very well:

|  | Li | Na | K | Rb | Cs |
|---|---|---|---|---|---|
| $\alpha_{th}(10^{-24}\,\text{cm}^3)$ | 32·6 | 25·2 | 43 | 45·5 | 57·2 |
| $\alpha_{exp}(10^{-24}\,\text{cm}^3)$ | 27 | 27 | 46 | 50 | 61 |

### 7. Refractive index of free electrons

From the equation of motion $m\ddot{r} = eE_0 e^{i\omega t} = eE$ we obtain $er = \alpha E$, where $\alpha = -e^2/m\omega^2$. Hence $n = \sqrt{\varepsilon} = \sqrt{(1 + 4\pi N\alpha)}$ is imaginary when

$$\frac{\omega^2}{c^2} = \left(\frac{2\pi}{\lambda}\right)^2 < \frac{4\pi e^2}{mc^2} N$$

For the alkali metals Li, Na, K, Rb, Cs: $\lambda = 1560, 2100, 2900, 3100, 3550\,\text{A.U.}$ Cf. the experimental values (R. W. Wood, *Physical Optics*, New York, 1934, p. 560) 2050, 2150, 3150, 3600, 4400 A.U.

$N \approx 10^6\,\text{cm}^{-3}$ for the ionosphere.

## Chapter DI (p. 271)

### 1. Energy of an ideal Fermi gas

The density of the k-values is $V/(2\pi)^3$. Hence the number of k-vectors within a sphere of radius $k_0$ is $(4\pi/3)k_0^3 \, V/(2\pi)^3$, which must be equal to $n/2$ for the ground state; thus $k_0^3 = 3\pi^2 n/V$. The smallest wavelength is therefore of the same order of magnitude as the mean distance between particles $(V/n)^{1/3}$. In addition:

$$E_0 = 2 \sum_{|\mathbf{k}|<k_0} \frac{\hbar^2}{2m} \mathbf{k}^2 = \frac{\hbar^2}{m} \frac{V}{(2\pi)^3} \int_{|\mathbf{k}|<k_0} k^4 \, dk = n \frac{3}{5} \frac{\hbar^2 k_0^2}{2m}$$

The mean energy per particle is 3/5 of the maximum energy.

### 2. Correlations for the ideal Fermi gas

(a) $$w = \frac{1}{n(n-1)V^2} \sum_{\substack{|\mathbf{k}|<k_0 \\ |\mathbf{k}'|<k_0}} \{1 - \delta_{s_1 s_2} \exp[i(\mathbf{k} - \mathbf{k}')(\mathbf{r}_1 - \mathbf{r}_2)]\}$$

When evaluating the summation it should be noted that $(\kappa_{s_1}, \kappa_{s_2}) = \delta_{s_1, s_2}, = \delta_{s_1, s_2}^2$, etc. Integrating over $\mathbf{k}$ and $\mathbf{k}'$ we obtain

$$w = \frac{n}{4(n-1)V^2} \{1 - \delta_{s_1 s_2} F(k_0 r)\}$$

where $$F(k_0 r) = \left\{3\frac{\sin k_0 r - k_0 r \cos k_0 r}{(k_0 r)^3}\right\}^2$$

(b) $$\rho_{s_1, s_2}(r) = \frac{n}{2V}\{1 - \delta_{s_1 s_2} F(k_0 r)\}$$

$$\rho_{11} = \frac{n}{2V}\{1 - F(k_0 r)\}; \qquad \rho_{12} = \frac{n}{2V}; \qquad \rho_{11} + \rho_{12} = \frac{n}{V}\{1 - \tfrac{1}{2}F(k_0 r)\}$$

When observed from a particle of spin $s_1 = 1$, the particles with $s_2 = -1$ possess a constant density, unaffected by the presence of the first particle; in contrast, the density of the particles with the same spin is reduced almost to zero in the neighbourhood of the point of observation, up to distances of the order of $1/k_0$

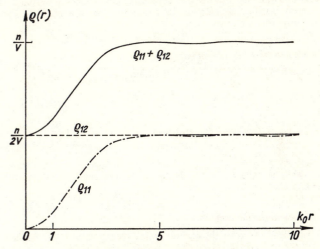

Fig. 68.—Densities in an ideal Fermi gas

(figure 68). (Note that $F(0) = 1$, corresponding to the "Fermi hole".) There is almost zero probability of finding two particles with the same spin at a distance less than $1/k_0$ from each other. Only half the "Fermi hole" occurs in the expression for the total density; the Fermi statistics therefore imply that the density in the neighbourhood of each particle is reduced by a factor of two, relative to the average density.

### 3. *Polarizability of helium*

From Exercise 5, p. 176:

$$\alpha = \frac{4}{2a_0} \{\overline{(x_1 + x_2)^2}\}2 = \frac{8}{a_0} a'^4, \text{ since } \overline{x_1^2} = \overline{x_2^2} = a'^2, \ \overline{x_1 x_2} = 0$$

Therefore
$$\alpha = \frac{8}{Z'^4} a_0^3 \approx a_0^3 = 0 \cdot 15 \times 10^{-24} \text{ cm}^3$$

The experimental value is $0 \cdot 2 \times 10^{-24}$ cm³. If we wish to make a more precise evaluation, using the method of variations and an initial function for the ground state with optimum $Z'$, the calculation follows the same lines as that in Exercise 5, p. 176. It should be noted, however, that the initial function is not an exact eigenfunction of the Hamiltonian operator in the absence of the electric field. The result is practically identical with that given above.

### 4. Rotation terms of the deuterium molecule

The complete wave function must be symmetric in the coordinates of the nuclei. Symmetric spin functions correspond to even values of $l$, antisymmetric spin functions to odd values. Six symmetric spin functions may be constructed from the two spins of value 1; one function with five-fold degeneracy corresponding to spin 2, and one function corresponding to spin 0. In addition, there are three antisymmetric functions corresponding to spin 1.

### 5. The exchange integral and spin interchange in the hydrogen molecule

The wave function may be put into the form

$$\phi(t) = 2c \exp\left\{-\frac{it}{\hbar}(K + \tfrac{1}{2}J)\right\} \left\{ [a(1)b(2)\,\alpha(1)\,\beta(2) - a(2)b(1)\,\alpha(2)\,\beta(1)] \cos\frac{Jt}{\hbar} \right.$$
$$\left. + i\,[a(1)b(2)\,\beta(1)\,\alpha(2) - a(2)b(1)\,\alpha(1)\,\beta(2)] \sin\frac{Jt}{\hbar}\right\}$$

$K$ and $J$ are given by (45.12). At the instant $t = 0$ atom $a$ has an $\alpha$-spin, atom $b$ a $\beta$-spin, while at time $t = \pi\hbar/2J$ the situation is reversed. From (45.12), $J \approx A$ when $S^4$ and $CS^2$ can be neglected.

### 6. Dirac's identity for the spin vector

It is sufficient to prove the identity when the operator $P^s$ is applied to the functions

$$\alpha(1)\,\alpha(2), \quad \alpha(1)\,\beta(2), \quad \beta(1)\,\alpha(2), \quad \beta(1)\,\beta(2)$$

because any function can be expressed in terms of them. Consider as an example the relation

$$\vec{\sigma}_1\,\vec{\sigma}_2 = 2(\sigma_1^+\,\sigma_2^- + \sigma_1^-\,\sigma_2^+) + \sigma_{1z}\,\sigma_{2z}$$

### 7. Eigenfunctions of $\mathrm{S}^2$

From the previous exercise:

$$(\vec{\sigma}_1 + \vec{\sigma}_2)^2 = \sigma_1^2 + \sigma_2^2 + 2\vec{\sigma}_1\,\vec{\sigma}_2 = 4(1 + P^s)$$

The eigenfunctions agree with those of $P^s$; they are

$$\alpha(1)\,\alpha(2), \quad \alpha(1)\,\beta(2) + \alpha(2)\,\beta(1), \quad \beta(1)\,\beta(2), \quad \alpha(1)\,\beta(2) - \alpha(2)\,\beta(1)$$

The eigenvalues are $2\hbar^2$ for the three symmetric functions and 0 for the antisymmetric one.

### 8. Van der Waals potential for the Thomson atomic model

(a)   $m\ddot{x}_1 = -fx_1 + 2\gamma x_2$      $m\ddot{y}_1 = -fy_1 - \gamma y_2$      $m\ddot{z}_1 = -fz_1 - \gamma z_2$

$m\ddot{x}_2 = -fx_2 + 2\gamma x_1$      $m\ddot{y}_2 = -fy_2 - \gamma y_1$      $m\ddot{z}_2 = -fz_2 - \gamma z_1$

where                     $f = e^2/R_A^3$ and $\gamma = e^2/R^3$

(b) The characteristic oscillations and eigenfrequencies are found by adding and subtracting the pairs of equations shown in (a). Only non-zero amplitudes are listed in the table below.

| Oscillation state | Eigenfrequency |
|---|---|
| $x_1 + x_2 \neq 0$ | $\omega_1 = \left\{ \dfrac{f - 2\gamma}{m} \right\}^{1/2}$ |
| $x_1 - x_2 \neq 0$ | $\omega_2 = \left\{ \dfrac{f + 2\gamma}{m} \right\}^{1/2}$ |
| $y_1 + y_2 \neq 0$ | $\omega_3 = \left\{ \dfrac{f + \gamma}{m} \right\}^{1/2}$ |
| $y_1 - y_2 \neq 0$ | $\omega_4 = \left\{ \dfrac{f - \gamma}{m} \right\}^{1/2}$ |
| $z_1 + z_2 \neq 0$ | $\omega_5 = \left\{ \dfrac{f + \gamma}{m} \right\}^{1/2}$ |
| $z_1 - z_2 \neq 0$ | $\omega_6 = \left\{ \dfrac{f - \gamma}{m} \right\}^{1/2}$ |

(c) $E(R) = \frac{1}{2}\hbar \sum\limits_{\nu=1}^{6} \omega_\nu$. Expanding in powers of $\gamma$, we obtain

$$E(R) = \tfrac{1}{2}\hbar\omega_0 \left\{ 6 - \frac{3}{2}\frac{\gamma^2}{f^2} \right\}$$

where $\omega_0 = \sqrt{(f/m)}$ is the frequency of the unperturbed Thomson atom. The energy of the ground state $E_0$ of the two unperturbed atoms is $3\hbar\omega_0$. If we now put $\hbar\omega_0$ equal to the excitation or ionization energy of the atom, $I$, and substitute the polarizability $\alpha$ of the isolated atom for $R_A{}^3$, then

$$E(R) = E_0 - \frac{3}{4}\frac{I\alpha^2}{R^6}$$

This formula is of much more general validity, as might be expected from its method of derivation. In the case of two atoms, 1 and 2, the correct perturbation calculation by the methods of quantum mechanics leads to the following approximate formula, which is suitable for estimates:

$$E(R) \approx E_0 - \frac{3}{2}\frac{I_1 I_2}{I_1 + I_2}\frac{\alpha_1 \alpha_2}{R^6}$$

In the classical model, $E(R)$ would be equal to 0 in the ground state since each electron would be at rest at the centre of its respective nucleus and no form of electrostatic interaction would occur. In quantum theory, on the other hand, the zero-point oscillations induce an attractive force, which can also be achieved according to classical mechanics by means of thermal motion.

(d) The ground-state function for a linear oscillator is $C \exp - \dfrac{m\omega}{2\hbar} x^2$. In the present case we obtain a product of six such functions:

$$C \exp \left\{ -\frac{m\omega_1}{4\hbar}(x_1 + x_2)^2 - \frac{m\omega_2}{4\hbar}(x_1 - x_2)^2 - \frac{m\omega_3}{4\hbar}[(y_1 + y_2)^2 + (z_1 + z_2)^2] \right.$$
$$\left. - \frac{m\omega_4}{4\hbar}[(y_1 - y_2)^2 + (z_1 - z_2)^2] \right\}$$

It should be noted that in the case of the oscillations for which $x_1 + x_2 \neq 0$, twice the mass and the coordinate $\frac{1}{2}(x_1 + x_2)$ are employed, while half the mass and the coordinate $x_1 - x_2$ enter into the oscillations for which $x_1 - x_2 \neq 0$.

*Proof*: If all $\omega_i$ are equal, as in the case of two separate atoms, the ground-state function must become $\exp\{(-m\omega_0/2\hbar)(\mathbf{r}_1{}^2 + \mathbf{r}_2{}^2)\}$.

We can immediately deduce from the above representation that all mean values of form $\overline{x_1}$, $\overline{x_2}$, $\overline{y_1}$ and $\overline{x_1y_1}$, $\overline{x_1y_2}$, . . . vanish.

The various correlation products such as $\overline{x_1 x_2}$ are found from $\overline{(x_1 + x_2)^2} = 2\hbar/m\omega_1$ and $\overline{(x_1 - x_2)^2} = 2\hbar/m\omega_2$, etc.:

$$\overline{x_1 x_2} = \frac{\hbar}{2m}\left(\frac{1}{\omega_1} - \frac{1}{\omega_2}\right) > 0; \quad \overline{y_1 y_2} = \overline{z_1 z_2} = \frac{\hbar}{2m}\left(\frac{1}{\omega_3} - \frac{1}{\omega_4}\right) < 0$$

The significance of the positive correlation $\overline{x_1 x_2} > 0$ is as follows: for a given $x_1 > 0$, the chance of finding electron 2 in the region $x_2 > 0$ is greater than the chance of encountering it in the region $x_2 < 0$; similarly for $x_1 < 0$ and $x_2 < 0$. The reason is clear: when $x_1 > 0$ the force on electron 2 is directed towards the right, and therefore tends to give preference to positive values of $x_2$. The opposite situation exists in the case of the y- and z-components.

### 9. *Forces between two helium atoms*

The result is $E_{el}(R) = 2\varepsilon_0 + 4C - 2A$, where $\varepsilon_0$ is the energy of the ground state of the helium atom when calculated with the optimum value of $Z$, and $C$ and $A$ have the same significance as in the case of the hydrogen molecule, but are calculated from the modified hydrogen functions appropriate to helium. Further integrals appear when the overlap integral $S$ cannot be neglected. The contribution of the Coulomb integral $4C$ is practically compensated by the electrostatic interaction $4e^2/R$, and the potential energy therefore consists almost entirely of the repulsion term $-2A$, where $A < 0$.

## Chapter EI (p. 296)

### 1. *Pressure and energy density in an ideal gas*

The number of particles in the interval $(\mathbf{p}, d\mathbf{p})$ incident on an element of surface $dS$ in time $dt$ is $f(\mathbf{p})d\mathbf{p}v_x dS dt$ (cf. figure 69). In the course of a collision (assumed to be elastic) each particle transfers momentum $2p_x$ to the wall. Therefore

$$P dS dt = \int_0^\infty dp_x \int\int_{-\infty}^{+\infty} dp_y dp_z 2p_x v_x f dS dt$$

since it is only permissible to integrate over those particles that actually strike the wall (i.e. $p_x$, $v_x > 0$). The general relations between the energy $E$, the momentum $\mathbf{p}$, and the velocity $\mathbf{v}$ are:

$$\mathbf{p} = \frac{m\mathbf{v}}{\sqrt{\left(1 - \dfrac{v^2}{c^2}\right)}} \qquad E = \frac{mc^2}{\sqrt{\left(1 - \dfrac{v^2}{c^2}\right)}}$$

$$E = \sqrt{(p^2 c^2 + m^2 c^4)} \qquad \mathbf{p}\mathbf{v} = pv = \frac{E^2 - m^2 c^4}{E}$$

Since the gas is assumed to be isotropic, the factor 2 may be replaced by integration over all values of $p_x$: hence $P = n\overline{p_x v_x}$. In addition,

$$\overline{p_x v_x} = \overline{p_y v_y} = \overline{p_z v_z} = \frac{1}{3}\,\overline{\mathbf{pv}} = \frac{1}{3}\left\{\frac{\overline{E^2 - m^2 c^4}}{E}\right\}$$

Therefore

$$P = \frac{n}{3}\left\{\frac{\overline{(E - mc^2)(E + mc^2)}}{E}\right\} = \begin{cases} \dfrac{n}{3}\bar{E} \ \text{for} \ E \gg mc^2, \ \text{or} \ m = 0 \\[2mm] \dfrac{2}{3}n\overline{(E - mc^2)} = \dfrac{2}{3}n\bar{E}_{\text{kin}} \ \text{for} \ E - mc^2 \ll mc^2 \end{cases}$$

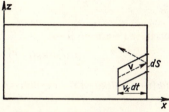

Fig. 69.—Calculation of the pressure of an ideal gas

## 2. *The solar constant*

From (46.4), the total energy emitted from the surface of the sun per unit time is $\sigma T^4 . 4\pi R_s^2$, where $R_s$ is the sun's radius. This energy passes through the surface of a sphere of radius $R_E$, where $R_E$ denotes the radius of the earth's orbit. At the earth, therefore, the incident energy per unit time and per unit area is

$$S = \sigma T^4\left(\frac{R_s}{R_E}\right)^2 = \sigma T^4\frac{\delta^2}{4}$$

The value of $\sigma$ is $1\cdot4 \times 10^{-12}\text{cal cm}^{-2}\text{s}^{-1}\text{deg}^{-4}$; putting $T = 6 \times 10^3$ and $\delta \approx 0\cdot009$ gives a value of about $2\text{cal cm}^{-2}\text{min}^{-1}$ for the solar constant.

# Chapter EII (p. 319)

## 1. *Fourier representation of Hertz's solution*

If the trial solution for $\phi$, say, is introduced into the initial equation, we obtain

$$\phi = \frac{1}{(2\pi)^4}\int 4\pi\rho(\mathbf{r}', t')\frac{\exp\,[i\mathbf{k}(\mathbf{r} - \mathbf{r}') - i\omega\,(t - t')]}{k^2 - \omega^2/c^2}\,d\mathbf{k}\,d\omega\,d\mathbf{r}'\,dt'$$

The path of integration in the plane of $\omega$ should be taken as shown in figure 70.

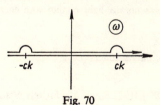

Fig. 70

When $t < t'$ the path of integration may be moved upwards into the positive imaginary region, and the integral round the contour is then zero. For $t > t'$ the path of integration must be moved downwards into the negative imaginary region, in which case it still includes the poles of the integrand at $\omega = -ck$ and $\omega = +ck$. Then from the theorem of residues:

$$\frac{4\pi}{(2\pi)^4} \int \frac{\exp[i\mathbf{k}(\mathbf{r}-\mathbf{r}')-i\omega(t-t')]}{k^2-\omega^2/c^2} d\mathbf{k}\,d\omega$$

$$= \frac{-ic}{(2\pi)^2} \int \exp i\mathbf{k}(\mathbf{r}-\mathbf{r}') \frac{\exp ick(t-t')-\exp -ick(t-t')}{k} d\mathbf{k}$$

$$= \frac{-c}{2\pi|\mathbf{r}-\mathbf{r}'|} \int_0^\infty \{\exp ik|\mathbf{r}-\mathbf{r}'|-\exp -ik|\mathbf{r}-\mathbf{r}'|\} \times$$
$$\times \{\exp ick(t-t')-\exp -ick(t-t')\}dk$$

$$= \frac{1}{|\mathbf{r}-\mathbf{r}'|} \int_{-\infty}^{+\infty} \Big\{ \exp\{ik|\mathbf{r}-\mathbf{r}'|-ick(t-t')\}-\exp\{ik|\mathbf{r}-\mathbf{r}'|+ick(t-t')\} d\frac{kc}{2\pi}$$

$$= \frac{\delta\left(t'-t+\frac{|\mathbf{r}-\mathbf{r}'|}{c}\right)}{|\mathbf{r}-\mathbf{r}'|}, \text{ since } t-t' > 0$$

When this result is introduced into the above integral and the integration is performed with respect to $t'$ we obtain the familiar formula for the retarded potential. The calculation in the case of $\mathbf{A}(\mathbf{r}, t)$ is similar.

### 2. *Total radiation momentum*

Using the relations (51.10) and (51.12), we obtain

$$\mathbf{G} = \sum_{\mathbf{k}, \lambda} \hbar\mathbf{k}b_{\mathbf{k}\lambda}^+ b_{\mathbf{k}\lambda}$$

Equation (52.2) shows that $\mathbf{G}$ commutes with $\mathscr{H}$, and that it is therefore constant with respect to time. The eigenvectors (52.6) are also eigenvectors of $\mathbf{G}$ with eigenvalues

$$\sum_{\mathbf{k}, \lambda} n_{\mathbf{k}\lambda}\hbar\mathbf{k}$$

This state therefore represents $n_{\mathbf{k}\lambda}$ light quanta with energy $\hbar\omega_{\mathbf{k}}$ and momentum $\hbar\mathbf{k}$.

### 3. *Angular momentum of the radiation*

We first obtain

$$\mathbf{M} = \sum_{\mathbf{k}} \sum_{\lambda,\lambda'} \tfrac{1}{2}i\hbar\{b_{\mathbf{k}\lambda}^+ b_{\mathbf{k}\lambda'} - b_{\mathbf{k}\lambda'}^+ b_{\mathbf{k}\lambda}\}(\mathbf{s}_{\mathbf{k}\lambda} \times \mathbf{s}_{\mathbf{k}\lambda'}) = \sum_{\mathbf{k}} \mathbf{M}^{(\mathbf{k})}$$

The vector product of the two mutually orthogonal polarization vectors is only non-zero if $\lambda \neq \lambda'$. $s_{k1} \times s_{k2} = e_k$ is a unit vector ($e_{-k} = e_k$) in the direction of $+\mathbf{k}$ if the vectors are chosen as in figure 71. Then

$$\mathbf{M}^{(k)} = e_k . i\hbar \{b_{k1}^+ b_{k2} - b_{k2}^+ b_{k1}\}$$

In order to find the eigenvalues of a Hermitian operator in the form of $i(b_+ b_2 - b_2^+ b_1)$ we may introduce the new operators

$$b_{(\sigma)} = \frac{1}{\sqrt{2}} (b_1 + i\sigma b_2), \text{ where } \sigma = \pm 1$$

(It should be verified that the commutation relations for the $b_{k(\sigma)}$ are identical with those for $b_{k\lambda}$.) Then

Eigenvalues

$$\mathbf{M}^{(k)} = e_k \hbar (b_{k(1)}^+ b_{k(1)} - b_{k(-1)}^+ b_{k(-1)}) \qquad e_k \hbar \quad (n_{k(1)} - n_{k(-1)})$$

$$\left. \begin{array}{l} \mathscr{H}_{(k)} = \hbar \omega_k \\ G^{(k)} = \hbar k \end{array} \right\} . (b_{k(1)}^+ b_{k(1)} + b_{k(-1)}^+ b_{k(-1)}) \qquad \left. \begin{array}{l} \hbar \omega_k \\ \hbar k \end{array} \right\} . (n_{k(1)} + n_{k(-1)})$$

Hence the eigenvalues of $\mathbf{M}^{(k)}$ are integral multiples of $\pm e_k \hbar$. The spin has the value $\hbar$ and is parallel or antiparallel to the propagation vector $\mathbf{k}$. The classical waves associated with $\sigma = 1$ clearly have corresponding amplitudes $b_{(1)} \neq 0$,

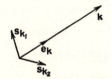

Fig. 71.—Polarization and propagation vectors

$b_{(-1)} = b_1 - ib_2 = 0$: i.e. the amplitudes $b_1$ and $b_2$ have the same magnitude and a phase difference of $\pi/2$. This is a circularly polarized wave (in the right-handed sense with respect to $\mathbf{k}$). It is evident from the above representation that $\mathscr{H}$, $\mathbf{G}$, and $\mathbf{M}$ commute. The Hilbert space is formed as in §52, except that the $b_{k(\sigma)}$ are employed in place of the $b_{k\lambda}$:

| | $\mathscr{H}$ with eigenvalue | $\sum\limits_{k,\sigma} \hbar \omega_k n_{k(\sigma)}$ |
|---|---|---|
| $\Phi \ldots, n_{k(\sigma)}, \ldots$ is an eigenfunction of the operator | $\mathbf{G}$ with eigenvalue | $\sum\limits_{k,\sigma} \hbar k n_{k(\sigma)}$ |
| | $\mathbf{M}$ with eigenvalue | $\sum\limits_{k,\sigma} \hbar e_k \sigma n_{k(\sigma)}$ |

The Hilbert space is spanned by these eigenfunctions.

## Chapter FI (p. 327)

**1.** *The energy spectrum of electrons in β-disintegration, according to Fermi*

$n(\mathbf{p}_e)\,d\mathbf{p}_e = C\,d\mathbf{p}_e \int d\mathbf{p}_\nu$ is first calculated, where $\mathbf{p}_\nu$ must be integrated subject to the condition $E_m - \mathcal{H}_e \leqslant \mathcal{H}_\nu \leqslant E_m - \mathcal{H}_e + \Delta E_m$. Since $\mathcal{H} = \{p^2c^2 + m^2c^4\}^{1/2}$, this is the same as integrating in $\mathbf{p}_\nu$-space over a spherical shell of radius

$$\bar{p}_\nu = \left\{ \frac{(E_m - \mathcal{H}_e)^2 - m_\nu^2 c^4}{c^2} \right\}^{1/2}$$

and thickness

$$\Delta \bar{p}_\nu = \frac{E_m - \mathcal{H}_e}{p_\nu c^2} \Delta E_m$$

Therefore

$$n(\mathbf{p}_e)\,d\mathbf{p}_e = C'\{(E_m - \mathcal{H}_e)^2 - m_\nu^2 c^4\}^{1/2}(E_m - \mathcal{H}_e)$$

where $C'$ is a new constant.

If this distribution is again integrated over $\mathbf{p}_e$, with $E_e \leqslant \mathcal{H}_e \leqslant E_e + dE_e$, the result is

$$n(E_e)\,dE_e = C''\{E_e^2 - m_e^2 c^4\}^{1/2} E_e\{(E_m - E_e)^2 - m_\nu^2 c^4\}^{1/2}(E_m - E_e)$$

In particular, for $m_\nu = 0$:

$$n(E_e)\,dE_e = C''\{E_e^2 - m_e^2 c^4\}^{1/2} E_e (E_m - E_e)^2$$

At values near the rest energy of the electron, $E_e = m_e c^2$, the distribution has the form $\sqrt{(E_e - m_e c^2)}$; at the maximum energy $E_m$ the slope is horizontal. The

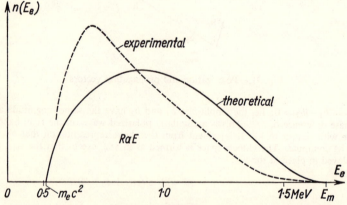

Fig. 72.—Experimental and theoretical distributions for $\beta$-disintegration of Ra E

theoretical distribution is compared with an experimental curve in figure 72. The theoretical curve is adjusted to $E_m$ and normalized to make the area under each curve the same.

**2.** *The relativistic Hamiltonian equations*

The treatment is the same as in §7.

## Chapter FII (p. 359)

### 1. *The spin operator of the Dirac equation*

From (57.10), $[\mathscr{H}, \vec{\sigma}'] = 2ic\vec{a} \times \mathbf{p}$, whence on forming the scalar product with $\mathbf{p}$ it follows that

$$[\mathscr{H}, (\vec{\sigma}'\mathbf{p})] = 2ic\vec{a}\mathbf{p} \times \mathbf{p} = 0$$

We can also see from the table on p. 356 that

$$\sigma'\mathbf{p} = (U\vec{\sigma}'U^+)\mathbf{p}$$

### 2. *The matrix trace as a unitary invariant*

For unitary transformations:

$$\tilde{A} = U^+AU \qquad \sum_m U^*{}_{nm} U_{lm} = \delta_{ln}$$

Therefore $\quad\sum\limits_m \tilde{A}_{mm} = \sum\limits_{l,m,n} U^*{}_{nm} A_{nl} U_{lm} = \sum\limits_n A_{nn}$

If $U$ is so chosen for Hermitian matrices that $A$ is diagonal, then $\sum\limits_n A_{nn} = \sum\limits_i a_i$.

In addition:

$$\operatorname{Tr} AB = \sum_{mn} A_{mn} B_{nm} = \sum_{mn} B_{nm} A_{mn} = \operatorname{Tr} BA$$

### 3. *Commutation with the Pauli matrices*

It follows from the condition

$$\left[\sigma_z, \begin{pmatrix} a & b \\ c & d \end{pmatrix}\right] = \begin{pmatrix} a & b \\ -c & -d \end{pmatrix} - \begin{pmatrix} a & -b \\ c & -d \end{pmatrix} = 0$$

that $b = c = 0$; further, $a = d$ because

$$\left[\sigma_x, \begin{pmatrix} a & 0 \\ 0 & d \end{pmatrix}\right] = \begin{pmatrix} 0 & d \\ a & 0 \end{pmatrix} - \begin{pmatrix} 0 & a \\ d & 0 \end{pmatrix} = 0$$

Since $2i\sigma_y = \sigma_z\sigma_x - \sigma_x\sigma_z$, the commutation relation between the matrix and $\sigma_y$ does not impose any further condition.

### 4. *Hilbert space for Fermi-Dirac statistics*

Since $\qquad\qquad b_k^+ b_k + b_k b_k^+ = 1 \quad$ and $\quad b_k^2 = 0$

it follows that $\qquad (b_k^+ b_k)^2 = b_k^+(1 - b_k^+ b_k)b_k = b_k^+ b_k$

Hence the equation for the eigenvalues $n_k$ of $b_k^+ b_k$ is $n_k^2 = n_k$, possessing the two solutions $n_k = 0$ or 1. If there is an eigenstate $\phi$ of $b_k^+ b_k$, then

$$(b_k \phi, b_k \phi) = n_k(\phi, \phi)$$

If $n_k = 0$, then $b_k \phi = 0$; if $n_k = 1$, $b_k \phi$ corresponds to the eigenvalue $n_k = 0$ because

$$b_k^+ b_k^2 \phi = (b_k^+ b_k) b_k \phi = 0 \quad \text{and} \quad b_k^2 = 0$$

$b_k$ therefore possesses the property of annihilating a particle of momentum $\hbar k$. Similarly, it may be verified that the operator $b_k^+$ generates a particle. Since $b_k^{+2} = 0$, a particle of type k can only be created once.

### 5. The spin and orbital current of the Dirac electron

$$\rho = \rho_b - \operatorname{div} \mathbf{P} \qquad \mathbf{j} = \mathbf{j}_b + e \operatorname{curl} \mathbf{M} + \dot{\mathbf{P}}$$

$$\rho_b = \frac{i\hbar}{2mc^2} (\psi^* \beta \dot{\psi} - \dot{\psi}^* \beta \psi) \qquad\qquad \mathbf{P} = \frac{i\hbar}{2mc} (\psi^* \vec{a} \beta \psi)$$

$$\mathbf{j}_b = \frac{i\hbar}{2m} (\operatorname{grad} \psi^* \beta \psi - \psi^* \beta \operatorname{grad} \psi) \qquad \mathbf{M} = \frac{\hbar}{2mc} (\psi^* \vec{\sigma}' \beta \psi)$$

$\mathbf{j}_b$ corresponds to the Schrödinger current, and $\rho_b$ is the associated relativistic generalization of the density. $\mathbf{M}$ is the "magnetization" and $\mathbf{P}$ the "polarization" of the electron. (See Vol. I, §§ 26 and 47, and p. 216.)

# H

---

## Index

845

# INDEX